The Handbook of Environmental Chemistry

Volume 108

Founding Editor: Otto Hutzinger

Series Editors: Damià Barceló • Andrey G. Kostianoy

In over four decades, *The Handbook of Environmental Chemistry* has established itself as the premier reference source, providing sound and solid knowledge about environmental topics from a chemical perspective. Written by leading experts with practical experience in the field, the series continues to be essential reading for environmental scientists as well as for environmental managers and decision-makers in industry, government, agencies and public-interest groups.

Two distinguished Series Editors, internationally renowned volume editors as well as a prestigious Editorial Board safeguard publication of volumes according to high scientific standards.

Presenting a wide spectrum of viewpoints and approaches in topical volumes, the scope of the series covers topics such as

- local and global changes of natural environment and climate
- anthropogenic impact on the environment
- water, air and soil pollution
- remediation and waste characterization
- environmental contaminants
- biogeochemistry and geoecology
- chemical reactions and processes
- chemical and biological transformations as well as physical transport of chemicals in the environment
- environmental modeling

A particular focus of the series lies on methodological advances in environmental analytical chemistry.

The Handbook of Environmental Chemistry is available both in print and online via https://link.springer.com/bookseries/698. Articles are published online as soon as they have been reviewed and approved for publication.

Meeting the needs of the scientific community, publication of volumes in subseries has been discontinued to achieve a broader scope for the series as a whole.

Removal and Degradation of Pharmaceutically Active Compounds in Wastewater Treatment

Volume Editors: Sara Rodriguez-Mozaz ·
Paqui Blánquez Cano ·
Montserrat Sarrà Adroguer

With contributions by

A. Agüera · L. Baptista-Pires · D. Barceló · M. Bilal · P. Blánquez ·
A. M. Botero-Coy · M. Brienza · G. Buttiglieri · M. Escolà Casas ·
M. Castaño-Trias · S. Esplugas · P. Gago-Ferrero · J. Giménez ·
A. Ginebreda · B. Guieysse · F. I. Hai · F. Hernández · A. Hom-Díaz ·
H. M. N. Iqbal · J. Lee · S. Malato · J. Marugán · V. Matamoros ·
J. A. Mir-Tutusaus · I. Oller · W. E. Price · J. Radjenovic ·
J. Rodríguez-Chueca · S. Rodríguez-Mozaz · J. A. Sánchez Pérez ·
C. Sans · M. Sarrà · E. A. Serna-Galvis · M. C. Tomei ·
R. A. Torres-Palma · P. Verlicchi · T. Vicent · F. Zietzschmann

Springer

Editors
Sara Rodriguez-Mozaz
Catalan Institute for Water Research (ICRA)
University of Girona
Girona, Spain

Paqui Blánquez Cano
Universitat Autònoma de Barcelona
Barcelona, Spain

Montserrat Sarrà Adroguer
Universitat Autònoma de Barcelona
Barcelona, Spain

ISSN 1867-979X ISSN 1616-864X (electronic)
The Handbook of Environmental Chemistry
ISBN 978-3-030-77508-7 ISBN 978-3-030-77509-4 (eBook)
https://doi.org/10.1007/978-3-030-77509-4

This Springer imprint is published by the registered company Springer Nature Switzerland AG.
The registered company address is: Gewerbestrasse 11, 6330 Cham, Switzerland

Series Preface

With remarkable vision, Prof. Otto Hutzinger initiated *The Handbook of Environmental Chemistry* in 1980 and became the founding Editor-in-Chief. At that time, environmental chemistry was an emerging field, aiming at a complete description of the Earth's environment, encompassing the physical, chemical, biological, and geological transformations of chemical substances occurring on a local as well as a global scale. Environmental chemistry was intended to provide an account of the impact of man's activities on the natural environment by describing observed changes.

While a considerable amount of knowledge has been accumulated over the last four decades, as reflected in the more than 150 volumes of *The Handbook of Environmental Chemistry*, there are still many scientific and policy challenges ahead due to the complexity and interdisciplinary nature of the field. The series will therefore continue to provide compilations of current knowledge. Contributions are written by leading experts with practical experience in their fields. *The Handbook of Environmental Chemistry* grows with the increases in our scientific understanding, and provides a valuable source not only for scientists but also for environmental managers and decision-makers. Today, the series covers a broad range of environmental topics from a chemical perspective, including methodological advances in environmental analytical chemistry.

In recent years, there has been a growing tendency to include subject matter of societal relevance in the broad view of environmental chemistry. Topics include life cycle analysis, environmental management, sustainable development, and socio-economic, legal and even political problems, among others. While these topics are of great importance for the development and acceptance of *The Handbook of Environmental Chemistry*, the publisher and Editors-in-Chief have decided to keep the handbook essentially a source of information on "hard sciences" with a particular emphasis on chemistry, but also covering biology, geology, hydrology and engineering as applied to environmental sciences.

The volumes of the series are written at an advanced level, addressing the needs of both researchers and graduate students, as well as of people outside the field of "pure" chemistry, including those in industry, business, government, research establishments, and public interest groups. It would be very satisfying to see these volumes used as a basis for graduate courses in environmental chemistry. With its high standards of scientific quality and clarity, *The Handbook of Environmental Chemistry* provides a solid basis from which scientists can share their knowledge on the different aspects of environmental problems, presenting a wide spectrum of viewpoints and approaches.

The Handbook of Environmental Chemistry is available both in print and online via https://link.springer.com/bookseries/698. Articles are published online as soon as they have been approved for publication. Authors, Volume Editors and Editors-in-Chief are rewarded by the broad acceptance of *The Handbook of Environmental Chemistry* by the scientific community, from whom suggestions for new topics to the Editors-in-Chief are always very welcome.

Damià Barceló
Andrey G. Kostianoy
Series Editors

Preface

Among organic micropollutants, pharmaceutically active compounds (PhACs) have been extensively investigated over the last few years due to their potential adverse ecological and/or human health effects. After administration, PhACs are excreted and delivered with wastewater to the wastewater treatment plants (WWTPs). Wastewater treatment plants thus become one of the first barriers in preventing them from reaching the ecosystems but also one of the main sources, since a significant proportion of PhACs is not properly eliminated in conventional WWTPs. Therefore, advances in wastewater treatment technologies are crucial to minimize the load of micropollutants, and thus to the optimal protection of both the environment and human health. In this context, this volume aims at providing a snapshot of selected wastewater treatment strategies, each of which is discussed in the corresponding chapter.

The efficiency of the wastewater treatment of choice can be evaluated by considering the overall removal of pollutants, although a better assessment is achieved if its environmental relevance is considered. Chapter One is specifically devoted to environmental risk assessment of pharmaceuticals in wastewater treatment, and different approaches are discussed and suggested as tools to evaluate the removal efficiency of a particular treatment. In Chapter Two, an overview of the fate of PhACs in conventional activated sludge (CAS) systems (the most standard practice in conventional WWTPs) is presented. Several parameters and constraints affecting PhAC removal in the water line of CAS systems are given and discussed, while for the sludge line different treatment options are also considered. Special attention is paid in these two chapters to the risks of pharmaceuticals in water reclamation practices. Even though most of the literature (and the chapters of this volume) deal with urban wastewater treatment, decentralized treatment of hospital effluents is becoming an interesting approach to reduce the impact on the aquatic environment of hazardous substances that are routinely administered or used in healthcare facilities. Chapter Three is entitled "New Insights into the Occurrence of Micropollutants and the Management and Treatment of Hospital Effluent," which deals with this type of effluents and reviews promising new technologies proven to improve the removal of emerging pollutants such as PhACs.

Another critical issue addressed in this volume is the development of appropriate analytical methods to determine the presence of PhACs and transformation products in the environment. This is of utmost importance for the proper evaluation of its disposal in the wastewater treatment process of choice. Target analytical methodologies are the established analytical approach for evaluating pollutant removal, but they can only cover a small proportion of organic pollutants. Consequently, some compounds remain uncharacterized despite their potential relevance. New analytical approaches are reported in Chapter Four (Suspect and Non-target Screening Methodologies for the Evaluation of the Behaviour of Polar Organic Micropollutants and Changes in the Molecule Fingerprint During Water Treatment). These strategies allow to provide information on a broader set of contaminants (including PhACs) to understand their behavior and their corresponding transformation products during wastewater treatment processes. The chapter also discusses the novel non-target workflows to characterize dissolved organic matter (DOM) during water treatment.

After these four chapters that are dedicated to general aspects, the core of the book is dedicated to report on different wastewater treatment strategies and their efficiency with respect to the removal of PhACs from the waste. Physical, chemical, and biological processes have been tested with varying degrees of success. Biological-based treatments considered in this book are enzymes, fungal reactors, nature-based solution (NBS) like wetlands and microalgae-based processes (Chapters Five to Eight). Chapter Nine reviews the application of membrane bioreactors, which combine biodegradation and membrane separation in a single step. Finally, the removal and degradation of PhACs in wastewater are considered using physicochemical treatment processes such as low-cost adsorbents, ozone and advanced oxidation processes (AOPs), and other advanced processes based on solar oxidation, electrochemical and sonochemical processes (Chapters Ten to Fourteen). Many of these treatments can be applied as single treatment steps but also as polishing treatment after conventional WWTP, or just as a combination of some of them in the treatment train of choice.

We hope that readers will be able to find in this volume the latest information on the topic that can help them to further develop water treatment and reuse projects. We would also like to acknowledge all the colleagues who agreed to join this project and share their knowledge, as well as the editor and publisher of the series for inviting us to prepare this book.

Girona, Spain — Sara Rodriguez-Mozaz
Barcelona, Spain — Paqui Blánquez Cano
Barcelona, Spain — Montserrat Sarrà Adroguer

Contents

Environmental Risk Assessment of Pharmaceuticals in Wastewater Treatment 1
Antoni Ginebreda, Damià Barceló, and Sara Rodríguez-Mozaz

Fate and Removal of Pharmaceuticals in CAS for Water and Sewage Sludge Reuse 23
Marc Castaño-Trias, Monica Brienza, Maria Concetta Tomei, and Gianluigi Buttiglieri

New Insights into the Occurrence of Micropollutants and the Management and Treatment of Hospital Effluent 53
Paola Verlicchi

Suspect and Non-target Screening Methodologies for the Evaluation of the Behaviour of Polar Organic Micropollutants and Changes in the Molecule Fingerprint During Water Treatment 97
Pablo Gago-Ferrero

Perspectives on the Feasibility of Using Enzymes for Pharmaceutical Removal in Wastewater 119
Muhammad Bilal, Hafiz M. N. Iqbal, and Damiá Barceló

Fungal Reactors: A Solution for the Removal of Pharmaceuticals in Urban and Hospital Wastewater 145
Josep Anton Mir-Tutusaus and Montserrat Sarrà

Novel Constructed Wetland Configurations for the Removal of Pharmaceuticals in Wastewater 163
Mònica Escolà Casas and Víctor Matamoros

Microalgae-Based Processes for the Removal of Pharmaceuticals in Wastewater 191
Paqui Blánquez, Andrea Hom-Díaz, Teresa Vicent, and Benoit Guieysse

Removal of Pharmaceuticals from Wastewater by Membrane Bioreactors: Factors, Mechanisms, and Perspectives 223
Faisal I. Hai and William E. Price

Adsorptive Removal of Pharmaceutically Active Compounds from Wastewater 239
Frederik Zietzschmann

Removal of Pharmaceutically Active Compounds (PhACs) in Wastewater by Ozone and Advanced Oxidation Processes 269
Javier Marugán, Jorge Rodríguez-Chueca, Santiago Esplugas, Carme Sans, and Sixto Malato

Removal and Degradation of Pharmaceutically Active Compounds (PhACs) in Wastewaters by Solar Advanced Oxidation Processes 299
Sixto Malato, Jaime Giménez, Isabel Oller, Ana Agüera, and José Antonio Sánchez Pérez

Advanced Electrochemical Processes for the Elimination of Pharmaceutical Compounds in Contaminated Waters 327
Jelena Radjenovic and Luis Baptista-Pires

Sonochemical Advanced Oxidation Processes for the Removal of Pharmaceuticals in Wastewater Effluents 349
Efraím A. Serna-Galvis, Judy Lee, Félix Hernández, Ana María Botero-Coy, and Ricardo A. Torres-Palma

Correction to: Advanced Electrochemical Processes for the Elimination of Pharmaceutical Compounds in Contaminated Waters . . . 383
Jelena Radjenovic and Luis Baptista-Pires

Environmental Risk Assessment of Pharmaceuticals in Wastewater Treatment

Antoni Ginebreda, Damià Barceló, and Sara Rodríguez-Mozaz

Contents

1 Introduction 2
  1.1 Why Pharmaceuticals? 2
2 Regulatory Aspects 4
  2.1 Environmental Risk Assessment Aspects in the Registration of Pharmaceuticals 4
  2.2 The Occurrence of Drugs in the Aquatic Environment 6
3 Overview of the Environmental Risk Assessment process 7
4 Environmental Risk Assessment of Wastewater Effluents 8
5 Environmental Risk Assessment as a Tool to Evaluate the Removal Efficiency of Wastewater Treatment 11
6 Discharge of Wastewater and Sludge in the Receiving Environment; Risk Assessment 13
7 Environmental and Human Health Risks of Reclaimed Water Irrigation 14
8 Future Perspectives in Environmental Risk Assessment of PhACs 15
  8.1 Emphasis on Environmental Mixtures 15
  8.2 Bioanalytical Tools to Capture Mixture Effects 16
  8.3 Toxicogenomics 17
  8.4 Antibiotic Resistance 17
References 18

A. Ginebreda
Department of Environmental Chemistry, IDAEA-CSIC, Barcelona, Spain

D. Barceló
Department of Environmental Chemistry, IDAEA-CSIC, Barcelona, Spain

Catalan Institute for Water Research (ICRA), Girona, Spain

University of Girona, Girona, Spain

S. Rodríguez-Mozaz (✉)
Catalan Institute for Water Research (ICRA), Girona, Spain

University of Girona, Girona, Spain
e-mail: srodriguez@icra.cat

Sara Rodriguez-Mozaz, Paqui Blánquez Cano, and Montserrat Sarrà Adroguer (eds.), *Removal and Degradation of Pharmaceutically Active Compounds in Wastewater Treatment*, Hdb Env Chem (2021) 108: 1–22, DOI 10.1007/698_2020_694,

Abstract The presence of pharmaceutically active compounds (PhACs) in the environment has gained increasing attention the past two decades. After administration, PhACs are excreted and thus delivered with wastewater to the wastewater treatment plants (WWTPs). WWTP effluents are usually discharged into receiving environment and, therefore, a high proportion of pollutants (including pharmaceuticals) can enter water bodies (surface or sea water), but also soils and eventually groundwater. The risk that pharmaceuticals can pose to the environment depends not only on compound concentration but also on its toxicity. In this chapter, different existing risk assessment approaches are presented, applied to the effluents from conventional WWTPs but also from alternative and advance wastewater treatment processes. Environmental Risk assessment (ERA) can also be applied as a tool to evaluate the removal efficiency of a wastewater treatment. Moreover, special focus is paid to assess the risk of pharmaceuticals in water reclamation practices such as crop irrigation. Finally, some research lines or topics that demand greater attention related to ERA in wastewater treatment are addressed such as toxicogenomics, antibiotic resistance, and evaluation of impact of chemical mixtures.

Keywords Environmental hazard, Pharmaceutically active compounds (PhACs), Toxicity, Wastewater effluent

1 Introduction

1.1 Why Pharmaceuticals?

Our continuously growing technological society has given rise to extensive and intensive use of chemicals on a global scale. More than 140,000 chemicals are routinely used by both industry and household [1]. Depending on their properties, the amounts produced, and the mode of use, a substantial fraction of these chemicals may potentially reach the environment. Pharmaceuticals are recognized as one of the pillars of our health system. More than 4,000 active compounds are thus globally and increasingly used worldwide, a fact that goes in parallel with their ubiquitous environmental occurrence. After the patient's medication with a pharmaceutically active compound (PhAC), it is metabolized and subsequently excreted through the body fluids as a mixture of the original parent compound and the metabolites generated in the treated organism. Finally, the complex mixture is evacuated through the sewage system and eventually treated in a wastewater plant, before the final effluents are released to the environment. Roughly, 80% of the world's wastewater is discharged without any treatment [1, 2]. Furthermore, even when it is properly treated, leaks from sewage pipes, storm overflows, and runoff during heavy rain events constitute non-negligible sources of untreated water as well [2]. Finally, because conventional WWTPs are not specifically designed for that purpose, some PhACs are simply not degraded and require advanced tertiary processes not easily affordable. Whereas excretion is the predominant source of PhACs into the environment, other pathways like inappropriate disposal, industrial spills, or manure spread (particularly for veterinary drugs) should not be overlooked. Once effluents reach the receiving waterbody PhACs and their metabolites may undergo further biotic and abiotic degradation processes such as sorption, photolysis, hydrolysis, and biodegradation potentially giving rise to new transformation products. Parent compounds plus the resulting transformation products (TPs) constitute a "chemical universe" whose long-term interactions with the different levels of biological organization (species, individuals, populations, and ecosystems) are still poorly known. The environmental concerns associated with PhACs are summarized in Box 1.

Box 1 Environmental Concerns Associated to Active Pharmaceutical Ingredients (APIs) and Their Transformation Products

- Drugs are regularly taken by the population with a clear growing trend. The process is ubiquitous on space and continuous along time and extends to some thousands of available APIs.
- Water treatment systems (if any) are not specifically designed to eliminate pharmaceuticals. Depending on each specific compound and characteristics of the treatment, their elimination maybe not complete. Further tertiary specific treatment steps may be necessary, but at a certain economic cost not always affordable.
- Even though many APIs are degraded, their continuous input into the environment makes them behave as "pseudo-persistent."
- Drugs are, by design, biologically active compounds, targeted to specific organisms (typically humans) and having precise modes of action. Possible side effects on other non-targeted unintentionally exposed organisms found in the receiving ecosystem biological communities are largely unknown.
- In addition to long-term ecotoxicity effects, there is a special human health concern for the specific case of antimicrobials (antibiotics, antifungals), whose environmental occurrence has been proved to promote the development of resistance genes (Antimicrobial Resistance Genes, ARG).
- The complex "cocktail" of many APIs, metabolites, and transformation may have unexpected and hardly predictable mixture interaction effects (i.e., synergistic, antagonistic, etc.).

Last but not least, the anthropogenic water cycle includes the safe supply of drinking water to the population. The occurrence of API residues in drinking water may pose a risk to human health, therefore their absence must be prevented in any case.

2 Regulatory Aspects

Altogether, it is recognized that the occurrence of drugs in the environment is not exempt from risk and there is a need of incorporating environmental aspects into the existing regulatory framework to assure the environmental safety of drugs. In that respect, two areas can be highlighted related to legislation (1) the registration process of drugs, and (2) the environmental occurrence of drugs in the environment, and more specifically in the aquatic environment. The current status of both aspects concerning drug legislation in the European Union is briefly reviewed below:

2.1 Environmental Risk Assessment Aspects in the Registration of Pharmaceuticals

The legal basis for the registration, production, distribution, and use of medicinal products in the EU member states is governed by the Directive 2001/83/EC. Following their provisions, any medical product should get an official authorization previous to its marketing. This is issued by the European Medicines Agency (EMA) after the proper completion of a registration process. According to article 8(3) of the directive, applicants must submit a dossier containing the necessary information of the pharmaceutical product to ensure its safety and therapeutic and clinical efficacy. As part of this authorization file, information regarding the potential risks of the PhAC to the environment is required. Therefore, companies wishing to register a new PhAC have to provide an Environmental Risk Assessment (ERA) [3, 4].

According to the EMA guideline [5], the overall process is depicted in Fig. 1 and includes a risk assessment and a specific hazard assessment. The risk assessment is focused on the environmental occurrence (exposure) and ecotoxic potential effects of the product on the exposed organisms. For some specific biological effects and substances (i.e., endocrine disruptors, antibiotics, etc.) additional aspects have to be considered. In turn, the hazard assessment refers to intrinsic properties of the products considered harmful for the living organisms exposed regardless of the concentration and specifically to the persistence, bioaccumulation, and toxicity (PBT) characteristics. Briefly, the procedure includes two phases (Fig. 1). In general, Phase I consists of a decision tree mostly addressed to differentiate among products that require a further assessment (Phase II) or those that not. This is done based on the Predicted Environmental Concentration in surface water (PEC) of the product estimated from its predicted use. If PEC $\geq$ 0.01 μg/L, the product enters Phase II, otherwise the process is finished. Phase II is a tiered process, starting with the study of physicochemical properties, environmental fate, and ecotoxicity and a predicted non-effect concentration (PNEC). Amongst other aspects like the potential risk to the groundwater and the soil, or the possibility of secondary poisoning in across the trophic chain, Tier A examines the risk ratio PEC/PNEC in surface water, and if it exceeds 1, a Tier B with PNEC refinement is performed. The PBT hazard

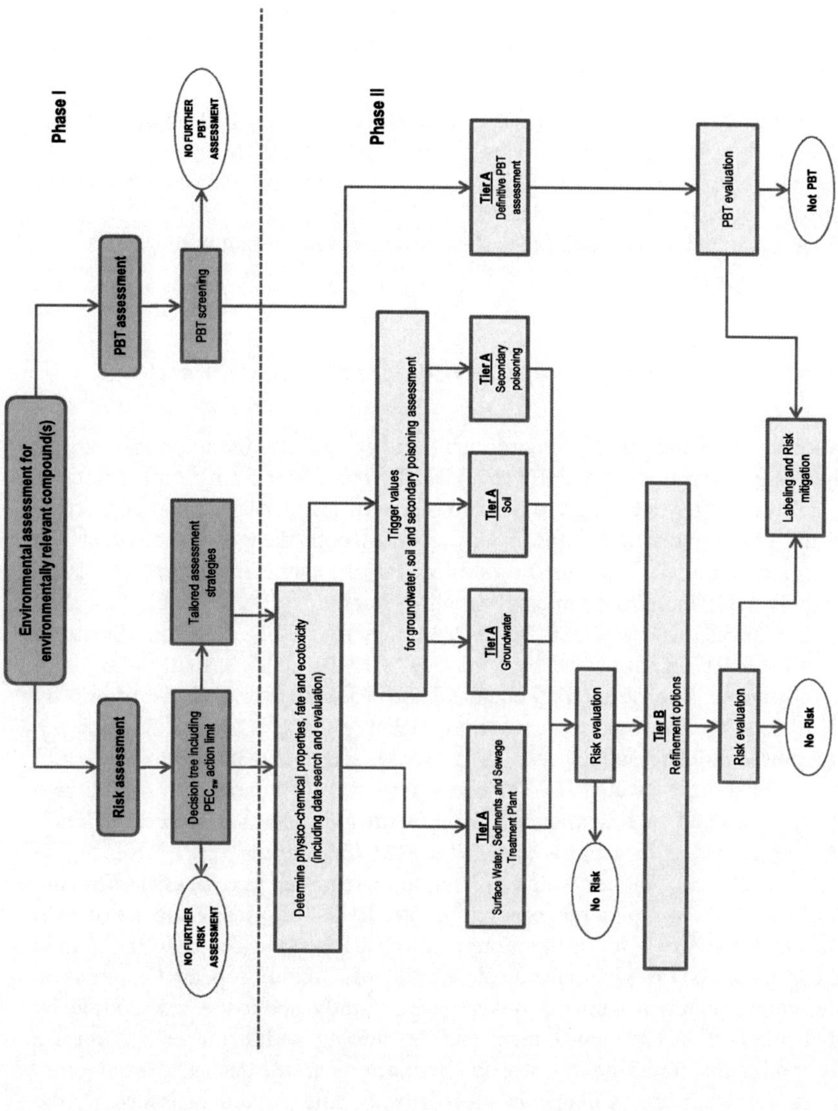

Fig. 1 Flow-chart of the environmental risk assessment process prescribed by the European Medicines Agency (EMA) to be used in the registration of new medicines (adapted from [5])

assessment, carried out (if necessary) in parallel to the risk assessment, aims at evaluating the potential long-term effects of the product in the environment, regardless of its environmental exposure concentration. Depending on the results obtained in Phase I (screening phase), a more detailed and definitive assessment is performed in Phase II.

Despite the implementation of the ERA procedures is a relevant advance to prevent the undesirable effects of pharmaceuticals in the environment, some limiting aspects are worth to be mentioned:

- Even though the inclusion of an ERA is mandatory in the registration of an PhAC, the final authorization or refusal does not depend on the ERA itself.
- ERAs are compulsory for new PhACs, but not for those authorized before the approval of the Directive 2001/83/EC.
- ERAs are conducted with pharmaceutical products rather than with PhACs.

2.2 *The Occurrence of Drugs in the Aquatic Environment*

The preservation of the aquatic environment, either surface (marine and fresh) or groundwater in Europe is essentially regulated by the Water Framework Directive (WFD) (Directive 2000/60/EC) and derived pieces of legislation. In that context, to achieve the good status of European water bodies both the good ecological and chemical status requirements must be fulfilled. For the latter, environmental levels must comply with the Environmental Quality Standards (EQS) of the so-called priority substances, whose EQS's were set up in the WFD daughter directives (Directive 2008/105/EC amended by Directive 2013/39/EU). Currently, the list of priority substances is constituted 45 chemical species. However, the "list of priority substances" is subjected to periodic revision, meaning that it is open to the incorporation of new candidate substances. To do so Directive 2013/39/EU (article 8b) foresees establishing a "watch list" of new substances of concern for which new monitoring data need to be gathered (sic) *"for the purpose of supporting future prioritization exercises in accordance with article 16(2) of the WFD."* Hence, the inclusion of a substance in the "watch list" might be regarded as a necessary (but not sufficient) condition to its final incorporation into the list of priority substances. On the other hand, the article 8c of the aforementioned Directive 2013/39/EU entitled "Specific provisions for pharmaceutical substances," the European Commission takes the commitment of carrying out a specific study about the risks posed by medicinal products in the environment and developing within the next 2 years a strategic approach to pollution of water by pharmaceutical substances. The outcomes of such an approach are available in the corresponding document issued by the European Commission [6].

Up to now the list of priority substances does not include any pharmaceutical substance. In contrast, several of them have been already added in the "watch list," in their three successive versions [7–9]. The compounds included in the latest are the antibiotics ciprofloxacin, amoxicillin, sulfamethoxazole, trimethoprim, and the psychiatric drugs venlafaxine and o-desmethylvenlafaxine (venlafaxine human metabolite).

3 Overview of the Environmental Risk Assessment process

Risk may be generally defined as the combination (i.e., product) of a probability of occurrence of a certain event by its hazard effects [10]:

$$\text{Risk} = \text{Occurrence} \times \text{Hazard effects} \tag{1}$$

Different existing risk assessment approaches have been developed in order to identify and rank compounds of environmental concern for both regulatory and monitoring purposes. Most of the existing schemes share the above risk assessment concept, however, they differ on how risk, occurrence and effects, are defined and hence quantified. Occurrence (exposure) data can be expressed in the form of measured environmental concentration (MEC) or predicted environmental concentrations (PEC) which can be indirectly estimated on the basis of, e.g. the data of annual production, sales rate, number of prescriptions, number of inhabitatants, etc. Hazard effects are usually based on ecotoxicity data of chemicals that can be either available from experimental measurements or predicted by the so-called in silico models (QSAR), such as ECOSAR.

Typically, for a single compound *i* risk is quantified using Eq. (2):

$$\text{Risk}\ (i) = \frac{C_i}{C_i(\text{ref})} \tag{2}$$

where c_i is the MEC or PEC of a compound in the water phase and $c_i(ref)$ is an ecotoxicity reference concentration for the same compound. The "reference concentration" can be an acute toxicity (EC50 or LC50), a chronic (long term) ecotoxicty (predicted no-effect concentration, PNEC) or even a legal value. In the former case Risk is expressed in "toxic units" (TU) [11], while the second is usually referred to as a "hazard or risk quotient" (HQ, RQ).

A review of different risk-based prioritization schemes for pharmaceuticals can be found at [12].

When dealing with multicomponent mixtures, three basic types of combined action models are typically taken into consideration [13, 14]: (a) concentration (or dose) addition (CA), for chemicals acting through a similar mode of action; (b) independent action (IA) for chemicals acting through a dissimilar mode of action; and (c) interactions between the chemicals present in the mixture, which includes any deviation from additivity either greater (synergistic) or lesser (antagonistic) than that predicted by CA or IA. It is worth noting that both CA and IA models rely on the assumption that the chemicals present in the mixture do not interact at the biological target site and hence do not influence each other's toxicity. On the other hand, at the environmental typically low concentrations and effect levels, the concentration-effect curves become linear and the two models converge. In turn, and for the same reasons, synergistic interactions are generally overlooked and their relevance is generally limited to mixtures with a small number of components at high concentrations [15, 16]. Altogether, the application of CA is generally an acceptable approach for mixture toxicity regardless of the components' similar or dissimilar mode of action. Hence, CA is usually the first tier approach used in several international recommendations and regulatory frameworks [13]. CA is usually expressed in Eq. (3):

$$\text{Risk} = \sum_{i=1}^{n} \frac{c_i}{c_i(\text{ref})} \tag{3}$$

where the index i extends over all the mixture components.

It is worth noting that both Eqs. (2) and (3) are referred to a specific ecotoxic assay or bioassay (organism, end-point, exposure time, etc.), and so it is the quantified risk. Therefore, it must be clearly specified.

4 Environmental Risk Assessment of Wastewater Effluents

There is a vast number of papers evaluating the risk of pharmaceuticals for the environment related to wastewater treatment and treated wastewater discharge. Treated wastewater effluents can be discharged in freshwater systems (lakes, rivers, and even groundwater through reclaimed water recharge) or directly in the ocean (in the case of coastal urban areas). In Table 1 some examples of different studies on the environmental risk of pharmaceuticals in different wastewater treatment scenarios are gathered. Due to water reuse activities, treated wastewater can also be applied for some activities, which can lead also to environmental and human impact, such as the use for irrigation of fields.

Most of the studies found in the literature are those focus on conventional activated sludge (CAS) which is the usual treatment in conventional WWTPs, and therefore most of publications are monitoring studies in full-scale WWTPs in different countries all over the world such as, South Africa, Kenya, Tunisia, Turkey, China, Iran, Brazil, Greece, Spain, Portugal, France, Sweden, Faroe Islands, Iceland, and Greenland as it is reported in Table 1 [19, 20, 22–35]. In these studies, risk or hazard quotients were calculated for the target pharmaceuticals found in the effluents of the selected WWTPs, whereas in some others the hazard posed in the receiving environment was directly calculated.

In addition, different types of alternative advanced technologies for treating emerging contaminants have been tested in the last years including physical-

Table 1 Selected environmental risk assessment studies related to the presence of PhACs in wastewater treatment processes

Ref.	Number of target compounds	Wastewater treatment	Receiving water body	Country	Highlighted compounds in WWTP effluents (high risk)
Guedes-Alonso et al. [17]	15 (steroid hormones)	Macrophyte pond-CW system	N.I.	Spain	No high risk
Lucas et al. [18]	81 PhACs	Fungal treatment	N.I.	Spain	N.I.
Afonso-Olivares et al. [19]	23 PhACs	CAS + reverse osmosis	N.I.	Spain	Ibuprofen, gemfibrozil, and ofloxacin
Díaz-Garduño et al. [20]	78 PhACs	CAS + photobioreactor	N.I.	Spain	No high risk
		CAS + multi-barrier treatment	N.I.		No high risk
Auvinen et al. [21]	12 PhACs	Constructed wetlands	River	France	No high risk
Ben et al. [22]	26 PhACs	14 WWTP (some tertiary treatment)	N.I.	China	Sulfamethoxazole clarithromycin erythromycin, ofloxacin, and ciprofloxacin
Kosma et al. [23]	16 PhACs	8 WWTP	N.I.	Greece	Diclofenac, trimethoprim, and sulfamethoxazole
Archer et al. [24]	31 PhACs	1 WWTP	River	South Africa	Carbamazepine, naproxen, diclofenac, and ibuprofen
Gros et al. [25]	73 PhACs	7 WWTP	River	Spain	No high risk
Kairigo et al. [26]	7 antibiotics	4 WWTP	River	Kenya	Amoxicillin, ciprofloxacin, trimethoprim, norfloxacin, and sulfamethoxazole
Mirzaei et al. [27]	7 antibiotics	2 WWTP	River	Iran	Cefixime, and azithromycin and erythromycin amoxicillin, penicillin, and ciprofloxacin
Pereira et al. [28]	10 psychiatric drugs	5 WWTP	River	Brazil	No high risk
Huber et al. [29]	36 PhACs	WWTP and untreated WW	Rivers, lakes, or the sea	Faroe Islands, Iceland, and Greenland	No high risk in water recipients

(continued)

Table 1 (continued)

Ref.	Number of target compounds	Wastewater treatment	Receiving water body	Country	Highlighted compounds in WWTP effluents (high risk)
Aydin et al. [30]	10 antibiotics	1 WWTP	Lake	Turkey	Azithromycin and clarithromycin
Afsa et al. [31]	40 PhACs	1 WWTP	Coastal seawater	Tunisia	No high risk in water recipients
Biel-Maeso et al. [32]	79 PhACs	1 WWTP	Coastal and oceanic water	Spain	Gemfibrozil and ofloxacine
Zhang et al. [33]	31 antibiotics	12 WWTP	NI	China	Erythromycin and clarithromycin
Pereira et al. [28]	11 PhACs	15 WWTP	NI	Portugal	Ciprofloxacin, bezafibrate, gemfibrozil, simvastatin, and diclofenac
Sörengård et al. [34]	78 PhACs	1 WWTP	River and Lake	Sweden	No high risk

N.I. Not Indicated

chemical and biological treatments, in order to find a proper alternative to conventional activated sludge (CAS) or even to be added as a tertiary treatment after CAS treatment. The use of this kind of technologies could be especially important in the treatment of wastewaters with a high PhACs concentration, such as hospital and to a lesser degree, veterinary hospital wastewaters [36]. Among all the publications reporting the presence, fate and removal of PhACs in these alternative wastewater systems, some have also considered the assessment of environmental risks. Lucas et al. [18] studied the impact of wastewater treatment based on fungi on the removal and environmental impact of a broad set of PhACs in different types of wastewater, whereas some other authors studied other biological treatments such as wetlands Guedes-Alonso et al. [17], who studied the occurrence of 15 steroid hormones along a waste-stabilization pond combined with a constructed wetlands (CWs). Final effluents showed a low ecological risk associated with steroid hormones in contrast to the medium-high ecological risks found in the influent samples.

5 Environmental Risk Assessment as a Tool to Evaluate the Removal Efficiency of Wastewater Treatment

Removal efficiency of wastewater treatment is usually assessed comparing the concentration of contaminants before and after the treatment under evaluation, i.e. as the percentage of reduction between the concentration in the initial untreated wastewater influent and the concentration in the treated wastewater effluent. However, as discussed in the previous section, actual risks of contaminants depend not only on their concentration but also on their hazardous effects. Therefore, some authors have proposed to evaluate the performance of water treatment in terms of reduction of total HQ. In this section some examples are discussed:

In the study by Verlicchi et al. [37], the environmental risk assessment was performed on the effluent from the hospital and the influent and effluent from the WWTP treating the effluents of the hospital among other urban wastewater contribution. The quotient between the maximum MEC and the PNEC (referring to acute toxicity) of each of the compounds detected was used as a marker of risk. 9 pharmaceuticals were contributing the most to the potential risk for the environment in all types of water although only 5 of them (the four antibiotics clarithromycin, erythromycin, ofloxacin, and sulfamethoxazole and the psychiatric drug fluoxetine) were posing a high risk for the environment both in the influent wastewater and the effluent of the municipal WWTP. The study points out antibiotic family as one of the most critical therapeutic classes, being highly recalcitrant to degradation and removal.

The elimination of broad set of pharmaceuticals (more than 80 compounds) by means of a biological treatment based on the fungus *Trametes versicolor* was evaluated in the study by Lucas et al. [18]. PhACs removal studied in different types of wastewaters (urban, reverse osmosis concentrate, hospital, and veterinary hospital wastewaters) were reviewed and compared with conventional activated sludge (CAS) treatment. Hazard indexes were calculated based on the ecotoxicity value for each compound and used for the evaluation of environmental risk. PhACs elimination achieved with the fungal treatment (mean value 76%) was similar to that achieved in the CAS treatment (85%). However, the fungal reactor was superior in removing, among all pharmaceutical compounds analyzed, those compounds that pose higher environmental risk (antibiotics and psychiatric drugs): concentration removal achieved with fungal treatment was 56% for antibiotics and 59% for psychiatric drugs whereas only 36% and 39% for antibiotics and psychiatric drugs, respectively, was achieved by CAS. This led to a reduction of environmental risk up to 93% in fungal treatment whereas only 53% of hazard reduction was achieved by CAS. Fungal treatment can thus be considered as a good alternative to conventional treatment technologies for the elimination of PhACs from wastewaters.

The monitoring of the health risk associated with the presence of pharmaceuticals along wastewater treatment was also performed by Ma et al. [38] in the frame of water reuse activities to assure reclaimed water safety. In their study, domestic wastewater was treated using an anaerobic-anoxic-oxic unit followed by a membrane bioreactor (A2O-MBR), and the reclaimed water was used for replenishing a landscape lake. A total of 58 organic micropollutants were detected in the system including 13 polycyclic aromatic hydrocarbons (PAHs), 16 phenols, 3 pesticides, and 26 pharmaceuticals and personal care products (PPCPs). In this case the authors used another approach using the models proposed by the US EPA [39] and calculated the health risks associated with the target micropollutants in the water reclamation and ecological reuse system, including carcinogenic risks (CRs) and noncarcinogenic risks (non-CRs) assuming that people are exposed to micropollutants in reclaimed water mainly by ingestion. Although carcinogenic and noncarcinogenic risks associated with all the detected micropollutants were at negligible levels, the HQ of PPCPs accounted for more than 90% of the total HQ. Authors conclude that this approach is useful for an effective evaluation of health risk reduction into the water reclamation and reuse systems and therefore to be used to improve strategies for guaranteeing the safety of reclaimed water reuse related to micropollutants.

Guedes-Alonso et al. [17] studied the occurrence of 15 steroid hormones along a waste-stabilization pond combined with constructed wetlands (CWs), which are natural wastewater treatment systems used in small communities because of their low cost and easy maintenance. The pond-CW system showed high elimination rates of the 8 steroid hormone residues detected with average removal efficiencies higher than 77%. This efficacy was confirmed in the ecological risk assessment performed. Final effluents showed a low ecological risk associated with steroid hormones in contrast to the medium-high ecological risks found in the influent samples.

The latest application of the hazard quotient (HQ) approach to evaluate wastewater treatment is the one carried out by Ramírez-Morales et al. [40], who monitored the occurrence of 70 pharmaceutical active compounds (PhACs) in 11 WWTPs across Costa Rica. Among the 33 PhACs found, 20 compounds in influents and 15 in the effluents showed high HQ. The top HQ values were obtained for risperidone, lovastatin, diphenhydramine and fluoxetine, caffeine and trimethoprim, which should be considered in prioritization lists of pharmaceuticals in the context of environmental hazard/risk. Likewise, almost all wastewater analyzed (both influents and effluents) seemed to pose a high hazard towards aquatic organisms.

6 Discharge of Wastewater and Sludge in the Receiving Environment; Risk Assessment

In their study Mirzaei et al. [27] pointed out six antibiotics cefixime, and azithromycin and erythromycin amoxicillin, penicillin and ciprofloxacin, as high risk to algae or bacteria in 2 WWTP effluents whereas only the 3 latest were posing a high risk in 2 rivers in Teheran (Iran). Kairigo et al. [26] also investigated 6 antibiotics for their occurrence and environmental risk assessment in 4 WWTPs and the receiving water bodies (river waters) in Kenya. The risk quotient for bacterial resistance selection in effluent and surface water ranged between <0.1 and 53, indicating a medium to high risk of antibiotic resistance developing within the study areas. The antibiotics posing medium to high risk in this respect were amoxicillin, ciprofloxacin, trimethoprim, norfloxacin, and sulfamethoxazole. In the work by Pivetta et al. [35], 10 psychotropic drugs widely consumed in Brazil were monitored at 5 WWTPs in the influents, at different stages of the treatments and in the surface waters from the Atibaia River and the Anhumas Creek, though he levels of the psychotropic compounds detected did not appear to present risks to the aquatic biota.

In the study by Afsa et al. [31], the occurrence of 40 pharmaceuticals belonging to several therapeutic groups was investigated in hospital effluent, wastewater treatment plant influent and effluent, and seawater in Mahdia, Tunisia. Measured environmental concentrations (MECs) detected in seawater samples seemed not pose a toxic effect to the aquatic organisms (except for salicylic acid, sulfamethoxazole, and fluoxetine). In another study in Spain Biel-Maeso et al. [32] also evaluated the impact of WWTP effluent discharges in sea water of the Gulf of Cadiz. Risk quotients were calculated for the 78 PhACs targeted and these were higher than 1 only for gemfibrozil and ofloxacin in one WWTP, whereas no high environmental risk was detected in both coastal and oceanic sampling areas.

To this respect it is interesting to highlight some studies such as the one by Rodriguez-Mozaz et al. [41] who were applying a specific dilution factor (the ratio between the volume of receiving water available and the domestic sewage discharge) to the PhACs concentration in the wastewater effluent of the corresponding country. This dilution factor or "National annual median dilution factor" was calculated for domestic effluents for approximately 100 countries by Keller et al. [42].

Sludge originated during wastewater treatment may contain a wide spectrum of organic and inorganic substances as well as microorganisms and viruses which are separated from the liquid phase during treatments. The main disposal routes are incineration, landfill, land application, and composting, with the specific percentages varying from country to country, although 53% of sludge in EU is estimated to be reused in agriculture either directly or after composting [43, 44]. Disposal routes of sewage sludge must fulfill specific regulations although no limits have been set for pharmaceuticals and other contaminants of emerging interest. However, there is ongoing debate within the scientific community in order to evaluate potential (environmental) risks in this kind of practice, due to the occurrence of toxic and persistent substances in sludge, such as aquifer contamination, the accumulation of pollutants in soil, and their transfer into the food chain [44]. Due to the lack of data regarding chronic and acute toxicity for terrestrial organisms with regard to PhACs, some authors [44, 45] evaluate the corresponding PNEC for sludge and soil on the basis of the PNEC for water and partition coefficient Kd of the selected compound according to Eq. (4).

$$\mathrm{PNEC_{sludge}} = \mathrm{PNEC_{water}} \times \mathrm{Kd} \times 1,000 \tag{4}$$

This approach was adopted by Verlicchi and Zambello [44], to evaluate the environmental risk posed by the spreading of treated sludge in agriculture. The authors first reviewed concentrations for 152 pharmaceuticals and 17 personal care products in sludge reported in 59 papers and further calculated the corresponding RQ. The most critical compounds found in the sludge-amended soil were the 3 antibiotics ciprofloxacin, ofloxacin, tetracycline, and other related compounds such as estradiol, caffeine, triclosan, and triclocarban.

7 Environmental and Human Health Risks of Reclaimed Water Irrigation

Uses of reclaimed water include the irrigation of landscapes, forests, and agricultural fields; the use by industries as an alternative to freshwater and the supplementation or augmentation of stream flows and groundwater aquifers [46]. Some reuse applications might imply a great potential for human exposure and some studies have assessed the corresponding risks for human health. In this section some examples of such studies are presented related to irrigation with reclaimed water. Reclaimed wastewater (RG) is increasingly used to irrigate agricultural land and to alleviate agricultural water shortages worldwide. This use has led to also growing concerns about soil contamination by pharmaceuticals and other contaminants and the human health risks associated with dietary crop intake [47]. As in the case of environmental risk assessment, specific hazard quotients are calculated to evaluate the risks of accumulated pharmaceutical residues to humans through the consumption of edible part of the plants irrigated by reclaimed water. An increasing number of studies are being published in the last few years to evaluate those risks and it is a topic of growing concern. Liu et al. [47] evaluated the occurrence and accumulation of 11 PPCPs and one active metabolite in soils and various crops (cucumber, eggplant, long bean, and wheat) from RW irrigation fields with different irrigation histories in Beijing and evaluated the human health risks associated with the consumption of these crops. The risk associated with the consumption of treated RW-irrigated crops was assessed using the threshold of toxicological concern (TTC) approach based on the Cramer classification tree [48, 49] and the hazard quotient (HQ) based on the description by Prosser and Sibley [50]. Cramer structural decision tree was constructed using Toxtree software (Toxtree V2.6.13) to classify substances as having genotoxic potential or into one of the three classes (I, II, III). The HQ was determined by dividing the estimated daily intake (EDI) by the acceptable daily intake (ADI), which is the amount that can be consumed daily over a person's lifespan without adverse effects. Humans are considered to be exposed to a potential hazard if the HQ value is >0.1 rather than >1, which is used when we recognize that humans may be exposed through other pathways. The estimated TTC and HQ values showed that the consumption of crops irrigated long term with RW presents a minimal risk to human health.

In another field study in Israel, Malchi et al. [51] evaluated the uptake of 15 PhACs by root crops (carrots and sweet potatoes) treated by wastewater and evaluated the potential risks for human health using the using the TTC approach. The TTC is a useful tool to estimate the safety of exposure to chemicals found at low concentration in foods and drinking water. Consumption of a PhACs above the corresponding TTC value indicates a possible risk of exposure. Authors show the extent of PhACs uptake is influenced by the physicochemical properties of the compound, the crop, and soil. The daily consumption of root crops irrigated with treated wastewater did not pose a health threat except for lamotrigine and 10,11-epoxycarbamazepine with concentrations above TTC value in edible organs.

Finally, Semerjian et al. characterized human health risks posed by 10 pharmaceuticals quantified in WWTP effluents used for landscape irrigation in United Arab Emirates, where wastewater reuse applications play an integral role in meeting water demands [46]. The human health risk was assessed and in none of the cases pharmaceutical exposure posed a high risk (RQ < 1) neither through dermal nor through ingestion exposure pathways. RQs were highest for the landscape worker followed by children playing in green areas and then adult using the athletic fields and golf courses irrigated by treated wastewater.

8 Future Perspectives in Environmental Risk Assessment of PhACs

8.1 Emphasis on Environmental Mixtures

Up to now, the risk assessment of chemicals mainly relies on the assessment of individual chemicals (Fig. 2). Methodologies and guidance for assessing risks from combined exposure to multiple chemicals have been developed for different

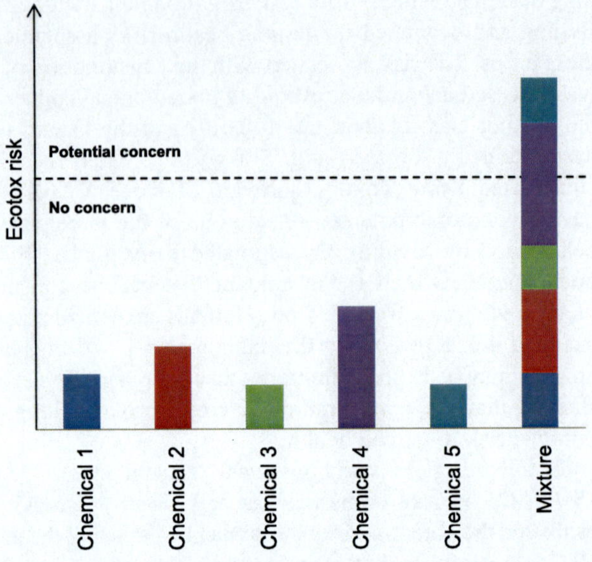

Fig. 2 Single compound vs. mixture risk, assuming concentration addition (CA) joint action. Even though every single component in the mixture is below the safety risk threshold, it may be exceeded by the whole mixture (Figure adapted from [52])

regulatory sectors, however, a harmonized, consistent approach for performing mixture risk assessments and management across different regulatory sectors is still lacking. Recent EU funded projects, such as SOLUTIONS (Grant no. 603437-2) proposed and developed an advanced methodological framework for identifying both priority pollutants and priority mixtures of chemicals posing the risk in European freshwaters [53]. The proposed framework aims at integrating different lines of evidence (LOE) regarding mixture risks. This encompasses evidence from different sources, namely, (1) ecological monitoring (field observations on so-called BQEs), (2) effect-based monitoring (in vitro or in vivo testing in the lab or on-site), (3) chemical monitoring in combination with component-based mixture risk assessment approaches, and (4) integrated modeling of co-exposure and resulting mixture risks. Where one or more lines of evidence identify groups of chemicals presenting a significant risk, these should be subject to prioritization for risk reduction measures. The above concerns strongly advocate for new legislation focusing on environmental mixtures rather than on single compounds [52].

8.2 *Bioanalytical Tools to Capture Mixture Effects*

Conventional ecotoxicological approaches make use of whole organism in in vivo bioassays to assess the toxicity of wastewater effluents. However, such bioassays exhibited some important drawbacks [14], as they are: (1) a limited sample throughput processing capability; (2) the inability to distinguish the effects of pollutants from those of the matrix (i.e., salinity, pH, etc.), and (3) they make unnecessary use of living organisms. This limitation was overcome with the development of cell-based in vitro bioassays with high-throughput processing capacity [54]. The application of such in vitro bioassays in ecological risk assessment is just at its beginnings but appears a promising tool for routine environmental surveys and reduces animal testing [55].

8.3 *Toxicogenomics*

The interrelation between chemical exposure (chemosphere) with the surrounding living organisms (ecosphere) and their adverse biological effects at different levels of biological complexity (including humans) compose altogether a chain of events known in the current environmental toxicology literature as Adverse Outcome Pathways (AOP) [56–58]. The investigation of the effects of the totality of exposure on all levels of biological organization and complexity, from the ecosystems to human health (i.e. the processes involved across the AOP chain) appears crucial.

Exposure and adverse effects can be regarded as the two extremes of the AOP causal chain of events [56, 58]. Whole organism assays provide useful information on adverse effects, but they cannot allow establishing a causal link between exposure and effects. Effect-driven cell assays can link exposure and effects but for a limited number of toxicological mechanisms and they also suffer from false positive or negative results. Alternatively, toxicogenomics – the combination of omic/system biology technologies with toxicity assays – offers the possibility to assess expected and unknown molecular toxicity mechanisms of individual and more complex chemical mixtures in exposed whole organisms or cells [59]. Furthermore, by combining toxicogenomics with AOPs it is feasible to progress on establishing causal links between exposure and adverse effects [60].

8.4 *Antibiotic Resistance*

Antibiotics were identified as the main contributors to the high environmental risk of wastewater effluents and ciprofloxacin, sulfamethoxazole, clarithromycin, and erythromycin are pointed out in several of the studies reviewed in this chapter as those posing a high risk for the environment (Table 1). In addition, antibiotics is the therapeutic group that provokes the greatest concern among pharmaceuticals due to their potential contribution to the development and spread of antibiotic resistance. Overuse and misuse of antibiotics has led to the emergence of antibiotic resistance bacteria, compromising the effectiveness of antimicrobial therapy since the infectious organisms are becoming resistant to commonly prescribed antibiotics. Emergence and spread of antibiotic resistance bacteria have indeed been classified by the World Health Organization [61] as one of the biggest threats to public health in the twenty-first century. In this line, the capacity of antibiotics to promote antimicrobial resistance spread has been considered in some ERA studies of antibiotics by using specific PNEC related to antibiotic resistance selection. [62, 63]. This approach was recently applied by Kairigo et al. [26] and Rodriguez-Mozaz et al. [41] to evaluate the risk posed by antibiotics present in wastewater effluents in the aquatic environment.

References

1. Kümmerer K, Clark JH, Zuin VG (2020) Rethinking chemistry for a circular economy. Science 367(6476):369–370. https://doi.org/10.1126/science.aba4979
2. Vermeulen R, Schymanski EL, Barabási A-L, Miller GW (2020) The exposome and health: where chemistry meets biology. Science 367:392–396
3. Lee D, Choi K (2019) Comparison of regulatory frameworks of environmental risk assessments for human pharmaceuticals in EU, USA, and Canada. Sci Total Environ 671:1026–1035
4. Walter S, Mitkidis K (2018) The risk assessment of Pharmaceuticals in the Environment: EU and US regulatory approach. Eur J Risk Regul 9:527–547
5. EMA (2018) Guideline on the environmental risk assessment of medicinal products for human use (draft). https://www.ema.europa.eu/en/documents/scientific-guideline/draft-guideline-environmental-risk-assessment-medicinal-products-human-use-revision-1_en.pdf. Accessed 20 Mar 2019
6. European Commission (2019) European Union strategic approach to pharmaceuticals in the environment. Brussels, 11.3.2019 COM(2019) 128 final
7. Commission Implementing Decision (EU) 2015/495 of 20 March 2015 establishing a watch list of substances for Union-wide monitoring in the field of water policy pursuant to Directive 2008/105/EC of the European Parliament and of the Council (notified under document C(2015) 1756) https://eur-lex.europa.eu/legal-content/EN/TXT/?uri=uriserv%3AOJ.L_.2015.078.01.0040.01.ENG
8. Commission Implementing Decision (EU) 2018/840 of 5 June 2018 establishing a watch list of substances for Union-wide monitoring in the field of water policy pursuant to Directive 2008/105/EC of the European Parliament and of the Council and repealing Commission Implementing Decision (EU) 2015/495 (notified under document C(2018) 3362) https://eur-lex.europa.eu/eli/dec_impl/2018/840/oj
9. Commission Implementing Decision (EU) 2020/1161 of 4 August 2020 Establishing a watch list of substances for union-wide monitoring in the field of water policy pursuant to directive 2008/105/EC of the European Parliament and of the council. Off J Eur Union L 257 (2020):32–35. https://eur-lex.europa.eu/legal-content/EN/TXT/?uri=uriserv:OJ.L_.2020.257.01.0032.01.ENG&toc=OJ:L:2020:257:TOC
10. Guillén D, Ginebreda A, Farré M, Darbra RM, Petrovic M, Barceló D (2012) Prioritization of chemicals in the aquatic environment analytical, modeling and regulatory perspective. Sci Total Environ 440:236–252
11. Sprague JB (1970) Measurement of pollutant toxicity to fish, II-utilizing and applying bioassay results. Water Res 4:3–32
12. Kuzmanovic M, Banjac Z, Ginebreda A, Petrovic M, Barcelo D (2013) Prioritization: selection of environmentally occurring pharmaceuticals to be monitored. Analysis, removal, effects and risk of pharmaceuticals in the water cycle occurrence and transformation in the environment. In: Petrovic M, Barcelo D, Pérez S (eds) Comprehensive analytical chemistry, vol 62, pp 71–90
13. Bopp SK, Barouki R, Brack W, Dalla Costa S, Dorne J-LCM, Drakvik PE, Faust M, Karjalainen TK, Kephalopoulos S, van Klaveren J, Kolossa-Gehring M, Kortenkamp A, Lettieri T, Nørager S, Rüegg J, Tarazona JV, van de Water B, van Gils J, Bergman A,

Lebret E, Trier X (2018) Current EU research activities on combined exposure to multiple chemicals. Environ Int 120:544–562
14. Escher BI, Stapleton HM, Schymanski EL (2020) Tracking complex mixtures of chemicals in our changing environment. Science 367:388–392
15. Boobis A, Budinsky R, Collie S, Crofton K, Embry M, Felter S, Hertzberg R, Kopp D, Mihlan G, Mumtaz M, Price P, Solomon K, Teuschler L, Yang R, Zaleski R (2011) Critical analysis of literature on low-dose synergy for use in screening chemical mixtures for risk assessment. Crit Rev Toxicol 41:369–383
16. Cedergreen N (2014) Quantifying synergy: a systematic review of mixture toxicity studies within environmental toxicology. PLoS One 9(5):e96580
17. Guedes-Alonso R, Montesdeoca-Esponda S, Herrera-Melián JA, Rodríguez-Rodríguez R, Ojeda-González Z, Landívar-Andrade V et al (2020) Pharmaceutical and personal care product residues in a macrophyte pond-constructed wetland treating wastewater from a university campus: presence, removal and ecological risk assessment. Sci Total Environ 703:135596
18. Lucas D, Barceló D, Rodriguez-Mozaz S (2016) Removal of pharmaceuticals from wastewater by fungal treatment and reduction of hazard quotients. Sci Total Environ 571:909–915
19. Afonso-Olivares C, Sosa-Ferrera Z, Santana-Rodríguez JJ (2017) Occurrence and environmental impact of pharmaceutical residues from conventional and natural wastewater treatment plants in gran Canaria (Spain). Sci Total Environ 599-600:934–943
20. Díaz-Garduño B, Pintado-Herrera MG, Biel-Maeso M, Rueda-Márquez JJ, Lara-Martín PA, Perales JA et al (2017) Environmental risk assessment of effluents as a whole emerging contaminant: efficiency of alternative tertiary treatments for wastewater depuration. Water Res 119:136–149
21. Auvinen H, Gebhardt W, Linnemann V, Du Laing G, Rousseau DPL (2017) Laboratory- and full-scale studies on the removal of pharmaceuticals in an aerated constructed wetland: effects of aeration and hydraulic retention time on the removal efficiency and assessment of the aquatic risk. Water Sci Technol 76(6):1457–1465
22. Ben W, Zhu B, Yuan X, Zhang Y, Yang M, Qiang Z (2018) Occurrence, removal and risk of organic micropollutants in wastewater treatment plants across China: comparison of wastewater treatment processes. Water Res 130:38–46
23. Kosma CI, Lambropoulou DA, Albanis TA (2014) Investigation of PPCPs in wastewater treatment plants in Greece: occurrence, removal and environmental risk assessment. Sci Total Environ 466-467:421–438
24. Archer E, Petrie B, Kasprzyk-Hordern B, Wolfaardt GM (2017) The fate of pharmaceuticals and personal care products (PPCPs), endocrine disrupting contaminants (EDCs), metabolites and illicit drugs in a WWTW and environmental waters. Chemosphere 174:437–446
25. Gros M, Petrović M, Ginebreda A, Barceló D (2010) Removal of pharmaceuticals during wastewater treatment and environmental risk assessment using hazard indexes. Environ Int 36(1):15–26
26. Kairigo P, Ngumba E, Sundberg LR, Gachanja A, Tuhkanen T (2020) Occurrence of antibiotics and risk of antibiotic resistance evolution in selected Kenyan wastewaters, surface waters and sediments. Sci Tot Environ 720
27. Mirzaei R, Mesdaghinia A, Hoseini SS, Yunesian M (2019) Antibiotics in urban wastewater and rivers of Tehran, Iran: consumption, mass load, occurrence, and ecological risk. Chemosphere 221:55–66
28. Pereira AMPT, Silva LJG, Meisel LM, Lino CM, Pena A (2015) Environmental impact of pharmaceuticals from Portuguese wastewaters: geographical and seasonal occurrence, removal and risk assessment. Environ Res 136:108–119
29. Huber S, Remberger M, Kaj L, Schlabach M, Jörundsdóttir HT, Vester J et al (2016) A first screening and risk assessment of pharmaceuticals and additives in personal care products in waste water, sludge, recipient water and sediment from Faroe Islands, Iceland and Greenland. Sci Total Environ 562:13–25

30. Aydin S, Aydin ME, Ulvi A, Kilic H (2019) Antibiotics in HOSPITAL effluents: occurrence, contribution to urban wastewater, removal in a wastewater treatment plant, and environmental risk assessment. Environ Sci Pollut Res 26(1):544–558
31. Afsa S, Hamden K, Lara Martin PA, Mansour HB (2020) Occurrence of 40 pharmaceutically active compounds in HOSPITAL and urban wastewaters and their contribution to Mahdia coastal seawater contamination. Environ Sci Pollut Res 27(2):1941–1955
32. Biel-Maeso M, Baena-Nogueras RM, Corada-Fernández C, Lara-Martín PA (2018) Occurrence, distribution and environmental risk of pharmaceutically active compounds (PhACs) in coastal and ocean waters from the Gulf of Cadiz (SW Spain). Sci Total Environ 612:649–659
33. Zhang X, Zhao H, Du J, Qu Y, Shen C, Tan F et al (2017) Occurrence, removal, and risk assessment of antibiotics in 12 wastewater treatment plants from Dalian, China. Environ Sci Pollut Res 24(19):16478–16487
34. Sörengård M, Campos-Pereira H, Ullberg M, Lai FY, Golovko O, Ahrens L (2019) Mass loads, source apportionment, and risk estimation of organic micropollutants from HOSPITAL and municipal wastewater in recipient catchments. Chemosphere 234:931–941
35. Pivetta RC, Rodrigues-Silva C, Ribeiro AR, Rath S (2020) Tracking the occurrence of psychotropic pharmaceuticals in Brazilian wastewater treatment plants and surface water, with assessment of environmental risks. Sci Total Environ 727:138661
36. Badia-Fabregat M, Lucas D, Gros M, Rodríguez-Mozaz S, Barceló D, Caminal G, Vicent T (2015) Identification of some factors affecting pharmaceutical active compounds (PhACs) removal in real wastewater. Case study of fungal treatment of reverse osmosis concentrate. J Hazard Mater 283:663–671
37. Verlicchi P, Al Aukidy M, Galletti A, Petrovic M, Barceló D (2012) HOSPITAL effluent: investigation of the concentrations and distribution of pharmaceuticals and environmental risk assessment. Sci Total Environ 430:109–118
38. Ma XY, Li Q, Wang XC, Wang Y, Wang D, Ngo HH (2018) Micropollutants removal and health risk reduction in a water reclamation and ecological reuse system. Water Res 138:272–281
39. U.S. EPA (1989) Risk assessment guidance for superfund volume I human health evaluation manual (part A)
40. Ramírez-Morales D, Masís-Mora M, Montiel-Mora JR, Cambronero-Heinrichs JC, Briceño-Guevara S, Rojas-Sánchez CE, Méndez-Rivera M, Arias-Mora V, Tormo-Budowski R, Brenes-Alfaro L, Rodríguez-Rodríguez CE (2020) Occurrence of pharmaceuticals, hazard assessment and ecotoxicological evaluation of wastewater treatment plants in Costa Rica. Sci Total Environ 746:141200
41. Rodriguez-Mozaz S et al (2020) Antibiotic residues in final effluents of European wastewater treatment plants and their impact on the aquatic environment. Environ Int 140:105733
42. Keller VDJ, Williams RJ, Lofthouse C, Johnson AC (2014) Worldwide estimation of river concentrations of any chemical originating from sewage treatment plants using dilution factors. Environ Toxicol Chem 33(2):447–452
43. Kelessidis A, Stasinakis AS (2012) Comparative study of the methods used for treatment and final disposal of sewage sludge in European countries. Waste Manag 32(6):1186–1195
44. Verlicchi P, Zambello E (2015) Pharmaceuticals and personal care products in untreated and treated sewage sludge: occurrence and environmental risk in the case of application on soil - a critical review. Sci Total Environ 538:750–767
45. Martín J, Camacho-Muñoz MD, Santos JL, Aparicio I, Alonso E (2012) Distribution and temporal evolution of pharmaceutically active compounds alongside sewage sludge treatment. Risk assessment of sludge application onto soils. J Environ Manag 102:18–25
46. Semerjian L, Shanableh A, Semreen MH, Samarai M (2018) Human health risk assessment of pharmaceuticals in treated wastewater reused for non-potable applications in Sharjah, United Arab Emirates. Environ Int 121:325–331

47. Liu X, Liang C, Liu X, Zhao F, Han C (2020) Occurrence and human health risk assessment of pharmaceuticals and personal care products in real agricultural systems with long-term reclaimed wastewater irrigation in Beijing, China. Ecotoxicol Environ Saf 190:110022
48. Cramer GM, Ford RA, Hall RL (1978) Estimation of toxic hazard – a decision tree approach. Food Cosmet Toxicol 16(3):255–276
49. Tomer M, Yehoshua M, Galit T, Moshe S, Benny C (2014) Irrigation of root vegetables with treated wastewater: evaluating uptake of pharmaceuticals and the associated human health risks. Environ Sci Technol 48(16):9325–9333
50. Prosser RS, Sibley PK (2015) Human health risk assessment of pharmaceuticals and personal care products in plant tissue due to biosolids and manure amendments, and wastewater irrigation. Environ Int 75:223–233
51. Malchi T, Maor Y, Tadmor G, Shenker M, Chefetz B (2014) Irrigation of root vegetables with treated wastewater: evaluating uptake of pharmaceuticals and the associated human health risks. Environ Sci Technol 48(16):9325–9333
52. Kortenkamp A, Faust M (2018) Regulate to reduce chemical mixture risk. Science 361 (6399):224–226
53. Faust M, Backhaus T, Altenburger R, Dulio V, van Gils J, Ginebreda A, Kortenkamp A, Munthe J, Posthuma L, Slobodnik J, Tollefsen KE, van Wezel A, Brack W (2019) Prioritisation of water pollutants: the EU Project SOLUTIONS proposes a methodological framework for the integration of mixture risk assessments into prioritisation procedures under the European Water Framework Directive. Environ Sci Eur 31:66
54. Attene-Ramos MS, Miller N, Huang R, Michael S, Robert M, Kavlock J, Austin CP, Shinn P, Simeonov A, Tice RR, Xia M (2013) The Tox21 robotic platform for the assessment of environmental chemicals – from vision to reality. Drug Discov Today 18:716–723
55. Villeneuve DL, Coady K, Escher BI, Mihaich E, Murphy CA, Schlekat T, Garcia-Reyero N (2019) High-throughput screening and environmental risk assessment: state of the science and emerging applications. Environ Toxicol Chem 38:12–26
56. Edwards SW, Tan Y-M, Villeneuve DL, Meek M, McQueen CA (2016) Adverse outcome pathways – organizing toxicological information to improve decision making. J Pharmacol Exp Ther 356(1):170–181
57. Escher BI, Hackermüller J, Polte T, Scholz S, Aigner A, Altenburger R, Böhme A, Bopp SK, Brack W, Busch W (2017) From the exposome to mechanistic understanding of chemical-induced adverse effects. Environ Int 99:97–106
58. Groh KJ, Carvalho RN, Chipman JK, Denslow ND, Halder M, Murphy CA, Roelofs D, Rolaki A, Schirmer K, Watanabe KH (2015) Development and application of the adverse outcome pathway framework for understanding and predicting chronic toxicity: I. Challenges and research needs in ecotoxicology. Chemosphere 120:764–777
59. Waters MD, Fostel JM (2004) Toxicogenomics and systems toxicology: aims and prospects. Nat Rev Genet 5(12):936
60. Ortiz-Villanueva E, Navarro-Martín L, Jaumot J, Benavente F, Sanz-Nebot V, Piña B, Tauler R (2017) Metabolic disruption of zebrafish (Danio rerio) embryos by bisphenol A. An integrated metabolomic and transcriptomic approach. Environ Pollut 231:22–36
61. WHO (2016) Antimicrobial resistance: global report on surveillance 2014. WHO
62. Bengtsson-Palme J, Larsson DGJ (2016) Concentrations of antibiotics predicted to select for resistant bacteria: proposed limits for environmental regulation. Environ Int 86:140–149
63. Tell J, Caldwell DJ, Häner A, Hellstern J, Hoeger B, Journel R et al (2019) Science-based targets for antibiotics in receiving waters from pharmaceutical manufacturing operations. Integr Environ Assess Manag 15(3):312–319

Fate and Removal of Pharmaceuticals in CAS for Water and Sewage Sludge Reuse

Marc Castaño-Trias, Monica Brienza, Maria Concetta Tomei, and Gianluigi Buttiglieri

Contents

1 Introduction 24
2 Bibliometric Study Regarding Micropollutants and PhACs 26
3 Influent PhAC Characterization in Municipal WWTP 29
3.1 Seasonal Variability in Influent WWTP PhAC Occurrence 30
4 Pharmaceutical Removal in Conventional Activated Sludge Systems 31
4.1 Compound Characteristics and Parameters Affecting PhAC Removal in CAS 32
4.2 Removal Rates of Selected Representative PhACs 34
5 Metabolites, Conjugates and Transformation Products 39
6 Regulation Regarding Water Reuse and PhACs 40
7 PhAC Fate in Sludge Treatment Line: Focus on Agricultural Reuse 41
7.1 Agricultural Sludge Reuse: Advantages and Environmental Impacts 41
7.2 Sludge Treatment Alternatives to Enhance PhAC Removal 42
7.3 EU Regulation for Agricultural Reuse 44
References 44

Abstract Among organic micropollutants, pharmaceutical active compounds (PhACs) have been extensively investigated in the last decades due to their potential adverse ecological and/or human health effects. Wastewater treatment plants are one of the first barriers in preventing them from reaching sensible ecosystems. In

M. Castaño-Trias and G. Buttiglieri (✉)
Catalan Institute for Water Research (ICRA), Girona, Spain

Universitat de Girona, Girona, Spain
e-mail: mcastano@icra.cat; gbuttiglieri@icra.cat

M. Brienza
Department of Science, University of Basilicata, Potenza, Italy
e-mail: monica.brienza@unibas.it

M. C. Tomei
Water Research Institute, C.N.R., Rome, Italy
e-mail: tomei@irsa.cnr.it

Sara Rodriguez-Mozaz, Paqui Blánquez Cano, and Montserrat Sarrà Adroguer (eds.), *Removal and Degradation of Pharmaceutically Active Compounds in Wastewater Treatment*, Hdb Env Chem (2021) 108: 23–52, DOI 10.1007/698_2020_666,

particular, conventional activated sludge (CAS) systems may be only partially effective in PhAC removal or degradation. The present work presents an overview of the fate of PhACs in CAS. Several parameters and constraints affecting PhAC removal are given and discussed including the physicochemical properties of the compounds and the operating parameters like the influent concentration, its seasonal variation, CAS hydraulic retention time and sludge retention time. A detailed data analysis is dedicated to selected representative compounds (i.e. carbamazepine, ibuprofen and sulfamethoxazole) in the water line of CAS systems, while for the sludge line, different treatment options are discussed, with special focus on the sequential anaerobic-aerobic digestion, a promising alternative to improve the removal of organic micropollutants, including PhACs, in the sludge matrix. Finally, the criteria for safe reuse of treated effluent and agricultural application of the produced sludge, related to PhACs, are discussed and analysed in the frame of EU and other regulations.

Keywords HRT, Regulation, Seasonal variability, Sludge treatment, SRT, Transformation products

1 Introduction

The presence of xenobiotics in the aquatic environment has raised environmental concern due to their high toxicity, at times even at low concentrations. These so-called micropollutants have been detected in all water matrices, and their removal in wastewater treatment plants (WWTPs) is known to be incomplete and challenging for several classes of compounds.

The focus of environmental research has extended in the last two decades beyond more established and known wastewater xenobiotics, such as some polychlorinated biphenyls, dioxins and pesticides, towards other classes of emerging organic micropollutants. Among the latter ones, pharmaceutical active compounds (PhACs) are of great interest, and they are getting more and more attention in the literature, as it can be seen in *bibliometric studies*. To give a figure of the relevance of these compounds, it is worth noting that in the European Union (EU), around 3,000 different PhACs are being used in medicine such as painkillers, antibiotics, contraceptives, beta-blockers, lipid regulators, tranquilizers and impotence drugs [1]. During and after medical treatment, humans and animals excrete a combination of intact and metabolized PhACs. Kümmerer and Henninger, for instance, reported that the metabolization rate of all antibiotics used in Germany was estimated to be around 30% [2]. In other words, approximately 70% of the total amount is excreted unchanged into raw wastewater. PhACs are continuously entering the environment via non-point sources (i.e. run-off or drainage from land surfaces) but mainly via point sources (i.e. both municipal and pharmaceutical factory WWTPs). Their harmful effects on the ecosystem and on the human health are well recognized and currently under extensive evaluation [3]. To this aim, the *characterization of WWTPs* in terms of PhAC concentration is mandatory.

Many advanced and more efficient wastewater treatment schemes are being proposed and tested, even at full scale. Nonetheless, the most common WWTPs in Europe, and worldwide, are anyway still based on *conventional activated sludge* (CAS) *systems*, to be noted that CAS are usually designed to remove organic carbon and nutrient loads but, generally, they are not designed to remove trace organics [4]. Consequently, the operative parameters that most affect *PhAC removal in CAS* need to be evaluated in order to improve the plant performance. Three main removal pathways are usually accounted in CAS: microbial processes (biodegradation, either metabolic or co-metabolic), sorption onto sludge flocs and volatilization (mainly during aeration). However, volatilization can be considered negligible for most PhACs, because of Henry's constant value of such molecules [5]. It must be generally remarked that extremely variable removal was reported in CAS treatments, even for the same compound. For instance, removal rates in the liquid fraction from 17% [6] to 89% [7] were reported for sulfamethoxazole (SMX), from 12% [8] to 80% [9] for tetracycline (TC), and from 4.3% to 72% [6] for erythromycin-H_2O. Other compounds, conversely, are usually reported as more recalcitrant: for example, less than 30% removal for carbamazepine (CBZ) [10] and 17% for lincomycin (LCM) [11]. Persistence has often been linked to their molecular properties, but also to process-specific factors such as sludge retention time (SRT), or hydraulic retention time (HRT), biomass concentration, cation-exchange properties as well as seasonal parameters such as temperature, precipitation rate and solar radiation [12].

For a comprehensive analysis of the fate of PhACs, it is vital to consider not only the parent compounds, or the metabolites that can be excreted, but also the so-called transformation products (TPs), generated from biotic/non-biotic processes. TPs and parent compounds in CAS can behave with a different pattern [13], and TPs can be less biodegradable, more toxic and inhibitory compared to the parent compound. Even though more and more information is now available on the occurrence and fate of TPs of PhACs in WWTPs, there are still many knowledge gaps on TP occurrence, fate, exposure, toxicity and risks.

PhACs that are inefficiently or not adequately eliminated may pass through CAS biological treatment processes. Consequently, they may enter the receiving waters as dissolved pollutants, via the WWTP effluent. Moreover, *water reuse*, e.g. agriculture reuse of treated wastewater, may be applied, also based on the recent EU law on minimum requirements for water reuse and/or other legislations and/or individual national regulations [14].

In parallel, also PhAC concentration in the solid phase should be considered. Pharmaceutical residues are, in fact, frequently detected in biosolids at $\mu g\ kg^{-1}$ to $mg\ kg^{-1}$ levels. For example, an average concentration of 68 $\mu g\ kg^{-1}$ dry weight (dw) for SMX in CAS was reported by Göbel et al. [15], and about 15 $\mu g\ kg^{-1}$ dw for TC in biosolids was shown by Spongberg and Witter [8]. Different *sludge treatment alternatives* are available. Among them, anaerobic digestion treatment has long been practiced and commonly applied as a renewable energy source in integrated waste management, but other ones may be available. Treated sludge may be used in agriculture, and there may be health constraints on the agricultural recycling of wastewater sludge [16]. Quite extensive information is available in terms of metals in sludge and sludge treatments. Less information is available in terms of micropollutants, and in particular of PhACs, thus requiring much more insights.

2 Bibliometric Study Regarding Micropollutants and PhACs

The interest of the scientific community on PhACs and the corresponding studies have been increasing in the last decades. In order to quantify the evolution and the past and current trends of investigations on the occurrence of pharmaceuticals compounds in wastewater, a bibliometric study has been carried out on Scopus database. This way both historical progress and quantitative trends of research publications on a specific topic can be effectively considered [17].

The analysis has been structured as follows: first, identification of research historical trends on micropollutants, analyses, fate and removal; second, country-specific distribution of paper productivity; and third, definition of most evaluated PhAC therapeutic classes.

The first analysis was performed in *SCOPUS* searching in the string "micropollutants" *OR* "emerging pollutants" *OR* "trace pollutants", with the applied filters "Chemistry" *OR* "Environmental Science" *OR* "Chemical Engineering", between 1980 and 2019. A total of 17,780 publications, in the whole period 1980–present, related to the micropollutants were identified, and the number of publications per year over the past 30 years is displayed in Fig. 1. Application of tandem mass analysers, in fact, started in the late 1970s, and they spotted a new possible challenge and corresponding treat for human health and the environment: the environmental occurrence of micropollutants, including PhACs.

In the decade 1980–1989, few publications over the total, only 576, are reported, also due to limitations in full commercialization of MS/MS equipment for micropollutants analyses. Subsequently, the increase in the attention of the research community was more and more sustained, and, for example, 266 research papers were remarkably published only in the year 2000. Then, 1,112 articles per year were published, on average, in the period 2010 to 2019. Most of the compounds themselves are not new in environmental matrices. Nonetheless, the increase in the number of publications is intrinsically linked to the extreme increase in the sensitivity provided by new analytical instruments and tools.

The second search aimed to evaluate if the paper title, abstract or keywords included also the words "fate", *OR* "monitoring", *OR* "removal", *OR* "analyses" *OR* "occurrence" that can be related, from different perspectives, to studies on

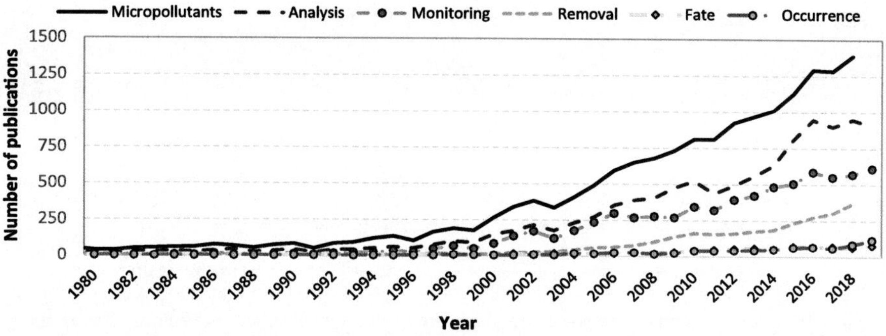

Fig. 1 Bibliometric search of publications about micropollutants and analysis, or monitoring, or removal, or fate or occurrence using *SCOPUS*. The series "micropollutants" was obtained searching in Title-abstract-keywords for the string "micropollutants" *OR* "trace pollutants" *OR* "emerging pollutants". The other series were obtained adding to the previous one the string *AND* "monitoring" *OR* "analyses" *OR* "removal" *OR* "occurrence" *OR* "fate". In all the series, also the filters "Chemistry" *OR* "Environmental Science" *OR* "Chemical Engineering" *AND* "PUBYEAR>1980" to "present" were added

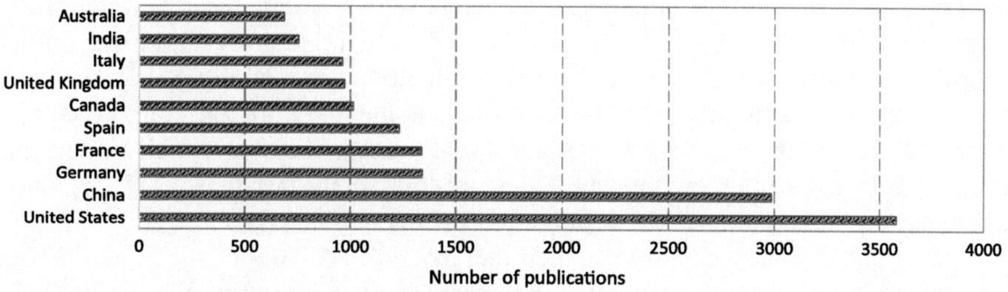

Fig. 2 Bibliometric search of number of publications on micropollutants, per country. The string is the same as the "micropollutants" series of the previous figure

micropollutants (Fig. 1). In 1980, only 1 publication was dealing with "fate", 2 with "removal" and 8 and 19 publications with "monitoring" and "analysis", respectively. A considerable increase can be observed, also in this case, in the last 20 years. In 2019 the published papers were 906, 611, 406, 118 and 81 for "analysis", "monitoring", "removal", "occurrence" and "fate", respectively. The increasing trends are similar, even though at different levels, demonstrating the interest of the scientific community in this topic, in particular as regards "analysis" and "monitoring". It might be related, on one hand, to the development of new methods of analysis and, on the other hand, to the research demand of reliable and accurate analytical and monitoring tools to evaluate contaminant removal and fate.

A complementary bibliometric search was performed to determine which country was leading the investigation. The result shows in Fig. 2 that the USA and China

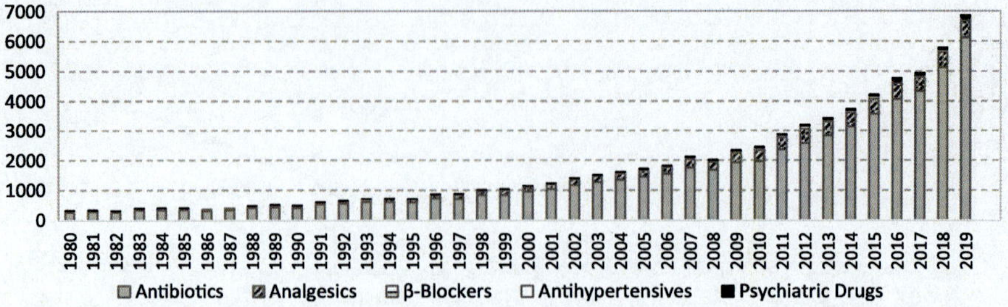

Fig. 3 Bibliometric search of publications about different pharmaceutical therapeutic classes using *SCOPUS* with the string "antibiotics" *OR* "analgesics" *OR* "β-blockers" *OR* "antihypertensives" or "psychiatric drugs" in Title-abstract-keywords with the filters Chemistry *OR* Environmental Science *OR* Chemical Engineering *AND* "*PUBYEAR*>1980" to "present"

have the highest research outputs on the topic on micropollutants with around 3000/3500 publications. Afterwards, Germany, France and Spain are the other countries in the top five. A noticeable difference is spotted between the first two and the other ones, with German manuscript production being only 37% and 45% with respect to the USA and China, while smaller difference is shown among the other countries, from Germany to Australia. Finally, five out of the ten countries with more publication on this topic are EU countries (39% of the total production, Fig. 2).

An even larger amount of studies is available in the literature reporting PhACs in their title, keywords, or abstract. Around 4,000 manuscripts were published in the decade 1980–1989 and more than 43,000 papers only in the last decade (Fig. 3), with an increasing trend comparable to previous Figures.

PhACs can be classified based on their therapeutic class. Among them, *antibiotics* are applied to stop bacterial growth in an organism (e.g. ciprofloxacin or amoxicillin); *analgesics* (usually extremely highly consumed) to relieve pain without affecting the conduction of nerve impulses, sensory perception or consciousness (e.g. naproxen); *β-blocking agents* to control abnormal heart beating rhythms and to control blood pressure (blocking epinephrine hormone, e.g. metoprolol, atenolol or sotalol); *antihypertensives* to prevent, control or treat hypertension (e.g. losartan); and *psychiatric drugs* to generate a change on consumers' mind, emotions and behaviours (to treat mental health diseases, e.g. venlafaxine to treat depression).

Bibliometric search highlights that antibiotics, out of the above-indicated categories, are the group with more scientific publications (from 3,301 in the decade 1980–1989 to 36,111 in the decade 2010–2019) with a percentage of around 83% over the total (Fig. 3). This large attention is also related to the thread that continuous exposure to antibiotics may spread antibiotic resistance in the environment and new antibiotics are eventually to be developed because old ones are not effective anymore [18].

3 Influent PhAC Characterization in Municipal WWTP

The development of effective strategies to remove organic micropollutants, and PhACs among them, from wastewater is critical. Whatever the applied strategy is, first it is mandatory to identify which compounds can be found in the influent, at which concentration they are and if significant variabilities are to be accounted for. The literature regarding PhACs in municipal WWTP influents is quite extensive. The data reported and treated in this section refer mainly to full-bodied reviews considering a large number of compounds and WWTPs (e.g. [9, 19–21]).

Among the PhACs, 18 compounds have been selected here, as representatives of different therapeutic classes, and their occurrences are compared. It is well known and reported that the concentration of these compounds in influent wastewater usually has a huge variability, as it can be observed in Fig. 4.

The mean concentration of these compounds in influent wastewater may be roughly calculated around 1 μg/L but heavily depending on the compound, the country and the WWTP. For most compounds, in fact, many orders of magnitude of difference are reported in real influent wastewater (e.g. ibuprofen can be found between 28 ng/L [21] and 372,000 ng/L [20]). Minimum and maximum concentrations may be representative of extremely low or high occurrence of these compounds in influent WWTP. Nonetheless, it is apparent that huge variability must be considered when evaluating the presence of PhACs in WWTPs.

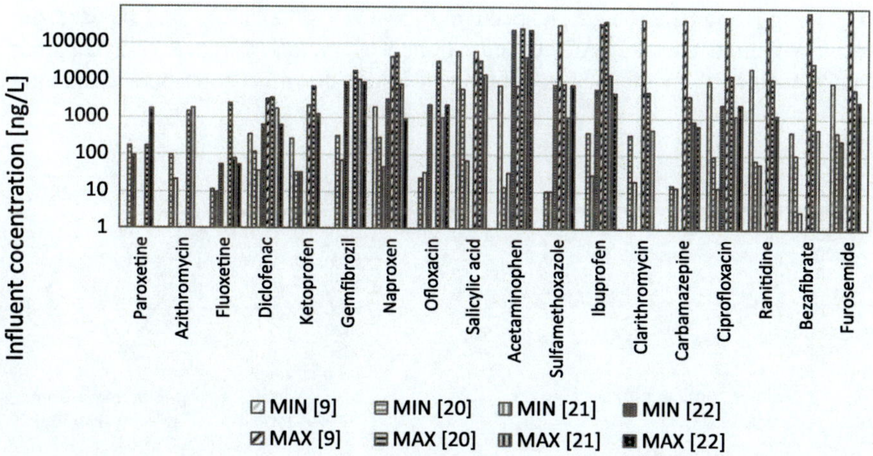

Fig. 4 Representation of the minimum and maximum influent concentrations to the WWTP included in [9, 20–22]

3.1 *Seasonal Variability in Influent WWTP PhAC Occurrence*

Seasonal/temporal variability is among the main variables that must be considered when evaluating micropollutant WWTP influent content. Some compounds may be present or not, based on the considered WWTP and/or season. For example, UV filter compounds are representative of extremely seasonal organic micropollutants because their use increases in the high touristic season [23]. Illicit drug use is reported to increase during the weekend, festivals and public holidays [24].

As regards PhACs, relevant seasonal variations were also reported for some therapeutic classes of compounds (i.e. analgesics/anti-inflammatory compounds or histamine receptor antagonists in the flowering period) based on their consumption [25]. Three representative PhACs of different therapeutic classes (the psychiatric drug venlafaxine, the analgesic ibuprofen and the antibiotic sulfamethoxazole) are here evaluated in more detail, reporting the average of WWTP influent PhAC concentrations in sampling campaigns performed in different seasons (Fig. 5; [21], info from ten papers).

Ibuprofen was confirmed as the compound reported at the highest concentration throughout the year. Moreover, a large seasonal variability was also confirmed between the winter-spring and the summer-autumn periods (on average 20,045 ng/L and 7,979 ng/L, respectively). Much smaller seasonal variability can be observed for the other two compounds, even though the minimum for both compounds was reported in autumn and the maximum in winter.

A strong seasonal effect can be confirmed also considering a larger list of 61 PhACs, from different therapeutic classes. It was confirmed, similarly to ibuprofen, a larger concentration in winter-spring than in summer-autumn (on average 181,311 ng/L and 62,256 ng/L, respectively), as it is visible in Fig. 5b. Likewise, the season with higher total PhAC concentrations was winter. Moreover, it is apparent that the analgesics are the analysed therapeutic class that most contribute to the total

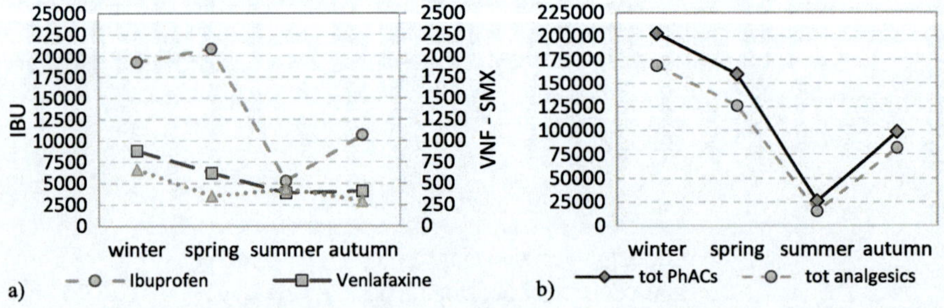

Fig. 5 (**a**) Average concentration, in ng/L, of three selected PhACs (IBU, ibuprofen; VNF, venlafaxine; SMX, sulfamethoxazole) throughout the year. (**b**) Total average concentrations of 61 PhACs and total average of analgesics, in ng/L. The reported data are obtained as average of ten papers [21]

influent PhAC concentration, with a much smaller contribution in summer (i.e. 83%, 79%, 59% and 83% of analgesics over the total PhAC concentration in winter, spring, summer and autumn, respectively). That the differences in season temperature may affect influent WWTP PhAC content for different reasons: from one side the consumption may change remarkably based on the season and the countries, and from the other one, higher temperature may lead to higher removal even in the sewer systems, before entering the WWTP [26].

4 Pharmaceutical Removal in Conventional Activated Sludge Systems

WWTPs were first designed only for biological oxygen demand removal. Through the decades they were upgraded with chemical phosphate precipitation, to reduce the phosphorus load discharge, to convert ammonia to nitrate and to partially convert nitrate to molecular nitrogen. CAS systems are still the most commonly applied in the world, but they are not designed to remove organic micropollutants, including PhACs [4]. Nonetheless, a full understanding of removal mechanisms can upgrade CAS capabilities and possibly reduce additional/alternative treatments and their cost of application.

PhACs comprehend an extremely large range of compounds with different chemical structures, persistence and effects in the environment. The most evaluated PhAC therapeutic classes in terms of occurrence and removal are usually the antibiotics, also due to their possible consequences in the environment, and the nonsteroidal anti-inflammatory drugs (NSAIDs), due to their high consumption [27]. These PhAC therapeutic classes exemplify different, and sometimes complementary, needs in terms of removal evaluation in treatment technologies. Some PhACs, in fact, may have influent WWTP concentrations lower than other compounds but with higher demonstrated or suspected effects on the environment [18] and/or more recalcitrant (e.g. some antibiotics). Some other PhACs, on the other hand, may usually exhibit higher removal rates, but as they have higher influent and effluent concentrations, they continuously enter the environment, e.g. NSAIDs [28]. Both classes of compounds, and the corresponding behaviours, are relevant. The topic of PhACs is clearly a remarkably interdisciplinary research area. The topic of PhACs is clearly a remarkably interdisciplinary research area *AND* IT poses obvious challenges for researchers across numerous disciplines (i.e. ecotoxicology, antimicrobial resistance genes, bioaccumulation, etc.) which are required in this field [29].

In the literature, there are several reviews presenting in detail and comparing the efficiency of CAS and of several advanced treatment technologies as regards the removal of several PhACs [30], among them the recent detailed review of Patel et al. [9] and Verlicchi et al. [20]. Analgesics and anti-inflammatory drug (e.g. naproxen, ibuprofen and acetaminophen) removal is generally high (over 75%) even though in some cases also lower removal rates are reported [9]. Among the broadly studied

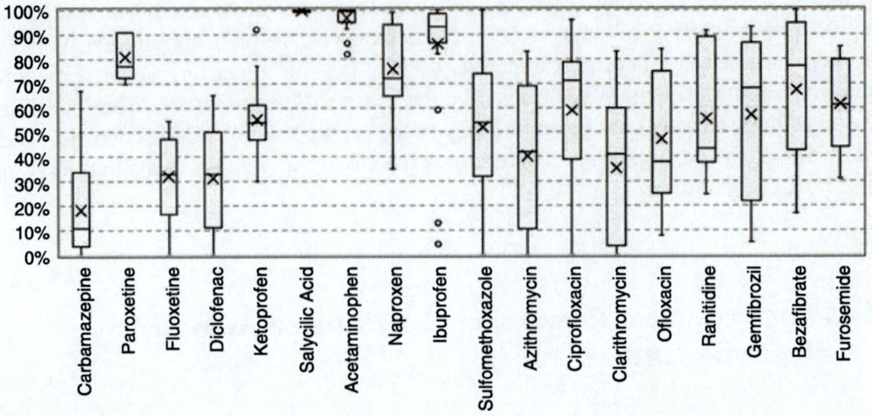

Fig. 6 Representation of reported PhAC removal in CAS (data from [9, 20, 36–38])

class of antibiotics, fluoroquinolones are a subclass of the most frequently investigated compounds. Similar results are reported about the removal rate of ciprofloxacin and ofloxacin in different countries such as Sweden [31], Italy [32], Finland [33] and Switzerland [34], with a higher percentage for ciprofloxacin than ofloxacin. The fluoroquinolone removal rate in CAS is associated also with the sorption on activated sludge, due to the relatively significant measured amount of these compounds in sludge, as reported by various authors [34, 35]. Other compounds, finally, are characterized by inconsistent results in terms of occurrence and fate like diclofenac, with removal rates ranging between 0% and 90% and, in some cases, with higher effluent concentration, compared with the influent.

In order to collect and combine the information from several reviews at the same time, the data from 40 papers are represented in Fig. 6. As discussed above, a large variability in terms of removal of the considered PhAC compounds can be observed.

4.1 *Compound Characteristics and Parameters Affecting PhAC Removal in CAS*

The removal of PhACs, like any other pollutant, is directly related to their chemical structure. For example, the presence of chloro, nitro and fluoro functional groups attached to aromatic rings may cause persistence and resistance to removal. On the other hand, electron-withdrawing groups, such as amines or alcohols, may enhance the removal in CAS [39–42].

As previously mentioned, the main removal pathways for PhACs in CAS are microbial processes (biodegradation, either metabolic or co-metabolic) and sorption onto sludge flocs and suspended material, being volatilization usually negligible. As regards sorption, different removal pathways can be observed if the compound has hydrophobic or hydrophilic molecular structure. High polarity is usually linked to higher water solubility, lowering adsorption, as it is reported for carbamazepine (7–10%). Compounds with medium polarity may have instead higher average removal percentages. As regards degradation, it is reported that PhACs with simpler molecular structure are more prone to biodegradation. Instead, the drugs with saturated aliphatic and aromatic rings (such as iomeprol and diatrizoate) exhibit resistance to degradation in CAS [43].

Nonetheless, whatever the chemical structure and characteristics are, also operative conditions of the chosen treatment technology, in this case CAS, must be considered and carefully evaluated case by case. As detailed in the next paragraph for selected PhACs, in fact, extremely different removal rates have been reported in the literature, even though CAS was the technology of choice in all the cases. A careful evaluation of both aspects, chemical and operative parameters, must be promoted in future studies. Among the main operative conditions and variables that can affect PhAC removal in CAS, there are the sludge retention time (SRT), the hydraulic retention time (HRT), the presence of aerobic/anoxic treatment, the seasonality effects and the influent concentration. These variables may lead to a much wider range of PhAC removals than those foreseen only based on their chemical properties only.

As regards the biology of the activated sludge, one of the most relevant parameters is SRT, representing the mean residence time of microorganisms in CAS systems. Several studies state that high SRT, allowing the development of slow-growing bacteria, is required for PhAC biodegradation/biotransformation. In addition, the enrichment of the microbial culture with new species may favour the development of a wider spectrum of metabolic pathways useful for PhAC degradation. This may explain the higher removal of some PhACs observed in WWTP removing nitrogen (i.e. at higher SRT and including slow-growing autotrophic species like nitrifiers) in comparison to the ones removing only organic matter [44–47]. Concerning the microbial diversity and the SRT effect, three hypotheses have been proposed: (1) microbial diversity is proportional to SRT; so higher SRT is related to higher microbial diversity; (2) activate sludge with high microbial diversity have more functional traits than those with low diversity; (3) high microbial diversity may result in higher PhAC removal potential [48]. It is reported that a SRT of 2–5 days is usually enough for the degradation of sulfamethoxazole, bezafibrate, aspirin and ibuprofen. Conversely, longer SRTs (5–15 days) were required for substantial degradation of iopromide, roxithromycin and diclofenac [49]. Finally, some other PhACs, such as carbamazepine, are usually not affected by SRT [50]. Nevertheless, even though SRT has been reported as influential for PhAC biodegradation, a clear correlation between SRT and biodegradation has not been completely elucidated.

In terms of hydraulics, the HRTs quantify the contact time between the microorganisms and the pollutants during the treatment. As it is for SRT, contrasting results are reported in the literature with similar HRT and other parameters, but dissimilar biodegradation efficiencies [51, 52]. Nonetheless, usually higher HRTs may result in longer contact/reaction time and higher PhAC removals.

Another important parameter that affects the biological activity is the temperature and, consequently, also PhAC degradation. In summer, for example, the increase in the temperature may favour the microbial activity, fasten bacterial metabolism and increase PhAC biodegradation. During the winter, on the other hand, a drop in the removal efficiencies can be sometimes anticipated. For instance, Vieno et al. [12] showed that low removal was most probably linked to the low water temperature in wintertime (~7°C). Adversely, Göbel et al. [46] reported no correlation between the water temperature and the elimination of sulphonamides, macrolides and trimethoprim in the range of 12 to 19°C. These data demonstrate that, as observed for SRT, the temperature effect is compound-specific and not of general validity.

Nonsteroidal anti-inflammatory drugs with low *pK*a, such as clofibric acid and bezafibrate, are ionized to their conjugate base at neutral pH *AND*, *THEREFORE*, they will have a *HIGH* affinity for the aqueous phase, reducing their adsorption on negatively charged sludge surface. *CONVERSELY*, zwitterionic and basic pharmaceuticals (e.g. the fluoroquinolone antibiotics) show significant adsorption [43].

Finally, anaerobic or aerobic conditions will also lead to different results in terms of removal. This factor is particularly critical for some compounds, such as the fluoroquinolone antibiotics [53]. These compounds are better removed under aerobic conditions, while their biodegradation under anoxic environments is almost negligible. This fact can be due to the higher oxidation potential of oxygen, compared to nitrate [43, 53].

4.2 Removal Rates of Selected Representative PhACs

As commented in the previous paragraph, operative parameters such as, HRT, SRT and temperature may affect directly or indirectly PhAC removal. In this paragraph a detailed analysis is reported on the effects of these parameters on the removal of carbamazepine, ibuprofen and sulfamethoxazole.

The psychiatric drug carbamazepine is generally scarcely or not removed in CAS systems (Table 1). This fact can be explained by its structure: the presence of heterocyclic N-containing aromatic rings makes difficult its biodegradation whatever the conditions are [54, 55].

Nonetheless, there are some exceptions (Fig. 7). In some cases, for sampling campaigns performed in winter times, a higher removal (i.e. 35%) was observed even though other parameters like HRT and SRT are in the same range (Fig. 7).

Based on the provided HRT and SRTs, when available, it can be confirmed the recalcitrance of the compound: the increase of HRT or SRT does not affect its removal in CAS (Fig. 8).

Table 1 Removal of carbamazepine in CAS systems

Season	Removal (%)	HRT (h)	SRT (days)	Country	Reference
Winter	−25	9–17	11–15	China	[56]
Winter	0	Non-specified	Non-specified	Italy	[32]
Winter	0	1.5–20	2–20	Finland	[33]
Winter	0	7.3	10	Switzerland	[5]
Winter	15	7.1–9.4	3.8–8.4	Japan	[57]
Winter	60	Non-specified	Non-specified	Spain	[58]
Winter	73	27	Non-specified	Spain	[59]
Winter	95	Non-specified	3–5	Turkey	[60]
Spring/autumn	−28	7.1–9.4	3.8–8.4	Japan	[57]
Spring/autumn	−21	Non-specified	Non-specified	Spain	[59]
Spring/autumn	−12	5	18	Germany	[61]
Spring/autumn	−1	27–48	Non-specified	Spain	[62]
Spring/autumn	10	12	3	Spain	[30]
Spring/autumn	10	11.5	10	Spain	[51]
Spring/autumn	10	23	7.5	Canada	[63]
Spring/autumn	32	Non-specified	Non-specified	Spain	[58]
Summer	−67	12–16	1.5–2.7	Spain	[64]
Summer	−4	12–16	1.5–2.7	Spain	[64]
Summer	0	16–24	15–25	USA	[65]
Summer	11	7.1–9.4	3.8–8.4	Japan	[57]
Summer	25	11	6–8	Canada	[66]
Summer	35	Non-specified	1.5–26	Austria	[67]
Mean year	11	12–17	1.5–5.1	Spain	[68]
Mean year	7	12–17	1.5–5.1	Spain	[68]
Non-specified	−47	24–48	1–26	Austria	[47]
Non-specified	−43	1.9–326	2–237	Austria	[47]
Non-specified	−43	24–48	1–26	Austria	[47]
Non-specified	−35	24–48	1–26	Austria	[47]
Non-specified	−11	1.9–326	2–237	Austria	[47]
Non-specified	−3	24–48	1–26	Austria	[47]
Non-specified	0	Non-specified	52–237	Austria	[69]
Non-specified	0	5.5	8	USA	[70]
Non-specified	10	24	60	Germany	[71]

The collected data regarding ibuprofen are shown in the following Table 2. Generally, high removals are reported in most of the studies, and biodegradation is usually accounted.

As it can be observed from Fig. 7, the removal of ibuprofen is generally not affected by the season, nor by operating parameters like SRT or HRT. Most of the considered papers reported high removal for this analgesic compound. Nonetheless, a lower removal was observed in few cases and most of them in the winter seasons.

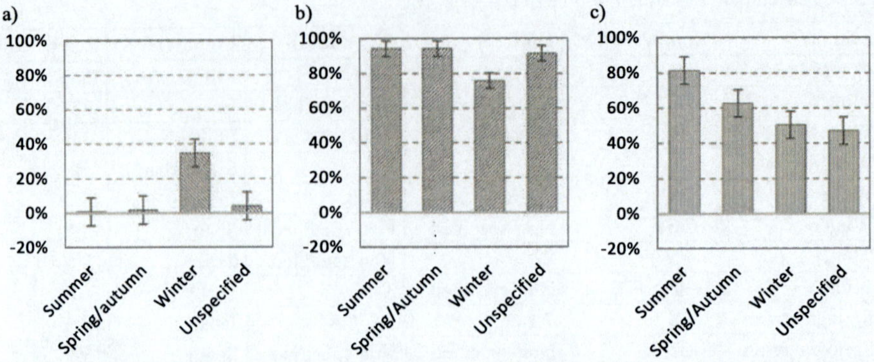

Fig. 7 CAS removal (%) in different seasons for carbamazepine (**a**), ibuprofen (**b**) and sulfamethoxazole (**c**). The reported data corresponds to averages of the references reported in Tables 1, 2 and 3

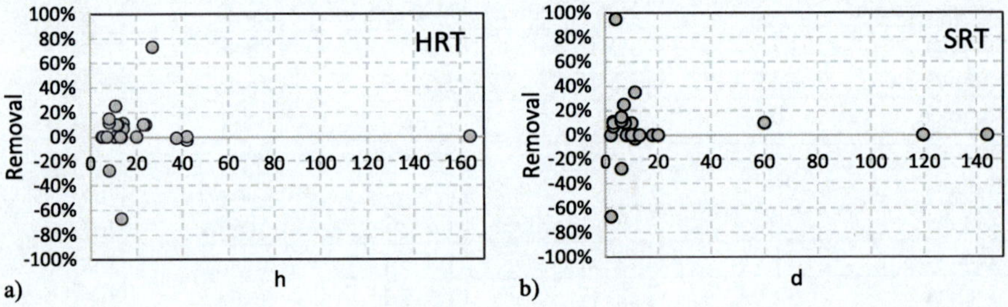

Fig. 8 CAS removal (%) of carbamazepine as a function of HRT (**a**) and SRT (**b**). The reported data corresponds to the references provided in Table 1

The lowest removal in winter, with the lowest temperature, jointly with the evaluation of its chemical structure, confirms that the main removal mechanism for ibuprofen in CAS is usually biodegradation, as also reported in the next paragraph. On the other hand, based on the provided HRT and SRTs, when available, it can be confirmed that ibuprofen is effectively removed also at low HRT and low SRT, thus confirming, even though with some exception, that it is one of the most easily removed analgesics in CAS (Fig. 9).

Finally, the collected data for sulfamethoxazole are reported in Table 3 and Fig. 7.

The removal of this compound is strongly variable in the sampled WWTPs. The variability is influenced by the season with higher removal in summer (on average 80%) and lower, on average, in winter. HRT and SRT confirmed variable removal results under a wide range of operating conditions (Fig. 10).

In the literature, HRT and SRT are recognized to be important parameters to be considered in this context [86]. Tailored adjustment of these parameters may have a

Table 2 Removal of ibuprofen in CAS systems

Season	Removal (%)	HRT (h)	SRT (days)	Country	Reference
Winter	48.5	Non-specified	Non-specified	Italy, France, Greece and Sweden	[72]
Winter	50	Non-specified	Non-specified	Italy, France, Greece and Sweden	[24]
Winter	65	24	Non-specified	Spain	[73]
Winter	66	Non-specified	Non-specified	Finland	[74]
Winter	84	27	Non-specified	Spain	[75]
Winter	95	8–15	4–21	China	[56]
Winter	98	Non-specified	Non-specified	Switzerland	[76]
Winter	100	Non-specified	Non-specified	Spain	[58]
Spring/autumn	83	12	3	Spain	[30]
Spring/autumn	86	13.5	1.5–2.7	UK	[77]
Spring/autumn	94	27	Non-specified	Spain	[75]
Spring/autumn	99	35	6–10	Sweden	[31]
Spring/autumn	99	11.5	10	Spain	[51]
Spring/autumn	100	36	20	Finland	[12]
Spring/autumn	100	Non-specified	Non-specified	Spain	[58]
Spring/autumn	100	27–48	Non-specified	Spain	[62]
Summer	−13	12.5	Non-specified	England	[78]
Summer	88	12–16	1.5–2.7	Spain	[64]
Summer	90	12–16	1.5–2.7	Spain	[64]
Summer	92	Non-specified	1.5–26	Austria	[67]
Summer	98	12	15–65	Japan	[79]
Summer	99	23	18	Greece	[5, 80]
Mean year	84	12–17	1.5–5.1	Spain	[68]
Mean year	87	12–17	1.5–5.1	Spain	[68]
Non-specified	−4.4	24–48	1–26	Austria	[47]
Non-specified	82	24	60	Germany	[71]
Non-specified	87	Non-specified	Non-specified	USA	[81]
Non-specified	92	24–48	1–26	Austria	[47]
Non-specified	97	Non-specified	52–237 days	Austria	[69]
Non-specified	98	1.9–326	2–237	Austria	[47]
Non-specified	98	24–48	1–26	Austria	[47]
Non-specified	99	24–48	1–26	Austria	[47]

beneficial influence on the removal of some PhACs and should be carefully considered. Moreover, prolonged SRT may favour nitrifying biomass and/or other slowly growing bacteria, which may have better xenobiotic removal capabilities.

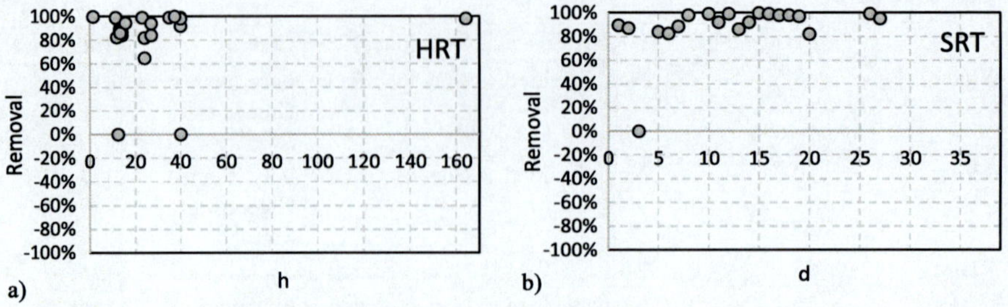

Fig. 9 CAS removal (%) of ibuprofen as a function of HRT (**a**) and SRT (**b**). The reported data corresponds to the references provided in Table 2

Table 3 Removal of sulfamethoxazole in CAS systems

Season	Removal (%)	HRT (h)	SRT (days)	Country	Reference
Winter	4.5	Non-specified	Non-specified	Switzerland	[46]
Winter	25	16.5	12.5	Australia	[82]
Winter	26	2.8–14	13–19	Japan	[83]
Winter	39	2.8–14	13–19	Japan	[83]
Winter	42	Non-specified	Non-specified	Italy	[32]
Winter	42	8–24	Non-specified	Sweden	[35]
Winter	61	15.8	21.4	China	[56]
Winter	74	Non-specified	Non-specified	Spain	[58]
Winter	94	27	Non-specified	Spain	[59]
Winter	97	10–17	7–12	China	[52]
Spring/autumn	10	11.5	10	Spain	[51]
Spring/autumn	56	12	3	Spain	[30]
Spring/autumn	62	Non-specified	Non-specified	Spain	[58]
Spring/autumn	71	27	Non-specified	Spain	[62]
Spring/autumn	71	27	Non-specified	Spain	[59]
Spring/autumn	75	10–17	7–12	China	[52]
Spring/autumn	94	10–17	7–12	China	[52]
Summer	51.9	7.9–15	16.8–43	South Korea	[84]
Summer	93	10–17	7–12	China	[52]
Summer	100	8–24	Non-specified	Sweden	[35]
Non-specified	32	Non-specified	52–237	Austria	[47]
Non-specified	62	14–16	2–4	Israel	[85]

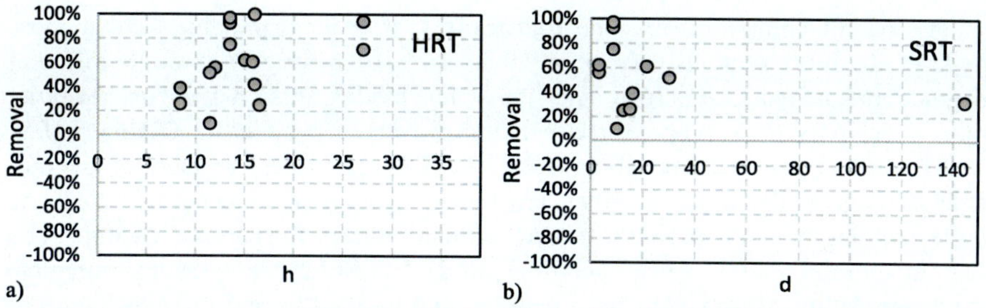

Fig. 10 CAS removal (%) of sulfamethoxazole as a function of HRT (**a**) and SRT (**b**). The reported data corresponds to the references provided in Table 3

5 Metabolites, Conjugates and Transformation Products

To fully understand the fate of PhACs, as for any other organic (micro)pollutant, in CAS it is crucial to consider also micropollutant metabolites (including conjugates) and transformation products.

Upon consumption, a variable amount of the ingested compound is excreted as a metabolite. Metabolites can be separated based on the undergone transformation: phase I metabolites and phase II metabolites. Phase I metabolites are originated through in vivo biochemical oxidation, reduction and hydrolysis reactions increasing aqueous solubility and facilitating the elimination from the body. On the other hand, phase II metabolites (or conjugated metabolites) are the result of the addition of a molecule to the parent compound (i.e. glucuronic acid). Phase II metabolites may deconjugate back to the parent compound in wastewater [87]. Afterwards, when PhACs enter WWTPs and/or in the environment, they can be exposed to biotic (biodegradation by bacteria and fungi)/non-biotic (hydrolysis or photolysis) processes [13].

The term "metabolite" is sometimes applied for compounds originated from both human, microbial and environmental sources. Nevertheless, in general it is recommended to only call "metabolites" those molecules that result from changes in human and animals and to call "transformation products" (TPs) to those generated from biotic/non-biotic processes [88]. In some cases, the same human metabolites can also be generated by microbial biodegradation (e.g. ibuprofen, [75]), but it is not the general case, and attention needs to be paid in this context.

The fraction of the parent compound to be excreted in an unchanged form depends on the structure of the micropollutant. For example, it may range from 20 to 75% for antibiotics. In the case of metoprolol, three main metabolites are formed by mammals (metoprolol acid, a-hydroxymetoprolol and O-desmethyl metoprolol), accounting for 85% of the urinary excretion [89]. Information on metabolites is increasing rapidly but still lacking for many compounds. The load of organic micropollutants entering CAS, hence, is greatly underestimated, if metabolites (and conjugates) are not accounted for.

The parent compound may be transformed, if it is degradable, into a more hydrophilic form (and usually of a lower molecular weight) through different degradation pathways. For example, ibuprofen biodegradation may lead, among other ones, to the formation of three main transformation products (1-OH-ibuprofen, 2-OH-ibuprofen and carboxy-ibuprofen [75]). TPs can be less or more biodegradable, and, generally speaking, they are found at a lower concentration than the corresponding parent compound in CAS effluent, except for persistent compounds (e.g. metoprolol acid, in some conditions, [89]). In just limited cases, the complete (bio)degradation pathway has been clarified and verified in real matrixes, concentrations (i.e. μg/L or ng/L levels) and conditions other than laboratory-controlled ones [13]. Each micropollutant can originate many TPs, and these TPs can also be transformed into different TPs; therefore their monitoring is one of the most challenging objectives.

Whatever the case, if they are not completely removed, TPs may pass through CAS and end up in the receiving waters. In some cases, the originated compound can even be more toxic than the parent compound [90]. On the other hand, some large organic micropollutants quickly lose their toxic features after the first degradation steps, making following treatment efforts counterproductive [21].

6 Regulation Regarding Water Reuse and PhACs

PhACs are ubiquitously detected in wastewater, but also freshwater, seawater and even drinking water. PhACs are among the compounds that can be referred to as "pseudo-persistent" contaminants as their transformation and removal are compensated by their continuous introduction into the environment [91]. Potential harmful effects both on the ecosystem and on human health are currently under extensive evaluation [3]. Continuous input of low-level pharmaceuticals into water bodies from WWTPs can adversely affect a variety of aquatic organisms [92] and lead to many kinds of anomalies [93]. PhACs can also enter the food chain, possibly resulting in biomagnification. Direct intake of these pollutants is also possible when human beings and animals drink already contaminated water. Additionally, antibiotics are not completely removed in WWTPs; microorganisms in the environment can be therefore continuously exposed to them, favouring the spread of antimicrobial resistance genes. According to the World Health Organization (WHO), "By 2050 – if effective interventions against antimicrobial resistance are not made – 10 million deaths and an economic loss of US$ 100 trillion may occur annually as the result of such resistance" [94].

The adoption of new legislation concerning water reuse, a better understanding of organic micropollutant impacts on the environment and public awareness campaigns about their hazards are paramount to encourage public and private initiatives to attain high-quality water and wastewater treatment standards in the next years [95]. Nonetheless the regulations at EU level that restrict PhAC presence in wastewater and the environment are still very limited. In 2020, the Council of the European Union proposed a new regulation to reduce the risks of water shortages by improving the availability of water and encourage its efficient reuse. The main focus was on water reuse in agricultural, while other possibilities (e.g. aquifer recharge) were finally disregarded. In this context, the removal of substances of emerging concern (mentioning in the risk assessment section also organic micropollutants, including PhACs) is fundamental for safe water reuse [14].

Moreover, micropollutants have been targeted with a Commission Implementing Decision in 2015 [96] and then updated through Decision 2018/480 [97]. The watch list is a list of micropollutants that may pose a significant risk to or via the aquatic environment but for which data are still insufficient to support their prioritization and must, therefore, be monitored Europe-wide by member states [13]. The PhACs included, for the first time in an EU Decision, were diclofenac in the 2015 watch List, amoxicillin and ciprofloxacin in the 2018 watch list and three macrolide antibiotics (erythromycin, clarithromycin and azithromycin) in both. Since the monitoring must be carried out in freshwater only, there is a missing link between the occurrence of such pollutants in wastewater and the risk associated with their presence when reclaimed wastewater is reused [98].

The situation outside the EU is different. Switzerland regulation established that WWTPs following certain criteria should be upgraded with advanced treatment by 2035 in order to ensure the removal of at least 80% of micropollutant loads (including amisulpride, carbamazepine, citalopram, clarithromycin, diclofenac, hydrochlorothiazide, metoprolol, venlafaxine, candesartan and irbesartan), tracking WWTPs efficiency by means of periodical monitoring campaigns [99]. Other countries (e.g. Australia, Singapore and the State of California) have implemented cutting water reuse policies, including specific regulations on grey water and direct or non-direct potable reuse. In some cases micropollutants, and PhACs among them, are also included to be monitored and/or considered in terms of potential stressors. The legislation of these countries might pave the way for implementation at EU levels [98].

Finally, the list of target compounds is typically created by selecting PhACs according to existing legislation, consumption, occurrence and available analytical methods. Nevertheless, the list of evaluated PhACs is frequently limited to previously studied compounds, while other PhACs are widely prescribed but overlooked. Moreover, new compounds are continuously introduced into the market and/or for emerging necessity (e.g. COVID-19). For this reason, other possibly relevant compounds should be also included in future legislation and studies.

7 PhAC Fate in Sludge Treatment Line: Focus on Agricultural Reuse

7.1 Agricultural Sludge Reuse: Advantages and Environmental Impacts

Nutrient contents and soil-conditioning properties of sewage sludge can be usefully exploited with its reuse in agriculture, especially for restoring overexploited land or for improving the humus content and water-holding capacity of light-textured sandy soil [100]. Agricultural reuse (directly or after composting) is a sustainable practice in the frame of the circular economy; it is extensively applied in several countries accounting 53% of the produced sludge in EU-27 countries [101], but the potential risk associated with the release of toxic and persistent substances in aquifers and soils has to be adequately considered.

PhACs constitute one of the broadest groups of compounds present in the urban wastewater; they are often characterized by hydrophobic-lipophilic properties which can favour their sorption (adsorption or absorption) onto the sludge. This prevalent removal mechanism causes an enrichment of the solid fraction at concentrations of several orders of magnitude higher than in the original wastewater [102].

Release of sorbed compounds onto the sludge in agricultural reuse can have a strong environmental impact because of the possible direct uptake by crops and consequent transfer to the food chain with a serious concern for human health and negative effects on plant growth [103]. In the specific case of PhACs, it has been observed that WWTPs concentrate antibiotics in the produced sludge, especially for hospitals and industrial wastewater, and the activated sludge systems provide a favourable environment with high microbial biomass concentration and selective process for horizontal gene transfer of antibiotic-resistant genes [102]. Moreover, the persistence and recalcitrance of many PhACs amplify negative impacts over time with consequences difficult to predict for long-term scenarios.

PhACs are the most investigated class of micropollutants in WWTPs, but published studies are mainly devoted to the characterization of the different types of sludge produced by different treatment options more than on the sludge treatment solutions able to provide a more effective removal of the accumulated pollutants. This aspect has been considered in recent publications, where the performance of different sludge treatment lines is analysed, not only in terms of solid reduction and stabilization but also taking into account the "quality" of the digested sludge, for several classes of micropollutants including PhACs [104, 105]. Results of these studies are summarized in the next section.

7.2 *Sludge Treatment Alternatives to Enhance PhAC Removal*

For WWTPs of medium/high potentiality (i.e. ≥30,000 P.E.), the preferred sludge treatment option is the anaerobic digestion, which provides good solid removal efficiencies and allows energy recovery though methane production. Mesophilic process ($T = 35°C$) is more commonly applied than the thermophilic one (T in the range of 55–60°C) given the lower energy demand, but the one-step mesophilic digestion is less efficient for organic compound removal (including PhACs), methane production and hygienization [106]. Enhancement of mesophilic sludge digestion can be achieved with appropriate pre- or post-treatments (both chemical-physical and biological) or with the combined anaerobic-aerobic sequential process.

Braguglia et al. analysed the fate of 11 selected PhACs in an extended study on the thermophilic digestion coupled with thermal hydrolysis pretreatment (THT) and of a dual-stage mesophilic/thermophilic digestion with (UMT) and without (MT) ultrasound pretreatment applied to secondary and mixed sludge [104]. PhAC concentration was in the range of 100–160 and 110–130 mg/kg$_{dw}$ in the secondary and mixed sludge, respectively. They observed that THT process was not effective in PhAC removal for all investigated conditions, while better results were achieved with MT process resulting in ~57% removal efficiency with no substantial beneficial effect exerted by the ultrasound pretreatment (UMT). Better performance of the combined MT process was detected for PhAC removal in secondary sludge than in mixed sludge digestion, where no statistically significant removal was observed.

Enhanced performance of sequential anaerobic-aerobic (AA) digestion with respect to the single-step anaerobic or aerobic process was demonstrated in previous studies: the motivation behind this approach is that the combined process provides differentiated reaction environments (anaerobic and aerobic) and thus optimal biodegradability conditions for the different volatile solid sludge fractions and pollutants present in the sludge matrix. A further improvement of the AA can be achieved by operating the post-aerobic step under mesophilic conditions [107] with a low external energy supply well compensated by the improved efficiency. Tomei et al. investigated the fate of a selected group of PhACs on secondary sludge treated in two AA schemes conducted with the same operating conditions for the anaerobic step (SRT 15 days $T = 37°C$), while the aerobic phase was operated at SRT = 12 days, $T = 20°C$ (AA1) and $T = 37°C$ (AA2) [105]. Table 4 shows the removal efficiencies for PhACs detected at the highest concentration among the investigated ones. Negative values indicate accumulation due to poor removal and volatile solid concentration decrease in the digestion process.

Table 4 highlights a high variability of the removal efficiencies which are attributable to the variability of the fed sludge, which remarkably affects the anaerobic removals in the two series, especially for the less biodegradable compounds, i.e. carbamazepine and diclofenac. Additional removal was achieved, for both compounds, in the post-aerobic step at both temperatures but with better

Table 4 Removal efficiencies (RE) of sequential anaerobic-aerobic (AA) digestion observed for representative PhACs (modified from [105]). Negative values indicate accumulation in the solid phase. The average PhAC concentration in the fed sludge was 140 ± 40 mg/kg$_{dw}$

	RE AA1 (%)			RE AA2 (%)		
Compound	Anaerobic	Aerobic	Total	Anaerobic	Aerobic	Total
Carbamazepine	34	3	36	−13	28	19
Dihydro-dihydroxy-carbamazepine	38	4.5	41	−46	13	−28
Dihydro-hydroxy-carbamazepine	56	25	67	−21	46	34
Clarithromycin	90	−271	63	85	−137	63
Diclofenac	41	2	43	−31	45	28
Fluconazole	51	−10	46	−3	24	22
Venlafaxine	45	8	49	19	27	40

performance in AA2 (at higher aerobic temperature). Similar results were observed for the other compounds with the only exception of clarithromycin, which was mainly removed in the anaerobic phase (85–90%).

The reported study cases are good examples of the possibility of producing sludge of better quality, in terms of PhAC content, with more efficient digestion lines, even if the complexity and variability of the sludge matrix require additional research in the field.

7.3 *EU Regulation for Agricultural Reuse*

The EU reference document for the regulation of land disposal of sewage sludge is the Sewage Sludge Directive (SSD) 86/278/EEC [108], which focuses essentially on heavy metals but does not set any threshold value for organic compounds for sludge application. National regulations in EU member states are established with the option to fix more restrictive limit values than the ones suggested in the SSD [108]. A more recent document is the "Working Document on Sludge and Biowaste" [109] with maximum proposed permissible levels for heavy metals and polycyclic aromatic hydrocarbons but not yet for PhACs.

This short overview highlights the need for revision and integration of the EU and national legislation to establish additional limit values for sludge reuse in agriculture, including a wider spectrum of organic contaminants exerting detrimental effects for both the human health and the environment. In the specific case of PhACs, a serious concern is that the sludge might be a reservoir of antibiotic resistance genes (ARGs) and, when reused in agriculture, could contribute to the spreading of the "antibiotic resistance", which is recognized as one of the most important challenges of contemporary medicine and a serious public health problem. Future regulations should also consider this phenomenon and define accurate parameters to be monitored.

Acknowledgements Marc Castaño acknowledges the FI fellowship (2020 FI_B 00711) from the Catalan Government. Gianluigi Buttiglieri acknowledges the Ramon y Cajal research fellowship RYC-2014-16754 and the CLEaN-TOUR project (CTM2017-85385-C2-1-R) from the Spanish Ministry of Economy and Competitiveness. The authors thank Generalitat de Catalunya through Consolidated Research Group 2017-SGR-1318. ICRA researchers thank funding from the CERCA programme. Monica Brienza acknowledges the "Attraction and International Mobility Programme" for her competitive contract funded by the European Union.

References

1. aus der Beek T, Weber FA, Bergmann A, Hickmann S, Ebert I, Hein A, Küster A (2016) Pharmaceuticals in the environment-global occurrences and perspectives. Environ Toxicol Chem 35(4):823–835. https://doi.org/10.1002/etc.3339
2. Kümmerer K, Henninger A (2003) Promoting resistance by the emission of antibiotics from hospitals and households into effluent. Clin Microbiol Infect 9(12):1203–1214. https://doi.org/10.1111/j.1469-0691.2003.00739.x

3. Majumder A, Gupta B, Gupta AK (2019) Pharmaceutically active compounds in aqueous environment: a status, toxicity and insights of remediation. Environ Res 176:108542. https://doi.org/10.1016/j.envres.2019.108542
4. Buttiglieri G (2008) Removal of emerging contaminants in wastewater treatment: conventional activated sludge treatment. In: Barceló D, Petrović M (eds) The handbook of environmental chemistry. Springer, Berlin, pp 1–35. https://doi.org/10.1007/698_5_098
5. Joss A, Zabczynski S, Göbel A, Hoffmann B, Löffler D, McArdell CS, Ternes TA, Thomsen A, Siegrist H (2006) Biological degradation of pharmaceuticals in municipal wastewater treatment: proposing a classification scheme. Water Res 40(8):1686–1696. https://doi.org/10.1016/j.watres.2006.02.014
6. Rosal R, Rodríguez A, Perdigón-Melón JA, Petre A, García-Calvo E, Gómez MJ, Agüera A, Fernández-Alba AR (2010) Occurrence of emerging pollutants in urban wastewater and their removal through biological treatment followed by ozonation. Water Res 44(2):578–588. https://doi.org/10.1016/j.watres.2009.07.004
7. Luo Y, Guo W, Ngo HH, Nghiem LD, Hai FI, Zhang J, Liang S, Wang XC (2014) A review on the occurrence of micropollutants in the aquatic environment and their fate and removal during wastewater treatment. Sci Total Environ 473–474:619–641. https://doi.org/10.1016/j.scitotenv.2013.12.065
8. Spongberg AL, Witter JD (2008) Pharmaceutical compounds in the wastewater process stream in Northwest Ohio. Sci Total Environ 397(1–3):148–157. https://doi.org/10.1016/j.scitotenv.2008.02.042
9. Patel M, Kumar R, Kishor K, Mlsna T, Pittman CU, Mohan D (2019) Pharmaceuticals of emerging concern in aquatic systems: chemistry, occurrence, effects, and removal methods. Chem Rev 119(6):3510–3673. https://doi.org/10.1021/acs.chemrev.8b00299
10. Miao XS, Yang JJ, Metcalfe CD (2005) Carbamazepine and its metabolites in wastewater and in biosolids in a municipal wastewater treatment plant. Environ Sci Technol 39 (19):7469–7475. https://doi.org/10.1021/es050261e
11. Karthikeyan KG, Meyer MT (2006) Occurrence of antibiotics in wastewater treatment facilities in Wisconsin, USA. Sci Total Environ 361(1–3):196–207. https://doi.org/10.1016/j.scitotenv.2005.06.030
12. Vieno NM, Tuhkanen T, Kronberg L (2005) Seasonal variation in the occurrence of pharmaceuticals in effluents from a sewage treatment plant and in the recipient water. Environ Sci Technol 39(21):8220–8226. https://doi.org/10.1021/es051124k
13. Carballa C, Alvarino T, Buttiglieri G, Choubert JM, Pons MN (2017) Innovative primary and secondary sewage treatment technologies for organic micropollutants abatement. In: Innovative wastewater treatment and resource recovery technologies: impacts on energy, economy and environment. https://doi.org/10.2166/9781780407876_0179
14. European Commission (2020) Regulation (EU) 2020/741 of the European Parliament and of the Council of 25 may 2020 on minimum requirements for water reuse
15. Göbel A, Thomsen A, McArdell CS, Joss A, Giger W (2005) Occurrence and sorption behavior of sulfonamides, macrolides, and trimethoprim in activated sludge treatment. Environ Sci Technol 39(11):3981–3989. https://doi.org/10.1021/es048550a
16. Godfree A (2003) Health constraints on the agricultural recycling of wastewater sludges. In: The handbook of water and wastewater microbiology, vol 1, pp 281–300. ISBN 0-12-470100-0
17. Chen H, Jiang W, Yang Y, Yang Y, Man X (2015) Global trends of municipal solid waste research from 1997 to 2014 using bibliometric analysis. J Air Waste Manage Assoc 65 (10):1161–1170. https://doi.org/10.1080/10962247.2015.1083913
18. Rodriguez-Mozaz S, Chamorro S, Marti E, Huerta B, Gros M, Sànchez-Melsió A, Borrego CM, Barceló D, Balcázar JL (2015) Occurrence of antibiotics and antibiotic resistance genes in hospital and urban wastewaters and their impact on the receiving river. Water Res 69:234–242. https://doi.org/10.1016/j.watres.2014.11.021

19. Couto CF, Lange LC, Amaral MCS (2019) Occurrence, fate and removal of pharmaceutically active compounds (PhACs) in water and wastewater treatment plants – a review. J Water Process Eng 32:100927. https://doi.org/10.1016/j.jwpe.2019.100927
20. Verlicchi P, Al Aukidy M, Zambello E (2012) Occurrence of pharmaceutical compounds in urban wastewater: removal, mass load and environmental risk after a secondary treatment-A review. Sci Total Environ. https://doi.org/10.1016/j.scitotenv.2012.04.028
21. Castaño-Trias M, Rodríguez-Mozaz S, Buttiglieri G (2021) 10 years in-house prioritization method development for organic micropollutants and its application to Spanish and Greece real samples of wastewater, river and sea. Submitted
22. Wang J, Wang S (2016) Removal of pharmaceuticals and personal care products (PPCPs) from wastewater: a review. J Environ Manage 182:620–640. https://doi.org/10.1016/j.jenvman.2016.07.049
23. Ekowati Y, Ferrero G, Farré MJ, Kennedy MD, Buttiglieri G (2019) Application of UVOX Redox® for swimming pool water treatment: microbial inactivation, disinfection byproduct formation and micropollutant removal. Chemosphere 220:176–184. https://doi.org/10.1016/j.chemosphere.2018.12.126
24. Castiglioni S, Borsotti A, Riva F, Zuccato E (2016) Illicit drug consumption estimated by wastewater analysis in different districts of Milan: a case study. Drug Alcohol Rev 35 (2):128–132. https://doi.org/10.1111/dar.12233
25. Bai X, Lutz A, Carroll R, Keteles K, Dahlin K, Murphy M, Nguyen D (2018) Occurrence, distribution, and seasonality of emerging contaminants in urban watersheds. Chemosphere 200:133–142. https://doi.org/10.1016/j.chemosphere.2018.02.106
26. Jelic A, Rodriguez-Mozaz S, Barceló D, Gutierrez O (2015) Impact of in-sewer transformation on 43 pharmaceuticals in a pressurized sewer under anaerobic conditions. Water Res 68:98–108. https://doi.org/10.1016/j.watres.2014.09.033
27. Correia A, Marcano L (2015) Presence and elimination of pharmaceutical compounds in wastewater treatment plants. Worldwide review and national perspective. Boletín de malariología y salud ambiental, pp 1–18
28. Kot-Wasik A, Jakimska A, Śliwka-Kaszyńska M (2016) Occurrence and seasonal variations of 25 pharmaceutical residues in wastewater and drinking water treatment plants. Environ Monit Assess 188(12). https://doi.org/10.1007/s10661-016-5637-0
29. Daughton CG (2016) Pharmaceuticals and the environment (PiE): evolution and impact of the published literature revealed by bibliometric analysis. Sci Total Environ 562:391–426. https://doi.org/10.1016/j.scitotenv.2016.03.109
30. Radjenovic J, Petrovic M, Barceló D (2007) Analysis of pharmaceuticals in wastewater and removal using a membrane bioreactor. Anal Bioanal Chem 387:1365–1377. https://doi.org/10.1007/s00216-006-0883-6
31. Zorita S, Mårtensson L, Mathiasson L (2009) Occurrence and removal of pharmaceuticals in a municipal sewage treatment system in the south of Sweden. Sci Total Environ 407 (8):2760–2770. https://doi.org/10.1016/j.scitotenv.2008.12.030
32. Castiglioni S, Bagnati R, Fanelli R, Pomati F, Calamari D, Zuccato E (2006) Removal of pharmaceuticals in sewage treatment plants in Italy. Environ Sci Technol 40(1):357–363. https://doi.org/10.1021/es050991m
33. Vieno N, Tuhkanen T, Kronberg L (2007) Elimination of pharmaceuticals in sewage treatment plants in Finland. Water Res 41(5):1001–1012. https://doi.org/10.1016/j.watres.2006.12.017
34. Golet EM, Alder AC, Giger W (2002) Environmental exposure and risk assessment of fluoroquinolone antibacterial agents in wastewater and river water of the Glatt Valley watershed, Switzerland. Environ Sci Technol 36(17):3645–3651. https://doi.org/10.1021/es0256212
35. Lindberg RH, Olofsson U, Rendahl P, Johansson MI, Tysklind M, Andersson BAV (2006) Behavior of fluoroquinolones and trimethoprim during mechanical, chemical, and active sludge treatment of sewage water and digestion of sludge. Environ Sci Technol 40 (3):1042–1048. https://doi.org/10.1021/es0516211

36. Sipma J, Osuna B, Collado N, Monclús H, Ferrero G, Comas J, Rodriguez-Roda I (2010) Comparison of removal of pharmaceuticals in MBR and activated sludge systems. Desalination 250(2):653–659. https://doi.org/10.1016/j.desal.2009.06.073
37. Kasprzyk-Hordern B, Dinsdale RM, Guwy AJ (2009) The removal of pharmaceuticals, personal care products, endocrine disruptors and illicit drugs during wastewater treatment and its impact on the quality of receiving waters. Water Res 43(2):363–380. https://doi.org/10.1016/j.watres.2008.10.047
38. Petrovic M, Verlicchi P (2014) Water treatment plants and pharmaceutical residues in Catalonia and Italy. Contrib Sci 10(2):135–150. https://doi.org/10.2436/cs.v10i2.139201
39. Dutta K, Lee MY, Lai WWP, Lee CH, Lin AYC, Lin CF, Lin JG (2014) Removal of pharmaceuticals and organic matter from municipal wastewater using two-stage anaerobic fluidized membrane bioreactor. Bioresour Technol 165(C):42–49. https://doi.org/10.1016/j.biortech.2014.03.054
40. Monsalvo VM, McDonald JA, Khan SJ, Le-Clech P (2014) Removal of trace organics by anaerobic membrane bioreactors. Water Res 49:103–112. https://doi.org/10.1016/j.watres.2013.11.026
41. Musson SE, Campo P, Tolaymat T, Suidan M, Townsend TG (2010) Assessment of the anaerobic degradation of six active pharmaceutical ingredients. Sci Total Environ 408 (9):2068–2074. https://doi.org/10.1016/j.scitotenv.2009.11.042
42. Adrian NR, Suflita JM (1994) Anaerobic biodegradation of halogenated and nonhalogenated heterocyclic compounds in aquifer slurries. Environ Toxicol Chem 13(10):1551–1557. https://doi.org/10.1002/etc.5620131002
43. Alvarino T, Lema J, Omil F, Suárez S (2018) Trends in organic micropollutants removal in secondary treatment of sewage. Rev Environ Sci Biotechnol 17:447–469. https://doi.org/10.1007/s11157-018-9472-3
44. Kassotaki E, Buttiglieri G, Ferrando-Climent L, Rodriguez-Roda I, Pijuan M (2016) Enhanced sulfamethoxazole degradation through ammonia oxidizing bacteria co-metabolism and fate of transformation products. Water Res 94:111–119. https://doi.org/10.1016/j.watres.2016.02.022
45. Miège C, Choubert JM, Ribeiro L, Eusèbe M, Coquery M (2009) Fate of pharmaceuticals and personal care products in wastewater treatment plants – conception of a database and first results. Environ Pollut 157(5):1721–1726. https://doi.org/10.1016/j.envpol.2008.11.045
46. Göbel A, McArdell CS, Joss A, Siegrist H, Giger W (2007) Fate of sulfonamides, macrolides, and trimethoprim in different wastewater treatment technologies. Sci Total Environ 372 (2–3):361–371. https://doi.org/10.1016/j.scitotenv.2006.07.039
47. Clara M, Kreuzinger N, Strenn B, Gans O, Kroiss H (2005) The solids retention time – a suitable design parameter to evaluate the capacity of wastewater treatment plants to remove micropollutants. Water Res 39(1):97–106. https://doi.org/10.1016/j.watres.2004.08.036
48. Bouju H, Buttiglieri G, Malpei F (2008) Perspectives of persistent organic pollutants (POPS) removal in an MBR pilot plant. Desalination 224(1–3):1–6. https://doi.org/10.1016/j.desal.2007.04.071
49. Ternes TA, Joss A, Siegrist H (2004) Scrutinizing pharmaceuticals and personal care products in wastewater treatment. Environ Sci Tech 38(20):392A–399A. https://doi.org/10.1021/es040639t
50. Metcalfe CD, Koenig BG, Bennie DT, Servos M, Ternes TA, Hirsch R (2003) Occurrence of neutral and acidic drugs in the effluents of Canadian sewage treatment plants. Environ Toxicol Chem 22(12):2872–2880. https://doi.org/10.1897/02-469
51. Radjenović J, Petrović M, Barceló D (2009) Fate and distribution of pharmaceuticals in wastewater and sewage sludge of the conventional activated sludge (CAS) and advanced membrane bioreactor (MBR) treatment. Water Res 43(3):831–841. https://doi.org/10.1016/j.watres.2008.11.043
52. Li B, Zhang T (2011) Mass flows and removal of antibiotics in two municipal wastewater treatment plants. Chemosphere 83(9):1284–1289. https://doi.org/10.1016/j.chemosphere.2011.03.002

53. Dorival-García N, Zafra-Gómez A, Navalón A, González-López J, Hontoria E, Vílchez JL (2013) Removal and degradation characteristics of quinolone antibiotics in laboratory-scale activated sludge reactors under aerobic, nitrifying and anoxic conditions. J Environ Manage 120:75–83. https://doi.org/10.1016/j.jenvman.2013.02.007
54. Boethling RS, Howard PH, Meylan W, Stiteler W, Beauman J, Tirado N (1994) Group contribution method for predicting probability and rate of aerobic biodegradation. Environ Sci Technol 28(3):459–465. https://doi.org/10.1021/es00052a018
55. Fernandez-Fontaina E, Gomes IB, Aga DS, Omil F, Lema JM, Carballa M (2016) Biotransformation of pharmaceuticals under nitrification, nitratation and heterotrophic conditions. Sci Total Environ 541:1439–1447. https://doi.org/10.1016/j.scitotenv.2015.10.010
56. Yan Q, Gao X, Huang L, Gan XM, Zhang YX, Chen YP, Peng X-Y, Guo JS (2014) Occurrence and fate of pharmaceutically active compounds in the largest municipal wastewater treatment plant in Southwest China: mass balance analysis and consumption back-calculated model. Chemosphere 99:160–170. https://doi.org/10.1016/j.chemosphere.2013.10.062
57. Nakada N, Tanishima T, Shinohara H, Kiri K, Takada H (2006) Pharmaceutical chemicals and endocrine disrupters in municipal wastewater in Tokyo and their removal during activated sludge treatment. Water Res 40(17):3297–3303. https://doi.org/10.1016/j.watres.2006.06.039
58. Čelić M, Gros M, Farré M, Barceló D, Petrović M (2019) Pharmaceuticals as chemical markers of wastewater contamination in the vulnerable area of the Ebro Delta (Spain). Sci Total Environ 652:952–963. https://doi.org/10.1016/j.scitotenv.2018.10.290
59. García-Galán MJ, Díaz-Cruz MS, Barceló D (2011) Occurrence of sulfonamide residues along the Ebro river basin. Removal in wastewater treatment plants and environmental impact assessment. Environ Int 37(2):462–473. https://doi.org/10.1016/j.envint.2010.11.011
60. Komesli OT, Muz M, Ak MS, Bakirdere S, Gokcay CF (2015) Occurrence, fate and removal of endocrine disrupting compounds (EDCs) in Turkish wastewater treatment plants. Chem Eng J 277:202–208. https://doi.org/10.1016/j.cej.2015.04.115
61. Wick A, Fink G, Joss A, Siegrist H, Ternes TA (2009) Fate of beta blockers and psycho-active drugs in conventional wastewater treatment. Water Res 43(4):1060–1074. https://doi.org/10.1016/j.watres.2008.11.031
62. Gros M, Rodríguez-Mozaz S, Barceló D (2012) Fast and comprehensive multi-residue analysis of a broad range of human and veterinary pharmaceuticals and some of their metabolites in surface and treated waters by ultra-high-performance liquid chromatography coupled to quadrupole-linear ion trap tandem mass spectrometry. J Chromatogr A 1248:104–121. https://doi.org/10.1016/j.chroma.2012.05.084
63. Lajeunesse A, Smyth SA, Barclay K, Sauvé S, Gagnon C (2012) Distribution of antidepressant residues in wastewater and biosolids following different treatment processes by municipal wastewater treatment plants in Canada. Water Res 46(17):5600–5612. https://doi.org/10.1016/j.watres.2012.07.042
64. Santos JL, Aparicio I, Alonso E (2007) Occurrence and risk assessment of pharmaceutically active compounds in wastewater treatment plants. A case study: Seville city (Spain). Environ Int 33(4):596–601. https://doi.org/10.1016/j.envint.2006.09.014
65. Subedi B, Kannan K (2015) Occurrence and fate of select psychoactive pharmaceuticals and antihypertensives in two wastewater treatment plants in New York state, USA. Sci Total Environ 514:273–280. https://doi.org/10.1016/j.scitotenv.2015.01.098
66. Kim M, Guerra P, Shah A, Parsa M, Alaee M, Smyth SA (2014) Removal of pharmaceuticals and personal care products in a membrane bioreactor wastewater treatment plant. Water Sci Technol 69(11):2221–2229. https://doi.org/10.2166/wst.2014.145
67. Kreuzinger N, Clara M, Strenn B, Kroiss H (2004) Relevance of the sludge retention time (SRT) as design criteria for wastewater treatment plants for the removal of endocrine disruptors and pharmaceuticals from wastewater. Water Sci Technol 50(5):149–156. https://doi.org/10.2166/wst.2004.0322

68. Santos JL, Aparicio I, Callejón M, Alonso E (2009) Occurrence of pharmaceutically active compounds during 1-year period in wastewaters from four wastewater treatment plants in Seville (Spain). J Hazard Mater 164(2–3):1509–1516. https://doi.org/10.1016/j.jhazmat.2008.09.073
69. Clara M, Strenn B, Ausserleitner M, Kreuzinger N (2004) Comparison of the behaviour of selected micropollutants in a membrane bioreactor and a conventional wastewater treatment plant. Water Sci Technol 50(5):29–36. https://doi.org/10.2166/wst.2004.0305
70. Inyang M, Flowers R, McAvoy D, Dickenson E (2016) Biotransformation of trace organic compounds by activated sludge from a biological nutrient removal treatment system. Bioresour Technol 216:778–784. https://doi.org/10.1016/j.biortech.2016.05.124
71. Suárez S, Ramil M, Omil F, Lema JM (2005) Removal of pharmaceutically active compounds in nitrifying-denitrifying plants. Water Sci Technol 52(8):9–14. https://doi.org/10.2166/wst.2005.0214
72. Andreozzi R, Marotta R, Paxéus N (2003) Pharmaceuticals in STP effluents and their solar photodegradation in aquatic environment. Chemosphere 50(10):1319–1330. https://doi.org/10.1016/S0045-6535(02)00769-5
73. Carballa M, Omil F, Lema JM, Llompart M, García-Jares C, Rodríguez I, Gómez M, Ternes T (2004) Behavior of pharmaceuticals, cosmetics and hormones in a sewage treatment plant. Water Res 38(12):2918–2926. https://doi.org/10.1016/j.watres.2004.03.029
74. Lindqvist N, Tuhkanen T, Kronberg L (2005) Occurrence of acidic pharmaceuticals in raw and treated sewages and in receiving waters. Water Res 39(11):2219–2228. https://doi.org/10.1016/j.watres.2005.04.003
75. Ferrando-Climent L, Collado N, Buttiglieri G, Gros M, Rodriguez-Roda I, Rodriguez-Mozaz-S, Barceló D (2012) Comprehensive study of ibuprofen and its metabolites in activated sludge batch experiments and aquatic environment. Sci Total Environ 438:404–413. https://doi.org/10.1016/j.scitotenv.2012.08.073
76. Tauxe-Wuersch A, De Alencastro LF, Grandjean D, Tarradellas J (2005) Occurrence of several acidic drugs in sewage treatment plants in Switzerland and risk assessment. Water Res 39(9):1761–1772. https://doi.org/10.1016/j.watres.2005.03.003
77. Jones OAH, Voulvoulis N, Lester JN (2007) The occurrence and removal of selected pharmaceutical compounds in a sewage treatment works utilising activated sludge treatment. Environ Pollut 145(3):738–744. https://doi.org/10.1016/j.envpol.2005.08.077
78. Roberts PH, Thomas KV (2006) The occurrence of selected pharmaceuticals in wastewater effluent and surface waters of the lower Tyne catchment. Sci Total Environ 356(1–3):143–153. https://doi.org/10.1016/j.scitotenv.2005.04.031
79. Kimura K, Hara H, Watanabe Y (2007) Elimination of selected acidic pharmaceuticals from municipal wastewater by an activated sludge system and membrane bioreactors. Environ Sci Technol 41(10):3708–3714. https://doi.org/10.1021/es061684z
80. Samaras VG, Stasinakis AS, Mamais D, Thomaidis NS, Lekkas TD (2013) Fate of selected pharmaceuticals and synthetic endocrine disrupting compounds during wastewater treatment and sludge anaerobic digestion. J Hazard Mater 244–245:259–267. https://doi.org/10.1016/j.jhazmat.2012.11.039
81. Yu JT, Bouwer EJ, Coelhan M (2006) Occurrence and biodegradability studies of selected pharmaceuticals and personal care products in sewage effluent. Agric Water Manag 86(1–2):72–80. https://doi.org/10.1016/j.agwat.2006.06.015
82. Watkinson AJ, Murby EJ, Costanzo SD (2007) Removal of antibiotics in conventional and advanced wastewater treatment: implications for environmental discharge and wastewater recycling. Water Res 41(18):4164–4176. https://doi.org/10.1016/j.watres.2007.04.005
83. Ghosh GC, Okuda T, Yamashita N, Tanaka H (2009) Occurrence and elimination of antibiotics at four sewage treatment plants in Japan and their effects on bacterial ammonia oxidation. Water Sci Technol 59(4):779–786. https://doi.org/10.2166/wst.2009.067
84. Behera SK, Kim HW, Oh JE, Park HS (2011) Occurrence and removal of antibiotics, hormones and several other pharmaceuticals in wastewater treatment plants of the largest

industrial city of Korea. Sci Total Environ 409(20):4351–4360. https://doi.org/10.1016/j.scitotenv.2011.07.015
85. Sahar E, Ernst M, Godehardt M, Hein A, Herr J, Kazner C, Melin T, Cikurel H, Aharoni A, Messalem R, Brenner A, Jekel M (2011) Comparison of two treatments for the removal of selected organic micropollutants and bulk organic matter: conventional activated sludge followed by ultrafiltration versus membrane bioreactor. Water Sci Technol 63(4):733–740. https://doi.org/10.2166/wst.2011.300
86. Kassotaki E, Pijuan M, Rodriguez-Roda I, Buttiglieri G (2019) Comparative assessment of endocrine disrupting compounds removal in heterotrophic and enriched nitrifying biomass. Chemosphere 217:659–668. https://doi.org/10.1016/j.chemosphere.2018.11.012
87. Evgenidou EN, Konstantinou IK, Lambropoulou DA (2015) Occurrence and removal of transformation products of PPCPs and illicit drugs in wastewaters: a review. Sci Total Environ 505:905–926. https://doi.org/10.1016/j.scitotenv.2014.10.021
88. Langin A, Schuster A, Kummerer K (2008) Chemicals in the environment – the need for a clear nomenclature: parent compounds, metabolites, transformation products and their elimination. Clean-Soil Air Water 36:349–350
89. Rubirola A, Llorca M, Rodriguez-Mozaz S, Casas N, Rodriguez-Roda I, Barceló D, Buttiglieri G (2014) Characterization of metoprolol biodegradation and its transformation products generated in activated sludge batch experiments and in full scale WWTPs. Water Res 63:21–32. https://doi.org/10.1016/j.watres.2014.05.031
90. Majewsky M, Glauner T, Horn H (2015) Systematic suspect screening and identification of sulfonamide antibiotic transformation products in the aquatic environment. Anal Bioanal Chem 407(19). https://doi.org/10.1007/s00216-015-8748-5
91. Barceló D, Petrovic M (2007) Pharmaceuticals and personal care products (PPCPs) in the environment. Anal Bioanal Chem 22(3):194–203. https://doi.org/10.1007/s00216-006-1012-2
92. Fent K, Weston AA, Caminada D (2006) Ecotoxicology of human pharmaceuticals. Aquat Toxicol 76(2):122–159. https://doi.org/10.1016/j.aquatox.2005.09.009
93. Zenker A, Cicero MR, Prestinaci F, Bottoni P, Carere M (2014) Bioaccumulation and biomagnification potential of pharmaceuticals with a focus to the aquatic environment. J Environ Manage 133:378–387. https://doi.org/10.1016/j.jenvman.2013.12.017
94. Tangcharoensathien V, Sattayawutthipong W, Kanjanapimai S, Kanpravidth W, Brown R, Sommanustweechai A (2017) Antimicrobial resistance: from global agenda to national strategic plan, Thailand. Bull World Health Organ 95(8):599–603. https://doi.org/10.2471/BLT.16.179648
95. Rizzo L, Malato S, Antakyali D, Beretsou VG, Đolić MB, Gernjak W, Heath E, Ivancev-Tumbas I, Karaolia P, Lado Ribeiro AR, Mascolo G, McArdell CS, Schaar H, Silva AMT, Fatta-Kassinos D (2019) Consolidated vs new advanced treatment methods for the removal of contaminants of emerging concern from urban wastewater. Sci Total Environ 655:986–1008. https://doi.org/10.1016/j.scitotenv.2018.11.265
96. European Commission (2015) EUR-Lex – 2015/495 of 20 March 2015 establishing a watch list of substances for Union-wide monitoring in the field of water policy pursuant to Directive 2008/105/EC of the European Parliament and of the Council. Off J Eur Communities L78:40–42
97. European Commission (2018) EUR-Lex – Commission implementing decision (EU) 2018/840 of 5 June 2018 establishing a watch list of substances for Union-wide monitoring in the field of water policy pursuant to Directive 2008/105/EC of the European Parliament and of the Council and repealing Commission Implementing Decision (EU) 2015/495. J Eur Union L141:9–12
98. Nika E, Gusmaroli L, Ghafourian M, Atanasova N, Buttiglieri G, Katsou E (2020) Nature-based solutions as enablers of circularity in water systems: a review on assessment methodologies, tools and indicators. Water Res 183:115988. https://doi.org/10.1016/j.watres.2020.115988

99. Swiss Confederation (2016) Ordinance of the Federal Law of introduction on water protection (Cantonal Ordinance on the water protection, OCPAc)
100. Verlicchi P, Zambello E (2015) Pharmaceuticals and personal care products in untreated and treated sewage sludge: occurrence and environmental risk in the case of application on soil – a critical review. Sci Total Environ 538:750–767. https://doi.org/10.1016/j.scitotenv.2015.08.108
101. Kelessidis A, Stasinakis AS (2012) Comparative study of the methods used for treatment and final disposal of sewage sludge in European countries. Waste Manag 32(6):1186–1195. https://doi.org/10.1016/j.wasman.2012.01.012
102. Mailler R, Gasperi J, Patureau D, Vulliet E, Delgenes N, Danel A, Deshayes S, Eudes V, Guerin S, Moilleron R, Chebbo G, Rocher V (2017) Fate of emerging and priority micropollutants during the sewage sludge treatment: case study of Paris conurbation. Part 1: contamination of the different types of sewage sludge. Waste Manag 59:379–393. https://doi.org/10.1016/j.wasman.2016.11.010
103. Smith SR (2009) Organic contaminants in sewage sludge (biosolids) and their significance for agricultural recycling. Phil Trans R Soc A Math Phys Eng Sci 367(1904):4005–4041. https://doi.org/10.1098/rsta.2009.0154
104. Braguglia CM, Coors A, Gallipoli A, Gianico A, Guillon E, Kunkel U, Mascolo G, Richter E, Ternes TA, Tomei MC, Mininni G (2015) Quality assessment of digested sludges produced by advanced stabilization processes. Environ Sci Pollut Res 22(10):7216–7235. https://doi.org/10.1007/s11356-014-3090-6
105. Tomei MC, Mosca Angelucci D, Mascolo G, Kunkel U (2019) Post-aerobic treatment to enhance the removal of conventional and emerging micropollutants in the digestion of waste sludge. Waste Manag 96
106. Riau V, De la Rubia MÁ, Pérez M (2010) Temperature-phased anaerobic digestion (TPAD) to obtain class A biosolids: a semi-continuous study. Bioresour Technol 101(8):2706–2712. https://doi.org/10.1016/j.biortech.2009.11.101
107. Tomei MC, Mosca Angelucci D, Levantesi C (2016) Two-stage anaerobic and post-aerobic mesophilic digestion of sewage sludge: analysis of process performance and hygienization potential. Sci Total Environ 545–546:453–464. https://doi.org/10.1016/j.scitotenv.2015.12.053
108. CEC (1986) Council Directive of 12 June 1986 on the Protection of the Environment, and in Particular of the Soil, when Sewage Sludge is used in Agriculture (86/278/EEC). Council of the European Communities. Off J Eur Communities No. L 181/6-12
109. EU Working Document on Sludge and Biowaste (2010) Retrieved from European Water Association: http://www.dwa.de/portale/ewa/ewa.nsf/home?readform&objectid=AC9B268C90BC9930C12577E00045E8E2; Last Accessed July 2020

New Insights into the Occurrence of Micropollutants and the Management and Treatment of Hospital Effluent

Paola Verlicchi

Contents

1 Introduction 55
2 Investigations into the Occurrence of Pharmaceuticals in Hospital Effluent 57
3 Investigations on Occurrence of Antibiotic-Resistant Genes and Bacteria (ARG and ARB) 65
4 Investigations on Hospital Effluent Management and Treatments 70
5 Prioritisation: Environmental Risk Assessment of PhCs in HWW 83
6 Strategies in HWW Management and Treatments: Lessons Learned from Singular Experiences 86
7 The Need for Further Debate and Research 89
References 90

Abstract This chapter deals with investigations carried out over the last 5 years on hospital effluent in terms of the occurrence of micropollutants; new and promising technologies tested to improve the removal of key compounds (including emerging contaminants); the environmental and health risk assessments of pharmaceutical residues and pathogens; and, finally, some of the strategies adopted in hospital effluent management and treatments through the discussion of some case studies. It emerges that the occurrence and treatment of hospital effluent are becoming issues of increasing concern also for countries such as Morocco, Tunisia, Iran and Colombia, whose research groups had not actively participated in the worldwide debate thus far. Their interest in these topics highlights the shared, global awareness of the need to adopt safe, economic and technically feasible technologies for the treatment of hospital effluent to reduce the impact on the aquatic environment of hazardous substances typically administered or used in healthcare facilities. The experiences reported and discussed herein demonstrate the worldwide efforts that have been

P. Verlicchi (✉)
Department of Engineering, University of Ferrara, Ferrara, Italy
e-mail: paola.verlicchi@unife.it

Sara Rodriguez-Mozaz, Paqui Blánquez Cano, and Montserrat Sarrà Adroguer (eds.), *Removal and Degradation of Pharmaceutically Active Compounds in Wastewater Treatment*, Hdb Env Chem (2021) 108: 53–96, DOI 10.1007/698_2020_677,

made and are still ongoing with the aim of reaching Sustainable Millennium Goal number 6 “Improve Clean Water and Sanitation” by 2030, as defined by the World Health Organization.

Keywords Environmental risk assessment, Hospital effluent, Management, Occurrence, Pharmaceuticals, Removal

Acronyms

AOP	Advanced oxidation process
AOX	Adsorbable organic halide
ARB	Antibiotic-resistant bacteria
ARG	Antibiotic-resistant gene
CPE	Carbapenemase-producing bacteria
CRE	Carbapenem-resistant *Enterobacteriaceae*
DOC	Dissolved organic carbon
DOX	Doxorubicin
EQS	Environmental quality standards
ESBL	Extended-spectrum β-lactam
FC	Faecal coliform
FIB	Fecal indicator bacteria
FS	Flat sheet
GAC	Granular activated carbon
HF	Hollow fibre
HWW	Hospital wastewater
LOD	Limit of detection
MBBR	Moving bed biofilm reactor
MBR	Membrane bioreactor
MDR	Multidrug resistance
MDRB	Multidrug-resistant bacteria
MEC	Measured environmental concentration
MLSS	Mixed liquor suspended solid
MLVSS	Mixed liquor volatile suspended solids
MTP	Metoprolol
MTPA	Metoprolol acid
NF	Nanofiltration
NSAID	Non-steroidal anti-inflammatory drug
O_3	Ozonation
OLR	Organic loading rate
PAC	Powder activated carbon
PNEC	Predicted no-effect concentration
RO	Reverse osmosis
RQ	Risk quotient
SMBR	Sponge membrane bioreactor

SVI	Sludge volume index
VRE	Vancomycin-resistant enterococci
WWTP	Wastewater treatment plant

1 Introduction

Over the last 10 years, an increasing number of studies have dealt with the chemical, physical and microbiological characteristics of hospital wastewater (HWW), their variability over time (day, month and year), its treatability, the removal of macro- and micropollutants and the risks related to their residues once released into the environment. There have also been discussions among the scientific community about the options to treat hospital wastewater separately or in combination with domestic wastewater. In 2010, an initial comprehensive review on hospital effluent as a source of micropollutants was published [1] and gave a snapshot into what was already known and also underlined the need for further research in a range of fields related to the management and treatment of this effluent. At the same time, some EU-funded projects have shed light on specific issues related to the treatment of hospital effluent. Of these, the PILLS project (https://cordis.europa.eu/project/id/214599/reporting) investigated wastewater quality in terms of selected micropollutants and researched and tested removal methods in order to identify and suggest better wastewater treatment at the source (hospitals and households) [2]. In 2015, another review was published focusing on the lessons learned on the management and treatment of hospital effluent [3], and, in 2018, a book by Springer published a collection of contributions from different expert research groups on hospital wastewater in terms of the occurrence of micropollutants, treatment options and risk assessment [4].

This chapter gives an update on the discussion and aims to highlight the main results achieved worldwide in the management and treatment of hospital effluent over the last 5 years. What emerges is that attention to these topics has involved new countries (including Morocco, Tunisia, Iran, Colombia), which were not among those reported in the previous review. In addition, the scientific debate tends to focus on antibiotic-resistant bacteria (ARB) and antibiotic-resistant genes (ARG) and their correlation with antibiotics and the risks related to the use of the receiving surface water for different needs (mainly for agricultural and drinking purposes). The main results refer to the studies reported on the map in Fig. 1. It clearly shows the countries where the studies took place as well as "knowledge spheres" (occurrence, ARB/ARG, risk, treatment, management, prioritisation) which have contributed to this update. On the same map, it is possible to see the countries where previous investigations on HWW treatments have been conducted, according to the study by [3].

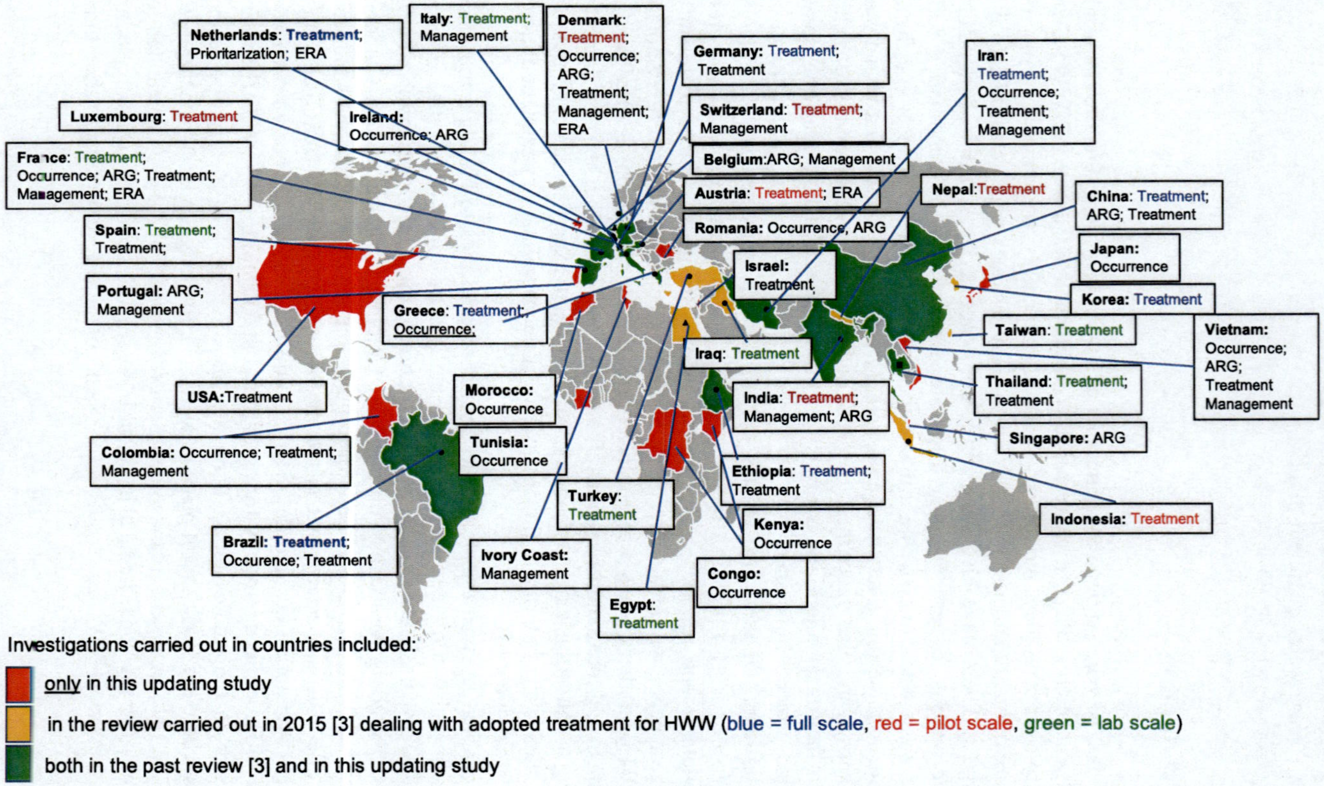

Fig. 1 Worldwide distribution of the studies on hospital effluent included in this chapter (red and green countries) and studies included in the review by [3] (yellow and green countries)

2 Investigations into the Occurrence of Pharmaceuticals in Hospital Effluent

Different classes of pollutants are present in HWW due to the different activities carried out within the healthcare structure. As noted in many studies, including [1, 5, 6], detergents and disinfectants are largely consumed in healthcare structures and represent classes of substances of concern. The occurrence of detergents and disinfectants in HWW was investigated in many cases in European countries [7]. Recently their consumption was analysed in the study conducted by [8] relating to the University Hospital of Marrakesh, Morocco (409 beds, 425 m^3/day). The study also analysed effluent in terms of common macropollutants (BOD, COD, nitrogen and phosphorus compounds), heavy metals and pathogens (bacteria and viruses) in order to evaluate the potential pollutant load and risk posed by the healthcare facility on the environment. It was found that, annually, around 17,100 kg of the detergents and disinfectants used included 25% sodium hypochlorite, 24% mild soap and 18% disinfectant glutaraldehyde. Administered pharmaceutical compounds (PhCs) over the same year exceeded 106,000 kg, with the class of anaesthetics at the top with more than 38,100 kg/year consumed, followed by the class of medicines for infectious diseases and parasitology which required 14,850 kg/year of PhCs. Water samples were taken at 12 different points within the hospital and underwent chemical and bacteriological analysis. Concentrations of copper (2.84 mg/L) and iron (3.38 mg/L) were higher than the standard limits in Morocco. This was due to the extensive consumption of drugs containing such minerals, but also to release from the old drainage pipes of the studied hospital, which was built in 1938. With reference to microorganisms, average concentrations were found of 3.9×10^7 CFU/100 mL for total coliforms, 2.6×10^7 CFU/100 mL for faecal coliform and 4.7×10^6 CFU/100 mL for intestinal enterococci. Regarding pathogens, the highest concentrations were found for *Staphylococcus aureus* (7.5×10^5 CFU/100 mL) and *Pseudomonas aeruginosa* (5.3×10^6 CFU/100 mL) in medical lab effluent and the presence of strains of *Salmonella* in intensive care unit, cardiology and gastroenterology surgery ward effluent, as well as in medical laboratory effluent and, finally, *Vibrio* in surgery and radiology ward effluent. They concluded that it is fundamental to know the characteristics of the hospital effluent in order to provide an adequate pre-treatment prior to its disposal into the public sewer system and to reduce the load directly discharged.

Investigations were carried out on the bacteriological contamination of hospitals and urban wastewaters in Kinshasa, which is the capital and largest city in the Democratic Republic of Congo [9, 10]. In this area, there are 20 hospitals, various medical centres and numerous polyclinics with different intrinsic characteristics. Big hospitals discharge their effluents in the urban drainage systems and then into the river without any treatment. Smaller healthcare structures and medical centres release their wastewater directly onto soil or directly into the river. Analysis carried out by [9] on the effluent of three hospitals showed that the concentrations of *E. coli* were in the range 4.5 × 10^5 UFC/100/L and 1.0 × 10^6 UFC/100 mL and of enterococci between 2.3 × 10^4 UFC/100 mL and 1.0 × 10^5 UFC/100 mL.

In the same area, [10] instead investigated the occurrence of toxic metals, ARGs and faecal indicator bacteria (FIB) in the sediments upstream and downstream of hospital effluent release points in the receiving rivers. They found that, depending on the size and activities within the healthcare structure, directly discharged raw hospital effluent leads to an increment in FIB, ARGs and heavy metal concentrations. In some cases, higher values of FIB and ARGs were found in sediment samples upstream of the hospital discharge point. This could be related to the fact that the studied rivers go through the City of Kinshasa and thus other sources could be responsible for the deterioration of the bacteriological quality in the rivers, including open defecation, uncontrolled landfills, unregulated effluent discharges and inadequate sewage collection near the studied sites. The authors remarked that the rivers act as a reservoir of these contaminants and, in tropical conditions like in Congo, they could potentially favour the transfer of mobile genetic elements carrying ARGs to susceptible bacterial pathogens.

Both studies remark the need for the adequate management and treatment of hospital effluent. In Congo, as in most other developing countries, river water is commonly used for both human and animal consumption and is also used for irrigation for fresh urban products. High values of metals, FIB and ARGs in rivers represent a risk to human health and the environment.

Ibrahim et al. [11] investigated the occurrence and frequency of sapoviruses in wastewater from three clinics located in the residential and business area of El Menzeh I, in the centre of Tunis (Tunisia), as well as their removal in the case of rotating biodisks (25 m^3/day, 2 m^3 the volume of the biodisk tank) and in a series of five oxidising ponds (average flow rate 47 m^3/day, overall HRT 13.3 days). As the analysed treatments were able to reduce the content of viruses of only 1 log unit (90 94%), they emphasised the inadequacy of the sanitary quality of these treated effluents for safe discharge into the natural receiving environment and also for indirect reuse purposes. In addition, they also underlined the urgent need for a (chemical or physical) disinfection step for hospital wastewater. The results highlighted that treated effluent may represent a possible route for the transmission of opportunistic pathogenic bacteria and sapoviruses. Lastly, they highlighted the high risk of infection related to this kind of water being released into the environment, if then to be used for agricultural and recreational purposes.

In another area in Africa, the Lake Victoria Basin (Kenya), the detection of the level of common antibiotics (ampicillin, amoxicillin, sulphamethoxazole, chloramphenicol and ciprofloxacin) in hospital lagoon effluents and sludge, as well as in the effluent and sludge of ten local wastewater treatment plants (WWTPs) and receiving rivers and lakes, was the focus of dedicated monitoring campaigns [12]. With regard to hospital lagoon effluent, it was found that selected antibiotics ranged between <0.05 and 0.79 ± 0.05 μg/L (the highest concentrations were due to ampicillin). Average concentrations in the ten sampled WWTP effluents were in the range 0.050.36 ± 0.04 μg/L, and the average values found in the receiving water bodies were in the range 0.050.29 ± 0.02 μg/L. Antibiotics were found in higher concentrations in sludges from hospital lagoons (<50 to 276 ng/g) rather than in WWTP sludge (<loq to 145 ng/g) and in sediments collected in the receiving streams (<50 to 88 ng/g).

These data show that the direct reuse of treated wastewater in agriculture could introduce antibiotics into the food chain after plant uptake, and the need to improve wastewater treatment in order to reduce the load of antibiotics released into surface water, sludge and soils is of urgent concern.

Composite samples of the raw hospital wastewater in Tumaco (Colombia) [13] were analysed in three different seasons (October 2016, April 2017 and July 2017). A high variability in concentrations was found, probably due to the different amount of medicines administered to patients during the three sampling campaigns. Acetaminophen was always the compound most present being in the range 10–78 μg/L; the other highest levels were found for five antibiotics, namely, azithromycin, ciprofloxacin, clarithromycin, clindamycin and norfloxacin, with concentrations above 10 μg/L in 8 out of 15 samples. The hospital effluent is directly discharged into surface water. The authors are well aware of the environmental risks related to this practice and comment on the urgent need to adopt adequate treatments for the removal of macro- and micropollutants. In this context, they suggest the adoption of adequate treatments which, according to Giannakis et al. [14], should be light-assisted advanced oxidation processes (AOPs): UV-based (UV, UV/H_2O_2) for developed countries and solar-supported ones for developing countries.

Research efforts have also led to the definition and validation of methods to analyse specific groups of compounds. In this context, Martins et al. [15] proposed an analytical method to analyse statin drugs and [16] proposed another method to analyse psychiatric drugs in water. Both studies then applied their developed methods to sample the effluent of the University Hospital (HUSM) of the Federal University of Santa Maria, RS Brazil. This is a medium-large hospital (403 beds, 475 m^3/day as the average flow rate), with a wide range of wards, including psychiatry, oncology, radiology, obstetrics, surgery, neurology and intensive care. The authors sampled the HWW after it had passed through a septic tank. The investigations by [15] on statins confirmed the presence of atorvastatin and simvastatin in all the (treated) hospital effluent samples: atorvastatin was in the range 18.8–35.3 μg/L, with an average of 28.8 μg/L, and simvastatin in concentrations between 30.3 and 38.5 μg/L, with an average of 36.7 μg/L. Compared to those reported by [17] in Norwegian hospitals, these ranges are much higher. This could be explained by the wide use of lipid regulator drugs not only by patients but also by hospital staff and visitors. The study outlined that an environmental risk assessment (on the basis of the risk quotient (RQ) approach that is the ratio between measured environmental concentration (MEC) and predicted no-effect concentration (PNEC)) highlights that atorvastatin and simvastatin concentrations in the treated effluent pose a high risk for the environment and further adequate treatment should be added to the existing septic tank.

Reichert et al. [16] compared the concentrations of six antipsychotic drugs (olanzapine, clozapine, haloperidol, risperidone, pimozide and chlorpromazine) at two sampling points within the hospital sewer network. Point A corresponds to the psychiatric wing effluent after its passage through an anaerobic/anaerobic septic tank, and point B corresponds to the emergency ward effluent after treatment in the dedicated septic tank. On the basis of the 2-h water samples collected on the first day of the investigation, haloperidol and clozapine occurred at the highest concentrations. Between 4 a.m. and 8 p.m., the largest number of antipsychotics was found in the treated HUSM wastewater. While from 2 a.m. to 4 a.m., no antipsychotic was detected ($\leq$ limit of detection, LOD) at either sampling point. Pimozide was never detected due to its limited use in the hospital. The composite water samples collected during the whole sampling week, from 8 a.m. to 6 p.m., showed that occurrence frequency was around 54% at point A and 40% at point B. The most present compounds were haloperidol and clozapine at both points, and the ranges of occurrence were 0.31–0.52 μg/L for olanzapine, 0.56–0.97 μg/L for clozapine, 1.43–2.73 μg/L for haloperidol and 0.92–0.98 μg/L for risperidone. Chlorpromazine was found only once at a concentration of 0.52 μg/L.

The study also carried out an environmental risk assessment following two approaches: the persistence-bioaccumulation-toxicity (PBT) approach [18] (http://www.pbtprofiler.net/) which assigns a mark from 1 to 3 (3 being the highest value) to each compound on the basis of its persistence, bioaccumulation and toxicity and the risk quotient (RQ) approach. The risk may be low if $RQ < 0.1$, moderate when $0.1 \leq RQ \leq 1$ and high if $RQ > 1$ [19]. The analysis showed that, based on PBT values, the highest risks should be related to the presence of clozapine, pimozide and haloperidol. On the basis of the risk quotient approach, a high risk is correlated to risperidone, clozapine and chlorpromazine. Both sampling points displayed a very high risk, which can be attributed to clozapine, but the risk posed at point A was much higher, owing to the additional occurrence of chlorpromazine and risperidone.

In the South of Iran, at the Bushehr coastline of the Persian Gulf, Kafaei et al. [20] investigated the occurrence of six antibiotics (tetracycline, norfloxacin, azithromycin, anhydroerythromycin, cephalexin and amoxicillin) in the raw effluents of two 85-bed hospitals (H1 and H2) and after being treated only in a septic tank (hospital H1, flow rate equal to 8 m^3/day, HRT 5 days) and in a conventional activated sludge system (hospital H2, with a flow rate of 14.5 m^3/day; HRT 2 days). Composite samples were collected during the dry and wet seasons, by mixing wastewater taken at 7 a.m., 3 p.m. and 11 p.m. Moreover, samples of seawater and marine sediments were analysed in order to evaluate the impact of the release of the treated effluent into the aquatic environment. The results showed that in the raw HWW, norfloxacin had the highest values of concentrations (150 and 198 ng/L). Tetracycline was the second most abundant antimicrobial agent in hospital effluents with its average concentrations being equal to 68 and 144 ng/L. With respect to cephalexin and amoxicillin in the H2 effluent, their mean concentrations were >11-fold and 5-fold higher, respectively, than the corresponding average concentrations in the H1 effluent. These differences could be related to different treatment practices in the two hospitals. Finally, macrolides showed close concentrations in the two hospital effluents. The comparison between the occurrence in the two wastewaters and in the raw domestic wastewater of the town highlighted very lower concentrations of the selected compounds in domestic wastewater. When investigating removal efficiency achieved by the two (modest) treatments, an average removal of 71% is shown in the septic tank, with concentrations of all selected antibiotics being in the range 2.9–42 ng/L and >84% in the conventional activated sludge system. They concluded that there should be further stages in the treatment of hospital effluent in order to enhance the removal of typical micropollutants.

Also the study carried out in two hospitals in the area of Hanoi (Vietnam) by Lien et al. [21] focused on antibiotics. Metronidazole, sulphamethoxazole, trimethoprim, ceftazidime, ciprofloxacin, ofloxacin and spiramycin were analysed in the raw effluents of a rural hospital (220 beds) and an urban hospital (520 beds). The investigations lasted 1 year and were based on continuous flow samples collected on weekdays of the last week of each month. These effluents are pre-treated on site and then subjected to a biological stage followed by a polishing step. With regard to raw wastewater, it was found that higher overall monthly concentrations of antibiotics were found for the rural hospital effluent where they were present in all the monthly samples. Concentrations ranged from 21.8 to 103.5 μg/L, with an average of 70.5 μg/L, resulting in a mean hospital bed contribution equal to 0.32 μg/(L bed). In the urban HWW, antibiotics were detected in 10 out of the 12 months with very high concentrations only in 3 months (September 151.1 μg/L, October 202.4 μg/L and December 337.3 μg/L). In the remaining months, concentrations were between 8.7 and 39.9 μg/L, with the average monthly value being 93.5 μg/L and a bed contribution of 0.18 μg/(L bed). The compounds most present were ciprofloxacin (on average 42.8 μg/L) in the raw rural hospital effluent and metronidazole (on average 36.5 μg/L) in the raw urban HWW.

The treatment applied to the two HWWs allowed an overall average removal of 49% in the rural and 67% in the urban site. The released amount of antibiotics was calculated for the rural hospital, and, on the basis of the provided water consumption within the structure, it was estimated at 61 g/month.

This result underlines the need to better investigate the risk posed by these residues in promoting antibiotic resistance due to the selective pressure placed on bacteria. Moreover, ARGs carried by bacterial contaminants can be transferred to other bacterial populations including pathogenic bacteria found in hospital wastewater.

In Japan, Azuma et al. [22] carried out an investigation into the occurrence of 41 compounds in HWW, including PhCs and their metabolites, and phytochemicals, of which the pollution status is unknown for the aquatic environment and, in particular, for hospital effluent. The selected target compounds belong to nine different classes: antivirals, antibacterials, anticancers, psychotropics, analgesic-antipyretics, bronchodilators, antipruritic, herbal medicines and phytoestrogens. The authors compared the effluent of a medium-large hospital (477 beds, average flow rate of 460 m^3/day) with the influent and effluent of the WWTP (420,000 people, flow rate of 134,000 m^3/day) to which the HWW is conveyed. From this study, it emerges that for many compounds, concentrations are in similar ranges in HWW and WWTP influent. Some results should be highlighted with regard to analgesics-antipyretics. In HWW, the top compounds are two conjugated metabolites of acetaminophen (acetaminophen glucuronide and acetaminophen sulphate), the average concentrations of which were as high as over 24 μg/L and about one order of magnitude higher than that of the mother compound. In WWTP influent, acetaminophen was detected at the same order of magnitude as in HWW, whereas in the case of its metabolites, only acetaminophen sulphate was found at a lesser concentration (on average 6.4 μg/L). Loxoprofen and its metabolite loxoprofen alcohol were found at higher concentrations in HWW than in WWTP influent: on average 15 μg/L and 14 μg/L, respectively, in HWW vs 3.5 and 2.3 μg/L in WWTP influent. Finally the study reported that in the ozonated effluent (8.6 mg O_3/L and 100 min as contact time), some recalcitrant compounds still remained above 10 ng/L (on average): 47 ng/L the anticancer bicalutamide, 17 ng/L carbamazepine, 13 ng/L indomethacin, 34 ng/L caffeine and 16 ng/L the antipruritic crotamiton.

Papageorgiou et al. [23] remarked that monitoring campaigns remain a useful tool that is able to provide up-to-date information on the current situation, mainly in those areas where PhC consumption data for the different users (in hospital and in urban settlements) are scarce or difficult to obtain. They carried out a 1-year monitoring campaign (24-h composite samples) on the raw effluent of two medium-sized hospitals in the city of Larissa, in the rural Thessaly region, Greece. Out of the 135 monitored compounds belonging to 37 therapeutic classes (including not frequently monitored substances such as thyroid hormones, antineoplastics, steroid hormones, antivirals, antiparkinsonian agents, anti-vertigo inotropic agents and cardiovascular agents), only 35 compounds were detected in the range < LOD 81.5 μg/L. The highest values were found for the stimulant caffeine (81.5 μg/L), the analgesic paracetamol (31,614 ng/L) and the antiseptic triclosan (10.5 μg/L). In addition, other compounds were detected at high concentration levels ranging from 1.2 to 9.8 μg/L. This is the case for the antibiotics levofloxacin and clindamycin, the antihypertensives irbesartan and valsartan, the non-steroidal anti-inflammatory drugs (NSAID) ibuprofen and diclofenac, the analgesic salicylic acid, the diuretic furosemide, the antidiabetic metformin and the anaesthetic lidocaine. They found that analgesics and anti-inflammatories, antibiotics and antihypertensives followed by antidiabetics, diuretics, psychiatrics, insect repellent and β-blockers seem to contribute more to the PPCP load in wastewaters. An investigation on the removal capacity of the existing WWTPs highlights that they are not able to remove all the key contaminants at a concentration corresponding to a low (acute and chronic) environmental risk.

Interesting results are reported by Szekeres et al. [24] who investigated the effluent of three hospitals, H1, H2 and H3 (with 535, 113 and 453 beds, respectively), in Cluj County in Romania. H2 and H3 currently release their raw effluent directly into the nearby river; H1 effluent instead is subjected to a dedicated treatment (conventional activated sludge followed by chlorination) and is then released into the public sewer network. The authors analysed the occurrence of 14 common antibiotics, 36 ARGs and the bacterial composition of the hospital effluents. They found that antibiotics occurred between 3.67 and 53.05 μg/L; the most abundant were β-lactams, glycopeptides and trimethoprim. Ampicillin was detected in all HWW samples, and the highest concentration was found in H1 (53.05 ± 0.08 μg/L), confirming the results found in other studies [12]. High concentrations of trimethoprim were also detected in H1 (13.06 ± 0.09 μg/L) and H3 (13.06 ± 0.09 μg/L) wastewater samples, together with vancomycin (13.98 ± 0.41), imipenem (14.42 ± 0.39 μg/L) and tazobactam (10.26 ± 0.08 μg/L) in H3 wastewater samples. Ciprofloxacin and norfloxacin were never detected.

As for ARGs, out of the 36 genes initially investigated during the prescreening process, 16 genes were detected in hospital wastewater samples. Fourteen genes confer resistance to β-lactams, aminoglycosides, chloramphenicol, macrolide-lincosamide-streptogramin B antibiotics, sulphonamides and tetracyclines. Genes encoding quaternary ammonium resistance and a transposon-related element were also detected. The *sulI* and qacED1 genes, which confer resistance to sulphonamides and quaternary ammonium, had the highest relative abundance with values ranging from 5.33×10^{-2} to 1.94×10^{-1} and 1.94×10^{-2} to 4.89×10^{-2} copies/16 rRNA gene copies, respectively. The abundance of the investigated ARG types versus classes of antibiotics administered in the three hospitals shows an apparent similar pattern. This leads to the consideration that the three hospital effluents may contain enough antibiotics to be able to exert selective pressure on bacteria species.

Regarding bacterial composition in the HWW, the dominant phyla detected were *Proteobacteria, Bacteroidetes, Firmicutes* and *Actinobacteria*. Finally, after the treatment (activated sludge process + chlorination) moderate removal efficiency of the studied pollutants was demonstrated, with a 55–81% decrease in antibiotic concentrations, 13 orders of magnitude lower than the abundance of ARGs, but with a slight increase in *Proteobacteria* phyla, from 97.5 to 98.7%, and *Bacteroidetes* phyla from 0.1 to 0.5%. It was also found that after the treatment, abundance of *Enterobacteriaceae* increased from 1.1 to 10%.

A monitoring investigation carried out by Lerat-Hardy et al. [25] between 2013 and 2016 on the effluent of different wards of the 1,500-bed Pellegrin hospital in the Bordeaux conurbation (France) which focused on rare-earth element concentrations showed that their concentrations may vary among the departments and, for most of them, the concentrations are much higher than in the surface water. As for gadolinium (Gd, a paramagnetic contrast agent for medical imaging methods such as magnetic resonance imaging), a maximum concentration of 98 μg/L was measured in the effluent from the surgical departments, being almost four orders of magnitude greater than the concentrations in the Garonne River. In addition, the study analysed the trend of Gd fluxes over a 15-year period in Garonne River water and found a continuous increment (from 413 ng/L in 2003 to 820 ng/L in 2017) which is probably related to the growing regional population and the increased number of MRI scanners. Other rare-earth elements Ce, Gd, Tb, Er and Tm were detected at relevant concentrations in the hospital effluent. These results lead to the consideration that, as hospitals contribute significantly to the emission of these elements, in particular Gd, further studies should be carried out on the chronic effects they have on the aquatic organisms living in the receiving water bodies.

A complete characterisation of the effluent of the large Danish Herlev Hospital, near Copenhagen (700 beds), is provided in the report by DHI [26]. This covered 40 PhCs, largely consumed in the healthcare facility, heavy metals, pathogens, antibiotic-resistant bacteria and oestrogenicity (expressed as 17β-oestradiol

Table 1 Hazard profile of the raw effluent of Herlev Hospital, near Copenhagen, Denmark

Parameters	Reasons
Ciprofloxacin, clarithromycin, sulphamethoxazole, diclofenac, capecitabine	Their concentration was 10–300 times higher than the corresponding PNEC in fresh water for water living organisms
Iomeprol	High concentration (2.5–7 mg/L)
ARG Cefotaxime-resistant *E. coli* Ciprofloxacin-resistant *E. coli* Vancomycin-resistant enterococci	High concentrations 1.0×10^5 9.8×10^5 (mean 4.4×10^5) MPN/100 mL 4.6×10^5 1.7×10^7 (mean 7.1×10^5) MPN/100 mL 1.5×10^5 1.1×10^6 (mean 4.8×10^5) MPN/100 mL
Norovirus	High concentrations genome copies/L (211 6.2×10^5, mean 1.7×10^5)
Zebra fish	100% mortality within 96 h
Daphnia	All test animals died
Estrogenic activity	Oestrogen effects

equivalents, EEQ). The highest mean concentrations were found for iomeprol (iodinated contrast medium) found at 2.9 mg/L, acetaminophen at 352 μg/L and ciprofloxacin at 13 μg/L. As for heavy metals, zinc and copper occurred at the highest concentrations, 100 and 110 μg/L, respectively, and lead and nickel at 4.0 and 4.8 μg/L, and as for estrogenic activity, it was found to be in the range 0.16–44 ng EEQ/L in the effluent. Regarding microorganisms, *E. coli* were in the range $3.1 \cdot 10^6$ $1.0 \cdot 10^7$ MPN/100 mL (on average 5.5×10^6), enterococci in the range 6.8×10^6 1.5×10^7 MPN/100 mL (on average 9.5×10^6) and norovirus between 211 6.2×10^5 genome copies/L (on average 1.7×10^5).

Finally, cefotaxime-resistant *E. coli* were in the range 1.0×10^5 9.8×10^5 MPN/100 mL (on average 4.4×10^5 MPN/100 mL), ciprofloxacin-resistant enterococci 4.6×10^5 1.5×10^7 MPN/100 mL (on average 7.1×10^5 MPN/100 mL) and vancomycin-resistant enterococci 1.5×10^5 1.1×10^6 MPN/100 mL (on average 4.8×10^5 MPN/100 mL).

The hazard profile of the raw Herlev Hospital effluent was related to the compounds reported in Table 1.

Lasek et al. [27] investigated the occurrence of biocides in the effluent of the large (1,900 beds) hospital in Poitiers in France at different points and in the final outfall. Large quantities of biocides are used daily in hospitals for hygiene purposes, to disinfect surfaces and medical instruments and to clean linen (if a laundry is present [1]), and also as antiseptics in healthcare procedures. They found that biocides are continuously discharged into the hospital sewage network over the 24 h, with higher quantities discharged during the day because of the greater number of healthcare and facility cleaning activities taking place during this period. The monitored biocides include chlorhexidine digluconate (CHD), bis(aminopropyl)laurylamine (BAPLA) and didecyldimethylammonium chloride (DDAC); only the latter was found in concentrations higher than its PNEC in the hospital sewage outfall. Due to the continuous release, biomagnification phenomena are envisaged.

Finally, some research works discussed the possibility of predicting hospital concentrations on the basis of the consumption of PhCs within the healthcare structure [28–30]. Of these, Verlicchi [29] compared, measured and predicted concentrations and loads in HWW with a focus on the strengths and weaknesses of the two approaches and discussed the main factors responsible for inaccurate predictions and uncertain measures on the basis of some case studies.

3 Investigations on Occurrence of Antibiotic-Resistant Genes and Bacteria (ARG and ARB)

Monitoring campaigns confirm the presence of different antibiotics in HWW which are generally directly discharged into the public sewer systems without any pre-treatment or, in some countries, are even discharged directly into the surface water body without treatment. Many pharmaceuticals, radionuclides and antimicrobial agents may contain heavy metals in their formula. The capability of a microorganism to stay alive and replicate in a heavy metal polluted environment depends on genetic/or physiological alteration. In particular, bacteria can promote resistance to an antimicrobial agent, for instance, by the production of extracellular enzymes which are able to eliminate antibiotic and metal toxicity. The interesting studies carried out by [10, 31, 32] note that there "has been a significant hypothesis regarding conceivable hereditary relationship among microbial *resilience* for these heavy metals and multidrug resistance". For instance, penicillins, carbapenems, cephalosporins and monobactams are recognised synthetically as beta-lactam anti-infection agents, and numerous microorganisms become resistant to these drugs by producing different beta-lactamases that are fit to inactivate a few types of these anti-infection agents.

The fact that antimicrobials exert a continuous selective pressure upon ARB is worrying the scientific community. The World Health Organization has recently included some ARB (including *Acinetobacter baumannii*, carbapenem-resistant; *Pseudomonas aeruginosa*, carbapenem-resistant; *Enterobacteriaceae*, carbapenem-resistant, third-generation cephalosporin-resistant with priority 1 and *Enterococcus faecium*, vancomycin-resistant with priority 2) on the critically important priority list of pathogens for which new antibiotics are necessary [33].

Antimicrobial selective pressure favours the intraspecies and interspecies horizontal transfer of resistance genes [34]. Antibiotic resistance causes prolonged illness, excess mortality and higher costs for patients and health systems. This is why recently the United Nations General Assembly remarked that antibiotic resistance is among the greatest global health risks, requiring urgent attention [35]. This risk could be higher in low and middle developing countries where HWWs are not properly managed and where they are often directly released into surface water bodies which are then used for irrigation purposes, promoting the spread of ARB and

ARG into the environment, but also for domestic uses, sometimes without any treatment [36].

Samples taken from the effluent of the King George's Medical University, in Lucknow, India, and analysed by Alam et al. [31], showed an elevated number of one and more antimicrobial tolerant coliform bacterial species. Selected isolates of these bacteria were recorded to have variable plasmids which encoded with the metal and drug resistance. The study concluded by remarking that the hospital aquatic environment may act as a possible source of resistance transfer among the pathogens as well as to humans. The finding of this investigation shows a potential medical issue as the prevalent coliform species has progressively been related to the outbreak of contamination at doctors' facility. It is prescribed that healthcare effluent must be treated before its discharge into the environment.

Faecal coliforms (FC), carbapenem-resistant *Enterobacteriaceae* (CRE) or carbapenemase-producing *Enterobacteriaceae* (CPE) producing $bla_{\mathrm{NDM\text{-}1}}$ and selected extended-spectrum β-lactam (ESBL)-resistant bacteria (bla_{CTX}, bla_{OXA}, bla_{TEM}) and genes (*intI*) were investigated in the effluent of 12 hospitals of different sizes and in five sewer drains (with no hospital) across New Delhi, India, over two seasons (winter and summer) [37]. A close association was found between faecal bacterial concentrations in HWW and concentrations of CRE and ESBL-resistant isolates and genes, which are released into the environment. Concentrations were higher in winter than in summer, with $bla_{\mathrm{MDN\text{-}1}}$ abundance exhibiting the highest differences (from three to nine orders of magnitude). Larger hospitals were found to release significantly higher concentrations of FC, CRE and $bla_{\mathrm{NDM\text{-}1}}$ on a mass-per-wastewater volume basis, confirming the findings by [10].

Finally, concentrations in HWW are higher than those found in the sewer network not receiving HWW, and the greatest difference refers to $bla_{\mathrm{NDM\text{-}1}}$: almost nine orders of magnitude.

The literature survey by [32] reports that major multidrug-resistant bacteria (MDRB) recovered in hospital effluents are extended-spectrum β-lactamase (ESBL)-producing *Escherichia coli*, vancomycin-resistant enterococci (VRE) and *Pseudomonas aeruginosa*. Moreover, they highlighted that conventional WWTPs are able to reduce the concentrations of *E. coli* but the proportion of ESBL-producing *E. coli* among *E. coli* increases significantly with treatment (0.30% in the inflow versus 0.61% in the outflow).

Attention should also be posed to the municipal treated sludge, often used as a fertiliser: it was found that a gram of the anaerobically digested sludge at the municipal WWTP in Besançon (France) contained on average 2.6×10^5 ESBL-producing *E. coli* [38].

The efficacy of wastewater treatment in reducing the content of multidrug-resistant bacteria is under discussion: some studies reported that the total load of enterococci decreases during wastewater treatment without a substantial change in species distribution, *E. faecium* being predominant [39]. Others found that there is an increment in the proportion of *E. faecalis* and *E. faecium* resistant to fluoroquinolone [40]. By contrast, the study carried out by Magiorakos et al. [41] reported that wastewater treatment efficiently reduced the concentration of VRE.

Table 2 Main results of the analysis of prevalence of antibiotic resistance in rural and urban raw and treated HWW [36] and of ARGs

Antibiotics	Raw rural HWW	Treated rural HWW	Raw urban HWW (%)	Treated urban HWW (%)	Overall, both hospitals
E. coli isolates (#)	84	74	60	47	265
Antibiotic with the highest prevalence of antibiotic resistance	Co-trimoxazole	Co-trimoxazole	Co-trimoxazole	Co-trimoxazole	Co-trimoxazole
Prevalence to at least 1 antibiotic (%)	94	74	88	68	
MDR (%)	44	26	32	21	32
ESBL-producing *E. coli* (#)	45	31	27	12	115
Ciprofloxacin-resistant *E. coli* (#)	21	26	14	8	69

Lien et al. [36] compared the antibiotic resistance of raw and treated effluents from a rural and an urban hospital in the area of Hanoi in Vietnam. They analysed the prevalence of resistance to at least one of the selected antibiotics (amoxicillin/clavulanic acid, ceftazidime, ceftriaxone, ciprofloxacin, co-trimoxazole (trimethoprim/sulphamethoxazole), fosfomycin, gentamicin and imipenem), to each studied antibiotic, and multidrug resistance (MDR) in *E. coli* isolates. The bacterial isolates that resulted as non-susceptible to at least one agent in three or more antibiotic categories (MDR $\geq$ 3) were considered multidrug-resistant, according to [42]. The collected results, summarised in Table 2, showed that in both HWWs (raw and treated), the highest prevalence of antibiotic resistance was for co-trimoxazole and the prevalence to at least one antibiotic was higher in percentage for rural HWW (raw as well as treated). MDR was found again in most cases for rural HWW.

Prevalence of resistance to the studied antibiotics in *E. coli* isolates from treated HWW was less common than raw HWW, with the only exception of ciprofloxacin and fosfomycin in the rural hospital, for which enrichment of resistant *E. coli* isolates after wastewater treatment was found. As for ARGs in the collected samples (ESBL-producing and ciprofloxacin-resistant *E. coli* strains), it was found that in the rural hospital, 76 *E. coli* isolates were ESBL-producing (48%) and, in particular, bla_{TEM} was detected in 97% of isolates and $bla_{CTX\text{-}M}$ in 76%. Both $bla_{CTX\text{-}M}$ and bla_{TEM} were detected in 75% of isolates. Quinolone-resistant gene (q_{epA}) was detected in 72% of ciprofloxacin-resistant isolates. All three genes (q_{epA}, $bla_{CTX\text{-}M}$ and bla_{TEM}) were detected in 51% of ciprofloxacin-resistant isolates. In the urban hospital, 39 *E. coli* isolates were ESBL-producing (36%), and, in particular, bla_{TEM} was detected in 95% of isolates and $bla_{CTX\text{-}M}$ in 41%. Both $bla_{CTX\text{-}M}$ and bla_{TEM} were detected in 41% of isolates. Quinolone-resistant gene (q_{epA}) was detected in 86% of ciprofloxacin-resistant isolates. All three genes (q_{epA}, $bla_{CTX\text{-}M}$ and bla_{TEM}) were detected in 36% of ciprofloxacin-resistant isolates. These results demonstrate the need to better investigate the presence of ARB and ARG in the treated effluent of the hospital and their occurrence in the receiving environment.

A recent investigation carried out in Sichuan province in China [43] highlighted the prevalence of ESBL- and carbapenem-resistant *Enterobacteriaceae* and non-*Enterobacteriaceae* species (such as *Acinetobacter* and *Aeromonas*) in hospital sewage systems and in the receiving waters and concluded that hospital WWTP effluent could be an important source of ARB propagation into the environment. The study suggested that efficient disinfection techniques are expected to improve the elimination of ARB during wastewater treatment, and closer monitoring of the quality of treated effluents is needed to minimise the dissemination of antimicrobial resistance into the receiving environment.

In Singapore the presence of densely urbanised areas in close proximity to ambient water sources might facilitate the spread of antibiotic resistance. In order to shed light on this issue, Haller et al. [44] analysed the prevalence of carbapenem-resistant bacteria (CRB) and ESBL-producing strains in the effluent of two hospitals: H1 (1,597 beds) and H2 (1,500 beds). They then characterised the isolated strains and identified which strains are responsible for the resistance. They sampled two blocks in H1, block H1a (corresponding to an isolation ward and clinical specialty wards) and block H1b (corresponding to general wards), whereas in H2, they analysed the mixed water of the whole healthcare structure.

The results showed the relatively high prevalence of *Enterobacteriaceae* as well as non-*Enterobacteriaceae* species (such as *Pseudomonas aeruginosa*, *Acinetobacter* spp. and *Aeromonas hydrophila*) resistant to modern extended-spectrum cephalosporins and carbapenems in the sampled hospital effluents. Moreover, it emerged that among the isolates resistant to these antibiotics, a high proportion of ESBL and carbapenemase producers were detected in all the hospital effluents and most of these strains exhibited a multidrug resistance phenotype. Non-disinfected hospital effluents carry significant quantities of antibiotic-resistant bacteria to municipal sewage. These findings lead to the identification of strategies able to control and limit the spread of antimicrobial resistance from hospital effluent and thus to the adequate treatment of this major source of antibiotic-resistant bacteria.

With regard to recent European investigations, there are interesting results within the SIPIBEL project at the pilot site of Bellecombe in France. In this site, in Scientrier, Haute-Savoie, France, the existing treatment plant receives and separately treats hospital effluent (146 m^3/day on average from a 450-bed hospital) and urban wastewater (5,355 m^3/day from a catchment area of 21,000 inhabitants) in two lines adopting the same conventional treatments (activated sludge processes). This configuration offers the possibility to carry on investigations regarding raw and treated HWW and urban wastewater characteristics, removal efficiencies in the two treatment trains and the environmental impact of the residual pollutant load in the two streams. In this context, Laquaz et al. [45] found that the relative abundance of the class 1 integrons was much higher in HWW than in the urban sample, from 0.5 to 4 in HWW and from 0.1 to 0.3 in UWW. The mixture of HWW and urban wastewater in the proportion of 1/3–2/3 did not show to have a synergetic or antagonist ecotoxic phenomena, nor does it promote the spread of antibiotic resistance.

In Portugal, Vaz-Moreira et al. [46] analysed the antibiotic resistance prevalence in the effluent of a large hospital (1,120 beds) and in raw and treated wastewater generated in the surrounding catchment area. They found that the prevalence of resistance to at least one antibiotic of three or more distinct classes (MDR $\geq$ 3) ranged from 91.7 to 72.4%, with the highest value being for hospital effluent followed by raw wastewater and then treated wastewater. Moreover, the resistance to antibiotics belonging to six or more different classes (MDR $\geq$ 6) was significantly higher in hospital effluent than in municipal wastewater.

As for the detection of carbapenemase-producing bacteria and ARGs conferring resistance to carbapenems, monitoring campaigns involved non-clinical environments including WWTP effluent. Among them, in Ireland, Cahill et al. [47] investigated the occurrence of CPEs in the effluent of a 693-bed hospital as well as in municipal sewer system before and after the emission of the hospital effluent. They were present in all the HWW samples. They found that hospital effluent represents the major source of CPEs in municipal wastewater and remarked that the need for the monitoring of hospital wastewater for CPEs could have important applications in the detection and risk management of unrecognised dissemination of CPEs in both the healthcare network and the environment. Moreover, according to a recent study by EPA [48], there is evidence that there should be regulations with regard to the release of hospital effluent into the municipal waste system in order to prevent the encouragement of antimicrobial resistance dissemination. In Belgium, Proia et al. [49] investigated the prevalence and spread of carbapenemase genes (CGs: bla_{KPC}, bla_{NDM} *and* $bla_{OXA\text{-}48}$) from its source in hospital wastewater to the river water after passing through the two WWTPs and made a distinction between particle-attached bacteria and free-living bacteria as bacteria lifestyle (attached to a particle or free in the river water) influences the behaviour in the environment. They found that the absolute abundances were the highest in HWW and they can be significantly reduced in WWTP effluents, but the relative abundances (normalised per 16S rRNA) were never lowered through wastewater treatment. Particularly, for the particle-attached bacteria, the relative abundances were significantly higher in the effluents with respect to the WWTP influent for all the genes. They concluded that the levels of carbapenemase genes are significantly lower than other genes conferring resistance to more widely used antibiotics, but they could rise to the levels of high prevalent resistance genes.

4 Investigations on Hospital Effluent Management and Treatments

Over these last 5 years, treatments of hospital effluent have been investigated in many countries, by different research groups, testing consolidated and modified technologies as well as new systems. What appears is that research has also been carried out in countries for which scarce information had been available on this matter. This is the case for Columbia [13], Iran [50] and Thailand [51] among the collection discussed herein. The investigations referred to the whole hospital effluent, the effluent from specific wards/departments inside the hospital or effluent from specific healthcare structures, such as nursing homes specialised in psychiatric diseases [52], as well as service units, for instance, laundry [53]. Technologies were tested with regard to different key compounds: from macropollutants to micropollutants and pathogens, from antibiotics to ARB and ARG.

Vo et al. [51] investigated the removal of acetaminophen from hospital effluent by a pilot vertical flow constructed wetland in Pathum Thani, Thailand. This compound was investigated as it frequently emerges in hospital effluent and its consumption is very high all over the world. The peculiarity of their study is the attempt to find a correlation between acetaminophen and peroxidase enzymes which seem to have an important role in the removal of (micro)contaminants, acting as catalysts for their degradation reactions. It was in fact found that these enzymes are produced within the plant species in the case of (micro)contaminants accumulating in their body and then causing stress and altering their biochemical system. In these circumstances, the plants induce the generation of reactive oxygen species, such as H_2O_2. But in the case of an overproduction of these species, the macromolecules, including nucleic acids, proteins and lipids, may be damaged. For these reasons, the plants trigger the antioxidant systems with the production of different peroxidase enzymes which are catalysts in the oxidation reactions between H_2O_2 and organic contaminants, thus reducing the stress of the plants. In their investigation, the hospital effluent fed to the constructed wetland (vegetated with *Scirpus validus,* size, 1.5 × 0.6 × 0.6; retention time, 5 days; flow rate, 7,585 L/day) was doped with a stock solution of acetaminophen up to 10 mg/L. On the basis of the 65-day investigation, it emerges that the peroxidase enzymes of *S. validus* planted in CW were feasible for monitoring acetaminophen. Further studies should evaluate if a correlation can be suggested for other micropollutants as this promising result could help reduce the costs for monitoring many micropollutants.

Other researchers [54] in South Ethiopia analysed the capacity of a subsurface flow bed filled with broken bricks in removing macropollutants (BOD, COD, nitrogen compounds and phosphorus) from a HWW. The bed was filled with broken bricks and fed with the effluent of Hawassa University Referral Hospital. The results showed that broken brick provides better removal of TKN, ammonia, nitrate and phosphate from wastewater than gravel and also provides better adsorption sites for ammonium, nitrate and phosphate. The results confirmed the possibility of using

Table 3 Main results collected by SeMPAC© technology

Compound	SBR	SBR + MF	SBR + MF/PAC (5 days after PAC addition)	SBR + MF/PAC (28 days after PAC addition)
Ibuprofen	>95%	>95%	>95%	>95%
Diclofenac	-45%	75%	>90%	87%
Carbamazepine	<0	<0	>90%	24%
Trimethoprim	<0	<0	>90%	75%
EE2	>95%	>95%	>90%	>90%

local materials (bricks) for filling the beds of constructed wetlands and to remove macropollutants from hospital effluent.

Apart from a few cases dealing with nature-based solutions for the treatment of HWW, most of the investigations referred to membrane technology: membrane bioreactor (MBR) coupled with other technologies acting as a polishing step; modified MBR, or MBR integrated with added materials; and pre-treatments to an MBR to improve the removal of target compounds. The main results are presented here and discussed.

Interesting results were achieved by the Spanish group of research [55] who tested the SeMPAC® process with HWW. SeMPAC® is a new hybrid system they developed (patents ES 2362298 B2, EP 2960214 A1) in 2012 and licensed to the Suez Group for the treatment of effluent containing recalcitrant compounds. It consists of a sequential batch reactor (SBR) followed by an external submerged microfiltration membrane compartment in which powdered activated carbon (PAC) is added to promote the removal of more persistent substances. The pilot plant used in the investigation included a homogenisation tank (to guarantee the homogenisation of the wastewater hydraulic and pollutant loads), an SBR (operating at 6 g VSS/L, HRT 16 h), a stand-by tank (to guarantee the continuous feed to the membrane compartment) and the chamber with the membrane module (pore size 0.2 μm, total surface 6.4 m^2, 7 min of permeation and 1 min of relaxation). Five PhCs were selected to investigate: ibuprofen, EE2, diclofenac, carbamazepine and trimethoprim.

The study lasted 150 days with the first 56 days with no PAC addition and the remaining 94 days with PAC (dosed at 1 g/L). Once added, the PAC particles integrated into the sludge flocks, strengthening their structure and increasing sludge settleability and filterability. In addition, they favour the development of a biofilm on the particle surface, and an enhancement of the microbial diversity was also found. At the same time, filamentous bacteria reduced.

An analysis of the biomass in the bioreactor with PAC revealed the presence of different species, including slowly growing species, such as worms and tardigrades, amoebae (related to nitrification), the protozoa *Carchesium polypinum* (indicators of high-quality effluent) and rotifers (able to improve flock growth because they can consume free cells, as well as small flocks).

As for the observed removal efficiencies for selected PhCs, the main results are summarised in Table 3. It emerges that without PAC addition, the readily biodegradable compound (ibuprofen) is efficiently removed in the biological reactor (SBR) and the following step of microfiltration does not significantly increase its removal (always >95%). For EE2, the removal was 80% in the SBR followed by an additional 15% in the membrane chamber; diclofenac achieved an important removal efficiency in the whole process (around 70%). These overall removal values are not generally observed in biological processes [56]. As for persistent compounds (carbamazepine and trimethoprim), no removal was found in these conditions. The presence of PAC enhances their removal: 5 days after PAC addition, high removals were observed for all the key compounds, but after 28 days, a rapid decrement was observed for carbamazepine (down to 24%) and trimethoprim (75%) and a slight worsening for diclofenac (87%). These values suggest that PAC was saturated of the different compounds after different time intervals and further sorbent should be added to maintain and guarantee the high removal observed with fresh PAC. Based on these tests, PAC should be added every 20 days in order to guarantee a removal of 50% at least for all the selected compounds.

Finally, the study remarked that the main operational conditions might be optimised on the basis of the behaviour of the different key compounds: (1) the recalcitrant carbamazepine defines PAC dosage; (2) the easily degradable ibuprofen sets duration of the aerobic cycle in the SBR; and (3) the denitrification efficiency defines the correct time length of the anoxic period.

Lima et al. [57] investigated the efficiency of a particular activated carbon in removing a target pharmaceutical (acetaminophen) from synthetic water, simulating hospital wastewater containing different pharmaceuticals (acetaminophen, enalapril, nicotinamide, diclofenac and metformin) and organic and inorganic salts. The tested activated carbon was obtained from the residual biomass of Brazil nut shells, chemically activated with $ZnCl_2$. The lab tests evaluated the sorption capacity with feeding containing up to 80 mg/L of acetaminophen and 30 mg/L of the other PhCs. It was found that the chemical activation with $ZnCl_2$ leads to activated carbons with better textural characteristics and much higher sorption capacities with regard to the target compound if compared to commercial and commonly used activated carbons.

In addition, it was also found that the proposed adsorbent can be regenerated up to 74% with a mixture of NaOH 0.1 M + 20% EtOH solution and reused up to four cycles.

In China, the most adopted treatment train for hospital effluent includes an MBR followed by a chemical disinfection. This sequence complies with the Chinese standards for the release of the hospital treated effluent in surface water bodies. Investigations are still ongoing to optimise the operational conditions with regard to the characteristics of local hospital effluent. Recently, great efforts have been made to find technologies able to better control the dissemination of resistant bacteria, including extended-spectrum β-lactamase-and carbapenemase-producing bacteria, from hospital sewage into the environment [43].

Combining treatments including MBR and a further filtration step (nanofiltration (NF), or reverse osmosis (RO)) is becoming a strategy of increasing concern. In this context, at Purpan Hospital in Toulouse (France). Lan et al. [58] studied the combination MBR + NF and identified the key characteristics of the MBR effluent which control fouling and flux decline. They also analysed the relationship between the content of salt (calcium and phosphate) in the permeate with the stabilised fluxes in the following NF step. In their 145-day experimental campaign, the lab scale MBR (500 mL), fed by raw hospital effluent, was equipped with immersed hollow fibre membranes (polysulfone 0.2 μm, surface of 7 m^2) and operated with an SRT of 40 days and an HRT of 24 h. The permeate was sent to a 10-L feeding tank which supplied the cross-flow nanofiltration unit where the flat feet membrane (NE 70 polyamide composite 1.4×10^{-2} m^2) was installed.

Filtrations were conducted at constant transmembrane pressures (535 bar) with a 0.3 m/s cross-flow velocity. Two types of fouling were observed depending on the pH range: silica and colloidal organic fouling when pH was in the range 6.05–6.8 and a combination of silica and colloidal organic fouling and calcium phosphate scaling when pH was between 7.20 and 8.27. The measured stable fluxes varied between 60 and 10 L/m^2 h depending on the type of compounds (calcium and phosphate) fouling the membrane and their distribution on the surface. Extremely low values appeared when calcium and phosphate scaling evenly occupied the entire membrane surface; greater stable fluxes (120–80 L/m^2 h) were obtained when only silica and organic fouling occurred. A correlation was found between the NF stable flux values for filtration of various MBR effluents on the basis of calcium and phosphate concentrations after judgement of the pH of the effluents.

At the Military Hospital in Ho Chi Minh (Vietnam), Tran et al. [59] investigated a similar combination. A pilot plant (MBR + NF) was in fact installed to treat the effluent (50 L/day) from the operating rooms of the healthcare facility. MBR characteristics: were: concentration of the mixed liquor volatile suspended solids (MLVSS) between 3.5 and 4.5 kg/m^3, hollow fibre microfiltration membrane equipped in the reactor, throughput of 12–18 L/m^2 h; pore size 0.1 mm, operating pressure $\leq$ 40 kPa, SRT of 40 days; as for NF modules: pore size equal to 1 nm, flux of 6–8 L/h. The system operated at three applied organic loading rates (OLR) of 0.5, 1.5 and 2.5 kgCOD/m^3 day. The excellent removal efficiencies observed for common organic pollutants (around 9,195% for COD), nutrients (9,098%for phosphorus, 9,598% for total iron) and microorganisms (4 log units for *E. coli* and 6 l.u. for coliforms) led the authors to think that the tested combination is a promising solution for the control of macropollutant release in the environment from hospital effluent. Nanofiltration was preferred to reverse osmosis due to the lower investment and operational costs, its capacity to retain high-valence ions while allowing monovalent ions to pass, the lower pressure required and the lower pressure requested and due to the fact that NF allows backwashing, which could significantly reduce the frequency of membrane fouling.

The combination MBR + ozonation (O_3) + granular activated carbon (GAC) filtration + UV irradiation was considered the best of the available technologies according to the results from the *Danish* project on hospital effluent treatment carried out between 2012 and 2015, which led to the construction of the full-scale treatment plant at Herlev Hospital (700 beds, 150 m^3/day). As stated in the report by DHI [26], the polishing sequence (O_3 + GAC + UV) was preferred to the GAC + O_3 + UV sequence, which was tested in parallel at the WWTP under construction, as it allowed greater removal of the key compounds from the permeate and lower operational and maintenance costs. With the adopted sequence, more than 99% of iodinated contract media – the most recalcitrant compounds, such as iomeprol, iohexol and iopamidol, which are largely administered at the Danish hospital – were also removed. After the treatment, oestrogenic activity in the range 0.16–44 ng EEQ/L in the influent samples was significantly reduced to <0.076–1.1 ng EEQ/L. As for heavy metals, their concentrations in the polished effluent were found below the environmental quality standards (EQS) for inland surface water. Disinfection was efficiently guaranteed by the MBR and ozonation steps. The ozonated effluent did not contain *E. coli* and coliforms at 37°C, but only culturable bacteria at 22 and 37°C, which were therefore assumed to be related to regrowth of bacteria in the polishing lines after membrane filtration. The UV reactor worked with a UV dose of 6–10 mJ/cm^2, which is lower than the common dose adopted in water works (around 40 mJ/cm^2). It was found that an applied dose of 6–10 mJ/cm^2 was sufficient to remove culturable bacteria at 22 and 37°C.

Paulus et al. [60] demonstrated that the O_3 + GAC + UV sequence provides an effective barrier effect against ARG contained in the HWW. They investigated the removal of ARG from hospital effluent pre-treated in an MBR by means of the patented Pharmafilter system technology (reproducing the same treatment technologies adopted for the full-scale WWTP receiving the effluent of Herlev Hospital in Denmark). They found the treatment reduced the ARG from 13 to 4 and the number of quantified antibiotics from 17 to 7. They agreed with the consideration that on-site treatment of a high-risk source of effluent was proved to be advantageous with regard to antibiotic as well as ARG reduction. Thus, legislative guidelines and suggested best practices should move in the direction of implementing on-site wastewater treatment.

The good performance reported by Ngo et al. [61] regarding the application of sponge media with high porosity (polyester-urethane sponge) within the aerobic compartment of an MBR (SMBR) in order to improve macropollutant removal and reduce membrane fouling resulted in testing the performance of this modified MBR with regard to micropollutants. In this context, the removal of antibiotics from the effluent of Trung Vuong Hospital (Vietnam) was investigated by Nguyen et al. [62]. They compared the performance of lab-scale MBRs (two reactors of 8 L, operating with a concentration of biomass of around 5,000 mg/L, with a ratio between volatile and total suspended solids (MLVSS/MLSS) equal to 0.79 and a sludge retention time of 20 days) equipped with hollow fibre (HF) and flat sheet (FS) membranes with the same pore size (0.4 μm) and surface area (0.1 m^2), operating at different high fluxes (10, 15 and 20 L/m^2 h). The analysed antibiotics and the range of concentrations in the raw HWW fed to the systems were norfloxacin (6.305–43.610 μg/L), ofloxacin (7.634–40.261 μg/L), ciprofloxacin (1.926–23.841 μg/L), sulphamethoxazole (0.378–2.078 μg/L), erythromycin (0.135–2.407 μg/L), tetracycline (0.036–1.612) μg/L and trimethoprim (0.676–2.911 μg/L). The results are reported in Table 4 in terms of ranges of concentration in the permeate from the two SMBRs and ranges of removal efficiencies observed in the two small reactors.

Table 4 Ranges of concentrations in the permeate and removal efficiencies achieved in SMBR equipped with HF and FS membranes and operating with a flux in the range of 10–20 L/m^2 h [62]

Antibiotic	SMBR-FS permeate (μg/L)	SMBR-HF permeate (μg/L)	SMBR-FS removal (%)	SMBR-HF removal (%)
Norfloxacin	0.07–0.10	0.08–0.09	93–99	62–86
Ofloxacin	0.2–2.1	0.22–6.73	73–93	68–93
Ciprofloxacin	0.75–8.52	0.69–8.12	76–93	54–70
Trimethoprim	0.049–0.494	0.058–0.809	60–97	47–93
Tetracycline	0.0001–0.106	Not detected		
Erythromycin	0.085–0.647	0.137–1.274	67–78	22–48

The slightly higher removal efficiency observed in flat sheet SMBR seems to be due to the higher concentration of MLVSS within the MBR equipped with FS membranes (4,546 ± 777 mg/L) compared to HF-SMBR (3,794 ± 1,243 mg/L). As for erythromycin, a higher removal was achieved with the flat sheet membranes, probably due to the better adsorption of erythromycin on the biomass and/or on the flat sheets as the operating conditions were the same in both reactors. Sulphamethoxazole was the only antibiotic which was not removed in these systems, and its concentration in the permeate increased mainly with the highest values of flux (15 and 20 L/m^2 h) due to back conversion of N4-acetylsulphamethazole to sulphamethoxazole during degradation [63].

The hybrid and modified MBR, consisting of the sponge MBR-HF proposed by Nguyen et al. [62] coupled with an ozonation step, was investigated by Vo et al. [64] in order to enhance the removal of the most recalcitrant antibiotics from the effluent of the hospital in Ho Chi Minh (Vietnam). The sponge MBR was operating at the same conditions as the previous investigators, and it was fed by real hospital effluent where the concentrations of the monitored antibiotics were those reported in Table 5. The chemical and physical properties of the investigated compounds influence the total removal that was achieved and is reported in the second-to-last column of Table 5. In general, an increment after ozonation was observed, and this can be attributed to the presence of ozone and hydroxyl radicals that are able to oxide double-bond carbons, activated aromatic groups and non-protonated amines [65]. A clear example is given by ciprofloxacin, which shows a higher ozone affinity due to a higher electron density at functional groups in the molecular structure.

Zotesso et al. [52] in Germany analysed the efficiency of ozonation on the effluent of a nursing home (271 beds, specialised in psychiatric diseases) that was pre-treated in an UF-MBR (UF membrane pore size, 0.04 μm, 10 m^2 area). The interesting study also analysed the effect of ozonation on the dissolved organic carbon (DOC, around 9 mg/L in the influent) and bromide (Br^- and $BrO_3{}^-$ were always below the limit of quantification of 0.1 mg/L) and the production of toxic substances (including transformation products). Selected micropollutants included analgesics (including ibuprofen, diclofenac and tramadol), psychiatric drugs (carbamazepine, bupropion), antidiabetics (metformin), antihypertensives (valsartan), beta-blockers (atenolol, metoprolol) and antibiotics (sulphamethoxazole). They tested different dosages of

Table 5 Data collected in the investigation by Vo et al. [64] in the system sponge MBR-HF + ozonation

Antibiotic	HWW influent (μg/L)	SMBR-HF permeate (μg/L)	SMBR-HF removal (%)	SMBR + O_3 effluent (μg/L)	O_3 removal (%)	SMBR-HF + O_3 removal (%)	Specific removal (μg/g O_3)
Norfloxacin	24.96 ± 18.65	3.51 ± 3.15	6,286	2.12 ± 0.31	46	92 ± 4	1.56
Ofloxacin	15.12 ± 7.49	4.77 ± 3.35	1,470	1.83 ± 0.25	1,321	88 ± 4	1.27
Ciprofloxacin	12.88 ± 10.96	3.82 ± 2.91	6,893	2.13 ± 0.29	1,321	83 ± 7	1.68
Trimethoprim	1.79 ± 1.12	0.13 ± 0.06	<0	0.05 ± 0.04	4–6	97 ± 2	0.11
Tetracycline	0.76 ± 0.73	n.d.±	2,460	n.d.		100	0.07
Erythromycin	2.15 ± 0.26	1.11 ± 0.17	4,793	0.21 ± 0.03	42	90 ± 1	0.46
Sulphamethoxazole	3.37 ± 0.54	1.86 ± 0.43	<0	1.14 ± 0.14	1,321	66 ± 1	0.37

ozone (2.5, 5, 10 and 15 mg O_3/L) and two HRTs (12.8 and 25 min). The results obtained showed that the removal was >90% for 12 out of 18 key compounds by applying 5 mg O_3/L and operating with HRT equal to 12.8 min. This dosage corresponds to a specific ozone consumption of 0.6 g O_3/(g DOC). As for DOC, only a slight (a few percentages) removal was found during the ozonation step. This fact confirmed the assumption by Bourgin et al. [66] that substances (micropollutants) are not completely mineralised after the ozonation step, but are turned into transformation products which need to be better investigated in terms of occurrence and toxicity, and a further treatment might be suggested. The study concluded that optimisation of the design of the ozonation stage might lead to a lower ozone dosage and lower specific ozone consumption, thus lower operational costs. Moreover, they highlighted that transformation products are expected to be generated as a low removal of organic content in the ozonation plant was found. These undesired compounds might require a further post-treatment. This issue was not considered and should be the subject of further research.

The pilot 100-L moving bed biofilm reactor (MBBR) investigated by Shokoohi et al. [67] fed with the effluent from Beasat Hospital in the town of Hamadan, Iran, showed a good removal of LAS (linear alkylbenzene sulfonic acid) anionic detergent. A removal of 92.3% of LAS was found with a filling rate (Kaldnes carriers) at 70% of the volume, biomass concentrations of 3,000 mg/L and HRT of 24 h.

In Denmark, interesting results were achieved by Tang et al. [68] who tested the removal of a selection of 22 PhCs at lab and pilot scale in the effluent of Aarhus University Hospital at Skejby, by means of the following two combinations: MBBR +ozonation and MBBR+ozonation+MBBR. In the MBBR, AnoxKaldnes™ K5 carriers (AnoxKaldnes, Lund, Sweden) were used with a filling ratio of 50%. The ozonation step was tested in order to reduce the concentrations of PhCs and toxicity and the second MBBR in order to remove biodegradable organic matter and toxicity generated from ozonation by-products. The ozone dosage able to guarantee 90% of the removal (DDO_3) of key compounds was normalised to the dissolved organic carbon (DOC) concentrations ($Z_{90} = DDO_3/DOC$). Propranolol, tramadol and trimethoprim had the lowest Z_{90}, between 0.72 and 0.88 (mg O_3/mg DOC), indicating that they are easily degraded by ozone. From the other side, iopromide had the highest Z_{90}, indicating that it reacted the slowest with ozone. They also found that ozone can reduce any toxicity by half (with regard to *Vibrio fischeri*) of MBBR effluent, and a subsequent polishing MBBR after ozonation can further reduce any residual toxicity in ozonated effluent.

In the large hospital of Valencia, Spain (1,000 beds, 6 oncology outpatient wards for adults and 2 for children), Ferre-Aracil et al. [69] evaluated and compared the efficacy of ozonation and ozonation combined with hydrogen peroxide as a pretreatment for the hospital effluent before mixing it with the municipal wastewaters and thus avoiding the dilution of pharmaceuticals arriving at the WWTPs. Analyses of composite samples of the hospital effluent (taken every 4 h from 8 a.m. to 8 p.m. for 5 consecutive days) were processed to determine some cytostatics administered in the structure. Out of the 17 compounds investigated, only four were detected: irinotecan hydrochloride trihydrate, ifosfamide, cyclophosphamide and capecitabine. Their measured concentrations decreased over the week with the highest values being on the Monday and the only exception of ifosfamide. This trend is correlated with the days on which chemotherapy treatment is administered. These treatments are usually applied on the day on which the experimental campaign started, and most patients returned to their homes, although some remain at the hospital and, consequently, the concentration diminishes as the week went by. The highest concentrations were found equal to 273 ng/L (irinotecan), 1,187 ng/L (cyclophosphamide) and 1,139 ng/L (capecitabine). They found in the tested operational conditions (under O_3 or O_3/H_2O_2) that cyclophosphamide was the most recalcitrant: it was still present when all the other 16 investigated cytostatics were completely removed. The results show that the removal efficiency was around 97% for cyclophosphamide and higher for the other cytostatics in relatively short reaction times (10 min) and for ozone gas concentrations in the order of 43.9 g/m^3. Moreover, they found that there was no apparent effect of hydrogen peroxide on ozone gas consumption.

Research has been carried out on the treatment of HWW with fungi which has highlighted promising results in the removal of specific compounds as reported in [3]. In this context Mir-Tutusaus et al. [70] evaluated the ability of *Trametes versicolor* in removing a wide spectrum of pharmaceuticals (PhCs) from real HWW as well as toxicity due to their residues occurring in the effluent. Samples taken from Sant Joan de Deu Hospital effluent (Barcelona, Spain) were first subjected to coagulation (by adding 95 mg/L or 190 mg/L of coagulant with 2 min of coagulation at 200 rpm), flocculation (by adding 10 mg/L and 20 mg/L of flocculant, respectively 15 min of flocculation at 20 rpm) and then settling (30 min). The pre-treated effluents were then subjected to biological treatment in the presence and absence of *T. versicolor*. They found that out of the 81 selected PhCs, 46 were detected in HWW. *Trametes versicolor* treatment consistently removed most of the detected PhCs, including the most recalcitrant ones (above all carbamazepine). In addition, the treated effluent did not exhibit any toxicity, and therefore the fungal treatment might have removed potential toxic metabolites.

Jaén-Gil et al. [71] instead investigated degradation, transformation and sorption of the β-blocker metoprolol MTP, and its recalcitrant metabolite metoprolol acid (MTPA), during a treatment based on the fungi *Ganoderma lucidum* – a batch aerobic fluidised bed reactor, fed with real HWW. They found a 33% removal of MTP and 64% of MTPA and a slight increase in toxicity, probably due to the presence of other contaminants in the HWW.

For the first time, Badia-Fabregat et al. [72] characterised *veterinary* hospital effluent. They found that it is quite similar to urban effluent in terms of macropollutants, whereas it presents inter-day variability for PhCs, with peaks of high concentrations for some antibiotics (metronidazole, cefalexin and ciprofloxacin), analgesics and anti-inflammatories (mainly salicylic acid). On the whole, antibiotics, analgesics and anti-inflammatories presented the highest concentrations. In general, the overall concentration levels of PhCs in veterinary hospital effluent were lower than those reported for hospital wastewaters [1, 73] probably due to the higher use of water (and therefore dilution of wastewater) for cleaning the veterinary facilities and the fact that urine from large animals (e.g. horses) was collected with straw and disposed separately. They concluded that the concentrations of pharmaceuticals are between those found in urban wastewater and hospital wastewater and thus some dedicated treatment at the source might be recommended. They investigated the efficacy of a fungal treatment with *Trametes versicolor*, which has already been tested for the treatment of hospital wastewater with promising results [71, 74]. They first investigated the removal of a wide spectrum of PhCs in veterinary effluent in batch mode in non-sterile conditions and at the same time analysed the evolution of the bacterial and fungal population. They found a removal of 66% for many PhCs and identified some microbial interactions which encouraged them to investigate the treatment of the same effluent in continuous mode. The corresponding removal efficiency was around 44%.

With the aim to better focus on the strategy to improve the removal of antibiotics and ARGs from veterinary hospital effluent, Lucas et al. [75] analysed the removal of 47 antibiotics belonging to 7 different groups (β-lactams, fluoroquinolones, macrolides, metronidazoles, sulphonamides, tetracyclines and trimethoprim) and a selection of ARGs bla_{TEM}, bla_{SHV}, *ermB, qnrS, sul*I, *tetW* and 16S rRNA. The observed removal was 77% of antibiotics after the fungal treatment, which is higher than the removal obtained in conventional treatment plants. As for the removal of ARGs, they found that the fungal treatment resulted in a good removal of the *sul*I and bla_{TEM} genes (removal rates of 56% and 100%, respectively), as compared with data obtained from a conventional WWTP, in which the removal rates were negative, 156 and 58% [76]. The *qnrS* gene showed a negative removal rate equal to 163% (in conventional WWTP its removal was even worse and was equal to 302%, [76]). As for *ermB* and *tetW* genes, fungal treatment showed a complete removal for both species. Very good results were obtained in conventional WWTPs which exhibited slightly lower values (82% and 97% for the two ARGs, [76]). They concluded that fungal treatment may represent an alternative solution to conventional WWTP as it showed a good removal of certain compounds such as ciprofloxacin, enrofloxacin, marbofloxacin and ampicillin, which are recalcitrant in conventional WWTPs. They also remarked that not only the occurrence of antibiotic compounds may influence the fate of antibiotic resistance but also other factors, including operational parameters of bioreactors, the wastewater-associated bacterial communities and their interaction with fungi.

Badia-Fabregat et al. [77] reported that many indigenous fungi rather than bacteria can successfully compete with the inoculated fungi and that wastewater origin and operational conditions have a stronger impact on the diversity of microbial communities developed in the bioreactors than the inoculation or not with *T. versicolor*.

As already reported, in some countries, HWWs are directly discharged into the receiving water, without specific treatments. This practice is followed by the hospital of Tumaco (Nariño), as well as from the city of Florencia (Caquetá), in Colombia, whose effluents are directly released in two rich biodiversity regions (the Pacific and Amazonian regions) [13]. The authors well know that the direct release of untreated hospital effluent into the sea may pose serious risk to aquatic life and underline the need to adopt adequate treatments to reduce such risk. With this aim, they investigated the efficacy in removing different pharmaceuticals from a real HWW by a combination of biological and sonochemical treatments (first sonochemical then sono-photo-Fenton). Fifteen pharmaceuticals (the analgesics acetaminophen and diclofenac; the psychiatric drugs carbamazepine and venlafaxine; the antihistamine loratadine; the antibiotics sulphamethoxazole (sulphonamide), trimethoprim, ciprofloxacin and norfloxacin (fluoroquinolones), erythromycin, azithromycin and clarithromycin (macrolides), and clindamycin (lincosamide); the β-blocker antihypertensives valsartan and irbesartan) were analysed in the raw effluent of the medium-sized (122 beds) hospital in Tumaco [78], after the two steps of the treatment trains. They found that, with the exception of acetaminophen and valsartan, all the other PhCs were not removed. If this effluent is then subjected to 90 min of sonochemical treatment, all the PhCs are removed with the exception of the two cited compounds. According to the authors, acetaminophen and valsartan could be sorbed onto the flocs formed during the biological step; ultrasonic waves would break them and cause the release of the two PhCs, thus justifying the observed increment of their concentration in the first 60 min of treatment. The modest decrement in the PhC concentration up to the end of the observation period (in the remaining 30 min) should be due to the chemical attack of the produced radicals ($HO^{\cdot}$) towards the PhCs, as well as to the chemical effects of ultrasound. In the sonochemical process, radicals are produced through acoustic cavitation phenomena. In order to enhance the degradation of the target pollutants, the authors also investigated a sono-photo-Fenton treatment that is an on-site photo-Fenton process where ferrous ions and UVC irradiation are added to the sonochemical component (volume of 350 mL; frequency equal to 375 kHz; power equal to 88 W/L; pH in the feeding equal to 7.9; [Fe^{2+}] equal to 90 μmol/L, and UVC of 4 W). The system showed a higher ability in removing recalcitrant compounds, and it was able to reach an average removal equal to 85% for the whole PhC load in the biologically treated effluent, while in the sonochemistry treatment, the average removal was 51%.

The Iranian study by Karami et al. [50] investigated the application of high-frequency ultrasound (1.8 MHz) to a conventional activated sludge process for the treatment of hospital effluent. They analysed the effect of the modified activated sludge process on the removal of COD and turbidity and on the sludge volume index (SVI) in a bioreactor characterised by a high biomass concentration (8,000 mg/L). It was found that with and without sonication, the removal of COD was similar, but in the sonicated system, SVI and effluent turbidity were lower than the non-sonicated system.

Giannakis et al. [14] discussed the need to adopt specific strategies for HWW management and treatment, distinguishing between developed and developing countries. They evaluated the application of light-assisted AOP solutions for both situations: UV-based AOPs in developed countries and solar-based AOPs in developing countries (focusing on the main characteristics of Ivory Coast and Colombia). They underlined that developed countries more or less have the problem of microorganisms under control and mainly focus on micropollutants, while the priority of developing countries should be the acute risk caused by microorganisms in the effluent which are often directly discharged in the surface water and thus indirectly reused for crop irrigation. In their study, they treated the effluent of the Vidy WWTP (Switzerland) by UV/H_2O_2 and photo-Fenton process and analysed the removal of microorganisms and a selection of micropollutants (carbamazepine, clarithromycin, diclofenac, metoprolol, venlafaxine, benzotriazole and mecoprop). They found UV-based treatment more efficient in the removal of microbiological and chemical contaminants: they achieved 100% removal after 10 min of exposure in the case of an effluent from an activated sludge system or an MBBR and after 30 min in the case of coagulation-flocculation treatment effluent. On the contrary, the photo-Fenton treatment achieved 20% of removal after 30 min of exposure. As for microorganisms, 5 min were sufficient for the UV treatments to complete their removal, whereas more than 3 h were needed for photo-Fenton. With regard to developing countries, they took as an example the conditions in Ivory Coast and Colombia and remarked that the application of solar photo-Fenton as a feasible AOP is suggested only after (the construction of) basic treatment (primary and secondary treatments), or as a possible implementation of a barrier of microbiological-related problems. They found photo-Fenton was able to achieve high bacterial removal, regrowth suppression and yeast and virus inactivation and was also a non-selective and effective system in removing micropollutants and organic matter. They concluded that light-assisted AOP application in hospitals wastewater is promising both in developed and developing countries.

The need to evaluate the health and environmental risks posed by reclaimed water in Israel led Petrovich et al. [79] to investigate (pre)treatments for the hospital effluent before its emission in the public sewer system. They tested the sequence, sedimentation + activated sludge process + secondary sedimentation on a pilot scale, and analysed microbial and viral communities and their antibiotic-resistant genes in raw hospital effluent and treated effluent. They found that ARGs, including many that confer resistance to antibiotics of high clinical relevance, were abundant and persistent throughout the pilot-scale system, including *aadA*, $bla_{\text{TEM-1}}$, $bla_{\text{NDM-1}}$, $bla_{\text{OXA-2}}$, bla_{OXA35}, $bla_{\text{OXA-10}}$, *mefA*, *mel*, *tetX* and *sulI*. They found an overall removal efficiency of ARGs throughout the pilot-scale system equal to 16% on a per genome equivalent basis. Moreover, based on their results, they did not identify strong linkages between ARGs and viruses in viral metagenomes, suggesting that the viral community may not play a significant role in promoting the horizontal gene transfer of ARGs in this hospital wastewater treatment system. They also commented on the lack of knowledge in this context, and the conflicting results from previous studies warrant further research on this topic.

The issue of the proper management and treatment of HWW is becoming of great concern for the scientific community of countries where attention has not yet been paid to this problem. For instance, in Iran, the recent study by Azizi et al. [80] remarks that in Kermanshah and Hamadan, there are 19 and 16 differently sized hospitals, respectively, with a capacity of 2,526 and 3,042 beds. The effluent of the healthcare facility is treated only in 7 out of 19 hospitals and 5 out of 16 hospitals in the 2 study areas. The adopted treatments generally include a biological process (activated sludge system). The study concludes with the need to urgently define policy for the construction of wastewater treatment in every single hospital and to demand teams of experts at the various phases: design, construction, exercise and control.

In Brazil, with the intention to reduce the impact of hospital effluent, which is generally discharged into the public sewage network without any treatment, at the Regional University Hospital of Maringá, Zotesso et al. [53] investigated the possibility of treating the hospital laundry wastewater on site and eventually to reuse the treated effluent for the requirements of internal services. The fact that, in this hospital, laundry water consumption represents 50% of the overall hospital water consumption and contains organic substances at high concentrations (COD in the range 211–1,162 mg/L), the research group was commissioned to evaluate an adequate treatment able to reduce the pollutant discharge load and also the demand for fresh water. Bearing this in mind, they tested the sequence:

- Coagulation-flocculation (by the addition of 40–140 mg/L of the organic cationic polymer Tanfloc in order to remove suspended solids and organic compounds, mainly colour and detergents)
- Anthracite filtration for the removal of turbidity
- UV/H_2O_2 in order to complete the removal of COD (the influence of H_2O_2 was evaluated by means of different [H_2O_2]/COD ratios in relation to the COD value of the pre-treated effluent, 0.5, 1, 1.5, 2)

The results show that it is possible to achieve a complete removal of COD. Moreover, the increase in [H_2O_2]/COD ratio favours COD removal; however, because H_2O_2 acts as an additional pollutant, its concentration cannot be indefinitely increased. However, the investigations did not determine if the tested system is efficient in both the disinfection of raw hospital laundry wastewater and the reduction of its toxicity, as well as if treated effluents meet the requirements for reuse.

Several attempts have been made to promote the degradation and mineralisation of specific compounds by specific technologies. Of these, Moreno et al. [81] investigated the electrochemical degradation of the chemotherapy drug doxorubicin (DOX) in hospital effluent (where it often occurs at concentrations greater than 1 μg/L) by means of nanostructured graphite electrodes with metallic oxides (graphite, TiO2@graphite and AuO-TiO2@graphite electrodes). The study was carried out on a laboratory scale with cells of a capacity of 5 mL. The assays were conducted using 3 mL of DOX 1.25 mg/L in tap water and in Milli-Q water (Millipore S.A) with NaCl 10 mmol/L and with an applied voltage of 1.5, 2.5 and 5.0 V. It was found that the technology seems promising as no undesired compounds are formed. The complete degradation was found with AuO-TiO2@graphite electrode fed with DOX 1.45 mg/L over 40 min. It is important to highlight that the addition of NaCl improves degradation due to the formation of active chlorine with high oxidation power.

With regard to ecotoxicological issues, it was found that DOX treatment by electro-oxidation caused no effect on embryo-larval development of zebrafish, but damage was found in the DNA after 96 h of exposure. This underlines the importance of an in-depth ecotoxicological evaluation during the development of electrochemical oxidation.

In Seifert et al. [82], a benchmark of the measures taken by 31 German hospitals registered in EMAS, regarding their environmental management, reports that there are no specific measures regarding hospital wastewater management. They generally are released into the public sewer systems and conveyed to a municipal WWTP. Some healthcare structures mention the renewal of certain devices: grease separators in the kitchen, light liquid separators, a water treatment plant or the rehabilitation of the drainage canals on the hospital site. However, no measures regarding the specific and dedicated treatment of wastewater or the removal of certain pollutants are mentioned.

Finally, it is also worth mentioning a recent study which had the main aim to suggest adequate treatments for HWW before its release into the public sewage in order to guarantee the safety to workers in the case of damage to the sewer network [83]. It was reported that in the USA, according to the guidance by the World Health Organization and Centers for Disease Control and Prevention, wastewater from patients undergoing treatment at hospitals for *Ebola* has to be managed as ordinary wastewater without specific pre-treatment. The authors underline that the potential for *Ebolavirus* transmission via liquid waste discharged into the wastewater environment is currently unknown. On the basis of this analysis, it was concluded that the results suggest that when sewer workers operate in a wastewater collection system downstream from a hospital treating patients with *Ebola*, they are exposed to a potential risk that requires more attention and that disposal into the sanitary sewer without further treatment may result in the insufficient protection of the safety of sewer workers.

To conclude, the issue of the treatment of emergency hospitals in zones affected by extreme conditions due to earthquakes, flooding and other natural disasters is addressed in Khan et al. [84]. In these areas, the first medical assistance is supplied by means of mobile hospitals where victims receive the initial medical care and can be transferred to other stationary hospitals. However, in many cases, these mobile hospitals can be transformed into field hospitals and have to provide definitive care right on the field [85]. The amount of antibiotics used for the treatment of patients in these healthcare structures is high, which in turn results in a high impact on the environment. For field hospital effluent, a treatment train is recommended that consists of pre-treatments followed by a membrane bioreactor and finally a photo-Fenton stage to complete the removal of the most recalcitrant micropollutants.

5 Prioritisation: Environmental Risk Assessment of PhCs in HWW

In Zillien et al. [86], an interesting modelling framework is presented and applied to two large *Dutch* academic hospitals in order to prioritise pharmaceuticals based on their relative risks for aquatic organisms, considering the three trophic levels, fish, daphnia and algae, using purchase and prescription data from the individual hospitals. The framework consists of an emission prediction module and a risk prioritisation module (based on the risk quotient, i.e. the ratio between predicted concentrations and corresponding predicted no-effect concentrations). The study was based on a selection of 16 pharmaceuticals, including antibiotics, antineoplastics, iodinated contrast media, non-steroidal anti-inflammatory drugs, anticonvulsants, antidepressants, lipid regulators, antidiabetics and beta-blockers. Three ranking strategies were considered in evaluating the most critical compounds: the first is based on the risks associated with the total release of pharmaceuticals from the hospital due to inpatients and staff personnel, the second strategy is based on the risk due to only hospitalised patients, and the third is based on risks related to the environment by comparing the total predicted emissions based on hospital prescriptions (regardless if excretion occurs within the hospital or outside) to the emissions of only hospitalised patients in order to estimate the urban impact of the hospital emission. Results from prioritising pharmaceuticals indicate that azithromycin and ciprofloxacin represent the highest risk to aquatic organisms in relation to the substance selection of this study. Similarly, NSAIDs scored relatively high even though the release of this pharmaceutical group is underestimated in both cases. Emission mitigation strategies by hospitals could therefore focus their efforts primarily on reducing antibiotic and NSAID emissions. Both modules of the presented framework require a limited amount of relatively simple input data. Therefore, this study provides healthcare managers with a useful and practical tool to acquire sufficient exploratory information on hospital emissions for prioritising potential emission reduction measures.

In the frame of the EU project CytoThreat, Mišík et al. [87] carried out an environmental risk assessment due to four anticancer drugs (5-fluorouracil (5FU), cisplatin (CDDP), imatinib mesylate (IM) and etoposide (ET)) in hospital effluent in terms of acute, subacute and chronic effects (growth inhibition, reproduction inhibition and mortality testing), as well as genotoxic effects (DNA damage, chromosomal damage and gene mutation) with regard to many species of bacteria, algae, cyanobacteria, plants, Crustacea, Rotifera, freshwater mussels, fish and cell lines derived from humans and fish. The risk was assessed by means of the risk quotient RQ between the predicted concentration and the corresponding PNEC found in the investigation for the different cases (toxicity or genotoxicity). The results showed that in terms of toxicity, the highest RQs were obtained with IM, followed by 5FU, CDDP and ET. All values are below 1.0 indicating that none of these compounds represents a significant environmental risk. As for the genotoxic effects, they found that the $RQ_{genotox}$ value for 5FU is between 10 and 100, indicating significant potential for adverse effects. The $RQ_{genotox}$ values obtained with IM and CDDP are between 1 and 10, which indicates a small potential for adverse effects, while ET represents no significant risk.

In conclusion, their study highlighted that acute, subacute and chronic toxic effects of cytostatics in the aquatic environment are unlikely but may occur when the compounds are not removed from hospital wastewaters by efficient treatments and where the measured concentrations were always found higher than those in urban wastewater. Furthermore, they indicate that certain species (e.g. Crustacea) are highly sensitive with regard to the induction of the genotoxic effects which were found with environmentally relevant concentrations of 5FU, CDDP and IM.

Within the SIPIBEL project, an investigation was carried out on ecotoxicity (*Daphnia magna* mobility, *Pseudokirchneriella subcapitata* growth, *Brachionus calyciflorus* reproduction and SOS chromotest) and antibiotic resistance (integron quantification) during the mixing of hospital and urban wastewater and the treatment steps they were subjected to Laquaz et al. [45]. The results showed that hospital effluent presents higher ecotoxicity than the mixture probably due to the high content of adsorbable organic halides (AOX, mainly including chloroform), ammonium, non-ionic detergents and certain pharmaceuticals. The observed variation in ecotoxicity over the day, week, month and year might be related to the specific activities at the local hospital and their variability over time. These include the pharmaceutical consumption pattern [28, 29] which depends on the treatment of various seasonal pathologies and some management factors such as disinfection campaigns that may also have led to the punctual release of high quantities of ecotoxic compounds [45]. The mixture of HWW and urban wastewater in the proportion of 1/3–2/3 did not show any promotion in the spread of antibiotic resistance.

Within the Danish project which led to the definition of guidelines for the management and treatment of HWW in Denmark [26], the health and environmental risk was assessed with reference to the treated effluent from Herlev Hospital in three different scenarios, which reflected the normal operating conditions of the WWTP receiving HWW, its partial failure (simulating the main failures occurring during the test periods of the plant) and its total breakdown. The WWTP effluent is discharged in the small Kangsa River, a tributary of the Harrestrup River which flows into the marine area of Kalveboderne, used for bathing and other recreational activities. As for the environmental risk, the study evaluated (modelled) the exceeding of the corresponding PNEC value in the marine area by the average concentrations of the most critical compounds (ciprofloxacin, clarithromycin, diclofenac and sulphamethoxazole), once the WWTP effluent has mixed with the receiving water and reached the sea (no decay rate of the key compounds is considered on the way to the final receptor). With regard to health risk, the risk of infection was assessed with norovirus, and the concentrations of FIB were estimated in the river and then in the final receptor (the sea). These values were then compared with the EU bathing water requirements (250 MPN/100 mL). The distribution of the expected concentrations of norovirus in the receiving environments were evaluated by means of Monte Carlo simulations; the risk of infection was calculated by applying the approximated β-Poisson norovirus dose-response model with immunity [88] and the distribution of the infection by using Monte Carlo simulations. The results of this complex risk assessment showed that even after a realistic incident leading to a decrease in treatment removal efficiency, the treated effluent will not affect safe bathing water quality in the marine area. A total breakdown of the WWTP over a period of 1 day will cause a breach of the bathing standards. In the same way, during normal operations, there will be no risk of infection with norovirus in the marine water.

The release of untreated wastewater will only have a small impact on the risk of norovirus infection in the final receptor. The discussion of the uncertainties reported in the report by DHI [26] is of great help in evaluating the reliability of the collected results and, if necessary, in which phase of the analysis an in depth evaluation could be carried out.

6 Strategies in HWW Management and Treatments: Lessons Learned from Singular Experiences

PhC concentrations in hospital effluents depend on the size of the hospital and wards/departments, the activities taking place inside the structure (laboratory, chemical analysis, kitchen, laundry, etc.) and water consumption. A reply to the question of how similar or different they are with respect to municipal wastewater involves a deep analysis of the characteristics of the healthcare structure and its surrounding area [17, 89, 90].

A quick look at the observed ranges of concentrations of the most common antibiotics in the two effluents reported in the review by Souza et al. [91] highlights that the concentration of antibiotics (including azithromycin, tetracycline, ofloxacin, ciprofloxacin, sulphamethoxazole, trimethoprim, metronidazole and lincomycin) in HWW is higher than in municipal wastewater, confirming the conclusion by Verlicchi et al. [1]. Moreover, they also found that small- and medium-sized hospitals tend to discharge higher mean concentrations than larger ones with regard to many analgesics and anti-inflammatories, antibiotics, lipid regulators and psychiatric drugs.

The occurrence of antibiotic-resistant bacteria in hospital effluent documented by the study by Haller et al. [44] suggests the need to improve their removal from this major source of immission in the environment and to discuss and share strategies to be put into practice. More effort should be made to monitor and further control the dissemination of resistant bacteria in the environment from such a major source.

The adoption of a separate treatment has benefits according to Alvarino et al. [55], for instance, the avoidance of dilution with domestic wastewater and the specific treatments of potential toxics. In 2014, in its updated report [92], the WHO remarked that although a large part of HWW could present similar characteristics to domestic wastewater, "a proportion of the generated wastewater from health-care facilities will pose a higher risk than domestic wastewater. Depending on the service level and tasks of the health-care facility, the wastewater might contain chemicals, pharmaceuticals and contagious biological agents, and might even contain radioisotopes". According to WHO [92], the effluent from a healthcare structure should be treated before its release into the environment. The best solution should be an on-site facility for the *pre-treatment* of the produced wastewater prior to its release into the public sewer network in order to remove the presence of hazardous components including microbiological pathogens, radioactive drugs, toxic chemicals and antibiotics. A discussion of the WHO guidelines and the specific regulations in force in some countries (France, Germany, Spain, the UK, Italy, India, China and Vietnam) is reported and discussed in [93] and in the book [4].

In a previous study, Verlicchi et al. [94] already presented and discussed four scenarios for the management and treatment of HWW, and, in another one [95], the main factors to consider in planning the proper management and treatment of HWW were proposed. They suggest different strategies in the case of developed and developing countries. For developed countries, they proposed options on the basis of hospital and catchment area size. Recently, Mousel and Pinnekamp [52] confirmed the adequateness of this approach and in particular the suggestion that for healthcare structures with large capacities connected to small WWTPs, dedicated treatments or at least (pre-)treating the wastewater could be useful in order to convey lower PhC loads to the municipal WWTP. Full-scale treatment plants for the on-site (pre-)treatment of HWW have already been built [96], and this barrier can effectively act against the spread of ARB and ARG in the environment.

Two interesting experiences deserve particular attention, the French and the Danish experiences, and lessons can be learned from them.

At the Bellecombe site, France, as already mentioned, the existing treatment plant receives and separately treats hospital effluent in two lines adopting the same conventional treatments (activated sludge processes). SIPIBEL (SIte PIlote de BELlocombe) is the name of the project, the preparation and implementation of which was coordinated by GRAIE (http://www.graie.org/Sipibel/index.html)) [97]. Ongoing monitoring and research activities highlighted that HWW, compared to urban wastewater, is characterised by a higher concentrations of organic carbon, AOX, phosphates, gadolinium, acetaminophen, ketoprofen, sulphamethoxazole and ciprofloxacin [98], higher ecotoxicity, as well as the presence of bacteria with potentially higher antibiotic resistance. A higher variability was also found in PhC removal efficiencies in urban WWTP than in hospital WWTP. This could be explained by a larger variation of contamination sources and stronger seasonal fluctuations in urban wastewater [99]. The results also showed higher removal efficiencies in HWW treatment train for antibiotics, analgesics and beta-blockers, while concentrations of antibiotics and analgesics in the hospital treated effluent were higher than in urban treated effluent because of their higher influent concentrations [99]. An analysis of the potential environmental impact of the two treated effluents highlighted that due to the different content of pharmaceutical residues and nutrients, the two treated effluents cause different environmental adaptations of bacterial communities. In fact, they found that biofilms exposed to hospital treated effluent exhibited less developed biomass and lower diversity. Lessons learned from the collected studies have led to remark that, on the basis of the removal efficiency of (analysed) PhCs, separate treatment is beneficial. However, high PhC concentrations in the hospital treated effluent and the following adaptations of biofilm communities note the importance of adopting adequate wastewater technologies for the control of specific hospital pollutants.

In Denmark, since 2009, hospital effluent is managed in the same manner as industrial effluent, and a permit is required before it can be discharged into the public sewer system. Around 10 years ago, the Danish authorities wanted to increase their knowledge on how HWW *can* be treated and whether suggested technologies are feasible from a technical and economic viewpoint. For this reason, in 2010–2011, pre-tests were carried out on a laboratory scale by DHI for the Danish Environmental Protection Agency on innovative technologies which were then followed by pilot and lab tests with the effluent from Herlev Hospital [26]. This first study phase showed that, with reference to the selected PhCs, the most adequate technologies were the membrane bio-reactor (MBR) technology combined with activated carbon, ozonation and UV. The most efficient sequence of treatment has not yet been clearly identified. It was during the private-public innovation project, which started in 2012, that this could be identified. The aim of the project was to acquire knowledge and define a "national" strategy in the management and treatment of HWW in the country. In this context, a full-scale WWTP was built (which has been in operation since 2014) for the effluent of Herlev Hospital (700 beds with a flow rate of 150,000 m^3/year, with plans for it to be increased to 900 beds and to 200,000 m^3/year by 2020). It includes an MBR (with nitrogen and phosphorus removal) followed by ozonation (2.5 and 3.4 mg O_3/mg DOC), GAC filtration and UV radiation (5–10 mJ/cm^2). This polishing sequence proved to be the most adequate after a period of tests on the field [26] which compared different sequences. The sequence O_3-GAC-UV was selected as it optimises the removal of key pharmaceuticals (a list of 40 compounds) and costs for GAC regeneration, most likely because the general organic matter is transformed into more water-soluble compounds by the ozonation. Moreover, no critical formation of ozone by-products, such as bromate or NDMA, was observed. The amount of GAC to be replaced yearly is around 3,000 kg. Based on the collected results, it was found that the treated effluent could be directly released into a small local stream (Kagså). It is also planned that part of the treated water will be reused as cooling water in the existing cooling tower at the hospital.

In the same time, a task group developed guidelines for the regulation of the management and treatment of hospital effluent. This document set a list of 40 pharmaceuticals and the corresponding guideline values (maximum values for discharge into the public sewer system) and a method to define a hospitals as a major, medium and minor point sources based on the amount of pharmaceuticals administered in the hospital and the hospital consumption of antibiotics with respect to the catchment area.

The study carried out a health and environmental risk assessment assuming (1) a complete breakdown of the hospital WWTP and the subsequent release of untreated wastewater, (2) a partially treated effluent and (3) normal operation of the WWTP. The main results and lessons learned during this investigation are well described and discussed in [26].

The French and the Danish strategies highlight how useful it can be to share the planning of research activities which involve decision-makers, administrators, technicians, practitioners, researchers and hospital administrative and technical staff in defining adequate strategies for the management and treatment of this particular effluent, also with a view to the ongoing European legislation, attention to the occurrence of micropollutants in water compartments and the control and reduction of their emission into the environment.

An environmental audit similar to that discussed in Vaccari et al. [100] which referred to pharmaceuticals consumed and released in a healthcare facility as well as the adopted treatment could help optimise management and minimise the impact of hospital (treated) effluent in the environment.

7 The Need for Further Debate and Research

The experiences herein reported and discussed demonstrate the efforts that have been carried out worldwide and that are still ongoing in reaching Sustainable Millennium Goal number 6 "Improve Clean Water and Sanitation" defined by the World Health Organization by 2030.

However, they also underline the need for further research. In particular, it emerges that:

- The direct disposal of HWW to a natural water body without any kind of treatment is the common practice for many countries [101].
- Adequate treatments are necessary to reduce the emission of micropollutants in the environment. National guidelines could help in this aim and should be defined by a team of expert involving scientists, technicians, hospital technical and administrative staff, as well as policy makers and decision-makers.
- HWW management and treatment should be dealt with in a holistic view, including all the water cycle, in particular in developing countries. An improvement in sanitation systems will reflect on the water sector, as discussed in [102].
- With regard to some critical compounds, households could be the main source of such compounds, and the municipal WWTP should be updated. This was the case for 94% of anticancer drugs which were due to outpatients in the hospital and catchment area investigated by [103].
- Investigations should be carried out on the efficacy of new treatments in removing key pharmaceuticals and not only the parent compounds but also the intermediates generated during treatment [71].
- Environmental risk assessment should also focus on the mixture of substances [86].
- Improving the elimination of antimicrobial-resistant microorganisms from HWW would appear to be a reasonable policy to put into practice, and more effort should be made to monitor and further control the dissemination of resistant bacteria in the environment from such a major source [44].

To conclude: much is already known but much more is still to be discovered and even more is to be done.

References

1. Verlicchi P, Galletti A, Petrovic M, BarcelÓ D (2010) Hospital effluents as a source of emerging pollutants: an overview of micropollutants and sustainable treatment options. J Hydrol 389:416–428
2. Pills_Report. Pharmaceutical input and elimination from local sources
3. Verlicchi P, Al Aukidy M, Zambello E (2015) What have we learned from worldwide experiences on the management and treatment of hospital effluent? – An overview and a discussion on perspectives. Sci Total Environ 514:467–491
4. Verlicchi P (2018) Hospital wastewaters. 60:17–32. http://link.springer.com/10.1007/978-3-319-62178-4
5. Kümmerer K (2001) Drugs in the environment: emission of drugs, diagnostic aids and disinfectants into wastewater by hospitals in relation to other sources – a review. Chemosphere 45(6–7):957–969
6. Weissbrodt D, Mcardell CS, Kovalova L, Hollender J, Siegrist H (2006) Hospital wastewater: mass flow analysis of pharmaceuticals and disinfectants. 1–2
7. Boillot C, Bazin C, Tissot-Guerraz F, Droguet J, Perraud M, Cetre JC et al (2008) Daily physicochemical, microbiological and ecotoxicological fluctuations of a hospital effluent according to technical and care activities. Sci Total Environ 403(1–3):113–129
8. El-Ogri F, Ouazzani N, Boraâm F, Mandi L (2016) A survey of wastewaters generated by a hospital in Marrakech city and their characterization. Desalin Water Treat 57 (36):17061–17074
9. Kilunga PI, Kayembe JM, Laffite A, Thevenon F, Devarajan N, Mulaji CK et al (2016) The impact of hospital and urban wastewaters on the bacteriological contamination of the water resources in Kinshasa, Democratic Republic of Congo. J Environ Sci Heal A Toxic/Hazardous Subst Environ Eng 51(12):1034–1042
10. Laffite A, Kilunga PI, Kayembe JM, Devarajan N, Mulaji CK, Giuliani G et al (2016) Hospital effluents are one of several sources of metal, antibiotic resistance genes, and bacterial markers disseminated in sub-saharan urban rivers. Front Microbiol 7:1–14
11. Ibrahim C, Hammami S, Chérif N, Mejri S, Pothier P, Hassen A (2019) Detection of Sapoviruses in two biological lines of Tunisian hospital wastewater treatment. Int J Environ Health Res 29(4):400–413. https://doi.org/10.1080/09603123.2018.1546835
12. Kimosop SJ, Getenga ZM, Orata F, Okello VA, Cheruiyot JK (2016) Residue levels and discharge loads of antibiotics in wastewater treatment plants (WWTPs), hospital lagoons, and rivers within Lake Victoria Basin, Kenya. Environ Monit Assess 188(9)
13. Botero-Coy AM, Martínez-Pachón D, Boix C, Rincón RJ, Castillo N, Arias-Marín LP et al (2018) An investigation into the occurrence and removal of pharmaceuticals in Colombian wastewater. Sci Total Environ 642:842–853. https://doi.org/10.1016/j.scitotenv.2018.06.088
14. Giannakis S, Rtimi S, Pulgarin C (2017) Light-assisted advanced oxidation processes for the elimination of chemical and microbiological pollution of wastewaters in developed and developing countries. Molecules 22(7)
15. Martins AF, Frank C, da S, Altissimo J, de Oliveira JA, da Silva DS, Reichert JF et al (2017) Determination of statin drugs in hospital effluent with dispersive liquid–liquid microextraction and quantification by liquid chromatography. J Environ Sci Heal A Toxic/Hazardous Subst Environ Eng 52(10):939–945. https://doi.org/10.1080/10934529.2017.1324706
16. Reichert JF, Souza DM, Martins AF (2019) Antipsychotic drugs in hospital wastewater and a preliminary risk assessment. Ecotoxicol Environ Saf 170:559–567. https://doi.org/10.1016/j.ecoenv.2018.12.021
17. Langford KH, Thomas KV (2009) Determination of pharmaceutical compounds in hospital effluents and their contribution to wastewater treatment works. Environ Int 35(5):766–770. https://doi.org/10.1016/j.envint.2009.02.007

18. Stockholm County Council (2014) Environmentally classified pharmaceuticals 2014-2015. 1–54. https://noharm-europe.org/documents/environmentally-classified-pharmaceuticals-2014-2015
19. Hernando MD, Mezcua M, Fernández-Alba AR, Barceló D (2006) Environmental risk assessment of pharmaceutical residues in wastewater effluents, surface waters and sediments. Talanta. 69(2 SPEC ISS):334–342
20. Kafaei R, Papari F, Seyedabadi M, Sahebi S, Tahmasebi R, Ahmadi M et al (2018) Occurrence, distribution, and potential sources of antibiotics pollution in the water-sediment of the northern coastline of the Persian Gulf, Iran. Sci Total Environ 627:703–712. https://doi.org/10.1016/j.scitotenv.2018.01.305
21. Lien LTQ, Hoa NQ, Chuc NTK, Thoa NTM, Phuc HD, Diwan V et al (2016) Antibiotics in wastewater of a rural and an urban hospital before and after wastewater treatment, and the relationship with antibiotic use-a one year study from Vietnam. Int J Environ Res Public Health 13(6):1–13
22. Azuma T, Arima N, Tsukada A, Hirami S, Matsuoka R, Moriwake R et al (2016) Detection of pharmaceuticals and phytochemicals together with their metabolites in hospital effluents in Japan, and their contribution to sewage treatment plant influents. Sci Total Environ 548–549:189–197. https://doi.org/10.1016/j.scitotenv.2015.12.157
23. Papageorgiou M, Zioris I, Danis T, Bikiaris D, Lambropoulou D (2019) Comprehensive investigation of a wide range of pharmaceuticals and personal care products in urban and hospital wastewaters in Greece. Sci Total Environ 694:133565. https://doi.org/10.1016/j.scitotenv.2019.07.371
24. Szekeres E, Baricz A, Chiriac CM, Farkas A, Opris O, Soran ML et al (2017) Abundance of antibiotics, antibiotic resistance genes and bacterial community composition in wastewater effluents from different Romanian hospitals. Environ Pollut 225:304–315
25. Lerat-Hardy A, Coynel A, Dutruch L, Pereto C, Bossy C, Gil-Diaz T et al (2019) Rare earth element fluxes over 15 years into a major European Estuary (Garonne-Gironde, SW France): Hospital effluents as a source of increasing gadolinium anomalies. Sci Total Environ 656:409–420. https://doi.org/10.1016/j.scitotenv.2018.11.343
26. DHI (2016) Full scale advanced wastewater treatment at Herlev Hospital – treatment performance and evaluation. 87
27. Lasek F, Karpel Vel Leitner N, Rauwel G, Blanchier L, Castel O, Ayraud-Thevenot S et al (2019) Discharge of biocidal products from healthcare activities into a sewage system – a case study at a French university hospital. Environ Sci Pollut Res 26(5):4938–4951
28. Verlicchi P, Zambello E (2016) Predicted and measured concentrations of pharmaceuticals in hospital effluents. Examination of the strengths and weaknesses of the two approaches through the analysis of a case study. Sci Total Environ. 565
29. Verlicchi P (2018) Pharmaceutical concentrations and loads in hospital effluents: Is a predictive model or direct measurement the most accurate approach? Handbook of environmental chemistry, vol 60
30. Daouk S, Chèvre N, Vernaz N, Widmer C, Daali Y, Fleury-Souverain S (2016) Dynamics of active pharmaceutical ingredients loads in a Swiss university hospital wastewaters and prediction of the related environmental risk for the aquatic ecosystems. Sci Total Environ 547:244–253. https://doi.org/10.1016/j.scitotenv.2015.12.117
31. Alam M, Imran M, Ahmad SS (2019) Screening of metal and antibiotic resistance in Beta-lactamase producing coliform Bacteria from Hospital wastewater of northern India. Recent Pat Biotechnol 14(1):63–77
32. Hocquet D, Muller A, Bertrand X (2016) What happens in hospitals does not stay in hospitals: antibiotic-resistant bacteria in hospital wastewater systems. J Hosp Infect 93(4):395–402. https://doi.org/10.1016/j.jhin.2016.01.010
33. WHO (2017) Global priority list of antibiotic-resistant bacteria to guide research, discovery, and development of new antibiotics. WHO. 7. https://www.who.int/news-room/detail/27-02-2017-who-publishes-list-of-bacteria-for-which-new-antibiotics-are-urgently-needed

34. Cantas L, Shah SQA, Cavaco LM, Manaia CM, Walsh F, Popowska M et al (2013) A brief multi-disciplinary review on antimicrobial resistance in medicine and its linkage to the global environmental microbiota. Front Microbiol 4:1–14
35. IAGC (2019) Report to the secretary general of the United Nations, no time to wait: securing the future from drug-resistant infections. Artforum Int 54(10):113–114. https://www.who.int/docs/default-source/documents/no-time-to-wait-securing-the-future-from-drug-resistant-infections-en.pdf?sfvrsn=5b424d7_6
36. Lien LTQ, Lan PT, Chuc NTK, Hoa NQ, Nhung PH, Thoa NTM et al (2017) Antibiotic resistance and antibiotic resistance genes in *Escherichia coli* isolates from hospital wastewater in Vietnam. Int J Environ Res Public Health 14(7):1–11
37. Lamba M, Graham DW, Ahammad SZ (2017) Hospital wastewater releases of Carbapenem-resistance pathogens and genes in Urban India. Environ Sci Technol 51(23):13906–13912
38. Bréchet C, Plantin J, Sauget M, Thouverez M, Talon D, Cholley P et al (2014) Wastewater treatment plants release large amounts of extended-spectrum β-lactamase-producing *Escherichia coli* into the environment. Clin Infect Dis 58(12):1658–1665
39. Leclercq R, Oberlé K, Galopin S, Cattoir V, Budzinski H, Petit F (2013) Changes in enterococcal populations and related antibiotic resistance along a medical center-wastewater treatment plant-river continuum. Appl Environ Microbiol 79(7):2428–2434
40. Łuczkiewicz A, Jankowska K, Fudala-Ksiazek S, Olańczuk-Neyman K (2010) Antimicrobial resistance of fecal indicators in municipal wastewater treatment plant. Water Res 44 (17):5089–5097
41. Rosenberg Goldstein RE, Micallef SA, Gibbs SG, George A, Claye E, Sapkota A et al (2014) Detection of vancomycin-resistant enterococci (VRE) at four U.S. wastewater treatment plants that provide effluent for reuse. Sci Total Environ 466–467:404–411. https://doi.org/10.1016/j.scitotenv.2013.07.039
42. Magiorakos AP, Srinivasan A, Carey RB, Carmeli Y, Falagas ME, Giske CG et al (2012) Multidrug-resistant, extensively drug-resistant and pandrug-resistant bacteria: an international expert proposal for interim standard definitions for acquired resistance. Clin Microbiol Infect 18(3):268–281. https://doi.org/10.1111/j.1469-0691.2011.03570.x
43. Zhang L, Ma X, Luo L, Hu N, Duan J, Tang Z et al (2020) The prevalence and characterization of extended-spectrum β-lactamase-and carbapenemase-producing bacteria from hospital sewage, treated effluents and receiving rivers. Int J Environ Res Public Health 17(4)
44. Haller L, Chen H, Ng C, Le TH, Koh TH, Barkham T et al (2018) Occurrence and characteristics of extended-spectrum β-lactamase- and carbapenemase-producing bacteria from hospital effluents in Singapore. Sci Total Environ 615:1119–1125. https://doi.org/10.1016/j.scitotenv.2017.09.217
45. Laquaz M, Dagot C, Bazin C, Bastide T, Gaschet M, Ploy MC et al (2018) Ecotoxicity and antibiotic resistance of a mixture of hospital and urban sewage in a wastewater treatment plant. Environ Sci Pollut Res 25(10):9243–9253
46. Vaz-Moreira I, Varela AR, Pereira TV, Fochat RC, Manaia CM (2016) Multidrug resistance in quinolone-resistant gram-negative Bacteria isolated from Hospital effluent and the municipal wastewater treatment plant. Microb Drug Resist 22(2):155–163
47. Cahill N, O'Connor L, Mahon B, Varley Á, McGrath E, Ryan P et al (2019) Hospital effluent: a reservoir for carbapenemase-producing Enterobacterales? Sci Total Environ 672:618–624. https://doi.org/10.1016/j.scitotenv.2019.03.428
48. Harris S, Cormican M (2016) Report No. 162 Hospital effluent: impact on the microbial environment and risk to human health Authors: Dearbháile Morris, Suvi Harris, Carol Morris, Enda Commins and Martin Cormican
49. Proia L, Anzil A, Borrego C, Farrè M, Llorca M, Sanchis J et al (2018) Occurrence and persistence of carbapenemases genes in hospital and wastewater treatment plants and propagation in the receiving river. J Hazard Mater 358:33–43. https://doi.org/10.1016/j.jhazmat.2018.06.058

50. Karami N, Mohammadi P, Zinatizadeh A, Falahi F, Aghamohammadi N (2018) High rate treatment of hospital wastewater using activated sludge process induced by high-frequency ultrasound. Ultrason Sonochem 46:89–98. https://doi.org/10.1016/j.ultsonch.2018.04.009
51. Vo HNP, Koottatep T, Chapagain SK, Panuvatvanich A, Polprasert C, Nguyen TMH et al (2019) Removal and monitoring acetaminophen-contaminated hospital wastewater by vertical flow constructed wetland and peroxidase enzymes. J Environ Manage 250:109526. https://doi.org/10.1016/j.jenvman.2019.109526
52. Mousel D, Pinnekamp J (2018) Ozonation of nursing home wastewater pretreated in a membrane bioreactor. Water Sci Technol 78(2):266–278
53. Zotesso JP, Cossich ES, Janeiro V, Tavares CRG (2017) Treatment of hospital laundry wastewater by UV/H2O2 process. Environ Sci Pollut Res 24(7):6278–6287. https://doi.org/10.1007/s11356-016-6860-5
54. Dires S, Birhanu T, Ambelu A (2019) Use of broken brick to enhance the removal of nutrients in subsurface flow constructed wetlands receiving hospital wastewater. Water Sci Technol 79 (1):156–164
55. Alvarino T, García-Sandá E, Gutiérrez-Prada I, Lema J, Omil F, Suárez S (2020) A new decentralized biological treatment process based on activated carbon targeting organic micropollutant removal from hospital wastewaters. Environ Sci Pollut Res 27(2):1214–1223
56. Verlicchi P, Al Aukidy M, Zambello E (2012) Occurrence of pharmaceutical compounds in urban wastewater: removal, mass load and environmental risk after a secondary treatment-a review. Sci Total Environ 429
57. Lima DR, Hosseini-Bandegharaei A, Thue PS, Lima EC, de Albuquerque YRT, dos Reis GS et al (2019) Efficient acetaminophen removal from water and hospital effluents treatment by activated carbons derived from Brazil nutshells. Colloids Surf A Physicochem Eng Asp 583:123966. https://doi.org/10.1016/j.colsurfa.2019.123966
58. Lan Y, Groenen-Serrano K, Coetsier C, Causserand C (2018) Nanofiltration performances after membrane bioreactor for hospital wastewater treatment: fouling mechanisms and the quantitative link between stable fluxes and the water matrix. Water Res 146:77–87
59. Tran T, Nguyen TB, Ho HL, Le DA, Lam TD, Nguyen DC et al (2019) Integration of membrane bioreactor and nanofiltration for the treatment process of real hospital wastewater in Ho Chi Minh City, Vietnam. Processes 7(3):1–13
60. Paulus GK, Hornstra LM, Alygizakis N, Slobodnik J, Thomaidis N, Medema G (2019) The impact of on-site hospital wastewater treatment on the downstream communal wastewater system in terms of antibiotics and antibiotic resistance genes. Int J Hyg Environ Health 222 (4):635–644. https://doi.org/10.1016/j.ijheh.2019.01.004
61. Ngo HH, Guo W, Xing W (2008) Evaluation of a novel sponge-submerged membrane bioreactor (SSMBR) for sustainable water reclamation. Bioresour Technol 99(7):2429–2435
62. Nguyen TT, Bui XT, Luu VP, Nguyen PD, Guo W, Ngo HH (2017) Removal of antibiotics in sponge membrane bioreactors treating hospital wastewater: comparison between hollow fiber and flat sheet membrane systems. Bioresour Technol 240:42–49. https://doi.org/10.1016/j.biortech.2017.02.118
63. García Galán MJ, Díaz-Cruz MS, Barceló D (2012) Removal of sulfonamide antibiotics upon conventional activated sludge and advanced membrane bioreactor treatment. Anal Bioanal Chem 404(5):1505–1515
64. Vo TKQ, Bui XT, Chen SS, Nguyen PD, Cao NDT, Vo TDH et al (2019) Hospital wastewater treatment by sponge membrane bioreactor coupled with ozonation process. Chemosphere 230:377–383. https://doi.org/10.1016/j.chemosphere.2019.05.009
65. Von Gunten U (2003) Ozonation of drinking water: part II. Disinfection and by-product formation in presence of bromide, iodide or chlorine. Water Res 37(7):1469–1487
66. Bourgin M, Beck B, Boehler M, Borowska E, Fleiner J, Salhi E et al (2018) Evaluation of a full-scale wastewater treatment plant upgraded with ozonation and biological post-treatments: abatement of micropollutants, formation of transformation products and oxidation by-products. Water Res 129:486–498. https://doi.org/10.1016/j.watres.2017.10.036

67. Shokoohi R, Torkshavand Z, Zolghadnasab H, Alikhani MY, Hemmat MS (2018) Study of the efficiency of moving bed biofilm reactor (MBBR) in LAS anionic detergent removal from hospital wastewater: determination of removing model according to response surface methodology (RSM). Water Sci Technol 2017(1):1–7
68. Tang K, Spiliotopoulou A, Chhetri RK, Ooi GTH, Kaarsholm KMS, Sundmark K et al (2019) Removal of pharmaceuticals, toxicity and natural fluorescence through the ozonation of biologically-treated hospital wastewater, with further polishing via a suspended biofilm. Chem Eng J 359:321–330. https://doi.org/10.1016/j.cej.2018.11.112
69. Ferre-Aracil J, Valcárcel Y, Negreira N, de Alda ML, Barceló D, Cardona SC et al (2016) Ozonation of hospital raw wastewaters for cytostatic compounds removal. Kinetic modelling and economic assessment of the process. Sci Total Environ 556:70–79. https://doi.org/10.1016/j.scitotenv.2016.02.202
70. Mir-Tutusaus JA, Parladé E, Llorca M, Villagrasa M, Barceló D, Rodriguez-Mozaz S et al (2017) Pharmaceuticals removal and microbial community assessment in a continuous fungal treatment of non-sterile real hospital wastewater after a coagulation-flocculation pretreatment. Water Res 116:65–75
71. Jaén-Gil A, Castellet-Rovira F, Llorca M, Villagrasa M, Sarrà M, Rodríguez-Mozaz S et al (2019) Fungal treatment of metoprolol and its recalcitrant metabolite metoprolol acid in hospital wastewater: biotransformation, sorption and ecotoxicological impact. Water Res 152:171–180
72. Badia-Fabregat M, Lucas D, Pereira MA, Alves M, Pennanen T, Fritze H et al (2016) Continuous fungal treatment of non-sterile veterinary hospital effluent: pharmaceuticals removal and microbial community assessment. Appl Microbiol Biotechnol 100(5):2401–2415
73. Kovalova L, Siegrist H, Singer H, Wittmer A, McArdell CS (2012) Hospital wastewater treatment by membrane bioreactor: performance and efficiency for organic micropollutant elimination. Environ Sci Technol 46(3):1536–1545
74. Cruz-Morató C, Lucas D, Llorca M, Rodriguez-Mozaz S, Gorga M, Petrovic M et al (2014) Hospital wastewater treatment by fungal bioreactor: removal efficiency for pharmaceuticals and endocrine disruptor compounds. Sci Total Environ 493:365–376. https://doi.org/10.1016/j.scitotenv.2014.05.117
75. Lucas D, Badia-Fabregat M, Vicent T, Caminal G, Rodríguez-Mozaz S, Balcázar JL et al (2016) Fungal treatment for the removal of antibiotics and antibiotic resistance genes in veterinary hospital wastewater. Chemosphere 152:301–308
76. Rodriguez-Mozaz S, Chamorro S, Marti E, Huerta B, Gros M, Sànchez-Melsió A et al (2015) Occurrence of antibiotics and antibiotic resistance genes in hospital and urban wastewaters and their impact on the receiving river. Water Res 69:234–242
77. Badia-Fabregat M, Lucas D, Tuomivirta T, Fritze H, Pennanen T, Rodríguez-Mozaz S et al (2017) Study of the effect of the bacterial and fungal communities present in real wastewater effluents on the performance of fungal treatments. Sci Total Environ 579:366–377. https://doi.org/10.1016/j.scitotenv.2016.11.088
78. Serna-Galvis EA, Silva-Agredo J, Botero-Coy AM, Moncayo-Lasso A, Hernández F, Torres-Palma RA (2019) Effective elimination of fifteen relevant pharmaceuticals in hospital wastewater from Colombia by combination of a biological system with a sonochemical process. Sci Total Environ 670:623–632. https://doi.org/10.1016/j.scitotenv.2019.03.153
79. Petrovich ML, Zilberman A, Kaplan A, Eliraz GR, Wang Y, Langenfeld K et al (2020) Microbial and viral communities and their antibiotic resistance genes throughout a Hospital wastewater treatment system. Front Microbiol 11:1–13
80. Azizi N, Kiomars S, Jila A, Kamaladdin K, Hooshmand S, Hossein A et al (2016) Management of collection, treatment and disposal of hospital wastewater. Case study: Hamadan and Kermanshah province hospitals. Int J Pharm Technol 8(2):12924–12933. www.ijptonline.com
81. Moreno EKG, Garcia LF, Lobón GS, Brito LB, Oliveira GAR, Luque R et al (2019) Ecotoxicological assessment and electrochemical remediation of doxorubicin. Ecotoxicol Environ Saf 179:143–150. https://doi.org/10.1016/j.ecoenv.2019.04.050

82. Seifert C, Guenther E (2019) Prevention is better than cure—environmental management measures in hospitals. Corp Soc Respon Environ Manag 26(4):781–790
83. Haas CN, Rycroft T, Bibby K, Casson L (2017) Risks from ebolavirus discharge from Hospitals to sewer workers. Water Environ Res 89(4):357–368
84. Khan NA, Ahmed S, Vambol S, Vambol V, Farooqi IH (2019) Field hospital wastewater treatment scenario. Ecol Quest 30(3):57–69
85. Bar-On E, Abargel A, Peleg K, Kreiss Y (2013) Coping with the challenges of early disaster response: 24 years of field hospital experience after earthquakes. Disaster Med Public Health Prep 7(5):491–498
86. Zillien C, van Loon C, Gülpen M, Tipatet K, Hanssen B, Beeltje H et al (2019) Risk-management tool for environmental prioritization of pharmaceuticals based on emissions from hospitals. Sci Total Environ 694:133733. https://doi.org/10.1016/j.scitotenv.2019.133733
87. Mišík M, Filipic M, Nersesyan A, Kundi M, Isidori M, Knasmueller S (2019) Environmental risk assessment of widely used anticancer drugs (5-fluorouracil, cisplatin, etoposide, imatinib mesylate). Water Res 164
88. Schmidt PJ (2015) Norovirus dose-response: are currently available data informative enough to determine how susceptible humans are to infection from a single virus? Risk Anal 35 (7):1364–1383
89. Verlicchi P, Galletti A, Petrovic M, Barceló D, Al Aukidy M, Zambello E (2013) Removal of selected pharmaceuticals from domestic wastewater in an activated sludge system followed by a horizontal subsurface flow bed – analysis of their respective contributions. Sci Total Environ 454–455:411–425
90. Santos LHMLM, Gros M, Rodriguez-Mozaz S, Delerue-Matos C, Pena A, Barceló D et al (2013) Contribution of hospital effluents to the load of pharmaceuticals in urban wastewaters: identification of ecologically relevant pharmaceuticals. Sci Total Environ 461–462:302–316. https://doi.org/10.1016/j.scitotenv.2013.04.077
91. Souza FS, Féris LA (2016) Hospital and municipal wastewater: identification of relevant pharmaceutical compounds. Water Environ Res 88(9):871–877
92. WHO (2014) In: Chartier Y et al. Safe management of wastes from health-care activities. World Heal Organ. p 329. http://www.searo.who.int
93. Carraro E, Bonetta S, Bertino C, Lorenzi E, Bonetta S, Gilli G (2016) Hospital effluents management: chemical, physical, microbiological risks and legislation in different countries. J Environ Manage 168:185–199. https://doi.org/10.1016/j.jenvman.2015.11.021
94. Verlicchi P, Galletti A, Masotti L (2010) Management of hospital wastewaters: the case of the effluent of a large hospital situated in a small town. Water Sci Technol 61(10)
95. Verlicchi P, Galletti A, Petrovic M, Barceló D (2012) Micro-pollutants in Hospital effluent: their fate, risk and treatment options. Handbook of environmental chemistry, vol 20
96. Beier S, Cramer C, Köster S, Mauer C, Palmowski L, Schröder HF et al (2011) Full scale membrane bioreactor treatment of hospital wastewater as forerunner for hot-spot wastewater treatment solutions in high density urban areas. Water Sci Technol 63(1):66–71
97. Chonova T, Lecomte V, Bertrand-Krajewski JL, Bouchez A, Labanowski J, Dagot C et al (2018) The SIPIBEL project: treatment of hospital and urban wastewater in a conventional urban wastewater treatment plant. Environ Sci Pollut Res 25(10):9197–9206
98. Wiest L, Chonova T, Bergé A, Baudot R, Bessueille-Barbier F, Ayouni-Derouiche L et al (2018) Two-year survey of specific hospital wastewater treatment and its impact on pharmaceutical discharges. Environ Sci Pollut Res 25(10):9207–9218
99. Chonova T, Keck F, Labanowski J, Montuelle B, Rimet F, Bouchez A (2016) Separate treatment of hospital and urban wastewaters: a real scale comparison of effluents and their effect on microbial communities. Sci Total Environ 542:965–975. https://doi.org/10.1016/j.scitotenv.2015.10.161

100. Vaccari M, Montasser W, Tudor T, Leone L (2017) Environmental audits and process flow mapping to assess management of solid waste and wastewater from a healthcare facility: an Italian case study. Environ Monit Assess 189(5)
101. Khan NA, Khan SU, Ahmed S, Farooqi IH, Yousefi M, Mohammadi AA et al (2020) Recent trends in disposal and treatment technologies of emerging-pollutants – a critical review. Trends Anal Chem 122
102. Verlicchi P, Grillini V (2020) Surface water and groundwater quality in South Africa and Mozambique-analysis of the most critical pollutants for drinking purposes and challenges in water treatment selection. Water 12(1)
103. Cristóvão MB, Torrejais J, Janssens R, Luis P, Van der Bruggen B, Dubey KK et al (2019) Treatment of anticancer drugs in hospital and wastewater effluents using nanofiltration. Sep Purif Technol 224:273–280. https://doi.org/10.1016/j.seppur.2019.05.016

Suspect and Non-target Screening Methodologies for the Evaluation of the Behaviour of Polar Organic Micropollutants and Changes in the Molecule Fingerprint During Water Treatment

Pablo Gago-Ferrero

Contents

1 Introduction 98
2 LC-HRMS-Based Analytical Methodologies 99
  2.1 Sampling Strategy and Sample Pretreatment 99
  2.2 Sample Extraction and Instrumental Analysis 100
3 Strategies for the Evaluation of the Behaviour of Organic Micropollutants During Wastewater Treatment 102
  3.1 Data Preprocessing 102
  3.2 Prioritisation Strategies 102
  3.3 Suspect/Non-target: Identification Approaches of Polar Organic Contaminants 107
  3.4 Novel Non-target Strategies to Assess Changes in the Molecular Fingerprints Without Structural Elucidation 109
References 111

Abstract The evaluation of water treatment systems in terms of presence, fate and removal of organic micropollutants is of paramount importance for the optimal protection of both the environment and human health. In this regard, there are significant gaps in the existing data for compounds present in treated water that remain uncharacterised but are still hazardous from both a toxicological and ecological point of view. This work aims to critically summarise the existing

---

P. Gago-Ferrero (✉)
Catalan Institute for Water Research (ICRA), Girona, Spain

University of Girona, Girona, Spain
e-mail: pgago@icra.cat

Sara Rodriguez-Mozaz, Paqui Blánquez Cano, and Montserrat Sarrà Adroguer (eds.), *Removal and Degradation of Pharmaceutically Active Compounds in Wastewater Treatment*, Hdb Env Chem (2021) 108: 97–118, DOI 10.1007/698_2020_662,

information about the application of suspect and non-target screening approaches based on liquid chromatography coupled to high-resolution mass spectrometry (LC-HRMS) to understand the behaviour of polar organic micropollutants and their corresponding transformation products during water treatment processes, focusing particularly on wastewater. Also, it includes a discussion about the novel non-target workflows whose objective is not the further structure elucidation of particular compounds but assessing changes in the molecular elemental compositions of the dissolved organic matter (DOM) during water treatment (without structural characterisation). These strategies can provide additional information and become a common step for a better understanding of treatment performance and transformation product formation.

Keywords Anthropogenic, Cheminformatics, Electrokinetic, Hydrophilic-lipophilic balance, Spectrometry

1 Introduction

Discharges from wastewater treatment plants (WWTPs) are one of the major sources of aquatic pollution. Commonly used WWTPs, primarily operating through biological processes, were developed and designed to protect natural aquatic systems and water resources mainly by removing loads of carbon, nitrogen and phosphorous present in the influent in the mg L^{-1} range. The increased detection of a wide range of emerging, not-regulated anthropogenic organic micropollutants (MPs) in the aquatic environment shows the limitations of conventional WWTPs in removing these compounds. Their continuous introduction into the environment, pseudo-persistence and intrinsic ability to interfere with organisms concern the scientific and public community because their potential toxic effects can threaten the ecological status of water bodies and human health via, e.g. the drinking water quality and/or long-term endocrine effects. Therefore, advances in wastewater treatment technologies are crucial in minimising the burden of wastewater-originated contaminants. In parallel to the technological developments, one of the major challenges remains the evaluation of novel and alternative treatment options regarding their potential to minimise the toxicological risks for both ecosystems and human health [1].

The global universe of chemicals is very complex and includes hundreds of thousands of substances in commercial use [2]. However, the effects of these compounds on the environment and human health and their combined effects are in most cases not at all or only partially investigated [3]. Within the last years, polar organic MPs (e.g. pharmaceuticals) became an increasing area of focus for environmental scientists and regulatory authorities encouraged by the advances in liquid chromatography coupled to mass spectrometry (LC-MS) technologies. The existing target screening methods are based on the preselection of certain chemicals and the use of reference standards and can only cover a small proportion of organic pollutants. However, it should be stressed that toxicity and related unwanted effects can be associated with chemicals that can be identified and those that cannot be identified as a result of limitations in the repertoire of chemical analytical techniques. Therefore, important site-specific and potentially ecotoxicologically relevant compounds are systematically missed [4], and the use of *suspect* and *non-target* screening analytical approaches, where no particular chemical is being searched for, is necessary to obtain a broader and more realistic picture. Thus, the improvements in the characterisation of water and the study of the fate of MPs during processes (e.g. WWTPs), where the entire bulk of the dissolved organic matter (DOM) and not few selected substances are considered, are of paramount importance. For a holistic risk assessment, target-based environmental monitoring should be accompanied by non-target analysis. This is particularly important in order to evaluate the behaviour and removal of organic MPs during water treatment processes, where a large number of transformation products (TPs) are continuously generated, which can pose a higher threat than the corresponding parent compounds to both the environment and human health.

This chapter aims to critically summarise the existing information about the application of suspect and non-target screening approaches based on liquid chromatography coupled to high-resolution mass spectrometry (LC-HRMS) to understand the behaviour of polar organic MPs and their corresponding TPs during water treatment processes, focusing particularly on wastewater. All this information is crucial to evaluate which are the potentially hazardous compounds that reach the environment through effluents and which improvements are urgently needed. Also, it is included a section describing novel non-target workflows whose objective is not the further structure elucidation of particular compounds but assessing changes in the molecular fingerprints of DOM during water treatment (without structural characterisation). These strategies can provide additional information and become a common step in research dealing with the evaluation of water treatment.

2 LC-HRMS-Based Analytical Methodologies

2.1 Sampling Strategy and Sample Pretreatment

The sampling strategy as well as the sample treatment procedure should be defined considering the research question that needs to be tackled as well as the resource availability. It can only be optimally designed when the objective and the methodology of a given study has been well formulated.

In order to evaluate the presence and behaviour of MPs during water treatment, it is necessary to collect different samples in time and space. For example, the studies focusing on assessing the removal of MPs in WWTPs usually need to collect at least influent and effluent (after secondary or tertiary treatment) water samples. Collection of grab samples is probably the most used approach due to its simplicity. However, it only allows the obtaining of the chemical profile of the exact time point when the samples were collected. Composite samples (e.g. 24-h integrated samples) allow the obtaining of a more comprehensive picture, and this approach is highly recommended (if not totally necessary) when fluctuating concentrations or episodic pollution events are expected (as in the case of WWTP effluents) and average concentrations or profiles need to be determined ([5]; Gago-Ferrero et al., [6]). Composite sampling should be carefully designed to avoid the loss of compounds due to degradation processes during the collection period. In this regard, collectors with thermostatic control are highly useful. For studies comparing the levels of MPs in different steps of the water treatment process (e.g. comparing influent and effluent water), it is of paramount importance to use time and flow proportional samples. In this manner, the water that is being analysed in the influent corresponds to the same one that will be analysed in other treatment steps a certain time later (e.g. secondary effluent or tertiary effluent). Otherwise, results can lead to wrong conclusions. The use of time-integrated passive samples is not common to evaluate wastewater treatments, but it is regularly used to evaluate waters affected by wastewater effluents [7–9].

Until now, the studies conducted using generic suspect and non-target screening approaches for the evaluation of wastewater treatment have primarily focused on aqueous samples (mainly influent and effluent). In these cases, pretreatment consisted basically of a filtration step through glass fibre and/or cellulose filters and adjusting the pH to a value optimum for the subsequent extraction. Filtration of the samples should be done as fast as possible to prevent bacterial degradation. This is the same sample pretreatment that it is usually applied to natural waters. In cases where the formation of disinfection by-products (DBPs) needs to be evaluated, it is necessary the previous addition of a quencher (e.g. ascorbic acid) in order to stop the disinfection reactions.

2.2 Sample Extraction and Instrumental Analysis

In the case of target methodologies, the choice of the sample extraction procedure is based on the properties of the analytes that are to be determined as well as the matrix of interest. Non-target methodologies follow a different conceptual approach since the aim normally consists of the determination of a wide range of still unknown substances with different physicochemical properties. All organic compounds present in the samples are potentially relevant, and therefore losses during the extraction process must be minimised. In this regard, strategies avoiding sample treatment, like a direct injection of filtrated liquid samples without any enrichment, are great options since they are fast and allow a comprehensive analysis when the sensitivity of the MS is high enough [10]. However, interferences with the matrix are common in wastewater, and their applicability is limited. Therefore, the most common strategy so far consists of conducting solid-phase extraction (SPE) by using different sorbent materials, depending on the specific objective of the study. There are studies using specific SPE sorbents for very specific compounds, like the use of coconut charcoal (Restek®) for the analysis of nitrosamines (in this case using GC-MS) [11, 12]. Other studies use stationary phases for a broader chemical enrichment, like the popular hydrophilic-lipophilic balance Oasis HLB which have been largely used for screening of pharmaceuticals, pesticides and other polar compounds [13]. For real wide-scope screening, the use of mixed-bed multilayered cartridges has become a very popular approach. These cartridges normally contain HLB (or equivalent) sorbents, anion- and cation-exchange resins (e.g. Strata XAW and Strata XCW, respectively) and the highly retentive nonpolar sorbent (e.g. Isolute ENV+), in order to enrich neutral, cationic and anionic species of a broad range of physicochemical properties [14, 15]. Using different cartridges stacked in series [16] or the combination of more specialised SPE sorbents to analyse substances with specific physicochemical properties (per- and polyfluoroalkyl substances (PFASs) or specific drugs) is also a valid option [16, 17]. Wide-scope SPE can lead to the extraction of a large number of substances that can significantly increase the matrix effect, causing serious difficulties in the analysis. Thus, it is always necessary to find a compromise to balance all the factors considering the analytical needs of the work. In some cases, procedures including more intense clean-up steps should be considered since the reduction of interferences can improve the outcome.

For wide-scope screening of water samples (both target and not target), most studies use LC coupled to HRMS, which provides high mass accuracy and high mass resolution. These features are necessary in order to identify suspected and unknown compounds. Hybrid instruments like linear ion trap/Orbitrap or quadrupole/TOF (qTOF) have shown high-performance detection capabilities, particularly for small molecules in complex matrices like wastewater. These technologies allow reliable identifications based on high-resolution accurate mass measurement of precursor and product ions. Electrospray (ESI) is by far the most used ionisation technique due to its good performance for polar organic MPs in aqueous matrices. However, it can be effectively complemented with other techniques suitable for less polar compounds like atmospheric pressure photoionisation (APPI) or atmospheric pressure chemical ionisation (APCI) [18, 19].

Ultra-high-performance liquid chromatography (UHPLC) increases the identification capability of MPs in complex matrices due to better separation power [20]. Also, it reduces the problem of co-eluting peaks and mass spectral overlap [21]. However, the shorter run times of this technique lead to less MS/MS data points per peak, which can negatively affect the identifications. Thus, the optimal solution involves finding a compromise, considering all the factors and the analytical needs for each specific study. The loss of MS/MS data points discouraged coupling UHPLC with Orbitrap type instruments in the past. However, the new generation of Orbitrap instruments, with shorter scan times, has solved this problem to an acceptable extent. Separation for HRMS-based screenings is mostly achieved using reverse-phase (RPLC) with C18-type columns, allowing good stability and performance for a very wide range of compounds. However, RPLC has important limitations, and highly polar compounds remain in many cases undetected. This is particularly very important for the identification of TPs (often with high polarity) during water treatment. In this regard, hydrophilic interaction liquid chromatography (HILIC) is an attractive complementary technique due to its ability to separate hydrophilic compounds that are poorly retained on RP columns as it has been demonstrated in different studies performed in wastewater [14, 22–25]. Recently, ion mobility separation coupled to HRMS has become available and promises significantly improved instrumental performance by adding one additional dimension of separation to the analytical system, i.e. separation in an electric field based on the shape-dependent velocity of the molecule [26, 27].

3 Strategies for the Evaluation of the Behaviour of Organic Micropollutants During Wastewater Treatment

3.1 Data Preprocessing

HRMS instruments such as qTOFs and Orbitraps acquire accurate mass and high-resolution MS and MS/MS full-scan spectra leading to a large amount of data. In the margins of the analytical conditions (e.g. sampling, enrichment method and solvents used) and instrumental limitations (e.g. ionizability, selectivity, sensitivity and resolution), the full mass spectral information of detectable compounds are stored in raw data files [28]. At this point, preprocessing is necessary to improve data quality and reduce complexity. This step normally includes chromatographic retention time alignment and mass correction, peak detection, subtraction of compounds present in blanks as well as "componentisation" via a grouping of isotopes, adducts, multi-charged ions and in-source fragments to define components (i.e. grouping all signals that likely belong to one unique molecular structure) [29, 30]. Differences in algorithms between different software influence significantly the outcome of the data preprocessing. Therefore, a good optimisation of the preprocessing step is crucial for obtaining high quality and reliable ready-to-use data. Different vendor software is available for preprocessing the data, although in many cases they do not allow full control of the process but only minor adjustments [31]. However, there are open-source possibilities that have been used in the preprocessing step of wastewater samples including RMassBank [32], enviMass (Martin [33]), XCMS [34] or MZmine 2 [35], among others.

3.2 Prioritisation Strategies

There are significant gaps in the existent data for compounds present in treated water that remain uncharacterised but are still hazardous from both a toxicological and ecological point of view. However, their prioritisation and further identification is not an easy task. Wastewater and other environmental samples are complex chemical mixtures containing tens of thousands of individual substances that produce a high number of peaks in LC-HRMS analysis. The suspect and non-target screening approaches are laborious, time-consuming and computationally challenging, especially when there is limited prior knowledge of the chemical classes likely to be present [36]. The complete elucidation of all the peaks present in the samples is not feasible since it would require extensive time and effort. Thus, prioritisation strategies (peak prioritisation) are necessary and a key step in any investigation involving non-target analysis to focus the efforts on relevant chemicals. Therefore, different prioritisation strategies should be applied to the set of obtained chromatographic peaks depending on the specific goals of the study and the research question to be answered [29, 30]. Prioritisation should also find an acceptable balance between the prioritised peaks and the number of false negatives caused by too strict procedures.

So far, many of the developed prioritisation strategies for environmental samples followed intensity-based criteria combined with the prioritisation of chemicals with a distinctive isotopic pattern (e.g. compounds containing halogens) [37–40], as these can be considered as relevant chemicals with reasonable identification changes. Other developed approaches used mass defect to focus the identification efforts on molecular formulas outside the matrix domain in complex samples [5, 10]. This strategy has also been proved useful when the objective was to find molecules with specific characteristics. An example can be found in the detection of perfluoroalkyl ether carboxylic acids and sulfonic acids in natural waters due to the negative mass defect of the multiple fluorine and oxygen atoms [41]. Peak prioritisation can be conducted based on effect-directed analysis (EDA), which is a valuable tool for understanding the link between the presence of mixtures of pollutants and their associated effects. EDA is useful for identifying predominant toxicants in complex environmental mixtures combining effect testing and fractionation [15, 18, 19, 42–46]. In EDA, sample fractionation includes adsorption, size exclusion and partition chromatography techniques, resulting in a significant simplification of the complex mixture, and facilitates linking the presence of a compound with their cause and effect. Other strategies that have been shown to be effective include time-series prioritisation (prioritising substances whose intensities varied substantially over a specific lapse of time in one sampling site [28, 47–50], are based on spatial variation [51] or use metabolomics logics [52]). In the field of trend analysis, Schlüsener et al. [47] used vendor software from SCIEX (MarkerView) to analyse long time-series LC-HRMS data coming from a specific site of Rhine river which was affected by effluents of WWTPs. Afterwards, they used open-source scripts to visualise the patterns and to perform autocorrelation to search and prioritise the features with high periodic variations. One of the main limitations in prioritisation efforts, particularly important for the evaluation of wastewater treatment, is that comparison between matrices (influent and effluent) is hampered by suppression of signals in matrix-rich samples [29, 30]. Isotopic-labelled internal standards are used for correction, but robust methods to correct for these influences for unknown compounds with various functional groups and thus varying ionisation efficiencies are needed [29, 30].

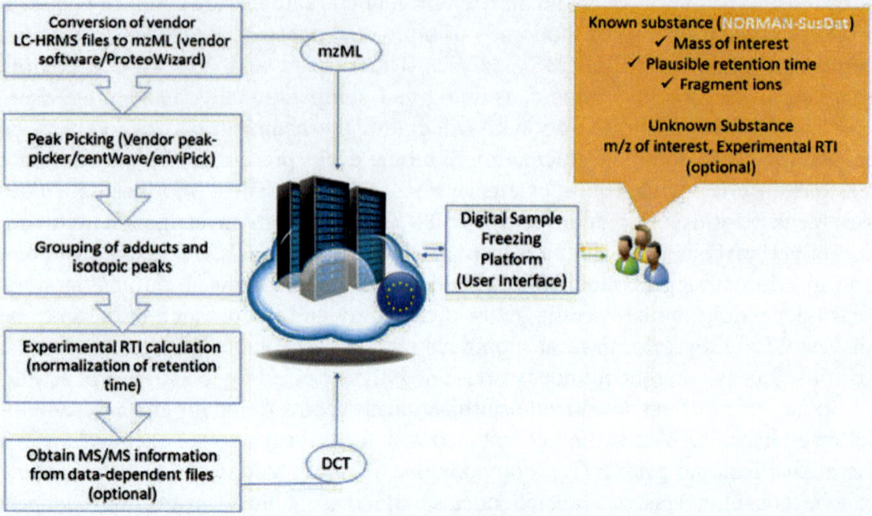

Fig. 1 Summary of the workflow used in the NORMAN Digital Sample Freezing Platform (DSFP) for automatic suspect screening. Adapted with permission from ref. Alygizakis et al. [28]. copyright (2019) Elsevier

Without a doubt the most commonly used conceptual approach to prioritise compounds is the so-called suspect screening that consists of the search for masses of compounds that are expected in the sample without the use of reference standards. Fragmentation information helps find structurally related compounds, and chromatographic retention time prediction models help in increasing the confidence in the identifications. Building the suspect list is a crucial step and should be done carefully in order to answer specific research questions. Suspect screening can include a large list of compounds (up to several thousands) for wide-scope screening. In this regard, a massive tool can be found in the NORMAN Digital Sample Freezing Platform (DSFP) [28], whose operation is summarised in Fig. 1. In this platform the layer-separated mzML files and their meta-data (instrumental, sample meta-data, matrix-specific meta-data and retention time of retention time index (RTI) calibration substances) are uploaded by the user. DSFP has integrated standard procedure to process the mzML files and store the files together with all meta-data for the generation of Data Collection Template (DCTs) databases. Uploading of files in DSFP enabled retrospective screening of >60,000 organic chemicals. Moreover, this will enable the possibility for future retrospective analysis when more advanced tools will be available. This tool also allows performing smaller and more specific suspect screening studies by selecting the compounds of interest or choosing smaller predefined suspect lists (e.g. pesticides, pharmaceuticals TPs, PFASs). In many cases when a specific research question has been clearly defined, it can be more efficient to create smaller smart suspect lists that allow a more time efficient management of the resources and efforts. Some applications of this type of more concise suspect consist of focusing on known toxic chemicals that have been previously detected [53, 54], focusing on specific lists of potential TPs [23] or using regulatory databases (market data) based on exposure index to focus on previously non-reported compounds with high chances of being present [55–57].

In the specific case of wastewater treatment, the study of the TPs that are generated during these processes is especially important. TPs encompass a large proportion of unregulated chemicals present in the environment and generated during water treatments that may be equally, if not more, persistent and toxic than their precursor chemicals. Other toxic TPs include metabolites of pesticides or pharmaceuticals [36]. From the non-target point of view, it is necessary to find the best approaches to determine this specific group of compounds considering their specificities. One simple and useful approach consists of prioritising components across related samples, from spatial and time trends through to "before and after" comparison of treatment technologies [29, 30]. In this approach, a sample is used to create a reference method, containing a comprehensive list of peaks with accompanying retention time and mass spectra. The reference sample serves as the basis of comparison for other samples. There are types of software that compute pairwise comparison to determine peaks present only in the samples and not in the references (e.g. generated during the treatment). Other interesting strategies include using metabolic logic in combination with multivariate statistics in order to find unknown metabolites of certain substances [52].

Obviously, the use of suspect screening is a powerful tool to detect TPs. The main limitation is that in several cases, the potential TPs that can be formed during treatment are not present in any available database. However, provided that there is some a priori knowledge of the classes of compounds present in a complex chemical mixture (e.g. wastewater), cheminformatics can be used to predict the structure of potential toxic TPs. Several powerful tools have been developed in this regard like the EAWAG path prediction system (EAWAG PPS), enviPath [58], PathPred [59], Meteor Nexus (Lhasa Ltd., Leeds, UK), ADMET Predictor software (Simulation Plus Inc., Lancaster, USA) or Catalogic (Laboratory of Mathematical Chemistry). An extensive explanation of these approaches can be found elsewhere [36, 60, 61].

A representative example for the study of the formation of TPs during water treatment can be found in the work carried out by Beretsou et al. [22]. This work aimed at understanding the fate and transformation of citalopram (a worldwide highly consumed antidepressant which has demonstrated incomplete removal by conventional wastewater treatment) during the biological treatment process. For this purpose, biodegradation batch experiments under aerobic conditions were carried out to investigate its behaviour and the formation of TPs during activated sludge treatment. One of the main objectives of the study was the identification of the formed TPs by applying suspect and non-target strategies based on LC-HRMS. As a first step, a suspect database of plausible TPs was compiled by using two different in silico prediction tools: (I) the EAWAG-Biocatalysis/Biodegradation Database Pathway Prediction System (EAWAG-BBD/PPS), an artificial intelligence system, which predicts microbial metabolic reactions based on biotransformation rules set

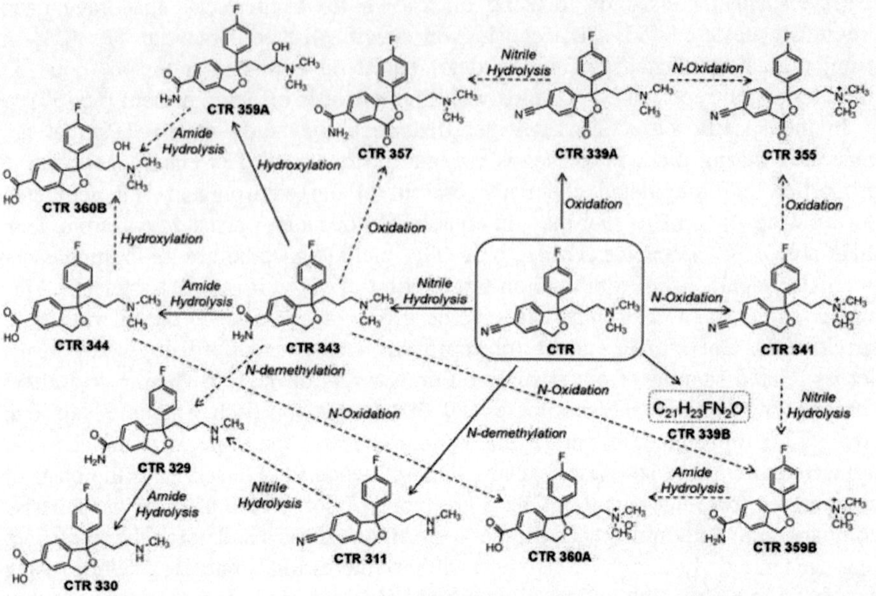

Fig. 2 Proposed biotransformation pathway for citalopram with transformation products identified following suspect and non-target screening strategies. Adapted with permission from Beretsou et al. [22]. copyright (2016) Elsevier

in the EAWAG-BBD and scientific literature and (II) the MetabolitePredict software (Metabolite Tools 2.0, Bruker Daltonics, Bremen, Germany), a rule-based expert system, which predicts metabolites from Phase I and Phase II and Cytochrome P450 reactions. As a second step, samples were also screened for additional TPs not present in the suspect database, following a non-target approach based on pairwise comparison to determine peaks present only in the biotic samples but absent in the control samples and that showed a meaningful time trend. After careful application of identification and confirmation strategies, 12 TPs of citalopram were identified through the use of suspect screening, and 2 additional TPs were detected by non-target screening. The pathway for the biotic transformation of citalopram is shown in Fig. 2. The performance of suspect screening and non-target screening as independent and complementary approaches resulted in comprehensive identification of the formed TPs within specific biotransformation systems. In this work, analyses were performed by both RPLC and hydrophilic interaction liquid chromatography (HILIC) to investigate their complementarity for the detection of additional compounds. In-house developed QSRR prediction models were also used to support identification in both chromatographic systems. The orthogonal confirmation by RPLC and HILIC analysis, along with chromatographic retention time prediction models, proved to be complementary strategies in the identification workflow. The importance of HILIC is emphasised in the detection of an additional TP (enabling the separation from its isomer), which could not be detected through RPLC analysis. In this regard, HILIC is becoming an attractive alternative (or complement) to the commonly used reversed-phase liquid chromatography (RPLC), due to its ability to separate hydrophilic compounds which are poorly retained on RPLC columns [14, 41].

3.3 *Suspect/Non-target: Identification Approaches of Polar Organic Contaminants*

After the crucial prioritisation step, efforts should be focused on the identification of the selected components. This step involves the evaluation of all information available from MS and MS/MS (molecular ion, isotope pattern and fragments), spectral and compound databases, chromatographic retention time information as well as meta-information such as the specific water treatment context. In general, for suspect screening, an exact mass match is not sufficient for identification alone [29, 30]. Suspect screening approaches, apart from a strict mass accuracy threshold, require a detailed study of the MS/MS spectra. A powerful option is the use of MS/MS libraries, which are in constant expansion and contain >1 million of MS/MS for more than 30,000 chemicals. Popular libraries include MassBank (www.massbank.eu (Europe), www.massbank.jp (Japan) and www.mona.scripps.edu (North America)), NIST MSMS (https://www.nist.gov/srd/nist-standard-reference-database-1a-v17), METLIN (www.metlin.scripps.edu) or mzCloud (www.mzcloud.org). However, in many cases, it is not possible to find the MS/MS spectra for the compounds of interest in available libraries. This is a very common situation for TPs, with still a very low presence in public databases. In these cases, in silico fragmentation techniques (such MetFrag [62], CFM-ID [63] or CSI-FingerID/SIRIUS [64]) can be used to obtain predicted fragments and compare with the experimental ones (as well as the manual expert inspection of the MS/MS spectra if there is sufficient structural information). Apart from this, other useful additional data to support identification include literature references, patent data or functional uses.

The identification of *known unknowns*, compounds without prior knowledge but present in large compound databases, is a more challenging but still doable task. In these cases, the first step consists of the determination of the molecular formula. For this purpose, there are various calculations based on mass accuracy, presence/absence of certain elements and using the presence of isotope signals for certain elements (e.g. sulphur), isotopic pattern fit and fragment sub-formula assignment. Most vendor software includes useful tools to perform this task. Other advance software also considers MS/MS data. MOLGEN-MS [65] suggests candidate structures based on the mass spectral fragmentation patterns, and it has shown excellent performance in the formula assignments. Some studies have also added steric energy to make the search more accurate [45]. Once the molecular formula has been determined, a combination of in silico fragmentation techniques with MS/MS libraries and expert knowledge are the most used tools to tentatively elucidate the structure of those unknown compounds. The most sophisticated in silico fragmentation approaches along with environmental context or metadata ranked the correct candidate to over 70%, showing a very good performance of these approaches [62, 66]. In water treatment, evaluation is often required to identify novel TPs and other unknown compounds. If these compounds have not been documented anywhere, the identification is much more challenging with no guarantees of success. However, some successful examples have been reported for small molecule and sufficient available structural information [67].

The evaluation of the chromatographic retention time (RT) plausibility for the compounds of interest using different developed models plays an important role in increasing confidence and eliminating less-plausible candidates in both suspect and non-target approaches. Tentatively identifications of the candidates are based on comparing how well the experimental RT (or retention index (RI)) matches the predicted RT (or RI) from computational models. Different RT prediction tools have been designed including quantitative structure retention relationships (QSRR) models [25, 68], artificial neural networks (ANNs) [69] or linear solvation energy relationship (LSER) [70]. It is noticeable that some of the platforms for automated suspect screening, like the NORMAN DSFP [28] or STOFFIDENT (https://www.lfu.bayern.de/stoffident), have implemented RT prediction models for a better identification performance. They include a step that calculates an experimental RTI for every feature detected based on the retention time observed and the calibration curve equation (RT ¼ f(RTI)), derived from the retention time of the standard calibration mixture [71].

Ion mobility (IM) separation coupled to HRMS (only QTOF systems) has recently been introduced to the market and promises significantly improved instrumental performance by adding one additional dimension of separation to the analytical system, i.e. separation in an electric field based on the shape-dependent velocity of the molecule [26, 27]. It significantly improves the identification of problematic compounds (e.g. isomers) and has been successfully applied to improve identification in both suspect and non-target methodologies [72].

The increasing interest in HRMS-based analysis has produced a very pronounced growth in the suspect and non-target identifications. This fact creates also a need to communicate the confidence in the identifications in a way that reflects all the evidence available. The varying levels of confidence are difficult to communicate concisely and accurately. So far, most of the studies report the confidence based on the hierarchical degrees of confidence described by Schymanski et al. [73]. In this classification, the compounds that have been confirmed with an available commercial standard are classified as Level 1 (maximum degree of confidence). The compounds identified without standard but by mass spectral library or database matching are reported with confidence Level 2 (probable structure). Level 3 (tentative candidate) describes a “grey zone”, where evidence exists for possible structure (s) but insufficient information for one exact structure only (e.g. positional isomers). Compounds for which structures cannot be elucidated are reported with Level 4 (if the molecular formula is unequivocally assigned) or Level 5, reporting only a mass of interest.

In many cases, although the aforementioned levels (published in 2014) are certainly useful, it is difficult to communicate the identification's confidence level in a concise manner. Therefore, a complementary system that allows the community to rapidly and concisely understand the identification proofs that have been obtained when a compound is reported is needed. In this regard, a system based on identification points (IPs), compatible between target and non-targeted approaches would be really valuable and would add accuracy to the field. Recently, a new system of identification points (IPs) based on the one described in the Commission Decision 2002/657/EC was applied to communicate the confidence level in the identification of the analytes for target analysis [74]. This system considers retention time, mass accuracy, isotopic fit and fragmentation, taking full advantage of the capacities of the HRMS instruments [74]. This system could be easily extended and encompass non-target approaches by including variables such as predicted retention time, determined MS/MS experimental ions and predicted fragment ions (both with mass accuracy) or ion mobility, assigning a specific value of IPs in each case.

3.4 Novel Non-target Strategies to Assess Changes in the Molecular Fingerprints Without Structural Elucidation

DOM is a heterogenic, complex mixture of proteins, lipids, polysaccharides, nucleic acids, soluble microbial products and also synthetic organic chemicals. In wastewater, these chemicals include thousands of compounds and related TPs, metabolites and, in effluents with tertiary treatment, also a wide range of disinfection by-products (DBPs). Wastewater is a matrix with high complexity, and HRMS analysis typically produces a large number of signals. Even after preprocessing steps, performing a manual structural identification of all the compounds is not a feasible task, and the same applies to other water treatment systems. Therefore, approaches are being developed to characterise the chemical composition of DOM in water, where previously most of the DOM components remained unidentified [75]. The objective is not the further structure elucidation of particular compounds but assessing changes in the molecular fingerprints during water treatments. It is possible to obtain relevant information from HRMS analysis on DOM if a large part of the detected features can be assigned a probable elemental composition (without focusing on structural elucidation of specific components). This is giving rise to a new field of study that allows the study of changes (e.g. the overall oxidation state, the number of sulfonated or chlorinated formed compounds or the average molecular weight) during wastewater treatment (as well as other treatments such as drinking water treatment and even soil remediation) by measuring shifts in the molecular fingerprint and other physicochemical changes.

The common protocol of a data-driven non-targeted analysis involves the same steps as described in the preprocessing section (Sect. 3.1) in order to obtain ready-to-use data. Data mining from HRMS data on DOM analysis yields to large lists of molecular features, understood as the integration of all the different MS features such as accuracy mass, isotopic information or average retention time (when chromatography is used). Based on that data, algorithms can be constructed to create an artificial list of DOM molecule elemental compositions (e.g. $C_nH_nO_nN_nS_nCl_nBr_n$) and therefore obtain DOM fingerprints. Such non-targeted workflow aims to gain insights on wastewater DOM transformation processes based on changes in this large number of detected molecular features. This new manner of evaluating HRMS data reduces manual data treatment present in the conventional suspect and non-target strategies, and it has been adapted from other fields like the characterisation of natural organic matter (NOM) [76–79] or petroleomics [80]. It has been widely applied to evaluate the OM in processed water [81–83], and there are also few studies on wastewater using different instrumentation. Back in 2011, Gonsior et al. performed a molecular characterisation of effluent OM using Fourier-transform ion cyclotron resonance MS (FTICR-MS) [84]. Also, by using FTICR-MS, Mesfioui et al. characterised effluent organic nitrogen (EON) from wastewater treatment plants and study its reactivity [85]. The results obtained from these non-target-based approaches for the molecular characterisation of DOM are highly influenced by the type of HRMS analyser, due to differences in resolution of masses, which affect the number of substances and how accurately the HRMS features can be assigned with an elemental composition. The differences obtained with different instrumentation systems are not that crucial when screening for specific substances [86] but are more important in the analysis of unknown DOM constituents and should be carefully considered [87]. It has been demonstrated that FTICR-MS is capable of unravelling complex DOM mixtures across aquatic systems [88, 89] and might allow a comprehensive evaluation of changes in the molecular fingerprints. However, the high cost of this technology made it inaccessible for most research laboratories. Recently, Orbitrap-MS (a much more accessible technology) has proven to be a valid alternative, despite its lower resolution, as it is able to detect subtle changes in DOM profiles. Verkh et al. use LC-HRMS with an LTQ-Orbitrap Velos™ to evaluate a real wastewater treatment system with secondary and tertiary treatment identifying significant chances in the DOM during these processes [90]. This study detected a reduction in the number of large molecules, lower average masses as well as an increase in saturated molecular features of the effluent OM.

In the field of drinking water treatment, these approaches have great potential. There are studies that have applied it to the study of the formation of DBPs and to investigate how DOM contributes to their generation where, i.e. Postigo et al. investigated the chemodiversity of DBPs in different drinking water plants [91]. In a similar way, Sanchis et al. investigated how DOM contributes to the formation of DBPs aiming to understand and minimise their formation [92]. This study showed that sample chlorination primarily involved a decrease in the concentration of non-nitrogenised lignin-like features, and it was also related with changes in the trihalomethanes and haloacetonitrile formation potential. The N-nitrosodimethylamine (NDMA) formation potential in natural waters has also been studied following a similar workflow, and a correlation was found with the number of compounds with high hydrogen saturation (H/C $\geq$ 1.5), which corresponded to reservoirs with higher background nutrient concentrations and wastewater indicators [93]. Other water processes like the KMnO4 oxidation have also been studied and linked with DOM [94]. These non-target approaches have shown to be also useful for other type of treatment processes. For example, a non-target workflow based on the evaluation of the molecular fingerprint without structural elucidation has been recently applied to an electrokinetic soil remediation process as a case study and allowed for an in-depth assessment of the chemical mechanistic processes, which has previously only been hypothesised and has contradictory results [95]. Many of these works use additional tools to evaluate changes to the DOM such as Van Krevelen diagrams, where the atomic ratio X/C, with X being an element of interest, is plotted against H/C. In other fields like NOM chemistry or petroleomics, the correlation between areas in these diagrams and functional classes of compounds provide important information to elucidate the chemical composition of OM [96–99]. In many studies, O/C ratio is plotted against H/C obtaining information about the oxidation of the DOM. Also, the monitoring of double bond equivalents (DBE) or DBE minus oxygen atoms (DBE-O) in wastewater treatment can be useful to estimate the quality of the process, to evaluate the hydrophobicity-altering reactions (e.g. oxidation or hydrolysis) and conceptually to even support the control of the process [90, 100].

These workflows for the assessment of changes in the molecular fingerprint can be very useful and become a common step in the research dealing with the evaluation of water treatments. They can be applied to tertiary treatments (e.g. UV irradiation, Fenton-like processes, ozonation, membranes or chlorination) and other innovative techniques. Overlooking potentially hazardous constituents in effluents might limit the understanding of the impact of the discharged organic matter on the environment. Overall changes in the elemental compositions can reveal changes in the degree of oxidation, the formation of TPs with specific elements like sulphur (e.g. using sulphate-based advanced oxidation processes) or halogens during disinfection, along with other valuable information. Therefore, their evaluation is crucial for a better understanding of the different treatments' performance and the TPs formation.

References

1. Prasse C, Stalter D, Schulte-Oehlmann U, Oehlmann J, Ternes TA (2015) Spoilt for choice: a critical review on the chemical and biological assessment of current wastewater treatment technologies. Water Res 87:237–270
2. Swedish EPA (2011) Swedish consumption and the global environment. http://www.naturvardsverket.se/Documents/publikationer/978-91-620-1285-4.pdf?pid=2687. Accessed 05 Apr 2020
3. EEA (2010) The European environment. State and outlook 2010

4. Hug C, Ulrich N, Schulze T, Brack W, Krauss M (2014) Identification of novel micropollutants in wastewater by a combination of suspect a nontarget screening. Environ Pollut 184:25–32
5. Moschet C, Piazzoli A, Singer H, Hollender J (2013) Alleviating the reference standard dilemma using a systematic exact mass suspect screening approach with liquid chromatography-high resolution mass spectrometry. Anal Chem 85:10312–10320
6. Menger F, Gago-Ferrero P, Wiberg K, Ahrens L (2021) Wide-scope screening of polar contaminants of concern in water: a critical review of liquid chromatography-high resolution mass spectrometry-based strategies. Trends Environ Anal Chem. (Online) https://www.sciencedirect.com/science/article/pii/S2214158820300301?via%3Dihub. https://doi.org/10.1016/j.teac.2020.e00102
7. Soulier C, Coureau C, Togola A (2016) Environmental forensics in groundwater coupling passive sampling and high-resolution mass spectrometry for screening. Sci Total Environ 563–564:845–854. https://doi.org/10.1016/j.scitotenv.2016.01.056
8. Gallen C, Heffernan AL, Kaserzon S, Dogruer G, Samanipour S, Gomez-Ramos MJ, Mueller JF (2019) Integrated chemical exposure assessment of coastal green turtle foraging grounds on the Great Barrier Reef. Sci Total Environ 657:401–409. https://doi.org/10.1016/j.scitotenv.2018.11.322
9. Menger F, Ahrens L, Wiberg K, Gago-Ferrero P (2021) Suspect screening based on market data of polar halogenated micropollutants in river water affected by wastewater. J Hazard Mat 401:123377. https://doi.org/10.1016/j.jhazmat.2020.123377
10. Vergeynst L, van Langenhove H, Joos P, Demeestere K (2014) Suspect screening and target quantification of multi-class pharmaceuticals in surface water based on large volume injection liquid chromatography and time-of-flight mass spectrometry. Anal Bioanal Chem 406 (11):2533–2547
11. Farré MJ, Insa S, Lamb A, Cojocariu C, Gernjak W (2020) Occurrence of: N -nitrosamines and their precursors in Spanish drinking water treatment plants and distribution systems. Environ Sci Water Res Technol 6(1):210–220
12. Farré MJ, Döderer K, Hearn L, Poussade Y, Keller J, Gernjak W (2011) Understanding the operational parameters affecting NDMA formation at advanced water treatment plants. J Hazard Mater 185(2–3):1575–1581
13. Leendert V, Van Langenhove H, Demeestere K (2015) Trends in liquid chromatography coupled to high-resolution mass spectrometry for multi-residue analysis of organic micropollutants in aquatic environments. TrAC Trends Anal Chem 67:192–208
14. Gago-Ferrero P, Schymanski EL, Bletsou AA, Aalizadeh R, Hollender J, Thomaidis NS (2015) Extended suspect and non-target strategies to characterize emerging polar organic contaminants in raw wastewater with LC-HRMS/MS. Environ Sci Technol 49 (20):12333–12341. https://doi.org/10.1021/acs.est.5b03454
15. Kern S, Fenner K, Singer HP, Schwarzenbach RP, Hollender J (2009) Identification of transformation products of organic contaminants in natural waters by computer-aided prediction and high-resolution mass spectrometry. Environ Sci Technol 43:7039–7046. https://doi.org/10.1021/es901979h
16. Wang Y, Yu N, Zhu X, Guo H, Jiang J, Wang X, Shi W, Wu J, Yu H, Wei S (2018) Suspect and nontarget screening of per- and Polyfluoroalkyl substances in wastewater from a fluorochemical manufacturing park. Environ Sci Technol 52:11007–11016. https://doi.org/10.1021/acs.est.8b03030
17. Casado J, Rodríguez I, Ramil M, Cela R (2014) Selective determination of antimycotic drugs in environmental water samples by mixed-mode solid-phase extraction and liquid chromatography quadrupole time-of-flight mass spectrometry. J Chromatogr A 1339:42–49. https://doi.org/10.1016/j.chroma.2014.02.087
18. Chiaia-Hernandez AC, Krauss M, Hollender J (2013) Environ Sci Technol 47:976e986
19. Gallampois CMJ, Schymanski EL, Krauss M, Ulrich N, Bataineh M, Brack W (2015) Multicriteria approach to select Polyaromatic River mutagen candidates. Environ Sci Technol 49(5):2959–2968
20. Terzic S, Ahel M (2011) Environ Pollut 159:557e566

21. Plumb R, Castro-Perez J, Granger J, Beattie I, Joncour K, Wright A (2004) Rapid Commun Mass Spectrom 18:2331e2337
22. Beretsou VG, Psoma AK, Gago-Ferrero P, Aalizadeh R, Fenner K, Thomaidis NS (2016) Identification of biotransformation products of citalopram formed in activated sludge. Water Res 103:205–214
23. Salas D, Borrull F, Fontanals N, Marcé RM (2017) Hydrophilic interaction liquid chromatography coupled to mass spectrometry-based detection to determine emerging organic contaminants in environmental samples. Trends Anal Chem 94:141–149. https://doi.org/10.1016/j.trac.2017.07.017
24. Bieber S, Greco G, Grosse S, Letzel T (2017) RPLC-HILIC and SFC with mass spectrometry: polarity-extended organic molecule screening in environmental (water) samples. Anal Chem 89(15):7907–7914. https://doi.org/10.1021/acs.analchem.7b00859
25. Aalizadeh R, Nika M-C, Thomaidis NS (2019) Development and application of retention time prediction models in the suspect and non-target screening of emerging contaminants. J Hazard Mater 363:277–285
26. Lanucara F, Holman SW, Gray CJ, Eyers CE (2014) The power of ion mobility-mass spectrometry for structural characterization and the study of conformational dynamics. Nat Chem 6:281–294. https://doi.org/10.1038/nchem.1889
27. Mollerup CB, Mardal M, Dalsgaard PW, Linnet K, Barron LP (2018) Prediction of collision cross section and retention time for broad scope screening in gradient reversed-phase liquid chromatography-ion mobility-high resolution accurate mass spectrometry. J Chromatogr A 1542:82–88. https://doi.org/10.1016/j.chroma.2018.02.025
28. Alygizakis NA, Oswald P, Thomaidis NS, Schymanski EL, Aalizadeh R, Schulze T, Oswaldova M, Slobodnik J (2019) NORMAN digital sample freezing platform: a European virtual platform to exchange liquid chromatography high resolution-mass spectrometry data and screen suspects in "digitally frozen" environmental samples. TrAC Trends Anal Chem 115:129–137. https://doi.org/10.1016/j.trac.2019.04.008
29. Hollender J, Schymanski EL, Singer HP, Ferguson PL (2017) Nontarget screening with high resolution mass spectrometry in the environment: ready to go? Environ Sci Technol 51:11505e11512
30. Hollender J, Schymanski EL, Singer HP, Ferguson PL (2017) Nontarget screening with high resolution mass spectrometry in the environment: ready to go? Environ Sci Technol 51:11505–11512. https://doi.org/10.1021/acs.est.7b02184
31. Rager JE, Strynar MJ, Liang S, McMahen RL, Richard AM, Grulke CM, Wambaugh JF, Isaacs KK, Judson R, Williams AJ, Sobus JR (2016) Linking high resolution mass spectrometry data with exposure and toxicity forecasts to advance high-throughput environmental monitoring. Environ Int 88:269–280. https://doi.org/10.1016/j.envint.2015.12.008
32. Stravs MA, Schymanski EL, Singer HP, Hollender J (2013) Automatic recalibration and processing of tandem mass spectra using formula annotation: recalibration and processing of MS/MS spectra. J Mass Spectrom 48:89–99. https://doi.org/10.1002/jms.3131
33. Loos M (2019) enviMass version 3.5 LC-HRMS trend detection workflow – R package. URL Zenodo. https://doi.org/10.5281/zenodo.1213098
34. Smith CA, Want EJ, O'Maille G, Abagyan R, Siuzdak G (2006) XCMS: processing mass spectrometry data for metabolite profiling using nonlinear peak alignment, matching, and identification. Anal Chem 78:779–787. https://doi.org/10.1021/ac051437y
35. Pluskal T, Castillo S, Villar-Briones A, Orešič M (2010) MZmine 2: modular framework for processing, visualizing, and analyzing mass spectrometry-based molecular profile data. BMC Bioinf 11:395. https://doi.org/10.1186/1471-2105-11-395
36. Chibwe L, Titaley IA, Hoh E, Simonich SL (2017) Integrated framework for identifying toxic transformation products in complex environmental mixtures. Environ Sci Technol Lett 4 (2):32–43. https://doi.org/10.1021/acs.estlett.6b00455
37. Bletsou AA, Jeon J, Hollender J, Archontaki E, Thomaidis NS (2015) Trends Anal Chem 66:32e44

38. Stasinakis AS, Gatidou G, Mamais D, Thomaidis NS, Lekkas TD (2008) Water Res 42:1796e1804
39. Schulze T, Weiss S, Schymanski EL, Carsten von der Ohe P, Schmitt-Jansen M, Altenburger R, Streck G, Brack W (2010) Environ Pollut 158(5):1461e1466
40. Escher BI, Fenner K (2011) Environ Sci Technol 45:3835e3847
41. Zonja B, Delgado A, Perez S, Barcelo D (2015) LC-HRMS suspect screening for detection-based prioritization of iodinated contrast media photodegradates in surface waters. Environ Sci Technol 49(6):3464–3472
42. Hernandez F, Ibanez M, Portoles T, Cervera MI, Sancho JV, Lopez FJ (2015) J Hazard Mater 282:86e95
43. Chibwe L, Geier MC, Nakamura J, Tanguay RL, Aitken MD, Simonich SLM (2015) Aerobic bioremediation of PAH contaminated soil results in increased genotoxicity and developmental toxicity. Environ Sci Technol 49(23):13889–13898
44. Lübcke-von Varel U, Streck G, Brack W (2008) Automated fractionation procedure for polycyclic aromatic compounds in sediment extracts on three coupled normal-phase high-performance liquid chromatography columns. J Chromatogr A 1185(1):31–42
45. Brack W, Ait-Aissa S, Burgess RM, Busch W, Creusot N, di Paolo C, Escher BI, Mark Hewitt L, Hilscherova K, Hollender J et al (2016) Effect-directed analysis supporting monitoring of aquatic environments an in-depth overview. Sci Total Environ 544:1073–1118
46. Fang M, Webster TF, Stapleton HM (2015) Effect-directed analysis of human peroxisome proliferator-activated nuclear receptors (PPARγ1) ligands in indoor dust. Environ Sci Technol 49(16):10065–10073
47. Schlusener MP, Kunkel U, Ternes TA (2015) Quaternary triphenylphosphonium compounds: a new class of environmental pollutants. Environ Sci Technol 49:14282–14291
48. Plassmann MM, Tengstrand E, Aberg KM, Benskin JP (2016) Non-target time trend screening: a data reduction strategy for detecting emerging contaminants in biological samples. Anal Bioanal Chem 408:4203–4208
49. Alygizakis NA, Gago-Ferrero P, Hollender J, Thomaidis NS (2019) Untargeted time-pattern analysis of LC-HRMS data to detect spills and compounds with high fluctuation in influent wastewater. J Hazard Mater 361:19–29. https://doi.org/10.1016/j.jhazmat.2018.08.073
50. Anliker S, Loos M, Comte R, Ruff M, Fenner K, Singer H (2020) Assessing emissions from pharmaceutical manufacturing based on temporal high-resolution mass spectrometry data. Environ Sci Technol. (in press). https://doi.org/10.1021/acs.est.9b07085
51. Ruff M, Mueller MS, Loos M, Singer HP (2015) Quantitative target and systematic nontarget analysis of polar organic micro-pollutants along the river Rhine using high resolution mass-spectrometry – identification of unknown sources and compounds. Water Res 87:145–154
52. Schollee JE, Schymanski EL, Avak SE, Loos M, Hollender J (2015) Prioritis zing unknown transformation products from biologically-treated wastewater using high resolution mass spectrometry, multivariate statistics, and metabolic logic. Anal Chem 87:12121–12129
53. Newton SR, McMahen RL, Sobus JR, Mansouri K, Williams AJ, McEachran AD, Strynar MJ (2018) Suspect screening and non-targeted analysis of drinking water using point-of-use filters. Environ Pollut 234:297–306. https://doi.org/10.1016/j.envpol.2017.11.033
54. Brunner AM, Dingemans MML, Baken KA, van Wezel AP (2019) Prioritising anthropogenic chemicals in drinking water and sources through combined use of mass spectrometry and ToxCast toxicity data. J Hazard Mater 364:332–338. https://doi.org/10.1016/j.jhazmat.2018.10.044
55. Gago-Ferrero P, Krettek A, Fischer S, Wiberg K (2018) Ahrens. Suspect screening and regulatory databases: a powerful combination to identify emerging micropollutants. Environ Sci Technol 52(12):6881–6894. https://doi.org/10.1021/acs.est.7b06598
56. Kiefer K, Müller A, Singer H, Hollender J (2019) New relevant pesticide transformation products in groundwater detected using target and suspect screening for agricultural and urban micropollutants with LC-HRMS. Water Res 165:114972. https://doi.org/10.1016/j.watres.2019.114972
57. Troger R, Kohler SJ, Franke V, Bergstedt O, Wiberg K (2020) Sci Total Environ, 706. http://eawag-bbd.ethz.ch/predict/aboutPPS.html. Accessed 05 Apr 2020

58. Wicker J, Lorsbach T, Gütlein M, Schmid E, Latino D, Kramer S, Fenner K (2016) enviPath – the environmental contaminant biotransformation pathway resource. Nucleic Acids Res 44 (D1):D502–D508
59. Moriya, Y.; Shigemizu, D.; Hattori, M.; Tokimatsu, T.; Kotera, M.; Goto, S.; Kanehisa, M. PathPred: an enzyme-catalyzed metabolic pathway prediction server. Nucleic Acids Res. 2010, 38, W138−W143. http://oasis-lmc.org/products/software/catalogic.aspx. Accessed 05 Apr 2020
60. Arora PK, Bae H (2014) Integration of bioinformatics to biodegradation. Biol Proced 16:8
61. Rucker C, Kummerer K (2012) Modeling and predicting aquatic aerobic biodegradation – a review from a user's perspective. Green Chem 14(4):875–887
62. Ruttkies C, Neumann S, Posch S (2019) Improving MetFrag with statistical learning of fragment annotations. BMC Bioinf 20. https://doi.org/10.1186/s12859-019-2954-7
63. Djoumbou-Feunang Y, Pon A, Karu N, Zheng J, Li C, Arndt D, Gautam M, Allen F, Wishart DS (2019) CFM-ID 3.0: significantly improved ESI-MS/MS prediction and compound identification. Metabolites 9. https://doi.org/10.3390/metabo9040072
64. Dührkop K, Fleischauer M, Ludwig M, Aksenov AA, Melnik AV, Meusel M, Dorrestein PC, Rousu J, Böcker S (2019) SIRIUS 4: a rapid tool for turning tandem mass spectra into metabolite structure information. Nat Methods 16:299–302. https://doi.org/10.1038/s41592-019-0344-8
65. Schymanski EL, Gerlich M, Ruttkies C, Neumann S (2014) Solving CASMI 2013 with MetFrag, MetFusion and MOLGEN-MS/MS. Mass Spectrom 3(SpecialIssue 2):S0036
66. Schymanski EL, Ruttkies C, Krauss M, Brouard C, Kind T, Duhrkop K, Allen F, Vaniya A, Verdegem D, Bocker S, Rousu J, Shen H, Tsugawa H, Sajed T, Fiehn O, Ghesquiere B, Neumann S (2017) Critical Assessment of Small Molecule Identification 2016: automated methods. J Cheminform 9(1):22
67. Huntscha S, Hofstetter TB, Schymanski EL, Spahr S, Hollender J (2014) Biotransformation of benzotriazoles: Insights from transformation product identification and compound-specific isotope analysis. Environ Sci Technol 48(8):4435–4443
68. Aalizadeh R, Thomaidis NS, Bletsou AA, Gago-Ferrero P (2016) Quantitative structure–retention relationship models to support nontarget high-resolution mass spectrometric screening of emerging contaminants in environmental samples. J Chem Inf Model 56:1384–1398. https://doi.org/10.1021/acs.jcim.5b00752
69. Bade R, Bijlsma L, Miller TH, Barron LP, Sancho JV, Hernández F (2015) Suspect screening of large numbers of emerging contaminants in environmental waters using artificial neural networks for chromatographic retention time prediction and high-resolution mass spectrometry data analysis. Sci Total Environ 538:934–941. https://doi.org/10.1016/j.scitotenv.2015.08.078
70. Ulrich N, Schüürmann G, Brack W (2011) Linear solvation energy relationships as classifiers in non-target analysis—a capillary liquid chromatography approach. J Chromatogr A 1218:8192–8196. https://doi.org/10.1016/j.chroma.2011.09.031
71. Gerlich M, Neumann S (2013) J Mass Spectrom 48:291e298
72. Escher B, Leusch F (2012) Bioanalytical tools in water quality assessment. IWA Publishing, London
73. Schymanski EL, Jeon J, Gulde R, Fenner K, Ruff M, Singer HP, Hollender J (2014) Identifying small molecules via high resolution mass spectrometry: communicating confidence. Environ Sci Technol 48:2097–2098. https://doi.org/10.1021/es5002105
74. Gago-Ferrero P, Bletsou AA, Damalas DE, Aalizadeh R, Alygizakis NA, Singer HP et al (2020) Wide-scope target screening of>2000 emerging contaminants in wastewater samples with UPLC-Q-ToF-HRMS/MS and smart evaluation of its performance through the validation of 195 selected representative analytes. J Hazard Mater 387:121712. https://doi.org/10.1016/j.jhazmat.2019.121712
75. Wode F, van Baar P, Dünnbier U, Hecht F, Taute T, Jekel M, Reemtsma T (2000) Current and legacy micropollutants on a wastewater infiltration site with a UPLC-high resolution MS target screening method. Water Res 2015(69):274–283

76. D'Andrilli J, Cooper WT, Foreman CM, Marshall AG (2015) An ultrahigh-resolution mass spectrometry index to estimate natural organic matter Lability: FTICRMS organic matter molecular Lability index. Rapid Commun Mass Spectrom 29(24):2385–2401
77. Lavonen EE, Kothawala DN, Tranvik LJ, Gonsior M, Schmitt-Kopplin P, Köhler SJ (2015) Tracking changes in the optical properties and molecular composition of dissolved organic matter during drinking water production. Water Res 85:286–294
78. Sleighter RL, Hatcher PG (2008) Molecular characterisation of dissolved organic matter (DOM) along a river to ocean transect of the lower Chesapeake Bay by ultrahigh resolution electrospray ionization Fourier transform ion cyclotron resonance mass spectrometry. Mar Chem 110(3–4):140–152
79. Hawkes JA, Dittmar T, Patriarca C, Tranvik L, Bergquist J (2016) Evaluation of the Orbitrap mass spectrometer for the molecular fingerprinting analysis of natural dissolved organic matter. Anal Chem 88:7698–7704
80. Wu Z, Rodgers RP, Marshall AG (2004) Two- and three-dimensional van Krevelen diagrams: a graphical analysis complementary to the Kendrick mass plot for sorting elemental compositions of complex organic mixtures based on ultrahigh-resolution broadband Fourier transform ion cyclotron resonance mass measurements. Anal Chem 76(9):2511–2516
81. Fang Z, He C, Li Y, Chung KH, Xu C, Shi Q (2017) Fractionation and characterisation of dissolved organic matter (DOM) in refinery wastewater by revised phase retention and ion-exchange adsorption solid phase extraction followed by ESI FT-ICR MS. Talanta 162:466–473
82. Cortés-Francisco N, Caixach J (2013) Molecular characterisation of dissolved organic matter through a desalination process by high resolution mass spectrometry. Environ Sci Technol 47 (17):9619–9627
83. Pereira AS, Islam MS, Gamal El-Din M, Martin JW (2013) Ozonation degrades all detectable organic compound classes in Oil Sands process-affected water; an application of high-performance liquid chromatography/Orbitrap mass spectrometry: ozonation degrades all detectable organic compound classes in OSPW. Rapid Commun Mass Spectrom 27 (21):2317–2326
84. Gonsior M, Zwartjes M, Cooper WJ, Song W, Ishida KP, Tseng LY, Jeung MK, Rosso D, Hertkorn N, Schmitt-Kopplin P (2011) Molecular characterisation of effluent organic matter identified by ultrahigh resolution mass spectrometry. Water Res 45(9):2943–2953
85. Mesfioui R, Love NG, Bronk DA, Mulholland MR, Hatcher PG (2012) Reactivity and chemical characterisation of effluent organic nitrogen from wastewater treatment plants determined by Fourier transform ion cyclotron resonance mass spectrometry. Water Res 46 (3):622–634
86. Bade R, Rousis NI, Bijlsma L, Gracia-Lor E, Castiglioni S, Sancho JV, Hernandez F (2015) Screening of pharmaceuticals and illicit drugs in wastewater and surface waters of Spain and Italy by high resolution mass spectrometry using UHPLC-QTOF MS and LCLTQ-Orbitrap MS. Anal Bioanal Chem 407(30):8979–8988
87. Remucal CK, Cory RM, Sander M, McNeill K (2012) Low molecular weight components in an aquatic humic substance as characterized by membrane Dialysis and Orbitrap mass spectrometry. Environ Sci Technol 46(17):9350–9359
88. Kellerman AM, Dittmar T, Kothawala DN, Tranvik LJ (2014) Chemodiversity of dissolved organic matter in lakes driven by climate and hydrology. Nat Commun 5:3804
89. Riedel T, Zark M, Vähätalo AV, Niggemann J, Spencer RGM, Hernes PJ, Dittmar T (2016) Molecular signatures of biogeochemical transformations in dissolved organic matter from ten world rivers. Front Earth Sci 4:85
90. Verkh Y, Rozman M, Petrovic M (2018) A non-targeted high-resolution mass spectrometry data analysis of dissolved organic matter in wastewater treatment. Chemosphere 200:397e404
91. Postigo C, Andersson A, Harir M, Bastviken D, Gonsior M, Schmitt-Kopplin P, Gago-Ferrero P, Ahrens L, Ahrens L, Wiberg K (2021) Unraveling chemodiversity of halogenated disinfection by-products formed during drinking water treatment using targeted and

non-targeted screening tools. J Hazard Mater. (Online) https://www.sciencedirect.com/science/article/pii/S2213343720307867. https://doi.org/10.1016/j.jece.2020.104437

92. Sanchís J, Jaén-Gil A, Gago-Ferrero P, Munthali E, Farré MJ (2020) Characterisation of organic matter by HRMS in surface waters: effects of chlorination on molecular fingerprints and correlation with DBP formation potential. Water Res 176:115743
93. Farré MJ, Jaén-Gil A, Hawkes J, Petrovic M, Catalán N (2019) Orbitrap molecular fingerprint of dissolved organic matter in natural waters and its relationship with NDMA formation potential. Sci Total Environ
94. Yuthawong V, Kasuga I, Kurisu F, Furumai H (2019) Molecular-level changes in dissolved organic matter compositions in Lake Inba water during KMnO4 oxidation: assessment by Orbitrap mass spectrometry. J Water Environ Technol 17:27–39
95. Sörengård M, Ahrens L, Alygizakis N, Jensen PE, Gago-Ferrero P (2021) Non-target and suspect screening strategies for electrodialytical soil remediation evaluation: assessing changes in the molecular fingerprints and per- and polyfluoroalkyl substances (PFASs). J Environ Chem Eng. (Under revision)
96. Zhang H, Zhang Y, Shi Q, Hu J, Chu M, Yu J, Yang M (2012) Study on transformation of natural organic matter in source water during chlorination and its chlorinated products using ultrahigh resolution mass spectrometry. Environ Sci Technol 46(8):4396–4402
97. Minor EC, Steinbring CJ, Longnecker K, Kujawinski EB (2012) Characterisation of dissolved organic matter in Lake Superior and its watershed using ultrahigh resolution mass spectrometry. Org Geochem 43:1–11
98. Lu M, Zhang Z, Qiao W, Wei X, Guan Y, Ma Q, Guan Y (2010) Remediation of petroleum-contaminated soil after composting by sequential treatment with Fenton-like oxidation and biodegradation. Bioresour Technol 101(7):2106–2113
99. Kim S, Kramer RW, Hatcher PG (2003) Graphical method for analysis of ultrahigh- resolution broadband mass spectra of natural organic matter, the Van Krevelen diagram. Anal Chem 75 (20):5336–5344
100. Marshall JW, Schmitt-Kopplin P, Schuetz N, Moritz F, Roullier-Gall C, Uhl J, Colyer A, Jones LL, Rychlik M, Taylor AJ (2018) Monitoring chemical changes during food sterilisation using ultrahigh resolution mass spectrometry. Food Chem 242:316–322

Perspectives on the Feasibility of Using Enzymes for Pharmaceutical Removal in Wastewater

Muhammad Bilal, Hafiz M. N. Iqbal, and Damiá Barceló

Contents

1 Introduction 120
2 Remediation of Pharmaceutically Active Compounds by Crude/Purified Enzymes 122
  2.1 Lignin Peroxidase: Catalytic Features and Pharmaceutical Degradation 123
  2.2 Manganese Peroxidase: Catalytic Features and Pharmaceutical Degradation 123
  2.3 Soybean Peroxidase and Chloroperoxidase: Catalytic Features and Pharmaceutical Degradation 124
  2.4 Horseradish Peroxidase: Catalytic Features and Pharmaceutical Degradation 126
  2.5 Laccases: Catalytic Features and Pharmaceutical Degradation 126
3 Enzyme Immobilization, Techniques, and Unique Advantages 130
4 Immobilized Peroxidases for Removal of Pharmaceutically Active Compounds 131
5 Immobilized Laccases for Removal of Pharmaceutically Active Compounds 134
6 Concluding Remarks and Outlook 138
References 139

Abstract This particular chapter spotlights the growing environmental concerns and hazardous consequences of numerous organic contaminants so-called emerging

---

M. Bilal
School of Life Science and Food Engineering, Huaiyin Institute of Technology, Huaian, China
e-mail: bilaluaf@hotmail.com

H. M. N. Iqbal (✉)
Tecnologico de Monterrey, School of Engineering and Sciences, Monterrey, NL, Mexico
e-mail: hafiz.iqbal@tec.mx

D. Barceló
Water and Soil Quality Research Group, Department of Environmental Chemistry, IDAEA-CSIC, Barcelona, Spain

Catalan Institute for Water Research (ICRA), Girona, Spain

College of Environmental and Resources Sciences, Zhejiang A&F University, Hangzhou, China
e-mail: dbcqam@cid.csic.es

Sara Rodriguez-Mozaz, Paqui Blánquez Cano, and Montserrat Sarrà Adroguer (eds.), *Removal and Degradation of Pharmaceutically Active Compounds in Wastewater Treatment*, Hdb Env Chem (2021) 108: 119–144, DOI 10.1007/698_2020_661,

contaminants (ECs). These ECs are being detected, though in different quantities, in different environmental matrices and wastewater treatment systems. With ever-increasing awareness, people are now more concerned about the wide-spread distribution of pharmaceutically related active compounds in water matrices. In turn, the free flow of ECs in water matrices poses notable adverse effects on human, aquatic animals, and naturally occurring plants, even at very small concentrations. Due to inadequacies and ineffectiveness of, in practice, physical and chemical-based remediation processes, robust treatment approaches, such as microorganisms and their novel enzyme-based degradation/removal of ECs, are of supreme interest. This chapter focuses on various pharmaceutically related ECs and their efficient mitigation from water matrices. Following a brief introduction, the focus is given to two main treatment approaches, i.e., (1) remediation of pharmaceutically active compounds by crude (pristine) and purified enzymes (i.e., lignin peroxidase, manganese peroxidase, soybean peroxidase, horseradish peroxidase, and laccases) and (2) immobilized enzyme-assisted degradation of pharmaceutically active compounds.

Keywords Biocatalysis, Biological risks, Enzymes, Hazardous compounds, Toxicity, Wastewater treatment

1 Introduction

A continuous rise in worldwide population and urbanization and their associated increase in the consumption of pharmaceuticals have redirected the researches' attention to the search for novel materials for rapid and efficient decontamination of both wastewaters and drinking waters. In addition to extensive heavy metal pollution problems, water contamination by pharmaceutical compounds has become a growing worldwide concern [1]. The concentrations of persistent organic pollutants like personal care products, pharmaceuticals, and their metabolites are continuously rising in the natural environment due to human activities [2–5]. The major source of pharmaceutical pollutants is the WWTP effluents. WWTPs are not designed to completely eliminate environmental pollutants, and thus they can percolate through WWTPs and incorporated into the aquatic systems (streams and rivers) [6]. For example, the contents of diclofenac (an anti-inflammatory drug) and carbamazepine reached 0.99 and 0.95 μg/L, respectively, in WWTP effluents [7]. Particularly, a detectable level of diclofenac has been identified in drinking waters, surface waters, and groundwaters in the range of ng/L to μg/L in Sweden, Spain, Switzerland, and the Baltic region [8–10]. Apart from this, other pharmaceuticals including tramadol, carbamazepine, ibuprofen, oxazepam, and naproxen have also recently been detected in drinking water supplies in some countries. These concentrations of pharmaceuticals can induce serious environmental threats such as

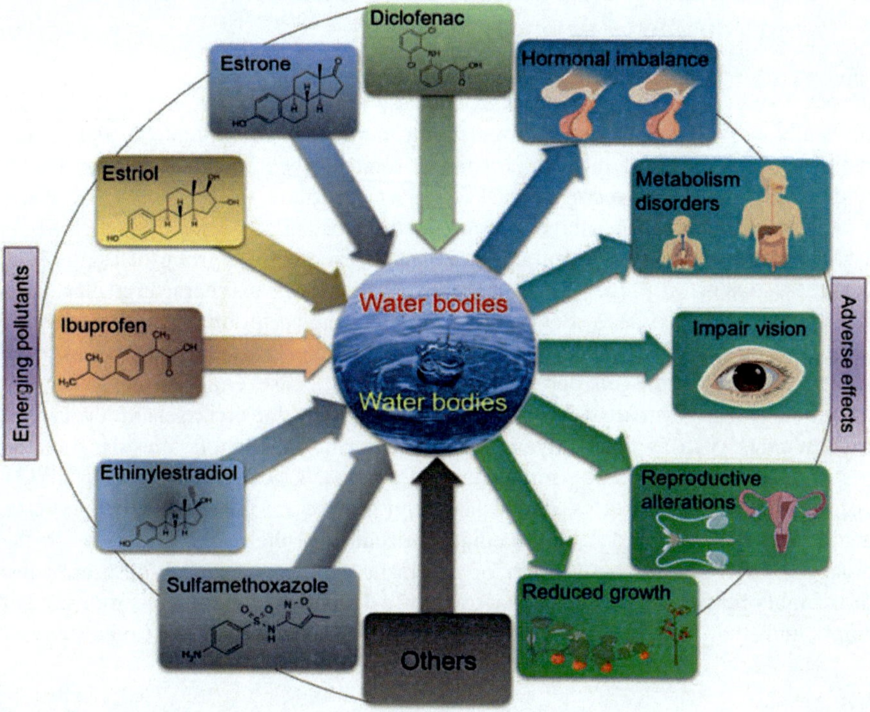

Fig. 1 Notable adverse effects of numerous ECs. Reprinted from Morsi et al. [12] with permission from Elsevier

congenital disorders, physical abnormalities, impairments of the endocrine and reproductive system, and feminization of some fish species [11, 12]. Due to the capability of micropollutants and pharmaceutically active compounds to cause adverse effects to the ecosystem and human health, they have attracted the principal research focus in recent days. Some notable adverse effects of numerous ECs are shown in Fig. 1 [12].

In this avenue, enzymatic treatment of water and wastewater has been advocated as a lucrative alternative due to their high biocatalytic performance and selectivity. Oxidoreductases such as laccase, MnP, LiP, and versatile peroxidase have garnered increasing research interest for potential aptitude to remove a large number of antibiotics and pharmaceutically active residues in wastewater samples.

2 Remediation of Pharmaceutically Active Compounds by Crude/Purified Enzymes

In recent years, enzymes have been extensively utilized for the removal of pharmaceutically active residues and antibiotics into innocuous compounds in wastewater samples. Application of microorganisms or their derived enzymes such as wood-rotting fungi has been recognized as a robust substitute for classical wastewater treatments. These microbes have demonstrated a substantial transformation ability to a vast number of environmental contaminants by secreting a group of ligninolytic enzymes, such as MnP, LiP, VP, and laccase. Peroxidases are characterized as heme and non-heme peroxidases. Peroxidases, in which heme groups serve as an active center, are an important class of oxidoreductases with pronounced ability to oxidize substrates using hydrogen peroxide (Fig. 2) [13]. These enzymes carry out the oxidation-reduction-mediated biotransformation of various classes of hazardous contaminants such as textile dyes, phenols, chlorinated phenols, cresols, dioxins, pesticides, herbicides, and personal care and pharmaceutically active products [13–16]. Among oxidoreductases, peroxidases and laccase are the most widely investigated class of enzymes for bioremediation because of their high capability in the biodegradation and decomposition of diverse organic pollutants. These enzymes form highly reactive free radicals that transform the parent contaminants into smaller compounds that are more decomposable and thus possess negligible toxicity.

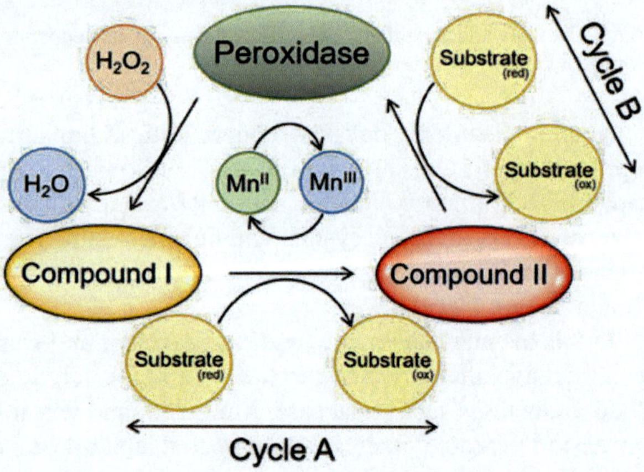

Fig. 2 A simplified schematic illustration of the catalytic cycle of peroxidases. Reprinted from Bilal et al. [13] with permission from Elsevier

2.1 Lignin Peroxidase: Catalytic Features and Pharmaceutical Degradation

LiP (EC 1.11.1.14) is a heme-containing enzyme that can catalyze the degradation of aromatic and halogenated phenolic substances with high redox potential. The range of isoelectric point and molecular weight of LiP has been documented as 3.3 to 4.7 and 38 kDa to 43 kDa, respectively [17]. This monomeric hemoprotein optimally executes its functions at acidic pH (around pH 3.0) in the presence of veratryl alcohol as a natural substrate [18]. As potent oxidants, microbial LiPs have demonstrated the aptitude for oxidative degradation of a wide spectrum of recalcitrant aromatic contaminants, including antibiotics and pharmaceutically active compounds. For instance, crude LiP extracted from the ligninolytic culture of *Phanerochaete chrysosporium* catalyzed the effective degradation of tetracycline and oxytetracycline [19]. Zhang and Geißen [20] markedly removed diclofenac and carbamazepine using crude LiP from *P. chrysosporium*. The high degradation properties of LiP toward persistent aromatic pollutants make it a versatile biocatalyst for diverse biotechnological applications. High redox potentialities of LiP to degrade lignin and other compounds of high redox potential have been ascribed to its exposed tryptophan residue (Trp171), which generates a reactive tryptophanyl radical on the enzyme's surface by relocating electron to the heme molecule. Differences in tryptophan milieu are associated with modifying the enzyme characteristics such as catalytic performance, substrate specificity, and functional stability [21]. The removal efficacy of carbamazepine remained stable in the range of 60–80% for 100 days in a fungal bioreactor that consists of three major biocatalysts including MnP, LiP, and laccase [22].

2.2 Manganese Peroxidase: Catalytic Features and Pharmaceutical Degradation

MnPs are known for their pronounced Mn^{2+}-mediated oxidative capacity for a large number of xenobiotic and aromatic compounds, including various industrial dyes, phenols, and non-phenols [23]. Though MnPs have been inspected for about three decades since their discovery with the fungal LiPs, very limited attention is devoted to MnPs [24, 25]. In recent times, there has been exceptional interest in the use of MnPs owing to their ubiquitous presence in almost all kinds of WRF and some soil-litter decomposers [26, 27]. MnPs originated from WRF are largely associated with the biotransformation of plant biomass and the removal of various refractory pollutants including antibiotics, endocrine disrupters, and pharmaceutically active residues. Moreover, these lignin-degrading enzymes are capable of complete bioconversion of recalcitrant lignin into carbon dioxide and water [27]. Consequently, MnPs have found prominent applications in various biotechnological sectors such as textiles, food, bioremediation, organic synthesis, and paper and pulp manufacturing [28, 29]. The Mn^{2+}-mediated catalytic mechanism of MnPs in oxidizing various aromatic compounds has been well elucidated. Mn^{2+} binds to MnP in a typical region formed by three conserved acidic amino acids, i.e., Glu35, Glu39, and Asp179, in MnP of *P. chrysosporium* [30]. MnP then oxidizes Mn^{2+} to Mn^{3+}, which subsequently is chelated with organic acids and serves as a diffusible oxidizing agent for a large spectrum of aromatic substances [31–33]. Two isoforms of MnPs, MnP1, and MnP2 and a laccase, Lac1, extracted from *T. polyzona* KU-RNW027 exhibited the potential for removing pharmaceutical compounds without a redox-mediated system. Under the optimized reaction environments of pH 4.5 and 50°C, the catalytic efficiencies of these enzymes led to effective remediation and removal of commonly consumed antibiotics ciprofloxacin, tetracycline, amoxicillin, and doxycycline [34]. Wen et al. [35] reported 84.3% and 72.5% removal of oxytetracycline and tetracycline within 4 h by applying 40 U/L of crude MnP extract from the *P. chrysosporium*. The tetracycline removal efficiency of MnP was markedly enhanced by the addition of H_2O_2 during the degradation reaction.

2.3 Soybean Peroxidase and Chloroperoxidase: Catalytic Features and Pharmaceutical Degradation

As a widely studied enzyme, SBP (EC. 1.11.1.7) is regarded as a lucrative biocatalyst and is present in considerable quantities in the soybean hulls [36, 37]. The SBP-assisted biocatalytic system presents a high biodegradation potential for a range of antibiotics, organic pollutants, and emerging contaminants by generating extremely reactive free radicals. Alneyadi et al. reported that both SBP and chloroperoxidase (CPO)-based treatment catalyzed the degradation of sulforhodamine B with the concomitant formation of at least one different degradation intermediate. Further, SBP-aided reaction results in degradation and detoxification of sulforhodamine B, whereas the CPO treatment does not affect the toxicity removal of this aromatic contaminant (Fig. 3) [38, 39]. In another study, Almaqdi et al. [40] conducted experiments for sulfamethoxazole degradation by a combinatorial approach using SBP enzymatic system as well as an advanced oxidation process (UV + H_2O_2). Initial experiments using the individual SBP + H_2O_2 system had no degradation effect; however, the sulfamethoxazole was profoundly transformed by the inclusion of 1-hydroxybenzotriazole as a redox mediator to the reaction system. Results revealed over 80% of the sulfamethoxazole was removed within an initial 10 min of reaction using a mixture of SBP + H_2O_2 system in the presence of HOBT mediators by producing various intermediates. To evaluate the remediation efficiencies of different peroxidases, a mixture containing 21 diverse arrays of emerging pollutants including antibiotics, textile dyes, analgesics, pesticides, herbicides, hormones, pharmaceutical, and personal care products and non-steroidal anti-inflammatory drugs was independently exposed to MnP, HRP, SBP, CPO, and lactoperoxidase (LPO). Experimental results showed that all of the tested

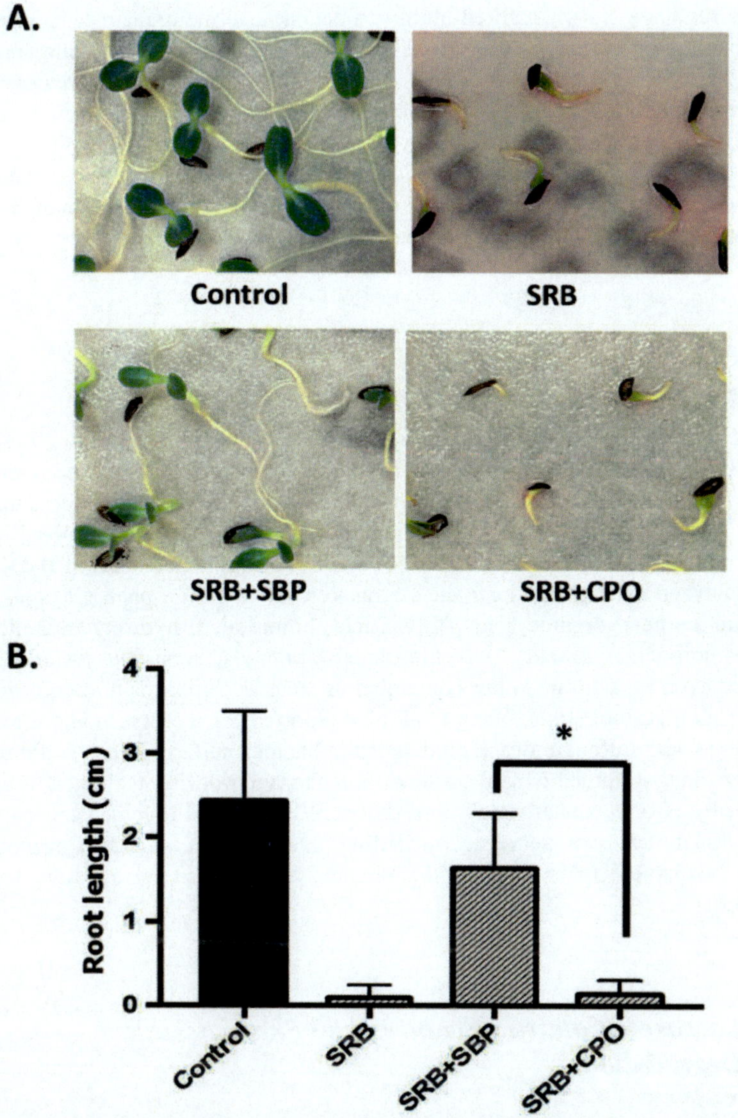

Fig. 3 SRB dye toxicity on *Lactuca sativa* seed. (**a**) A representative sample of *L. sativa* seed treated by SRB dye (10 ppm), SRB dye decolorized by SBP (SRB + SBP) and SRB dye treated by CPO (SRB + CPO); (**b**) root length (cm) after treating the seeds with SRB dye samples. Statistical analysis was performed using unpaired t-test ($n = 80$). The asterisk (*) shows a significant difference ($p < 0.05$). Reprinted from Alneyadi et al. [38], an open-access article distributed under the Creative Commons Attribution (CC BY) license (http://creativecommons.org/licenses/by/4.0/). Copyright (2017) the authors. Licensee MDPI, Basel, Switzerland

peroxidases have shown marked ability to decompose meloxicam (a nonsteroidal anti-inflammatory drug), with SBP as a slightly superior catalyst compared with others. Nevertheless, it appears that all of the organic contaminants were not equally bio-transformed and degraded by the catalytic actions of peroxidases. For illustration, SBP was not capable of degrading roxithromycin (a semi-synthetic macrolide antibiotic), and it also found to recalcitrant to MnP and CPO treatments exhibiting only about 25% transformation. On the other hand, HRP and LPO treatment induced efficient degradation with about 95% and 80% degradation, respectively [40].

2.4 *Horseradish Peroxidase: Catalytic Features and Pharmaceutical Degradation*

Plant HRP (EC 1.11.1.7) is a Class III peroxidase that has attracted incredible interest as a widely studied enzyme to oxidizing a variety of phenolic substrates such as alkylphenols, phenols, halophenols, bisphenol A, triclosan, lignosulfonates, and hormones by one-electron transfer [41–43]. The supplementation of H_2O_2 in the HRP-catalyzed reaction results in the biotransformation of non-phenolic compounds (aliphatic amines), organic acids (fulvic acid, humic acid, hydroxycinnamic acid, and perfluorooctanoic acid), polychlorinated biphenyls, persistent organic pollutants, and even recalcitrant water contaminants such as swine manure and azo dyes [44, 45]. In a current study, Yang et al. [46] reported for the first time the efficient removal of six sulfonamides including sulfadiazine, sulfapyridine, sulfathiazole, sulfamerazine, sulfamethoxazole, and sulfamethoxypyridazine from aqueous environment by HRP-mediated catalytic reaction. When spiked together, the concomitant sulfonamides can serve as mediating agents for accelerated removal of sulfamethoxazole. Up to 47% of diclofenac was removed by a native form of HRP [47].

2.5 *Laccases: Catalytic Features and Pharmaceutical Degradation*

As trendy enzymes, laccase is a multi-copper containing blue oxidase enzyme that belongs to oxidoreductases. The induction, formation, and secretion of extracellular laccase varies in the presence of different physical and nutrient growth conditions like pH, temperature, moisture, aeration, nutrients, mediators, and inhibitors. It is revealed that copper plays a role in the induction of laccases by most of the laccase producers [48]. Laccases typically catalyze their substrates (phenolic as well as non-phenolics, i.e., aromatic amines, phenol, or its derivatives) oxidation with a simultaneous four-electron reduction of molecular oxygen to form a corresponding radical, with water as a by-product (Fig. 4) [13]. Though laccases can be extracted

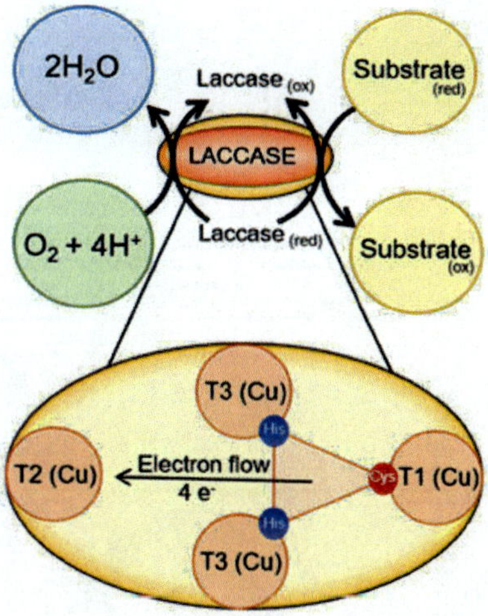

Fig. 4 A simplified schematic illustration of the catalytic cycle of laccase. Reprinted from Bilal et al. [13] with permission from Elsevier

from fungi, bacteria, plants, and insects [49], microbial-derived laccases, in particular WRF, have garnered increasing interest because of broad substrate specificity and high oxidation capability. It drives the oxidative cleavage of a diversity of phenolic compounds, diamines, aromatic amines, and electron-rich substrates accompanied by simultaneous mono-electronic reduction of molecular oxygen to water. Laccases accelerate the conversion of phenoxyl radicals to the ketone, demethoxylation, and cleavage of carbon in phenolic structures. Oxidation and biotransformation of a spectrum of non-phenolic structures of lignin by laccase can also be possible through the inclusion of different natural/synthetic redox mediators including 2,2′-azino-bis(3-ethylbenzthiazoline-6-sulfonic acid) (ABTS), syringaldehyde, acetosyringone, vanillin, and 1-hydroxybenzotriazole (HOBT). During the last few years, the application of microbial laccases has been widely explored for the bioremediation of pharmaceuticals, xenobiotics, organic pollutants, endocrine-disrupting compounds, dyes, polychlorinated biphenyls, organochlorines, polyaromatic hydrocarbons, and 2,4,6-trinitrotoluene [13–15, 50–52].

Pharmaceutical products released openly by pharmaceutical manufacturers or hospital activities have substantial health-related apprehensions. In a very recent study, laccase treatment drives the dehydrochlorination of an insecticide, lindane, to form non-toxic and environmentally friendlier by-products (Fig. 5) [53]. A promising aspect of laccases functions is the cleavage of different substrates without the necessity of external source of Mn^{2+}, H_2O_2, or co-factors that are generally included in the case of other peroxidases such as MnP or LiP. Guardado et al. [54] appraised the degradation capability of a novel *Pycnoporus sanguineus CS43* laccase for decomposing various pharmaceutical micropollutants. Degradation profile indicated a removal efficacy of 40%, 80%, and 100% for ciprofloxacin, amoxicillin, and sulfamethoxazole, respectively, within 3 h of syringaldehyde-mediated catalytic reaction time. The recombinant laccase from *T. versicolor* evinced to be a highly proficient catalyst for the decomposition of pharmaceuticals-based environmental

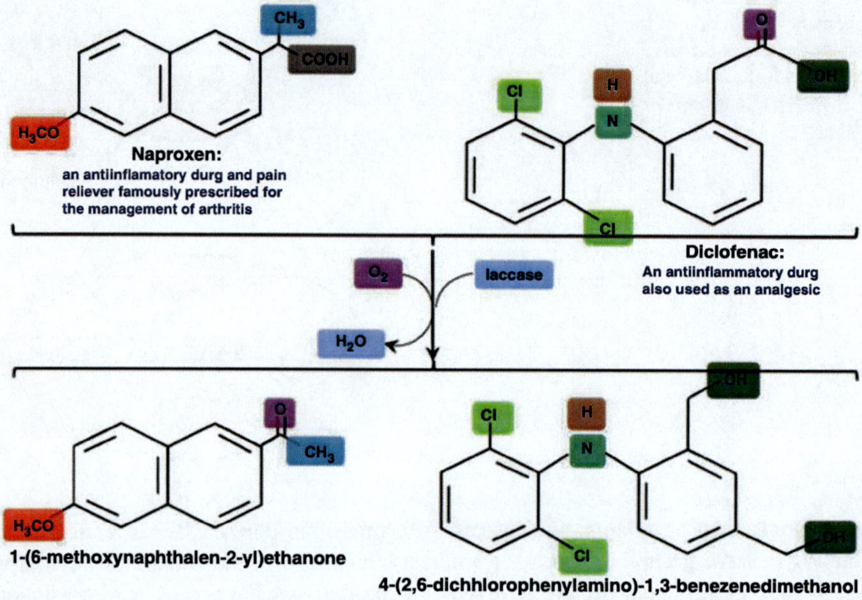

Fig. 5 The oxidative cleavage of some pharmaceuticals by laccase enzyme. Reprinted from Unuofin et al. [53] with permission under Creative Commons Attribution (CC BY) license (http://creativecommons.org/licenses/by/4.0/). Copyright (2019) the authors. Licensee MDPI, Basel, Switzerland

pollutants. After 24 h of catalytic treatment with a recombinant enzyme, the sulfamethoxazole and diclofenac were removed to 51.1% and 46.8%, respectively. A complete degradation efficiency was obtained within 1 h by the supplementation of 1 mM ABTS as a redox mediator [55]. Alharbi et al. [56] investigated the removal of four pharmaceutical compounds, i.e., sulfamethoxazole (SMX), diclofenac (DCF), carbamazepine (CBZ), and trimethoprim (TMP), separately as well as a mixture at various concentrations, by laccase from *T. versicolor*. In contrast to mixtures, the pharmaceuticals tested individually were more efficiently degraded under identical reaction environments. For instance, diclofenac (5 mg/L) was degraded to an undetectable level within 8 h when treated in the individual experiment, whereas the removal time was increased to 24 h following exposure to the catalytic system in the form of a mixture with other pharmaceutical compounds. The comparable tendency was noted with other tested pharmaceuticals, where 82 vs. 34%, 95 vs. 39%, and 56 vs. 49% degradation occurred after 48 h of biocatalytic treatment with carbamazepine, trimethoprim, and sulfamethoxazole ascertained separately or as in mixture forms, respectively. Toxicity analysis revealed the non-toxicity of all the laccase-treated pharmaceutical and effluent samples. Table 1 summarizes recent degradation studies of pharmaceutically active compounds by free laccase enzymes.

Table 1 Removal studies of pharmaceutical compounds by free laccases under various operating conditions

Enzyme source	Pharmaceutical compound	Reaction conditions	Removal (%)	Reference
Aspergillus oryzae	Ciprofloxacin hydrochloride	Enzyme loading of 0.02% (w/v), temperature 60°C, power input 75 W, frequency 22 kHz, and agitation 200 rpm	51	Sutar and Rathod [57]
Aspergillus oryzae	Cetirizine dihydrochloride	Optimal experimental parameters were enzyme loading of 0.02%, temperature 50°C, power input 100 W, frequency 25 kHz, and agitation 200 rpm	91	Sutar and Rathod [58]
Pycnoporus sanguineus CS43	Amoxicillin	Pollutant concentration, 20 mg L^{-1} Redox mediator syringaldehyde Incubation time, 3 h	80	Guardado et al. [54]
Pycnoporus sanguineus CS43	Sulfamethoxazole	Pollutant concentration, 20 mg L^{-1} Redox mediator syringaldehyde Incubation time, 3 h	100	Guardado et al. [54]
Pycnoporus sanguineus CS43	Ciprofloxacin	Pollutant concentration, 20 mg L^{-1} Redox mediator syringaldehyde Incubation time, 3 h	40	Guardado et al. [54]
Trametes versicolor	Doxycycline	Reaction mixture contained 10^{-4} M doxycycline, partially purified laccase (10 nkat mL^{-1}), malonate buffer (50 mM, pH 4.5), and 0.2 mM HBT as a redox mediator	100	Suda et al. [59]
Trametes versicolor	Chlortetracycline	Reaction mixture contained 10^{-4} M chlortetracycline, partially purified laccase (10 nkat mL^{-1}), malonate buffer (50 mM, pH 4.5), and 0.2 mM HBT as a redox mediator	100	Suda et al. [59]
Trametes versicolor	Tetracycline	Reaction mixture contained 10^{-4} M tetracycline, partially purified laccase (10 nkat mL^{-1}), malonate buffer (50 mM, pH 4.5), and 0.2 mM HBT as a redox mediator	100	Suda et al. [59]
Trametes versicolor	Oxytetracycline	Reaction mixture contained 10^{-4} M oxytetracycline, partially purified laccase (10 nkat mL^{-1}), malonate buffer (50 mM, pH 4.5), and 0.2 mM HBT as a redox mediator	100	Suda et al. [59]
Trametes versicolor	Tetracycline	Antibiotic of 100 μg/mL, laccase 17.5 μg/mL, pH 7.0, temperature 20°C, incubation time 18 h	78%	Llorca et al. [60]

3 Enzyme Immobilization, Techniques, and Unique Advantages

Enzymes work perfectly under normal physiological environments, and their functionalities are highly based on their conformations. Harsh and adverse environmental conditions that are usually experienced in effluent streams such as high ionic strength, extreme temperatures, presence of inhibitors, and very low or high pH can affect the conformation of the enzyme, thus resulting in its denaturation [61, 62]. These inadequacies can be overcome by immobilizing enzyme on a solid support. Immobilization is a process in which the enzyme is attached to an insoluble support carrier, where it is held in a proper geometrical conformation resulting in increased stability and reusability of the enzyme [14, 63]. The immobilization process converts the enzyme from its homogenous form to a heterogeneous catalyst (immobilized enzyme) to give an immobilized biocatalyst [64]. The immobilized enzymes can be used for the continuous bioremediation of great volumes of effluent effectively [61].

The immobilization of enzymes to different biocompatible supports can be achieved using three major techniques. The first method is binding to a carrier, and it can be further subdivided into the physical binding, also known as adsorption, and chemical binding via covalent linkages [65]. In adsorption, the catalyst is adsorbed to the outside surface of inert support, which can be a glass, matrix, or alginate beads. This technique is not very efficient, and the coupling between the enzyme and support is feeble to keep the enzyme attached to its place. Covalent binding entails the attachment of the biocatalyst to support by covalent bonds directly or using a cross-linking reagent such as glutaraldehyde, which will be attached to the enzyme at one side and the support from its other side. Covalent binding is more effective and stronger than adsorption. The second method is entrapment or encapsulation in which the enzyme is either trapped in a polymeric matrix network or encapsulated within a solid carrier. The last technique is the development of cross-linking of enzyme aggregates (CLEAs) or crystals (CLECs), which are considered as carrier-free immobilized biocatalytic systems [63, 65]. It is important to mention that the solid support used in immobilization should be inexpensive, ecologically friendly, and non-toxic and does not exhibit any undesirable effect on the biodegraded solution [14].

The use of carrier-immobilized enzymes can address most of the issues that are faced when using an enzyme in the free form. Non-immobilized forms of enzymes cannot be reused and recycled after the treatment and are considered highly expensive. Immobilization overcomes part of these issues as it allows for the repeated usability of enzymes and increases their recycling efficiencies in multiple continuous cycles [13, 66]. Additionally, immobilization allows the long-term and durable stability of enzymes as they become more resistant to degradation and denaturation and stabilized against harsh temperatures, pH, and pressure conditions [13, 67]. Since the enzyme immobilization process offers numerous advantages compared to the free state of the enzymes, it can be considered as an easy and effective way to boost up the catalytic potentialities of enzymes.

4 Immobilized Peroxidases for Removal of Pharmaceutically Active Compounds

Though the free form of biocatalysts may be efficient in removing pharmaceutically active compounds, their scale-up exploitation is associated with some drawbacks. The soluble nature of free enzymes in the aqueous media renders them challenging to retain in the reactor system. Moreover, instability toward thermal or chemical denaturation diminishes their applicability to establish enzyme-driven processes, which in turn profoundly increase the processing cost. The deployment of carrier-immobilized enzyme enables the easy retrieval of both the products and enzyme molecules facilitating their multiple-time reutilization in consecutive biocatalytic operations, controlled reactions termination, and wider bioreactor configurations. In the last decade, immobilized peroxidases have been effectively used in many studies for the removal of different pharmaceutically active compounds due to their elevated functional stability and repeated usability (Table 2). Touahar et al. [68] developed a versatile and multifunctional combined cross-linked enzyme aggregate (combi-CLEA) to remove an array of different pharmaceuticals. To this end, three oxidative enzymes including versatile peroxidase from *B. adusta*, laccase from *T. versicolor*, and glucose oxidase from *A. niger* were concurrently cross-linked to prepare combi-CLEAs. The as-synthesized combi-CLEAs presented the better capability for resisting denaturing environments such as low pH 3.0 and elevated temperature of 60°C. The immobilized biocatalytic system exhibited high removal efficiency to synthetic wastewater and pharmaceutical cocktail consisting of diazepam, fenofibrate, caffeine, naproxen, acetaminophen, indomethacin, ketoprofen, diclofenac, mefenamic acid, ciprofloxacin, bezafibrate, trimethoprim, and carbamazepine. The MnP-Tween 80 system comprising unsaturated fatty acid has shown the potential to eliminate the antidepressant sertraline and miconazole resulting in 85% and 88% removal of sertraline and miconazole after enzyme exposure to 24 h. The developed enzyme system was also able to remove intermediate metabolites of sertraline such as desmethylsertraline [71].

In a recent report, Pylypchuk et al. [1] engineered novel enzyme-based nanocomposites by adsorbing LiP and HRP onto magnetite nanoparticles followed by sol-gel encapsulation in a silica matrix, and they used these for the removal of common pharmaceutical pollutants in a green process (Fig. 6a). Encapsulation results in an improvement in the thermal stability profile of the biocatalysts. The encapsulated enzyme exhibited notable selectivity in oxidative removal of the organic contaminants carbamazepine, paracetamol, and diclofenac with high activity. Particularly, sol-gel immobilized HRP and LiP-based nanostructured composites showed catalytic activity over 20 successive removal cycles for 20 days at a high

Table 2 Removal studies of pharmaceutical compounds by peroxidases under various operating conditions

Enzyme source	Enzyme name	Immobilization support	Pharmaceutical compound	Reaction conditions	Removal (%)	Reference
Commercial	Horseradish peroxidase and lignin peroxidase	Magnetic sol-gel encapsulated horseradish peroxidase and lignin peroxidase composite	Carbamazepine	Carbamazepine solution (1 mL) was added to 30 mg of sol-gel composites sample under vigorous shaking in the presence of 3 μL of H_2O_2 (3.6%)	68	Pylypchuk et al. [1]
Commercial	Horseradish peroxidase and lignin peroxidase	Magnetic sol-gel encapsulated horseradish peroxidase and lignin peroxidase composite	Diclofenac	Diclofenac solution (1 mL) was added to 30 mg of sol-gel composites sample under vigorous shaking in the presence of 3 μL of H_2O_2 (3.6%)	64	Pylypchuk et al. [1]
Versatile peroxidase from *Bjerkandera adusta*, laccase *Trametes versicolor*, and glucose oxidase *Aspergillus niger*	Laccase, versatile peroxidase, and glucose oxidase	Combined cross-linked enzyme aggregates from laccase, versatile peroxidase, and glucose oxidase	Acetaminophen, naproxen, mefenamic acid, indomethacin, diclofenac, ketoprofen, caffeine, diazepam, ciprofloxacin, trimethoprim, fenofibrate, bezafibrate, and carbamazepine	Treatment was carried out at pH 5.0, 20°C in 125 mL Erlenmeyer flask at 150 rpm for 5 h. Free form of enzymes or combi-CLEA was included to achieve final Lac and Mn-oxidizing activities of 750 U/L and 250 U/L, respectively	>80	Touahar et al. [68]
Caldaromyces fumago	Chloroperoxidase	Chitosan microspheres	Sulfamethoxazole	The removal was determined in a volume of 1 mL containing 20 μM of the substrate, 20 mM	>80	García-Zamora et al. [69]

				of KCl, and 1 macrosphere in a 60 mM phosphate buffer at pH 3.0 under gentle agitation for 10 min		
Caldaromyces fumago	Chloroperoxidase	Chitosan macrospheres	Naproxen	–	>80	García-Zamora et al. [69]
Caldaromyces fumago	Chloroperoxidase	Chitosan macrospheres	Tetracycline	–	>80	García-Zamora et al. [69]
Pichia methanolica	Lignin peroxidase	Fe_3O_4@SiO_2@polydopamine nanoparticles	Tetracycline	1.0 g of immobilized LiP was employed to degrade pollutants with 5 mg/L of original concentration in 200 mL of sodium tartrate buffer for different periods	100	Guo et al. [70]
Pichia methanolica	Lignin peroxidase	Fe_3O_4@SiO_2@polydopamine nanoparticles	Dibutyl phthalate	–	100	Guo et al. [70]
Pichia methanolica	Lignin peroxidase	Fe_3O_4@SiO_2@polydopamine nanoparticles	5-Chlorophenol	–	100	Guo et al. [70]
Pichia methanolica	Lignin peroxidase	Fe_3O_4@SiO_2@polydopamine nanoparticles	Phenol	–	100	Guo et al. [70]
Pichia methanolica	Lignin peroxidase	Fe_3O_4@SiO_2@polydopamine nanoparticles	Phenanthrene	–	79	Guo et al. [70]
Pichia methanolica	Lignin peroxidase	Fe_3O_4@SiO_2@polydopamine nanoparticles	Fluoranthene	–	73	Guo et al. [70]
Pichia methanolica	Lignin peroxidase	Fe_3O_4@SiO_2@polydopamine nanoparticles	Benzo(a)pyrene	–	65	Guo et al. [70]

temperature of 55°C. In contrast to native enzymes, the encapsulated biocatalysts also displayed improved stability in acidic medium. Under the optimal conditions, sol-gel-assisted LiP composites showed 64% and 68% degradation of diclofenac and carbamazepine, respectively. Both diclofenac and carbamazepine were completely degraded at pH 3.0 and 55°C within 3 days by treatment with both immobilized LiP and HRP. After characterizing decomposedproducts of the drug through NMR spectroscopy, a decomposition pathway was proposed by the peroxidase enzymes (Fig. 6b). It was noted that the degradation of diclofenac by oxidases initiates from hydroxylation of its structural molecule. For instance, two major degradation metabolites, viz., 4′-hydroxy diclofenac and 5-hydroxy diclofenac, were detected by laccase-assisted catalytic treatment. Quinones like diclofenac-2,5-iminoquinone might also be generated as diclofenac degradation intermediates, but only 5–10% of the degraded diclofenac was converted into quinones, indicating that a major portion of this drug (90–95%) was probably removed via other oxidation routes [72].

5 Immobilized Laccases for Removal of Pharmaceutically Active Compounds

Immobilized form of laccases from many strains of WRF has been demonstrated for the degradation and removal of various pharmaceutically active compounds in wastewater effluents. Yang et al. [73] exploited an immobilized form of laccase and tested its ability to degrading different antibiotics. The method that they used for the immobilization process is CLEAs as they prepared magnetic-CLEAs for laccase and used it for the biodegradation of antibiotics. Results showed that laccase M-CLEAs were capable of degrading more than 80% of tetracycline within 12 h treatment. Three degradation products of tetracycline were detected using LC-TOF-MS designated as TP 459, TP 431, and TP 396 that have elution time at 2.69, 6.01, and 6.35 min, respectively. In a very recent study, Bilal et al. [13–15] demonstrated the ability of free and immobilized laccase for the degradation of bisphenol A, which is an organic synthetic compound [14]. In their work, laccase was covalently attached to chitosan beads using glutaraldehyde as a coupling agent. The immobilized biocatalyst showed good stability and preserved 71.24% of its initial activity after ten cycles of treatments. Bisphenol A was almost completely degraded (more than 99%) by the immobilized laccase after 150 min. Likewise, there are a plethora of studies that have documented the ability of immobilized biocatalytic systems to degrade different contaminants of emerging concerns efficiently. A summary of recent degradation studies of different pharmaceutical pollutants by immobilized laccase-based biocatalytic systems is summarized in Table 3. Immobilization of biocatalysts on different supports is a promising and ecologically friendly technique for the elimination of different pollutants from wastewater. The role of three natural phenolic substances, including acetosyringone, syringaldehyde, and p-coumaric acid, was evaluated as mediators in eliminating a routinely used

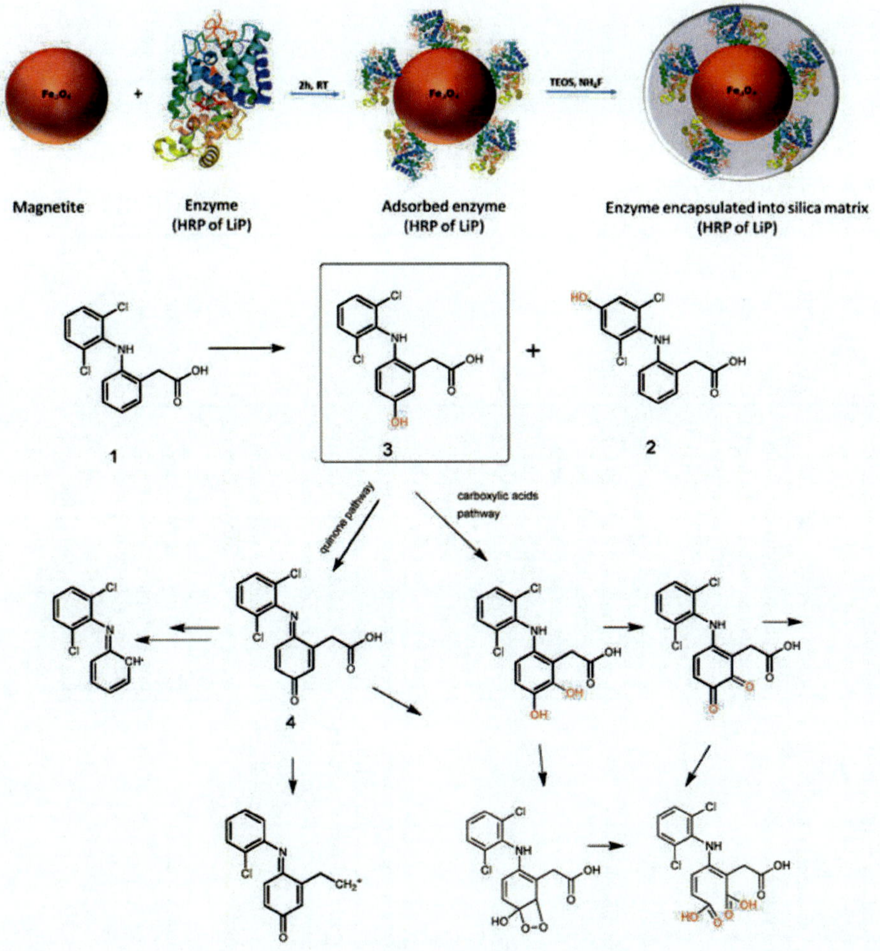

Fig. 6 (**a**) Sol-gel encapsulation of HRP and LiP peroxidases adsorbed onto magnetite nanoparticles. (**b**) Possible pathways in diclofenac enzymatic oxidative decomposition. Reprinted from Pylypchuk et al. [1] an open-access article distributed under the terms and conditions of the Creative Commons Attribution (CC BY) license (http://creativecommons.org/licenses/by/4.0/)

recalcitrant pharmaceutical carbamazepine by the native as well as insolubilized laccases. Among these mediators, p-coumaric acid facilitated the optimal biocatalytic performance for removing carbamazepine with over 60% elimination efficiency after 96 h of enzymatic treatment [79].

Enzyme immobilization onto porous membranes, so-called the development of enzyme-based membrane reactors, is an attractive method to solve the issue associated with the free enzymes due to the provision of high surface area by membranes for biocatalytic reactions. Noticeable advantages of laccases immobilized on membranes are reusability, durable shelf life, and steadiness against pH and temperature

Table 3 Removal studies of pharmaceutical compounds by immobilized laccases under various operating conditions

Enzyme source	Immobilization support	Pharmaceutical compound	Reaction conditions	Removal (%)	Reference
Trametes versicolor	Bentonite-derived mesoporous materials	Tetracycline	Reaction mixture containing tetracycline solution (10 mg/L) and the immobilized enzyme was placed at 303 K for 2 h in the presence of 1-hydroxybenzotriazole	60	Wen et al. [74]
Phoma sp. UHH 5-1-03	Cross-linked to polyvinylidene fluoride membranes	Acetaminophen	Concentration, 10 μM; 20 U L-1 of immobilized enzyme; incubation at room temperature for 24 h using 180 rpm	>85	Jahangiri et al. [75]
Phoma sp. UHH 5–1-03	Cross-linked to polyvinylidene fluoride membranes	Mefenamic acid	Concentration, 10 μM; 20 U L-1 of immobilized enzyme; incubation at room temperature for 24 h using 180 rpm	>85	Jahangiri et al. [75]
Trametes versicolor	Hybrid bioreactor consisting of a combined cross-linked laccase and tyrosinase aggregates	Acetaminophen, naproxen, mefenamic acid, ibuprofen, ketoprofen, indomethacin, trimethoprim, ciprofloxacin, ofloxacin, caffeine, carbamazepine, bezafibrate, fenofibrate, atenolol	Pharmaceutical solution (10 mg/L) was mixed with the combi-CLEAs to a final activity of 80 U/L of Lac and 50 U/L of Tyr The experiment was continuously operated 5 days at room temperature	100	Ba et al. [76]
Trametes versicolor	Polyacrylonitrile-biochar composite nanofibrous membrane	Chlortetracycline	Chlortetracycline concentration (200 ppb) in water was pumped at three different fluxes (1, 2, and 3 $mL/cm^2{\cdot}h$) into the test setup in the dead-end configuration	58.3%, 40.7%, and 22.6% chlortetracycline removal efficiency at flux rates of 1, 2, and 3 $mL/h{\cdot}cm^2$	Taheran et al. [77]

Trametes versicolor	Oxygen functionalized nanobiochars	Carbamazepine	50 mg immobilized laccase was dispersed in 20 mL of carbamazepine solution (20 ng/mL) and allowed to react at 25°C and 200 rpm for 24 h	83% and 86% degradation in spiked water and secondary effluent, respectively	Naghdi et al. [78]
Trametes versicolor	Membrane hybrid reactor	Carbamazepine	Carbamazepine solution with an initial concentration of 20 μM was incubated with immobilized laccase after 96 h	60	Ji et al. [79]
Trametes versicolor	Cross-linked enzyme aggregates on amine-functionalized magnetic nanoparticles	Acetaminophen, atenolol, diclofenac, epoxy carbamazepine, and mefenamic acid	Reaction mixture containing 100 μg/L of each of pharmaceuticals was reacted with 1,000 U/L of the immobilized enzyme for 6 h using 0.1 mM ABTS at pH 7.0 and 125 rpm	100	Kumar and Cabana [80]
Aspergillus oryzae	Enzymatic membrane reactor	Diclofenac	Loading rate 480 mg/L d (diclofenac); redox-mediator syringaldehyde (5 mM) The reactor was operated at a flux of 1.1 L/m2 h via a peristaltic pump with an 8 min on and 1 min off cycle	>80	Nguyen et al. [81]

fluctuations [82]. Laccase enzyme produced by *T. versicolor* was subjected to immobilization on an electrospun polyacrylonitrile-biochar composite membrane and applied for the degradation of a commonly used chlortetracycline antibiotic from aqueous solution. The newly developed immobilized biocatalyst catalyzed 58.3%, 40.7%, and 22.6% degradation of chlortetracycline continuously at flux rates of 1, 2, and 3 mL/h·cm^2 [77]. Cross-linking of laccase from *Phoma* sp. UHH 5-1-03 to polyvinylidene fluoride membranes by electron beam irradiation was employed to remove a mixture of pharmaceutical pollutants including indomethacin, naproxen, bezafibrate, acetaminophen, ketoprofen, and mefenamic acid in actual municipal wastewater samples. Degradation profile of the immobilized laccase showed greater removal efficiency (more than 80%) of acetaminophen and mefenamic acid in wastewater in continuously eliminating operational mode.

6 Concluding Remarks and Outlook

In conclusion, based on the above-discussed literature with suitable examples, environmental contamination with a range of anthropogenic pollutants of emerging concern has become a global problem. This chapter is of particular interest, which highlights the significant potential of numerous enzymes in pristine or immobilized forms for efficient removal/degradation of ECs. The growing contamination of water matrices with a controlled or uncontrolled discharge of incompletely or inadequately treated hazardous industrial wastes harshly affects the whole living ecosystem. Considering the above-discussed scenarios, there is a dire need to develop highly efficient bioremediation strategies that are clean, green, sustainable, and environmental-friendly and can replace the in-practice inefficient remediation approaches. Various wastewater treatment technologies using crude (pristine), purified enzymes, and immobilized enzyme systems are discussed with a particular emphasis on the degradation and detoxification of pharmaceutically active compounds. This study emphasized that the enzyme-based biocatalytic treatment processes have gained unprecedented importance in mitigating an array of ECs typically present in wastewater effluents. Despite an effective technology with proven bioremediation potential, enzymes exploitability is currently far from being and has not yet been incorporated in large-scale water treatment systems. Future studies should be attentive to increase the application of enzymes under the actual treatment environments. Moreover, an in-depth understanding of the extensive environmental and commercial influences regarding the practical applicability of enzyme-based biocatalytic agents to achieve efficient remediation of numerous contaminants of emerging concern is critical to attaining the sustainability of immobilized enzyme technology.

Acknowledgments Authors are grateful to their representative universities/institutes for providing literature facilities.

Conflict of Interests Authors declare no conflict of interest in any capacity, including financial and competing.

References

1. Pylypchuk IV, Daniel G, Kessler VG, Seisenbaeva GA (2020) Removal of diclofenac, paracetamol, and carbamazepine from model aqueous solutions by magnetic Sol–Gel encapsulated horseradish peroxidase and lignin peroxidase composites. Nano 10(2):282
2. Ali N, Khan A, Bilal M, Malik S, Badshah S, Iqbal H (2020) Chitosan-based bio-composite modified with Thiocarbamate moiety for decontamination of cations from the aqueous media. Molecules 25(1):226
3. Ali N, Said A, Ali F, Raziq F, Ali Z, Bilal M et al (2020) Photocatalytic degradation of congo red dye from aqueous environment using cobalt ferrite nanostructures: development, characterization, and photocatalytic performance. Water Air Soil Pollut *231*(2):50
4. Ebele AJ, Abdallah MAE, Harrad S (2017) Pharmaceuticals and personal care products (PPCPs) in the freshwater aquatic environment. Emerg Contam 3(1):1–16
5. Khan A, Ali N, Bilal M, Malik S, Badshah S, Iqbal H (2019) Engineering functionalized chitosan-based sorbent material: characterization and sorption of toxic elements. Appl Sci 9 (23):5138
6. Petrović M, Gonzalez S, Barceló D (2003) Analysis and removal of emerging contaminants in wastewater and drinking water. TrAC Trends Anal Chem 22(10):685–696
7. Tixier C, Singer HP, Oellers S, Müller SR (2003) Occurrence and fate of carbamazepine, clofibric acid, diclofenac, ibuprofen, ketoprofen, and naproxen in surface waters. Environ Sci Technol 37(6):1061–1068
8. Aldekoa J, Medici C, Osorio V, Pérez S, Marcé R, Barceló D, Francés F (2013) Modelling the emerging pollutant diclofenac with the GREAT-ER model: application to the Llobregat River basin. J Hazard Mater 263:207–213
9. Buser HR, Poiger T, Müller MD (1998) Occurrence and fate of the pharmaceutical drug diclofenac in surface waters: rapid photodegradation in a lake. Environ Sci Technol 32 (22):3449–3456
10. Hallgren P, Wallberg P (2015) Background report on pharmaceutical concentrations and effects in the Baltic Sea. In: Policy area hazards of the EU strategy for the Baltic Sea Region. Swedish Environmental Protection Agency, Stockholm
11. Belhaj D, Baccar R, Jaabiri I, Bouzid J, Kallel M, Ayadi H, Zhou JL (2015) Fate of selected estrogenic hormones in an urban sewage treatment plant in Tunisia (North Africa). Sci Total Environ 505:154–160
12. Morsi R, Bilal M, Iqbal HMN, Ashraf SS (2020) Laccases and peroxidases: the smart, greener and futuristic biocatalytic tools to mitigate recalcitrant emerging pollutants. Sci Total Environ:136572
13. Bilal M, Adeel M, Rasheed T, Zhao Y, Iqbal HMN (2019) Emerging contaminants of high concern and their enzyme-assisted biodegradation – a review. Environ Int 124:336–353
14. Bilal M, Jing Z, Zhao Y, Iqbal HM (2019) Immobilization of fungal laccase on glutaraldehyde cross-linked chitosan beads and its bio-catalytic potential to degrade bisphenol A. Biocatal Agric Biotechnol 19:101174
15. Bilal M, Ashraf SS, Barceló D, Iqbal HM (2019) Biocatalytic degradation/redefining "removal" fate of pharmaceutically active compounds and antibiotics in the aquatic environment. Sci Total Environ 691:1190–1211
16. Bilal M, Rasheed T, Iqbal HMN, Yan Y (2018) Peroxidases-assisted removal of environmentally-related hazardous pollutants with reference to the reaction mechanisms of industrial dyes. Sci Total Environ 644:1–13

17. Glumoff T, Harvey P, Molinari S, Goble M, Frank G, Palmer JM et al (1990) Lignin peroxidase from *Phanerochaete chrysosporium* molecular and kinetic characterization of isozymes. Eur J Biochem 187:515–520
18. Furukawa T, Bello FO, Horsfall L (2014) Microbial enzyme systems for lignin degradation and their transcriptional regulation. Front Biol 9:448–471
19. Wen X, Jia Y, Li J (2009) Degradation of tetracycline and oxytetracycline by crude lignin peroxidase prepared from *Phanerochaete chrysosporium*–a white rot fungus. Chemosphere 75 (8):1003–1007
20. Zhang Y, Geißen SU (2010) In vitro degradation of carbamazepine and diclofenac by crude lignin peroxidase. J Hazard Mater 176(1–3):1089–1092
21. Ivancich A, Mazza G, Desbois A (2001) Comparative electron paramagnetic resonance study of radical intermediates in turnip peroxidase isozymes. Biochemistry 40:6860–6866
22. Zhang Y, Geißen SU (2012) Elimination of carbamazepine in a non-sterile fungal bioreactor. Bioresour Technol 112:221–227
23. Heinfling A, Ruiz-Dueñas FJ, Martínez MJ, Bergbauer M, Szewzyk U, Martínez AT (1998) A study on reducing substrates of manganese-oxidizing peroxidases from Pleurotus eryngii and Bjerkandera adusta. FEBS Lett 428(3):141–146
24. Glenn JK, Gold MH (1985) Purification and characterization of an extracellular Mn(II)-dependent peroxidase from the lignin-degrading basidiomycete, Phanerochaete chrysosporium. Arch Biochem Biophys 242(2):329–341
25. Paszczyński A, Huynh VB, Crawford R (1985) Enzymatic activities of an extracellular, manganese-dependent peroxidase from Phanerochaete chrysosporium. FEMS Microbiol Lett 29(1–2):37–41
26. Bilal M, Asgher M, Parra-Saldivar R, Hu H, Wang W, Zhang X, Iqbal HM (2017) Immobilized ligninolytic enzymes: an innovative and environmental responsive technology to tackle dye-based industrial pollutants–a review. Sci Total Environ 576:646–659
27. Kellner H, Luis P, Pecyna MJ, Barbi F, Kapturska D, Krüger D et al (2014) Widespread occurrence of expressed fungal secretory peroxidases in forest soils. PLoS One 9(4):e95557
28. Champagne PP, Ramsay JA (2005) Contribution of manganese peroxidase and laccase to dye decoloration by Trametes versicolor. Appl Microbiol Biotechnol 69(3):276
29. Maciel MJM, Ribeiro HCT (2010) Industrial and biotechnological applications of ligninolytic enzymes of the basidiomycota: a review. Electron J Biotechnol 13(6):14–15
30. Sundaramoorthy M, Kishi K, Gold MH, Poulos TL (1994) The crystal structure of manganese peroxidase from Phanerochaete chrysosporium at 2.06-A resolution. J Biol Chem 269 (52):32759–32767
31. Hildén K, Mäkelä MR, Steffen KT, Hofrichter M, Hatakka A, Archer DB, Lundell TK (2014) Biochemical and molecular characterization of an atypical manganese peroxidase of the litter-decomposing fungus Agrocybe praecox. Fungal Genet Biol 72:131–136
32. Li L, Liu B, Yang J, Zhang Q, He C, Jia R (2019) Catalytic properties of a short manganese peroxidase from Irpex lacteus F17 and the role of Glu166 in the Mn2+-independent activity. Int J Biol Macromol 136:859–869
33. Wong DW (2009) Structure and action mechanism of ligninolytic enzymes. Appl Biochem Biotechnol 157(2):174–209
34. Lueangjaroenkit P, Teerapatsakul C, Sakka K, Sakka M, Kimura T, Kunitake E, Chitradon L (2019) Two manganese peroxidases and a laccase of *Trametes polyzona* KU-RNW027 with novel properties for dye and pharmaceutical product degradation in redox mediator-free system. Mycobiology 47:217–229
35. Wen X, Jia Y, Li J (2010) Enzymatic degradation of tetracycline and oxytetracycline by crude manganese peroxidase prepared from *Phanerochaete chrysosporium.* J Hazard Mater 177 (1–3):924–928
36. Bódalo A, Gomez JL, Gomez E, Bastida J, Maximo MF (2006) Comparison of commercial peroxidases for removing phenol from water solutions. Chemosphere 63(4):626–632

37. Fernandes M, Souza DH, Henriques RO, Alves MV, Skoronski E, Junior AF (2020) Obtaining soybean peroxidase from soybean hulls and its application for detoxification of 2, 4-dichlorophenol contaminated water. J Environ Chem Eng 8:103786
38. Alneyadi A, Shah I, AbuQamar S, Ashraf S (2017) Differential degradation and detoxification of an aromatic pollutant by two different peroxidases. Biomol Ther 7(1):31
39. Alneyadi AH, Ashraf SS (2016) Differential enzymatic degradation of thiazole pollutants by two different peroxidases–a comparative study. Chem Eng J 303:529–538
40. Almaqdi KA, Morsi R, Alhayuti B, Alharthi F, Ashraf SS (2019) LC-MSMS based screening of emerging pollutant degradation by different peroxidases. BMC Biotechnol 19(1):83
41. Arnoldsson K, Andersson PL, Haglund P (2012) Formation of environmentally relevant brominated dioxins from 2,4,6,-tribromophenol via bromoperoxidase-catalyzed dimerization. Environ Sci Technol 46(13):7239–7244
42. Baumer JD, Valério A, de Souza SMGU, Erzinger GS, Furigo Jr A, de Souza AAU (2018) Toxicity of enzymatically decolored textile dyes solution by horseradish peroxidase. J Hazard Mater 360:82–88
43. Zhou H, Yang D, Qiu X, Wu X, Li Y (2013) A novel and efficient polymerization of lignosulfonates by horseradish peroxidase/H2O2 incubation. Appl Microbiol Biotechnol 97 (24):10309–10320
44. Colosi LM, Pinto RA, Huang Q, Weber WJJ (2009) Peroxidase-mediated degradation of perfluorooctanoic acid. Environ Toxicol Chem Int J 28(2):264–271
45. Ling KQ, Li WS, Sayre LM (2008) Oxidations of N-(3-indoleethyl) cyclic aliphatic amines by horseradish peroxidase: the indole ring binds to the enzyme and mediates electron-transfer amine oxidation. J Am Chem Soc 130(3):933–944
46. Yang L, Shi Y, Li J, Fang L, Luan T (2018) Transformation of aqueous sulfonamides under horseradish peroxidase and characterization of sulfur dioxide extrusion products from sulfadiazine. Chemosphere 200:164–172
47. Xu R, Si Y, Li F, Zhang B (2015) Enzymatic removal of paracetamol from aqueous phase: horseradish peroxidase immobilized on nanofibrous membranes. Environ Sci Pollut Res 22 (5):3838–3846
48. Berry TD, Filley TR, Blanchette RA (2014) Oxidative enzymatic response of white-rot fungi to single-walled carbon nanotubes. Environ Pollut 193:197–204
49. Giardina P, Faraco V, Pezzella C, Piscitelli A, Vanhulle S, Sannia G (2010) Laccases: a never-ending story. Cell Mol Life Sci 67(3):369–385
50. Asgher M, Noreen S, Bilal M (2017) Enhancing catalytic functionality of Trametes versicolor IBL-04 laccase by immobilization on chitosan microspheres. Chem Eng Res Des 119:1–11
51. Asgher M, Noreen S, Bilal M (2017) Enhancement of catalytic, reusability, and long-term stability features of Trametes versicolor IBL-04 laccase immobilized on different polymers. Int J Biol Macromol 95:54–62
52. Moreno AD, Ibarra D, Eugenio ME, Tomás-Pejó E (2019) Laccases as versatile enzymes: from industrial uses to novel applications. J Chem Technol Biotechnol 95:481. https://doi.org/10.1002/jctb.6224
53. Unuofin JO, Okoh AI, Nwodo UU (2019) Aptitude of oxidative enzymes for treatment of wastewater pollutants: a laccase perspective. Molecules (Basel, Switzerland) 24(11):2064
54. Guardado ALP, Belleville MP, Alanis MDJR, Saldivar RP, Sanchez-Marcano J (2019) Effect of redox mediators in pharmaceuticals degradation by laccase: a comparative study. Process Biochem 78:123–131
55. Litwińska K, Bischoff F, Matthes F, Bode R, Rutten T, Kunze G (2019) Characterization of recombinant laccase from Trametes versicolor synthesized by Arxula adeninivorans and its application in the degradation of pharmaceuticals. AMB Express 9(1):102
56. Alharbi SK, Nghiem LD, Van De Merwe JP, Leusch FD, Asif MB, Hai FI, Price WE (2019) Degradation of diclofenac, trimethoprim, carbamazepine, and sulfamethoxazole by laccase from Trametes versicolor: transformation products and toxicity of treated effluent. Biocatal Biotransformation 37(6):399–408

57. Sutar RS, Rathod VK (2015) Ultrasound assisted laccase catalyzed degradation of ciprofloxacin hydrochloride. J Ind Eng Chem 31:276–282
58. Sutar RS, Rathod VK (2015) Ultrasound assisted enzyme catalyzed degradation of cetirizine dihydrochloride. Ultrason Sonochem 24:80–86
59. Suda T, Hata T, Kawai S, Okamura H, Nishida T (2012) Treatment of tetracycline antibiotics by laccase in the presence of 1-hydroxybenzotriazole. Bioresour Technol 103(1):498–501
60. Llorca M, Rodríguez-Mozaz S, Couillerot O, Panigoni K, de Gunzburg J, Bayer S et al (2015) Identification of new transformation products during enzymatic treatment of tetracycline and erythromycin antibiotics at laboratory scale by an on-line turbulent flow liquid-chromatography coupled to a high resolution mass spectrometer LTQ-Orbitrap. Chemosphere 119:90–98
61. Mugdha A, Usha M (2012) Enzymatic treatment of wastewater containing dyestuffs using different delivery systems. Sci Rev Chem Commun 2(1):31–40
62. Yang RL, Zhao XJ, Wu TT, Bilal M, Wang ZY, Luo HZ, Yang WJ (2019) A novel and highly regioselective biocatalytic approach to acetylation of helicid by using whole-cell biocatalysts in organic solvents. Catal Commun 128:105707
63. Asgher M, Shahid M, Kamal S, Iqbal HMN (2014) Recent trends and valorization of immobilization strategies and ligninolytic enzymes by industrial biotechnology. J Mol Catal B Enzym 101:56–66
64. Zdarta J, Meyer AS, Jesionowski T, Pinelo M (2018) Developments in support materials for immobilization of oxidoreductases: a comprehensive review. Adv Colloid Interf Sci 258:1–20
65. Sheldon RA, van Pelt S (2013) Enzyme immobilisation in biocatalysis: why, what and how. Chem Soc Rev 42(15):6223–6235
66. Ren S, Li C, Jiao X, Jia S, Jiang Y, Bilal M, Cui J (2019) Recent progress in multienzymes co-immobilization and multienzyme system applications. Chem Eng J 373:1254–1278
67. Bilal M, Asgher M (2015) Dye decolorization and detoxification potential of Ca-alginate beads immobilized manganese peroxidase. BMC Biotechnol 15(1):111
68. Touahar IE, Haroune L, Ba S, Bellenger JP, Cabana H (2014) Characterization of combined cross-linked enzyme aggregates from laccase, versatile peroxidase and glucose oxidase, and their utilization for the elimination of pharmaceuticals. Sci Total Environ 481:90–99
69. García-Zamora JL, León-Aguirre K, Quiroz-Morales R, Parra-Saldívar R, Gómez-Patiño MB, Arrieta-Baez D et al (2018) Chloroperoxidase-mediated halogenation of selected pharmaceutical micropollutants. Catalysts 8(1):32
70. Guo J, Liu X, Zhang X, Wu J, Chai C, Ma D et al (2019) Immobilized lignin peroxidase on Fe3O4@ SiO2@ polydopamine nanoparticles for degradation of organic pollutants. Int J Biol Macromol 138:433–440
71. Inoue S, Igarashi Y, Yoneda Y, Kawai S, Okamura H, Nishida T (2015) Elimination and detoxification of fungicide miconazole and antidepressant sertraline by manganese peroxidase-dependent lipid peroxidation system. Int Biodeterior Biodegradation 100:79–84
72. Forrez I, Carballa M, Verbeken K, Vanhaecke L, Ternes T, Boon N, Verstraete W (2010) Diclofenac oxidation by biogenic manganese oxides. Environ Sci Technol 44(9):3449–3454
73. Yang J, Lin Y, Yang X, Ng TB, Ye X, Lin J (2017) Degradation of tetracycline by immobilized laccase and the proposed transformation pathway. J Hazard Mater 322:525–531
74. Wen X, Zeng Z, Du C, Huang D, Zeng G, Xiao R et al (2019) Immobilized laccase on bentonite-derived mesoporous materials for removal of tetracycline. Chemosphere 222:865–871
75. Jahangiri E, Thomas I, Schulze A, Seiwert B, Cabana H, Schlosser D (2018) Characterisation of electron beam irradiation-immobilised laccase for application in wastewater treatment. Sci Total Environ 624:309–322
76. Ba S, Haroune L, Soumano L, Bellenger JP, Jones JP, Cabana H (2018) A hybrid bioreactor based on insolubilized tyrosinase and laccase catalysis and microfiltration membrane remove pharmaceuticals from wastewater. Chemosphere 201:749–755

77. Taheran M, Naghdi M, Brar SK, Knystautas EJ, Verma M, Surampalli RY (2017) Degradation of chlortetracycline using immobilized laccase on Polyacrylonitrile-biochar composite nanofibrous membrane. Sci Total Environ 605:315–321
78. Naghdi M, Taheran M, Brar SK, Kermanshahi-pour A, Verma M, Surampalli RY (2017) Immobilized laccase on oxygen functionalized nanobiochars through mineral acids treatment for removal of carbamazepine. Sci Total Environ 584:393–401
79. Ji C, Hou J, Wang K, Zhang Y, Chen V (2016) Biocatalytic degradation of carbamazepine with immobilized laccase-mediator membrane hybrid reactor. J Membr Sci 502:11–20
80. Kumar VV, Cabana H (2016) Towards high potential magnetic biocatalysts for on-demand elimination of pharmaceuticals. Bioresour Technol 200:81–89
81. Nguyen LN, Hai FI, Price WE, Leusch FD, Roddick F, McAdam EJ et al (2014) Continuous biotransformation of bisphenol A and diclofenac by laccase in an enzymatic membrane reactor. Int Biodeterior Biodegradation 95:25–32
82. Wang Q, Cui J, Li G, Zhang J, Huang F, Wei Q (2014) Laccase immobilization by chelated metal ion coordination chemistry. Polymers 6(9):2357–2370

Fungal Reactors: A Solution for the Removal of Pharmaceuticals in Urban and Hospital Wastewater

Josep Anton Mir-Tutusaus and Montserrat Sarrà

Contents

1 Introduction 146
2 White-Rot Fungi 146
2.1 Enzymatic System of the Fungi 147
2.2 Bio-Oxidation/Redox-Mediated System 149
3 Type of Reactors with Fungi 150
3.1 Retained Fungi 150
3.2 Immobilized Fungi 151
4 Treatments of UWW and HWW with Fungi 152
4.1 Batch System 152
4.2 Continuous Treatment 154
5 Conclusions 158
References 158

Abstract The pharmaceutical occurrence in surface and ground water bodies is due to low efficiency of the wastewater treatment plants in removing those pollutants. Their biological transformation using white-rot fungi has been proposed due to their unspecific intracellular and extracellular oxidoreductase enzymatic systems. This chapter summarizes and analyzes the studies performed on pharmaceuticals removal from urban and hospital wastewater using fungal reactors operating in batch or continuous mode. Due to low fungal growth rate, all reactors are based on the biomass reuse through the retention through the pellet morphology or membrane, or by immobilization on a support. The treatment of real wastewater in non-sterile conditions requires the assessment of effect of the native microorganism on the fungal treatment. The chapter also offers an insight into fungal enzymatic systems,

---

J. A. Mir-Tutusaus and M. Sarrà (✉)
Departament d'Enginyeria Química Biològica i Ambiental, Escola d'Enginyeria, Universitat Autònoma de Barcelona, Barcelona, Spain
e-mail: jantonz@gmail.com; Montserrat.Sarra@uab.cat

Sara Rodriguez-Mozaz, Paqui Blánquez Cano, and Montserrat Sarrà Adroguer (eds.), *Removal and Degradation of Pharmaceutically Active Compounds in Wastewater Treatment*, Hdb Env Chem (2021) 108: 145–162, DOI 10.1007/698_2020_660,

types of reactors, and strategies to reduce bacteria effect and consequently maintaining fungal activity during a continuous treatment during long periods.

Keywords Biomass retention, Immobilization, Non-sterile conditions, Real wastewater

1 Introduction

Current wastewater treatment plants are not designed to remove pharmaceuticals; as a consequence, their effluents are one of the main sources of this pollution in surface and ground water bodies [1]. Although the concentration of those compounds is usually low, ranging from ng/L to μg/L, there is an increasing concern due their high activity even at a very low concentration. Among the many alternatives studied to remove emerging pollutants, biological treatments are preferred because they are regarded as a more environmentally friendly process. Nevertheless, not all biological systems have evidenced the same efficiency in removing pharmaceuticals compounds from wastewater. Among these eco-friendly treatments, the use of white-rot fungi (WRF) is an interesting option because these fungi have demonstrated the capacity to degrade a wide range of recalcitrant organic compounds, including emerging pollutants, due to their unspecific enzymatic system. Despite their high potential, an application at full scale does not exist. One of the drawbacks of the technology is the difficulty in maintaining the fungal activity for a long period of time since bacteria exert competitive pressure in fungal survival. So, the reactor design and the operational strategies are key aspects to consider in achieving a scale-up process to remove emerging pollutants using WRF.

This chapter shows some of the latest research efforts at improving fungal bioreactors for removing pharmaceuticals from urban and hospital wastewater. The first section presents some general aspects of the WRF and their enzymatic systems, focusing on pollutant biodegradation. The second section is devoted to describing the type of reactors using fungi as retained biomass or as immobilized biomass. Finally, the third section reviews treatments of urban and hospital wastewater with fungi while focusing on the removal of pharmaceutically active compounds.

2 White-Rot Fungi

It is not uncommon in nature to find among fungal species the ability to break down cellulose and hemicellulose, polysaccharides that together with lignin form the main components of lignocellulose. This process is part of the primary metabolism and as such, it yields a positive energy gain. However, the ability to depolymerize the third component, lignin, is much more scarce [2]. Lignin is a complex and heterogeneous polyphenolic polymer, the degradation of which does not result in a net energy gain; rather, it is a step used by few organisms to access polysaccharides inside lignin-carbohydrate complexes and exploit this ecological niche. White-rot fungi (WRF) are a group of fungal species that are able to do just that. In nature they achieve depolymerization of lignin through a mixture of extracellular ligninolytic enzymes and mediators and intracellular monooxygenases. This enzymatic system happens to be very unspecific and has been studied for transforming a wide range of micropollutants [3, 4].

WRF owe their name, *white-rot*, to the whitish aspect that a wood takes after being attacked by these fungi. Most known WRF are basidiomycetes, but also certain ascomycetes are able to produce white rot. Some relevant fungal species that have been studied for their ability to remove pharmaceuticals include *Ganoderma* spp., *Phanerochaete chrysosporium*, *Pleurotus ostreatus*, and *Trametes versicolor* [5].

2.1 Enzymatic System of the Fungi

The enzymatic system of fungi is very complex and diverse. In regard to xenobiotic transformation, it can be classified into extracellular and intracellular. However, metabolites generated by either method may be catabolized extra- and intracellularly, metabolites produced by intracellular enzymes may be excreted, and metabolites produced by extracellular enzymes may be internalized. Additionally, transformation products (TPs) may form conjugates, be further transformed, or result in mineralization. TPs may be excreted at different oxidation stages or be transformed by fungal, non-fungal, or abiotic oxidation (Fig. 1).

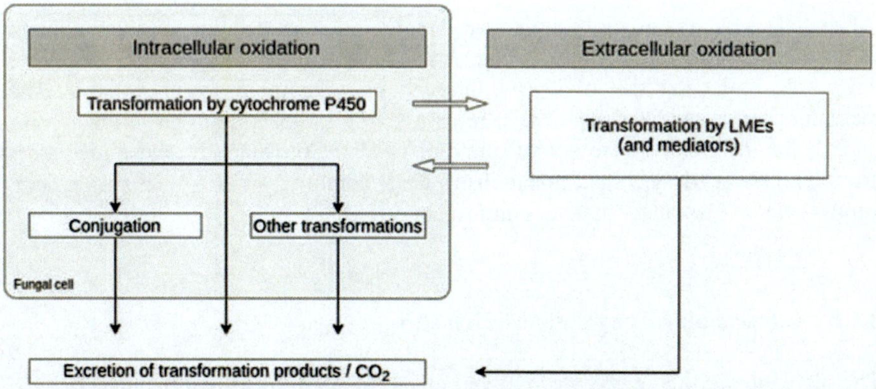

Fig. 1 Simplified schema of micropollutant transformation pathways in white-rot fungi

2.1.1 Extracellular Enzymatic Mechanism

The extracellular enzymatic mechanism of WRF is composed mainly of oxidoreductases called lignin-modifying enzymes (LMEs) and mediators, all of which are mainly active in the acidic pH range. Several LMEs have been described in the literature, four of which are described below; however, they are not the only enzymes involved in lignin transformation [6]:

- Lignin peroxidase (LiP). Monomeric glycosylated hemoproteins with relatively low specificity, although their preferred substrates are non-phenolic compounds [7]. Like most LMEs, LiP is involved in lignin transformation, but it is not essential, as some fungi capable of lignin transformation do not produce this enzyme. Their reaction mechanism is H_2O_2-dependent one-electron oxidation and is usually mediated by redox mediator veratryl alcohol [8, 9].
- Manganese peroxidase (MnP). Heme glycosylated enzyme, its reaction mechanism is H_2O_2-dependent, via one-electron oxidation of Mn^{2+} to Mn^{3+}, which subsequently oxidates the organic substrate [9, 10]. Mn^{3+} can diffuse into the lignified cell wall and attack it from inside, sometimes with the help of some organic acids such as oxalic acid that chelates Mn^{3+} and acts as redox mediator.
- Laccase (Lac). Laccases are multicopper enzymes and are produced by several wood-transforming basidiomycetes and ascomycetes. Their active site contains four copper atoms, and their reaction mechanism involves the reduction of oxygen to water [11, 12]. Small organic compounds, known as mediators, can be oxidized by laccases, and these molecules may further attack other substrates, thus widening the range of compounds laccase can transform.
- Versatile peroxidase (VP). These enzymes, first described in *Pleurotus* and *Bjerkandera* species, are peroxidases with MnP-LiP hybrid properties, capable of transforming typical substrates of both previous peroxidases (Mn^{2+} and veratryl alcohol, respectively) [7, 13]. Thanks to their distinct multiple binding sites, they can also transform aromatic compounds not efficiently attacked by MnP and LiP [14].

Mediators are small, low-molecular-weight redox molecules that are able to diffuse through the lignocellulosic complex. They are especially important because due to the molecular size of LMEs, interactions with lignin are improbable. The mediators, once oxidized by LMEs, remain stable radicals that can oxidize compounds that originally were not substrate of LMEs. Conveniently, these mediators are often produced by fungal metabolism itself, notably 3-hydroxyanthranilic acid, veratryl alcohol, oxalate, malate, and fumarate [15].

2.1.2 Intracellular Enzymatic Mechanism

The main intracellular enzymes involved in xenobiotic transformation are related to the cytochrome P450 superfamily. They are intracellular monooxygenases typically found in mammals that are involved in the metabolism of a wide range of compounds. The reaction mechanism involves the reduction of O_2 to H_2O, while oxidizing the substrate. Cytochrome P450 can, therefore, transform aromatic or aliphatic structures of many molecules, including organic pollutants.

In fungus, cytochrome P450 has been shown to play a role in the metabolism of anti-inflammatory drugs, lipid regulators and analgesics, among other compounds [16–19].

2.2 *Bio-Oxidation/Redox-Mediated System*

In addition to their extracellular oxidation potential, white-rot fungi show the ability to reduce lignin depolymerization products, such as aromatic aldehydes, acids, and quinones, through different intracellular and membrane-bound systems. So, the simultaneous actions of both systems, extracellular oxidative enzymes and intracellular reduction mechanism, lead to the establishment of redox cycles. One of the functions of these cycles is the production of reactive oxygen species (ROS), i.e., superoxide anion radicals ($O_2\cdot^-$), H_2O_2 and hydroxyl radicals ($\cdot OH$) depending on the fungus, the depolymerization product, and the oxidizing enzyme [20]. Among the ROS produced by fungi, hydroxyl radical has the highest redox potential, and it can be chemically generated by the Fenton reaction ($H_2O_2 + Fe^{2+} \rightarrow OH^- + \cdot OH + Fe^{3+}$). But the biological generation of the hydroxyl radical involves lignin-oxidizing enzymes to produce H_2O_2, fungal mycelia, and its metabolic products as metal chelating agents. Therefore, the use of mediators in lignin and/or environmental pollutants degradation by ligninolytic enzymes performs similarly to the advanced oxidation process in the generation of highly reactive radicals. The process applied to evidence the degradation of several selected pharmaceuticals through the quinone redox cycling mechanism can be achieved from molecular oxygen, incubating fungi and chelated ferric ion. The advanced bio-oxidation process can be achieved by incubating the fungi with a lignin-derived quinone (2,6-dimethoxy-1,4-benzoquinone, DBQ) and chelated ferric ion (Fe^{3+}-oxalate) [21]. Under these conditions, DBQ are converted to hydroquinone ($DBQH_2$) by an intracellular quinone reductase; subsequently, extracellular lignin-modifying enzyme (laccase or peroxidase) oxidizes $DBQH_2$ to semiquinone radicals ($DBQ\cdot^-$). Then the Fenton's reagent is formed by these radical's autoxidation catalyzed by Fe^{3+} ($DBQ\cdot^- + Fe^{3+}$-oxalate $\rightarrow DBQ + Fe^{2+}$-oxalate and Fe^{2+}-oxalate $+ O_2 \leftrightarrow Fe^{3+}$-oxalate $+ O_2\cdot^-$). Finally, the $O_2\cdot^-$ dismutation produces H_2O_2 precursor of hydroxyl radicals. This process was first evidenced by Gómez-Toribio et al. [20], who established the conditions to enhance the radical production.

The unspecificity, the short reaction time required, and the high oxidation potential of $^\bullet OH$ radicals are the main advantages of the process in removing recalcitrant organic pollutants. Despite these advantages, reports of effluent treatment are scarce, even at lab scale, due to quinone depletion probably due to radical attack. Recent studies are focused in using gallic acid as a mediator because it is also a transformation product of the lignin and it can generate semiquinone compounds through the fungal redox cycle [22].

3 Type of Reactors with Fungi

3.1 Retained Fungi

Several strategies exist to retain fungal biomass inside the bioreactor. A prominent one is the auto-immobilization of fungal mass in the form of pellets. This immobilization allows for decoupling the solids retention time and the hydraulic retention time (i.e., better separation of fungal biomass and liquid matrix). In addition to retain the fungus inside the reactor, other advantages can be observed for pelleted biomass versus free hyphae: improved settling, possibility of biomass reuse, low clogging effect, high volumetric productivity, and lower adhesion to parts of the reactor [23–25]. The fact that pellet formation is often separated from the treatment stage is also an advantage. Different approaches for pellet growth can be found in the literature [26–28]. In general, it can be acknowledged that pellet diameter can be controlled by shear stress, often controlled by agitation/aeration speed, and inoculum concentration [29, 30]. However, pelleted biomass comes with some limitations, such as lower nutrient and oxygen transfer inside the pellet, which can result in autolysis and difficulty of maintaining fungal pellets during long-term operations [23, 29, 31].

Some representative reactors with pelleted biomass comprise stirred tanks, fluidized beds, and airlift/bubble columns:

- Stirred tank reactor (STR). It consists of a tank with mechanical agitation that provides both agitation and a mechanism to disperse air, which is usually supplied at the bottom of the reactor. Agitation speed is a main variable in these systems, which controls the diameter of the pellets, the stress of the cells, and the mass and oxygen transfer. Increased shear stress and agitation may lead to the rupture of pellets [32].
- Airlift reactor and bubble column reactor. The airlift reactor incorporates a cylindrical baffle that separates the liquid moving down from the liquid rising. This baffle provides an engineered path that air bubbles follow, therefore ensuring proper aeration and mixing by recirculation. It does not need any mechanical agitation. The bubble diameter and the flow rate of aeration control the diameter and fluffiness of the pellets, the mass, and oxygen transfer. These systems produce less shear stress than their mechanically stirred counterparts and require less energy. Bubble column reactors similarly rely on aeration for oxygen transfer and mixing purposes, but they do not use any baffle; instead, a gas distributor on the bottom ensures (non-uniform) agitation [33].
- Fluidized bed reactor (FBR). The fungal pellets are suspended in the stream of air and influent, both of which flow from the bottom of the reactor. The mixing does not require mechanical agitation and similarly to airflit reactors; the airflow controls pellet diameter and mass/oxygen transfer. Poor fluidization is a problem in fluidized bed reactors using pelleted biomass due to pellet aggregation; partial biomass purging helps in reducing this difficulty. Some improvements in the system involve (1) fluidization by air pulses, which improves fluidization quality, avoids conglomerates of mycelia, and maintains more uniform pellets [29, 34],

and (2) reactor head with a diameter larger than the reactor column, which aids in reducing foam formation, facilitates gas-liquid phase separation, and facilitates pellet formation by allowing pellets to spin before descending through the column [35].

The use of membrane bioreactors (MBRs) constitutes another important retention strategy. MBRs have long been used both in urban and industrial wastewater treatment [36–38], and there has been an increased interest in micropollutant removal on these systems [39]. Fungal MBRs tend to take the form of a bubble reactor coupled with a membrane stage and rely on air flow for mixing [40]. The membrane enables the retention of fungal biomass while allowing the flow of medium, thus allowing for shorter hydraulic retention times than other fungal reactors. Additionally, depending on the membrane type and its pore size, the retention of enzymes and other microorganisms is also possible [3, 41, 42]. Membrane fouling and durability of the membrane are a concern in membrane reactors (especially in fungal MBRs), which require more maintenance and monitoring [43, 44].

3.2 *Immobilized Fungi*

A lot of diversity can be found in the literature about fungal reactors with immobilized biomass. Some diversity comes from the fact that fungi can be immobilized onto a wide variety of carriers, both inert, stainless steel mesh, polyamide fiber net, fiberglass net, polyurethane foam, polyurethane cubes, foam glass beads, etc. [45, 46], and non-inert, some of which can also be partially used as a carbon source: pine chips, pallet wood, pine bark, hazelnut shell, nutshell, etc. [47, 48].

- Trickling packed bed reactors. Some particularly successful reactors using immobilized fungi are trickling packed-bed reactors. They consist of a packed bed full of packing material that may be already colonized by the fungus or rely on the reactor operation to grow the fungus onto the material. The air is usually pumped from the bottom of the reactor and the influent is fed from the top. Some setups also include a recycling loop to recirculate the effluent [47–49].
- Packed bed reactor. Another alternative is a packed bed reactor. The operation of such reactors may not require aeration (oxygen is supplied dissolved in the influent), and they tend to be operated with an upflow of medium [50, 51].
- Moving packed bed reactor. A moving packed bed reactor is another option for WRF, in which the packing material with the immobilized fungus is placed inside a holed cylinder cage rotating inside the reactor container. The air is pumped from the bottom of the reactor with an air diffuser [45].
- Rotating biological contactors (RBCs). These reactors consist of spaced disks mounted on a rotating shaft, which is half submerged on an open container that contains the influent. The reactor is first operated with growth medium until a

biofilm of the target fungus is grown onto the disks [52]. A *rotating suspension cartridge reactor* is a variation of the RBC that use holed cartridges filled with the fungus immobilized onto packing material [53]. Another reactor inspired by the RBC is the rotating drum reactor, which consists of an inner tube, filled with the immobilized fungus, located in the center of the longitudinal axis of an open polyvinylchloride channel [54]. These types of reactor do not need aeration because the rotating shaft ensures that biomass is intermittently submerged in the liquid and in contact with air.

4 Treatments of UWW and HWW with Fungi

Some prominent wastewater treatments are reviewed in this section. It is significant to note that at the time of writing only a few studies can be found that use real wastewater. And of the ones that do treat real wastewater, only some of them do so mimicking real-world variables: environmentally significant micropollutant concentration and non-sterile conditions. In addition, most studies are run at laboratory scale, being few the experiments found at pilot-scale plants and none, to the best of the authors' knowledge, at full-scale WWTPs.

4.1 Batch System

Most of the white-rot fungal bioreactor batch treatments reviewed in this section used fluidized bed bioreactors. Those studies can be now regarded as a stepping stone needed to demonstrate the feasibility of white-rot fungal reactor treatments of real wastewater, and put some light into the effect of sterility in the removal of pharmaceutical active compounds from that matrix (Table 1).

Cruz-Morató et al. [56] demonstrated for the first time the degradation of pharmaceuticals at pre-existent concentrations in a WRF bioreactor treating real urban wastewater under non-sterile conditions. In subsequent experiments, they demonstrated the possibility of removing a mixture of pharmaceuticals and endocrine disruptor compounds in non-sterile hospital wastewater (HWW), with presumably a higher micropollutant initial concentration [58]. They performed parallel experiments with heat-sterilized and non-sterile hospital wastewater, and both operations reduced the concentration and estrogenic activity of the influent. The non-sterile treatment, however, performed substantially worse with a 53.3% removal of the initial micropollutant load while the sterile operation showed a 83.2% removal. *T. versicolor*, the WRF used in the study, played a major role in the pollutant degradation.

A 10 L FBR was used to remove spiked micropollutants from non-sterile HWW. The setup removed 87% of X-ray contrast agent iopromide and 98.5% antibiotic ofloxacin in sterile HWW, while similar results were obtained for the non-sterile

Table 1 Batch treatments based on white-rot fungi

Source	Reactor	Biomass morphology	Matrix	Spiked Matrix	Sterility	Duration of the treatment (day)	Pharmaceuticals
Gros et al. [55]	Fluidized bed reactor	Pellets	Hospital wastewater	No	No	8	Iopromide; ofloxacin
Gros et al. [55]	Fluidized bed reactor	Pellets	Hospital wastewater	No	Yes	8	Iopromide; ofloxacin
Cruz-Morató et al. [56]	Fluidized bed reactor	Pellets	Urban wastewater	No	No	8	Naproxen; ibuprofen; acetaminophen; salicylic acid; ketoprofen; codeine; erythromycin; metronidazole; ciprofloxacin; azithromycin; cephalexin; propranolol; carbamazepine; 10,11-epoxyCBZ; 2-hydroxyCBZ; acridone; citalopram
Cruz-Morató et al. [56]	Fluidized bed reactor	Pellets	Urban wastewater	No	Yes	8	Naproxen; ibuprofen; acetaminophen; salicylic acid; ketoprofen; codeine; erythromycin; metronidazole; ciprofloxacin; azithromycin; cephalexin; propranolol; carbamazepine; 10,11-epoxyCBZ; 2-hydroxyCBZ; acridone; citalopram
Badia-Fabregat et al. [57]	Fluidized bed reactor	Pellets	Veterinary hospital wastewater	No	No	15	Furosemide; salicylic acid; ketoprofen; ibuprofen; piroxicam; diclofenac; naproxen; indomethacin; metronidazole; ciprofloxacin; diazepam; fluvastatin; gemfibrozil; atorvastatin; clopidogrel; albendazole; ranitidine
Cruz-Morató et al. [58]	Fluidized bed reactor	Pellets	Hospital wastewater	No	No	8	51 PhACs
Cruz-Morató et al. [58]	Fluidized bed reactor	Pellets	Hospital wastewater	No	Yes	8	49 PhACs
Ferrando-Climent et al. [59]	Fluidized bed reactor	Pellets	Hospital wastewater	No	No	8	Ciprofloxacin; tamoxifen; ifosfamide; cyclophosphamide; vincristine; docetaxel; paclitaxel; etoposide; methotrexate; azathioprine
Ferrando-Climent et al. [59]	Fluidized bed reactor	Pellets	Hospital wastewater	No	Yes	8	Ciprofloxacin; tamoxifen; ifosfamide; cyclophosphamide; vincristine; docetaxel; paclitaxel; etoposide; methotrexate; azathioprine

operation: 65.4% and 99%, respectively [55]. A similar pilot-scale reactor was used to test the ability of *T. versicolor* to degrade recalcitrant anticancer drugs from HWW, both raw and sterilized [59]. This paper highlighted the interesting case of ciprofloxacin, which was more efficiently eliminated under non-sterile conditions. This fact was attributed to a synergistic contribution between fungi and wastewater-native bacteria, although no further discussion was offered.

That "synergistic contribution" was evaluated in a different study using DGGE for microbial community assessment [57]. The experiment achieved a longer batch operation, 15 days, using *T. versicolor* in a 1.5 L lab-scale fluidized bed bioreactor treating veterinary hospital wastewater. Under non-sterile conditions, the treatment showed a 66% removal of the micropollutants present at pre-existent concentrations, while no removal was observed in the uninoculated control reactor. It was demonstrated that the presence of WRF was mandatory for degradation to occur, and they hypothesized that the synergistic effect was due to the native wastewater microorganisms, especially bacteria, degrading fungal transformation products and metabolites. However, no direct relationship between detected non-WRF microorganisms and degraded pollutants could be established. Finally, the study also highlighted the necessity of adjusting nutrient addition in non-sterile treatments taking into account the nutrient consumption of native microorganisms.

4.2 *Continuous Treatment*

It is known that a continuous wastewater treatment is preferred over batch treatment unless the volume treated is very low. As a result, more continuous white-rot fungal treatments can be found in the literature, although one could argue that they could not have existed without prior reactor batch studies, seeing that all of them are more recent (Table 2).

All operations reviewed in this section were conducted in non-sterile conditions and can be classified by the reactors they used: fluidized bed reactors (FBR), rotating biological contactors (RBC), and trickling bed reactors.

The *fluidized bed reactor* operations reviewed in this section all used pelleted biomass. As discussed in Sect. 3, the use of pellets enables the retention of fungal biomass while allowing the flow of liquid. However, attention must be paid to oxygen and mass transfer inside the pellets. The need for optimizing the nutrients addition in non-sterile continuous treatments, both carbon-to-nitrogen ratio and quantity to reduce the cost of operation and the COD of the effluent, was highlighted by Badia-Fabregat et al. [57]. It also identified some WRF-bacteria and WRF-other fungi interactions and described the inactivation of the fungus *T. versicolor*, which could be confirmed with DGGE. The operation lasted 26 days and removed a 44% of overall pharmaceutical load, but the fungus was inactivated during some periods of time. An experiment with a similar FBR, yet with spiked anti-inflammatories (at a much higher concentration than the concentration found in the environment), included the pre-treatment of influent through UV light, coagulation-flocculation

Table 2 Continuous treatments based on white-rot fungi

Source	Reactor	Biomass morphology	Matrix	Spiked matrix	Sterility	Duration of the treatment (day)	HRT	Pharmaceuticals
Kresinová et al. [60]	Trickling bed reactor	Grown in straw	Effluent from secondary treatment	No	No	28	46 h–8 h	Bisphenol A; estrone (E1); 17-α-ethynylestradiol (EE2); triclosan; 4-n-nonylphenol
Zhang and Geißen [61]	Plate bioreactor	Immobilization in polyether foam	Effluent from WWTP	Yes	No	100	36 h	Carbamazepine
Badia-Fabregat et al. [57]	Fluidized bed reactor	Pellets	Veterinary hospital wastewater	No	No	26	3.3 d	Furosemide; salicylic acid; ketoprofen; ibuprofen; piroxicam; diclofenac; naproxen; indomethacin; metronidazole; ciprofloxacin; diazepam; fluvastatin; gemfibrozil; atorvastatin; clopidogrel; albendazole; ranitidine
Mir-Tutusaus et al. [62]	Fluidized bed reactor	Pellets	Hospital wastewater (flocculated)	Yes	No	28	3 d	Ibuprofen; ketoprofen
Mir-Tutusaus et al. [63]	Fluidized bed reactor	Pellets	Hospital wastewater (flocculated)	No	No	56	3 d	Acetaminophen; diclofenac; ibuprofen; ketoprofen; thiabendazole; ciprofloxacin; ronidazole; sulfamethoxazole; trimethoprim; ofloxacin; warfarin; valsartan; atenolol; sotalol; furosemide; hydrochlorothiazide; ranitidine ; loratadine; atorvastatin; gemfibrozil; 10.11-epoxycarbamazepine; 2-hydroxycarbamazepine; acridone; carbamazepine; citalopram; diazepam; norfluoxetine; olanzapine; sertraline; trazodone; venlafaxine; dexamethasone

(continued)

Table 2 (continued)

Source	Reactor	Biomass morphology	Matrix	Spiked matrix	Sterility	Duration of the treatment (day)	HRT	Pharmaceuticals
Cruz del Álamo et al. [52]	Rotating biological contactor	Immobilized on rotating biological contactors	Effluent from primary treatment	No	No	22	24 h	Amoxicillin; metronidazole; sulfamethoxazole; carbamazepine; 4-acetamidoantipyrine; gemfibrozil; hydrochlorothiazide; iohexol
Torán et al. [48]	Fluidized bed reactor	Pellets with wood core	Hospital wastewater (flocculated)	Yes	No	28	3 d	Ibuprofen; ketoprofen; naproxen
Torán et al. [48]	Trickling packed-bed reactor	Immobilized on pallet wood	Hospital wastewater (flocculated)	Yes	No	49	3 d	Ibuprofen; ketoprofen; naproxen
Mir-Tutusaus et al. [29]	Fluidized bed reactor	Pellets	Hospital wastewater (flocculated)	Yes	No	21	3 d	Ibuprofen; ketoprofen; naproxen
Mir-Tutusaus et al. [64]	Fluidized bed reactor	Pellets	Hospital wastewater (flocculated)	No	No	91	3 d	74 pharmaceutically active compounds.
Cruz del Álamo et al. [22]	Rotating biological contactor	Immobilized on rotating biological contactors	Effluent from primary treatment of a pilot WWTP	Yes	No	40	1 d	Antipyrine; clofibric acid; atenolol; caffeine; carbamazepine; diclofenac; gemfibrozil; hydrochlorothiazide; ibuprofen; ranitidine; sulfamethoxazole; sulpiride

and both strategies, and the operation of the same reactor in sequential batch and continuous mode [62]. The study's objective was to decrease the influent microbial concentration in an attempt to increase the operation time. Specifically, the main objective was to increase the time the white-rot fungus was active inside the reactor. The best-performing strategy was the coagulation-flocculation pre-treatment step coupled to a continuous fungal treatment, which led to an operation that lasted 28 days with an average removal of 80% of the spiked concentration. A similar study with spiked anti-inflammatories was conducted by Torán et al. [48] using fungal pellets with a wooden core, so as to avoid the addition of an external carbon and nitrogen source. The operation ensured fungal viability up to day 21, when removal performance degraded.

A 2-L FBR inoculated with pelleted *T. versicolor* was operated for 56 days treating flocculated HWW at environmentally relevant concentration of pharmaceutically active compounds [63]. A DGGE and sequencing approach was used to evaluate the microbial community arisen in inoculated and uninoculated reactors, and it revealed a possible inhibition of *T. versicolor* activity due to *Candida*. Further studies with similar reactors but spiked pharmaceuticals identified an optimal carbon-to-nitrogen ratio of 7.5 in regard to nutrient addition (i.e., glucose and ammonium nitrate) and found that an aeration of 0.8 L·min^{-1} and a pellet diameter of around 2 mm helped in extending the treatment length and fungal activity [29]. A long-term continuous treatment of hospital wastewater was performed using the nutrient addition, aeration, and pellet diameter values found previously [64]. The study removed a 70% of influent pharmaceutical concentration for 91 days and demonstrated the presence during the whole treatment of *T. versicolor* by means of qPCR.

Del Álamo et al. [22] operated a 10 L *rotating biological contactor* with real wastewater (collected after the primary treatment of a pilot scale WWTP) for 40 days. The white-rot fungus *T. versicolor* was grown in the form of biofilm into the disks in the start-up period prior to the wastewater treatment. During the treatment phase, the operation achieved a 40–95% removal of 12 micropollutants spiked at 50 μg·L^{-1} each without the need of addition of nutrients (which can be regarded as a mandatory requirement in FBRs). Additionally, RBCs do not typically need an external aeration source, as the aeration is attained through rotation of the reactor itself. However, the study does report the addition of promoters of advanced bio-oxidation: metallic species in the form of Fe^{3+}-oxalate and Mn^{2+}-nitrate and quinone-like compounds such as gallic acid. A similar experiment was conducted with UWW at environmentally relevant micropollutant concentration, without spiking of pharmaceutical compounds [52]. The operation also used promoters of bio-oxidation and lasted 22 days when using the primary effluent of a full-scale WWTP. A 2 L *plate bioreactor* was used for the treatment of the effluent of urban WWTP with spiked carbamazepine [61]. The WRF *P. chrysosporium* was immobilized in a sheet of polyether foam submerged in a recipient with the WWTP effluent, placed in a water bath at 34–37°C. In this case no bio-oxidation promoters were added, but the addition of carbon and nitrogen sources was needed to maintain the removal of the spiked carbamazepine around 60%.

The *trickle bed reactors* reviewed in this section used WRF immobilized or grown on lignocellulosic substrates. The first experiment reviewed is a pilot-scale study that used a reactor of 30 L working volume and treated the secondary effluent of a WWTP [60]. The operation used the fungus *Pleurotus ostreatus* (an edible industrial mushroom that can be cultivated in farms), and it decreased by 76% of the concentration of the endocrine-disrupting compounds (EDCs) analyzed. The operation lasted 28 days, and it demonstrated the possibility of a fungal-based tertiary treatment for the degradation of EDCs. Another study evaluated the ability of *T. versicolor* immobilized in pallet wood of removing wastewater with spiked pharmaceutical compounds in a lab-scale trickle packed-bed reactor [48]. This study was not intended as a tertiary treatment but as an onsite treatment for hospital wastewater in order to reduce pharmaceutical load. The experiment successfully removed pharmaceutical compounds for 49 days in non-sterile flocculated wastewater without the addition of carbon or nitrogen source (as it is needed in FBRs).

5 Conclusions

Biotechnology processes based on fungal systems at industrial scale have been developed along the history, for example, antibiotic production. But, regarding wastewater treatment, although the high degrading capacity of the fungi, particularly to antibiotics and pharmaceutical compounds, treatment processes are still scarce even at lab scale. In general, it seems that bacterial concentration in raw wastewater hindered the robust operation of the batch and continuous treatments reviewed. Nevertheless, the reduction of this microbiological load, either using a pre-treatment (e.g., coagulation-flocculation) or by feeding the reactor a more processed influent (e.g., effluent from primary or secondary treatment), helped maintaining the fungus active for a longer period. The use of WRF immobilized onto lignocellulosic substrates avoided the need of nutrients addition, although systems with this type of biomass tend to need longer start-up periods for biomass growth. So, the technology is mature enough for different pilot-scale reactor setups to flourish soon, and such studies should be encouraged.

References

1. Verlicchi P, Al Aukidy M, Zambello E (2012) Occurrence of pharmaceutical compounds in urban wastewater: removal, mass load and environmental risk after a secondary treatment – a review. Sci Total Environ 429:123–155. https://doi.org/10.1016/j.scitotenv.2012.04.028
2. Cragg SM, Beckham GT, Bruce NC, Bugg TDH, Distel DL, Dupree P, Etxabe AG, Goodell BS, Jellison J, McGeehan JE, McQueen-Mason SJ, Schnorr K, Walton PH, Watts JEM, Zimmer M (2015) Lignocellulose degradation mechanisms across the tree of life. Curr Opin Chem Biol. https://doi.org/10.1016/j.cbpa.2015.10.018

3. Nguyen LN, Hai FI, Yang S, Kang J, Leusch FDL, Roddick F, Price WE, Nghiem LD (2013) Removal of trace organic contaminants by an MBR comprising a mixed culture of bacteria and white-rot fungi. Bioresour Technol 148:234–241. https://doi.org/10.1016/j.biortech.2013.08.142
4. Tuomela M, Hatakka A (2019) 6.17 – oxidative fungal enzymes for bioremediation. In: Moo-Young M (ed) Comprehensive biotechnology3rd edn. Pergamon, Oxford, pp 224–239. https://doi.org/10.1016/B978-0-444-64046-8.00349-9
5. Schwarze FWMR (2007) Wood decay under the microscope. Fungal Biol Rev. https://doi.org/10.1016/j.fbr.2007.09.001
6. Pointing SB (2001) Feasibility of bioremediation by white-rot fungi. Appl Microbiol Biotechnol 57:20–33. https://doi.org/10.1007/s002530100745
7. Martínez AT (2002) Molecular biology and structure-function of lignin-degrading heme peroxidases. Enzym Microb Technol. https://doi.org/10.1016/S0141-0229(01)00521-X
8. Hammel KE (1995) Mechanisms for polycyclic aromatic hydrocarbon degradation by ligninolytic fungi. Environ Health Perspect 105:41–43. https://doi.org/10.2307/3432477
9. Hofrichter M, Ullrich R, Pecyna MJ, Liers C, Lundell T (2010) New and classic families of secreted fungal heme peroxidases. Appl Microbiol Biotechnol. https://doi.org/10.1007/s00253-010-2633-0
10. Hofrichter M (2002) Review: lignin conversion by manganese peroxidase (MnP). Enzym Microb Technol 30:454–466. https://doi.org/10.1016/S0141-0229(01)00528-2
11. Gianfreda L, Xu F, Bollag J-M (1999) Laccases: a useful group of oxidoreductive enzymes. Biorem J 3:1–26. https://doi.org/10.1080/10889869991219163
12. Tuor U, Winterhalter K, Fiechter A (1995) Enzymes of white-rot fungi involved in lignin degradation and ecological determinants for wood decay. J Biotechnol. https://doi.org/10.1016/0168-1656(95)00042-O
13. Camarero S, Sarkar S, Ruiz-Dueñas FJ, Martínez MJ, Martínez ÁT (1999) Description of a versatile peroxidase involved in the natural degradation of lignin that has both manganese peroxidase and lignin peroxidase substrate interaction sites. J Biol Chem 274:10324–10330. https://doi.org/10.1074/jbc.274.15.10324
14. Carabajal M, Kellner H, Levin L, Jehmlich N, Hofrichter M, Ullrich R (2013) The secretome of Trametes versicolor grown on tomato juice medium and purification of the secreted oxidoreductases including a versatile peroxidase. J Biotechnol 168:15–23. https://doi.org/10.1016/j.jbiotec.2013.08.007
15. Leonowicz A, Matuszewska A, Luterek J, Ziegenhagen D, Wojtaś-Wasilewska M, Cho N-S, Hofrichter M, Rogalski J (1999) Biodegradation of lignin by white rot fungi. Fungal Genet Biol 27:175–185. https://doi.org/10.1006/fgbi.1999.1150
16. Cerniglia CE (1997) Fungal metabolism of polycyclic aromatic hydrocarbons: past, present and future applications in bioremediation. J Ind Microbiol Biotechnol. https://doi.org/10.1038/sj.jim.2900459
17. Golan-rozen N, Chefetz B, Ben-Ari J, Geva J, Hadar Y (2011) Transformation of the recalcitrant pharmaceutical compound carbamazepine by pleurotus ostreatus: role of cytochrome P450 monooxygenase Anda manganese peroxidase. Environ Sci Technol 45:6800–6805. https://doi.org/10.1021/es200298t
18. Marco-Urrea E, Pérez-Trujillo M, Cruz-Morató C, Caminal G, Vicent T (2010) White-rot fungus-mediated degradation of the analgesic ketoprofen and identification of intermediates by HPLC-DAD-MS and NMR. Chemosphere 78:474–481. https://doi.org/10.1016/j.chemosphere.2009.10.009
19. Marco-Urrea E, Pérez-Trujillo M, Vicent T, Caminal G (2009) Ability of white-rot fungi to remove selected pharmaceuticals and identification of degradation products of ibuprofen by Trametes versicolor. Chemosphere 74:765–772. https://doi.org/10.1016/j.chemosphere.2008.10.040
20. Gómez-Toribio V, García-Martín AB, Martínez MJ, Martínez ÁT, Guillén F (2009) Enhancing the production of hydroxyl radicals by Pleurotus eryngii via quinone redox cycling for pollutant removal. Appl Environ Microbiol 75:3954–3962. https://doi.org/10.1128/AEM.02138-08

21. Marco-Urrea E, Radjenović J, Caminal G, Petrović M, Vicent T, Barceló D (2010) Oxidation of atenolol, propranolol, carbamazepine and clofibric acid by a biological Fenton-like system mediated by the white-rot fungus Trametes versicolor. Water Res 44:521–532. https://doi.org/10.1016/j.watres.2009.09.049
22. Cruz del Álamo A, Pariente MI, Vasiliadou I, Padrino B, Puyol D, Molina R, Martínez F (2018) Removal of pharmaceutical compounds from urban wastewater by an advanced bio-oxidation process based on fungi Trametes versicolor immobilized in a continuous RBC system. Environ Sci Pollut Res 25:34884–34892. https://doi.org/10.1007/s11356-017-1053-4
23. El-Enshasy HA (2007) Filamentous fungal cultures – process characteristics, products, and applications. In: Bioprocessing for value-added products from renewable resources. Elsevier, Amsterdam, pp 225–261. https://doi.org/10.1016/B978-044452114-9/50010-4
24. Rodarte-Morales AI, Feijoo G, Moreira MT, Lema JM (2012) Biotransformation of three pharmaceutical active compounds by the fungus Phanerochaete chrysosporium in a fed batch stirred reactor under air and oxygen supply. Biodegradation 23:145–156. https://doi.org/10.1007/s10532-011-9494-9
25. Xin B, Xia Y, Zhang Y, Aslam H, Liu C, Chen S (2012) A feasible method for growing fungal pellets in a column reactor inoculated with mycelium fragments and their application for dye bioaccumulation from aqueous solution. Bioresour Technol 105:100–105. https://doi.org/10.1016/j.biortech.2011.11.062
26. Borràs E, Blánquez P, Sarrà M, Caminal G, Vicent T (2008) Trametes versicolor pellets production: low-cost medium and scale-up. Biochem Eng J 42:61–66. https://doi.org/10.1016/j.bej.2008.05.014
27. Espinosa-Ortiz EJ, Rene ER, Pakshirajan K, van Hullebusch ED, Lens PNLL (2015) Fungal pelleted reactors in wastewater treatment: applications and perspectives. Chem Eng J 283:553–571. https://doi.org/10.1016/j.cej.2015.07.068
28. Nair RB, Lennartsson PR, Taherzadeh MJ (2016) Mycelial pellet formation by edible ascomycete filamentous fungi, Neurospora intermedia. AMB Express 6:31. https://doi.org/10.1186/s13568-016-0203-2
29. Mir-Tutusaus JAJA, Caminal G, Sarrà M (2018) Influence of process variables in a continuous treatment of non-sterile hospital wastewater by Trametes versicolor and novel method for inoculum production. J Environ Manag 212:415–423. https://doi.org/10.1016/j.jenvman.2018.02.018
30. Sitanggang AB, Wu H-S, Wang SS, Ho Y-C (2010) Effect of pellet size and stimulating factor on the glucosamine production using Aspergillus sp. BCRC 31742. Bioresour Technol 101:3595–3601. https://doi.org/10.1016/j.biortech.2009.12.084
31. Pazouki M, Panda T (2000) Understanding the morphology of fungi. Bioprocess Eng 22:127–143. https://doi.org/10.1007/s004490050022
32. Moreira MT, Feijoo G, Lema JM (2003) Fungal bioreactors: applications to white-rot fungi. Rev Environ Sci Biotechnol 2:247–259. https://doi.org/10.1023/B:RESB.0000040463.80808.dc
33. Kantarci N, Borak F, Ulgen KO (2005) Bubble column reactors. Process Biochem. https://doi.org/10.1016/j.procbio.2004.10.004
34. Moreira MT, Sanromán A, Feijoo G, Lema JM (1996) Control of pellet morphology of filamentous fungi in fluidized bed bioreactors by means of a pulsing flow. Application to Aspergillus Niger and Phanerochaete chrysosporium. Enzym Microb Technol 19:261–266. https://doi.org/10.1016/0141-0229(95)00244-8
35. Blánquez P, Sarrà M, Vicent T (2008) Development of a continuous process to adapt the textile wastewater treatment by fungi to industrial conditions. Process Biochem 43:1–7. https://doi.org/10.1016/j.procbio.2007.10.002
36. Grandclément C, Seyssiecq I, Piram A, Wong-Wah-Chung P, Vanot G, Tiliacos N, Roche N, Doumenq P (2017) From the conventional biological wastewater treatment to hybrid processes, the evaluation of organic micropollutant removal: a review. Water Res 111:297–317. https://doi.org/10.1016/j.watres.2017.01.005

37. Song W, Xie B, Huang S, Zhao F, Shi X (2020) Aerobic membrane bioreactors for industrial wastewater treatment. In: Current developments in biotechnology and bioengineering. Elsevier, Amsterdam, pp 129–145. https://doi.org/10.1016/b978-0-12-819809-4.00006-1
38. Thomas O, Thomas MF (2017) Urban wastewater. In: UV-visible spectrophotometry of water and wastewater. Elsevier, Amsterdam, pp 281–315. https://doi.org/10.1016/B978-0-444-63897-7.00009-3
39. Goswami L, Vinoth Kumar R, Borah SN, Arul Manikandan N, Pakshirajan K, Pugazhenthi G (2018) Membrane bioreactor and integrated membrane bioreactor systems for micropollutant removal from wastewater: a review. J Water Process Eng. https://doi.org/10.1016/j.jwpe.2018.10.024
40. Yang S, Hai FI, Nghiem LD, Nguyen LN, Roddick F, Price WE (2013) Removal of bisphenol A and diclofenac by a novel fungal membrane bioreactor operated under non-sterile conditions. Int Biodeterior Biodegrad 85:483–490. https://doi.org/10.1016/j.ibiod.2013.03.012
41. Asif MB, Hai FI, Singh L, Price WE, Nghiem LD (2017) Degradation of pharmaceuticals and personal care products by white-rot fungi—a critical review. Curr Pollut Rep 3:88–103. https://doi.org/10.1007/s40726-017-0049-5
42. Nguyen LN, Hai FI, Yang S, Kang J, Leusch FDL, Roddick F, Price WE, Nghiem LD (2014) Removal of pharmaceuticals, steroid hormones, phytoestrogens, UV-filters, industrial chemicals and pesticides by Trametes versicolor: role of biosorption and biodegradation. Int Biodeterior Biodegrad 88:169–175. https://doi.org/10.1016/j.ibiod.2013.12.017
43. Asif MB, Hai FI, Jegatheesan V, Price WE, Nghiem LD, Yamamoto K (2018) Applications of membrane bioreactors in biotechnology processes. In: Current trends and future developments on (bio-) membranes: membrane processes in the pharmaceutical and biotechnological field. Elsevier, Amsterdam, pp 223–257. https://doi.org/10.1016/B978-0-12-813606-5.00008-7
44. Iorhemen O, Hamza R, Tay J (2016) Membrane bioreactor (MBR) technology for wastewater treatment and reclamation: membrane fouling. Membranes (Basel) 6:33. https://doi.org/10.3390/membranes6020033
45. Bardi A, Yuan Q, Siracusa G, Becarelli S, Di Gregorio S, Tigini V, Levin DB, Petroni G, Munz G (2019) Stability of fungal biomass continuously fed with tannic acid in a non-sterile moving-packed bed reactor. J Environ Manag 247:67–77. https://doi.org/10.1016/j.jenvman.2019.06.036
46. Gao D, Zeng Y, Wen X, Qian Y (2008) Competition strategies for the incubation of white rot fungi under non-sterile conditions. Process Biochem 43:937–944. https://doi.org/10.1016/j.procbio.2008.04.026
47. Ehlers GA, Rose PD (2005) Immobilized white-rot fungal biodegradation of phenol and chlorinated phenol in trickling packed-bed reactors by employing sequencing batch operation. Bioresour Technol 96:1264–1275. https://doi.org/10.1016/j.biortech.2004.10.015
48. Torán J, Blánquez P, Caminal G (2017) Comparison between several reactors with Trametes versicolor immobilized on lignocellulosic support for the continuous treatments of hospital wastewater. Bioresour Technol 243:966–974. https://doi.org/10.1016/j.biortech.2017.07.055
49. Li X, Xu J, de Toledo RA, Shim H, De Toledo RA, Shim H (2015) Enhanced removal of naproxen and carbamazepine from wastewater using a novel countercurrent seepage bioreactor immobilized with Phanerochaete chrysosporium under non-sterile conditions. Bioresour Technol 197:465–474. https://doi.org/10.1016/j.biortech.2015.08.118
50. Basak B, Jeon B-HH, Kurade MB, Saratale GD, Bhunia B, Chatterjee PK, Dey A (2019) Biodegradation of high concentration phenol using sugarcane bagasse immobilized *Candida tropicalis* PHB5 in a packed-bed column reactor. Ecotoxicol Environ Saf 180:317–325. https://doi.org/10.1016/j.ecoenv.2019.05.020
51. Kurade MB, Waghmode TR, Xiong JQ, Govindwar SP, Jeon BH (2019) Decolorization of textile industry effluent using immobilized consortium cells in upflow fixed bed reactor. J Clean Prod 213:884–891. https://doi.org/10.1016/j.jclepro.2018.12.218
52. Cruz del Álamo A, Pariente MI, Martínez F, Molina R, del Álamo AC, Pariente MI, Martínez F, Molina R (2020) Trametes versicolor immobilized on rotating biological contactors as

alternative biological treatment for the removal of emerging concern micropollutants. Water Res 170:115313. https://doi.org/10.1016/j.watres.2019.115313
53. Li X, Xu J, de Toledo RA, Shim H (2016) Enhanced carbamazepine removal by immobilized Phanerochaete chrysosporium in a novel rotating suspension cartridge reactor under non-sterile condition. Int Biodeterior Biodegradation 115:102–109. https://doi.org/10.1016/j.ibiod.2016.08.003
54. Beltrán-Flores E, Torán J, Caminal G, Blánquez P, Sarrà M (2020) The removal of diuron from agricultural wastewaters by Trametes versicolor immobilized on pinewood in simple channel reactors. Sci Total Environ 728:138414. https://doi.org/10.1016/j.scitotenv.2020.138414
55. Gros M, Cruz-Morato C, Marco-Urrea E, Longrée P, Singer H, Sarrà M, Hollender J, Vicent T, Rodriguez-Mozaz S, Barceló D (2014) Biodegradation of the X-ray contrast agent iopromide and the fluoroquinolone antibiotic ofloxacin by the white rot fungus Trametes versicolor in hospital wastewaters and identification of degradation products. Water Res 60:228–241. https://doi.org/10.1016/j.watres.2014.04.042
56. Cruz-Morató C, Ferrando-Climent L, Rodriguez-Mozaz S, Barceló D, Marco-Urrea E, Vicent T, Sarrà M (2013) Degradation of pharmaceuticals in non-sterile urban wastewater by Trametes versicolor in a fluidized bed bioreactor. Water Res 47:5200–5210. https://doi.org/10.1016/j.watres.2013.06.007
57. Badia-Fabregat M, Lucas D, Pereira MA, Alves M, Pennanen T, Fritze H, Rodríguez-Mozaz S, Barceló D, Vicent T, Caminal G (2015) Continuous fungal treatment of non-sterile veterinary hospital effluent: pharmaceuticals removal and microbial community assessment. Appl Microbiol Biotechnol 100:2401–2415. https://doi.org/10.1007/s00253-015-7105-0
58. Cruz-Morató C, Lucas D, Llorca M, Rodriguez-Mozaz S, Gorga M, Petrovic M, Barceló D, Vicent T, Sarrà M, Marco-Urrea E, Petrović M, Barceló D, Vicent T, Sarrà M, Marco-Urrea E, Petrovic M, Barceló D, Vicent T, Sarrà M, Marco-Urrea E (2014) Hospital wastewater treatment by fungal bioreactor: removal efficiency for pharmaceuticals and endocrine disruptor compounds. Sci Total Environ 493:365–376. https://doi.org/10.1016/j.scitotenv.2014.05.117
59. Ferrando-Climent L, Cruz-Morató C, Marco-Urrea E, Vicent T, Sarrà M, Rodriguez-Mozaz S, Barceló D (2015) Non conventional biological treatment based on Trametes versicolor for the elimination of recalcitrant anticancer drugs in hospital wastewater. Chemosphere 136:9–19. https://doi.org/10.1016/j.chemosphere.2015.03.051
60. Kresinová Z, Linhartová L, Filipová A, Ezechiáš M, Mašín P, Cajthaml T, Křesinová Z, Linhartová L, Filipová A, Ezechiáš M, Mašín P, Cajthaml T (2018) Biodegradation of endocrine disruptors in urban wastewater using Pleurotus ostreatus bioreactor. New Biotechnol 43:53–61. https://doi.org/10.1016/j.nbt.2017.05.004
61. Zhang Y, Geißen S-UU (2012) Elimination of carbamazepine in a non-sterile fungal bioreactor. Bioresour Technol 112:221–227. https://doi.org/10.1016/j.biortech.2012.02.073
62. Mir-Tutusaus JA, Sarrà M, Caminal G (2016) Continuous treatment of non-sterile hospital wastewater by Trametes versicolor: how to increase fungal viability by means of operational strategies and pretreatments. J Hazard Mater 318:561–570. https://doi.org/10.1016/j.jhazmat.2016.07.036
63. Mir-Tutusaus JA, Parladé E, Llorca M, Villagrasa M, Barceló D, Rodriguez-Mozaz S, Martinez-Alonso M, Gaju N, Caminal G, Sarrà M (2017) Pharmaceuticals removal and microbial community assessment in a continuous fungal treatment of non-sterile real hospital wastewater after a coagulation-flocculation pretreatment. Water Res 116:65–75. https://doi.org/10.1016/j.watres.2017.03.005
64. Mir-Tutusaus JA, Parladé E, Villagrasa M, Barceló D, Rodríguez-Mozaz S, Martínez-Alonso M, Gaju N, Sarrà M, Caminal G (2019) Long-term continuous treatment of non-sterile real hospital wastewater by Trametes versicolor. J Biol Eng 13:47. https://doi.org/10.1186/s13036-019-0179-y

Novel Constructed Wetland Configurations for the Removal of Pharmaceuticals in Wastewater

Mònica Escolà Casas and Víctor Matamoros

Contents

1 Introduction ... 164
2 Common CW Configurations Used for the Removal of PhACs ... 165
2.1 Removal of PhACs by SSFCWs ... 166
2.2 Removal of PhACs by SFCWs ... 166
2.3 Removal of PhACs with Hybrid CW Systems ... 167
3 Novel CW Approaches for Pharmaceutical Removal ... 167
3.1 Modifying the Filling Material of CWs ... 167
3.2 Enhancing the Biodegradation Capacity of CWs ... 174
3.3 CWs Combined with Other Wastewater Treatment Technologies ... 180
4 Challenges and Future Trends ... 182
5 Concluding Remarks ... 184
References ... 185

Abstract Nature-based solutions such as constructed wetlands (CWs) are a cost-effective option to perform wastewater treatment. Multiple studies have already shown that common CW configurations can remove pharmaceutically active compounds (PhACs) from wastewater mainly due to sorption, biodegradation, and photodegradation processes. Even so, recent research has proposed novel CW configurations to improve PhAC removal. This chapter will review existing studies on the three following novel CW approaches: (1) CWs' filling material modification, (2) biodegradation enhancement, and (3) integration of CWs with intensive wastewater treatment technologies. Waste-to-product filling materials such as biochar and cork can be used to enhance CWs' adsorption capacities. On the other side, biodegradation can be improved by incorporating microbial fuel cells, forced aeration, or

M. Escolà Casas (✉) and V. Matamoros
Department of Environmental Chemistry, IDAEA-CSIC, Barcelona, Spain
e-mail: mecgam@cid.csic.es

Sara Rodriguez-Mozaz, Paqui Blánquez Cano, and Montserrat Sarrà Adroguer (eds.), *Removal and Degradation of Pharmaceutically Active Compounds in Wastewater Treatment*, Hdb Env Chem (2021) 108: 163–190, DOI 10.1007/698_2020_681,

bioaugmentation technologies into CWs. Finally, combinations of CWs with intensive wastewater treatments may offer a wider range of biodegradation pathways for PhACs and a reduction of the resulting transformation products (TPs). Future research on CWs should include the monitoring of TPs, a higher number of PhACs, ecotoxicological effects, and antimicrobial resistance. In addition, one of the main limitations of CW technology is still the high surface area required, which may be solved by future studies combining novel CW technologies.

Keywords Bioaugmentation, Constructed wetlands, Filling material, Forced aeration, Microbial fuel cells, Pharmaceuticals

1 Introduction

Constructed wetlands (CWs) are probably one of the clearest examples of a nature-based solution for wastewater treatment. Wetlands have been involved in the treatment of the wastewater since humans started to discharge sewage into the environment [1]. However, it was not until the early 1950 that artificial wetlands started to be built to treat wastewater [2]. During the last decades, CWs were developed into fully engineered systems, and nowadays they are defined as a sustainable, low-cost, robust, and efficient wastewater treatment technology which, unlike other technologies, provides ecosystem services [1]. Among these ecosystem services, CWs support animal biodiversity, which contributes to dampen the occurrence of zoonosis [3] and therefore can minimize the risk of future disease outbreaks.

CWs have been typically designed to remove biochemical oxygen demand (BOD), nutrients (nitrogen and phosphorus), total suspended solids, and bacteria [4–6]. Nevertheless, different studies have also reported their ability to remove the so-called contaminants of emerging concern (CEC) from urban and domestic wastewaters, which include pharmaceutically active compounds (PhACs) [7–11]. Furthermore, CWs are also capable of removing PhACs from livestock, swine [12], hospital [13], and aquaculture [14] wastewaters. The removal of PhACs in CWs involves different processes such as adsorption to the filling material, biodegradation by the soil's or rhizosphere's microbial community, and plant uptake [11] but also photodegradation, hydrolysis, and volatilization [15, 16]. According to this, the efficiency of CWs to degrade PhACs from wastewater depends on their physicochemical properties, the characteristics of the filling material, the planted species, the microbial populations, the wastewater characteristics, and the environmental conditions [17]. As PhACs belong to different chemical families with very diverse physicochemical properties, the removal mechanisms are complex and almost compound-specific.

The main aim of this chapter is to provide a review of novel CW configurations for the attenuation of PhACs from wastewater. Nevertheless, to understand the attenuation processes of PhACs in CWs, a general overview on common CW configurations will also be included. Novel configurations will be classified on three strategies: (1) modifying the filling material, (2) enhancing the biodegradation mechanisms, and (3) combining CWs with other wastewater treatment technologies. This chapter will also explain the challenges that novel CW configurations need to overcome regarding the attenuation of PhACs and the derived future research trends. Finally, conclusions about the removal of PhACs in novel CWs will be stated.

2 Common CW Configurations Used for the Removal of PhACs

Typically, CWs can be designed as subsurface flow constructed wetlands (SSFCWs), where the water level is maintained below the surface of the medium placed in the beds [18], or as surface flow constructed wetlands (SFCWs), which are flooded systems that expose the water to the atmosphere (Fig. 1) [2]. Based on water flow direction, SSFCWs are subdivided into horizontal flow (HF) and vertical flow (VF) systems [19] (Fig. 1). In addition, there are hybrid systems, which consist of various types of those CW designs combined according to the wastewater treatment needs (Fig. 1). Despite their design, all CWs work in a similar way: solids are removed by filtration and settling, BOD declines due to consumption of organic compounds by the microbial community, ammonia is oxidized by microorganisms under aerobic conditions, nitrate is reduced under anaerobic conditions, and

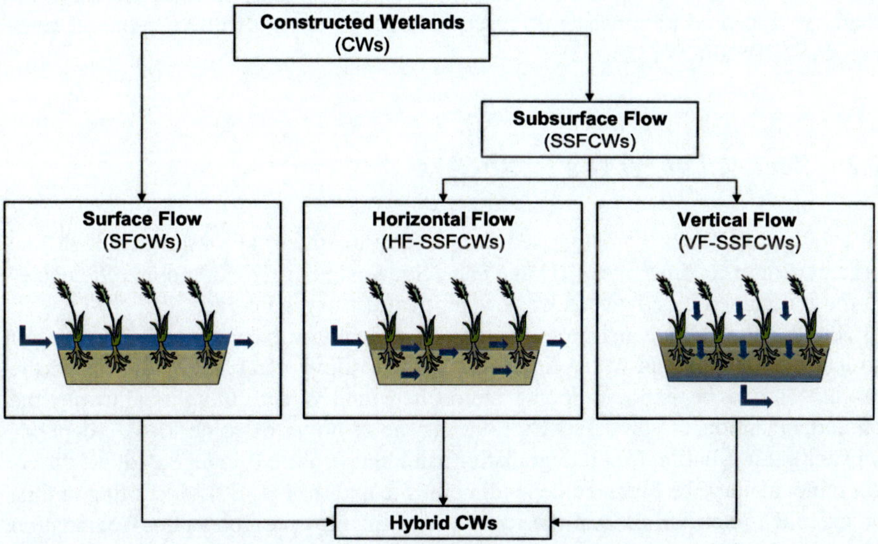

Fig. 1 Scheme of most common CW configurations used for the removal of PhACs

phosphorus may be absorbed by sediments [4]. All these processes are facilitated by plants, which maintain a favorable environment for microbial communities [20]. The removal of these pollutants has been the main focus of research in CW configurations [1], and several thorough reviews about this topic have been already written [4–6].

2.1 Removal of PhACs by SSFCWs

SSFCWs are mainly performing as a secondary treatment with relatively high organic matter loading rates [21]. However, HF-SSFCWs have successfully been used for both secondary and tertiary treatment [6].

HF-SSFCWs can degrade organic matter by sedimentation and filtration as well as by aerobic and anaerobic microbial processes [6]. In such systems, redox potential measurements have confirmed that the environment is anoxic-anaerobic, but aerobic microenvironments may exist around the plant roots, around algae blooms, or near the surface [11]. It has been seen that aerobic conditions lead to more efficient biodegradation pathways of PhACs than anoxic-anaerobic conditions [11]. Therefore, shallower (and thus more aerobic) HF-SSFCWs have shown to be better at degrading most PhACs [22]. However, not all PhACs are degraded under aerobic conditions [11], and the presence of oxic, anoxic, and anaerobic environments offers the possibility to degrade a wider range of PhACs. For example, the biodegradation of diclofenac has been linked to anaerobic conditions or even to the combination of oxic-anoxic conditions [23].

VF-SSFCWs, with unsaturated flow and therefore mainly with an aerobic environment, appeared to consistently perform better in terms of PhAC removal compared to HF-SSFCWs [24, 25].

2.2 Removal of PhACs by SFCWs

Generally, SFCWs are mostly used as a tertiary treatment of wastewater with low organic matter loading rates [21] and can achieve similar CEC removal efficiencies to those obtained by advanced wastewater treatment systems [26, 27].

SFCWs typically consist of a sequence of shallow basins and water control structures that maintain water depth [2]. SCFWs allow water to flow on top of the ground and to be exposed to the atmosphere and direct sunlight, allowing the photodegradation of some PhACs [28]. As the removal of some PhACs depends on sunlight irradiation (photodegradation) and temperature (biodegradation), different removals may be obtained depending on site and season [26]. According to this, an extended retention time of the water (i.e., 1 month) promotes photodegradation and biodegradation reactions so removal efficiencies of PhACs can reach as high values as those reported for HF-SSFCWs and conventional WWTPs [26]. In particular, SFCWs have shown to have the same capacity to remove antibiotics as conventional activated sludge wastewater systems [29]. In addition, the short exposure (25 days) to low levels of antibiotics (ng/L to μg/L) in SFCWs did not promote the proliferation of antibiotic resistance genes (ARG), which were as abundant as in natural wetlands [29]. However, the increase of ARG after longer exposure periods should be assessed [29].

2.3 *Removal of PhACs with Hybrid CW Systems*

In the last years, the combined use of the abovementioned CW configurations has been tested for removing PhACs. Depending on the nature of the compounds to be degraded, some CW designs are better suited than others; so the combination of different CW systems can considerably reduce the total discharge of PhACs [30]. For example, three different real-scale hybrid treatment combinations (one 3-pond system and two different pond-SFCW-HF-SSFCW systems) obtained PhAC removals mainly over 70% [31]. These systems had highest removal efficiency in the first treatment step (pond, SFCW, or SSFCW) probably due to dependency on concentration [31]. Another example is a study testing a full-scale hybrid CW, consisting of a VF-SSFCW, a HF-SSFCW, and a SFCW [32]. This hybrid configuration achieved removals over 80% of all PhACs and other CEC, thanks to the diversity of physicochemical conditions occurring in the different CW configurations, which allowed synergies between biotic and abiotic removal mechanisms [32].

3 Novel CW Approaches for Pharmaceutical Removal

Multiple studies are exploring novel CW designs to boost the attenuation of PhACs from wastewater (Fig. 2). These studies follow three strategies: (1) modification of the filling material or support matrix, (2) enhancement of the biodegradation mechanisms, and (3) combination of CWs with other wastewater treatment technologies. The use of these strategies for removing PhACs from wastewater will be the main focus of this chapter.

3.1 *Modifying the Filling Material of CWs*

The choice of the filling material mainly influences the adsorption capacity of the CWs for removing PhACs, but it can also promote biofilm growth [33], increase residence time, and shape the microbial community, which will ultimately determine the biodegradation of pollutants [34]. The latest years, alternative CW approaches based on the replacement of the filling material (e.g., clays and biochar) as well as

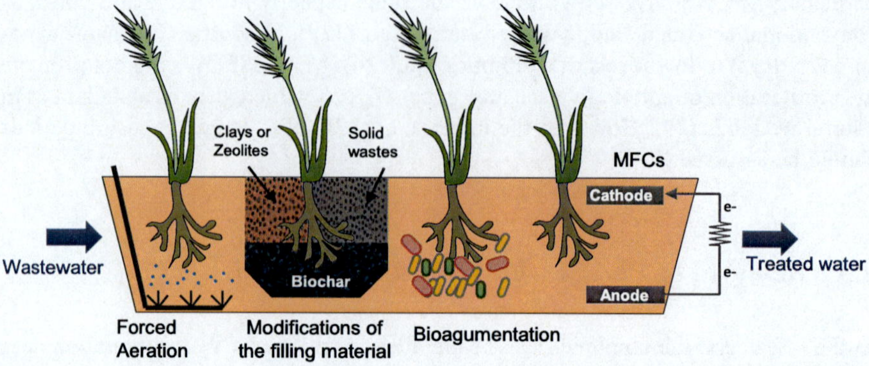

Fig. 2 Overview of novel CW configurations discussed in this chapter. Modifications of the filling material (clays or zeolites, solid wastes, and biochar) are discussed in Sect. 3.1. Strategies for enhancing biodegradation (microbial fuel cells (MFCs), forced aeration, and bioaugmentation) are discussed in Sect. 3.2. A HF-SSFCW has been represented here, but novel CW configurations may be applied to other types of CWs

giving new usages to materials (e.g., waste products) have been studied for the removal of carbon, nitrogen, and phosphorus content. However, these new approaches have also shown to improve the removal of xenobiotic compounds such as PhACs.

3.1.1 Clays and Zeolites

Clays can be used to enhance the removal of PhACs in CWs. Thermally processed clay materials like lightweight expanded clay aggregate (LECA) (300 US dollars per m^3, 0.28 kg/L density[1]) or exfoliated vermiculite (117 US dollars per m^3, 0.060–0.12 kg/L density[2]) have been successfully used for regular wastewater treatment as filters or as filling material in CWs [35]. These materials have also lately been tested for the removal of PhACs, and promising results have been obtained in several studies. Among such materials, the use of LECA has shown excellent results with the attenuation of several PhACs (Table 1) [35, 36, 38, 39, 42]. For example, the removals of ibuprofen and carbamazepine in planted LECA were both over 82% in two separate studies [38, 39] (Table 1). Also, exfoliated vermiculite may have some specific applications in CWs as it has shown higher and faster adsorption capacities per unit mass than LECA, but, in general, its low density makes it less efficient than LECA per unit of volume [35].

With the same composition as clays but with different crystallographic structure, zeolites are another appealing material that can be applied in CWs. Zeolites are

[1]Data for bulk acquisition of Leca® LWA in https://www.specialistaggregates.com/ in May 2020.
[2]Data for bulk acquisition of exfoliated vermiculite in https://www.perlipol.com.pl/ in May 2020.

Table 1 Assessment of different filling materials (LECA, exfoliated vermiculite, zeolites, cork, and biochar) which have been or could be applied in CWs for the attenuation of different PhACs

Compound and concentration tested	Technology	Dimensions	Type of water	Removal efficiency (%)	Hydraulic parameters	Reference
Atenolol 1 mg/L	SSFCW, LECA (unplanted and planted)	Microcosm batch	Real wastewater after secondary treatment	Unplanted: 82% Planted: 92–95%	HRT = 4 days	[36]
Caffeine 3.8–4.7 μg/L	Biochar (unplanted, active biofilm/inactive biofilm/no biofilm)	Laboratory scale	Real autoclaved wastewater	97–99% (range active biofilm/inactive biofilm/no biofilm)	HLR: 50 Lm^2/days	[37]
Carbamazepine 6.9–2.3 μg/L	Biochar (unplanted active biofilm/inactive biofilm/no biofilm)	Laboratory scale	Real autoclaved wastewater	Active biofilm: 99% Inactive biofilm: 73% No biofilm: 99%	HLR: 50 Lm^2/days	[37]
Carbamazepine 58 μg/L	SSFCW, LECA (planted and unplanted)	Laboratory scale	Real wastewater	Planted: 89% Unplanted: 95%	HRT = 3 days	[38]
Carbamazepine 1,000 μg/L	SSFCW, LECA (planted and unplanted, winter and summer)	Microcosm batch	Real wastewater after secondary treatment	Planted: 88–96% Unplanted: 87–86% (winter and summer, respectively)	HRT = 7 days	[39]
Carbamazepine 35 mg/L	Granulated cork sorption	Laboratory-scale batch	Real wastewater after secondary treatment	63%	Contact time = 6 days	[40]
Clofibric acid 1 mg/L	SSFCW, LECA (planted and unplanted, winter and summer)	Microcosm batch	Real wastewater after secondary treatment	Planted: 48–75% Unplanted: 40–43% (winter and summer, respectively)	HRT = 7 days	[39]
Clofibric acid 35 mg/L	Granulated cork sorption	Laboratory-scale batch	Real wastewater after secondary treatment	8%	Contact time = 6 days	[40]

(continued)

Table 1 (continued)

Compound and concentration tested	Technology	Dimensions	Type of water	Removal efficiency (%)	Hydraulic parameters	Reference
Erythromycin 11.6 μg/L	HF-SSFCW (planted zeolite)	Mesocosm scale	Raw wastewater	83%	HLR: 96 Lm^2/days	[41]
Furosemide 1–11 mg/L	Cork and LECA, adsorption (unplanted)	Laboratory scale	Ultra-pure water	LECA: >89%, concentration dependent Cork: 73–77%	HRT = 7 days	[42]
Gemfibrozil 1–30 mg/L	LECA and exfoliated vermiculite adsorption (unplanted)	Laboratory scale	Ultra-pure water	LECA: 95–100% Exfoliated vermiculite: 35–40%	LECA: HRT = 4 days Exfoliated vermiculite: HRT = 2 days	[35]
Ibuprofen 33 μg/L	SSFCW, LECA (planted and unplanted)	Laboratory scale	Real wastewater	Planted: 89% Unplanted: 94%	HRT = 3 days	[38]
Ibuprofen 1 mg/L	SSFCW, LECA (planted and unplanted, winter and summer)	Microcosm batch	Real wastewater after secondary treatment	Planted: 82–96% Unplanted: 73–91% (winter and summer, respectively)	HRT = 7 days	[39]
Ibuprofen 35 mg/L	Granulated cork sorption (unplanted)	Laboratory-scale batch	Real wastewater after secondary treatment	95–100%	Contact time = 6 days	[40]
Lincomycin 46 ng/L	HF-SSFCW (planted zeolite)	Mesocosm scale	Raw wastewater	55%	HLR: 96 Lm^2/days	[41]
Mefenamic acid 1–30 mg/L	LECA and exfoliated vermiculite adsorption (unplanted)	Laboratory scale	Ultra-pure water	LECA: 95–100% Exfoliated vermiculite: 35–40%	LECA: HRT = 4 days Exfoliated vermiculite: HRT = 2 days	[35]

Metoprolol 1.9–3.1 μg/L	Biochar (unplanted active biofilm/inactive biofilm/no biofilm)	Laboratory scale	Real autoclaved wastewater	95–99% (range active biofilm/inactive biofilm/no biofilm)	HLR: 50 Lm^2/days	[37]
Monensin 144 ng/L	HF-SSFCW (planted zeolite)	Mesocosm scale	Raw wastewater	95%	HLR: 96 Lm^2/days	[41]
Naproxen 1–30 mg/L	LECA and exfoliated vermiculite adsorption (unplanted)	Laboratory scale	Ultra-pure water	LECA: 95–100% Exfoliated vermiculite: 80%	LECA: HRT = 4 days Exfoliated vermiculite: HRT = 2 days	[35]
Novobiocin 6 ng/L	HF-SSFCW (planted zeolite)	Mesocosm scale	Raw wastewater	100%	HLR: 96 Lm^2/d	[41]
Ofloxacin 7 ng/L	HF-SSFCW (planted zeolite)	Mesocosm scale	Raw wastewater	100%	HLR: 96 Lm^2/days	[41]
Oxytetracycline 233 ng/L	IVCW (zeolites vs. quartz)	Laboratory scale	Swine wastewater aerobically digested	Planted zeolite: 90% Planted quartz: 70–80%	HRT = 7 days	[43]
Ranitidine 3.2–7.2 μg/L	Biochar (unplanted active biofilm/inactive biofilm/no biofilm)	Laboratory scale	Real autoclaved wastewater	97–99% (range active biofilm/inactive biofilm/no biofilm)	HLR: 50 Lm^2/days	[37]
Sulfadiazine 33 μg/L	SSFCW, LECA (planted and unplanted)	Laboratory scale	Real wastewater	Planted: 67% Unplanted: 94%	HRT = 3 days	[38]
Sulfadiazine 22 ng/L	IVCW (zeolites vs. quartz)	Laboratory scale	Swine wastewater aerobically digested	Planted zeolite: 60–70% Planted quartz: 30–40%	HRT = 7 days	[43]
Sulfamethazine 10 ng/L	IVCW (zeolites vs. quartz)	Laboratory scale	Swine wastewater aerobically digested	Planted zeolite: 60–70% Planted quartz: 50–60%	HRT = 7 days	[43]
Sulfamethazine 8 ng/L	HF-SSFCW (planted zeolite)	Mesocosm scale	Raw wastewater	52%	HLR: 96 Lm^2/days	[41]
Sulfamethoxazole 0.6 ng/L	IVCW (zeolites vs. quartz)	Laboratory scale	Swine wastewater aerobically digested	Planted zeolite: 45–50% Planted quartz: 45–50%	HRT = 7 days	[43]

(continued)

Table 1 (continued)

Compound and concentration tested	Technology	Dimensions	Type of water	Removal efficiency (%)	Hydraulic parameters	Reference
Sulfamerazine 96 ng/L	HF-SSFCW (planted zeolite)	Mesocosm scale	Raw wastewater	100%	HLR: 96 Lm^2/ days	[41]
Tetracycline 539 ng/L	IVCW (zeolites vs. quartz)	Laboratory scale	Swine wastewater aerobically digested	Planted zeolite: 80–90% Planted quartz: 70–80%	HRT = 7 days	[43]

HRT hydraulic retention time, *HLR* hydraulic loading rate natural or artificially produced aluminosilicate minerals which are normally used for their adsorption, ion exchange, molecular sieve, and catalytic properties [44]. In CWs, zeolites alone or applied together with other materials have shown to have a successful removal of nitrogen from domestic, urban, and livestock wastewaters, with a less satisfactory removal of chemical oxygen demand (COD) and phosphorus [45]. Interestingly, the usage of zeolites in CWs has shown to improve the removal efficiency of PhACs such as antibiotics as well as ARGs [41, 43] (Table 1). Although natural zeolites are relatively affordable (1,000 US dollar per ton [46]) and abundant, they are still a limited natural resource and producing them artificially implies energetic costs. This is a disadvantage compared to other available materials that can be cheaply produced like LECA or obtained from waste products (e.g., cork and biochar), but their efficiency in removing antibiotics and ARGs may pay off the costs.

3.1.2 Solid Wastes

The use of solid wastes from agriculture, industry, and construction to fill CWs has been explored as a way to reduce landfill disposal of those materials in the context of circular economy [47]. Among the different tested materials, cork granulates showed the ability to adsorb organic compounds [47]. Despite their selective adsorption potential, cork-filled treatment wetlands appeared to be specifically an appropriate technology to treat nitrate-polluted groundwater [48] and to adsorb some PhACs [40, 42] (Table 1). It was also noted that cork allowed shorter equilibration times and larger removal of PhACs per weight of sorbent than LECA materials [40]. In another study, LECA reached greater removals of furosemide than cork (Table 1), but cork granulates achieved higher adsorption in the first 24 h [42]. Therefore, cork granulates were suggested as preliminary filter to retain most of the PhAC concentration with a low hydraulic retention time (HRT), followed by a LECA filter [42]. In addition, cork sorbents have become a potential low-cost choice (240–260 US dollars per m^3; 0.06–0.22 kg/L density[3]) as they can be obtained from cork industrial wastes [49]. Nevertheless, no pilot or full-scale CW system based on cork material has been built to demonstrate the real effectiveness of this material. Other solid wastes like oyster shells were able to remove 17–60% of antibiotics, but COD and nitrogen removals were not as satisfactory [41]. Similar results have also been reported with other comparable materials (snail shells) [47].

3.1.3 Biochar

Biochar used as filling material in CWs seems to be a promising option to enhance PhAC removal. Biochar is an interesting charcoal-like product obtained when

[3]Data for bulk acquisition of natural cork granulates in https://www.corklink.com/ in May 2020.

biomass (which can be a waste product) is heated at elevated temperatures with little or no available oxygen (pyrolysis). Biochar price varies between 90 and 3,000 US dollars per ton, mostly depending on the country of production [50]. When applied to the soil, biochar enhances soil's fertility and quality by increasing the moisture retention capacity, attracting beneficial fungi and microbes, and retaining nutrients in the soil [51]. Due to this, biochar is used as filling material in CWs, mainly to remove nitrogen [52] but also COD, phosphorus [53], and even phenols and metals [54]. In spite of this, there are currently no available studies on the effect of biochar applied in CWs towards PhAC removal from wastewater. Nonetheless, trials in unplanted biofilters based on biochar suggest that its usage may enhance degradation of PhACs in CWs. Dalahmeh et al. [37] observed that filters based on biochar resulted in the simultaneous biodegradation and adsorption of recalcitrant pharmaceuticals, like carbamazepine (73–99%), and less recalcitrant compounds like metoprolol, caffeine, and ranitidine (Table 1). In the case of the recalcitrant carbamazepine, the obtained removals were achieved mainly by adsorption processes, but some biodegradation still occurred (Table 1). Same authors suggested that as biofilm contributed to biodegradation, biochar with biofilm could extend adsorbent lifetime.

All in all, the usage of different filling materials in CWs seems to improve the removal of PhACs and seems especially useful to remove recalcitrant compounds, which can only be removed by sorption. However, all reported studies (Table 1) were conducted at laboratory, microcosm, or mesocosm scale. Therefore, further studies should move towards testing such materials in pilot- and real-scale CWs, with longer operation times. Future studies need to evaluate the applicability of these materials under real wastewater conditions because clogging, material degradation, and exhaustion of the sorption capacity (the main removal mechanism for some PhACs) may eventually happen.

3.2 *Enhancing the Biodegradation Capacity of CWs*

Besides modifying the filling material, another approach to enhance the elimination of PhACs from wastewater is to enhance the biodegradation capacity of CWs. This is done by stimulating the microbiological activity through different mechanisms.

3.2.1 Microbial Fuel Cell

A new research line based on the integration of microbial fuel cells (MCFs) (Fig. 3) into CWs (MFC-CWs) has recently been developed (Fig. 2). In a MFC, exoelectrogenic bacteria oxidize organic and inorganic substrates, and electrons are shifted from the anode, via a conductive material and a resistor, to a higher redox electron acceptor, like oxygen, at the cathode [56]. These bioelectrochemical systems and the related microbial electrolysis cell use the electric current to improve pollutant degradation [1].

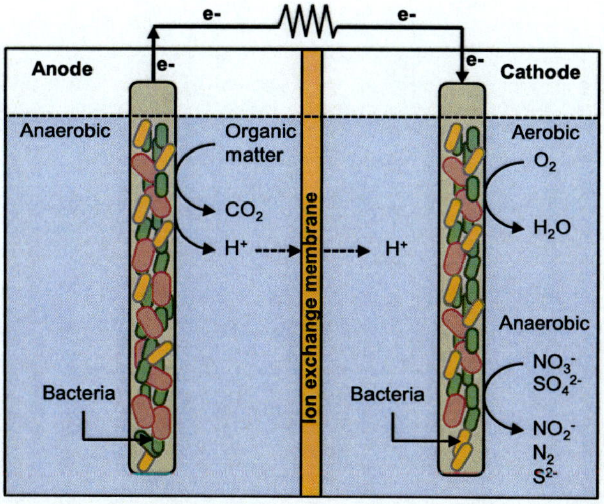

Fig. 3 Scheme of a typical two-chambered MFC separated by an ion exchange membrane. Electrons and protons are generated from organic matter oxidation via anaerobic respiration or fermentation by microorganisms in the anode chamber. Electrons are transferred to the cathode through an external circuit and produce electricity. Protons are transferred to the cathode through the ion exchange membrane. Electrons and protons react with oxygen (aerobic cathode), nitrate, or sulfate (anaerobic cathode). Adapted from [55]

Limited studies have been conducted applying MFC-CWs towards CEC degradation. In a recent study, MFCs have been integrated into an upper-flow CW (MFC-UCW) [57] (Table 2). In that study, at least 60% of 10 mg/L of ibuprofen and bisphenol-A were removed from wastewater with a HRT of 0.7 days [57]. In an even more recent study, MFC-CWs with different configurations were tested for the removal of sulfadiazine, carbamazepine, naproxen, and ibuprofen [60] (Table 2). These studies demonstrated that the operation of the MFC-CW system increased the accumulation of PhACs (sulfadiazine, carbamazepine, naproxen, and ibuprofen) on the electrode layers by electrosorption, which ended up selecting bacterial communities with enhanced xenobiotics metabolism. This resulted in an enhanced removal of PhACs but also with an enriched number of sulfonamide ARGs [60]. Such findings suggested that MFC-CW can be a promising technology for the removal of PhACs, but potential risks due to the accumulation of ARGs should be assessed.

3.2.2 Aerated CWs

Oxygen availability in CWs results in improved biodegradation, reduced clogging, enhanced removal of organic nutrients, and reduced land area requirement [8]. Therefore, another way to enhance biodegradation of PhACs in CWs is the incorporation of forced aeration in their design (Fig. 2). Forced aeration has been implemented in both, HF- and VF-SSFCW to treat wastewater with very high organic loads, which

Table 2 Overview of articles that tested the removal of PhACs from wastewater in CWs with enhanced biodegradation mechanisms (MCFs, forced aeration, and bioaugmentation)

Compound and concentration tested	Technology	Dimensions	Type of water	Removal efficiency (%)	Hydraulic parameters	Reference
Caffeine 64 μg/L	HF-SSFCW with forced aeration VF-SSFCW with forced aeration Two-cell reciprocating system	Pilot-scale	Real wastewater after primary treatment (septic tank)	HF-SSFCW with forced aeration: 100% VF-SSFCW with forced aeration: 99% Two-cell reciprocating system: 99% (93–100% with non-aerated VF and HF-SSFCW)	HRT > 3 days	[58]
Caffeine 25 μg/L	SFCW SFCW-stabilization tank	Laboratory-scale	Synthetic wastewater	SFCW: 98% SFCW-stabilization tank: 98.2% Both systems: 13 days aerated +17days non-aerated, removal rate did not vary	HRT = 7 days	[59]
Carbamazepine 4 mg/L	Up-flow CW-MFC with activated carbon	Laboratory-scale	Synthetic wastewater	>99% (both HRTs) Accumulation on the anode	HRT=3 and 1.5 days	[60]
Carbamazepine 3 μg/L	HF-SSFCW with forced aeration VF-SSFCW with forced aeration Two-cell reciprocating system	Pilot-scale	Real wastewater after primary treatment (septic tank)	HF-SSFCW with forced aeration: −5% VF-SSFCW with forced aeration: −4% Two-cell reciprocating system: −5% (−8 to −4% with non-aerated VF-SSFCW and 2% with HF-SSFCW)	HRT > 3 days	[58]
Diclofenac μg/L 5.58 ± 2.95	VF-SSFCW With aeration, planted	Pilot-scale	Real wastewater after primary treatment	58% (same achieved in non-aerated, unsaturated system)	HLR = 0.095 m/days (intermittent loading)	[61]

Diclofenac 6 μg/L	HF-SSFCW with forced aeration VF-SSFCW with forced aeration Two-cell reciprocating system	Pilot-scale	Real wastewater after primary treatment (septic tank)	HF-SSFCW with forced aeration: 34% VF-SSFCW with forced aeration: 74% Two-cell reciprocating system: 75% (53–78% with non-aerated VF-SSFCW and 17% with HF-SSFCW)	HRT > 3 days	[58]
Ibuprofen 50 mg/L	HF-SSFCWs, bioaugmentation in planted system	Laboratory-scale	Synthetic wastewater	75–80% (5% higher than without bioaugmentation) Biodegradation 80 days earlier in bioaugmented CWs	HLR=40 Lm^2/days	[62]
Ibuprofen 10 mg/L	Up-flow CW-MFC with activated carbon	Laboratory-scale	Synthetic wastewater	96% (9% higher than control)	HRT = 0.7 days	[57]
Ibuprofen 4 mg/L	Up-flow CW-MFC with activated carbon	Laboratory-scale	Synthetic wastewater	>99% (both HRTs) Accumulation on anode	HRTs=3 and 1.5 days	[60]
Ibuprofen 100 μg/L	Forced aeration, planted and unplanted CWs	Mesocosm	Synthetic wastewater	65% for unplanted (higher than non-aerated unsaturated systems (45%)) 68–97% for planted (same as non-aerated saturated and unsaturated systems)	HLR=0.34 m/days	[63]
Ibuprofen 44.5 ± 13.0 μg/L	VF-SSFCW With aeration, planted	Pilot-scale	Real wastewater after primary treatment	99% (95% achieved in non-aerated, saturated systems)	HLR of 0.095 m/days (intermittent loading)	[61]
Ibuprofen 24 μg/L	HF-SSFCW with forced aeration VF-SSFCW with forced aeration	Pilot-scale	Real wastewater after primary treatment (septic tank)	HF-SSFCW with forced aeration: 95% VF-SSFCW with forced aeration: 98%	HRT > 3 days	[58]

(continued)

Table 2 (continued)

Compound and concentration tested	Technology	Dimensions	Type of water	Removal efficiency (%)	Hydraulic parameters	Reference
	Two-cell reciprocating system			Two-cell reciprocating system: 99% (95–98% with non-aerated VF-SSFCWs, 9% with non-aerated HF-SSFCW)		
Metformin 910 ng/L	VF-SSFCW with forced aeration, planted	Pilot-scale	Real wastewater after primary treatment (direct from sewer, after settling)	99% (30–40% non-aerated)	HRT = 1 day	[64]
Naproxen 4 mg/L	Up-flow CW-MFC with activated carbon	Laboratory-scale	Synthetic wastewater	>99% (both HRTs) Accumulation on the anode	HRTs = 3 and 1.5 days	[60]
Naproxen 24 μg/L	HF-SSFCW with forced aeration VF-SSFCW with forced aeration Two-cell reciprocating system	Pilot-scale	Real wastewater after primary treatment (septic tank)	HF-SSFCW with forced aeration: 84% VF-SSFCW with forced aeration: 94% Two-cell reciprocating system: 99% (89–97% with non-aerated VF-SSFCWs and 23% with HF-SSFCW)	HRT > 3 days	[58]
Sulfadiazine 4 mg/L	Up-flow CW-MFC with activated carbon	Laboratory-scale	Synthetic wastewater	>99% (both HRTs) Accumulation on anode	HRTs=3 and 1.5 days	[60]
Valsartan 2,270 ng/L	VF-SSFCW Forced aeration, planted	Pilot-scale	Real wastewater after primary treatment (direct from sewer, after settling)	99% (30–40% non-aerated)	HRT = 1 day	[64]

HRT hydraulic retention time, *HLR* hydraulic loading rate are oxygen limited and therefore achieve poor removal of nitrogen [52, 65]. HF-SSFCWs particularly are often oxygen limited due to permanent water clogging of the reed bed [66]. For regular pollutants, aeration can increase the nitrification rate of such CWs, but the economic costs of this might not pay off [66].

Regarding PhACs, it has been observed that aerated saturated CWs have a higher microbial diversity than unsaturated systems [63], which might result in the biodegradation of a wider range of compounds. Forced aeration has only been tested for a few PhACs (Table 2). For ibuprofen, forced aeration achieved 65% removal in unplanted HF-SSFCWs, a 20% higher than that found for equivalent saturated or unsaturated systems; however, aeration had not much influence when comparing equivalent planted systems [63]. Similar results were achieved in planted VF-SSFCW with forced aeration, where ibuprofen and diclofenac removals (99% and 58%, respectively) were not different from VF-SSFCWs operating with non-aerated unsaturated flows [61]. Also, aeration in VF-SSFCWs has been tested during municipal and hospital wastewater treatment [64]. In this study, aeration did not enhance the removal of sotalol, carbamazepine, and diclofenac (less than 50% removal in aerated and non-aerated systems); but it resulted in a complete removal (99%) of valsartan and metformin, which were poorly removed in non-aerated systems (17% and 68%, respectively). Finally, another study tested aerated and non-aerated HF- and VF-SSFCWs and, while there were only small differences in VF-SSFCWs (0–16% greater removals in aerated systems), HF-SSFCW aeration improved the removal efficiencies of ibuprofen, naproxen, and diclofenac from 19 to 95%, from 23 to 84%, and from 17 to 24%, respectively [58]. According to those studies, forced aeration seems to be useful to increase the biodegradation of PhACs in systems with low dissolved oxygen levels such as HF-SSFCWs or unplanted systems. Nonetheless, this is mostly valid for compounds that undergo aerobic degradation or degradation in aerobic-anaerobic environments, and, depending on the circumstances, the installation and maintenance costs might not pay off the increased removals.

3.2.3 Bioaugmentation

Bioaugmentation consists in adding biological material to improve pollutant biodegradation and ecosystem function [67]. In wastewater treatment, this is usually translated into adding the required cultured microorganisms into the media to accelerate the degradation of contaminants (Fig. 2). To do so, microorganisms or uncharacterized microbial consortiums isolated from the same media to be treated are usually preferred, as they may persist longer and perform better than externally obtained microorganisms [67].

Bioaugmentation has shown to be able to treat wastewater from different origins (pharmaceutical factories, pulp mills, and other factories) containing different organic and inorganic chemicals, but in CWs it has mainly been studied to enhance nitrogen and phosphorus removal [68–70]. Regarding organic contaminants, one study demonstrated that bioaugmentation enhanced atrazine removal in CWs from 74.6 to 87.1% [71]. For PhACs, there is only one available study that tested the addition of ibuprofen-degrading bacteria into CWs [62] (Table 2). In this study, wastewater containing 11–92 μg/L of ibuprofen was incubated with 500 mg/L of ibuprofen to isolate a bacterium living solely on ibuprofen. After that, a biofilm of the selected bacteria (*Sphingobium yanoikuyae*) was artificially grown on gravel, which was later on applied to a HF-SSFCW. The inoculated gravel improved the removal efficiency of ibuprofen in the CW only by 5%. However, the key finding of the study was that the addition of these bacteria reduced with more than 80 days the stabilization time of the CW compared with the non-inoculated CWs [62]. In addition, this study found that plants were important to maintain the ibuprofen-degrading bacteria community because roots provided aerobic environments.

The limitation of this novel approach comes from the fact that inoculated microorganisms might be washed out, predated, or killed by environmental conditions; periodic inoculation must be considered to maintain the biomass, which is translated into costs and complex operation [68]. To avoid continuous inoculation, microorganism immobilization has been pointed as a solution [72]. In this case, CWs are especially well suited for it as bacteria could be immobilized in the filling material.

As it happens with the filling material approaches, most processes to enhance biodegradation have been mainly tested at laboratory scale, and further testing in pilot and real scale should be the next research step. This will be important to evaluate the applicability of these technologies in terms of costs and efficiency. Forced aeration is the only approach that has been tested on pilot-scale systems, and it has shown to enhance PhAC removal. According to the results, forced aeration seems to improve the attenuation of PhACs by 20–60% in comparison to conventional HF-SSFCWs, but limited improvements have been observed in comparison to conventional VF-SSFCWs.

3.3 CWs Combined with Other Wastewater Treatment Technologies

Most of the times, solutions to remove PhACs from wastewater are thought in terms of using one specific type of technology, and therefore limitations have to be accepted. However, the combination of complementary treatment processes could overcome some of those limitations. In this sense, intensive technologies such as membrane bioreactors (MBRs), moving bed biofilm reactors (MBBRs), or advanced oxidation processes (AOPs) combined with CWs should be taken into consideration. The specific combination of these intensive technologies with CWs has barely been explored in terms of PhACs elimination.

MBRs and MBBRs are intensive biological wastewater treatment techniques. The combination of these techniques with CWs is not common, and, to the knowledge of the authors, these combinations have not been tested to remove PhACs. So far, only common wastewater treatment studies have been conducted by using these combinations. For example, a submerged MBR (SMBR) acting as a secondary treatment step was coupled with an integrated VF-SSFCW which acted as a tertiary treatment step and reached 97.5, 87.5, and 83.5% removals for COD, total nitrogen, and total phosphorus, respectively, with HRTs of 7.5 h in the SMBR and 11.5 h in the VF-SSFCW [73]. A CW combined with a MBBR has also been tested for the removal of COD, total nitrogen, and total phosphorus and achieved removals of 85.9–97.5%, 46.1–84.5%, and 27.9–69.5%, respectively (HRT: 8–15 h) [74]. Even though these combination systems were not tested for PhACs yet, they could be an interesting solution to eliminate PhACs due to their capacity to develop different microbial communities in each of the technologies and therefore diversifying the possible degradation pathways of these compounds. In fact, MBBRs are able to eliminate certain organic micropollutants in several stages of the wastewater treatment [75–77].

The AOPs are techniques aimed at the production of hydroxyl radicals, one of the most powerful oxidant species, capable of oxidizing a range of organic contaminants [78]. Due to the oxidation process, these techniques generate transformation products (TPs) that can be more toxic than the parent compounds [79]. For this reason, CWs could be added after AOPs to eliminate such products that, moreover, could be more biodegradable than the parent compounds.

Among AOPs, photocatalysis is based on aqueous-phase hydroxyl radical chemistry that combines low-energy UV-A light with semiconductors like TiO_2 that act as photocatalysts [80]. As photocatalysis can be conducted with sunlight, it matches the low-cost approach of CWs. In fact, Melián et al. [80] observed that the combination of photocatalysis with CWs was capable of achieving a complete removal of 100 mg/L of phenol and the generated TPs over 1 day. This was also tested by conducting the photocatalysis of several pesticides spiked into ultra-pure water, mixing the photocatalyzed products with wastewater, and applying it to CWs [81]. Like this, complete removals of most compounds and a reduction of at least 78% of toxicity were achieved, achievements that photodegradation or CWs could not accomplish independently [81]. In the same way, mild photocatalysis combined with sludge-based biological treatment achieved complete removal of recalcitrant PhACs such as diclofenac and gemfibrozil [82]. Additionally, it has been shown that natural organic matter released from CW plants allowed a higher photodegradation kinetic rates of recalcitrant compounds like carbamazepine and paraxanthine (ca. 0.001 min^{-1}, both) than organic matter from a WWTP effluent (ca. 0.00016 min^{-1}, both) [83]. Therefore, photocatalysis combined with CWs could be an interesting sustainable and low-cost solution to enhance the removal of PhACs from wastewater.

Besides photocatalysis, ozonation has also been tested in combination with CWs. In ozonation, oxidation occurs via the ozone itself or by the generated hydroxyl radicals. The combination of HF-SSFCW followed by a polishing ozonation step enhanced the removals of PhACs from real WWTP influent during a laboratory-scale experiment, and ibuprofen and naproxen removals increased from 92 to 94% and 83% to 96%, respectively [84]. In this case, the HF-SSFCW conducted a secondary wastewater treatment, while ozonation was established as a tertiary wastewater treatment. Nonetheless, this CW-ozone combination could be also used as a whole tertiary treatment unit. In this case, ozonation could be used as a first polishing step, and afterwards CWs could biodegrade the remaining PhACs and TPs. In any case, ozonation is an expensive technology, with a potential explosion hazard and toxicity related to its generation [85]. Therefore, before using ozone, other options to eliminate PhACs should be evaluated.

To summarize, the combined use of intensive biological wastewater treatment (MBR and MBBR) or AOPs techniques with CWs is a promising solution to eliminate PhACs. Until now, those combinations have not been yet investigated or only tested at laboratory scale. Particularly, the combination of AOPs with CWs may help to reduce the amount of energy needed to oxidize PhACs as well as to minimize the presence of oxidized TPs generated by the former technology. Furthermore, the water effluent from CWs has a greater ecological richness than that from AOPs, comparable to that found in continental surface waters. Therefore further investigation is needed in this field.

4 Challenges and Future Trends

Despite CWs being a low-cost alternative to conventional WWTPs with a remarkable capacity to remove PhACs, there are still some challenges to be overcome to become a more competitive technology (Fig. 4). Therefore, future research trends should address those challenges that, in several cases, are shared with other wastewater treatment technologies.

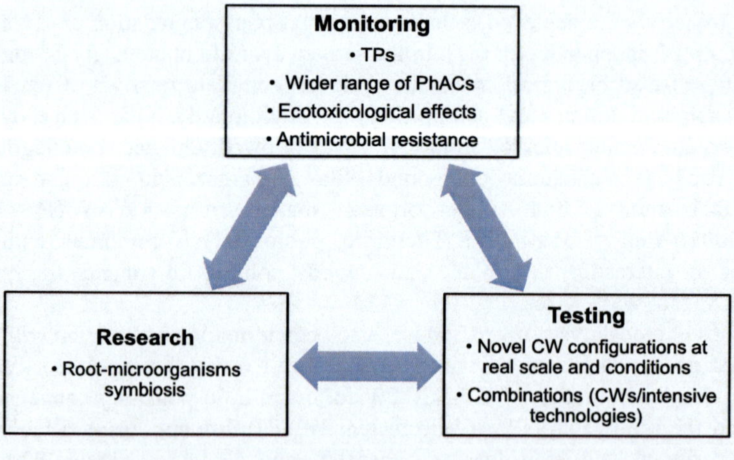

Fig. 4 Overview of future research strategies to overcome the challenges of wastewater treatment with CWs. Monitoring, research, and testing actions also relate to each other

Monitoring: Firstly, most studies about PhAC degradation by CWs or other wastewater treatment technologies have focused on the removal of the parent compounds. However, PhAC degradation may result in TPs which may be more or less toxic than the original compound. As an attempt to solve this, TP formation assessment via toxicity tests has been proposed to complement chemical analysis [86]. On the other side, limited studies in CWs, and only for a few compounds, have demonstrated that several TPs are left after water treatment [8]. Further studies should address the formation and toxicity of TPs [87]. For example, one TP of diclofenac was identified after passing through filtration-UV-chlorination [88], but it was not detected after a CW treatment [89, 90]. Thereafter, the same authors found out that the presence of aquatic plants aided to enhance the removal of this TP from water [91]. Second, PhACs include a large list of compounds, but only a few of them have been studied repeatedly in CWs, hindering a deeper understanding of treatment processes. This phenomenon, known as Matthew Effect, is generally occurring in the study of PhACs in the environment [92]. Taking into account the recent advances in analytical chemistry, future research should analyze a wider range and larger number of compounds. In relation to that, the ecotoxicological effects of PhACs and their TPs should also be monitored and assessed. While conventional toxicological studies of a few compounds can be performed to complement chemical analysis [86, 93], new in silico tools using big data, machine learning, and artificial intelligence seem to be the future way to conduct ecotoxicological assessment studies in CWs [94]. Finally, antimicrobial resistance is a raising threat in global health, and wastewater treatment has been pointed out as one of the main sources of ARGs [95]. Several authors have claimed that CWs are a potential spot for horizontal gene transfer that could diffuse antimicrobial resistant bacteria and ARG to the aquatic environment [96, 97]. This could be particularly significant when bacteria are inoculated into CWs to enhance nutrient and PhAC removal (bioaugmentation). In this case, it would be important to start characterizing the antimicrobial resistance profile of the inoculated bacteria [96]. In contrast, some studies point towards the potential of CWs to attenuate antimicrobial resistance [96, 98]. For example, SFCWs have shown removals of 56–100% for 12 antibiotics without increasing the amount of ARGs [29]. In addition, the usage of zeolites in a SSFCW mesocosm achieved good ARG removal (85% for HLR = 10 cm/days) [41]. Future research should verify and explore the utility of CWs to fight antimicrobial resistance.

Research: An important challenge that CW technology needs to overcome is the high surface area required for wastewater treatment. Besides adsorption, biodegradation is the most important process in CWs, which is remarkably enhanced by the symbiotic interactions between the plants and the microbial community [96]. These interactions are mainly conducted through the release of chemical root exudates as it has been observed in recent publications [99]. Therefore, this interaction is predominantly happening within the root zone (rhizosphere), reducing the effective volume of the CW. By enhancing the symbiotic interactions between plants and the surrounding microbial communities, the surface area/volume demand of the CW could be reduced [100]. This could be performed by the addition of synthetic chemical root exudates to the CWs, as it has already been employed in the enhancement of the bioremediation of contaminated soils [101, 102]. Therefore, future research trends on CWs will certainly need to focus on the understanding of the role of root exudates on the biodegradation enhancement of PhACs.

Testing: Novel CW configurations have only been tested at laboratory scale. Some novel CW configurations have shown promising results removing PhACs but only on a laboratory scale and under controlled conditions (e.g., temperature and synthetic wastewater). Future research should put efforts into testing these novel configurations at pilot or full scale under real conditions. This will be essential to evaluate the applicability of these technologies in terms of costs and efficiency. On the other hand, nearly all novel CW configurations enhance only one removal mechanism: sorption or biodegradation. This approach is favorable to study and understand the efficiency of the technology, but it limits the type of PhACs that can be removed. As a solution, recent studies have started to test combinations of the available novel CW configurations. For example, a full-scale CW has been constructed using LECA, which enhances sorption and biofilm growth, while it operates with forced aeration, which enhances biodegradation [103]. With such configuration, the system achieved removals over 78% of 10 out of 11 PhACs, with high removal of recalcitrant compounds such as carbamazepine and diclofenac. These types of approaches, as well as the combination of novel CWs with intensive wastewater treatment technologies, should be further studied. Additionally, synergies between different approaches have to be considered. For example, the nature of the filling material of the CW determines the associated microbial communities, but it could also be used to enhance the immobilization of bacteria when applying bioaugmentation.

5 Concluding Remarks

CWs are sustainable, low-cost, robust, and efficient wastewater treatment systems that can remove PhACs at least as effectively as conventional WWTPs. It is known that most PhACs undergo biodegradation under aerobic conditions and only a few of them are degraded under anoxic-anaerobic conditions. Nevertheless, lately, different novel CW configurations have been developed to improve the attenuation of PhACs from wastewater. First, novel filling materials offer enhanced adsorption capacities which are, up to now, the only mechanism to remove recalcitrant compounds. Besides PhACs, zeolites have shown capacity to remove ARGs. Interestingly, novel filling materials can also be part of circular economy, as they can be obtained from waste materials. Second, biodegradation processes can be boosted in CWs. MCF-CWs have shown to increase removal of PhACs, but the accumulation of antibiotics on the anode layer could trigger antimicrobial resistance processes. Forced aeration enhanced the biodegradation rates of PhACs by 20–60% in comparison to HF-SSFCWs, but limited improvements were obtained in VF-SSFCWs. Bioaugmentation for PhACs has only been tested for ibuprofen but has shown to be effective for other types of organic contaminants like atrazine. Also, bioaugmentation may shorten the CW adaptation phase, starting to degrade PhACs earlier. Third, the combination of CWs with intensive wastewater treatment technologies such as MBBR and MBR may offer a broader range of biodegradation pathways, while CWs placed after AOPs may allow the biodegradation of generated TPs. Finally, future research should assess TPs, expand the number of studied PhACs, assess ecotoxicity, evaluate antimicrobial resistance processes, investigate root-microorganism symbiosis, and test additional CW combinations to improve CW wastewater treatment technology.

Acknowledgments The authors gratefully acknowledge the financial support of the Spanish Ministry of Science, Innovation and Universities under Project CTM2017-91355-EXP and Generalitat de Catalunya for the Beatriu de Pinós grant-programme.

References

1. Carvalho PN, Arias CA, Brix H (2017) Constructed wetlands for water treatment: new developments. Water. https://doi.org/10.3390/w9060397
2. Vymazal J (2010) Constructed wetlands for wastewater treatment. Water. https://doi.org/10.3390/w2030530
3. Ostfeld RS (2009) Biodiversity loss and the rise of zoonotic pathogens. Clin Microbiol Infect. https://doi.org/10.1111/j.1469-0691.2008.02691.x
4. Fitch MW (2014) Constructed wetlands. In: Comprehensive water quality and purification. https://doi.org/10.1016/B978-0-12-382182-9.00053-0
5. Scholz M, Lee B (2005) Constructed wetlands: a review. Int J Environ Stud 62:421–447. https://doi.org/10.1080/00207230500119783
6. Vymazal J, Kröpfelová L (2009) Removal of organics in constructed wetlands with horizontal sub-surface flow: a review of the field experience. Sci Total Environ. https://doi.org/10.1016/j.scitotenv.2008.08.032
7. Hijosa-Valsero M, Matamoros V, Sidrach-Cardona R et al (2010) Comprehensive assessment of the design configuration of constructed wetlands for the removal of pharmaceuticals and personal care products from urban wastewaters. Water Res:44. https://doi.org/10.1016/j.watres.2010.04.022
8. Ilyas H, van Hullebusch ED (2020) Performance comparison of different types of constructed wetlands for the removal of pharmaceuticals and their transformation products: a review. Environ Sci Pollut Res Int. https://doi.org/10.1007/s11356-020-08165-w
9. Li Y, Zhu G, Ng WJ, Tan SK (2014) A review on removing pharmaceutical contaminants from wastewater by constructed wetlands: design, performance and mechanism. Sci Total Environ. https://doi.org/10.1016/j.scitotenv.2013.09.018
10. Matamoros V, Bayona JM (2013) Removal of pharmaceutical compounds from wastewater and surface water by natural treatments. Compr Anal Chem 62:409–433. https://doi.org/10.1016/B978-0-444-62657-8.00012-4
11. Verlicchi P, Zambello E (2014) How efficient are constructed wetlands in removing pharmaceuticals from untreated and treated urban wastewaters? A review. Sci Total Environ. https://doi.org/10.1016/j.scitotenv.2013.10.085
12. Liu L, Hong LY, Wang Z et al (2014) Behavior of tetracycline and sulfamethazine with corresponding resistance genes from swine wastewater in pilot-scale constructed wetlands. J Hazard Mater. https://doi.org/10.1016/j.jhazmat.2014.06.015

13. Vo HNP, Koottatep T, Chapagain SK et al (2019) Removal and monitoring acetaminophen-contaminated hospital wastewater by vertical flow constructed wetland and peroxidase enzymes. J Environ Manage. https://doi.org/10.1016/j.jenvman.2019.109526
14. Bôto M, Almeida CMR, Mucha AP (2016) Potential of constructed wetlands for removal of antibiotics from saline aquaculture effluents. Water. https://doi.org/10.3390/w8100465
15. Loffler D, Rombke J, Meller M, Ternes TA (2005) Environmental fate of pharmaceuticals in water/sediment systems. Environ Sci Technol 39:5209–5218. https://doi.org/10.1021/es0484146
16. Verlicchi P, Al Aukidy M, Jelic A et al (2014) Comparison of measured and predicted concentrations of selected pharmaceuticals in wastewater and surface water: a case study of a catchment area in the Po Valley (Italy). Sci Total Environ. https://doi.org/10.1016/j.scitotenv.2013.10.026
17. Dordio AV, Carvalho AJP (2013) Organic xenobiotics removal in constructed wetlands, with emphasis on the importance of the support matrix. J Hazard Mater. https://doi.org/10.1016/j.jhazmat.2013.03.008
18. Brix H (1994) Use of constructed wetlands in water pollution control: historical development, present status, and future perspectives. In: Water science and technology
19. Ávila C, Matamoros V, Reyes-Contreras C et al (2014) Attenuation of emerging organic contaminants in a hybrid constructed wetland system under different hydraulic loading rates and their associated toxicological effects in wastewater. Sci Total Environ. https://doi.org/10.1016/j.scitotenv.2013.10.065
20. Brix H (2003) Plants used in constructed wetlands and their functions. Proceedings of 1st international seminar on the use of aquatic macrophytes for wastewater treatment in constructed wetlands
21. Kadlec RH, Wallace S (2008) Treatment wetlands. CRC Press, Boca Raton
22. Matamoros V, Bayona JM (2006) Elimination of pharmaceuticals and personal care products in subsurface flow constructed wetlands. Environ Sci Technol. https://doi.org/10.1021/es0607741
23. Ávila C, Pedescoll A, Matamoros V et al (2010) Capacity of a horizontal subsurface flow constructed wetland system for the removal of emerging pollutants: an injection experiment. Chemosphere. https://doi.org/10.1016/j.chemosphere.2010.08.006
24. Matamoros V, Arias C, Brix H, Bayona JM (2009) Preliminary screening of small-scale domestic wastewater treatment systems for removal of pharmaceutical and personal care products. Water Res 43. https://doi.org/10.1016/j.watres.2008.10.005
25. Matamoros V, Arias C, Brix H, Bayona JM (2007) Removal of pharmaceuticals and personal care products (PPCPs) from urban wastewater in a pilot vertical flow constructed wetland and a sand filter. Environ Sci Technol. https://doi.org/10.1021/es071594+
26. Matamoros V, García J, Bayona JM (2008) Organic micropollutant removal in a full-scale surface flow constructed wetland fed with secondary effluent. Water Res:42. https://doi.org/10.1016/j.watres.2007.08.016
27. Matamoros V, Salvadó V (2012) Evaluation of the seasonal performance of a water reclamation pond-constructed wetland system for removing emerging contaminants. Chemosphere. https://doi.org/10.1016/j.chemosphere.2011.09.020
28. Matamoros V, Duhec A, Albaigés J, Bayona JM (2009) Photodegradation of carbamazepine, ibuprofen, ketoprofen and 17α-ethinylestradiol in fresh and seawater. Water Air Soil Pollut. https://doi.org/10.1007/s11270-008-9765-1
29. Berglund B, Khan GA, Weisner SEB et al (2014) Efficient removal of antibiotics in surface-flow constructed wetlands, with no observed impact on antibiotic resistance genes. Sci Total Environ. https://doi.org/10.1016/j.scitotenv.2013.12.128
30. Rühmland S, Wick A, Ternes TA, Barjenbruch M (2015) Fate of pharmaceuticals in a subsurface flow constructed wetland and two ponds. Ecol Eng. https://doi.org/10.1016/j.ecoleng.2015.01.036

31. Hijosa-Valsero M, Matamoros V, Martín-Villacorta J et al (2010) Assessment of full-scale natural systems for the removal of PPCPs from wastewater in small communities. Water Res 44. https://doi.org/10.1016/j.watres.2009.10.032
32. Ávila C, Bayona JM, Martín I et al (2015) Emerging organic contaminant removal in a full-scale hybrid constructed wetland system for wastewater treatment and reuse. Ecol Eng. https://doi.org/10.1016/j.ecoleng.2014.07.056
33. Hadjiev D, Dimitrov D, Martinov M, Sire O (2007) Enhancement of the biofilm formation on polymeric supports by surface conditioning. Enzyme Microb Technol. https://doi.org/10.1016/j.enzmictec.2006.06.022
34. Zhang L, Lyu T, Ramírez Vargas CA et al (2018) New insights into the effects of support matrix on the removal of organic micro-pollutants and the microbial community in constructed wetlands. Environ Pollut. https://doi.org/10.1016/j.envpol.2018.05.028
35. Dordio AV, Miranda S, Prates Ramalho JP, Carvalho AJP (2017) Mechanisms of removal of three widespread pharmaceuticals by two clay materials. J Hazard Mater. https://doi.org/10.1016/j.jhazmat.2016.05.091
36. Dordio A, Pinto J, Dias CB et al (2009) Atenolol removal in microcosm constructed wetlands. Int J Environ Anal Chem. https://doi.org/10.1080/03067310902962502
37. Dalahmeh S, Ahrens L, Gros M et al (2018) Potential of biochar filters for onsite sewage treatment: adsorption and biological degradation of pharmaceuticals in laboratory filters with active, inactive and no biofilm. Sci Total Environ. https://doi.org/10.1016/j.scitotenv.2017.08.178
38. Özengin N, Elmaci A (2016) Removal of pharmaceutical products in a constructed wetland. Iran J Biotechnol. https://doi.org/10.15171/ijb.1223
39. Dordio A, Carvalho AJP, Teixeira DM et al (2010) Removal of pharmaceuticals in microcosm constructed wetlands using Typha spp. and LECA. Bioresour Technol. https://doi.org/10.1016/j.biortech.2009.09.001
40. Dordio AV, Gonçalves P, Texeira D et al (2011) Pharmaceuticals sorption behaviour in granulated cork for the selection of a support matrix for a constructed wetlands system. Int J Environ Anal Chem. https://doi.org/10.1080/03067319.2010.510605
41. Chen J, Wei XD, Liu YS et al (2016) Removal of antibiotics and antibiotic resistance genes from domestic sewage by constructed wetlands: optimization of wetland substrates and hydraulic loading. Sci Total Environ. https://doi.org/10.1016/j.scitotenv.2016.04.176
42. Machado AI, Dordio A, Fragoso R et al (2017) Furosemide removal in constructed wetlands: comparative efficiency of LECA and Cork granulates as support matrix. J Environ Manage. https://doi.org/10.1016/j.jenvman.2017.08.002
43. Du L, Zhao Y, Wang C et al (2020) Removal performance of antibiotics and antibiotic resistance genes in swine wastewater by integrated vertical-flow constructed wetlands with zeolite substrate. Sci Total Environ. https://doi.org/10.1016/j.scitotenv.2020.137765
44. USGS (2020) Zeolites statistics and information. Zeolites Stat Inf.: https://www.usgs.gov/centers/nmic/zeolites-statistics-and-information. Accessed May 2020
45. Wang H, Xu J, Sheng L (2020) Purification mechanism of sewage from constructed wetlands with zeolite substrates: a review. J Clean Prod. https://doi.org/10.1016/j.jclepro.2020.120760
46. CM Zeolite (2013) Zeolite CM discussion note: Castle Mountain Zeolite or polyacrylamide as water saver crystals. http://www.cmzeolites.com.au/_literature_149098/Zeolite_Compared_to_PAM_as_a_Water_Saver_Cystal
47. Mateus DMR, Pinho HJO (2018) Screening of solid waste as filler material for constructed wetlands. Paper presented in IOP Conference Series: Earth and Environmental Science
48. Aguilar L, Gallegos Á, Arias CA et al (2019) Microbial nitrate removal efficiency in groundwater polluted from agricultural activities with hybrid cork treatment wetlands. Sci Total Environ. https://doi.org/10.1016/j.scitotenv.2018.10.426
49. Mateus DMR, Pinho HJO (2020) Evaluation of solid waste stratified mixtures as constructed wetland fillers under different operation modes. J Clean Prod. https://doi.org/10.1016/j.jclepro.2020.119986

50. Maroušek J, Strunecký O, Stehel V (2019) Biochar farming: defining economically perspective applications. Clean Techn Environ Policy 21:1389. https://doi.org/10.1007/s10098-019-01728-7
51. Lehmann J (2007) Bio-energy in the black. Front Ecol Environ. https://doi.org/10.1890/1540-9295(2007)5[381:BITB]2.0.CO;2
52. Zhou X, Jia L, Liang C et al (2018) Simultaneous enhancement of nitrogen removal and nitrous oxide reduction by a saturated biochar-based intermittent aeration vertical flow constructed wetland: effects of influent strength. Chem Eng J. https://doi.org/10.1016/j.cej.2017.11.066
53. Gupta P, Ann TW, Lee SM (2016) Use of biochar to enhance constructed wetland performance in wastewater reclamation. Environ Eng Res. https://doi.org/10.4491/eer.2015.067
54. Abedi T, Mojiri A (2019) Constructed wetland modified by biochar/zeolite addition for enhanced wastewater treatment. Environ Technol Innov. https://doi.org/10.1016/j.eti.2019.100472
55. Clark DP, Pazdernik NJ (2016) Chapter 12 – environmental biotechnology. In: Clark DP, Pazdernik NJBT-B, Second E (eds). Academic Cell, Boston, pp 393–418
56. Corbella C, Puigagut J (2015) Microbial fuel cells implemented in constructed wetlands: fundamentals, current research and future perspectives. Contrib Sci. https://doi.org/10.2436/20.7010.01.220
57. Li H, Zhang S, Yang XL et al (2019) Enhanced degradation of bisphenol A and ibuprofen by an up-flow microbial fuel cell-coupled constructed wetland and analysis of bacterial community structure. Chemosphere. https://doi.org/10.1016/j.chemosphere.2018.11.022
58. Kahl S, Nivala J, van Afferden M et al (2017) Effect of design and operational conditions on the performance of subsurface flow treatment wetlands: emerging organic contaminants as indicators. Water Res:125. https://doi.org/10.1016/j.watres.2017.09.004
59. Li J, Zhou Q, Campos LC (2017) Removal of selected emerging PPCP compounds using greater duckweed (*Spirodela polyrhiza*) based lab-scale free water constructed wetland. Water Res. https://doi.org/10.1016/j.watres.2017.09.002
60. Li H, Cai Y, Gu Z et al (2020) Accumulation of sulfonamide resistance genes and bacterial community function prediction in microbial fuel cell-constructed wetland treating pharmaceutical wastewater. Chemosphere. https://doi.org/10.1016/j.chemosphere.2020.126014
61. Ávila C, Nivala J, Olsson L et al (2014) Emerging organic contaminants in vertical subsurface flow constructed wetlands: influence of media size, loading frequency and use of active aeration. Sci Total Environ. https://doi.org/10.1016/j.scitotenv.2014.06.128
62. Balciunas EM, Kappelmeyer U, Harms H, Heipieper HJ (2020) Increasing ibuprofen degradation in constructed wetlands by bioaugmentation with gravel containing biofilms of an ibuprofen-degrading Sphingobium yanoikuyae. Eng Life Sci. https://doi.org/10.1002/elsc.201900097
63. Zhang L, Lyu T, Zhang Y et al (2018) Impacts of design configuration and plants on the functionality of the microbial community of mesocosm-scale constructed wetlands treating ibuprofen. Water Res:131. https://doi.org/10.1016/j.watres.2017.12.050
64. Auvinen H, Havran I, Hubau L et al (2017) Removal of pharmaceuticals by a pilot aerated sub-surface flow constructed wetland treating municipal and hospital wastewater. Ecol Eng. https://doi.org/10.1016/j.ecoleng.2016.12.031
65. Uggetti E, Hughes-Riley T, Morris RH et al (2016) Intermittent aeration to improve wastewater treatment efficiency in pilot-scale constructed wetland. Sci Total Environ. https://doi.org/10.1016/j.scitotenv.2016.03.195
66. Rous V, Vymazal J, Hnátková T (2019) Treatment wetlands aeration efficiency: a review. Ecol Eng. https://doi.org/10.1016/j.ecoleng.2019.06.006
67. Free A, McDonald MA, Pagaling E (2018) Diversity-function relationships in natural, applied, and engineered microbial ecosystems. Adv Appl Microbiol. https://doi.org/10.1016/bs.aambs.2018.07.002

68. Meng P, Pei H, Hu W et al (2014) How to increase microbial degradation in constructed wetlands: influencing factors and improvement measures. Bioresour Technol. https://doi.org/10.1016/j.biortech.2014.01.095
69. Shao Y, Pei H, Hu W et al (2014) Bioaugmentation in lab scale constructed wetland microcosms for treating polluted river water and domestic wastewater in northern China. Int Biodeter Biodegr. https://doi.org/10.1016/j.ibiod.2014.05.027
70. Zhao X, Yang J, Bai S et al (2016) Microbial population dynamics in response to bioaugmentation in a constructed wetland system under 10°C. Bioresour Technol. https://doi.org/10.1016/j.biortech.2016.01.043
71. Zhao X, Bai S, Li C et al (2019) Bioaugmentation of atrazine removal in constructed wetland: performance, microbial dynamics, and environmental impacts. Bioresour Technol. https://doi.org/10.1016/j.biortech.2019.121618
72. Suzana Cudia SM, Claudia M, Fiúza LMCG et al (2013) Immobilization of microbial cells: a promising tool for treatment of toxic pollutants in industrial wastewater. Afr J Biotechnol. https://doi.org/10.5897/ajb12.2677
73. Xiao ER, Liang W, He F et al (2010) Performance of the combined SMBR-IVCW system for wastewater treatment. Desalination. https://doi.org/10.1016/j.desal.2008.11.041
74. Lai C, Sun Y, Guo Y et al (2020) A novel integrated bio-reactor of moving bed and constructed wetland (MBCW) for domestic wastewater treatment and its microbial community diversity. Environ Technol. https://doi.org/10.1080/09593330.2019.1709904
75. Escolà Casas M, Chhetri RK, Ooi G et al (2015) Biodegradation of pharmaceuticals in hospital wastewater by staged moving bed biofilm reactors (MBBR). Water Res:83. https://doi.org/10.1016/j.watres.2015.06.042
76. Torresi E (2017) Removal of micropollutants in moving bed biofilm reactors (MBBRs). Dissertation, Danish Technical University
77. Torresi E, Escolà Casas M, Polesel F et al (2017) Impact of external carbon dose on the removal of micropollutants using methanol and ethanol in post-denitrifying moving bed biofilm reactors. Water Res:108. https://doi.org/10.1016/j.watres.2016.10.068
78. Liu R, Zhao Y, Doherty L et al (2015) A review of incorporation of constructed wetland with other treatment processes. Chem Eng J. https://doi.org/10.1016/j.cej.2015.05.023
79. Reungoat J, Escher BI, Macova M et al (2012) Ozonation and biological activated carbon filtration of wastewater treatment plant effluents. Water Res:46. https://doi.org/10.1016/j.watres.2011.11.064
80. Melián JAH, Araña J, Ortega JA et al (2008) Comparative study of phenolics degradation between biological and photocatalytic systems. J Sol Energy Eng. https://doi.org/10.1115/1.2969800
81. Araña J, Garriga i Cabo C, Fernández Rodríguez C et al (2008) Combining TiO2-photocatalysis and wetland reactors for the efficient treatment of pesticides. Chemosphere. https://doi.org/10.1016/j.chemosphere.2007.10.008
82. de Wilt HA (2018) Pharmaceutical removal: synergy between biological and chemical processes for wastewater treatment. Dissertation, Wageningen University
83. Lee E, Shon HK, Cho J (2014) Role of wetland organic matters as photosensitizer for degradation of micropollutants and metabolites. J Hazard Mater. https://doi.org/10.1016/j.jhazmat.2014.05.001
84. Lancheros JC, Madera-Parra CA, Caselles-Osorio A et al (2019) Ibuprofen and naproxen removal from domestic wastewater using a horizontal subsurface flow constructed wetland coupled to ozonation. Ecol Eng. https://doi.org/10.1016/j.ecoleng.2019.05.007
85. Ozonetech (2020) Ozone safety. https://www.ozonetech.com/ozone-safety. Accessed May 2020
86. Escher BI, Fenner K (2011) Recent advances in environmental risk assessment of transformation products. Environ Sci Technol. https://doi.org/10.1021/es1030799
87. Yin L, Wang B, Yuan H et al (2017) Pay special attention to the transformation products of PPCPs in environment. Emerg Contam. https://doi.org/10.1016/j.emcon.2017.04.001

88. Matamoros V, Salvadó V (2013) Evaluation of a coagulation/flocculation-lamellar clarifier and filtration-UV-chlorination reactor for removing emerging contaminants at full-scale wastewater treatment plants in Spain. J Environ Manage 117:96–102. https://doi.org/10.1016/j.jenvman.2012.12.021
89. Matamoros V (2010) Attenuation of emerging contaminants by wetland systems. Cases from Catalonia and Denmark. http://www.phytosudoe.eu/wp-content/uploads/2018/10/10oct_2_Matamoros_emerging-contaminants.pdf
90. Matamoros V, Sala L, Salvadó V (2012) Evaluation of a biologically-based filtration water reclamation plant for removing emerging contaminants: a pilot plant study. Bioresour Technol. https://doi.org/10.1016/j.biortech.2011.11.036
91. Matamoros V, Nguyen LX, Arias CA et al (2012) Evaluation of aquatic plants for removing polar microcontaminants: a microcosm experiment. Chemosphere. https://doi.org/10.1016/j.chemosphere.2012.04.004
92. Daughton CG (2014) The Matthew effect and widely prescribed pharmaceuticals lacking environmental monitoring: case study of an exposure-assessment vulnerability. Sci Total Environ. https://doi.org/10.1016/j.scitotenv.2013.06.111
93. Han GH, Hur HG, Kim SD (2006) Ecotoxicological risk of pharmaceuticals from wastewater treatment plants in Korea: occurrence and toxicity to *Daphnia magna*. Environ Toxicol Chem 25:265–271. https://doi.org/10.1897/05-193r.1
94. Kar S, Sanderson H, Roy K et al (2020) Ecotoxicological assessment of pharmaceuticals and personal care products using predictive toxicology approaches. Green Chem. https://doi.org/10.1039/C9GC03265G
95. Pazda M, Kumirska J, Stepnowski P, Mulkiewicz E (2019) Antibiotic resistance genes identified in wastewater treatment plant systems – a review. Sci Total Environ. https://doi.org/10.1016/j.scitotenv.2019.134023
96. Riva V, Riva F, Vergani L et al (2020) Microbial assisted phytodepuration for water reclamation: environmental benefits and threats. Chemosphere. https://doi.org/10.1016/j.chemosphere.2019.124843
97. Zhang S, Lu YX, Zhang JJ et al (2020) Constructed wetland revealed efficient sulfamethoxazole removal but enhanced the spread of antibiotic resistance genes. Molecules. https://doi.org/10.3390/molecules25040834
98. Chen QL, Cui HL, Su JQ et al (2019) Antibiotic Resistomes in plant microbiomes. Trends Plant Sci. https://doi.org/10.1016/j.tplants.2019.02.010
99. de la Paz A, Salinas N, Matamoros V (2019) Unravelling the role of vegetation in the attenuation of contaminants of emerging concern from wetland systems: preliminary results from column studies. Water Res 166. https://doi.org/10.1016/j.watres.2019.115031
100. SIMAGUA Project (2019) CTM2017-91355-EXP. https://simaguasite123me/. Accessed May 2020
101. Lu H, Sun J, Zhu L (2017) The role of artificial root exudate components in facilitating the degradation of pyrene in soil. Sci Rep. https://doi.org/10.1038/s41598-017-07413-3
102. Urbaniak M, Mierzejewska E, Tankiewicz M (2019) The stimulating role of syringic acid, a plant secondary metabolite, in the microbial degradation of structurally-related herbicide. MCPA PeerJ. https://doi.org/10.7717/peerj.6745
103. Auvinen H, Gebhardt W, Linnemann V et al (2017) Laboratory- and full-scale studies on the removal of pharmaceuticals in an aerated constructed wetland: effects of aeration and hydraulic retention time on the removal efficiency and assessment of the aquatic risk. Water Sci Technol. https://doi.org/10.2166/wst.2017.328

Microalgae-Based Processes for the Removal of Pharmaceuticals in Wastewater

Paqui Blánquez, Andrea Hom-Díaz, Teresa Vicent, and Benoit Guieysse

Contents

1 An Introduction to Microalgae and Microalgae-Based Wastewater Treatment 192
  1.1 Microalgae Biochemistry and Ecological Significance 192
  1.2 Microalgae Biotechnology 193
2 Pharmaceutical Removal Under Controlled Conditions During Simulated Microalgae-Based Wastewater Treatment 195
  2.1 General Mechanisms of Pharmaceutical Removal During Wastewater Treatment . 195
  2.2 Relevance of Individual Mechanisms to Microalgae-Based Wastewater Treatment 196
3 Algae Reactor Design for Wastewater Treatment 201
  3.1 Photobioreactor Design 201
4 Removal of Pharmaceutical Compounds from Wastewaters in Microalgae Reactors 206
  4.1 Pharmaceutical Removal in Synthetic WW 206
  4.2 Pharmaceutical Removal in Real WW 213
5 Concluding Remarks 217
References 218

Abstract Microalgae-based technologies deliver economic and effective treatment for municipal and industrial effluents while also providing capacity for nutrient recovery and the development of algae-derived products. In order to assess the potential of microalgae-based technologies for the removal of pharmaceuticals (PhACs) found in wastewater effluents, this chapter aims to (1) introduce microalgae and their ecological significance; (2) describe the principle of pollutant removal in

P. Blánquez (✉), A. Hom-Díaz, and T. Vicent
Chemical, Biological and Environmental Engineering Department, Escola d'Enginyeria, Universitat Autònoma de Barcelona, Barcelona, Spain
e-mail: paqui.blanquez@uab.cat; andreahomdiaz@gmail.com; teresa.vicent@uab.cat

B. Guieysse
School of Engineering and Advanced Technology, Massey University, Palmerston North, New Zealand
e-mail: B.J.Guieysse@massey.ac.nz

Sara Rodriguez-Mozaz, Paqui Blánquez Cano, and Montserrat Sarrà Adroguer (eds.), *Removal and Degradation of Pharmaceutically Active Compounds in Wastewater Treatment*, Hdb Env Chem (2021) 108: 191–222, DOI 10.1007/698_2020_682,

microalgae-based systems; (3) review the general mechanisms of pharmaceutical removal occurring during wastewater treatment and discuss their potential significance in microalgae-based systems; (4) describe the main photobioreactors used for microalgae-based wastewater treatment; and (5) summarize recent progress on microalgae-based PhACs removal with particular focus on efficiency and mechanisms at pilot and full-scale systems. Overall, the data available demonstrate microalgae based-systems providing capacities for PhACs removal via biodegradation, biosorption and bioaccumulation that are similar or higher than the capacities commonly reported during conventional biological treatment. In addition, microalgae-based wastewater system also provides the capacity for pollutant photodegradation and volatilization. Full-scale demonstration during long-term treatment remains needed.

Keywords Microalgae, Pharmaceuticals (PhACs), Photobioreactors, Wastewater treatment

1 An Introduction to Microalgae and Microalgae-Based Wastewater Treatment

1.1 Microalgae Biochemistry and Ecological Significance

Microalgae represent a phylogenetically diverse group of unicellular eukaryotes that can be found grouped into chains. These microorganisms are found in almost any environment (e.g. lakes, rivers, oceans, soil), including in extreme habitats (e.g. ice, hypersaline environments), and can develop symbiotic relationships with other organisms (e.g. with fungi in lichen). Prokaryotic cyanobacteria have similar abilities and functions, and consequently, eukaryotic and prokaryotic microbial phototrophs will henceforth be commonly referred to as microalgae for simplicity.

Microalgae carry out the photosynthetic reduction of carbon dioxide (CO_2) into biomass using sunlight as energy source. Briefly, light-dependent reactions photolyse water into O_2 and convert sunlight energy onto energy carrier ATP (adenosine triphosphate) and NADPH (nicotinamide adenine dinucleotide phosphate). Light-independent reactions then use these ATP and NAPDH to reduce CO_2 into 3-carbon sugars that are then further processed into cellular components. The conversion of sunlight energy into chemical energy is typically carried out by chlorophyll, but other pigments can be involved. Beside carbon fixation, microalgae must also access the minerals needed to build cell components and enzymes, especially nitrogen (N) and phosphorus (P). N is found in amino acids (the component of proteins) and nucleotides (the component of nucleic acids) and can represent up to 10% of the total dry cell weight. N is generally obtained from ammonium (NH_4^+) or nitrate (NO_3^-), and the latter must reduce the former before incorporation into amino acids. P is typically found in phospholipids, nucleotides and ATP and represents approximately 1% of the total dry cell weight, although various microalgae can intracellularly accumulate P as polyphosphate up to 3–4% of the cell weight [1]. Interestingly, nutrient starvation (especially N) can trigger lipid accumulation in microalgae cells, with interesting application for bioenergy generation.

Microalgae have critical global ecological functions as primary producer (converting carbon dioxide into organic matter) and oxygen producer and for nitrogen fixation. Microalgae can however also generate ecological issues when they have been introduced in new environments, or when nutrient pollution (N and/or P) cause excessive algae growth leading to the eutrophication of water bodies. As anthropogenic pollution causes massive eutrophication issues worldwide, microalgae may also contribute to global anthropogenic N_2O emissions [2] because these microorganisms can synthetize this potent greenhouse gas, including during wastewater treatment [3]. Fortunately, the ability of microalgae to effectively scavenge N and P from aquatic bodies also provides means of water treatment and nutrient recycle [1], as further discussed below.

1.2 Microalgae Biotechnology

Industrial applications of microalgae currently focus on food production using a few species (e.g. *Chlorella vulgaris* and *Arthrospira platensis*) cultivated in raceway ponds, but there is considerable interest in also using microalgae to produce high-value bioactive compounds such as pharmaceuticals, nutraceuticals or cosmetics. In addition, and while microalgae cultivation for bioenergy (e.g. biodiesel) generation has generated considerable interest in the past, this application is nowadays considered to be neither economic nor sustainable, and research in this area is now focusing on combining microalgae-based wastewater treatment with bioenergy generation.

Microalgae are commonly found (at least periodically) in maturation and stabilization ponds during wastewater treatment (WWT), but their growth in these systems is opportunistic (and often problematic) rather than engineered. In contrast, the high rate algal ponds (HRAPs) developed by Oswald and co-workers in the 1960s were specifically designed and operated to enhance microalgae activity by optimizing light supply to these microorganisms (see Sect. 3). These systems rely on the exchange of materials between microalgae and heterotrophic microorganisms to remove pollutants: the microalgae carry out the photosynthetic fixation of CO_2 into biomass, thereby releasing O_2 from the photolytic oxidation of water. This O_2 is then used as a final electron acceptor by heterotrophs converting organic pollutants into biomass and CO_2, the latter becoming available as a carbon source for microalgae photosynthetic growth (Fig. 1). This in situ biological oxygen supply can significantly reduce energy consumption in comparison to mechanically aerated treatment alternatives (Box 1). Many microalgae can also carry out heterotrophic growth at

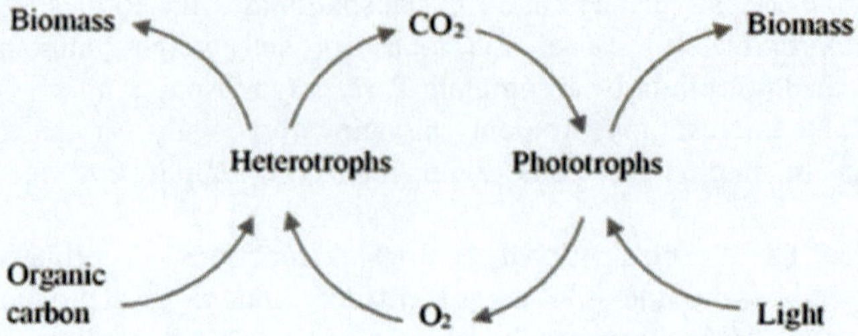

Fig. 1 Principle of organic pollutant removal in HRAP (adapted from [4])

night and even during daytime and therefore provide additional capacity (to strict heterotrophs) for removing organic pollutants.

In addition to providing O_2, photosynthesis occurring during microalgae-based WWT produces additional biomass, which improves N and P removal by assimilation. This additional biomass must however be harvested to achieve wastewater discharge compliance, and sludge management can represent up to 50% of WWT costs. Sludge can be anaerobically digested, and the additional algal biomass means more biogas is produced per unit of wastewater treated than during 'conventional' treatment [5]. However, the algal biomass can be hard to digest anaerobically, and digestion releases N and P into a stream (digestate) that must be further managed. The digestate and/or undigested biosolids may be used as fertilizer [5], but this practice is practically limited by logistic issues (fertilizers are only needed in certain places at certain times) and the risks of pollutant spreading (e.g. pathogens or heavy metals) onto agricultural products (this also limits the use of raw biomass as animal feed). Sludge management is therefore especially critical in the context of hazardous pollutants such as pharmaceuticals because sorption onto the sludge can cause the potential release (or transfer) of these pollutants during sludge disposal. There is currently considerable interest in developing biorefineries based on algae-based WWT, but major risks remain when the end products are intended for animal or human consumption, and these processes are difficult to control considering the variability of wastewater.

Box 1 Microalgae-Based Wastewater Treatment: A Case Study
Posadas et al. [6] sized a HRAP for the treatment of secondary domestic wastewater with a daily average flow rate of 920 m^3/d representative of 2000 people equivalents (PE). The treatment system required to achieve year-round bCOD (biodegradable chemical oxygen demand) and TN (total nitrogen) compliance under temperate climate was estimated to 12.9 m^2 per PE, for an energy requirement for pond-mixing of 0.039 kWh per m^3 of wastewater treated. In comparison, treating the same effluent with an activated sludge process would only require around 0.04 m^2 per PE, but at an energy

(continued)

Box 1 (continued)
input of 0.23 kWh/m^3 for combined bCOD/TN removal [7]. As can be seen, biological oxygen supply can significantly reduce energy consumption compared to mechanically aerated treatment alternatives, but the relatively low photosynthetic efficiency of microalgae means large areas are required to supply enough light to the microalgae. This large land requirement can be limiting but land value generally appreciates over time, and the large treatment volume thus provided buffers flow variations. Critically, Posadas et al. [6] also showed that sizing an HRAP for N removals de facto provides enough oxygenation capacity for bCOD removal, even without prior primary treatment (preliminary treatment is still needed). Altogether, Posadas et al.'s [6] analysis demonstrated HRAPs are best used when combined bCOD and TN is needed at relatively small scale. This does not, however, exclude their integration within existing treatment systems to, as examples, provide nutrient removal following secondary treatment [8], centrate treatment [9] or upgrade biogas [10].

2 Pharmaceutical Removal Under Controlled Conditions During Simulated Microalgae-Based Wastewater Treatment

2.1 General Mechanisms of Pharmaceutical Removal During Wastewater Treatment

The mechanisms driving pharmaceuticals (PhACs) removal during conventional biological WWT are well-established and include hydrolysis, volatilization, biodegradation, biosorption and bioaccumulation. During advanced WWT, the mechanisms targeted include sorption (e.g. onto activated carbon) and chemical oxidation with radicals generated chemically or photochemically. A distinction should be made between 'destructive' mechanisms that (should) convert pollutants into harmless substances (e.g. hydrolysis, biodegradation and chemical oxidation) and mechanisms that only transfer these pollutants into another medium (e.g. a sorbent that must be regenerated or safely disposed of). Destructive mechanisms may however lead to the formation toxic and/or recalcitrant products: this must be assessed and avoided.

The field efficiency of the mechanisms listed above depends on the chemistry of pollutants (some being intrinsically more biodegradable or reactive than others), their concentrations (the rate of removal generally increases with concentration), wastewater composition and environmental conditions (e.g. temperature). Removal is thus especially influenced by the presence of substances that can either enhance (e.g. radical precursors or co-substrates for co-metabolism) or reduce (e.g. substances competing for radicals or sorption sites) removal. The presence of interfering substances is especially challenging because emerging pollutants such as PhACs are typically found at very low concentrations in wastewater, meaning they are easily 'outcompeted' by other substances.

2.2 *Relevance of Individual Mechanisms to Microalgae-Based Wastewater Treatment*

To establish the potential significance of a removal mechanism, pollutant removal must often be assessed under specific conditions that 'isolate' this mechanism against a negative control where no significant removal occurs. In the context of microalgae-based WWT, hydrolytic removal is therefore typically assessed in darkness in clean water (this test is often the negative control) and/or chemically sterilized wastewater. Photodegradation is tested under light irradiation in either clean (sometimes buffered) water, defined media (where known interfering substances are added at known concentrations) or sterilized wastewater (that can also be spiked with interfering substances). Biodegradation can be tested under darkness (under artificial aeration, if needed) or under light illumination (using UV filters and/or controls to avoid/assess photodegradation) in defined media or wastewater inoculated with active microalgae or microalgae consortium. Biosorption and bioaccumulation can be tested with dead or live biomass (generally under darkness). It can be difficult to distinguish between mechanisms (e.g. biosorption from bioaccumulation), so several controls and advanced analytics will be needed. Various levels of reactants and/or combinations of conditions should also be tested (e.g. incubation times, light intensity, biomass concentration), especially when kinetics are quantified, and experiments must be replicated under controlled conditions to demonstrate statistical significance. The need to conduct many experiments under rigorously controlled conditions generally means that relevance must be sacrificed for data quality (e.g. the pollutants are often spiked at relatively high concentrations, and microcosms are incubated under constant artificial light, temperature, etc.). The results can therefore statistically demonstrate the significance of a mechanism under the conditions tested, but they may not evidence that the mechanism investigated is also significant under real-life conditions (e.g. low pollutant concentration, variable conditions, etc.). The following sections describe and demonstrated mechanisms and discuss their full-scale potential during microalgae-based treatment. Please refer to the reviews by Wang et al. [11], Leng et al. [12], [13] and Norvill et al. [14] for more examples and in-depth discussion.

Biodegradation by heterotrophs is a major mechanism of PhACs removal during conventional biological WWT and is generally best carried out under aerobic conditions. This potential is also established for microalgae and microalgal consortia for a broad range of PhACs. For example, Peng et al. [15] studied the biotransformation of progesterone and norgestrel by the microalgae *Scenedesmus obliquus* and *Chlorella pyrenoidosa*. Using controls, these authors showed that hydrolysis and volatilization were negligible and that biosorption to dead algae cells or bioaccumulation to live cells had little impact under the conditions studied. Instead, they reported pollutant-dependent and species-dependent biodegradation and, based on the transformation products detected, showed this biodegradation likely involved hydroxylation, reduction and oxidation as main pathways.

While many PhACs can inhibit microalgae at very high concentrations found in certain effluents (ppm), toxicity will typically not prevent biodegradation at the concentration found in domestic wastewater (ppt). Instead, low pollutant concentrations may limit biodegradation when the degrading microorganisms do not 'recognize' a pollutant as a potential substrate below a certain threshold. For this reason, co-metabolism is suspected to be the main mechanism for trace pollutant biodegradation during biological WWT.

Biodegradation by microalgae and other microorganisms generally involves a series of sequential steps catalysed by enzymatic systems such as cytochrome P450 [15, 16]. As most pharmaceuticals are relatively soluble in comparison to persistent pollutants such as dioxins and PCBs, lack of catabolic ability and/or low bioavailability is unlikely to prevent their biodegradation. Each reaction step also typically yields products that are more soluble and readily biodegradable, meaning there is little risk of metabolite accumulation. A high biodiversity is generally seen as beneficial for the removal of hazardous pollutants as it increases the chance that the required metabolic abilities are available and as it can provide alternative pathways via synergetic interactions (although competition also occurs). In this regard, microalgae-based WWT systems harbour microalgae as well as diverse heterotrophic communities that are also likely to possess a high metabolic versatility [6]. These systems can also harbour ammonia oxidizers with well-established pollutants' co-metabolic abilities.

Biodegradation efficiency typically increases with the mean cell retention time (MCRT) during conventional biological WWT. A high MCRT is practically achieved by recycling (e.g. activated sludge process), retaining (e.g. membrane bioreactors) or attaching (e.g. trickling biofilter) cells during conventional WWT. However, cells are typically not recycled, retained or attached in photobioreactors, although these configurations have been successfully tested [17–19]. Instead, the MCRT is increased via operation at high average hydraulic retention times (HRTs) as the HRT equals the MCRT when cells are not retained or recycled. Hence, biodegradation should also be considered as the major removal mechanism during full-scale microalgae-based WWT, and the microalgae may either biodegrade the pollutants or support other biodegrading species. Finally, HRAP broth conditions can vary from supersaturated in O_2 during daytime (when photosynthetic O_2 output $>$ O_2 demand) to anaerobic at night (if atmospheric O_2 transfer $<$ demand). This variability may enhance biodiversity and enable anaerobic-aerobic degradation pathways [20].

Adsorption onto bacterial and microalgal biomass is a well-established passive mechanism for the removal of relatively hydrophobic PhACs. Biosorption efficiency is highly dependent on pollutant structure (hydrophobicity and functional groups) and sorbent properties (i.e. the hydrophobicity and functional groups found on algal cell walls and extracellular substances). For example, Santaeufemia et al. [21] reported oxytetracycline maximum sorption capacities of 29.18 mg/d and 4.54 mg/g for live and dead biomass of the microalga *Phaeodactylum tricornutum*, respectively, while de Godos et al. [22] evidenced tetracycline sorption onto biomass dominated by *Chlorella vulgaris*. At full scale, the significance of this mechanism will also depend on interfering substances competing for binding sites and environmental conditions impacting sorption equilibrium and/or kinetics (e.g. temperature, water hardness, pH). A high biomass productivity is also critical, meaning photosynthetic growth should provide enhanced capabilities for pollutant removal via adsorption.

Bioaccumulation involves the active (energy-demanding) transfer and storage of pollutants inside the cells. Bioaccumulation efficiency is highly dependent on pollutant structure and algae biology. For example, Stravs et al. [16] studied the bioaccumulation of 15 fungicides and 9 PhACs (in various mixtures) in the cyanobacteria *Microcystis aeruginosa* and *Synechococcus* sp. as well as the green alga *Chlamydomonas reinhardtii*. These authors reported bioaccumulation factors up to 1,000 and above and species-dependant correlations between the hydrophobicity of a pollutant and its bioconcentration factor. As for biosorption, enhanced biomass productivity can increase the significance of bioaccumulation during microalgae-based WWT.

Photodegradation is a well-established removal mechanism for many PhACs in aquatic environments [23] and during certain advanced oxidation processes (AOPs). This mechanism can involve direct photolysis, when the pollutant absorbs light energy (generally UV-B or low UV-A) and is then converted into a new product, and photooxidation when light irradiation causes the formation of radicals that then oxidize the pollutants (these radicals can be formed from the irradiations of photosensitizers, or they can be released by photosynthesizing microalgae). The potential of these mechanisms is well demonstrated in laboratory studies: for example, Bai and Acharya [24] evaluated the removal of seven endocrine-disrupting chemicals (EDCs) in wastewater (WW) inoculated with the green microalga *Nannochloris* sp. Using various inoculums, these authors demonstrated triclosan was removed by photolysis, even in the presence of the microalgae, but this mechanism has little impact on the other pollutants tested. Efficiency is therefore dependent on pollutant structure (e.g. ability to absorb UV radiation) and broth characteristics (e.g. presence of photosensitizers or competing compounds). Given that photobioreactors (PBs) are specifically designed to optimize light supply (see Sect. 3), PhACs photodegradation may be significant during microalgae-based WWT (unlike during conventional WWT). It is however difficult to assess the full-scale potential of photodegradation because the microalgae can 'shade' pollutants (or photosensitizers) from direct light exposure. This effect may however be counteracted by mixing, enabling frequent (but brief) exposure to high-light intensities at or near the reactor surface. Finally, photodegradation can be incomplete but photodegradation products can be subsequently biodegraded [25].

Hydrolysis is generally not seen as a main route of PhACs removal during WWT [26], although certain compounds such as β-lactam antibiotics can hydrolyse rapidly in microalgae microcosms [27]. In these cases, hydrolysis may initiate biodegradation. To our knowledge, this mechanism is often not tested under the combination of high temperature and high pH that can occur during microalgae-based WWT, meaning it may be underestimated during laboratory studies. Hydrolysis can be enzymatically catalysed, but this mechanism should then be considered as biodegradation.

Volatilization of pollutants has been evidenced during laboratory assays, but only in the case of volatile substances with high Henry's law constant (H) values. For example, Matamoros et al. [28] reported the volatilization of non-ionized pollutants with H values of 0.1234–0.323 Pa m^3 mol^{-1} (4-octylphenol, galaxolide and tributyl phosphate) in aerated 2.5 L microcosms containing green microalgae, whereas ionized (e.g. ibuprofen) and low H pollutants (e.g. caffeine) were not removed by this mechanism. The high surface-to-volume ratio of HRAPs may support significant volatilization for high H-values of PhACs, but low concentration will strongly limit mass transfer from the aqueous broth to the atmospheric phase.

To conclude, microalgae-based WWT design and operation enables the development of diverse populations of phototrophic and heterotrophic microorganisms, many of which harbour heterotrophic biodegradation capacities. Microalgae activity may also support bacterial biodegradation via synergistic associations, and phototrophic biomass production likely benefits PhACs removal via biosorption and bioaccumulation (following biomass harvesting and disposal). Microalgae-based WWT systems are also specifically designed to optimize light supply to microalgae, which may enable PhACs photolysis and photodegradation and, in specific cases, pollutant volatilization from open ponds. Photosynthetic activity also drives daily variations in temperature, pH and dissolved oxygen (DO) concentrations that may impact the kinetics of removal mechanisms, including hydrolysis. High HRT operation will finally improve the quantitative significance of all removal mechanisms considered.

Overall, the design and operation of microalgae-based WWT appears to support the removal potential of all known mechanisms. This is especially important given that the efficiency of individual mechanism depends on pollutant structure, meaning microalgae-based WWT should be able to remove a broad range of contaminants (from volatile to biodegradable and hydrophobic). Different removal mechanisms may compete for pollutant removal and thus provide redundancy, and several mechanisms may even be required for complete treatment: for example, bioaccumulation may be required before cells can initiate biodegradation, and biosorption/bioaccumulation may act as a buffer to maintain efficiency under rapidly changing influent pollutant concentration or when the main removal process ceases. This effect was demonstrated by Hom-Diaz et al. [29] who showed the removal of the antibiotic ciprofloxacin was driven by photodegradation during daytime and biosorption at night. Similar results were reported by Norvill et al. [30] for tetracycline. Table 1 provides a summary of the removal mechanisms and their potential significance.

Table 1 Potential significance of main pharmaceutical removal mechanisms in HRAPs (see [14] for further discussion and original references; volatilization is not considered as broadly significant and is therefore not included)

Mechanism	Critical parameters	Relevance to HRAP
Pharmaceuticals are mainly removed via biodegradation and biosorption during conventional WWT. Biodegradation is generally more efficient for soluble pollutants, while biosorption is more efficient for hydrophobic pollutants, but many other factors must be taken into consideration	High biodiversity and aerobic conditions are generally beneficial[a] A high MCRT can be achieved by recycling, retaining, or attaching cells, or by using a large reactor volume (high HRT) A high MCRT increases biodiversity, which increases biodegradation potential	Microalgae can biodegrade numerous pharmaceuticals and indirectly support other degrading species The variability of HRAP broth conditions may enable new biodegradation pathways[b] HRAP are generally operated at high MCRT (=HRT) of 7–10 days, which is comparable to most WWT processes[c]
Sorption onto activated carbon is commonly used for hazardous pollutant removal. Alternative materials can be used, including active biomass, but costs and other issues have hitherto prevented commercialization	Biosorption and bioaccumulation efficiency depends on pollutant structure, cell biology and biochemistry. At constant biosorption affinity/bioaccumulation ratio, biomass productivity determines the amount of pollutant removed when biomass is harvested	Microalgae can enhance pharmaceutical removal via biosorption and bioaccumulation. Algae activity also boosts the overall biomass productivity
Photolysis and photodegradation (direct or indirect) are powerful mechanisms of pollutant removal in many aquatic environments	Removal kinetics and efficiency are highly dependent on wastewater chemistry (interfering substance or sensitizers), light attenuation and individual pollutant chemistry	High HRT operation and mixing increase the probability of pollutant exposure to UV and/or radicals. Algal biomass causes rapid light attenuation, but algal activity may enhance photodegradation via the production of radicals and photosensitizers and/or by increasing pH, DO and temperature during daytime
Hydrolysis enables the breakdown of large biopolymers prior to their biodegradation	Hydrolysis is highly dependent on wastewater composition, pollutant chemistry and HRT	Hydrolysis is only significant for a few pharmaceuticals. The combination of high HRT operation, high pH and high temperature may increase hydrolysis

[a]Reduced pollutants, such as pentachlorophenol, are often best biodegraded anaerobically by serving as electron acceptor. Anaerobic degradation is however often incomplete and must be completed by aerobic treatment

[b]Facultative ponds are stratified (from anaerobic in the sedimentation zone to aerobic near the surface) and other stabilization ponds can be anaerobic or aerobic. Most 'intensive' WWT processes are operated aerobically, but certain activated sludge process configuration supports cell recycle between anaerobic, anoxic and aerobic zones. Biofilm bioreactors are generally operated aerobically but anaerobic micro-zones can be found in deep biofilm (or biogranules)

[c]Higher MCRT are used in membrane and biofilm bioreactors targeting hazardous pollutants

3 Algae Reactor Design for Wastewater Treatment

The geometry and operation of photobioreactors (PBs) determine how light is supplied to the microalgae and, therefore, how efficient the microalgae use this light during photosynthesis. The light source and its intensity are critical factors for microalgal growth, and light must often be 'diluted' throughout the entire culture, while excessive shading should be reduced by mixing. Several external factors affect microalgae activity during cultivation outdoors, such as day/night cycles and seasonal or geographic changes in climatic conditions. The pH of the culture medium affects the biochemical reactions taking place and is often optimal around pH 7–9. During daytime, pH can increase if the photosynthetic uptake of CO_2 exceeds its replenishment rate via atmospheric diffusion and microbial respiration, and in these cases, the pH also decreases overnight as photosynthesis stops [31].

3.1 Photobioreactor Design

Works on large-scale algal cultures began in the United States in 1948–1950, and the first significant outdoor pilot plant studies on the production of *Chlorella* were carried out in 1951 (Massachusetts, USA). It was not until the 1950s that commercial farming started in Germany. The potential of microalgae to provide oxygen and capabilities for biomass valorization during wastewater treatment was also demonstrated in the USA in the 1960s. Microalgae-based WWT indeed provides additional capacities for nutrient and organic matter removal at a much lower energy requirement compared to conventional activated sludge systems relying on mechanical aeration [32, 33]. Microalgae-based WWT has therefore been demonstrated to deliver economic and efficient treatment [34, 35], and, from the perspective of algae cultivation, domestic wastewater streams are readily available and cost-effective substrates for microalgal biomass production [36–39]. Many bioreactors can be used for the dual purpose of removing nutrients from wastewaters and growing biomass, and several reviews have discussed these aspects in detail [4, 40–44]. Algae cultivation systems are commonly classified as open and closed systems as further discussed below.

3.1.1 Closed Systems

One of the first closed photobioreactors was built by Gudin [45] as a pilot plant with a total working volume of 6.5 m^3. This reactor was operated for several years to cultivate *Porphyridium cruentum* (productivity: 20–25 g/ (m^2·d)) for the production of an exocellular sulphated polysaccharide used as a thickening agent. Many designs have been tested at the laboratory or pilot scale since this pioneer work (Fig. 2), including vertical reactors, flat plate, plastic bags, green wall panel and annular and

Fig. 2 Microalgal closed photoreactors. Top: vertical column; Bottom left: flat panel; Bottom right: horizontal tubular

tubular reactors. The use of closed photobioreactors prevents direct evaporation, minimizes contamination risks and CO_2 losses and potentially enables high light utilization efficiencies boosting biomass productivity; all these attributes also lead to more reproducible conditions. The disadvantages of closed PBs include high construction costs, biomass fouling, sedimentation, leakage [46] and difficulties in economically controlling temperature during outdoor cultivation. In addition, the scale-up of closed systems can be limited by high respiration losses at high cell density operation [47]. Consequently, closed PBs are currently best used for the cultivation of pure strains and high-value algal products [48], and their design seeks to optimize light supply, mixing and mass transfer according to the unique physiological and growth characteristics of the strain cultivated.

In order to reduce equipment costs, transparent bags (i.e. plastic) have been located on rigid frames [49]. Tubular systems are widely used and made of transparent pipes with small internal diameters (the working volume can be increased by extending the tube length). The main disadvantage of tubular PBs is the high mixing energy consumption in long continuous tubes [44]. Vertical column PBs are made of vertical tubing equipped with a gas sparging system installed at the bottom of the reactor as agitation system. Bubble column and airlift reactors are examples of vertical column PBs [44]. The configuration of column PBs is simple as these reactors contain no moving parts. Hence, microalgae only experience mild shear stresses. This configuration also support efficient gas-liquid mass transfer, and bubble behaviour plays an important role in CO_2 transfer in the microalgae solution, which in turn affects microalgal growth and CO_2 uptake efficiency. Flat panel PBs are simple and advantageous in terms of biomass production and energy consumption as large illumination surface areas can improve light supply. Mixing is achieved using rising air bubbles.

3.1.2 Open Systems

Open ponds are the most common platform used for large-scale outdoor microalgae cultivation due to their relatively low capital and operational costs. They are also the most suitable for WWT, where they are usually known as high rate algal ponds (HRAPs). In these systems, microalgae are cultivated in 0.2–0.5 m deep open channels stirred with a paddle wheel, which is economic and easy to build and operate. On the downside, open ponds are prone to contamination, due to constant exposure to atmospheric pollutants, and prone to low productivity resulting from poor mixing, low light penetration and difficult control. Evaporation losses can also be considerable.

Microalgae can be found in three main types of ponds used for wastewater treatment: facultative ponds, maturation ponds and HRAPs (Fig. 3). Facultative

Fig. 3 Microalgal open system: high rate algal pond

and maturation ponds are often collectively referred to as waste stabilization ponds (WSP) or WWT lagoons and are widely used by farms and small communities [50–52]. While facultative ponds are 1–2 m deep and receive raw WW or anaerobic waste effluent, maturation ponds are 1–1.5 m deep and receive secondary effluents from conventional WWT: maturation ponds are therefore used for tertiary WWT with the main objective of pathogen removal [14]. A major disadvantage of WWT ponds is the relatively large land requirement compared with electromechanical treatment systems [5]. HRAPs retain most of the advantages of conventional ponds (simplicity and economy) but overcome many of their drawbacks (poor and highly variable effluent quality, limited nutrient removal). Critically, HRAPs also recovers wastewater nutrients as a harvestable algal/bacterial biomass that can be used as fertilizer or biofuel [5]. The effluent can be partially recycled in order to compensate evaporation losses [53]. HRAPs require low power input and are relatively easy to maintain [47]. They can be constructed of concrete, glass fibre or membrane [54] and can receive raw wastewater, primary effluent or effluent from a previous waste treatment pond [14]. A HRAP combined with gravity settling pretreatment (to remove suspended solids) and followed by secondary clarification (to remove algal/bacterial biomass) would fit within the footprint of an existing two-pond oxidation pond system [5]. HRAPs are operated under long HRT (8–14 d) to enable sufficient phototroph growth to provide oxygen for aerobic biodegradation. Daily variations in pH and dissolved oxygen (DO) occur due to photosynthetic activity, and pond temperature has a direct influence on cell density [55]. Pond performance and ecology (e.g. algal/bacterial dominance) can also be affected by environmental conditions.

HRAPs have now also been studied for emerging contaminant (EC) removal [55, 56]. Although single-species microalgae cultures have been applied for nutrient and EC removal [15, 57–59], it is difficult to maintain a microalgal monoculture in open processes. Accordingly, several studies have reported the advantages of co-cultivated photosynthetic microorganisms during WWT. Koreivienė et al. [60] thus reported high nutrient removal efficiencies from municipal WW using a non-native consortium composed by *Chlorella* sp. and *Scenedesmus* sp. Renuka et al. [61] described the potential of using a consortium of non-native and native species to obtain a self-flocculating culture in the treatment of a primary-treated sewage effluent. Most of the consortia reported involve *Chlorella* and *Scenedesmus* species [60–63] (Box 2).

Box 2 Full-Scale Wastewater Treatment Plant: A Case Study

An important example of full-scale wastewater treatment plant is the All-Gas project in the El Torno treatment plant in Chiclana, supported by the European Union. It was initiated to demonstrate the sustainable large-scale production of biofuels based on the low-cost cultivation of microalgae. It is the largest site in the world for the production of biofuel from algae harvested using wastewater. Pretreated wastewater is transformed into an algae biofuel which provides four times the distance of conventional biofuels such as sugar ethanol or palm oil

(continued)

Box 2 (continued)
diesel per L. The complete process chain is designed for a cultivation area of up to 10 ha, making wastewater treatment energy self-sufficient, and recycling the nitrogen and phosphorus from wastewater into microalgal biomass. The full process chain at demonstration scale has been implemented, consisting of:

- A cultivation area composed of four raceway ponds with a surface area of 200 m^2 each in which an average of 100 ton/ha-year biomass can be produced (Fig. 4).
- A separation and thickening system based on dissolved air flotation (3 units). At this stage, the algae broth is concentrated 100 times at very low energy requirement.
- A 750 m^3 anaerobic digester in which more than 200 L of biomethane per kg of added volatile solid can be produced.
- The demo operation showed that each hectare of algae culture has a wastewater treatment capacity of around 1,000 m^3/d, producing biomethane above 13,000 kg CH4/year – enough to fuel 20 cars (at 4.5 kg/100 km and 15,000 km/year). This represents four times more than what is achieved with conventional biofuels, such as bioethanol from sugarcane or biodiesel from palm oil, both yielding about 5,000 L/ha/year.

All-Gas project is an example of circular economy, since a waste is converted sustainably into raw materials with added value, providing an innovative and environment-friendly process.

Fig. 4 Raceway ponds of All-Gas project in the El Torno treatment plant in Chiclana, Spain (EU FP7 All-Gas project). Source: AQUALIA

4 Removal of Pharmaceutical Compounds from Wastewaters in Microalgae Reactors

The mechanisms for pollutant removal involved during conventional or advanced treatments cannot be directly extrapolated to algae-based wastewater (WW) treatment because algae photosynthesis causes specific environmental and ecological conditions to occur [14]. Unfortunately, PhACs removal has not yet been extensively studied during microalgae-based WWT, and Table 2 summarizes published results on the elimination of PhACs in algae photobioreactors in synthetic WW and real WW. Some of these studies are also discussed in recently published reviews on organic micropollutant removal [13, 44, 72, 73].

4.1 *Pharmaceutical Removal in Synthetic WW*

The first specific study on PhACs removal during microalgal WWT treatment was published by de Godos et al. [22] who reported the removal of the antibiotic tetracycline from synthetic WW using a pilot-scale HRAP (14 L). Tetracycline was added after 46 days of operation and reached a stable value of 0.79 ± 0.02 mg/L. Active removal was confirmed with an efficiency increasing from 60 ± 1% on day 61 to 69 ± 1% on day 76. The main mechanisms involved were photodegradation and biosorption.

Another antibiotic was selected by Hom-Diaz et al. [29] who evaluated the removal of ciprofloxacin (CPX) in indoor and outdoor HRAPs. These microalgae systems were fed with primary wastewater effluent. When spiked at 2 mg/L, CPX was efficiently removed from photobioreactors continuously operated under various conditions of artificial illumination and hydraulic residence times (7–3 d). These findings were confirmed in a 1,000 L pilot HRAP operated outdoors (Fig. 5) and laboratory batch assays performed using HRAP microcosms incubated under the same outdoor conditions. Spiked experiments were conducted by stopping the wastewater influent supply during 24 h and immediately adding CPX to an initial concentration of 2 mg/L. HRAP was also monitored throughout the day to assess photodegradation and sorption. CPX removal occurred immediately after spiking in all assays, independent of the time of the day. There was a pollutant removal of 15% during daytime and 18–22% at night time. The final CPX removal percentage was 38% after 24 h. Photodegradation also took place during continuous CPX removal, and it was not strongly influenced by the suspended biomass concentration. Hom-Diaz et al. [29] concluded that the main CPX removal mechanisms were sorption and photodegradation, although biodegradation may also occur.

Some ECs can be transformed into metabolites that are even more detrimental to human health and natural ecosystems than their parent compounds. Ismail et al. [64] therefore studied the degradation of a mixture of the analgesics ketoprofen, paracetamol and aspirin by focusing on toxicity reduction and the elimination of the

Table 2 Studies of PhACs removal in algae photobioreactors treating synthetic or real WW

Reactor type	PhAC	WW	Operational description	Removal (%)	References
HRAP (14 L)	Tetracycline 0.79 ± 0.02 mg/L	Synthetic	Addition at the steady state	60–69 ± 1%	de Godos et al., [22]
HRAP (1,000 L) (Fig. 5)	Ciprofloxacin 2 mg/L	Urban WW primary influent	Batch experiments a) Sunrise spike (24 h) b) Sunset spike (night) c) Sunset spike (24 h)	a) 15% during light hours, up to 23% at the end of the experiment b) 18% c) 22% during night hours, up to 38% at the end of the experiment	Hom-Diaz et al., [29]
Stirred tank (5 L)	Ketoprofen 0.5 mM, paracetamol 0.25–0.5 mM, aspirin 0.5 mM	Synthetic	HRT 3d–4d	95% removal (HRT 4d)	Ismail et al., [64]
Tubular periphyton PB (30 L)	Carbamazepine ibuprofen, hydro-chlorothiazide, gemfibrozil 10 μg/L each one	Synthetic	HRT 2d–4d	Carbamazepine 6.45%, ibuprofen 48.7%, hydrochlorothiazide 26.2%, gemfibrozil 20.6%	Kang et al., [65]
Semiopen multitubular PB (1,200 L) (Fig. 6)	17β-Estradiol (2 mg/L)	Toilet WW	Batch operation	Night 81% Day 77–55%	Parladé et al., [66]
HRAP (470 L) (Fig. 7)	26 emerging contaminants 0–24 μg/L	Urban WW primary influent	HRT 4d-8d	84–59% (warm and cold season, respectively) Ibuprofen, ketoprofen, naproxen, diclofenac, carbamazepine: average 40% and 62% at 4d and 8d, respectively	Matamoros et al., [56]
Open PB (650 L)	52 PhACs 0.16–2.9 μg/L	Urban WW and CO_2	Repeated (7 d)	a) >90% for atenolol, bisoprolol, metoprolol, clarithromycin, bupropion, atracurium, diltiazem and terbutaline b) 50–90% for 14 PhACs c) Very low or nonquantifiable for 18 PhACs,	Gentili and Fick, [67]

(continued)

Table 2 (continued)

Reactor type	PhAC	WW	Operational description	Removal (%)	References
				11% Carbamazepine	
Semiopen multitubular PB (1,200 L) (Fig. 6)	16 PhACs anti-inflammatories 50–59 μg/L, antibiotics 65–5,662 ng/L	Toilet WW	HRT 8d-12d	Anti-inflammatories >98% Antibiotics >48%	Hom-Diaz et al., [68]
HRAP (9,600 L)	64 PhACs average concentration 223 mg/L	Urban WW	HRT 6d	Analgesics and anti-inflammatories >85% Diclofenac 55–71% Antibiotics 63% Other 77%	Villar-Navarro et al., [69]
Semiopen multitubular PB (8,500 L) (Fig. 8)	PhACs and other micropollutants The highest for carbamazepine 510 ng/L	Agricultural *	HRT 16 d	Ibuprofen 50% Diclofenac 61% Carbamazepine <40%	García-Galán et al. [70]
HRAP (470 L) (Fig. 7)	12 PhACs and 26 TPs 166–23,811 ng/L	Urban WW	HRT 4.5 d	Acetaminophen 100% Ibuprofen 79% Sulfamethoxazole 80% Diclofenac 53% Carbamazepine 0%	García-Galán et al. [71]

Fig. 5 Concrete 1,000 L HRAP located in Palmerston North, New Zealand. Courtesy of Dr. Andrea Hom-Diaz

Fig. 6 1,200 L semiopen multitubular PB located on the roof of the Chemical, Biological and Environmental Engineering Department at the Universitat Autònoma de Barcelona, Spain. Courtesy of Dr. Andrea Hom-Diaz

Fig. 7 470 L HRAPs located outdoors at the laboratory of the GEMMA research group (Universitat Politècnica de Catalunya – BarcelonaTech, Spain). Courtesy of Dr. Ivet Ferrer

Fig. 8 Full-scale semi-closed tubular PBR located in Agropolis, Barcelona, Spain. Courtesy of Dr. Ivet Ferrer

metabolites formed. An artificial microalgal-bacterial consortium was used in a stirred-tank photobioreactor. The toxic paracetamol metabolite p-aminophenol was formed at a HRT of 3 days, and increasing the HRT to 4 days enhanced bioremediation efficiency. After acclimatization under continuous illumination, both the analgesics and p-aminophenol were removed by 95%.

Kang et al. [65] employed periphyton to treat a synthetic wastewater spiked with the pharmaceuticals and personal care products (PPCPs) carbamazepine (CBZ), ibuprofen (IBU), hydrochlorothiazide (HCT) and gemfibrozil (GEM) as well as the industrial chemical bisphenol A (BPA). The photobioreactor was made of acrylic glass and divided into five channels. The reactors were illuminated using LED lamps, and four treatments with different illuminances (7,000–5,000 lx) and HRT (4-2d) were carried out. After 10 d of pollutant-free operation, standard solutions of the PPCPs were added to the storage tanks resulting in a concentration of 10 μg/L for each compound. CBZ and GEM were poorly removed (6.45–20.6%), HCT and IBU were moderately removed (26.2–48.7%), and BPA was effectively removed (72–86.4%). High HRT operation enhanced nutrient and PPCP removal, but change in illumination period had contrasting effects on the nutrients and PPCP removals, suggesting that dark/light illumination cycles supported the development of microbial communities with high PPCP removal capacity. These results agree with the findings by Hom-Diaz et al. [68] who reported that there is no correlation between the removal of PhACs and biomass concentration in the photobioreactor. Kang et al. [65] thus suggested that changes of dark/light illumination periods could be a strategy to improve micropollutant removal in periphyton photobioreactors.

Few publications have studied the evolution of bacteria and microalgae populations and their influence on pollutant removal in photobioreactors. Kang et al. [65] identified algae belonging to 4 phyla and 11 genera in their study. The cyanophyta *Phormidium* accounted for over 94% of the total algae DNA extracted. As *Phormidium* was therefore likely the dominant algal genus in all treatments, the differences in nutrient and PPCP removal efficiencies observed should be more related to the bacterial composition. Kang et al. [65] reported that a high bacterial diversity favoured nutrient removal but that a lower bacterial diversity was associated with better PPCP removal.

The hormone 17β-estradiol (E2) is a widespread emerging contaminant (EC) detected in urban and industrial WW [74]. Due to their environmental fate and toxicological effects, the EU commission has proposed to list E2, diclofenac and 17-α-ethinylestradiol (EE2) as priority substances. Although the concentrations currently detected in WW are in the range of ng to μg/L, E2 and EE2 can induce adverse ecological effects and can act as endocrine disruptors [75]. High E2 removal has been achieved in microalgal monocultures in spiked experiments [59, 76]. Parladé et al. [66] later aimed to assess how E2 was removed and how microbial community changed during 1 year of photobioreactor operation for treating toilet wastewater. Briefly, the system was designed as a 1,200 L semiopen multitubular PR (Fig. 6). Two distribution chambers were placed at each end of the tubes to transfer and distribute the culture evenly between the tubes. The PB was operated at HRTs of 8 and 12 days. E2 spiking experiments were conducted during steady-state operation corresponding to different seasons of the year.

Complete E2 removal was achieved during the first assay (8d HRT) and 81% of the removal took place during the night. Two assays conducted at 12 d HRT showed lower final removal (77% and 55%), but the behaviour was similar as rapid removal took place during the first 10–13 h followed by a slower removal rate. Complete removal of the hormone could thus be attained under favourable seasonal conditions at temperatures between 10 and 25°C, and more than 50% of the pollutant was still removed in 24 h at low temperature and under low solar irradiation. Estrone (E1) has been reported as an E2 biodegradation product from the microalgae *Chlorella vulgaris* [77], although it can also be formed during E2 oxidation processes. In the study of Parladé et al. [66], E1 was formed at the same rate as E2 disappeared, and a rapid conversion occurred at the beginning of the experiment. The transformation rate decreased after 5 h, and E1 formation was still taking place at the end of the experiment since E2 was not completely removed. E1 was formed in the presence of active algal biomass but was not detected in the killed controls, under either light or dark conditions, demonstrating that E1 was a product from biodegradation. Sorption ranged from 8% to 50%, suggesting that E2 sorption varied with biomass composition. As earlier reported by Hom-Diaz et al. [59], algal biomass composition has a direct effect on the adsorption of pollutants. Parladé et al. [66] concluded that both adsorption and biodegradation were fast both in light and in darkness.

Parladé et al. [66] also evaluated the effect of bioaugmentation with a pure *Scenedesmus* culture on E2 removal. Eukaryotic diversity was assessed using denaturing gradient gel electrophoresis, and samples were taken immediately before and after 24 h of each E2 spike. No changes occurred after E2 addition, suggesting that the oestrogenic compound did not have any effect on populations. The analysis revealed the presence of five different phyla within the Eukarya domain and two unclassified sequences. Briefly, the PB microalgae flora was dominated by *Chlorella* species which remained present at all stages, followed by *Pseudospongiococcum* in less abundance. Other abundant phylotypes, which were only present episodically, included protozoa, rotifer or parasitic chytrid species that fed on the algae. It is likely that changes in the microbial community were partially responsible for the differences in E2 removal reported. Bioaugmentation with *Scenedesmus* could not outmatch the native PB community in terms of biomass yield. The PB community was composed of transient species that appeared and disappeared over time and stable microalgae like *Chlorella*, which remained unaffected by seasonal variations. Parladé et al. [66] concluded that a high biodiversity with some well-established species was preferable to a high biomass concentration for E2 removal.

In addition to supporting satisfactory WWT efficiency, the biomass accumulated during microalgae-based WWT can be valorized as, for example, animal feed, soil conditioners, plant fertilizers or biofuel [72]. Ismail et al. [64] extensively analysed the biomass harvested and reported total crude protein, carbohydrate and lipid levels of 50%, 16% and 7%, respectively. According to Brown et al. [78], the levels of protein, carbohydrate and lipid can range from 6% to52%, 5% to 23% and 7% to 23%, respectively, in microalgal biomass. The high level of crude protein reported by Ismail et al. [64] is likely a consequence of the high nitrogen supply. Moreover, the total protein concentration of the harvested biomass was rich in essential (65.5%), nonessential (29%) and conditional amino acids (5.5%), which suggests a 'high-quality protein-rich' biomass suitable for use as animal feed. Interestingly, the harvested biomass contained a good blend of medium and long-chain fatty acids that make it a good biodiesel-producing candidate. The water-extractable fraction of the biomass possessed a high phenolic content and antioxidant capacity. Ismail et al. [64] concluded that microalgae-based WWT could provide an integrated eco-friendly and cost-efficient strategy for remediating pharmaceutical wastewater.

4.2 Pharmaceutical Removal in Real WW

Matamoros et al. [56] studied the effect of HRT and meteorological conditions on the removal efficiencies of 26 ECs (pharmaceuticals, fragrances, fire retardants and pesticides, among others) in two HRAP pilot plants fed with urban wastewater. The pilot plant was located outdoors at the laboratory of the GEMMA research group (Universitat Politècnica de Catalunya – BarcelonaTech, Spain). Urban wastewater was pumped from a municipal sewer to a homogenization tank. From there, the wastewater was screened and conveyed to two identical parallel lines, each equipped with a primary settler, a pilot high-rate algal pond and a secondary clarifier for biomass separation (Fig. 7). The experimental HRAPs were made of PVC raceway ponds equipped with a paddle wheel for stirring the mixed liquor. Each HRAP had a nominal volume of 0.47 m^3 at a water depth of 0.3 m and a surface area of 1.54 m^2.

The concentration of ECs in the HRAP influent ranged from undetected to 24 μg/L. Caffeine, acetaminophen and ibuprofen were usually detected at >9 μg/L in the influent, and these pollutants were efficiently removed (>90%). Ketoprofen and naproxen were removed by 60–90%, diclofenac and diazinon were removed by 40–60%, while carbamazepine was poorly removed by <40%. The results showed removal efficiencies varied greatly with pollutant chemistry. The most abundant pollutants in the biomass were the most hydrophobic ones, such as musk fragrances, but adsorption did not seem to be a quantitatively significant removal mechanism for any PhACs. No significant differences in removal performance were observed at the two HRTs tested during the warm season, but significant differences were found during the cold season for those compounds previously known to be affected by biodegradation (i.e. caffeine, ibuprofen and naproxen), photodegradation (i.e. ketoprofen and triclosan) and sorption or volatilization (i.e. musk fragrances). Most of the compounds were nevertheless removed at 4 d HRT in both seasons. In contrast to other authors, Matamoros et al. [56] reported that the effect of seasonality on pollutant removal performance was low or null (i.e. for caffeine, acetaminophen and ibuprofen) for the most abundant compounds. To assess the potential for detoxification, Matamoros et al. [56] computed the hazard quotients (HQs) based on the chemical composition of water samples and predicted non-effect concentrations (PNECs) for different aquatic organisms. Acetaminophen, ibuprofen and oxybenzone exhibited high HQs in influent wastewater samples ($HQ > 1$), mainly due to their high concentrations. After treatment in the HRAP system, all the studied ECs had an $HQ < 1$. Matamoros et al. [56] thus concluded that HRAPs were at least as efficient as conventional activated sludge, constructed wetlands and waste stabilization ponds for EC removal and detoxification.

While algae can be simultaneously used for WWT treatment, CO_2 abatement and bioenergy production, few studies have considered the potential use of cultivating algae to simultaneously treat flue gases and municipal wastewater containing pharmaceuticals. Gentili and Fick [67] investigated the removal of pharmaceuticals in urban WW in a 650 L open outdoor photobioreactor supplied with flue gas. The photobioreactor was constructed following the open-pond principle, where water flow was generated by a mechanical device. It was supported by a metal frame and constructed of thin fibreglass to allow light penetration from the top (as in a traditional open pond), sides and bottom. The reactor also had an empty space in its centre to improve illumination. Municipal wastewater was treated during 1 week in several batches. Flue gas (9% CO_2) from a combined heat and power plant incinerating municipal and industrial solid wastes (Umeå Energi, Umeå, Sweden) was bubbled through the algal culture at a flow rate of approximately 3 L/min. The pH in all batches was 8.3 ± 0.9. Bubbling was stopped at night to prevent medium acidification in the absence of light. Experiments were conducted from April to May 2012 to study the performance of the algae under different environmental conditions of light and temperature (10–32°C). The concentrations of the 52 pharmaceuticals detected ranged from 0.16 to 2.9 μg/L. Removal efficiencies were very high (>90%) for atenolol, bisoprolol, metoprolol, clarithromycin, bupropion, atracurium, diltiazem and terbutaline, moderate (50–90%) for 14 PhACs and very low or nonquantifiable for 18 PhACs. These results agreed with past studies on different biological WW treatment processes [79, 80]. Many pharmaceuticals were metabolized and excreted as glucuronides or other conjugated metabolites that can be converted back to their parent compounds by enzymatic processes. Deconjugation can indeed occur in sewage treatment processes and fungal WW treatment, showing negative removal efficiencies for macrolide antibiotics, carbamazepine and other pharmaceuticals. Several pharmaceuticals were degraded more efficiently in the algal photobioreactor than in the influent treated without microalgae. On average, only 8 pharmaceuticals were not removed nor had negative removal rates in the presence of microalgae compared to 20 without microalgae. The reduction in pharmaceuticals content was positively correlated with light intensity inside the culture, dissolved oxygen concentration and the reduction in nitrogen content. Some authors have suggested that photolysis is a critical removal pathway for several micropollutants including pharmaceuticals [56, 81]. However, and although Gentili and Fick [67] recognized the importance of this mechanism in the elimination of PhACs, the removal of PhACs was lower in a control PB free-of-algae than in the inoculated system.

Hom-Diaz et al. [68] studied the removal of PhACs in an outdoor pilot tubular PB treating toilet WW at HRTs of 8 days and 12 days (Period I and Period II, respectively). The experimental set-up was previously described (Fig. 6). Toilet WW was collected from the toilet drain of the Chemical, Biological and Environmental Engineering Department (Universitat Autònoma de Barcelona) and settled. The supernatant was conducted to a second settler from which the WW was pumped into the microalgal PB. The HRT of the settlers (also used as homogenization tanks) was 48 h. The PB treated 150 L/day (8 d HRT) in Period I and 100 L/day (12 d HRT) in Period II. The PBR performance was monitored from September to December 2015. Nutrient removal efficiency was very high but slightly different for the two periods. No large differences in the influent PhAC levels were observed between October and December. Five anti-inflammatory compounds (acetaminophen, ibuprofen, naproxen, salicylic acid and ketoprofen) were detected: acetaminophen (paracetamol) was found at the highest concentrations (50.2–58.7 μg/L) in both sampling periods and was removed by >98%. Ibuprofen, the second most abundant compound detected in the influent WW during both operating periods (39.0–52.8 μg/L), was also removed by >98%, evidencing again a high biodegradability noted in past studies. The antibiotics ciprofloxacin and ofloxacin were detected at concentrations of 2,629–294 ng/L and 5,662–65 ng/L, respectively. Azithromycin was only detected in Period I (385 ng/L), and erythromycin was only detected in Period II (661 ng/L). Antibiotic removal efficiencies varied between periods, but remained >48%. Higher removal efficiencies were reported during Period II despite the lower temperature and light irradiation. Hom-Diaz et al. [68] stated this could be a consequence of the HRT increase from 8 to 12 days, in agreement with the study by Matamoros et al. [56]. Hom-Diaz et al. [68] also studied the harvesting of the algal biomass since its separation from the treated effluent is essential for discharge compliance and biomass valorization.

Villar-Navarro et al. [69] studied the efficacy of PhACs removal from urban wastewater by comparing a conventional treatment using activated sludge against a nonconventional treatment scheme involving HRAPs. This research was performed at the WWTP of Cadiz (Spain). The nonconventional treatment line was a pilot-scale installation within the WWTP (Fig. 4) consisting of three upflow anaerobic sludge blanket (UASB) reactors followed by six HRAPs. These were open shallow ponds (9.6 m^3 each) equipped with a paddle wheel. After grit and grease removal, the WWTP influent was diverted from the conventional to the nonconventional line. The wastewater was either treated through the UASB reactors before entering the HRAPs (February to April 2015) or directly directed onto the HRAPs (May to July 2015). The HRT in the HRAPs was between 72 and 216 h (144 h average). The HRAPs removed 74% and 92% of dissolved nitrogen and phosphorous respectively, while the conventional treatment only satisfactorily removed phosphorus, probably due to insufficient HRT in the biological reactor. Sixty-four out of 81 target PhACs were detected in the WW influent, at an average total concentration of 223 mg/L. Analgesics and anti-inflammatories were the predominant group in the influent (>100 mg/L) and were removed above 85% during both conventional and nonconventional treatments. The average removal efficiency of diclofenac in the conventional line was 19%, where it was never exceeding 30%. In comparison, this pollutant was removed between 55 and 71% following HRAP treatment. These results agreed with past demonstrations of the influence of light on diclofenac removal. Removal from the HRAP was also higher for antibiotics (63%) and other PhACs (77%). The average removal efficiencies were however similar (94% vs. 92%) for both treatment lines when comparing total PhACs concentrations.

The research by García-Galán et al. [70] focused on the treatment of agricultural runoff water for biomass production. Agricultural runoff can contain a wide variety of hazardous contaminants including pesticides, heavy metals and numerous veterinary PhACs (mainly antibiotics and their metabolites). This study was performed over two different periods (December and April) in batch and continuous (16 d HRT) modes. The PB was a full-scale (8.5 m^3) semi-closed hybrid tubular horizontal photobioreactor (HTH-PB, Fig. 8), similar to the reactor used by Hom-Diaz et al. [68], located at the Universitat Politècnica de Catalunya facilities (Barcelona, Spain). The feed for the HTH-PB was pumped from a drainage channel that runs through different cropped fields in the area, and it also received the discharge from an urban WWTP.

PhACs removal was evaluated during four consecutive days at the end of the continuous experiment. Carbamazepine was detected at the highest concentration in the influent (average 510 ng/L) followed by diclofenac (average 150 ng/L). The poor removal efficiency obtained for carbamazepine (<40%) agrees with prior results in microalgae systems [56, 67]. The best removal efficiency was achieved with diclofenac (61%). Diclofenac removal rates can range from 0% to 90% during municipal wastewater treatment, but Villar-Navarro et al. [69] concluded diclofenac removal was higher in algae systems (55–71%) than during conventional treatment with activated sludge. These results agree with findings from Zhang et al. [82] who reported that the removal of diclofenac can be mostly attributed to photodegradation, while carbamazepine could not be photodegraded, biodegraded or adsorbed. The removal of ibuprofen achieved by García-Galán et al. [70] was low when compared to removals obtained in HRAP studies [56, 69] and in a similar tubular PB [68].

García-Galán et al. [71] recently published a new research related to PhACs removal in microalgal systems using the set-up described by Matamoros et al. [56]. Two HRAP systems (Fig. 7) were tested with and without primary treatment during the beginning of summer (high solar irradiation). The aim was to assess the effect of primary treatment on the removal of 12 PhACs and 26 of their main metabolites in urban wastewater. Each HRAP had a volume of 470 L and was operated continuously at 4.5 d HRT. The concentration of the 12 PhACs and their transformation products detected (15) before and after primary treatment was not statistically different. The analgesic ibuprofen (IBU) was detected at the highest concentrations (330–23,811 ng/L). Diclofenac (DCF) and acetylsalicylic acid (AcSAc) were detected in all influent samples at 271–2126 ng/L and 166–5,428 ng/L, respectively. Acetaminophen (ACM) was detected up to 7,800 ng/L, and the antibiotic sulfamethoxazole (SMX) was detected in all influent samples at 70–900 ng/L. This compound is considered ubiquitous in natural systems and has been identified as the main responsible for the increase of sulfonamide resistance genes in WWTPs [83]. The antidepressant venlafaxine (VFX) and the antiepileptic carbamazepine (CBZ) were present in all the influent samples at average concentrations ranging between 664 and 702 ng/L and at concentrations <33 ng/L, respectively. The very low concentration of CBZ detected is not surprising since CBZ can be found in conjugate form in wastewater and therefore its concentration is often underestimated. Some PhACs metabolites were already detected in the influents, including metabolites of IBU, CBZ and SMX, as reported by Jelic et al. [84] who noted degradation may occur in sewer networks. The results showed pretreatment had no impact on PhACs removal from the HRAPs. SMX and metronidazole (MTZ) removal averaged 80% and > 89% respectively, but these pollutants were not detected in biomass algae, which suggests that sorption did not cause their removal. García-Galán et al. [71] therefore concluded that biodegradation should be considered as the main elimination route of the antibiotics during microalgae treatment. High removals were also obtained for ACM (100%) and IBU (79%), in agreement with past studies early cited. DCF was removed around 53%, which is much lower than reported by Matamoros et al. [56] using the same experimental HRAPs during the warm/summer season. DCF was however detected at quite high concentrations in biomass samples, so bioadsorption/bioaccumulation could partly account for its elimination. CBZ was not removed, confirming the low biodegradability of this drug, although other authors obtained removals of 23–60% in HRAPs [56, 69]. García-Galán et al. [71] concluded that HRAPs could be operated without primary treatment and provided high removal efficiencies for a broad range of PhACs, including metabolites and transformation products. Bioadsorption/bioaccumulation to microalgal biomass was needed to remove non-biodegradable compounds such as venlafaxine.

5 Concluding Remarks

Overall, nearly all studies conducted at laboratory to full scale and using synthetic media or real wastewater converge in demonstrating that microalgae-based wastewater technologies support efficient removal of pharmaceutical compounds during WWT, with the added advantage of removing more nutrients, producing potentially valuable biomass and reducing energy demand in comparison to mechanically aerated treatment systems. Biodegradation is often suspected as the main removal mechanism, but biosorption/bioaccumulation and photodegradation can also be important. Because pollutant removal by specific removal mechanisms is highly dependent on pollutant chemistry, the broader range of pollutant-removing mechanism potentially taking place during microalgae-based WWT should provide more efficient and resilient pharmaceutical removal.

The HRT is a key design parameter for the removal of microcontaminants during microalgae-based WW treatment (as in the absence of cell retention or recycling, the HRT is also the mean cell residence time). The relative long HRTs used during microalgal-based treatment systems should therefore favour all removal mechanisms, including especially photodegradation and hydrolysis. Experimental data indeed confirm high HRT operation is generally beneficial to pollutant removal.

References

1. Shilton AN, Powell N, Guieysse B (2012) Plant based phosphorus recovery from wastewater via algae and macrophytes. Curr Opin Biotechnol 23(6):884–889. https://doi.org/10.1016/j.copbio.2012.07.002
2. Plouviez M, Shilton A, Packer MA, Guieysse B (2019) Nitrous oxide emissions from microalgae: potential pathways and significance. J Appl Phycol 31:1–8. https://doi.org/10.1007/s10811-018-1531-1
3. Plouviez M, Chambonniere P, Shilton A, Packer M, Guieysse B (2019) Nitrous oxide (N2O) emissions during real domestic wastewater treatment in an outdoor pilot-scale high rate algae pond. Algal Res 44. https://doi.org/10.1016/j.algal.2019.101670
4. Muñoz R, Guieysse B (2006) Algal-bacterial processes for the treatment of hazardous contaminants: a review. Water Res 40:2799–2815. https://doi.org/10.1016/j.watres.2006.06.011
5. Craggs R, Sutherland D, Campbell H (2012) Hectare-scale demonstration of high rate algal ponds for enhanced wastewater treatment and biofuel production. J Appl Phycol 24:329–337. https://doi.org/10.1007/s10811-012-9810-8
6. Posadas E, Alcántara C, García-Encina PA, Gouveia L, Guieysse B, Norvill Z, Acién FG, Markou G, Congestri R, Koreiviene J, Muñoz R (2017) Microalgae-Based Biofuels and Bioproducts. 3 - Microalgae cultivation in wastewater, 67–91. From Feedstock Cultivation to End-products. Woodhead Publishing Series in Energy. https://doi.org/10.1016/B978-0-08-101023-5.00003-0
7. Metcalf and Eddy, Inc (2011) In: Franklin LM, David Stensel H (eds) Wastewater engineering. Treatment and reuse, 4th edn. George Tchobanoglous
8. Arbib Z, Ruiz J, Álvarez-Díaz P, Garrido-Pérez C, Barragán J, Perales JA (2013) Effect of pH control by means of flue gas addition on three different photo-bioreactors treating urban wastewater in long-term operation. Ecol Eng 57:226–235. https://doi.org/10.1016/j.ecoleng.2013.04.040
9. Ledda C, Romero Villegas GI, Adani F, Acién Fernández FG, Molina Grima E (2015) Utilization of centrate from wastewater treatment for the outdoor production of *Nannochloropsis gaditana* biomass at pilot scale. Algal Res 12:17–25. https://doi.org/10.1016/j.algal.2015.08.002
10. Bahr M, Díaz I, Domínguez A, González-Sánchez A, Muñoz R (2014) Microalgal-biotechnology as a platform for an integral biogas upgrading and nutrient removal from anaerobic effluents. Environ Sci Technol 48:573–581. https://doi.org/10.1021/es403596m
11. Wang Y, Liu J, Kang D, Wu C, Wu Y (2017) Removal of pharmaceuticals and personal care products from wastewater using algae-based technologies: a review. Rev Environ Sci Biotechnol 16:717–735. https://doi.org/10.1007/s11157-017-9446-x
12. Leng L, Wei L, Xiong Q, Xu S, Li W, Lv S, Lu Q, Wan L, Wen Z, Zhou W (2020) Use of microalgae based technology for the removal of antibiotics from wastewater: a review. Chemosphere 238:124680. https://doi.org/10.1016/j.chemosphere.2019.124680
13. Xiong J-Q, Kurade MB, Jeon B-H (2018) Can microalgae remove pharmaceutical contaminants from water? Curr Opin Biotechnol 36(1):30–44. https://doi.org/10.1016/j.tibtech.2017.09.003
14. Norvill ZN, Shilton A, Guieysse B (2016) Emerging contaminant degradation and removal in algal wastewater treatment ponds: identifying the research gaps. J Hazard Mater 313:291–309. https://doi.org/10.1016/j.hazmat.2016.03.085
15. Peng F-Q, Ying G-G, Yang B, Liu S, Lai H-J, Liu Y-S, Chen Z-F, Zhou G-J (2014) Biotransformation of progesterone and norgestrel by two freshwater microalgae (Scenedesmus obliquus and Chlorella pyrenoidosa): transformation kinetics and products identification. Chemosphere 95:581–588. https://doi.org/10.1016/j.chemosphere.2013.10.013
16. Stravs MA, Pomati F, Hollender J (2017) Exploring micropollutant biotransformation in three freshwater phytoplankton species. Environ Sci Process Impacts 19(6):822–832. https://doi.org/10.1039/c7em00100b

17. Ferrando L, Matamoros V (2020) Attenuation of nitrates, antibiotics and pesticides from groundwater using immobilised microalgae-based systems. Sci Total Environ 703 (10):134740. https://doi.org/10.1016/j.scitotenv.2019.134740
18. Muñoz R, Köllnera C, Guieysse B (2009) Biofilm photobioreactors for the treatment of industrial wastewaters. J Hazard Mater 161(1):29–34. https://doi.org/10.1016/j.jhazmat.2008.03.018
19. Park JBK, Craggs RJ, Shilton AN (2011) Recycling algae to improve species control and harvest efficiency from a high rate algal pond. Water Res 45:6637–6649. https://doi.org/10.1016/j.watres.2011.09.042
20. Ronen Z, Abeliovich A (2000) Anaerobic-aerobic process for microbial degradation of tetrabromobisphenol A. Appl Environ Microbiol 66:2372–2377. https://doi.org/10.1128/AEM.66.6.2372-2377.2000
21. Santaeufemia S, Torres E, Mera R, Abalde J (2016) Bioremediation of oxytetracycline in seawater by living and dead biomass of the microalga Phaeodactylum tricornutum. J Hazard Mat 320:315–332. https://doi.org/10.1016/j.jhazmat.2016.08.042
22. de Godos I, Muñoz R, Guieysse B (2012) Tetracycline removal during wastewater treatment in high-rate algal ponds. J Hazard Mater 229-230:446–449. https://doi.org/10.1016/j.jhazmat.2012.05.106
23. Challis JK, Hanson ML, Friesen KJ, Wong CS (2014) A critical assessment of the photodegradation of pharmaceuticals in aquatic environments: defining our current understanding and identifying knowledge gaps. Environ Sci Process Impacts 16(4):672–696. https://doi.org/10.1039/c3em00615h
24. Bai X, Acharya K (2019) Removal of seven endocrine disrupting chemicals (EDCs) from municipal wastewater effluents by a freshwater green alga. Environ Pollut 247:534–540. https://doi.org/10.1016/j.envpol.2019.01.075
25. Poirier-Larabie S, Segura PA, Gagnona C (2016) Degradation of the pharmaceuticals diclofenac and sulfamethoxazole and their transformation products under controlled environmental conditions. Sci Total Environ 557–558:257–267. https://doi.org/10.1016/j.scitotenv.2016.03.057
26. Monteiro SC, Boxal ABA (2010) Occurrence and fate of human pharmaceuticals in the environment in: reviews of environmental contamination and toxicology. In: Reviews of Environmental Contamination and Toxicology, vol 202. Springer, New York. https://doi.org/10.1007/978-1-4419-1157-5_2
27. Längin A, Alexy R, König A, Kümmerer K (2009) Deactivation and transformation products in biodegradability testing of ß-lactams amoxicillin and piperacillin. Chemosphere 75 (3):347–354. https://doi.org/10.1016/j.chemosphere.2008.12.032
28. Matamoros V, Uggetti E, Garcia J, Bayona JM (2016) Assessment of the mechanisms involved in the removal of emerging contaminants by microalgae from wastewater: a laboratory scale study. J Hazard Mater 301:197–205. https://doi.org/10.1016/j.jhazmat.2015.08.050
29. Hom-Diaz A, Norvill ZN, Blanquez P, Vicent T, Guieysse B (2017) Ciprofloxacin removal during secondary domestic wastewater treatment in high rate algal ponds. Chemosphere 180:33–41. https://doi.org/10.1016/j.chemosphere.2017.03.125
30. Norvill ZN, Toledo-Cervantes A, Blanco S, Shilton A, Guieysse B, Muñoz R (2017) Photodegradation and sorption govern tetracycline removal during wastewater treatment in algal ponds. Bioresour Technol 232:35–43. https://doi.org/10.1016/j.biortech.2017.02.011
31. Sutherland DL, Howard-Williams C, Turnbull MH, Paul A, Craggs RJ (2015) Enhancing microalgal photosynthesis and productivity in wastewater treatment high rate algal ponds for biofuel production. Bioresour Technol 184:222–229. https://doi.org/10.1016/j.biortech.2014.10.074
32. Park JBK, Craggs RJ, Shilton AN (2011) Wastewater treatment high rate algal ponds for biofuel production. Bioresour Technol 102:35–42. https://doi.org/10.1016/j.biortech.2010.06.158
33. Romero-Villegas GI, Fiamengo M, Acién-Fernández FG, Molina-Grima E (2018) Utilization of centrate for the outdoor production of marine microalgae at the pilot-scale in raceway

photobioreactors. J Environ Manag 228:506–516. https://doi.org/10.1016/j.jenvman.2018.08.020
34. de la Noüe J, Laliberté G, Proulx D (1992) Algae and waste water. J Appl Phycol 4:247–254. https://doi.org/10.1007/BF02161210
35. Oswald WJ (1988) Micro-algae and wastewater treatment. In: Borowitzka MA (ed) Microalgal biotechnology. Cambridge University Press, Cambridge, pp 691–707
36. Batstone DJ, Hülsen T, Mehta CM, Keller J (2015) Platforms for energy and nutrient recovery from domestic wastewater: a review. Chemosphere 140:2–11. https://doi.org/10.1016/j.chemosphere.2014.10.021
37. Kong QX, Li L, Martinez B, Chen P, Ruan R (2010) Culture of microalgae *Chlamydomonas reinhardtii* in wastewater for biomass feedstock production. Appl Biochem Biotechnol 160:9–18. https://doi.org/10.1007/s12010-009-8670-4
38. Seco A, Aparicio S, González-Camejo J, Jiménez-Benítez A, Mateo O, Mora JF, Noriega-Hevia G, Sanchis-Perucho P, Serna-García R, Zamorano-López N, Giménez JB, Ruiz-Martinez A, Aguado D, Barat R, Borrás L, Bouzas A, Martí N, Pachés M, Ribes J, Robles A, Ruano MV, Serralta J, Ferrer J (2018) Resource recovery from sulphate rich sewage through an innovative anaerobic-based water resource recovery facility (WRRF). Water Sci Technol 78 (9):1925–1936. https://doi.org/10.2166/wst.2018.492
39. Wang L, Min M, Li Y, Chen P, Chen Y, Liu Y, Wang Y, Ruan R (2010) Cultivation of green algae Chlorella sp. in different wastewaters from municipal wastewater treatment plant. Appl Biochem Biotechnol 162:1174–1186. https://doi.org/10.1007/s12010-009-8866-7
40. Carvalho AP, Meireles LA, Malcata FX (2006) Microalgal reactors: a review of enclosed system designs and performances. Biotechnol Prog 22:1490–1506. https://doi.org/10.1021/bp060065r
41. Chisti Y (2007) Biodiesel from microalgae. Biotechnol Adv 25:294–306. https://doi.org/10.1016/j.biotechadv.2007.02.001
42. Rawat I, Kumar R, Mutanda T, Bux F (2011) Dual role of microalgae: phycoremediation of domestic wastewater and biomass production for sustainable biofuels production. Appl Energy 88:3411–3424. https://doi.org/10.1016/j.apenergy.2010.11.025
43. Shen Y, Yuan W, Pei ZJ, Wu Q, Mao E (2009) Microalgae mass production methods. Trans ASABE 52:1275–1287
44. Vo HNP, Ngo HH, Guo W, Minh T, Nguyen H, Liu Y, Liu Y, Nguyen DD, Chang SW (2019) A critical review on designs and applications of microalgae-based photobioreactors for pollutants treatment. Sci Total Environ 651(1):1549–1568. http://dx.doi.org/10.1016 /j.scitotenv.2018.09.282
45. Gudin C (1976) Method of growing plant cells. US Patent 3,955,317
46. Huang Q, Jiang F, Wang L, Yang C (2017) Design of photobioreactors for mass cultivation of photosynthetic organisms. Engineering 3(3):318–329. https://doi.org/10.1016/J.ENG.2017.03.020
47. Ugwu CU, Aoyagi H, Uchiyama H (2008) Photobioreactors for mass cultivation of algae. Bioresour Technol 99:4021–4028. https://doi.org/10.1016/j.biortech.2007.01.046
48. Chen CY, Yeh KL, Aisyah R, Lee DJ, Chang JS (2011) Cultivation, photobioreactor design and harvesting of microalgae for biodiesel production: a critical review. Bioresour Technol 102:71–81. https://doi.org/10.1016/j.biortech.2010.06.159
49. Tredici MR, Rodolf L (2004) Reactor for industrial culture of photosynthetic microorganisms 2004/074423 A2
50. Crites RW, Middlebrooks EJ, Reed SC (2006) Natural systems for wastewater treatment methods, vol 96. https://doi.org/10.1016/S0026-0576(98)80320-6
51. Heaven S, Lock AC, Pak LN, Rspaev MK (2003) Waste stabilisation ponds in extreme continental climates: a comparison of design methods from the USA, Canada, northern Europe and the former Soviet Union. Water Sci Technol 48:25–33. https://doi.org/10.2166/wst.2003.0078

52. Shilton A, Walmsley N (2005) Introduction to pond treatment technology. In: Shilton A (ed) Pond treatment technology. IWA Publishing, London, p 479
53. Handler RM, Canter CE, Kalnes TN, Lupton FS, Kholiqov O, Shonnard DR, Blowers P (2012) Evaluation of environmental impacts from microalgae cultivation in open-air raceway ponds: analysis of the prior literature and investigation of wide variance in predicted impacts. Algal Res 1:83–92. https://doi.org/10.1016/j.algal.2012.02.003
54. Brennan L, Owende P (2010) Biofuels from microalgae-a review of technologies for production, processing, and extractions of biofuels and co-products. Renew Sustain Energy Rev 14:557–577. https://doi.org/10.1016/j.rser.2009.10.009
55. García J, Green BF, Lundquist T, Mujeriego R, Hernández-Mariné M, Oswald WJ (2006) Long term diurnal variations in contaminant removal in high rate ponds treating urban wastewater. Bioresour Technol 97:1709–1715. https://doi.org/10.1016/j.biortech.2005.07.019
56. Matamoros V, Gutierrez R, Ferrer I, García J, Bayona JM (2015) Capability of microalgae-based wastewater treatment systems to remove emerging organic contaminants: a pilot-scale study. J Hazard Mater 288:34–42. https://doi.org/10.1016/j.jhazmat.2015.02.002
57. Escapa C, Coimbra RN, Paniagua S, García AI, Otero M (2015) Nutrients and pharmaceuticals removal from wastewater by culture and harvesting of Chlorella sorokiniana. Bioresour Technol 185:276–284. https://doi.org/10.1016/j.biortech.2015.03.004
58. Escapa C, Coimbra RN, Paniagua S, García AI, Otero M (2016) Comparative assessment of diclofenac removal from water by different microalgae strains. Algal Res 18:127–134. https://doi.org/10.1016/j.algal.2016.06.008
59. Hom-Diaz A, Llorca M, Rodríguez-Mozaz S, Vicent T, Barceló D, Blánquez P (2015) Microalgae cultivation on wastewater digestate: b-estradiol and 17a ethynylestradiol degradation and transformation products identification. J Environ Manag 155:106–113. https://doi.org/10.1016/j.jenvman.2015.03.003
60. Koreivienė J, Valčiukas R, Karosienė J, Baltrėnas P (2014) Testing of Chlorella/Scenedesmus microalgae consortia for remediation of wastewater, CO 2 mitigation and algae biomass feasibility for lipid production. J Environ Eng Landsc Manag 22:105–114. https://doi.org/10.3846/16486897.2013.911182
61. Renuka N, Sood A, Ratha SK, Prasanna R, Ahluwalia AS (2013) Evaluation of microalgal consortia for treatment of primary treated sewage effluent and biomass production. J Appl Phycol 25:1529–1537. https://doi.org/10.1007/s10811-013-9982-x
62. Qin L, Wang Z, Sun Y, Shu Q, Feng P, Zhu L, Xu J, Yuan Z (2016) Microalgae consortia cultivation in dairy wastewater to improve the potential of nutrient removal and biodiesel feedstock production. Environ Sci Pollut Res 23:8379–8387. https://doi.org/10.1007/s11356-015-6004-3
63. Su Y, Mennerich A, Urban B (2012) Coupled nutrient removal and biomass production with mixed algal culture: impact of biotic and abiotic factors. Bioresour Technol 118:469–476. https://doi.org/10.1016/j.biortech.2012.05.093
64. Ismail MM, Essam TM, Ragab YM, El-Sayed AE-kB, Mourad FE (2017) Remediation of a mixture of analgesics in a stirred-tank photobioreactor using microalgal-bacterial consortium coupled with attempt to valorise the harvested biomass. Bioresour Technol 232:364–371. https://doi.org/10.1016/j.biortech.2017.02.062
65. Kang D, Zhao Q, Wu Y, Wu C, Xiang W (2018) Removal of nutrients and pharmaceuticals and personal care products from wastewater using periphyton photobioreactors. Bioresour Technol 248:113–119. https://doi.org/10.1016/j.biortech.2017.06.153
66. Parladé E, Hom-Diaz A, Blánquez P, Martínez-Alonso M, Vicent T, Gaju N (2018) Effect of cultivation conditions on ß-estradiol removal in laboratory and pilot-plant photobioreactors by an algal-bacterial consortium treating urban wastewater. Water Res 137:86–96. https://doi.org/10.1016/j.watres.2018.02.060
67. Gentili FG, Fick J (2017) Algal cultivation in urban wastewater: an efficient way to reduce pharmaceutical pollutants. J Appl Phycol 29:255–262. https://doi.org/10.1007/s10811-016-0950-0

68. Hom-Diaz A, Jaén-Gil A, Bello-Laserna I, Rodríguez-Mozaz S, Vicent T, Barceló D, Blánquez P (2017) Performance of a microalgal photobioreactor treating toilet wastewater: pharmaceutically active compound removal and biomass harvesting. Sci Total Environ 592:1–11. https://doi.org/10.1016/j.scitotenv.2017.02.224
69. Villar-Navarro E, Baena-Nogueras RM, Paniw M, Perales JA (2018) Removal of pharmaceuticals in urban wastewater: high rate algae pond (HRAP) based technologies as an alternative to activated sludge based processes. Wat Res 139:19–29. https://doi.org/10.1016/j.watres.2018.03.072
70. García-Galán MJ, Gutiérrez R, Uggetti E, Matamoros V, García J, Ferrer I (2018) Use of full-scale hybrid horizontal tubular photobioreactors to process agricultural runoff. Biosyst Eng 166:138–149. https://doi.org/10.1016/j.biosystemseng.2017.11.016
71. García-Galán MJ, Arashiroa L, Santos LHMLM, Insa S, Rodríguez-Mozaz S, Barceló D, Ferrer I, Garfí M (2020) Fate of priority pharmaceuticals and their main metabolites and transformation products in microalgae-based wastewater treatment systems. J Hazard Mater 390:121771. https://doi.org/10.1016/j.jhazmat.2019.121771
72. Garrido-Cardenas JA, Manzano-Agugliaro F, Acien-Fernandez FG, Molina-Grima E (2018) Microalgae research worldwide. Algal Res 35:50–60. https://doi.org/10.1016/j.algal.2018.08.005
73. Tolboom SN, Carrillo-Nieves D, Rostro-Alanis M, Quiroz R, Barceló D, Iqbal HMN, Parra-Saldivar R (2019) Algal-based removal strategies for hazardous contaminants from the environment – a review. Sci Total Environ 665:358–366. https://doi.org/10.1016/j.scitotenv.2019.02.129
74. Santos LHMLM, Araújo AN, Fachini A, Pena A, Delerue-Matos C, Montenegro MCBSM (2010) Ecotoxicological aspects related to the presence of pharmaceuticals in the aquatic environment. J Hazard Mater 175:45–95. https://doi.org/10.1016/j.jhazmat.2009.10.100
75. Jobling S, Nolam M, Tyler CR, Brighty G, Sumpter JP (1998) Widespread sexual disruption in wild fish. Environ Sci Technol 32:2498–2506. https://doi.org/10.1021/es9710870
76. Shi W, Wang L, Rousseau DPL, Lens PNL (2010) Removal of estrone, 17a ethinylestradiol, and 17b-estradiol in algae and duckweed-based wastewater treatment systems. Environ Sci Pollut Res Int 17:824–833. https://doi.org/10.1007/s11356-010-0301-7
77. Lai KM, Scrimshaw MD, Lester JN (2002) Biotransformation and bioconcentration of steroid estrogens by *Chlorella vulgaris*. Appl Environ Microbiol 68:859–864. https://doi.org/10.1128/AEM.68.2.859
78. Brown MR, Jeffrey SW, Volkman JK, Dunstan GA (1997) Nutritional properties of microalgae for mariculture. Aquaculture 151:315–331. https://doi.org/10.1016/S0044-8486(96)01501-3
79. Gros M, Petrović M, Ginebreda A, Barceló D (2010) Removal of pharmaceuticals during wastewater treatment and environmental risk assessment using hazard indexes. Environ Int 36:15–26. https://doi.org/10.1016/j.envint.2009.09.002
80. Mir-Tutusaus JA, Parladé E, Villagrasa M, Barceló D, Rodriguez-Mozaz S, Marinez-Alonso M, Gaju N, Sarrà M, Caminal G (2019) Long-term continuous treatment of non sterile real hospital wastewater by Trametes versicolor. J. Biol Eng 13:47. https://doi.org/10.1186/s13036-019-0179-y
81. de Wilt A, Butkovskyi A, Tuantet K, Hernandez Leal L, Fernandes TV, Langenhoff A, Zeeman G (2016) Micropollutant removal in an algal treatment system fed with source separated wastewater streams. J Hazard Mater 304:84–92. https://doi.org/10.1016/j.jhazmat.2015.10.033
82. Zhang N, Liu G, Liu H, Wang Y, He Z, Wang G (2011) Diclofenac photodegradation under simulated sunlight: effect of different forms of nitrogen and kinetics. J Hazard Mater 192 (1):411–418. https://doi.org/10.1016/j.jhazmat.2011.05.038
83. Guo X, Pang W, Dou C, Yin D (2017) Sulfamethoxazole and COD increase abundance of sulfonamide resistance genes and change bacterial community structures within sequencing batch reactors. Chemosphere 175:21–27. https://doi.org/10.1016/j.chemosphere.2017.01.134
84. Jelic A, Rodriguez-Mozaz S, Barceló D, Gutierrez O (2015) Impact of in-sewer transformation on 43 pharmaceuticals in a pressurized sewer under anaerobic conditions. Water Res 68:98–108. https://doi.org/10.1016/j.watres.2014.09.033

Removal of Pharmaceuticals from Wastewater by Membrane Bioreactors: Factors, Mechanisms, and Perspectives

Faisal I. Hai and William E. Price

Contents

1 Introduction 224
2 Pharmaceuticals Removal by MBR 224
3 Factors and Mechanisms Affecting Pharmaceuticals Removal by MBR 226
3.1 Characteristics of the Pharmaceuticals 227
3.2 Operating Parameters 229
4 Bioaugmented MBR for Pharmaceutical Removal 232
5 Conclusion and Perspective 233
References 234

Abstract This chapter reviews the performance of membrane bioreactors (MBR) for the removal of pharmaceuticals from wastewater. Many factors affect pharmaceuticals removal by MBR. The factors include physicochemical characteristics of the pharmaceuticals and MBR operational parameters. The presence of a membrane allows various adsorption mechanisms for pharmaceuticals. Conceptually MBRs can deliver better removal of pharmaceuticals than conventional activated sludge (CAS) processes because of operation under longer solids retention times (SRTs), higher mixed liquor suspended solids (MLSS) concentrations, and complete retention of suspended solids. However, the available reports from lab and full-scale plants are sometimes conflicting. Further work is required to improve our understanding of the overall fate and degradation pathways of pharmaceuticals in MBRs.

Keywords Membrane bioreactor, Pharmaceuticals, Biodegradation, Adsorption, Sludge retention time

---

F. I. Hai (✉) and W. E. Price
Strategic Water Infrastructure Laboratory, The University of Wollongong, Wollongong, NSW, Australia
e-mail: faisal@uow.edu.au; wprice@uow.edu.au

Sara Rodriguez-Mozaz, Paqui Blánquez Cano, and Montserrat Sarrà Adroguer (eds.), *Removal and Degradation of Pharmaceutically Active Compounds in Wastewater Treatment*, Hdb Env Chem (2021) 108: 223–238, DOI 10.1007/698_2020_676,

1 Introduction

The conventional wastewater treatment plants were not designed to remove emerging contaminants such as pharmaceuticals. The ineffective removal of pharmaceuticals from wastewater is a significant concern because pharmaceuticals are biologically active and can present detrimental effects to the aquatic environment. The effects include acute and chronic toxicity to aquatic organisms, accumulation in the ecosystem, and various adverse effects on human health [1]. Membrane bioreactors (MBR) are an advancement over the conventional activated sludge processes. It combines biodegradation and membrane separation in a single step. Details on the basic principle, design, and operation of MBRs are available elsewhere [2]. This chapter provides a brief overview of the factors and mechanisms of pharmaceuticals removal by MBR.

2 Pharmaceuticals Removal by MBR

Compared to the conventional activated sludge (CAS) processes, MBRs can operate under longer solids retention times (SRTs), higher mixed liquor suspended solids (MLSS) concentration, and complete retention of suspended solids. Hai et al. [3] noted that MBRs may offer additional removal mechanisms for pharmaceuticals as follows: (1) the membrane is an effective barrier to solids, onto which pharmaceuticals may adsorb [4–6]; (2) pharmaceuticals may also adsorb on membrane surface [7]; and (3) the longer SRT in MBR may enhance pharmaceuticals biodegradation.

Compared to CAS, better and/or more stable removal by MBR has been reported for pharmaceuticals which have moderate to high biodegradability. On the other hand, similar to CAS, low and unstable removal performance by MBR has been observed for biologically persistent hydrophilic pharmaceuticals [8–10]. In some cases, similar removal efficiency has been observed for CAS and MBR. Examples of such pharmaceuticals include naproxen, acetaminophen, ibuprofen, and paroxetine [11, 12]. Some pharmaceuticals, namely, carbamazepine and hydrochlorothiazide have been widely reported to be poorly removed by both CAS and MBR [13, 14]. Thus based on the extent of their removal, pharmaceuticals can be categorized into (1) better removal by MBR (e.g., ketoprofen, mefenamic acid, and naproxen); (2) easy removal by both CAS and MBR (e.g. ibuprofen); and (3) poor removal by either of them (e.g., carbamazepine, clofibric acid, dichlorprop, and diclofenac) [3].

Gonzalez et al. [15] observed similar high removal (over 95%) of linear alkylbenzene sulfonates compounds by CAS and MBR processes. However, MBR performance was more stable and independent of the influent concentrations. Furthermore, MBR effluent quality in terms of chemical oxygen demand (COD), NH_4^+ concentration, and total suspended solids (TSS) was consistently better. Radjenovic et al. [13] reported better removal of the pharmaceutical compounds investigated, for example, bezafibrate, diclofenac, ketoprofen, gemfibrozil, pravastatin, ranitidine, and ofloxacin. Better or more uniform removal of pharmaceutical compounds by MBR may be due to the smaller sludge flocs which improve mass transfer, and thus removal efficiency increases [13]. Gobel et al. [16] reported 20% better removal of sulfamethoxazole by MBR. Compared to unstable removal of trimethoprim, clarithromycin, and dehydro-erythromycin by CAS, MBR achieved up to 90% removal of these pharmaceuticals at SRTs over 60 days [16]. Radjenovic et al. [13] reported better removal by MBR of a range of pharmaceuticals including β-blockers (atenolol and metoprolol), lipid regulators, and cholesterol-lowering medicine (gemfibrozil, bezafibrate, clofibric acid, and pravastatin), antibiotics (ofloxacin and erythromycin), anti-ulcer agent (ranitidine), and some analgesics and anti-inflammatory drugs (propyphenazone, mefenamic acid, and diclofenac). In a pilot scale study by Radjenovic et al. [14] compounds such as mefenamic acid, diclofenac, indomethacin, pravastatin, propyphenazone, and gemfibrozil were poorly removed by CAS. However, they were better removed by MBR.

Post-treatment by ultrafiltration following CAS treatment was reported to increase antibiotics removal by around 30% [6]. This improvement was attributed to adsorption of the antibiotics on suspended biomass and on the biomass attached to membrane surface. Boonyaroj et al. [4] reported a similar observation. Urase et al. [17] note that in MBRs additional removal of hydrophobic pharmaceuticals may occur due to the extra retention of the pharmaceuticals on suspended solids deposited on the membrane. Li et al. [18] reported that dosing of powdered activated carbon to MBR can lead to significant additional removal of pharmaceuticals. However, for hydrophilic pharmaceuticals, such additional removal by MBR may not occur [19].

It appears that MBR performs better than CAS for pharmaceuticals with moderate biodegradability, while the removal performance of CAS and MBR may be similar for easily biodegradable and resistant compounds. For example, in batch experiments with MBR and CAS sludge, similar kinetic biodegradation constant (k_{biol}) of $\leq$ 0.1 L/(g $_{SS}$· d) for diclofenac was observed by Joss et al. [20]. Clara et al. [21] reported on the importance of SRT on pharmaceuticals removal by both CAS and MBR. They noted that when operated at comparable SRT, no significant differences in the treatment efficiency between CAS and MBR may be observed. It is, however, noteworthy that pharmaceuticals removal by MBR is more stable (Figs. 1 and 2) [22]. While MBR may not lead to higher removal efficiencies, it shows a more consistent performance and stable response to fluctuating pharmaceuticals load in wastewater. Additionally, MBR installations are more compact and offer more flexibility to fine-tune biological performance.

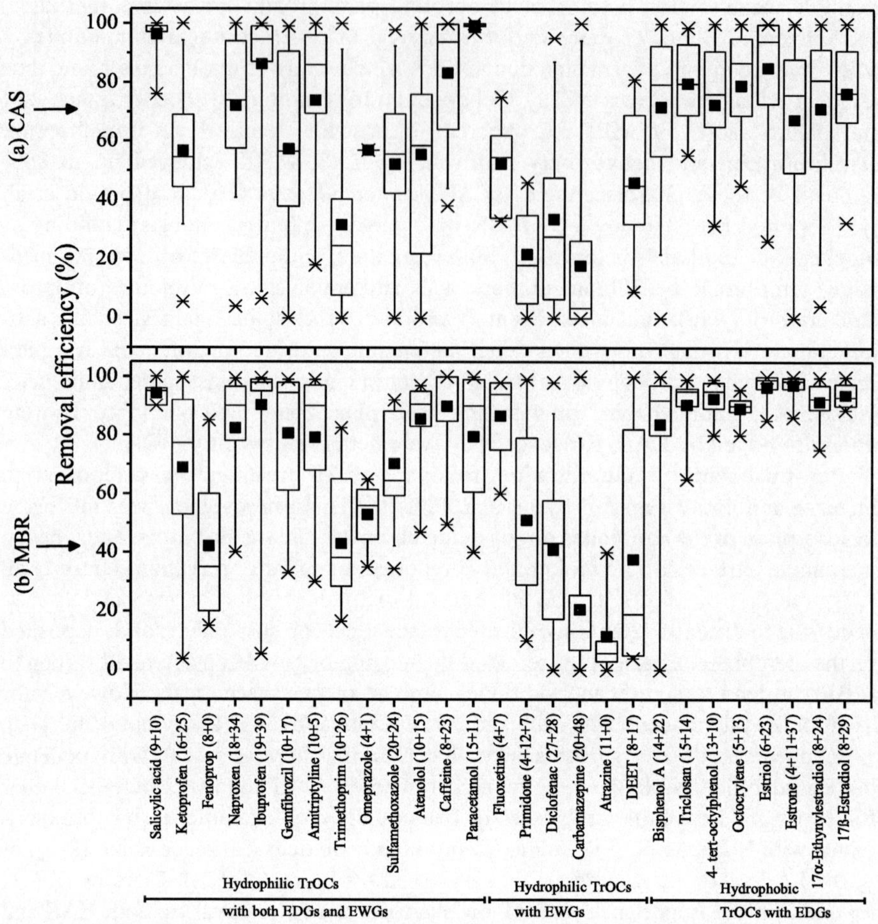

Fig. 1 Variations in the removal of emerging contaminants including pharmaceuticals by (**a**) CAS and (**b**) MBR. Box-and-whisker plots show information about the interquartile range, median (horizontal line in the box), min and max (whiskers), and average (block square in the box). Numbers in parenthesis on the X-axis represent the no. of data points (MBR + CAS). Adapted by permission from Asif, Ansari, Chen, Nghiem, Price, and Hai [22]

3 Factors and Mechanisms Affecting Pharmaceuticals Removal by MBR

The mechanisms of pharmaceuticals removal do not follow a general rule since their relative contribution depends on the physicochemical properties of the pharmaceuticals, the origin and composition of the wastewater, and the operating parameters of the wastewater treatment facility [11].

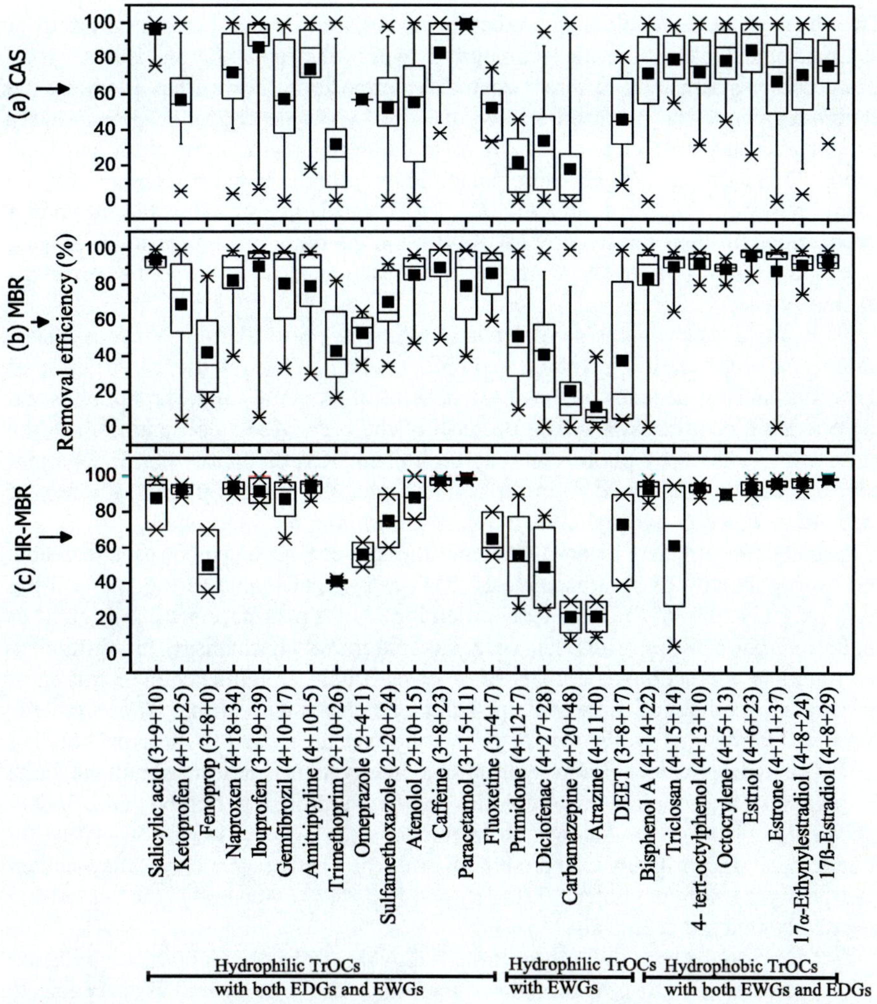

Fig. 2 Box-and-whisker plot showing the biodegradation of emerging trace organic contaminants (TrOCs) including pharmaceuticals in (**a**) CAS, (**b**) conventional MBR, and (**c**) high retention MBR. This chapter discusses the case of conventional MBR. For discussion on high retention MBR, which is an emerging format, see the Ref. [3]. Figure reproduced with permission from Asif, Ansari, Chen, Nghiem, Price, and Hai [22]

3.1 Characteristics of the Pharmaceuticals

Removal of pharmaceuticals even from the same therapeutic group may be different [23]. Indeed Tadkaew et al. [10] observed no general trend for any of the six therapeutic classes of pharmaceuticals investigated: their removal efficiencies by MBR treatment varied widely (10–98%) even within the same class of compounds.

Thus it may not be possible to predict removal based on the therapeutic group of pharmaceutical. This is because compounds within the same pharmaceutical group can possess significantly different chemical composition. For example, among the antibiotics, while the macrolides have one ring with side chains or sugars, every sulfonamide has two relatively small rings connected by a sulfur atom and C–N bonds. Thus sulfonamides are more polar and hydrophilic in nature compared to the macrolides (log $K_{ow} < 1$ and log $K_{ow} > 3$, respectively). This has important implications for their removal by MBRs [6]. On the other hand, Onesios et al. [23] observed antiseptic removals of greater than 70% in majority of the studies reviewed in their article.

Four broad mechanisms govern pharmaceuticals removal from aqueous phase: photolysis, volatilization, sorption, and biodegradation. The relative contribution of these mechanisms depends on the physicochemical properties of the compounds and the operating condition of MBR. Because of the high MLSS concentration in the bioreactor, removal by photolysis may be limited. If sufficiently volatile, pharmaceuticals can be transferred from the aqueous phase to the gas phase. Similarly, if they are sufficiently hydrophobic, they can partition to the solid phase.

Log D or the effective octanol-water partitioning coefficient can be used to denote the hydrophobicity of pharmaceuticals. This can be calculated as $\log D_{(\mathrm{pH})} = \log K_{ow} - \log(1+10^{(\mathrm{pH}-\mathrm{p}Ka)})$. In general, when log D of a pharmaceutical is about 3 or higher, sorption to biosolids can be a major removal mechanism. In addition to hydrophobic interactions, a number of other reactions can influence the partition of the organic compound between the solid and the liquid phase. These include complex formations with metal ions, ion exchange, and hydrogen bonding [24]. The hydrophobic and non-hydrophobic interactions can occur simultaneously. For example, fluoroquinolone antibiotics are very hydrophilic, but they adsorb onto sludge significantly via electrostatic interactions [25]. Stevens-Garmon et al. [26] reported that the positively charged pharmaceutically active compounds they studied had the highest sorption potential (log $K_d = 2.8$–3.8), as compared to the neutral and negatively charged chemicals.

For compounds with low biosorption, removal mainly occurs by biodegradation [27]. Suarez et al. [28] proposed that compounds could be classified according to their k_{biol} into very highly ($k_{biol} > 5$ L/(g $_{SS}$· d)), highly ($1 < k_{biol} < 5$ L/(g $_{SS}$· d)), moderately ($0.5 < k_{biol} < 1$ L/(g $_{SS}$· d)), and hardly ($k_{biol} < 0.5$ L/(g $_{SS}$· d)) biodegradable. Available reports show the relationship between the molecular structure and/or functional moieties of pharmaceutical and their biodegradability during MBR treatment [10, 19, 29]. For example, compounds with complex structure (e.g., alkyl chain branch) [30] and toxic groups (e.g., halogens groups [31]) show higher resistance to complete biodegradation. Tadkaew et al. [10] systematically studied the influence of electron withdrawing groups (EWGs) and electron donating groups (EDGs) on pharmaceuticals removal: hydrophilic and moderately hydrophobic (log $D < 3.2$) compounds possessing strong electron withdrawing functional groups showed removal efficiency of less than 20%. By contrast, high removal efficiencies were observed with most compounds bearing electron donating functional groups such as hydroxyl and primary amine groups.

3.2 Operating Parameters

3.2.1 Mixed Liquor Suspended Solids Properties

When biosorption is significant, higher MLSS concentration can result in better removal of pharmaceuticals from the aqueous phase [32]. On the other hand, a minimum level of MLSS must be maintained to sustain biodegradation. For example, Li et al. [33] found that the biodegradation of hormones depended significantly on the level of mixed liquor volatile suspended solids (MLVSS) concentration which varied between the low levels of 0.435 and 1.75 g/L. In a study by Shariati et al. [34], the removal of acetaminophen improved from 20 to 40% when the MLSS concentration increased from 2 to 15 g/L. This performance improvement can be attributed to biodegradation improvement as acetaminophen is a hydrophilic compound. Better acclimatization of the microorganisms to the influent can improve biodegradation. For example, naproxen removal performance of a pilot plant increased from 27 to 99% during the first 300 days, possibly due to acclimation of bacteria to this pharmaceutical [35].

3.2.2 Solids Retention Time (SRT)

With increasing SRT, the microbial population in an MBR may become more diversified (e.g., slow-growing bacteria eventually reaching effective population numbers). The microorganisms may also diversify their metabolic activity in response to the lower food to microorganism (F/M) ratio in MBR under long SRT [36].

A survey conducted in some full-scale CAS plants revealed better removal of gemfibrozil and diclofenac under SRTs over 30 days than below 15 days [37]. Similarly Kimura et al. [38] reported better removal of ketoprofen and diclofenac by MBR at an SRT of 65 days compared to that at an SRT of 15 days. Tambosi et al. [39] also observed roxithromycin removal to increase when SRT was doubled from the initial SRT of 15 days.

The minimum SRT required for effective removal depends on the compound. Caffeine and oxybenzone are well removed from the aqueous phase by MBR. They showed a critical SRT of less than 5 d [40]. On the other hand, compounds categorized as showing poor removal (e.g., galaxolide and tris(2-chloroethyl) phosphate) were reported to require an SRT greater than 15 d. For compounds such as macrolide antibiotics azithromycin, erythromycin, and clarithromycin, a clear increase in removal was observed when SRT exceeded 60 days, whereas, in the same study, higher reduction of roxithromycin was observed already at a SRT of 33 days [16].

Figure 3 indicates that there is no universal relationship between SRT and pharmaceuticals removal by MBR. Nevertheless, the removal of a few pharmaceuticals such as diclofenac, sulfamethoxazole, and mefenamic acid could improve

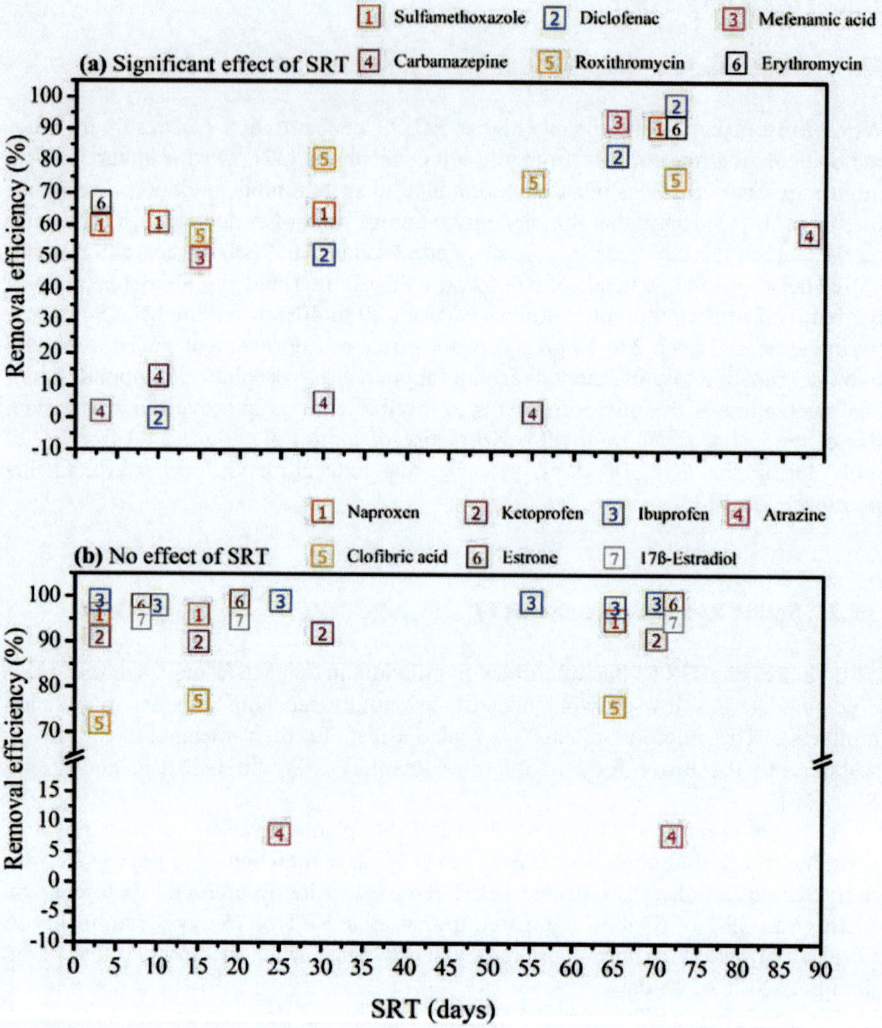

Fig. 3 Effect of SRT on the aqueous phase removal of the selected compounds by MBRs. Reproduced with permission from Asif, Ansari, Chen, Nghiem, Price, and Hai [22]

significantly with the increase in SRT. On the other hand, SRT did not influence the removal of easily biodegradable as well as resistant hydrophilic compounds [22]. Notably, in MBRs, SRT is independent of HRT. Therefore most MBR studies report no obvious effect of HRT on pharmaceuticals [3]. For example, no discernible influence of HRT (3.9–8 h) on BPA removal was observed in an MBR [41].

3.2.3 Pharmaceutical Loading

The concentration of pharmaceuticals in wastewater is usually in microgram to nanogram per liter level. At such low concentrations, biodegradation of pharmaceuticals can happen only if a primary substrate is available for the corresponding bacteria to grow on. Thus cometabolism may be the main removal mechanism of pharmaceuticals, meaning that in presence of other substrate, bacteria break down the pharmaceuticals but do not use them as carbon source. For instance, in batch biodegradation tests, bezafibrate, naproxen, and ibuprofen were degraded only via cometabolism [42]. On the other hand, in a lab-scale sequencing batch reactor study, sulfamethoxazole was biodegraded in presence of acetate, but there was an indication that the microorganisms used the degraded antibiotic as a nitrogen source [43].

3.2.4 Mixed Liquor pH

Wastewater pH can influence pharmaceuticals removal by influencing the physiology of the microorganisms (i.e., optimum pH for enzymatic activity) and the solubility of pharmaceuticals in wastewater [11]. Depending on the pH, pharmaceuticals can exist in different protonation states. This can affect the hydrophobicity (log *D*) at different pH values [11, 44]. For example, during sludge dewatering and conditioning with lime, the pH is increased to over 9. This can cause desorption of pharmaceuticals. On the other hand, compared to neutral pH, at an acidic pH, Urase et al. [17] observed improved removal of several pharmaceuticals, such as ibuprofen, fenoprofen, naproxen, and diclofenac. The removal of the neutral compounds remained unchanged. Despite the possibility of enhanced adsorption of certain ionizable pharmaceuticals on sludge at acidic pH, it may have detrimental impact on microbial groups that are involved in removal of TOC, TN, and/or total phosphorous (TP) removal [45, 46].

3.2.5 Mixed Liquor Temperature

Temperature fluctuation can cause deterioration of treated water quality and disrupt the treatment system [11, 47]. This is because temperature conditions significantly affect microbial growth and activity as well as solubility and other physicochemical properties of organics [11]. Relatively high effluent concentrations of certain pharmaceuticals during low winter temperature are observed [48, 49]. On the other hand, in lab-scale MBR experiments, Hai et al. [50] observed that compared to the operation in the temperature range of 10–35°C, operation at 10 and 45°C led to significantly reduced removal efficiency of the hydrophilic pharmaceuticals investigated.

3.2.6 Mixed Liquor Dissolved Oxygen Concentration

Biodegradation of organics can occur under aerobic (molecular oxygen available), denitrifying (no molecular oxygen available, nitrate available), or anaerobic (neither molecular oxygen nor nitrate available) conditions. Different removal efficiencies of pharmaceuticals have been observed for anaerobic, anoxic, and aerobic conditions [51]. For example, when DO level was high, Thompson et al. [52] observed biodegradation to be responsible for the majority of triclosan removal, while sorption became the dominant removal mechanism when the DO level was low. Suarez et al. [28] classified pharmaceuticals as follows: (1) highly biodegradable under aerobic and anoxic conditions, e.g., ibuprofen, fluoxetine; (2) highly biodegradable under aerobic conditions but persistent in the anoxic reactor, e.g., diclofenac, naproxen, roxithromycin, and erythromycin; (3) moderately biodegradable under aerobic and anoxic conditions, e.g., citalopram; and (4) resistant to biological transformation, for example, sulfamethoxazole, trimethoprim, carbamazepine, and diazepam.

A few studies have recommended combining aerobic/anoxic/anaerobic treatment sequences. Zwiener and Frimmel [53] observed better degradation of diclofenac in an anoxic biofilm reactor. Park et al. [54] reported high removal of carbamazepine removal by a constructed wetland under anoxic conditions. In a study by Suarez et al. [35], several pharmaceuticals such as fluoxetine, trimethoprim, and erythromycin showed significant biodegradation during anoxic treatment, while the remaining pharmaceuticals were only removed during the aerobic treatment [35].

In addition to the studies on the effect of anoxic/aerobic operation regimes on pharmaceuticals removal by MBR, some studies have highlighted the relationship between achievement of nitrification and/or denitrification and pharmaceuticals removal. There is circumstantial evidence that micropollutant biodegradation can be cometabolically mediated under operating conditions that allow for enrichment of nitrifiers [55, 56].

4 Bioaugmented MBR for Pharmaceutical Removal

Bioaugmentation can be implemented by adding special microbes which either use micropollutant as the main carbon source or can degrade micropollutants due to "cometabolism." Only a few bioaugmentation studies involving MBR are available. Pure culture of white-rot fungi (WRF) can degrade micropollutants effectively in absence of contamination from other microbes [57]. WRF secrete enzymes such as laccase and different kinds of peroxidases [58–62]. However, in non-sterile environment, which is common for wastewater conditions, the performance of WRF usually diminishes over time. Yang et al. [63] tested a fungal MBR. The MBR was first inoculated with the WRF *Trametes versicolor*, but over time a mixed bacteria-fungi community developed within the bioreactor. A synthetic wastewater comprising bisphenol A and diclofenac was treated by the MBR. It is noteworthy that diclofenac is recalcitrant to bacterial degradation. In batch pure cultures, the fungus degraded both micropollutants effectively. However, during MBR operation, diclofenac removal was not as effective, possibly because of continuous loss of the extracellular enzymes with treated effluent and contamination by bacteria. Nevertheless at a micropollutant loading of around 400 μg/L d, steady removal of bisphenol A (80–90%) and diclofenac (~55%) was observed. Nguyen, Hai, Yang, Kang, Leusch, Roddick, Price, and Nghiem [64] investigated the performance of a fungal MBR for the removal of 30 micropollutants from synthetic wastewater. When compared to a conventional bacterial MBR, the bioaugmented MBR showed 30–55% better removal for some micropollutants, namely, pentachlorophenol, fenoprop, naproxen, clofibric acid, ketoprofen, and diclofenac. Importantly, contamination by bacteria significantly reduces the performance of fungal MBR during operation over long term [65–70]. Such contamination can (1) reduce the growth and enzyme secretion by WRF and (2) consume/destabilize the produced enzymes [63, 71, 72]. Bacteria are fast-growing prokaryotes. Thus they can outperform WRF and dominate in fungal MBR [72, 73]. Some strategies to mitigate bacterial contamination have been reviewed by Asif, Hai, Singh, Price, and Nghiem [74]. The strategies include biomass replacement, immobilized fungal growth, influent pretreatment, and use of micro-screen for fungal retention within the bioreactor, while allowing passage of bacteria. An alternative MBR is enzymatic membrane bioreactor (EMBR) which combines an enzymatic bioreactor with a membrane which can retain the enzyme [75–78].

5 Conclusion and Perspective

This chapter critically discusses the potential advantages of the MBR process over conventional CAS process for the biodegradation and removal of pharmaceuticals. MBR offers some advantages for a range of pharmaceuticals because of its operation under longer SRT, higher biomass concentrations, and complete retention of suspended solids. Additionally, pharmaceuticals are adsorbed either by the membrane directly or onto the retained solids. The longer SRT in MBR may enhance biodegradation of pharmaceuticals. Nevertheless, in reality, performance of MBR over CAS depends on the operating conditions and the particular pharmaceutical being investigated. Additionally, there are a number of probable reasons why laboratory-scale studies do not exactly match removals reported from their full-scale counterparts. Such reasons include the use of arbitrarily chosen pharmaceuticals concentrations that are much higher than those encountered in wastewater and in the environment.

References

1. Deblonde T, Cossu-Leguille C, Hartemann P (2011) Emerging pollutants in wastewater: a review of the literature. Int J Hyg Environ Health 214:442–448
2. Hai FI, Yamamoto K, Lee C-H (2019) Membrane biological reactors: theory, modeling, design, management and applications to wastewater reuse, 2nd edn. IWA Publishing, London, ISBN electronic: 9781780409177. https://doi.org/10.2166/9781780409177
3. Hai FI, Nghiem LD, Khan SJ, Asif MB, Price WE, Yamamoto K (2019) Removal of emerging trace organic contaminants (TrOC) by MBR. In: Hai FI, Yamamoto K, Lee C-H (eds) Membrane biological reactors: theory, modeling, design, management and applications to wastewater reuse. IWA Publishing, London, pp 413–468. https://doi.org/10.2166/9781780409177_0413
4. Boonyaroj V, Chiemchaisri C, Chiemchaisri W, Theepharaksapan S, Yamamoto K (2012) Toxic organic micro-pollutants removal mechanisms in long-term operated membrane bioreactor treating municipal solid waste leachate. Bioresour Technol 113:174–180
5. Johnson AC, Sumpter JP (2001) Removal of endocrine-disrupting chemicals in activated sludge treatment works. Environ Sci Technol 35:4697–4703
6. Sahar E, Messalem R, Cikurel H, Aharoni A, Brenner A, Godehardt M, Jekel M, Ernst M (2011) Fate of antibiotics in activated sludge followed by ultrafiltration (CAS-UF) and in a membrane bioreactor (MBR). Water Res 45:4827–4836
7. Chang S, Waite TD, Schäfer AI, Fane AG (2003) Adsorption of the endocrine-active compound estrone on microfiltration hollow fiber membranes. Environ Sci Technol 37:3158–3163
8. Clara M, Strenn B, Gans O, Martinez E, Kreuzinger N, Kroiss H (2005) Removal of selected pharmaceuticals, fragrances and endocrine disrupting compounds in a membrane bioreactor and conventional wastewater treatment plants. Water Res 39:4797–4807
9. Joss A, Keller E, Alder AC, Göbel A, McArdell CS, Ternes T, Siegrist H (2005) Removal of pharmaceuticals and fragrances in biological wastewater treatment. Water Res 39:3139–3152
10. Tadkaew N, Hai FI, McDonald JA, Khan SJ, Nghiem LD (2011) Removal of trace organics by MBR treatment: the role of molecular properties. Water Res 45:2439–2451
11. Cirja M, Ivashechkin P, Schäffer A, Corvini P (2008) Factors affecting the removal of organic micropollutants from wastewater in conventional treatment plants (CTP) and membrane bioreactors (MBR). Rev Environ Sci Biotechnol 7:61–78
12. Oulton RL, Kohn T, Cwiertny DM (2010) Pharmaceuticals and personal care products in effluent matrices: a survey of transformation and removal during wastewater treatment and implications for wastewater management. J Environ Monit 12:1956–1978
13. Radjenovic J, Petrovic M, Barceló D (2007) Analysis of pharmaceuticals in wastewater and removal using a membrane bioreactor. Anal Bioanal Chem 387:1365–1377
14. Radjenovic J, Petroviic M, Barcelo D (2009) Fate and distribution of pharmaceuticals in wastewater and sewage sludge of the conventional activated sludge (CAS) and advanced membrane bioreactor (MBR) treatment. Water Res 43:831–841
15. Gonzalez S, Petrovic M, Barcelo D (2007) Removal of a broad range of surfactants from municipal wastewater – comparison between membrane bioreactor and conventional activated sludge treatment. Chemosphere 67:335–343
16. Gobel A, McArdell CS, Joss A, Siegrist H, Giger W (2007) Fate of sulfonamides, macrolides, and trimethoprim in different wastewater treatment technologies. Sci Total Environ 372:361–371
17. Urase T, Kagawa C, Kikuta T (2005) Factors affecting removal of pharmaceutical substances and estrogens in membrane separation bioreactors. Desalination 178:107–113
18. Li X, Hai FI, Nghiem LD (2011) Simultaneous activated carbon adsorption within a membrane bioreactor for an enhanced micropollutant removal. Bioresour Technol 102:5319–5324
19. Reemtsma T, Zywicki B, Stueber M, Kloepfer A, Jekel M (2002) Removal of sulfur-organic polar micropollutants in a membrane bioreactor treating industrial wastewater. Environ Sci Technol 36:1102–1106

20. Joss A, Zabczynski S, Gobel A, Hoffmann B, Loffler D, McArdell CS, Ternes TA, Thomsen A, Siegrist H (2006) Biological degradation of pharmaceuticals in municipal wastewater treatment: proposing a classification scheme. Water Res 40:1686–1696
21. Clara M, Kreuzinger N, Strenn B, Gans O, Kroiss H (2005) The solids retention time – a suitable design parameter to evaluate the capacity of wastewater treatment plants to remove micropollutants. Water Res 39:97–106
22. Asif MB, Ansari AJ, Chen S-S, Nghiem LD, Price WE, Hai FI (2018) Understanding the mechanisms of trace organic contaminant removal by high retention membrane bioreactors: a critical review. Environ Sci Pollut Res 26:34085–34100
23. Onesios K, Yu J, Bouwer E (2009) Biodegradation and removal of pharmaceuticals and personal care products in treatment systems: a review. Biodegradation 20:441–466
24. Boxall ABA, Blackwell P, Cavallo R, Kay P, Tolls J (2002) The sorption and transport of a sulphonamide antibiotic in soil systems. Toxicol Lett 131:19–28
25. Göbel A, McArdell CS, Joss A, Siegrist H, Giger W (2007) Fate of sulfonamides, macrolides, and trimethoprim in different wastewater treatment technologies. Sci Total Environ 372:361–371
26. Stevens-Garmon J, Drewes JE, Khan SJ, McDonald JA, Dickenson ERV (2011) Sorption of emerging trace organic compounds onto wastewater sludge solids. Water Res 45:3417–3426
27. Joss A, Keller E, Alder AC, Gobel A, McArdell CS, Ternes T, Siegrist H (2005) Removal of pharmaceuticals and fragrances in biological wastewater treatment. Water Res 39:3139–3152
28. Suarez S, Lema JM, Omil F (2010) Removal of pharmaceutical and personal care products (PPCPs) under nitrifying and denitrifying conditions. Water Res 44:3214–3224
29. Kümmerer K, Al-Ahmad A (1997) Biodegradability of the anti-tumour agents 5-fluorouracil, cytarabine, and gemcitabine: impact of the chemical structure and synergistic toxicity with hospital effluent. Acta Hydrochim Hydrobiol 25:166–172
30. Corvini P, Schäffer A, Schlosser D (2006) Microbial degradation of nonylphenol and other alkylphenols – our evolving view. Appl Microbiol Biotechnol 72:223–243
31. Hai FI, Tadkaew N, McDonald JA, Khan SJ, Nghiem LD (2011) Is halogen content the most important factor in the removal of halogenated trace organics by MBR treatment? Bioresour Technol 102:6299–6303
32. Auriol M, Filali-Meknassi Y, Tyagi RD, Adams CD, Surampalli RY (2006) Endocrine disrupting compounds removal from wastewater, a new challenge. Process Biochem 41:525–539
33. Li F, Yuasa A, Obara A, Mathews AP (2005) Aerobic batch degradation of 17-β estradiol (E2) by activated sludge: effects of spiking E2 concentrations, MLVSS and temperatures. Water Res 39:2065–2075
34. Shariati FP, Mehrnia MR, Salmasi BM, Heran M, Wisniewski C, Sarrafzadeh MH (2010) Membrane bioreactor for treatment of pharmaceutical wastewater containing acetaminophen. Desalination 250:798–800
35. Suarez S, Reif R, Lema JM, Omil F (2012) Mass balance of pharmaceutical and personal care products in a pilot-scale single-sludge system: influence of T, SRT and recirculation ratio. Chemosphere 89:164–171
36. Ternes TA, Joss A, Siegrist H (2004) Peer reviewed: scrutinizing pharmaceuticals and personal care products in wastewater treatment. Environ Sci Technol 38:392A–399A
37. Lishman L, Smyth SA, Sarafin K, Kleywegt S, Toito J, Peart T, Lee B, Servos M, Beland M, Seto P (2006) Occurrence and reductions of pharmaceuticals and personal care products and estrogens by municipal wastewater treatment plants in Ontario, Canada. Sci Total Environ 367:544–558
38. Kimura K, Hara H, Watanabe Y (2007) Elimination of selected acidic pharmaceuticals from municipal wastewater by an activated sludge system and membrane bioreactors. Environ Sci Technol 41:3708–3714

39. Tambosi JL, de Sena RF, Favier M, Gebhardt W, Jose HJ, Schroder HF, Moreira RdFPM (2010) Removal of pharmaceutical compounds in membrane bioreactors (MBR) applying submerged membranes. Desalination 261:148–156
40. Oppenheimer J, Stephenson R, Burbano A, Liu L (2007) Characterizing the passage of personal care products through wastewater treatment processes. Water Environ Res 79:2564–2577
41. Chen J, Huang X, Lee D (2008) Bisphenol A removal by a membrane bioreactor. Process Biochem 43:451–456
42. Quintana JB, Weiss S, Reemtsma T (2005) Pathways and metabolites of microbial degradation of selected acidic pharmaceutical and their occurrence in municipal wastewater treated by a membrane bioreactor. Water Res 39:2654–2664
43. Drillia P, Dokianakis SN, Fountoulakis MS, Kornaros M, Stamatelatou K, Lyberatos G (2005) On the occasional biodegradation of pharmaceuticals in the activated sludge process: the example of the antibiotic sulfamethoxazole. J Hazard Mater 122:259–265
44. Tadkaew N, Sivakumar M, Khan SJ, McDonald JA, Nghiem LD (2010) Effect of mixed liquor pH on the removal of trace organic contaminants in a membrane bioreactor. Bioresour Technol 101:1494–1500
45. Zhang T, Liu Y, Fang HHP (2005) Effect of pH change on the performance and microbial community of enhanced biological phosphate removal process. Biotechnol Bioeng 92:173–182
46. Baldwin DD, Campbell CE (2001) Short-term effects of low pH on the microfauna of an activated sludge wastewater treatment system. Water Qual Res J Can 36:519–535
47. Morgan-Sagastume F, Allen DG (2003) Effects of temperature transient conditions on aerobic biological treatment of wastewater. Water Res 37:3590–3601
48. Sui Q, Huang J, Deng S, Chen W, Yu G (2011) Seasonal variation in the occurrence and removal of pharmaceuticals and personal care products in different biological wastewater treatment processes. Environ Sci Tech 45:3341–3348
49. Vieno NM, Tuhkanen T, Kronberg L (2005) Seasonal variation in the occurrence of pharmaceuticals in effluents from a sewage treatment plant and in the recipient water. Environ Sci Technol 39:8220–8226
50. Hai FI, Tessmer K, Nguyen LN, Kang J, Price WE, Nghiem LD (2011) Removal of micropollutants by membrane bioreactor under temperature variation. J Membr Sci 383:144–151
51. Joss A, Andersen H, Ternes T, Richle PR, Siegrist H (2004) Removal of estrogens in municipal wastewater treatment under aerobic and anaerobic conditions: consequences for plant optimization. Environ Sci Technol 38:3047–3055
52. Thompson A, Griffin P, Stuetz R, Cartmell E (2005) The fate and removal of triclosan during wastewater treatment. Water Environ Res 77:63–67
53. Zwiener C, Frimmel FH (2003) Short-term tests with a pilot sewage plant and biofilm reactors for the biological degradation of the pharmaceutical compounds clofibric acid, ibuprofen, and diclofenac. Sci Total Environ 309:201–211
54. Park N, Vanderford BJ, Snyder SA, Sarp S, Kim SD, Cho J (2009) Effective controls of micropollutants included in wastewater effluent using constructed wetlands under anoxic condition. Ecol Eng 35:418–423
55. Yi T, Harper Jr WF (2007) The link between nitrification and biotransformation of 17-α-ethinylestradiol. Environ Sci Tech 41:4311–4316
56. Vader JS, van Ginkel CG, Sperling FMGM, de Jong J, de Boer W, de Graaf JS, van der Most M, Stokman PGW (2000) Degradation of ethinyl estradiol by nitrifying activated sludge. Chemosphere 41:1239–1243
57. Yang S, Hai FI, Nghiem LD, Roddick F, Price WE, Moreira MT, Magram SF (2013) Understanding the factors controlling the removal of trace organic contaminants by white-rot fungi and their lignin modifying enzymes: a critical review. Bioresour Technol 141:97–108

58. Marco-Urrea E, Pérez-Trujillo M, Blánquez P, Vicent T, Caminal G (2010) Biodegradation of the analgesic naproxen by Trametes versicolor and identification of intermediates using HPLC-DAD-MS and NMR. Bioresour Technol 101:2159–2166
59. Kang A-R, Choi H-T, Song H-G (2008) Optimization of bisphenol A biodegradation by Trametes versicolor. Korean J Microbiol 44:37–42
60. Marco-Urrea E, Gabarrell X, Sarrà M, Caminal G, Vicent T, Reddy CA (2006) Novel aerobic perchloroethylene degradation by the white-rot fungus Trametes versicolor. Environ Sci Technol 40:7796–7802
61. Tran NH, Urase T, Kusakabe O (2010) Biodegradation characteristics of pharmaceutical substances by whole fungal culture Trametes versicolor and its laccase. J Water Environ Technol 8:125–140
62. Nguyen LN, Hai FI, Yang S, Kang J, Leusch FDL, Roddick F, Price WE, Nghiem LD (2014) Removal of pharmaceuticals, steroid hormones, phytoestrogens, UV-filters, industrial chemicals and pesticides by Trametes versicolor: role of biosorption and biodegradation. Int Biodeter Biodegr 88:169–175
63. Yang S, Hai FI, Nghiem LD, Nguyen LN, Roddick F, Price WE (2013) Removal of bisphenol A and diclofenac by a novel fungal membrane bioreactor operated under non-sterile conditions. Int Biodeterior Biodegradation 85:483–490
64. Nguyen LN, Hai FI, Yang S, Kang J, Leusch FD, Roddick F, Price WE, Nghiem LD (2013) Removal of trace organic contaminants by an MBR comprising a mixed culture of bacteria and white-rot fungi. Bioresour Technol 148:234–241
65. Jelic A, Cruz-Morató C, Marco-Urrea E, Sarrà M, Perez S, Vicent T, Petrović M, Barcelo D (2012) Degradation of carbamazepine by Trametes versicolor in an air pulsed fluidized bed bioreactor and identification of intermediates. Water Res 46:955–964
66. Zhang Y, Geißen S-U (2012) Elimination of carbamazepine in a non-sterile fungal bioreactor. Bioresour Technol 112:221–227
67. Ferrando-Climent L, Cruz-Morató C, Marco-Urrea E, Vicent T, Sarrà M, Rodriguez-Mozaz S, Barceló D (2015) Non conventional biological treatment based on Trametes versicolor for the elimination of recalcitrant anticancer drugs in hospital wastewater. Chemosphere 136:9–19
68. Cruz-Morató C, Ferrando-Climent L, Rodriguez-Mozaz S, Barceló D, Marco-Urrea E, Vicent T, Sarrà M (2013) Degradation of pharmaceuticals in non-sterile urban wastewater by Trametes versicolor in a fluidized bed bioreactor. Water Res 47:5200–5210
69. Cruz-Morató C, Lucas D, Llorca M, Rodriguez-Mozaz S, Gorga M, Petrovic M, Barceló D, Vicent T, Sarrà M, Marco-Urrea E (2014) Hospital wastewater treatment by fungal bioreactor: removal efficiency for pharmaceuticals and endocrine disruptor compounds. Sci Total Environ 493:365–376
70. Badia-Fabregat M, Lucas D, Tuomivirta T, Fritze H, Pennanen T, Rodríguez-Mozaz S, Barceló D, Caminal G, Vicent T (2017) Study of the effect of the bacterial and fungal communities present in real wastewater effluents on the performance of fungal treatments. Sci Total Environ 579:366–377
71. Espinosa-Ortiz EJ, Rene ER, Pakshirajan K, van Hullebusch ED, Lens PN (2016) Fungal pelleted reactors in wastewater treatment: applications and perspectives. Chem Eng J 283:553–571
72. Libra JA, Borchert M, Banit S (2003) Competition strategies for the decolorization of a textile-reactive dye with the white-rot fungi Trametes versicolor under non-sterile conditions. Biotechnol Bioeng 82:736–744
73. Hai FI, Yamamoto K, Nakajima F, Fukushi K (2009) Factors governing performance of continuous fungal reactor during non-sterile operation – the case of a membrane bioreactor treating textile wastewater. Chemosphere 74:810–817
74. Asif MB, Hai FI, Singh L, Price WE, Nghiem LD (2017) Degradation of pharmaceuticals and personal care products by white-rot fungi – a critical review. Curr Pollut Rep 3:88–103

75. Nguyen LN, Hai FI, Dosseto A, Richardson C, Price WE, Nghiem LD (2016) Continuous adsorption and biotransformation of micropollutants by granular activated carbon-bound laccase in a packed-bed enzyme reactor. Bioresour Technol 210:108–116
76. Nguyen LN, Hai FI, Price WE, Kang J, Leusch FD, Roddick F, van de Merwe JP, Magram SF, Nghiem LD (2015) Degradation of a broad spectrum of trace organic contaminants by an enzymatic membrane reactor: complementary role of membrane retention and enzymatic degradation. Int Biodeter Biodegr 99:115–122
77. Lloret L, Eibes G, Feijoo G, Moreira MT, Lema JM (2013) Application of response surface methodology to study the removal of estrogens in a laccase-mediated continuous membrane reactor. Biocatal Biotransformation 31:197–207
78. Nguyen LN, Hai FI, Price WE, Leusch FD, Roddick F, McAdam EJ, Magram SF, Nghiem LD (2014) Continuous biotransformation of bisphenol A and diclofenac by laccase in an enzymatic membrane reactor. Int Biodeter Biodegr 95:25–32

Adsorptive Removal of Pharmaceutically Active Compounds from Wastewater

Frederik Zietzschmann

Contents

1 Introduction 240
2 Challenges of PhAC Adsorption in Wastewaters 241
2.1 Properties of PhACs and Adsorbents 241
2.2 Competitive Adsorption Phenomena 243
2.3 Inorganic Water Composition 247
3 Adsorbent Materials 247
3.1 Carbon-Based Materials 248
3.2 Other Materials 249
3.3 Comparison 250
4 Practical Aspects 251
4.1 Powdered Applications 251
4.2 Granular Applications 253
4.3 Applicability, Costs, and Environmental Impacts 258
5 Conclusion and Outlook 259
References 260

Abstract This chapter provides some fundamental aspects on adsorption of pharmaceutically active compounds in wastewater treatment scenarios, with a strong focus on actual practicality and the most recent scientific output. It discusses the key properties of adsorbing substances and adsorbent materials impacting the associated removals. The text demonstrates how adverse competitive effects by dissolved organic matter substantially impact adsorption of targeted compounds. This chapter further highlights possibilities for efficiently exploiting the available capacities of powdered and granular adsorbents. It introduces several practical and easily applicable means for prediction of the occurring adsorptive processes. In addition, the text

F. Zietzschmann (✉)
Berliner Wasserbetriebe, Berlin, Germany

Delft University of Technology, Delft, The Netherlands

Technische Universität Berlin, Berlin, Germany
e-mail: frederik.zietzschmann@bwb.de

Sara Rodriguez-Mozaz, Paqui Blánquez Cano, and Montserrat Sarrà Adroguer (eds.), *Removal and Degradation of Pharmaceutically Active Compounds in Wastewater Treatment*, Hdb Env Chem (2021) 108: 239–268, DOI 10.1007/698_2020_687,

provides some insights into available adsorbents and new material developments, comparing their advantages, downsides, and overall potentials for successful application. The last section gives a short overview on the associated environmental impacts and costs of adsorptive stages when installed on wastewater treatment plants. The conclusion summarizes the main findings and relates them to open research questions and additional future needs beyond strictly scientific disciplines.

Keywords Activated carbon, Adsorbent, Adsorption, Advanced wastewater treatment, Dissolved organic matter, Pharmaceutically active compound

1 Introduction

To address the problem of pharmaceutically active compounds (PhACs) in the water cycle (e.g., [1, 2]), authorities are exploring managerial and regulative measures [3, 4], ultimately implying advanced technological solutions [5, 6]. A major goal of all approaches is to diminish PhAC emissions into the aquatic environment from point sources – of which wastewater treatment plants (WWTPs) are by far the most relevant [7]. In comparison to PhAC treatment in drinking water works, removing PhACs in WWTPs protects the entire water cycle (rivers, lakes, groundwaters) and thus also drinking water resources [8].

The variety of available techniques in advanced wastewater treatment for PhAC removal includes, among others, improved biological stages, oxidative-reductive approaches, membrane processes, and *adsorption*. When compared to the other techniques, adsorption stands out due to a multitude of features leading to several advantages:

- *Broad effect*, since many adsorbent materials are highly heterogeneous and can adsorb compounds with largely varying chemical properties
- *True removal of contaminants*, since contaminant molecules attach onto the adsorbent and therefore literally exit the aqueous phase (in contrast to chemical transformation in oxidative-reductive processes and brine production in membrane processes)
- *Low-tech and robust*, since reactors must primarily assure good contact between water and adsorbent (no temperature or chemical reaction control, etc.)
- *Renewable resources-based*, since manifold organic carbon-based raw materials are suitable as adsorbents (e.g., plant detritus)
- *Cheap*, since adsorbents can be produced from otherwise unused materials (e.g., organic waste)
- *Safe*, since many adsorbent materials are not harmful for the environment or for humans

However – despite these advantages – adsorption of PhACs in wastewaters or WWTP effluents involves several peculiarities and difficulties. In addition, the substantiality of the above-listed advantages strongly depends on the adsorbent material considered and the application settings (water type, PhAC, etc.). Acknowledging these specificities of adsorption in wastewater treatment for PhAC removal, this chapter gives an overview on current knowledge, applications, and assessment tools.

2 Challenges of PhAC Adsorption in Wastewaters

2.1 Properties of PhACs and Adsorbents

2.1.1 PhACs

Refractory PhACs that remain in the liquid phase during wastewater treatment are often relatively polar, hydrophilic molecules [1, 2, 9]. This is undesirable because many adsorbent materials such as activated carbon generally target compounds with overall low polarity and low hydrophilicity (e.g., [10]). Accordingly, efficient adsorptive removal of polar/hydrophilic compounds often requires comparatively high adsorbent doses [11–13]. Increased hydrophilicity or decreased hydrophobicity of a PhAC, as indicated by low distribution coefficients logD, can lead to dramatically reduced loadings on activated carbons; polarizability similarly can affect adsorptive PhAC removals [14]. Poor removability generally occurs for extremely polar compounds such as charged ions like gabapentin (cf. Table 1) which typically occurs as a zwitterion in aqueous phases (e.g., [15]).

However, hydrophobic or aromatic systems within PhAC molecules can overcompensate adverse charge effects; such PhACs are still removed very well – as in the case of diclofenac or bezafibrate [16, 17]. Moreover, certain charge effects can even increase PhAC removal, for example, in the case of metoprolol [18]. This substance carries a positive charge around typical real water pHs (6–8), allowing it to interact very well with negatively charged adsorbent surfaces as those of most activated carbons [19]. The overall affinity of aromatic structures toward many adsorbents is underlined by the typically high adsorptive removals of PhACs carrying aromatic electron systems, like carbamazepine (cf. Table 1).

In addition to hydrophobicity/charge/aromaticity, molecular size of target PhACs can strongly affect their removal by different adsorbents. Many materials used for aqueous adsorption treatment are extremely porous (e.g., [20]), and targeted compounds must be able to enter the pore structure. Otherwise, they cannot adequately access adsorption sites and would suffer from size-exclusion effects [21–23]. An example for a comparatively large PhAC is clarithromycin with a molecular weight of 748 g/mol, shown in Table 1, right. Besides the large molecular size of clarithromycin, its low aromaticity, high polarizability, and proton-donating groups

Table 1 Examples for typical PhACs found in wastewaters and WWTP effluents, with names, medical application, and some major molecular properties [25]

Name	Gabapentin	Carbamazepine	Clarithromycin
Medical application	Analgesic, antiepileptic, veterinary drug	Anticonvulsant/ antiepileptic	Antibiotic
Molecular structure	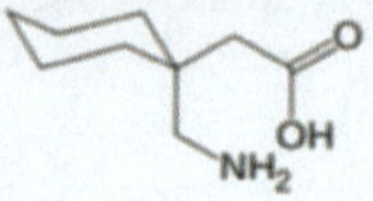	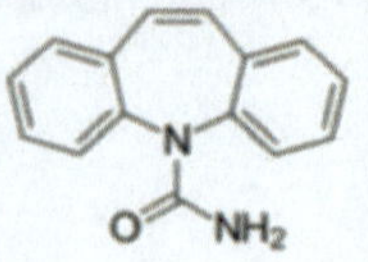	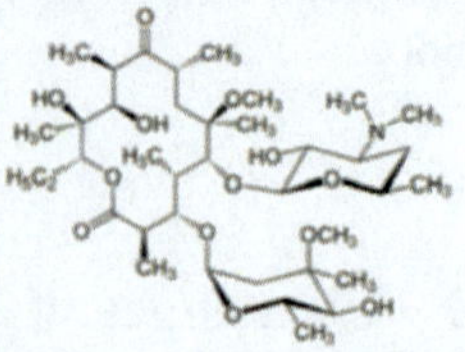
Molar mass (g/mol)	171.2	236.3	748.0
Solubility (mg/L)	4,490	17.7	0.342
Log D at pH 7	−1.1	2.45	1.25
Charge at pH 7	+-(zwitterion)	Neutral	+
Aromatic rings	0	2	0

are good examples for manifold molecular properties of PhACs, ultimately determining a PhAC's overall adsorbability by intricate interactions.

Despite adverse effects of very large molecular sizes, PhACs which are *too small* in relation to the average pore size of an adsorbent may not be removed well, either. For example, very small water contaminants like nitrobenzene adsorb very poorly onto activated carbons with a relatively large average pore size and otherwise extremely high specific internal surfaces [11, 19]. This may appear unexpected but probably results from poor overall interactions with the adsorbent surface inside pores that are too large. Despite such potentially detrimental effects, PhACs of low molecular size have the advantage of rapid diffusion within an adsorbent's porous system, allowing them to reach adsorption sites rapidly [24].

2.1.2 Adsorbents

The pore size of adsorbents logically influences the involved transport phenomena as well [24, 26, 27]. Large average pore sizes of adsorbents allow for rapid establishment of near-equilibrium conditions, whereas this takes longer for adsorbents of low average pore size [19]. Similarly, the particle size of adsorbents strongly influences the velocity of the overall process. Small particle sizes help in reducing the travel distances of PhACs within adsorbent particles and often lead to substantially improved performance [28]. Since intraparticle diffusion is generally the slowest mass transfer step in adsorptive systems, super-fine adsorbents can improve the overall process dramatically [29]. However, reducing the average particle size of adsorbents can also have detrimental effects on adsorbent capacities [30] and particle removal (cf. Sect. 4.1.1).

As indicated, the surface chemistry of adsorbents plays a crucial role in removing PhACs from aqueous solutions. In particular, charged PhACs can be repulsed from identical charges on adsorbent surfaces, or attracted by opposite charges [20]. In addition, a large variety of charge interactions, like dispersive/van der Waals forces, H-bonds, electron donor-acceptor interactions, and electrostatic attraction/repulsion, largely impact the adsorption of PhACs [31]. It is noteworthy that the surface chemistry of adsorbents strongly depends on, and can be adjusted by, the respective production approach [32]. An often unnoticed effect of carbonaceous adsorbents is their ability to induce redox reactions of PhACs, depending on the chemical surface composition of the adsorbent [33].

2.2 *Competitive Adsorption Phenomena*

2.2.1 Concentration Aspects

Most organic contaminants in raw wastewater normally occur in concentrations in the low to medium μg/L range. This concentration range is a combined result of (1) use and discharge of PhACs by polluters (households, industry) and (2) dilution and degradation/transformation processes in sewers. In WWTP effluents, the concentrations are lower, usually in the range of ng/L or μg/L [34]. Thus, the concentrations of PhACs are some 1,000 to >100,000 times lower than those of bulk dissolved organic matter (DOM), whose concentration usually ranges within 10–30 mg/L in WWTP effluents (~5–15 mg/L dissolved organic carbon, DOC). Since both PhACs and DOM are attracted toward adsorbent surfaces, they compete for the available adsorption sites, with substantial adverse effects on PhAC removal. The vast concentration difference of PhACs and DOM makes competitive adsorption phenomena appear quite unsurprising. The severity of competitive adsorption is related to DOM concentrations of treated waters in most cases [35], which makes wastewaters and WWTP effluents particularly unfavorable, due to overall very high DOM concentrations [36].

An example for adverse competitive DOM impacts on PhAC adsorption is shown in Fig. 1 via isotherms of carbamazepine adsorbing onto activated carbon from different waters. For pure water, the carbamazepine loading amounts to ~22 μg/mg at a liquid-phase concentration of 1 μg/L. At the same liquid-phase concentration, the loading declines to ~2.6 μg/mg in WWTP effluent at an initial carbamazepine concentration $c_0 = 58$ μg/L and ~0.27 μg/mg at $c_0 = 2.5$ μg/L, respectively. Given that typical WWTP effluents contain carbamazepine concentrations around 1–3 μg/L, loadings are thus ~100-fold decreased compared to pure water. Note that in real waters like the WWTP effluent shown in Fig. 1, the PhAC initial concentration has a direct impact on the position of the isotherm [37]. At increased initial PhAC concentrations, more PhAC molecules are present and can compete against DOM,

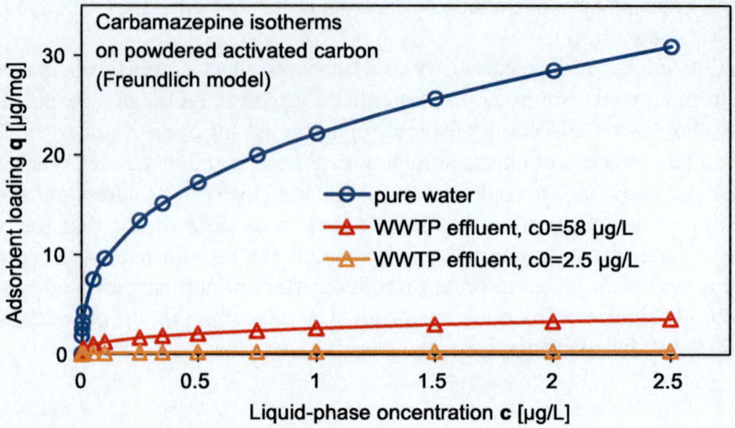

Fig. 1 Isotherms of carbamazepine (loadings *q* vs liquid-phase concentrations *c* after 48 h contact time, modeled as pseudo single-solute Freundlich isotherms based on real data) on powdered activated carbon (*Norit SAE Super*) in pure water and WWTP effluent (Berlin, Germany, DOC = 10–12 mg/L) at two different initial concentrations (c_0), depicted concentrations limited to ranges typically found in WWTP effluents for carbamazepine; unpublished data by the author

leading to overall higher loadings. In contrast, isotherms in pure water – without competing DOM – are mostly independent of the initial PhAC concentration.

2.2.2 Properties of Competitive DOM

DOM encompasses ubiquitously present background organic water constituents, originating from (1) microbes/plants/fungi/animals, (2) various anthropogenic applications, and (3) physical/chemical/biological transformation/degradation of substances from the former two groups. For these origins, DOM can have vastly variable molecular compositions, and the diversity of DOM constituents in a single water sample is enormous. The adsorption of DOM and therefore the competitiveness of DOM against PhAC adsorption relate to the large variability of DOM constituents [36–40].

To understand the underlying physical-chemical phenomena of DOM competitive adsorption, model substances can be helpful [19, 31, 41–46]. Nevertheless, despite the important fundamental insights of such studies, the actual mechanisms of competitive DOM adsorption in real waters remain partly unknown: The major complexity of DOM cannot be comprehensively simulated by model compounds. The only way to obtain manageable analytical information on DOM is to use "overall" characteristics from sum parameters like DOC or optical properties (e.g., ultraviolet/visual (UV-vis) light absorption). More advanced analytical techniques provide vastly more information but simultaneously lead to increased complexity of the data that we have to understand (e.g., [47, 48]).

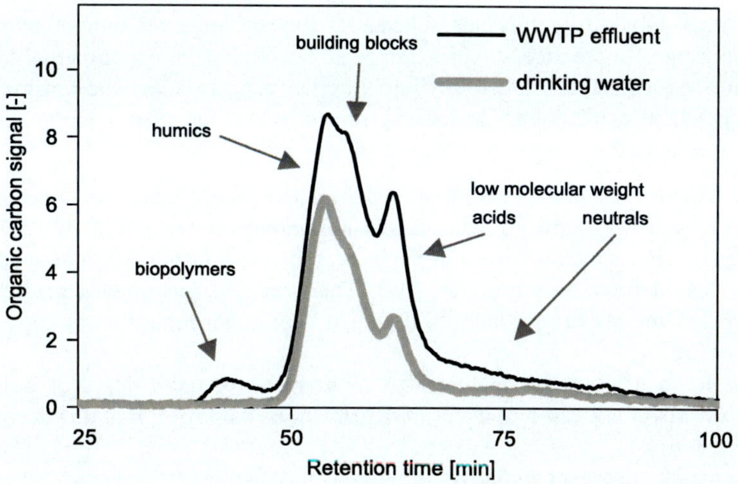

Fig. 2 LC-OCD chromatograms for Berlin (Germany) WWTP effluent (DOC = 10.9 mg/L) and drinking water (DOC = 5.1 mg/L) with indicators for typical DOM fractions according to [57]; unpublished data by the author

Advanced DOM characterization by liquid size-exclusion chromatography with online organic carbon and UV_{254} detection (LC-OCD-UV or LC-OCD) appears to provide a balance of increased analytical information and data manageability with respect to competitive adsorption. LC-OCD-UV demonstrates that different DOM fractions impose variable competitive stress on PhACs [49–53]. In particular, low molecular weight (LMW) constituents of DOM exert strong adverse impacts on PhAC adsorption [54–56]. These LMW DOM compounds have molecular weights of ~100–1,000 g/mol, thus in the range of most PhACs. Obviously, PhACs and LMW DOM target similar adsorption sites on many adsorbent materials and therefore compete with each other, leading to negative outcomes on PhAC removal. Exemplary LC-OCD chromatograms for Berlin WWTP effluent and drinking water (for comparison) are given in Fig. 2, with indicators for the typical DOM fractions encountered in real waters. LC-OCD bases upon size-exclusion chromatography which is why the largest DOM fractions (biopolymers/polysaccharides) elute the earliest and the smallest fractions (LMW acids and neutrals) elute the latest. Regarding competitive adsorption, the LMW acids and neutrals are the fractions that cause the strongest adverse effects against PhAC adsorption [52, 54, 55]. Relative to the total DOM, WWTP effluent contains higher amounts of these LMW compounds than drinking water (cf. Fig. 2).

2.2.3 Practical Implications

For the large majority of adsorbents, competitive adsorption is the most important phenomenon when considering PhAC removal in real waters, particularly in wastewaters. Obviously, this has substantial impacts on operation of adsorptive treatment stages in practice, e.g., in reactor dimensioning, or experimental tests for performance assessments. The following important aspects have to be considered for correct practical application, particularly for carbonaceous adsorbents as the most widely used material:

(a) The position of PhAC adsorption isotherms in typical loading-vs-concentration diagrams moves upward with increasing initial concentration of the PhAC (Fig. 1). This is a direct result of the increased amount of PhAC able to compete with DOM for adsorption sites [37]. Therefore, the adsorbent capacity for a PhAC is linearly proportional to the PhAC initial concentration in a given real water [58].
(b) This leads to constant relative PhAC removals for fixed doses of powdered adsorbents in a given real water, independent of the PhAC initial concentration [35, 58–61].
(c) In granular adsorbent applications, adsorbent capacities are linearly correlated to PhAC initial concentrations, meaning that relative breakthrough curves (depicted over specific throughputs, e.g., bed volumes) are independent of the PhAC initial concentration [52, 62, 63].
(d) The linear proportionality between adsorbent capacities and initial PhAC concentration in real waters is sometimes *falsely* described as "linear isotherm." This becomes obvious when considering the ideal adsorbed solution theory (IAST) equivalent background compound (EBC) model for variable initial PhAC concentration at constant EBC concentrations [36, 37, 58].
(e) Besides impacts on adsorption isotherms, competitive DOM constituents hamper the mass transfer/kinetics, resulting in increased contact times in real waters compared to pure/synthetic waters [51]. The slower mass transfer occurs in powdered and granular applications and has substantial effects on reactor design, residence times, and invest as well as operational costs.
(f) Competitive adsorption affects weak adsorbates more severely than strong adsorbates, and operators/practitioners should meticulously check if adsorptive removals of weak adsorbates are sufficient [15, 28, 64].
(g) Different types of wastewaters induce varying adverse effects on PhAC adsorption due to the variable competitiveness of different DOM fractions; WWTP effluents are generally preferable as compared to raw wastewaters or influents [40].
(h) The importance of the above-listed aspects for PhAC adsorption from real waters can strongly depend on the type of adsorbent. Non-carbonaceous adsorbents may behave differently to a large extent. For example, zeolites usually suffer far less from competitive adsorption but usually are only effective for specific PhACs [65].

2.3 *Inorganic Water Composition*

The pH and salt composition of a solution can have strong effects on adsorptive mechanisms. It is normally not feasible to adjust these parameters in practical applications; but knowing their influences can help in understanding the occurring phenomena. This can lead to improved adsorbent choices, e.g., by increasing the point of zero charge (pH_{PZC}) aiming at predominantly protonated adsorbent surfaces, alleviating adsorption of negatively charged PhACs [19, 31]. Since pH or ionic composition normally does not vary strongly for most wastewaters, potential favorable or adverse effects are usually relatively stable for a given treatment location/wastewater. However, some WWTPs are partially or strongly influenced by dilution, intermittent industrial discharges, or other factors inducing much variation to the treated waters. In such cases, variable inorganic wastewater composition should be considered regarding potential effects on PhAC adsorption.

3 Adsorbent Materials

The variety of adsorbent materials for PhAC removal is very large. A rough overview on research output related to various adsorbent materials for wastewater purification is given in Fig. 3, comparing yearly scientific publications, as obtained by searching for a fixed set of terms relating to PhACs, wastewater, and adsorption, combined with different material classifications. The search for "activated carbon"

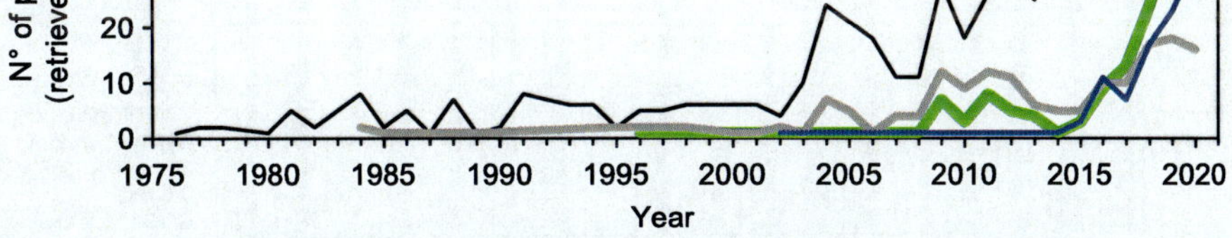

Fig. 3 Number of publications per year from a Scopus® literature search on 28 May 2020 using the command "[XYZ] AND wastewater AND adsorption AND (pharmaceutical OR antibiotic OR (organic AND (pollutant OR contaminant OR compound OR substance))) AND NOT (desalination OR (heavy metal) OR (ion exchange) OR salt OR inorganic)" in the search fields "article title" and "keyword," with *[XYZ]* referring to the search items in the legend above [66]

clearly returns the highest numbers of hits each year, but publications on new adsorbents are certainly on the rise during the last 5 years or so.

3.1 *Carbon-Based Materials*

3.1.1 Activated Carbon

Activated carbon (AC) is by far the most commonly used adsorbent for the removal of organic contaminants (including PhACs) from impacted waters. Its ratio of performance to costs is still unrivalled, and besides water purification, AC has additional applications in hundreds of industries. Thus, the overall availability is vast, resulting in a rapidly growing global market projected to pass 3.3×10^9 kg in 2021 [67]. Theoretically, this amount would allow for the treatment of ~1.7×10^{14} L of water at a typical AC dose of 20 mg/L, thus enough for treating 50 L of water per person per day year-round at a world population of 9×10^9. With manifold different production techniques [20, 68], the product diversity is enormous.

This diversity results in strongly variable PhAC removals, as demonstrated in Fig. 4 which compares the removals of 2 PhACs and a metabolite, as achieved with 18 different powdered activated carbons (PACs). Almost all PACs demonstrate the highest removal for carbamazepine, followed by 4-formylaminoantipyrine and gabapentin. The results in Fig. 4 further underline that there are large differences between the performances achieved with the various products tested here. For example, product "B" reaches >95% carbamazepine, ~70% 4-formylaminoantipyrine, and ~25% gabapentin removal, respectively. In contrast, product "K" achieves <5%, 5–10%, and ~10%, respectively. These data highlight

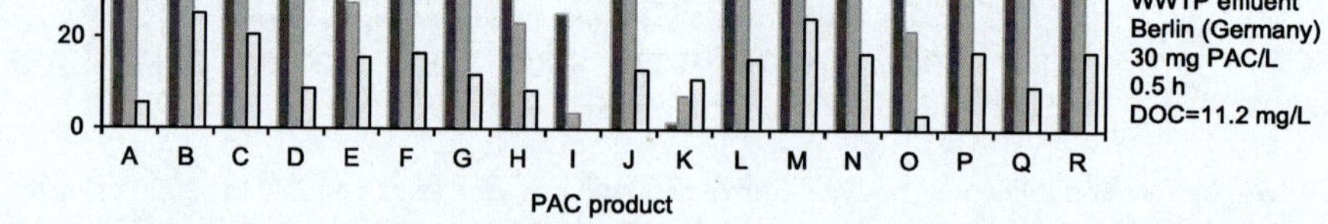

Fig. 4 Removal of carbamazepine, 4-formylaminoantipyrine (metabolite of analgesic metamizole), and gabapentin by 18 different PACs from WWTP effluent (Berlin, Germany) at 30 mg PAC/L and 30 min contact time, initial DOC = 11.2 mg/L; unpublished data by the author

the importance to assess the respective scenario prior to application, especially regarding targeted PhACs and available products.

3.1.2 Low-Cost Carbonaceous Adsorbents

An extensive amount of studies examines low-cost adsorbents for PhAC removal; several reviews provide detailed overviews on the tested materials [69–71] and production techniques [72, 73]. Despite the plenty of studies on alternative adsorbents, two fundamental shortcomings often jeopardize their practical applicability: (1) use of de-ionized/pure/synthetic water instead of real (waste)water and (2) unrealistically high PhAC concentrations in mg/L range (instead of ng/L or μg/L range) – leading to substantially overestimated potentials (cf. Sect. 2.2.1). Under more realistic conditions, low-cost adsorbents may only reach similar performance as activated carbons of moderate quality [73, 74]. Nonetheless, from environmental and economical perspectives, less efficient low-cost adsorbents like biochar can still be a better choice than activated carbons due to their reduced global warming and/or health effects [75] and reduced costs [76].

3.1.3 Special Carbonaceous Materials

Special carbonaceous adsorbent materials cover a niche for applications where the required amounts are low or where the costs play a subordinate role. In wastewater treatment both conditions are seldomly met, and such adsorbents are normally not applied unless production techniques can be transferred to a larger scale and thus be reduced in costs. The materials often have special properties, e.g., they are partly magnetic, allowing for alleviated removal from treated waters and reuse [77], or highly activated with specific internal surfaces up to >3,000 m^2/g [78], or contain doping elements such as nitrogen leading to modified internal surface chemistries [79].

3.2 Other Materials

3.2.1 Zeolites and Clays

Clays are silicate-based inorganic minerals of highly variable composition, and zeolites are a subgroup of silicates, with aluminum replacing silicon in defined positions within the mineral framework. The overall costs are usually cheap as compared to commercially available high-grade activated carbons. Clay/zeolitic minerals have defined symmetrical pores, resulting in highly specific adsorptive behavior. For example, the cage-like pore structure of zeolites allows for specific removal of substances that fit into the cages, reducing the proneness to competitive adsorption [65]. However, zeolites may be less effective when a multitude of PhACs shall be removed simultaneously [59]. For some PhACs like the antibiotic sulfamethoxazole, zeolites can reach relatively good removals from real WWTP effluent in the range 43% to 76% with 10 mg/L dosages [80], thus comparable to removals achieved with ACs [12]. A general peculiarity of adsorption onto zeolites can be S-shaped isotherms because of increased adsorbate-adsorbate interactions at high concentrations/loadings [81]. Despite these examples for ongoing research activity, there is still a lack of widespread insights into PhAC adsorption onto clay-based materials under realistic conditions [82].

3.2.2 Oligo- and Polysaccharides, Metal-Organic Frameworks, and Molecularly Imprinted Polymers

Cyclodextrins are natural circular oligosaccharides whose application as adsorbents for PhAC removal from contaminated waters is relatively new [83, 84]. The application on practical scales is still challenging since cyclodextrins swell upon contact with water, making controlled usage delicate. β-Cyclodextrins appear to be less prone to DOM adsorption competition than ACs but also demonstrate less overall adsorption capacity; doped onto cellulose microcrystals, they achieved some promising granular applications on the lab scale [18]. Other saccharide-based adsorbents include chitosans [85] and various chitosan-containing composite materials. Metal-organic frameworks (MOFs) are a new class of highly tailored synthetic material combinations of inorganic building blocks and organic linkers. Their application for (waste)water purification appears promising [86] but is strongly limited due to the elaborate synthesis of MOFs. Similarly, molecularly imprinted polymers are a new class of potential adsorbents [87], but selectivity for certain PhACs and elaborate production are bottlenecks for wider application.

3.3 Comparison

Despite the above-given examples for new developments, the choice of adsorbent base material is still relatively straightforward, since carbonaceous adsorbents are clearly the most widespread. A fundamental factor is the related costs, which range from 1 €/kg to 5 €/kg for bulk activated carbons, e.g., 1.60 €/kg for GAC and 1.80 €/kg for PAC [88]. Most "high-end" adsorbent materials cannot easily compete with such prices. Furthermore, the capacities for production need to be in the range of several tons per day for supplying a *single* large-scale WWTP (e.g., 250,000 m^3/day × 20 g/m^3 = 5 t/day).

In addition, the choice strongly depends on the respective goal. In most cases, broad effects are sought with the aim of removing a multitude of PhACs, some of which might not even be in the scope of analytical monitoring. This aspect, again, favors carbonaceous adsorbents over more specialized materials like zeolites or cyclodextrins. Due to their highly defined internal structure, especially zeolites have strong affinities for very specific PhACs, e.g., regarding size or charge, while blocking out non-suitable PhACs. Logically, a wide range of PhACs with largely variable structural/chemical properties suffers from such exclusive material characteristics [59].

4 Practical Aspects

4.1 Powdered Applications

Powdered adsorbents such as powdered activated carbon (PAC) are fine particles in the low to medium µm range. They are mixed to the treated water either in dry form or as concentrated slurries, at variable locations within WWTPs. After the desired contact time, powdered adsorbent and water are separated by flocculation, sedimentation, filtration, centrifugation, or combinations thereof.

4.1.1 Process Options

The primary goal in powdered adsorption process optimization is to exploit the available adsorbent capacity to the fullest extent possible while minimizing the required contact time. To achieve these goals, contact between adsorbents and treated water should be intensified where possible. However, competitive adsorption by DOM impedes rapid exploitation of adsorbent capacities and decelerates mass transfer (cf. Sect. 2.2.3). Therefore, several means for improved/prolonged adsorbent-water contact have been developed, resulting in different options for powdered adsorbent processes:

(a) *Direct dosage into existing biological stages* on WWTP, with continuous integration of the powdered adsorbent into sludge flocs, subsequent separation of the adsorbent with the sludge, and recirculation of the sludge-adsorbent slurry into the biology influent [88, 89]. Realization of this design is relatively straightforward since WWTPs have to be modified only marginally (installation of adsorbent dosage). However, increased DOM concentrations and competitiveness in early WWTP stages can reduce the overall efficiency to a considerable extent [40].

(b) *Direct dosage into advanced wastewater treatment stages* like clarifiers or filtration. Separation of the powdered adsorbent is usually achieved by coagulation-sedimentation and/or rapid/deep-bed filtration. If existing filtration stages are used for adsorbent separation, the filtration regimes (i.e., backwash intervals) are usually affected due to the increased loads to the filter material [90, 91]. Even distribution of the powdered adsorbent over the filter bed can substantially increase the performance [92].

(c) *Implementation of separate advanced slurry contactors/reactors,* followed by separation by sedimentation/filtration/centrifugation with recirculation of the adsorbent slurry for repeated usage of the adsorbent and increasing adsorbent concentrations in the system. This setup requires extensive additional modifications to wastewater treatment trains but normally rewards the efforts with superior adsorbent exploitation [15, 93–95].

Independent of the kind of powdered adsorbent process, separation of the adsorbent from the slurry is always necessary to avoid discharge of loaded adsorbents into receiving waters [96]. This is particularly important for very fine adsorbent materials like super-fine PAC and therefore demands for well-adjusted coagulation [97]. Nonetheless, super-fine PAC can have substantial improvements regarding adsorption capacities and required contact times as compared to conventional PAC [98], resulting in reduced slurry adsorber sizes and shorter contact times [29]. Additional improvements regarding adsorbent-water separation can be achieved by membrane filtration which usually allows for complete retention of adsorbent particles [99].

A general drawback of PAC as compared to GAC is that PAC normally cannot be reactivated equally well because of the following factors: (1) the overall burn-off during thermal reactivation is higher for PAC, (2) used PAC is usually mixed with various wastewater sludge components which cause interferences, and (3) the adsorptive qualities of PAC decrease more severely at the relatively high temperatures needed [100]. However, examples exist of pilot-scale thermal reactivation and vacuum regeneration [100, 101]. For adsorbents like zeolites, which can withstand strongly oxidizing agents, regeneration by in situ oxidation, e.g., using ozone, of adsorbed PhACs appears to be an interesting option [102].

4.1.2 Performance Prediction

Predicting the adsorptive performance is important for efficient operation and reduced costs; however, various factors influence the process efficiency. For example, variable incoming flows can increase/reduce contact times of adsorbent and water, and variable water composition, e.g., regarding competitive DOM concentrations, can influence PhAC removal. In addition, different grades of delivered adsorbent batches can induce variations of the overall adsorptive removal as well. Last but not least, it is important to realize that the relationship between PhAC removal and adsorbent dosage is strongly nonlinear in most cases and therefore the effects of variable dosages are not always easy to estimate [103].

Theoretical predictions of adsorption performance involve various concepts, e.g., quantitative structure-activity relationships (QSARs) relating adsorbent, PhAC, and water characteristics to performance indicators like PhAC removal [14, 104]. An important shortcoming of such semi-descriptive methods for performance predictions is that they often lack the integration of water quality parameters. Thus, they do not account for competitive adsorption, and the associated models usually have to be empirically corrected for influences of DOM [105]. This can be overcome by initializing competitive adsorption models in an integrated fashion, additionally using water quality characteristics such as the concentration of LMW DOM [36].

Despite the usefulness of the above-given examples of performance prediction for powdered adsorbents in lab-based studies, these approaches are seldom useful on truly practical scales. The problem is that they require a relatively large amount of analytical information on the adsorbent, the PhAC, and the water composition – all of which are not always easily available, particularly not on operating full-scale plants. Therefore, strictly operational performance assessments, which should be available without substantial loss of time, have to be simplified. However, direct online measurements of PhAC removals are not possible with the currently available analytical techniques, since they require trained personnel on advanced analytical equipment and sophisticated sample preparation (high-performance liquid chromatography-mass spectrometry).

Therefore, several approaches exist to project adsorptive PhAC removals by using surrogate parameters' adsorptive removal, as was initially conceived for oxidative stages in a similar way [106]. The AC-induced removals of PhACs, either as individual PhAC or averaged over several PhACs, can be correlated to easily measurable sum parameters, e.g., spectrometric or DOC. For example, ultraviolet absorption at 254 nm (UV_{254}) allows for relatively precise performance assessments, mostly independent of:

- PAC product and concentration [11, 107]
- PhAC initial concentration [36]
- WWTP effluent [13, 36]
- PAC recirculation stage [108]
- Coagulation, if adequately corrected for [109]

An example for such a correlation between the removal of the lipid-lowering agent bezafibrate and the removal of UV_{254} is given in Fig. 5, showing a mostly linear relation of the two parameters. Note that in Fig. 5, the linearity covers the entire range of the y-axis, which is a result of the quite similar PAC doses of 20 mg/L and 30 mg/L used here. In cases of higher variability of the PAC doses, such correlations are normally L-shaped since the PhAC removals approach 100% while the UV_{254} removals continuously increase [11]. This concept is also practically applicable, and PAC dosages can be empirically adjusted toward the desired performances using online measurements on pilot or full scales [110, 111].

4.2 Granular Applications

Granular adsorbent particles typically lie in the range of 0.5 mm to 3 mm and are applied as loose bulk fillings in fixed-bed filters. The treated water passes through the filter bed, coming into contact with the adsorbent particles and thus promoting adsorption of PhACs. In contrast to powdered adsorbent applications, granular applications contain a large excess adsorbent during the initial phase of operation.

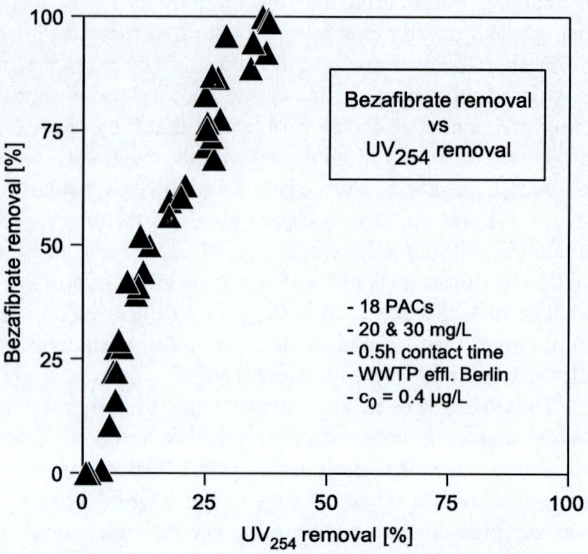

Fig. 5 Removals of the lipid-lowering agent bezafibrate vs removals of UV_{254} for various PAC products, two PAC doses, 0.5 h contact time in Berlin WWTP effluent; unpublished data by the author

During continuous operation, more and more volume of water is treated, and the ratio of adsorbent mass to treated water approaches values similar to powdered applications. Similar to "classical" granular filter materials, granular adsorbent media have to be backwashed regularly, especially when treating turbid waters – like wastewaters.

4.2.1 Operation Fundamentals

Mass transfer resistances, particularly within adsorbent pores of larger granular media, are the reason that not all PhAC molecules can enter the same adsorbent grains at the same time. Some PhAC molecules travel farer down the filter bed and, at a certain point, reach the filter effluent – without having been adsorbed. This marks the beginning of breakthrough, and PhAC effluent concentrations continuously rise thereafter, resulting in the typical occurrence of breakthrough curves (BTCs). Examples for BTCs, depicted as the dimensionless relative effluent concentration (effluent concentration divided by influent concentration, c/c_0) over the specific throughput in bed volumes fed to the filter, are shown in Fig. 6. Note that the only difference between the exemplary BTCs in Fig. 6 originates from differently fast film and/or internal mass transfer rates; the overall adsorption capacities are the same in all three cases.

Ideally, BTCs are very steep, indicating fast external and internal mass transfer and near-complete exploitation of the adsorber capacity (cf. black line in Fig. 6). Such BTCs can be expected in well-designed adsorbers without competitive effects by DOM. More realistically, BTCs in strongly competitive waters like WWTP

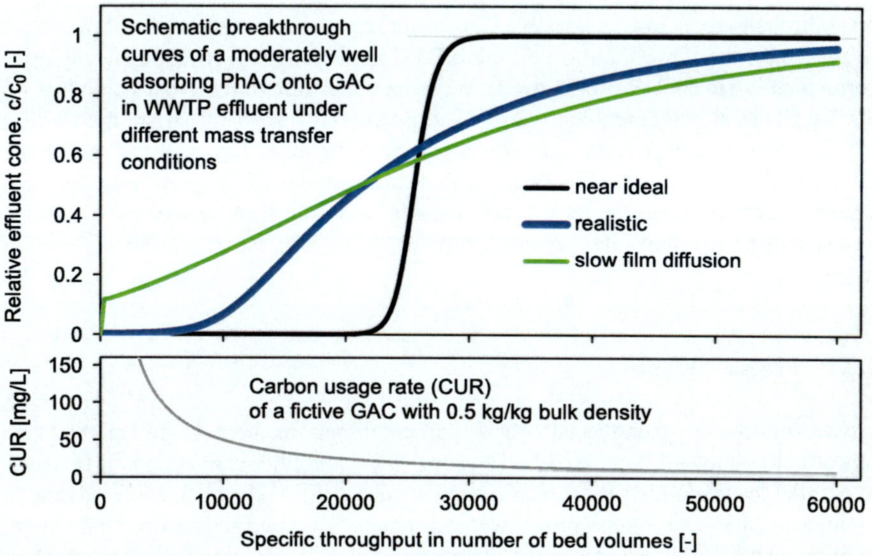

Fig. 6 *Top:* Examples for BTCs of a fictive moderately well-adsorbing PhAC onto GAC from WWTP effluent in three scenarios with variable film and internal mass transfer resistances; thin gray line indicates $c/c_0 = 1$; modeled as pseudo single-solute BTCs using the FAST 2.1 software [116]. *Bottom:* Carbon usage rate (CUR) corresponding to the specific throughput in number of bed volumes for a GAC with bulk density 0.5 kg/L

effluents are far from ideal, usually with early increases of c/c_0 and overall flat appearance (cf. blue line in Fig. 6), indicating slow usage of the available adsorption capacity [52, 62]. This shape of BTC is predominantly caused by pore-blocking DOM, which hampers PhAC mass transfer toward and within the adsorbent pores [103]. Increased water-adsorbent contact times, usually expressed via the empty bed contact time (EBCT), can help in improving the BTC shape [112]. However, this effect cannot be arbitrarily exploited and levels off at a certain filter height/EBCT [55, 113]. Moreover, practical considerations limit the EBCT since it directly relates to the adsorbent amount and thus, the adsorber dimensioning, which logically should be kept reasonably small for economic reasons.

Unfortunately, flat and stretched appearance of BTCs in strongly competitive waters can be profoundly aggravated if the advective mass transfer toward the adsorbent grains is inefficient (cf. green line in Fig. 6), characterized by almost immediate breakthrough. Such BTCs indicate poorly conceived adsorbers, either characterized by very low Reynolds numbers $\ll 1$ or grains being too big (e.g., ~2.5 mm) [114]. In addition to DOM interference by pore blocking, strong direct site competition by DOM can also occur, particularly for weakly adsorbing PhACs. In such cases, competing DOM can displace previously adsorbed PhAC, and this can lead to temporary concentration overshoots with $c/c_0 > 1$ [115].

The so-called carbon usage rate (CUR, Fig. 6 bottom) serves as a measure for the efficiency of granular adsorber operation. Ideally, CURs at specific levels of breakthrough are as low as possible. For example, at a minimum removal of ≥70% ($c/c_0 \leq 0.3$), the throughput on the blue BTC is maximally ~16,000 bed volumes, corresponding to a CUR of ~31 mg/L. With the dimension of mass per volume (e.g., mg/L), CURs are comparable to dosages of powdered adsorbents. Also, the CUR is helpful since many granular adsorbents are sold by mass (not volume) and have different bulk densities. Note that the definition of the CUR as a "rate" is very disadvantageous, since the term "rate" usually relates to time-dependent quantities like speeds or reaction rates. A more suitable expression would simply be *carbon usage*.

4.2.2 Process Options

Process options for granular adsorbers, and combinations thereof, are various, e.g., upward or downward flow fixed-bed systems [117] and filter connections in series or in parallel for profound performance improvements [88, 118, 119]. More advanced options include continuously backwashed/moving-bed or fluidized-bed adsorbers with advantages due to uninterrupted operation [120]. Continuously backwashed filters with conventional media like sand can often be easily upgraded to adsorbers [121]. Such modifications are particularly attractive for small- and medium-sized WWTPs, as they are feasible by simple adjustments and an exchange of the filter media. Setups with fluidized beds, held in place with filtration membranes, allow for usage of micro-grain granular adsorbents with superb kinetic properties [64]. A relatively widespread application of granular adsorbers is as polishing step for the effluents of oxidative stages, to reduce potential harmful toxic effects from by-products [122].

Despite the above-given examples for new approaches to granular adsorption, most setups are still relatively "classic" fixed-bed filters with granular activated carbon. Due to the vast knowledge on rapid filtration, granular adsorbers are relatively straightforward to operate, and to date, manifold examples of successful application exist [113]. Granular adsorbers are only slightly different in their operation than typical rapid filters, e.g., filter velocities are lower and, therefore, contact times are longer. Very generally, the treated water passes through a bed of 1–2.5 m height, at velocities of, e.g., 5–12 m/h, resulting in EBCTs of 5–30 min. Shorter EBCTs cause rapid and inefficient PhAC breakthrough curves, whereas longer EBCTs are normally not feasible as of space constraints (cf. Sect. 4.2.1). Also, the advantageous effects of increasing EBCTs for reducing adverse kinetic impacts by DOM tend to flatten at a certain point [52].

Due to the comparatively slow continuous usage of adsorbent in granular applications, operation times often exceed several months or years. Therefore, substantial additional biological effects occur, increasing the overall removal of various PhACs beyond what could be expected of solely adsorptive removals [117]. The long operational times of granular adsorbers also require regular backwashing due to particulate deposits within the filter media. Since wastewaters are comparatively turbid waters, the overall frequency of backwashes in treating such waters is high, e.g., >1/day, thus much higher than in, e.g., drinking water treatment. The high turbidity is also the reason why granular adsorption is normally limited to WWTP effluents, thus wastewaters whose loads of suspended solids do not require interrupting filter operation too frequently for backwashing [123].

For many granular adsorbents, regeneration and reactivation are attractive options to increase the operation times of filter media and therefore raise the overall efficiency. Considering that reactivated GAC products cost about 1/5 of virgin GAC, the reuse of spent filter materials can also result in profound economic benefits [113]. Of course, adsorbent reuse is also more environmentally friendly, given that many adsorbents are resource-consuming during their production [88]. There are manifold possibilities for adsorbent regeneration and reactivation, ranging from thermal [124] to chemical, microbial, and vacuum processes [101].

4.2.3 Performance Prediction

The interactions in granular adsorbers between the adsorbent internal (specific internal surface, pore volume distribution) and external properties (particle size), water composition (PhACs, DOM), and operational parameters (EBCT, flow conditions, long operation times) make performance prediction generally more delicate than in powdered applications. In addition, regular backwash can partly disturb the filter media filling and lead to stratification [125]. A very straightforward tool for robust and online performance prediction is via the use of easily measurable surrogate parameters which allow for inferring PhAC removals, e.g., UV_{254} or fluorescence [126].

An elaborate but mostly accurate method for adsorber performance prediction involves rapid small-scale column tests (RSSCTs), miniature duplicates of large-scale filters with identical flow conditions [115]. RSSCTs allow for shortening filter run times and decreasing necessary volumes up to several orders of magnitude. Despite their overall successful application for WWTP effluent scenarios [62], caution must be kept regarding blockage of the filter bed and associated operational difficulties [127]. Using DOC concentrations and different RSSCTs allows for the determination of fouling parameters to account for pore-blocking effects [128].

A general finding from various RSSCTs (and larger-scale tests) is that BTCs on GAC are mostly independent of the PhAC influent concentration in a given water [62, 129]. This translates into a linear relationship between the GAC capacity and the influent concentration [62]. Furthermore, the analytical information on LMW DOM (cf. Fig. 2) proves useful for the alignment of BTCs in different types of waters [62]. Granular adsorber performance can further be predicted via deterministic models, however requiring manifold simplifications as in Fig. 6 [116], or BTC data for prior fitting [55]. More complex models include, thus, direct adsorption site competitive effects by DOM [130] and pore-blocking effects [131], yet difficulties remain regarding initialization. Regression-based models like quantitative structure-activity relationships (QSARs) use adsorbent or PhAC properties, among others, and BTC data to obtain best-fit parameters [132, 133].

4.3 Applicability, Costs, and Environmental Impacts

The decision if adsorption is a feasible option for PhAC removal in a specific wastewater treatment scenario depends on various factors. Profound practical experience has been gained in full-scale advanced wastewater treatment using adsorption (and oxidative techniques) during the last years in a variety of countries, e.g., Switzerland [134] and Germany [135]. The available resources of information from various case studies and fully operational plants contain different treatment designs/approaches, sizes, and locations, among others. In addition, many full-scale research activities are still ongoing, providing more details on special adsorption stages designs or adsorbent materials (e.g., [119]). Despite such plenty of knowledge, decision-makers need to assess each case individually for monetary, societal, and environmental benefits and expenditures.

The costs for advanced wastewater treatment cover relatively wide ranges, depending on the targeted PhACs' removability/adsorbability, the respective WWTP size, construction and operational design, and applied adsorbents and chemicals, among other factors. For German adsorptive treatment stages using activated carbon, numbers of 0.05 €/m^3 up to 0.30 €/m^3 in WWTP effluent are reported, translating to 2 €/a/inhabitant up to 12 €/a/inhabitant [88]. Note, however, that such cost calculations can lose their validity rapidly due to increasing capital costs as a consequence of augmented demand, inflation, and other factors.

Since adsorbent production and operation of treatment stages both consume resources and energy, potential adverse environmental impacts have to be crucially assessed. For example, the energy demand for the production of virgin activated carbon is comparatively high and becomes more feasible if reuse/reactivation of spent adsorbent is considered, too [136]. Reactivation is normally only available for GAC, not for PAC. However, PAC increases the heating value of the accruing sludge, thus partly balancing the drawback of not being reactivatable. Moreover, the source material for activated carbon production can increase the cumulated energy consumption of advanced wastewater treatment stages by >100%, e.g., when switching from coconut husk to lignite. Similarly, the global warming potential profoundly depends on source materials, possibility for reactivation, adsorbent usage, as well as the location of treatment [88].

Examples for the cumulated energy consumption and global warming potential, including all indirect effects (e.g., production), for different advanced wastewater treatment scenarios with activated carbon, are given in Fig. 7. The error margins outline the high variability as induced by different material sources for the produced activated carbon: renewable (left, coconut husk) versus fossil (right, lignite). Note that the PAC and GAC scenarios were chosen for similar expectable PhAC removals [88] – and differ substantially considering the GAC-related CURs corresponding to the shown bed volumes of throughput (cf. Fig. 6). The reason for this higher GAC consumption as compared to PAC is adverse kinetic DOM impacts, resulting in unfavorable BTC shape. Remarkably, the ecologic benefits from reactivation,

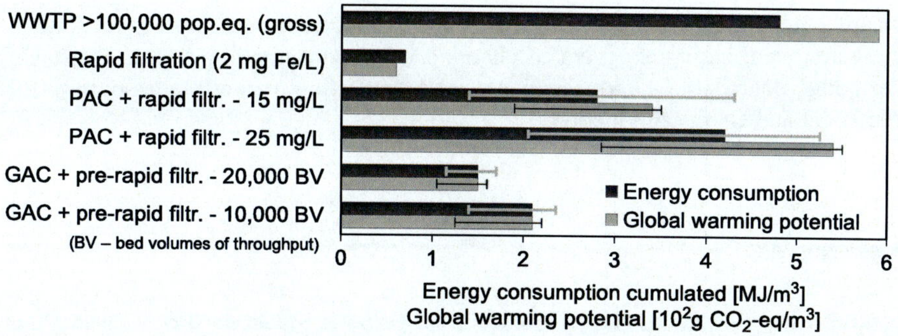

Fig. 7 Energy consumption (cumulated) and global warming potential per m^3 WWTP effluent for two PAC (dosages) and two GAC (throughput) treatment scenarios, with error bars indicating min/max values (see text for more information) and average values for typical large WWTPs (without advanced treatment) and rapid filtration for comparison (data adapted from [88])

incorporated within the GAC data in Fig. 7, clearly outweigh the drawback from unfavorable BTCs.

5 Conclusion and Outlook

Adsorption is an increasingly viable option for the removal of pharmaceutically active compounds from wastewaters, with several countries installing full-scale capacities in various locations. The insights gained on efficient adsorbent materials, installation, operation, performance prediction, and adsorptive mechanisms are substantial. The variety of process designs comprises simple upgrades of existing plants, e.g., dosage of powdered adsorbents into biological basins, as well as more sophisticated installations, like reaction chambers with sludge recirculation. Similarly, granular adsorbents can serve as exchange material for media in existing rapid filters – or as filling of newly conceived downstream adsorbers. Future research needs to optimize processes regarding high and continuous removal of pharmaceutically active compounds, with a particular focus on understanding and alleviating adverse competitive effects of dissolved organic matter.

Besides profound developments in adsorber design and operation, demands for improved adsorbent materials push manufacturers toward increasing efficiency and reducing environmental impacts. Carbonaceous adsorbents are still by far the most widespread, but various new structures such as zeolites, cyclodextrins, and metal-organic frameworks create growing interests – at least from a scientific perspective. The competitiveness of these new adsorbents will depend on the achievability of broad removal of pharmaceuticals, large production volumes, aggressive prices, toxicological harmlessness, and low impacts during production.

Last but not least, improvements of water quality by advanced treatments should not be pitted against external environmental costs. To avoid impacts from adsorbent manufacture and its excessive usage, we must exploit additional means, too. Among these are overthinking exaggerated pharmaceutical consumption, e.g., in livestock, designing degradable pharmaceuticals, and exploiting renewable resources for adsorbent and energy production.

References

1. Reemtsma T, Weiss S, Mueller J, Petrovic M, Gonzalez S, Barcelo D et al (2006) Polar pollutants entry into the water cycle by municipal wastewater: a European perspective. Environ Sci Technol 40(17):5451–5458
2. Schulze S, Zahn D, Montes R, Rodil R, Quintana JB, Knepper TP et al (2019) Occurrence of emerging persistent and mobile organic contaminants in European water samples. Water Res 153:80–90
3. Swiss Confederation (2016) Verordnung des UVEK zur Überprüfung des Reinigungseffekts von Massnahmen zur elimination von organischen Spurenstoffen bei Abwasserreinigungsanlagen (ordinance of the UVEK on the review of purification effects by measures for the elimiation of organic micro-pollutants in wastewater treatment plants). Document number 814.201.231. https://www.admin.ch/opc/de/classified-compilation/20160123/index.html
4. European Commission (2018) Watch list of substances for Union-wide monitoring in the field of water policy. Document number 32018D0840. http://data.europa.eu/eli/dec_impl/2018/840/oj
5. Blair B, Nikolaus A, Hedman C, Klaper R, Grundl T (2015) Evaluating the degradation, sorption, and negative mass balances of pharmaceuticals and personal care products during wastewater treatment. Chemosphere 134:395–401
6. Eggen RIL, Hollender J, Joss A, Scharer M, Stamm C (2014) Reducing the discharge of micropollutants in the aquatic environment: the benefits of upgrading wastewater treatment plants. Environ Sci Technol 48(14):7683–7689
7. Ruff M, Mueller MS, Loos M, Singer HP (2015) Quantitative target and systematic non-target analysis of polar organic micro-pollutants along the river Rhine using high-resolution mass-spectrometry – identification of unknown sources and compounds. Water Res 87:145–154
8. McCance W, Jones OAH, Edwards M, Surapaneni A, Chadalavada S, Currell M (2018) Contaminants of emerging concern as novel groundwater tracers for delineating wastewater impacts in urban and peri-urban areas. Water Res 146:118–133
9. Reemtsma T, Berger U, Arp HPH, Gallard H, Knepper TP, Neumann M et al (2016) Mind the gap: persistent and Mobile organic compounds water contaminants that slip through. Environ Sci Technol 50(19):10308–10315
10. Worch E (2012) Adsorption Technology in Water Treatment – fundamentals, processes, and modeling. De Gruyter, Berlin
11. Zietzschmann F, Altmann J, Ruhl AS, Dünnbier U, Dommisch I, Sperlich A et al (2014) Estimating organic micro-pollutant removal potential of activated carbons using UV absorption and carbon characteristics. Water Res 56:48–55
12. Mailler R, Gasperi J, Coquet Y, Deshayes S, Zedek S, Cren-Olive C et al (2015) Study of a large scale powdered activated carbon pilot: removals of a wide range of emerging and priority micropollutants from wastewater treatment plant effluents. Water Res 72:315–330
13. Altmann J, Ruhl AS, Zietzschmann F, Jekel M (2014) Direct comparison of ozonation and adsorption onto powdered activated carbon for micropollutant removal in advanced wastewater treatment. Water Res 55:185–193

14. de Ridder DJ, Villacorte L, Verliefde AR, Verberk JQ, Heijman SG, Amy GL, van Dijk JC (2010) Modeling equilibrium adsorption of organic micropollutants onto activated carbon. Water Res 44(10):3077–3086
15. Zietzschmann F, Dittmar S, Splettstößer L, Hunsicker J, Dittmann D, Meinel F et al (2019) Fast empirical lab method for performance projections of large-scale powdered activated carbon re-circulation plants. Chemosphere 215:563–573
16. Wang J, Wang S (2016) Removal of pharmaceuticals and personal care products (PPCPs) from wastewater: a review. J Environ Manag 182:620–640
17. Radovic LR, Silva IF, Ume JI, Menéndez JA, Leon CALY, Scaroni AW (1997) An experimental and theoretical study of the adsorption of aromatics possessing electron-withdrawing and electron-donating functional groups by chemically modified activated carbons. Carbon 35 (9):1339–1348
18. Ling Y, Alzate-Sánchez DM, Klemes MJ, Dichtel WR, Helbling DE (2020) Evaluating the effects of water matrix constituents on micropollutant removal by activated carbon and β-cyclodextrin polymer adsorbents. Water Res 173:115551
19. Dittmar S, Zietzschmann F, Mai M, Worch E, Jekel M, Ruhl AS (2018) Simulating effluent organic matter competition in micropollutant adsorption onto activated carbon using a surrogate competitor. Environ Sci Technol 52(14):7859–7866
20. Bandosz TJ, Ania CO, Dfaz JAMN, Beguin F, Choma J, Ingaki M et al (2006) Activated carbon surfaces in environmental remediation. Elsevier, Amsterdam
21. Kilduff JE, Karanfil T, Chin YP, Weber WJ (1996) Adsorption of natural organic polyelectrolytes by activated carbon: a size-exclusion chromatography study. Environ Sci Technol 30(4):1336–1343
22. Ding L, Snoeyink VL, Marinas BJ, Yue ZR, Economy J (2008) Effects of powdered activated carbon pore size distribution on the competitive adsorption of aqueous atrazine and natural organic matter. Environ Sci Technol 42(4):1227–1231
23. Ji LL, Liu FL, Xu ZY, Zheng SR, Zhu DQ (2010) Adsorption of pharmaceutical antibiotics on template-synthesized ordered micro- and mesoporous carbons. Environ Sci Technol 44 (8):3116–3122
24. Piai L, Dykstra JE, Adishakti MG, Blokland M, Langenhoff AAM, van der Wal A (2019) Diffusion of hydrophilic organic micropollutants in granular activated carbon with different pore sizes. Water Res 162:518–527
25. U.S. National Library of Medicine (2020) ChemIDplus advanced. https://chem.nlm.nih.gov/chemidplus/. Accessed 28 May 2020
26. Aschermann G, Schröder C, Zietzschmann F, Jekel M (2019) Organic micropollutant desorption in various water matrices – activated carbon pore characteristics determine the reversibility of adsorption. Chemosphere 237:124415
27. Pelekani C, Snoeyink VL (1999) Competitive adsorption in natural water: role of activated carbon pore size. Water Res 33(5):1209–1219
28. Nowotny N, Epp B, von Sonntag C, Fahlenkamp H (2007) Quantification and modeling of the elimination behavior of ecologically problematic wastewater micropollutants by adsorption on powdered and granulated activated carbon. Environ Sci Technol 41(6):2050–2055
29. Bonvin F, Jost L, Randin L, Bonvin E, Kohn T (2016) Super-fine powdered activated carbon (SPAC) for efficient removal of micropollutants from wastewater treatment plant effluent. Water Res 90:90–99
30. Takaesu H, Matsui Y, Nishimura Y, Matsushita T, Shirasaki N (2019) Micro-milling superfine powdered activated carbon decreases adsorption capacity by introducing oxygen/hydrogen-containing functional groups on carbon surface from water. Water Res 155:66–75
31. Kah M, Sigmund G, Xiao F, Hofmann T (2017) Sorption of ionizable and ionic organic compounds to biochar, activated carbon and other carbonaceous materials. Water Res 124:673–692

32. Salame II, Bandosz TJ (2001) Surface chemistry of activated carbons: combining the results of temperature-programmed desorption, boehm, and potentiometric titrations. J Colloid Interf Sci 240(1):252–258
33. Nielsen L, Biggs MJ, Skinner W, Bandosz TJ (2014) The effects of activated carbon surface features on the reactive adsorption of carbamazepine and sulfamethoxazole. Carbon 80:419–432
34. Tran NH, Reinhard M, Gin KY-H (2018) Occurrence and fate of emerging contaminants in municipal wastewater treatment plants from different geographical regions-a review. Water Res 133:182–207
35. Graham MR, Summers RS, Simpson MR, MacLeod BW (2000) Modeling equilibrium adsorption of 2-methylisoborneol and geosmin in natural waters. Water Res 34(8):2291–2300
36. Zietzschmann F, Aschermann G, Jekel M (2016) Comparing and modeling organic micro-pollutant adsorption onto powdered activated carbon in different drinking waters and WWTP effluents. Water Res 102:190–201
37. Najm IN, Snoeyink VL, Richard Y (1991) Effect of initial concentration of a Soc in natural-water on its adsorption by activated carbon. J Am Water Works Ass 83(8):57–63
38. Summers RS, Haist B, Koehler J, Ritz J, Zimmer G, Sontheimer H (1989) The influence of background organic-matter on Gac adsorption. J Am Water Works Ass 81(5):66–74
39. Rabolt B (1988) Untersuchungen zur konkurrierenden adsorption von Mikroverunreinigungen und natürlichen organischen Wasserinhaltsstoffen. Doctoral Dissertation, Technische Universität Dresden, Dresden
40. Streicher J, Ruhl AS, Gnirss R, Jekel M (2016) Where to dose powdered activated carbon in a wastewater treatment plant for organic micro-pollutant removal. Chemosphere 156:88–94
41. Kilduff JE, Karanfil T, Weber WJ (1996) Competitive interactions among components of humic acids in granular activated carbon adsorption systems: effects of solution chemistry. Environ Sci Technol 30(4):1344–1351
42. Li QL, Snoeyink VL, Mariaas BJ, Campos C (2003) Elucidating competitive adsorption mechanisms of atrazine and NOM using model compounds. Water Res 37(4):773–784
43. Pignatello JJ, Kwon S, Lu YF (2006) Effect of natural organic substances on the surface and adsorptive properties of environmental black carbon (char): attenuation of surface activity by humic and fulvic acids. Environ Sci Technol 40(24):7757–7763
44. Yu ZR, Peldszus S, Huck PM (2008) Adsorption characteristics of selected pharmaceuticals and an endocrine disrupting compound – naproxen, carbamazepine and nonylphenol – on activated carbon. Water Res 42(12):2873–2882
45. Nam SW, Choi DJ, Kim SK, Her N, Zoh KD (2014) Adsorption characteristics of selected hydrophilic and hydrophobic micropollutants in water using activated carbon. J Hazard Mater 270:144–152
46. Dittmar S, Ruhl AS, Jekel M (2017) Untersuchung von Modellstoffen zur simulation der Adsorptionskonkurrenz in gereinigten Abwässern (Model substances for simulating competitive adsorption in treated wastewaters). Vom Wasser 115(2):45–47. In German
47. Hertkorn N, Frommberger M, Witt M, Koch BP, Schmitt-Kopplin P, Perdue EM (2008) Natural organic matter and the event horizon of mass spectrometry. Anal Chem 80 (23):8908–8919
48. Matilainen A, Gjessing ET, Lahtinen T, Hed L, Bhatnagar A, Sillanpaa M (2011) An overview of the methods used in the characterisation of natural organic matter (NOM) in relation to drinking water treatment. Chemosphere 83(11):1431–1442
49. Zietzschmann F, Mitchell R-L, Jekel M (2015) Impacts of ozonation on the competition between organic micro-pollutants and effluent organic matter in adsorption onto powdered activated carbon. Water Res 84:153–160
50. Aschermann G, Neubert L, Zietzschmann F, Jekel M (2019) Impact of different DOM size fractions on the desorption of organic micropollutants from activated carbon. Water Res 161:161–170

51. Li QL, Snoeyink VL, Marinas BJ, Campos C (2003) Pore blockage effect of NOM on atrazine adsorption kinetics of PAC: the roles of PAC pore size distribution and NOM molecular weight. Water Res 37(20):4863–4872
52. Kennedy AM, Summers RS (2015) Effect of DOM size on organic micropollutant adsorption by GAC. Environ Sci Technol 49(11):6617–6624
53. Wang Q, Zietzschmann F, Yu J, Hofman R, An W, Yang M, Rietveld LC (2020) Projecting competition between 2-methylisoborneol and natural organic matter in adsorption onto activated carbon from ozonated source waters. Water Res 173:115574
54. Zietzschmann F, Worch E, Altmann J, Ruhl AS, Sperlich A, Meinel F, Jekel M (2014) Impact of EfOM size on competition in activated carbon adsorption of organic micro-pollutants from treated wastewater. Water Res 65:297–306
55. Zietzschmann F, Müller J, Sperlich A, Ruhl AS, Meinel F, Altmann J, Jekel M (2014) Rapid small-scale column testing of granular activated carbon for organic micro-pollutant removal in treated domestic wastewater. Water Sci Technol 70(7):1271–1278
56. Al Mardini F, Legube B (2010) Effect of the adsorbate (Bromacil) equilibrium concentration in water on its adsorption on powdered activated carbon. Part 3: competition with natural organic matter. J Hazard Mater 182(1–3):10–17
57. Huber SA, Balz A, Abert M, Pronk W (2011) Characterisation of aquatic humic and non-humic matter with size-exclusion chromatography – organic carbon detection – organic nitrogen detection (LC-OCD-OND). Water Res 45(2):879–885
58. Knappe DRU, Matsui Y, Snoeyink VL, Roche P, Prados MJ, Bourbigot MM (1998) Predicting the capacity of powdered activated carbon for trace organic compounds in natural waters. Environ Sci Technol 32(11):1694–1698
59. Rossner A, Snyder SA, Knappe DRU (2009) Removal of emerging contaminants of concern by alternative adsorbents. Water Res 43(15):3787–3796
60. Kovalova L, Knappe DRU, Lehnberg K, Kazner C, Hollender J (2013) Removal of highly polar micropollutants from wastewater by powdered activated carbon. Environ Sci Pollut R 20 (6):3607–3615
61. Matsui Y, Fukuda Y, Inoue T, Matsushita T (2003) Effect of natural organic matter on powdered activated carbon adsorption of trace contaminants: characteristics and mechanism of competitive adsorption. Water Res 37(18):4413–4424
62. Zietzschmann F, Stützer C, Jekel M (2016) Granular activated carbon adsorption of organic micro-pollutants in drinking water and treated wastewater – aligning breakthrough curves and capacities. Water Res 92:180–187
63. Corwin CJ, Summers RS (2011) Adsorption and desorption of trace organic contaminants from granular activated carbon adsorbers after intermittent loading and throughout backwash cycles. Water Res 45(2):417–426
64. Mailler R, Gasperi J, Coquet Y, Bulete A, Vulliet E, Deshayes S et al (2016) Removal of a wide range of emerging pollutants from wastewater treatment plant discharges by micro-grain activated carbon in fluidized bed as tertiary treatment at large pilot scale. Sci Total Environ 542:983–996
65. Jiang N, Shang R, Heijman SGJ, Rietveld LC (2018) High-silica zeolites for adsorption of organic micro-pollutants in water treatment: a review. Water Res 144:145–161
66. Elsevier (2020) Scopus document search. https://www.scopus.com. Accessed 28 May 2020
67. Statista (2020) Market volume of activated carbon worldwide in 2016 and 2021. Statista GmbH. https://www.statista.com/statistics/963555/global-market-volume-activated-carbon/. Accessed 16 May 2020
68. Danish M, Ahmad T (2018) A review on utilization of wood biomass as a sustainable precursor for activated carbon production and application. Renew Sust Energ Rev 87:1–21
69. Silva CP, Jaria G, Otero M, Esteves VI, Calisto V (2018) Waste-based alternative adsorbents for the remediation of pharmaceutical contaminated waters: has a step forward already been taken? Bioresour Technol 250:888–901

70. Quesada HB, Baptista ATA, Cusioli LF, Seibert D, de Oliveira BC, Bergamasco R (2019) Surface water pollution by pharmaceuticals and an alternative of removal by low-cost adsorbents: a review. Chemosphere 222:766–780
71. Tran HN, Nguyen HC, Woo SH, Nguyen TV, Vigneswaran S, Hosseini-Bandegharaei A et al (2019) Removal of various contaminants from water by renewable lignocellulose-derived biosorbents: a comprehensive and critical review. Crit Rev Environ Sci Technol 49 (23):2155–2219
72. Tan XF, Liu SB, Liu YG, Gu YL, Zeng GM, Hua XJ et al (2017) Biochar as potential sustainable precursors for activated carbon production: multiple applications in environmental protection and energy storage. Bioresour Technol 227:359–372
73. Benstoem F, Becker G, Firk J, Kaless M, Wuest D, Pinnekamp J, Kruse A (2018) Elimination of micropollutants by activated carbon produced from fibers taken from wastewater screenings using hydrothermal carbonization. J Environ Manag 211:278–286
74. Shimabuku KK, Kearns JP, Martinez JE, Mahoney RB, Moreno-Vasquez L, Summers RS (2016) Biochar sorbents for sulfamethoxazole removal from surface water, stormwater, and wastewater effluent. Water Res 96:236–245
75. Thompson KA, Shimabuku KK, Kearns JP, Knappe DRU, Summers RS, Cook SM (2016) Environmental comparison of biochar and activated carbon for tertiary wastewater treatment. Environ Sci Technol 50(20):11253–11262
76. Kearns JP, Wellborn LS, Summers RS, Knappe DRU (2014) 2,4-D adsorption to biochars: effect of preparation conditions on equilibrium adsorption capacity and comparison with commercial activated carbon literature data. Water Res 62:20–28
77. Rocha LS, Pereira D, Sousa É, Otero M, Esteves VI, Calisto V (2020) Recent advances on the development and application of magnetic activated carbon and char for the removal of pharmaceutical compounds from waters: a review. Sci Total Environ 718:137272
78. Li SJ, Han KH, Li JX, Li M, Lu CM (2017) Preparation and characterization of super activated carbon produced from gulfweed by KOH activation. Microporous Mesoporous Mater 243:291–300
79. Schneidermann C, Jackel N, Oswald S, Giebeler L, Presser V, Borchardt L (2017) Solvent-free Mechanochemical synthesis of nitrogen-doped Nanoporous carbon for electrochemical energy storage. ChemSusChem 10(11):2416–2424
80. de Sousa DNR, Insa S, Mozeto AA, Petrovic M, Chaves TF, Fadini PS (2018) Equilibrium and kinetic studies of the adsorption of antibiotics from aqueous solutions onto powdered zeolites. Chemosphere 205:137–146
81. Jiang N, Erdős M, Moultos OA, Shang R, Vlugt TJH, Heijman SGJ, Rietveld LC (2020) The adsorption mechanisms of organic micropollutants on high-silica zeolites causing S-shaped adsorption isotherms: an experimental and Monte Carlo simulation study. Chem Eng J 389:123968
82. Thiebault T (2020) Raw and modified clays and clay minerals for the removal of pharmaceutical products from aqueous solutions: state of the art and future perspectives. Crit Rev Environ Sci Technol 50(14):1451–1514
83. Morin-Crini N, Crini G (2013) Environmental applications of water-insoluble beta-cyclodextrin-epichlorohydrin polymers. Prog Polym Sci 38(2):344–368
84. Sikder MT, Rahman MM, Jakariya M, Hosokawa T, Kurasaki M, Saito T (2019) Remediation of water pollution with native cyclodextrins and modified cyclodextrins: a comparative overview and perspectives. Chem Eng J 355:920–941
85. Zhang X, Bai RB (2003) Mechanisms and kinetics of humic acid adsorption onto chitosan-coated granules. J Colloid Interface Sci 264(1):30–38
86. Tian C, Zhao J, Ou X, Wan J, Cai Y, Lin Z et al (2018) Enhanced adsorption of p-Arsanilic acid from water by amine-modified UiO-67 as examined using extended X-ray absorption fine structure, X-ray photoelectron spectroscopy, and density functional theory calculations. Environ Sci Technol 52(6):3466–3475

87. Meng Z, Chen W, Mulchandani A (2005) Removal of estrogenic pollutants from contaminated water using molecularly imprinted polymers. Environ Sci Technol 39(22):8958–8962
88. Metzger S, Alt K, Benstoem F, Biebersdorf N, Boehler M, Bornemann C et al (2019) Aktivkohleeinsatz auf kommunalen Kläranlagen zur Spurenstoffentfernung (Activated carbon application on municipal wastewater treatment plants for trace compound removal). Deutsche Vereinigung für Wasserwirtschaft, Abwasser und Abfall e.V. (DWA), Hennef (Germany). (In German)
89. Boehler M, Zwickenpflug B, Hollender J, Ternes T, Joss A, Siegrist H (2012) Removal of micropollutants in municipal wastewater treatment plants by powder-activated carbon. Water Sci Technol 66(10):2115–2121
90. Altmann J, Sperlich A, Jekel M (2015) Integrating organic micropollutant removal into tertiary filtration: combining PAC adsorption with advanced phosphorus removal. Water Res 84:58–65
91. Loewenberg J, Zenker A, Krahnstover T, Boehler M, Baggenstos M, Koch G, Wintgens T (2016) Upgrade of deep bed filtration with activated carbon dosage for compact micropollutant removal from wastewater in technical scale. Water Res 94:246–256
92. Altmann J, Ruhl AS, Sauter D, Pohl J, Jekel M (2015) How to dose powdered activated carbon in deep bed filtration for efficient micropollutant removal. Water Res 78:9–17
93. Meinel F, Zietzschmann F, Ruhl AS, Sperlich A, Jekel M (2016) The benefits of powdered activated carbon recirculation for micropollutant removal in advanced wastewater treatment. Water Res 91:97–103
94. Nicolet L, Rott U (1999) Recirculation of powdered activated carbon for the adsorption of dyes in municipal wastewater treatment plants. Water Sci Technol 40(1):191–198
95. Karelid V, Larsson G, Bjorlenius B (2017) Effects of recirculation in a three-tank pilot-scale system for pharmaceutical removal with powdered activated carbon. J Environ Manag 193:163–171
96. Krahnstöver T, Wintgens T (2018) Separating powdered activated carbon (PAC) from wastewater – technical process options and assessment of removal efficiency. J Environ Chem Eng 6(5):5744–5762
97. Nakazawa Y, Matsui Y, Hanamura Y, Shinno K, Shirasaki N, Matsushita T (2018) Minimizing residual black particles in sand filtrate when applying super-fine powdered activated carbon: coagulants and coagulation conditions. Water Res 147:311–320
98. Ando N, Matsui Y, Kurotobi R, Nakano Y, Matsushita T, Ohno K (2010) Comparison of natural organic matter adsorption capacities of super-powdered activated carbon and powdered activated carbon. Water Res 44(14):4127–4136
99. Loewenberg J, Zenker A, Baggenstos M, Koch G, Kazner C, Wintgens T (2014) Comparison of two PAC/UF processes for the removal of micropollutants from wastewater treatment plant effluent: process performance and removal efficiency. Water Res 56:26–36
100. Hernandez LA, Harriott P (1976) Regeneration of powdered active carbon in fluidized bed. Environ Sci Technol 10(5):454–456
101. Salvador F, Martin-Sanchez N, Sanchez-Hernandez R, Sanchez-Montero MJ, Izquierdo C (2015) Regeneration of carbonaceous adsorbents. Part II: chemical, microbiological and vacuum regeneration. Microporous Mesoporous Mater 202:277–296
102. Reungoat J, Pic JS, Manéro MH, Debellefontaine H (2007) Adsorption of nitrobenzene from water onto high silica zeolites and regeneration by ozone. Sep Sci Technol 42(7):1447–1463
103. Quinlivan PA, Li L, Knappe DRU (2005) Effects of activated carbon characteristics on the simultaneous adsorption of aqueous organic micropollutants and natural organic matter. Water Res 39(8):1663–1673
104. Li L, Quinlivan PA, Knappe DRU (2005) Predicting adsorption isotherms for aqueous organic micropollutants from activated carbon and pollutant properties. Environ Sci Technol 39 (9):3393–3400
105. Worch E (2010) Competitive adsorption of micropollutants and NOM onto activated carbon: comparison of different model approaches. J Water Supply Res Technol 59(5):285–297

106. Bahr C, Schumacher J, Ernst M, Luck F, Heinzmann B, Jekel M (2007) SUVA as control parameter for the effective ozonation of organic pollutants in secondary effluent. Water Sci Technol 55(12):267–274
107. Ziska AD, Park M, Anumol T, Snyder SA (2016) Predicting trace organic compound attenuation with spectroscopic parameters in powdered activated carbon processes. Chemosphere 156:163–171
108. Zietzschmann F, Altmann J, Hannemann C, Jekel M (2015) Lab-testing, predicting, and modeling multi-stage activated carbon adsorption of organic micro-pollutants from treated wastewater. Water Res 83:52–60
109. Altmann J, Zietzschmann F, Geiling EL, Ruhl AS, Sperlich A, Jekel M (2015) Impacts of coagulation on the adsorption of organic micropollutants onto powdered activated carbon in treated domestic wastewater. Chemosphere 125:198–204
110. Rößler A, Metzger S (2016) Application of SAC(254) measurement for the assessment of micropollutant removal in the adsorptive treatment stage of a municipal wastewater treatment plant. Water Pract Technol 11(2):503–515
111. Altmann J, Massa L, Sperlich A, Gnirss R, Jekel M (2016) UV254 absorbance as real-time monitoring and control parameter for micropollutant removal in advanced wastewater treatment with powdered activated carbon. Water Res 94:240–245
112. Kennedy AM, Reinert AM, Knappe DRU, Ferrer I, Summers RS (2015) Full- and pilot-scale GAC adsorption of organic micropollutants. Water Res 68:238–248
113. Benstoem F, Nahrstedt A, Boehler M, Knopp G, Montag D, Siegrist H, Pinnekamp J (2017) Performance of granular activated carbon to remove micropollutants from municipal wastewater-A meta-analysis of pilot- and large-scale studies. Chemosphere 185:105–118
114. Meinel F, Ruhl AS, Sperlich A, Zietzschmann E, Jekel M (2015) Pilot-scale investigation of micropollutant removal with granular and powdered activated carbon. Water Air Soil Pollut 226(1)
115. Crittenden JC, Reddy PS, Arora H, Trynoski J, Hand DW, Perram DL, Summers RS (1991) Predicting Gac performance with rapid small-scale column tests. J Am Water Works Ass 83 (1):77–87
116. Schimmelpfennig S, Sperlich A (2011) FAST – fixed-bed adsorption simulation tool 2.1beta. Berlin, Germany. http://www.fast-software.de/
117. Altmann J, Rehfeld D, Träder K, Sperlich A, Jekel M (2016) Combination of granular activated carbon adsorption and deep-bed filtration as a single advanced wastewater treatment step for organic micropollutant and phosphorus removal. Water Res 92:131–139
118. Crittenden JC, Trussell RR, Hand DW, Howe KJ, Tchobanoglous G (2012) MWH's water treatment – principles and design, 3rd edn. Wiley, Hoboken (acid-free paper)
119. Nahrstedt A, Rohn A, Alt K, Wu X, Schlösser F, Schröder K-H (2016) Mikroschadstoffelimination mittels granulierter Aktivkohle im Ablauf der Kläranlage Gütersloh-Putzhagen – Kurzbericht (micropollutant elimination by granular activated carbon in the effluent of WWTP Gütersloh-Putzhagen – short report). IWW Rheinisch-Westfälisches Institut für Wasserforschung gemeinnützige GmbH. In German
120. Wunderlin P, Meier A, Grelot J (2017) Elimination von Mikroverunreinigungen auf ARA: Aktueller Stand der Verfahren und künftige Entwicklungen. Aqua Gas 11:60–70. In German
121. Hertel M, Maurer P, Steinmetz H (2014) Auswahl und Überprüfung granulierter Aktivkohlen (GAK) für den Einsatz in kontinuierlich gespülten Filtern (selection and screening of granular activated carbons for application in continuously backwashed filters). Wasser Abfall 16 (11):40–45. In German
122. Bourgin M, Beck B, Boehler M, Borowska E, Fleiner J, Salhi E et al (2018) Evaluation of a full-scale wastewater, treatment plant upgraded with ozonation and biological post-treatments: abatement of micropollutants, formation of transformation products and oxidation by-products. Water Res 129:486–498
123. Bornemann C, Hachenberg M, Kazner C, Herr J, Jagemann P, Lyko S et al (2012) Ertüchtigung kommunaler Kläranlagen, insbesondere kommunaler

Flockungsfiltrationsanlagen durch den Einsatz von Aktivkohle – MICROFlock; Abschlussbericht (upgrading municipal wastewater treatment plants, especially flocculation filters, by activated carbon; final report). RWTH Aachen. In German
124. Salvador F, Martin-Sanchez N, Sanchez-Hernandez R, Sanchez-Montero MJ, Izquierdo C (2015) Regeneration of carbonaceous adsorbents. Part I: thermal regeneration. Microporous Mesoporous Mater 202:259–276
125. Ruhl AS, Zietzschmann F, Altmann J, Meinel F, Sperlich A, Jekel M (2015) Stratification of granular activated carbon filters for advanced wastewater treatment. Water Air Soil Pollut 226 (11)
126. Anumol T, Sgroi M, Park M, Roccaro P, Snyder SA (2015) Predicting trace organic compound breakthrough in granular activated carbon using fluorescence and UV absorbance as surrogates. Water Res 76:76–87
127. Freihardt J, Jekel M, Ruhl AS (2017) Comparing test methods for granular activated carbon for organic micropollutant elimination. J Environ Chem Eng 5(3):2542–2551
128. Kennedy AM, Reinert AM, Knappe DRU, Summers RS (2017) Prediction of full-scale GAC adsorption of organic micropollutants. Environ Eng Sci 34(7):496–507
129. Summers RS, Kim SM, Shimabuku K, Chae SH, Corwin CJ (2013) Granular activated carbon adsorption of MIB in the presence of dissolved organic matter. Water Res 47(10):3507–3513
130. Worch E (2008) Fixed-bed adsorption in drinking water treatment: a critical review on models and parameter estimation. J Water Supply Res Technol 57(3):171–183
131. Schideman LC, Marinas BJ, Snoeyink VL, Campos C (2006) Three-component competitive adsorption model for fixed-bed and moving-bed granular activated carbon adsorbers. Part I. Model development. Environ Sci Technol 40(21):6805–6811
132. de Ridder DJ, McConville M, Verliefde ARD, van der Aa LTJ, Heijman SGJ, Verberk JQJC et al (2009) Development of a predictive model to determine micropollutant removal using granular activated carbon. Drink Water Eng Sci 2(2):57–62
133. Liang CH, Chiang PC, Chang EE (2007) Modeling the behaviors of adsorption and biodegradation in biological activated carbon filters. Water Res 41(15):3241–3250
134. VSA (2020) VSA-plattform "Verfahrenstechnik Mikroverunreinigungen". Verband Schweizer Abwasser- und Gewässerschutzfachleute. https://www.micropoll.ch. Accessed 30 May 2020
135. KomS (2020) Micropollutants Competence Centre BW. Deutsche Vereinigung für Wasserwirtschaft, Abwasser und Abfall e. V. (DWA). https://koms-bw.de/en. Accessed 30 May 2020
136. Mousel D, Palmowski L, Pinnekamp J (2017) Energy demand for elimination of organic micropollutants in municipal wastewater treatment plants. Sci Total Environ 575:1139–1149

Removal of Pharmaceutically Active Compounds (PhACs) in Wastewater by Ozone and Advanced Oxidation Processes

Javier Marugán, Jorge Rodríguez-Chueca, Santiago Esplugas, Carme Sans, and Sixto Malato

Contents

1 Introduction 270
2 Ozone-Based Processes 272
  2.1 Fundamentals 272
  2.2 Removal of PhACs from Wastewaters by Ozonation 276
3 Fenton and Fenton-Like Processes 279
  3.1 Fundamentals 279
  3.2 Removal of PhACs from Wastewaters by Fenton and Fenton-Like Processes 280
4 UV/Peroxide Processes 284
  4.1 Fundamentals 284
  4.2 Removal of PhACs from Wastewaters by UV/H_2O_2 286
5 Concluding Remarks 292
References 293

J. Marugán (✉)
Department of Chemical and Environmental Technology, ESCET, Universidad Rey Juan Carlos, Madrid, Spain
e-mail: javier.marugan@urjc.es

J. Rodríguez-Chueca
Department of Chemical and Environmental Engineering Escuela Técnica Superior de Ingenieros Industriales, Universidad Politécnica de Madrid, Madrid, Spain
e-mail: jorge.rodriguez.chueca@upm.es

S. Esplugas and C. Sans
Department of Chemical Engineering and Analytical Chemistry, Faculty of Chemistry, Universitat de Barcelona, Barcelona, Spain
e-mail: santi.esplugas@ub.edu; carmesans@ub.edu

S. Malato
Plataforma Solar de Almería (CIEMAT), Carretera de Senés, Almería, Spain
e-mail: smalato@psa.es

Sara Rodriguez-Mozaz, Paqui Blánquez Cano, and Montserrat Sarrà Adroguer (eds.), *Removal and Degradation of Pharmaceutically Active Compounds in Wastewater Treatment*, Hdb Env Chem (2021) 108: 269–298, DOI 10.1007/698_2020_664,

Abstract During the last decades, many water treatment processes have been proposed to deal with the increasing water quality requirements demanded for the urban wastewater effluents (UWWE). Among them, the so-called Advanced Oxidation Processes (AOPs) include a wide range of technologies based on the generation of very reactive and non-selective radical species with a strong oxidation potential that can readily remove not only PhACs but also other organic micropollutants. Together with ozone-based processes, these technologies constitute promising alternatives to be used as final barrier to remove these contaminants before the discharge of the effluents to the environment. This chapter focuses on the processes with higher potential to be implemented at large scale for the removal of PhACs in urban wastewater treatment plants: ozone-based, Fenton, and UV/H_2O_2 processes. Along the chapter, the fundamentals of these processes are introduced, including some of the more relevant modifications from the basic processes, and the more recent studies about their application for the removal of PhACs from UWWE at pilot and full scale are reviewed. From a scientific and technical point of view, the results from these researches confirm the successful removal of a wide range of PhACs and other micropollutants, although the absolute efficiency depends on the water matrix and the specific substances monitored along the treatment. As final remarks, other aspects are also included, such as the need to control the potential formation of by-products during the process and the required optimization of the processes to become competitive in economic terms.

Keywords Fenton, Hydrogen peroxide, Hydroxyl radicals, Ozone, UV

1 Introduction

The available technologies to remove contaminants in aqueous solution are very diverse. Physical treatments just concentrate the contaminants, whereas chemical and biochemical processes degrade these compounds, mainly oxidizing the contaminants. Glaze et al. [1] defined "the Advanced Oxidation Processes (AOPs) as those involved in the generation of hydroxyl radicals in sufficient quantity to affect water purification." Glaze referred to water treatments involving UV/H_2O_2/O_3 used in "light conditions," that is, room temperature and atmospheric pressure. Later on, the definition of AOPs was extended to the role of other radicals generated, as hydroperoxyl, persulfate, or peroxymonosulfate [2–4], and other treatments like Fenton, TiO_2 photocatalysis, electrochemistry, ultrasounds, non-thermal plasma, and wet oxidation were included in AOPs (Fig. 1). The distinction of applications at light conditions is nowadays questionable. The name AOPs is also doubtful because some of the named processes are used in the reduction instead of the oxidation attack to the contaminants. Nowadays, the applications to the AOPs have been extended from water to wastewaters, air, and soil, mainly at laboratory

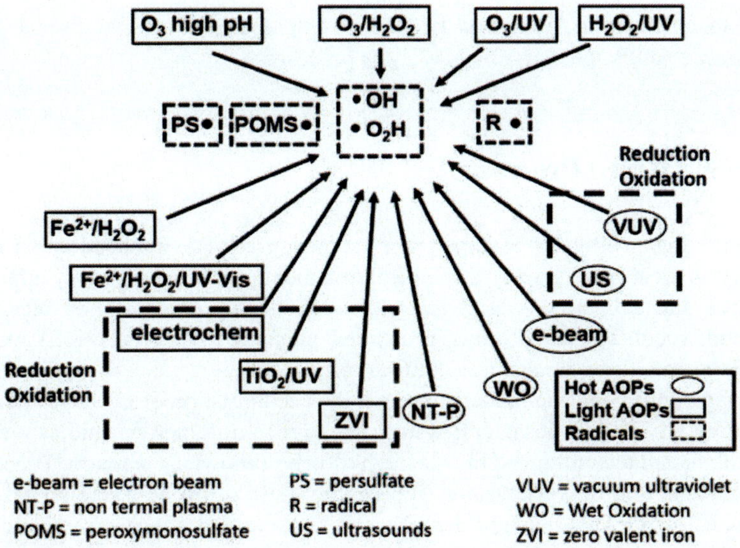

Fig. 1 Advanced oxidation processes

and pilot scales. Only ozone-based AOPs, Fenton, and UV/H_2O_2 are implemented at industrial scale for water and wastewater treatment.

AOPs involve chemical reactions promoting the degradation of the contaminants, and it is more a transformation of compounds than a simple fading. In addition, it is important to consider the possibility to generate by-products more harmful than the parent compounds. In consequence, biodegradability and toxicology tests of treated waters turn as important as the chemical analysis of the contaminant. The water matrix (pure, groundwater, wastewater) also influences in the efficiency and the removal rate of these contaminants [5], mainly because the oxidant is also consumed by other compounds in the water matrix (organic matter, ions, particulate matter). For most wastewaters, contaminant concentration and effluent toxicity decrease, and biodegradability increases as AOP treatment is extended. Because of this and the cost of chemicals and energy involved in AOP treatments, the combination of these technologies with other less costly treatments as biodegradation, adsorption, and membrane separation has been extensively studied [6].

Nowadays, there is major concern about the presence of PhACs in wastewater treatment plant effluents. These compounds are continuously released to the environment, and, in many cases, natural degradation is not enough to remove them efficiently. Consequently, PhACs remain in natural waters affecting the water quality and human health. Many reviews on the applications of AOPs for the treatment of water and wastewater containing PhACs compounds can be found in the literature [7–10]. Today it is possible to find extensive information on removal of PhACs in wastewater by AOPs, mainly at lab and pilot plant scale. Accordingly, this section aims to present a general vision of the removal of PhACs in wastewater by

AOPs. Specific care is devoted to the AOPs implemented at industrial scale, that is, ozone-based AOPs, Fenton, and UV/H_2O_2 processes.

2 Ozone-Based Processes

After more than 100 years of experience on industrial-scale applications of ozone, nowadays ozonation is already a chemical treatment process extensively applied on municipal and industrial wastewater disinfection and oxidation of recalcitrant organic microcontaminants (drugs, pesticides, phenolic compounds, etc.) and inorganic chemicals (iron, manganese, sulfites, etc.).

Among current environmental concerns, the accumulation of microcontaminants and particularly PhACs in natural waters has triggered numerous studies on ozonation and related technologies, like ozone/hydrogen (known as peroxone process) or catalytic ozonation of wastewaters. The effectiveness of ozone-based processes on the removal of PhACs strongly depends on chemical and engineering parameters, like ozone generation and transfer from the gas phase to the wastewater, PhACs reactivity with ozone, and, particularly, composition of the water matrix. This section summarizes the current knowledge on PhACs remediation in wastewaters by ozone-based processes.

2.1 Fundamentals

Ozone (O_3) is a pale blue gas with a robust oxidizing capacity ($E_0 = 2.07$ eV), slightly soluble in water. It easily decomposes to oxygen atoms and molecules; that is why in water treatment, ozone must be generated "in situ" [11]. This gas with high oxidation potential is produced by submitting oxygen molecules to high electrical voltages. There are different methods for producing ozone being the corona discharge one of the most used by far. This one consists of passing oxygen between two electrodes separated by a dielectric material and connected to a source of high voltage producing a corona discharge. The oxygen in the discharge gap is converted to ozone.

Once ozone is produced, it is bubbled into a reactor containing the solution to be treated. The ozone transfer efficiency from gas to liquid phase is fundamental to achieve optimal reaction kinetics rates between ozone and dissolved compounds. This transfer is mainly controlled by physical parameters, such as temperature, gas flow rate, bubble size, ozone partial pressure, and reactor geometry, and also chemical factors like pH, ionic strength, composition of aqueous solutions, etc. [12]. Therefore, it is important to distinguish between applied ozone dose and transferred ozone dose (TOD), which is defined by Eq. (1) and represents the accumulated amount of ozone that is transferred to the water sample per unit of volume and time:

$$TOD = \int_0^t \frac{F_{\text{gas}}}{V_{\text{liq}}} \cdot \left([O_3]_{\text{gas,in}} - [O_3]_{\text{gas,out}}\right) \cdot \mathrm{d}t \tag{1}$$

Ozone in the liquid phase can react with organic compounds in two ways, direct or indirect. These two reaction pathways lead to different oxidation products, and they are controlled by different reaction kinetics. Direct ozone reactions with organic water constituents (comprising PhACs) are quite selective mechanisms including cycloaddition, electrophilic substitution, and electron transfer reactions. Thus, molecular ozone preferably reacts with organic matter with double bonds, activated aromatic groups, or amines moieties [13, 14]. The indirect reaction pathway involves hydroxyl radicals (•OH), mainly generated by radical chain decomposition of ozone initiated by hydroxide ions (OH^-) in water. Reaction between O_3 and OH^- is very slow ($k = 70\ M^{-1}\ s^{-1}$); therefore indirect ozone reactions through hydroxyl radicals only contribute to pollutants removal at neutral or alkaline pH conditions. Hydroxyl radicals are extremely reactive transient species, more reactive, and significantly less selective than molecular ozone. Hydroxyl radicals' reactivity involves three main mechanisms: addition reactions, hydrogen abstraction reactions, and electron transfer reactions, being addition reactions to double bonds (like C=C, C=N, S=O) and aromatic rings the most common and fast [14].

Other •OH generation mechanism is ozone reactions with the dissolved organic matter (DOM) contained in wastewater matrices, particularly with DOM aromatic functional groups. The electron-rich aromatic components of DOM can undergo electron transfer with O_3 generating the intermediate ozonide anion ($O_3^{\bullet -}$) which can give rise to •OH generation. Similarly, phenolic moieties generated by hydroxylation of aromatic DOM moieties can also lead to •OH production by electron transfer with O_3. However, DOM present in water also acts as radical chain inhibitor, scavenging •OH or consuming O_3 without further •OH or chain carrier production [15]. Even with this dual behavior of DOM, organic matter content in wastewaters dramatically increases ozone consumption for pharmaceuticals depletion.

Moreover, some inorganic anions present in the wastewater can interfere in the ozonation process. Carbonate and bicarbonate anions are well-known •OH scavengers and radical chain inhibitors, since they contribute to the termination of the ozone decomposition mechanism [16]. Other important species is nitrite, contributing to ozone depletion of 3.4 g O_3 per g NO_2-N. Finally, organic particulate matter content of water consumes both O_3 and •OH, increasing overall ozone requirements.

Considering all the issues exposed, ozonation performance for the removal of PhACs in wastewaters hardly depends on the quantity and type of organic matter and inorganic constituents of the water matrix.

2.1.1 Ozonation Kinetics

There are many experimental shreds of evidence, at laboratory, pilot plant, and industrial scale, demonstrating the effectiveness of ozonation for the oxidation of

Table 1 Micropollutants classification and some PhACs with their second-order reaction rate constant with ozone and hydroxyl radical (adapted from [17])

Group of MPs	Reactivity	k_{O3} [M^{-1} s^{-1}]	$k_{\bullet OH}$ [M^{-1} s^{-1}]
I	*High with O_3 and $^{\bullet}OH$*	*>1 × 10^5*	*>5 × 10^9*
Triclosan	Antimicrobials	3.8 × 10^7	9.6 × 10^9
Diclofenac	NSAID	1.0 × 10^6	7.5 × 10^9
Sulfamethoxazole	Antibiotics	5.7 × 10^5	5.5 × 10^9
Carbamazepine	Anticonvulsant	3.0 × 10^5	8.8 × 10^9
Trimethoprim	Antibiotic	2.7 × 10^5	6.9 × 10^9
Naproxen	NSAID	2.0 × 10^5	9.6 × 10^9
II	*Moderate with O_3 and high with $^{\bullet}OH$*	*10^{-1} × 10^5*	*>5 × 10^9*
Gemfibrozil	Lipid regulator	5.0 × 10^4	10 × 10^9
Atenolol	B-blocker	1.7 × 10^3	8.0 × 10^9
III	*Low with O_3 and high with $^{\bullet}OH$*	*<10*	*>5 × 10^9*
Ibuprofen	NSAID	9.6	7.4 × 10^9
Phenytoin	Antiepileptic	<10	6.3 × 10^9
Primidone	Anticonvulsant	<10	6.7 × 10^9
IV	*Low with O_3 and moderate with $^{\bullet}OH$*	*<10*	*1 × 10^9–5 × 10^9*
Meprobamate	Antianxiety drug	<1	3.7 × 10^9
V	*Low with both O_3 and $^{\bullet}OH$*	*<10*	*<1 × 10^9*

PhACs in wastewaters. Since common municipal wastewater pH is neutral or slightly alkaline, ozonation efficiency directly depends on the second-order rate constants of reactions between the pharmaceutical (PhAC) and both ozone and hydroxyl radicals, as well as the ozone and •OH exposures ($\int[O_3]dt$ and $\int[\bullet OH]dt$) as described by Eq. (2):

$$-\ln\left(\frac{[\mathrm{PhAC}]}{[\mathrm{PhAC}]_0}\right) = k_{O3}\cdot\int[O_3]\mathrm{d}t + k_{\bullet OH}\cdot\int\left[{}^{\bullet}OH\right]\mathrm{d}t \qquad (2)$$

Most organic micropollutants, including PhACs, present high reactivity with •OH, with second-order rate constants higher than 10^9 M^{-1} s^{-1}. However, kinetics in direct ozone reactions largely depends on the structure of the target chemicals due to the specific reaction pathways of molecular ozone. Consequently, the second-order rate constants for ozone oxidation of organic compounds vary more than eight orders of magnitude, between 0.1 M^{-1} s^{-1} and 10^7 M^{-1} s^{-1}. Lee and coworkers classified micropollutants depending on their reactivity with ozone and •OH [17]. Classification goes for group I, including the most reactive compounds with both oxidants down to group V, containing the most resistant to ozonation treatment. The second-order kinetic constant of most common PhACs, both with ozone and •OH, is already available in the literature. Table 1 includes micropollutants classification and some examples of PhACs, together with their second-order reaction rate constant with ozone and hydroxyl radical.

The acid-base dissociation constant (pK_a) of a PhAC is also a key physicochemical parameter influencing ozonation performance. As the majority of drugs are weak acids and/or bases, the knowledge of their dissociation constant helps in understanding the ionic form a molecule will take across a range of pH values. This is particularly important since the reactivity of ozone with some organic dissociating compounds extraordinarily changes with pH by several orders of magnitude [13].

2.1.2 Peroxone

The promotion of in situ generation of •OH by combining ozone with hydrogen peroxide (H_2O_2) is known as peroxone process. The generation mechanism is based on the favored reaction of ozone with HO_2^-, the equilibrium deprotonated form of H_2O_2 [18]. On the other hand, H_2O_2 reaction with ozone is very slow ($k < 0.01\ M^{-1}\ s^{-1}$), while its reactivity with •OH is substantially faster ($k = 2.7 \times 10^7\ M^{-1}\ s^{-1}$). Consequently, increasing H_2O_2 concentration may result in an overall process efficiency decline, due to the reduction of the available amount of •OH in the reaction media [18].

There are many experimental pieces of evidence on the benefits of peroxone in terms of PhACs abatement and bromates production reduction, compared with single ozonation, for drinking water applications [19]. However, this enhancement in micropollutants abatement is not evident for urban wastewaters presenting a higher pollution load, at the current industrial ozone doses applied [20].

2.1.3 Catalytic Ozonation

Ozone combined with homogeneous and heterogeneous catalysts can also enhance •OH production. Homogeneous process, using transition metals in solution, and heterogeneous catalysts, like metal oxides, supported metals, and even minerals or metallic residues, have been widely studied demonstrating the benefits associated to the potential generation of •OH [21, 22].

However, several problems arise related to the use of catalysts that makes industrial application quite difficult. Thus, the homogeneous process requires metals removal from water effluents, increasing treatment costs. Heterogeneous systems deal with problems like catalyst fouling and deactivation, metals leakage into water media, and synthesis and/or preparation costs. All these drawbacks, together with the objective to improve the overall efficiency, are the reasons that, nowadays, catalytic ozonation is one of the main research topics in water treatment.

2.2 Removal of PhACs from Wastewaters by Ozonation

Although PhACs were not within the 45 substances/groups of substances included in the EU priority substances list (established in Directive 2013/39/EU), 5 out of 15 contaminants of emerging concern (CECs) in the latest watch list are antibiotics (Decision 2018/810/EU). Moreover, the Swiss Federal Office for the Environment (FOEN) established a list of 12 indicator substances to be removed by at least 80% in the upgraded wastewater treatment plants, among them are 11 PhACs (amisulpride, carbamazepine, citalopram, clarithromycin, diclofenac, metoprolol, hydrochlorothiazide, venlafaxine, candesartan, irbesartan, and mecoprop).

Although ozonation efficiency directly depends on the second-order rate constants of PhACs with ozone and •OH, urban wastewater effluents (UWWE) composition, mainly its effluent organic matter (EfOM) content, is the main parameter controlling overall process efficiency. This is the reason why ozone dosage for UWWE is commonly referred as specific ozone dose, that is, mg of ozone supplied per mg of dissolved organic carbon [mg O_3 mg C^{-1}]. Under the current applied specific ozone doses between 0.4 and 0.6 mg O_3 mg C^{-1}, there has been reported acceptable (50–80%) to good (>80%) efficiencies removal for PhACs belonging to groups I and II (i.e., triclosan, naproxen, trimethoprim, ciprofloxacin, clarithromycin, diclofenac, erythromycin, metoprolol, sulfamethoxazole, among others). PhACs with rate constants with ozone lower than 10 M^{-1} s^{-1} (from group III to V) are considered ozone-resistant compounds (like meprobamate, X-ray contrast agent iopromide, etc.). They are partially or poorly abated when applying ozone doses up to about 1.0 mg O_3 mg C^{-1} [23, 24]. Higher ozone doses would be necessary to improve their depletion, increasing ozone treatment cost. Figure 2 shows some PhACs removal obtained from concentrated municipal wastewaters at different applied specific ozone dose [25].

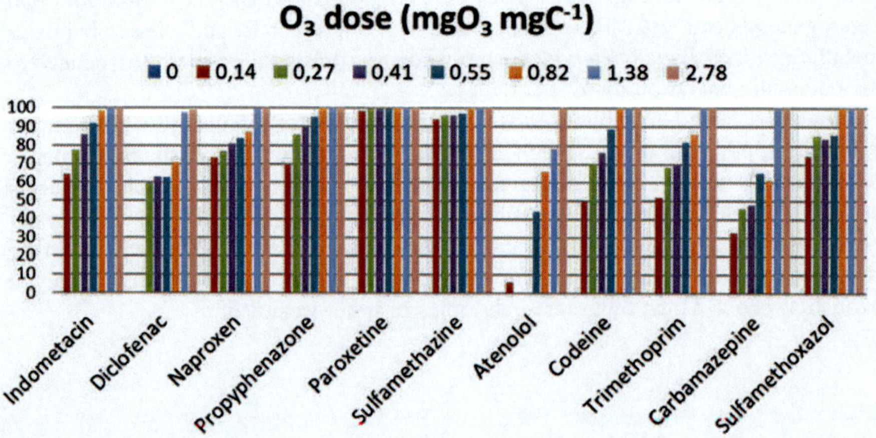

Fig. 2 PhACs removal at different ozone doses [25]

In UWWE ozonation, there is a competition for ozone between the PhACs and the wastewater constituents, that is, EfOM and some inorganic ions. These compounds rapidly consume transferred ozone in a process called primary ozonation or instantaneous ozone demand (IOD) stage. Wastewater organic and inorganic constituents also strongly scavenge the hydroxyl radicals formed through ozone decomposition reactions, resulting in poor availability of this oxidant in the reaction medium. After IOD completion, a secondary ozonation stage characterized by a slower ozone consumption and significantly lower •OH scavenging rate starts. However, UWWE ozonation typically employs ozone dosages below IOD fulfillment [26]; thus a satisfactory removal of recalcitrant PhACs is not accomplished. Application of peroxone (O_3/H_2O_2) process for the intensification of •OH oxidation neither appears to contribute significantly to recalcitrant PhACs removal in wastewaters, probably because the ozone decomposition process, at the current applied doses, is strongly controlled by reactions with EfOM [27].

As every year water resources become scarcer, wastewater reclamation and reuse gains prominence as an alternative to increasing water supplies in many areas of the world. Therefore, obtaining high-quality reclaimed water is already imperative. Ozonation is a water treatment technology with a proven capability to remove microcontaminants, including PhACs, from UWWE. However, efforts to further remove both highly reactive and also ozone-resistant micropollutants during ozonation and ozone-based processes are still necessary, especially for water entering the reuse cycle either in potable (drinkable) form or in non-potable (not drinkable) form.

2.2.1 Control of Ozonation By-products

Ozonation rarely leads to UWWE mineralization but to the so-called transformation products (TPs). A combination of advanced analytical techniques and the knowledge of mechanisms of ozone and •OH reactions with PhACs allows the elucidation of many of these products. Although the existence or not of the functional pharmacophores in the TPs structures can be an indicator of their potential hazard, their potential risks to humans and the environment associated to their presence in water matrices, particularly for water reuse, should be assessed. Protocols comprising a wide range of specific and unspecific toxicity bioassays have been developed over the last years, as well as computer-aided methods based on the existing knowledge for a number of model compounds [28].

The generation of bromate, an oxidation by-product from bromide, also needs to be carefully evaluated. Bromate is a highly carcinogenic compound, and its generation can be significant in the ozonation of industrial wastewaters and sewer systems with saline intrusion. Even though the process is still not fully understood due to its highly complicated chemistry, efficient strategies for its mitigation during wastewater ozonation are nowadays fairly well-known, such as low ozone dosage and the use of the peroxone process [29].

The generation of other oxidation by-products from water pollutants and EfOM transformations during ozonation can also bring some toxicity concerns – like

N-nitrosodimethylamine (NDMA) that originated, for example, from the oxidation of sulfonamide moiety of some antibiotics—and others, like aldehydes, ketones, and carboxylic acids, although in general less toxic, that are formed in significant concentrations. These compounds contribute to the increase of the biodegradability of ozonized effluents, which may trigger some technical and additional environmental problems, like formation of toxic disinfection by-products, bacterial growth in distribution systems, or eutrophication in receiving water bodies [30]. In order to mitigate problems with oxidation products, ozonation is integrated with post-treatments with other wastewater technologies, such as granular activated carbon and biofiltration strategies.

2.2.2 Municipal Wastewater Ozonation at Industrial Scale

UWWE have been proven as the main source of PhACs in natural water compartments. Protection of sensitive waters and drinking water resources together with load reduction for downstream water use has become a priority for some countries, like Switzerland. Moreover, the reclamation and reuse of municipal wastewater is a well-established practice in many areas of the world with water scarcity, like California and the southeast of Australia, and these activities will continue to increase.

High-quality effluents and water supplies can be obtained by further treatment of UWWE. Ozone is nowadays recognized as one of the best chemical oxidants for reducing the amount of micropollutants discharged with UWWE. This strategy is the one adopted by Switzerland and it is currently under development. By 2040, circa 100 wastewater treatment plants (out of a total of 650) will be upgraded with a tertiary treatment either with ozonation or powdered activated carbon [31]. In other European countries such as France, the Netherlands, or Germany, similar practices are currently under evaluation.

Ozonation is also one of the short- or medium-term future prospect technology in wastewater reclamation for water reuse. This technology can be employed after activated sludge when reuse is intended for agricultural and municipal applications or as part of a multibarrier advanced treatment in the case of indirect and direct potable reuse. Many cities have integrated reuse plans into their water management strategies, for example, to irrigate parks, playgrounds, and golf courses and for other non-potable uses such as toilet flushing, fire protection, street cleaning, and fountains. For example, Los Angeles has set the goal of reusing 40% of its municipal wastewaters within 20 years. St. Petersburg in Florida completely reuses all of its wastewater through a specific distribution system, at a cost about one-third less than the drinking-quality water. Many studies are being conducted on the feasibility of reclaiming wastewater to drinking quality. Although implementation results are still scarce, areas facing limited water availabilities and high costs of developing new supplies are considering reclaimed water as a potential source of potable water in the near future.

3 Fenton and Fenton-Like Processes

3.1 Fundamentals

Fenton reaction was described for the first time by Henry John Horstman Fenton, who discovered that H_2O_2 could be activated in the presence of ferrous ions to oxidize tartaric acid [32]. The mechanism of Fenton oxidation is quite complex [33]. Briefly, it relies on the reaction between ferrous iron (as catalyst) and H_2O_2 (as oxidant) generating •OH, as stated in the simplified sequence of reactions 3 and 4. The regeneration of ferrous iron from ferric iron by reaction 4 is the rate-limiting step in the catalytic iron cycle, if iron is added in small amounts. The values of the kinetic constants have been reported by Sychev and Isaak [34]. It has only been in 1968 that it has been first suggested as means for wastewater treatment by Bishop et al. [35]. In the presence of H_2O_2 and at $pH < 3$, the reaction system is autocatalytic, because Fe (III) reacts with H_2O_2 giving Fe(II), which is generated at a slow rate:

$$Fe^{2+} + H_2O_2 \rightarrow Fe^{3+} + {}^{\bullet}OH + OH^-, k = 53 - 76\ M^{-1}s^{-1} \quad (3)$$

$$Fe^{3+} + H_2O_2 \rightarrow Fe^{2+} + H^+ + HO_2^{\bullet}, k = 1 - 2 \times 10^{-2}\ M^{-1}s^{-1} \quad (4)$$

Many Fenton-based processes have risen in the last years, suggesting the future intensification of the use of the classical Fenton process coupled to radiation or electrochemistry, and/or involving heterogeneous catalysts. The optimization of the catalyst and oxidant doses makes the process capable of treating complex water matrices. It has long been demonstrated that the increase of H_2O_2 concentration results in a higher generation of •OH, which in turn leads to the increase of the degradation rate. Nevertheless, the use of excessive oxidant concentration is not encouraged either, since massive amount of H_2O_2 can induce antagonistic reactions (i.e., reaction of the oxidant with the produced hydroxyl radicals) and thus the formation of radicals that are less reactive than the hydroxyl radicals.

ZVI-based in situ chemical oxidation has been proposed as a potential alternative for the traditional Fenton reaction because ZVI would leach dissolved iron species regularly and activate molecular oxygen. ZVI can be used to oxidize organic and inorganic compounds in the presence of oxygen. The important oxidants are hydroxyl radicals, ferryl ions, and superoxide radicals [36]. In the oxidation of ZVI in aerobic water, hydrogen peroxide is produced as an intermediate product (reaction 5). Further reaction for the formation of H_2O_2 is the oxidation of ferrous iron by O_2 (reaction 6). H_2O_2 can react with ferrous iron by Fenton (reaction 3). Acidic conditions promote the production of highly reactive •OH (reaction 7), whereas neutral and basic conditions lead to the formation of oxidizing species such as ferryl iron (reaction 8). Moreover, fast recycling of ferric iron into ferrous species at the metal surface can take place (reaction 9). It has been also recently proposed [37] an analogous process using persulfate (reaction 10) for PhACs removal. Acidic conditions will be not discussed here, as an application for

UWWE containing PhACs would not be a consistent application, as stated in the following paragraphs.

$$Fe^0 + 2H_2O + O_2 \rightarrow Fe^{2+} + H_2O_2 + 2OH^- \tag{5}$$

$$Fe^{2+} + O_2^{\bullet -} + 2H^+ \rightarrow Fe^{3+} + H_2O_2 \tag{6}$$

$$Fe^{2+} + H_2O_2 \rightarrow Fe^{3+} + {}^{\bullet}OH + OH^- \tag{7}$$

$$Fe^{2+} + H_2O_2 \rightarrow FeO^{2+} + H_2O \tag{8}$$

$$2Fe^{3+} + Fe^0 \rightarrow 3Fe^{2+} \tag{9}$$

$$Fe^0 + S_2O_8^{2-} \rightarrow Fe^{2+} + 2SO_4^{2-} \tag{10}$$

Chelating agents are substances with several coordinating sites and can form bonds with metal ions. They can be used at pH higher than 3 to form complexes with Fe(III)/Fe(II), and keep it soluble, thus enhancing the production of oxidative species by the reaction of H_2O_2 with Fe(III)/Fe(II). The criteria to successfully achieve such goal have been recently defined by Clarizia et al. [38]: lowering costs, minimal increase of total organic carbon (TOC), low toxicity, and high biodegradability. Different chelating agents have been tested (most common have been humic acids, fulvic acids, citric acid, oxalic acid, aminopolycarboxylic acids).

3.2 *Removal of PhACs from Wastewaters by Fenton and Fenton-Like Processes*

Treating UWWE resulting in rapid and complete removal of pharmaceutically active compounds (PhACs) should take also into account that the process should be mild (to remove microcontaminants from huge streams of water not large quantities of catalyst, oxidant, or energy, keeping the environmental advantage of PhACs removal) and competitive with best available technologies (BAT) as ozonation and adsorption on active carbon (PAC and GAC) [24]. The subsequent use or reuse of water should be taken into account, for large quantities of iron could impede the reuse in irrigation of crops, the main application of treated water. Recommended maximum concentration in irrigation water is Fe <5.0 mg/L, as it is not toxic to plants in aerated soils [39]. However, FAO established the limit in Fe from 5 mg/L for long-term use until 20 mg/L for short-term use (periods of up to 20 years on fine-textured neutral or alkaline soil) [40]. Wastewater reuse for irrigation in agriculture is by far the most established end use in low-income countries as well as in arid and semi-arid ones [41]. Another important factor to take into account is the legislation that establishes environmental limits of iron emissions. Specifically, in Andalucia (Spain, Decree 109/2015), a limit of 3.3 and 2.2 mg/L is established for wastewater discharge in coastal or surface waters, respectively. High conductivity would not be also recommendable for reuse for irrigation.

BAT already applied at real scale in a few countries for PhACs removal in UWWE have been reported to range in 0.04–0.10 €/m^3 investment costs for ~50,000 population equivalent (PE) and 0.03–0.09 €/m^3 operation costs. Therefore, for mid-scale plants (~50,000 PE), the costs range from 0.10 to 0.15 €/m^3. Capitalized investment-related and operation cost yield to the annual costs for the given life cycle are stated in Rizzo et al. [24], decreasing further with the increasing plant size. Therefore, Fenton for PhACs removal in UWWE should be applied at low iron concentration (in the range of a few mg/L) and minimizing the use of additional chemicals to improve the efficiency, as it would dramatically increase costs up to the available BAT.

Because of the low solubility product of ferric iron hydroxide ($K_S(Fe(OH)_3) \approx 10^{-37}$), precipitation starts at pH 2.5–3.5 depending on the iron concentration and the temperature. The precipitation process starts with the formation of dimers and oligomers, which at continuation gradually polymerize further and lose water until forming finally insoluble iron hydroxides (e.g., goethite or hematite). At pH above 4, the formation and precipitation of ferric oxyhydroxides inhibit both the production of •OH radicals and the regeneration of Fe(II) [42]. The precipitation and aging processes are also temperature-dependent, and more and faster precipitation takes place at higher temperatures [43]. The precipitate is difficult to re-dissolve through acidification and insoluble above pH 1–1.5 [44]. Therefore, conventional Fenton should be applied at acidic pH.

However, urban wastewater effluents containing PhACs also contain inorganic carbon as carbonates, mainly as hydrogen carbonate (HCO_3^-), in the range of hundreds of mg/L. Roughly, HCO_3^- 250 mg/L needs around 200 mg/L H_2SO_4 for being balanced and 50 mg/L H_2SO_4 for attaining pH < 3. Therefore, 250 kg H_2SO_4 are needed per 1,000 m^3 of urban wastewater effluent. Before water disposal or reuse, pH 7–8 should be recovered, using at least NaOH 40 mg/L, or other neutralizing agents. The salinity of effluent would increase >0.5 mS/cm, leading also to considerable loads of iron-containing sludge that will require further management. Applying conventional Fenton to urban wastewater effluents would dramatically increase effluent salinity, environmental impact, and treatment costs. Therefore, from the point of view of the authors, it could be disregarded for PhACs removal.

It has also been demonstrated that by adding iron at different steps (i.e., sequential iron dosage), it is possible to operate Fenton at initial neutral pH but substantially increasing iron dosing and decreasing the reaction rate compared to Fenton at acidic pH [45]. Using Fenton at neutral pH is not viable without adding large quantities of iron and long treatment times, as Fe(II) 30 mg/L and 90 min or 72 min of reaction for eliminating 51 or 32 microcontaminants (ranging from tens of nanogram per liter to micrograms per liter) detected in urban wastewater effluents from two different treatment plants [46]. Similar results were obtained by Soriano-Molina et al. [47] as only 35% of PhACs and other microcontaminants were removed from UWWE with $FeSO_4$ (Fe 20 mg/L) at circumneutral pH. Increasing iron concentration >5 mg/

L would impede proper reuse of treated water in many crops, as stated earlier. It can contribute to soil acidification and loss of availability of essential phosphorus and molybdenum and may result in unsightly deposits on plants, equipment, and buildings [48]. But when using Fe <5.0 mg/L, Fenton could not remove almost any PhACs [49].

Another important factor that strongly influences Fenton performance concerning the removal of PhACs is the complex chemical composition of organic matter present (OM) in UWWE. OM components react readily with hydroxyl radicals, thus reducing the radical concentration and the direct attack of HO• toward the PhACs. Under the inherent wastewater pH, Fe(III) can be complexed by DOM resulting in the formation of stable and soluble complexes that can participate in further reactions [50]. Humic substances are the largest fraction of dissolved organic matter in natural water, and an important fraction of UWWE, as they result from microbiological and chemical transformation of organic debris. They contain carboxylic acids, phenolic compounds, alcoholic quinine, and amino and amido groups which enable them to support ion exchange and redox processes and to form complexes. They can also facilitate the reduction of Fe(III) to Fe(II) due to the presence of quinone moiety [51].

The occurrence of inorganic anions (i.e., chlorides, sulfates, carbonates) in wastewater can influence the degradation rate of PhACs during the Fenton treatment by consuming HO•. They can also complex the iron species ($FeCl_2^+$, $FeCl_2^+$, and Fe $(SO_4)^{2-}$) interfering with the reduction of Fe(III) to Fe(II) that is required to sustain Fenton mechanism. The inorganic anions scavenge the hydroxyl radicals to generate inorganic radicals ($Cl^•$, $SO_4^{-•}$, and $CO_3^{•-}$) which, in turn, react with the organic contaminants, albeit at a slower rate [5]. $SO_4^{-•}$ have recently gained great attention due to the high redox potential [52], reacting preferentially through an electron transfer mechanism. Phosphate is a specific case as it can precipitate the iron catalyst as iron(III) phosphate in a wide range of pH.

In conclusion, conventional Fenton or Fenton at circumneutral pH seems to be not suitable for PhACs removal in UWWE. Other available processes based on electrochemical processes [53] would be also disregarded due to a lack of efficiency (and therefore high costs) with usual UWWE due to the low salt content (0.9–2.2 mS/cm conductivity). Therefore, other research line has been explored and are now hot topics in Fenton and Fenton-like processes: zero-valent iron (ZVI) and application of chelating agents to form complexes with Fe and permit Fenton reaction at high pH.

The use of waste-metallic iron shavings instead of commercial ZVI powders showed promising results for the elimination of PhACs [54] but always after activation by immersion in acid solution to remove the surface layer of the oxidized iron (Fig. 3).

It has been demonstrated the negligible effect of the air/ZVI system for the treatment of the wastewater, unlike the promising results reported in the degradation of model pollutants. It is clear the significant role of both ZVI and H_2O_2 together in the reaction system. The formation of iron oxides supported on the metallic iron

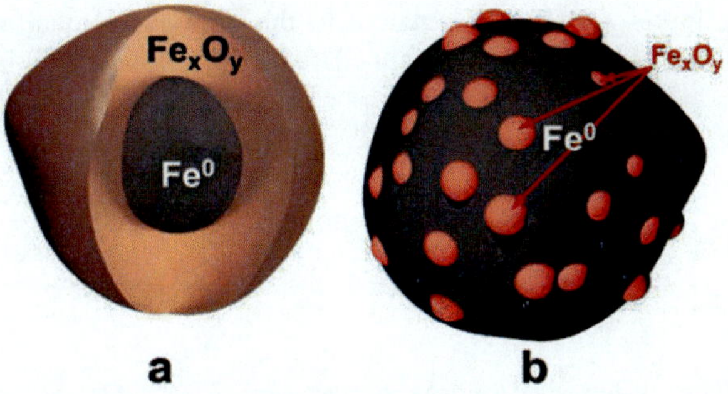

Fig. 3 Waste-metallic iron with Fe^0 inside (**a**). Formation of iron oxides supported on the metallic Fe^0 surface at circumneutral pH (**b**)

surface (Fig. 3b) is plausible at circumneutral pH, which would also reduce the presence of the iron species in the treated aqueous solution [55].

Many chelating agents presented important drawbacks to be applied for PhACs removal in UWWE. Carboxylic acids (citric acid, oxalic acid) add an important extra dissolved organic carbon and reduced final pH; humic substances resulted in low degradation rates due to the quite low amount (few mg/L) that can be found in UWWE. Therefore, it is required extracting and concentrating actions from water which is not easy and does not fulfill the "lowest-cost" principle. Aminopolycarboxylic acids (APCAs) act similarly to polycarboxylic acids, and they are used widely in domestic products, industrial applications, and soil remediation. APCAs contain several carboxylate groups linked to one or more nitrogen atoms and can chelate metal ions by forming one or more heteroatomic rings around the ions [56].

EDTA, as one of the most common APCAs, has been used widely. However, the concern about the potential of toxicity and low biodegradability makes it a persistent organic pollutant and limits its applicability as iron chelate [57] for treatment of UWWE by Fenton processes. EDDS (ethylenediaminedisuccinic acid) is a structural isomer of EDTA and is reported to have metal-complexing properties similar to EDTA [58]. Complete and rapid biodegradation of [S,S]-EDDS (SS-ethylenediaminedisuccinic acid) has been described in many environmental matrices as water [59] and soil [60], being easily biodegraded by activated sludge [61]. Therefore, [S,S]-EDDS is safe when used as Fe-chelating agent during Fenton treatment of UWWE. Currently, 80% of fertilizers used in agriculture are synthetic iron chelates. In addition, the effect of the sub-products generated by photodecomposition has been previously investigated resulting non-toxic for crops [62]. Therefore, its use as a possible iron chelate for wastewater treatment and further reuse in agriculture seems to be a plausible option.

The removal of PhACs in the EDDS-driven Fenton reaction was successful at pH between 8 and 9 [63]. Therefore, that of EDDS application may be limited by its rapid reaction with •OH that involves the decomposition of EDDS and the decrease in efficiency of Fenton process. It is important to highlight that the treatment of several grams of PhACs per cubic meter of UWWE should be achieved in just a few minutes to be an alternative to the previously mentioned BAT. In conclusion, it would be possible to apply EDDS-iron complex only in these circumstances.

4 UV/Peroxide Processes

4.1 *Fundamentals*

The system UV/hydrogen peroxide (UV/H_2O_2) is one of the most extensively studied AOPs for the removal of water pollutants, both chemical and biological [64, 65]. In this case, the generation of the •OH comes from the photolysis of the hydrogen peroxide, specifically through the cleavage of the peroxide bond (O-O). Baxendale and Wilson [66] reported the formation of two hydroxyl radicals per absorbed photon with a quantum yield close to 1.0 at 254 nm according to reaction 11:

$$H_2O_2 \overset{h\nu}{\rightarrow} 2\,{}^{\bullet}OH, \phi_{254nm} = 1.0 \text{ mol Einstein}^{-1} \quad (11)$$

The photolysis of H_2O_2 is controlled by certain parameters such as H_2O_2 concentration, UV-C dose, UV wavelength, pH, and alkalinity of water. The optimal control of these parameters allows maximizing the efficiency of the treatment. Obviously, the increase of the UV dose leads to an increase in the efficiency of the generation of •OH. In contrast, the concentration of H_2O_2 is a critical parameter. Since the molar absorption coefficient of H_2O_2 is low ($\varepsilon_{254nm} = 18.6\ M^{-1}\ cm^{-1}$), a high concentration is required to absorb a significant fraction of UV in the range of 200–300 nm (the maximum absorbance of hydrogen peroxide is around 220 nm). However, excessive dosage can cause scavenging of •OH according to the reactions (12)–(14):

$${}^{\bullet}OH + H_2O_2 \rightarrow HO_2^{\bullet} + H_2O \quad (12)$$

$$HO_2^{\bullet} + H_2O_2 \rightarrow {}^{\bullet}OH + H_2O + O_2 \quad (13)$$

$$HO_2^{\bullet} + HO_2^{\bullet} \rightarrow H_2O_2 + O_2 \quad (14)$$

Besides, the activation of H_2O_2 is affected by the UV transmittance of the medium, or in other words, the effectiveness of UV/H_2O_2 process is susceptible to water matrix. Alkaline pH favored the process, as the molar absorption coefficient of the HO_2^- conjugate base is much higher ($\varepsilon_{254nm} = 240\ M^{-1}\ cm^{-1}$), improving significantly the efficiency of the photon absorption step. However, water alkalinity strongly affects the yield of the process, as carbonates and bicarbonates present act as •OH scavengers (reactions 15 and 16), reducing the net availability of •OH for the degradation of pollutants:

$$^{\bullet}OH + CO_3^{2-} \rightarrow OH^- + CO_3^{\bullet -} \ k = 2 \times 10^8 \ M^{-1} \ s^{-1} \tag{15}$$

$$^{\bullet}OH + HCO_3^- \rightarrow OH^- + HCO_3^{\bullet} \ k = 1.5 \times 10^7 \ M^{-1} \ s^{-1} \tag{16}$$

Among the advantages of the UV/H_2O_2 processes for the removal of chemical pollutants, it can be mentioned that their degradation might occur also via direct photolysis and the possibility of carrying out the simultaneous disinfection of water by the damages produced by the UV-C on the DNA of microorganisms [67]. In addition, the design of the treatment facilities is simple [68], and existing facilities can be adapted [69], taking advantage of the high degree of maturity and widely full-scale application of UV disinfection processes. Moreover, UV/H_2O_2 minimizes the formation of by-products, as trihalomethanes (THMs) and haloacetic acids (HAAs) generated as a consequence of reaction of chlorine used as oxidant with natural organic matter (NOM) [70]. Finally, UV/H_2O_2 is an AOP with a relative low cost if it is compared with other advanced treatments. This low cost gives it great versatility, being useful as a single or complementary treatment to other treatments (coupled with filtration processes, adsorption, etc.) [70–72]. The cost of the UV/H_2O_2 process is associated with the intensity required from the UV lamps and their low energy efficiency [69]. The introduction of UV-LED lamps promises to reduce the associated costs, although now they are not competitive because of the associated cooling costs [73].

As main limitations of the process, it can be mentioned the strong sensitivity of the efficiency to water turbidity, reducing UV transmittance, and water matrix effects, especially radical scavenging substances. For this reason, the application of this technology is especially of interest as a tail-end technology treating the final effluent before discharge. It is also worth noting that H_2O_2 is a powerful oxidizing substance, so its handling and storage require some specific precautions.

Recently, the scientific community has focused their interest in the application of a modification of UV/H_2O_2 system, based on substituting the H_2O_2 by persulfate salts as sodium or potassium persulfate and potassium peroxymonosulfate [74]. The substantial difference with UV/H_2O_2 system is the additional generation of sulfate radicals, with a comparable redox potential to hydroxyl radicals (2.5–3.1 V vs. 1.8–2.7 V for $SO_4^{\bullet -}$ and •OH, respectively) [75]. The reactions 17 and 18 show the generation of hydroxyl and/or sulfate radicals by cleavage of the O-O bond of persulfate (PS) and peroxymonosulfate (PMS) anion [76]:

$$S_2O_8^{2-} \xrightarrow{h\nu} 2SO_4^{\bullet -}, \ \phi_{254nm} = 1.4 \ \text{mol/Einstein} \tag{17}$$

$$HSO_5^- / SO_5^{2-} \xrightarrow{h\nu} SO_4^{\bullet -} + {}^{\bullet}OH, \phi_{254nm} = 1.04 \ \text{mol/Einstein} \tag{18}$$

These treatment systems are as simple to apply as UV/H_2O_2, with the only difference that persulfate salts are solids and hydrogen peroxide is liquid. In both cases, the oxidizing character of both substances forces precautions for storage and handling.

4.2 *Removal of PhACs from Wastewaters by UV/H_2O_2*

As it has been previously mentioned, the UV/H_2O_2 process is a promising technology with a high level of maturity thanks to the extensive bench-scale studies carried out during the last decades. However, the study of the process continues nowadays due to the current challenges and associated human risks, such as the formation of by-products [77], the presence of substances of emerging concern [78], and antibiotic resistance genes [79].

The high level of bench-scale tests of the UV/H_2O_2 system has helped through the scaling up of pilot- and even full-scale application, as a feasible technology for the removal of PhACs. Along this section, it will be shown the state of the art about pilot- and full-scale application of UV/H_2O_2 system for the removal of PhACs and other micropollutants or emerging concern. Table 2 summarizes recent reports about the application of UV/H_2O_2, UV/PS, and UV/PMS processes at pilot scale, whereas Table 3 summarizes some recent studies at full scale. As it can be observed, there is an important variety of working conditions in the presented researches, such as different water matrices, working pH, UV dose, concentration of hydrogen peroxide, flow rates, kind of reactor, or even the studied PhACs, making difficult the direct comparison of the results. Nevertheless, there is a common trend, that is, the use of UV/H_2O_2 system as a final step in a tertiary treatment in a wastewater treatment plant, in some cases treating directly the effluent after the secondary treatment and in other cases treating the effluent after some previous stages as microfiltration, ultrafiltration, reverse osmosis, adsorption, etc. Besides, there is a common pattern in most of the cases, which is the comparison of the process with other AOPs, including UV/Cl_2 system.

Miklos et al. [68] reported the viability of applying UV-C/H_2O_2 as a tertiary treatment for municipal wastewater, studying the removal efficiency of this technology over 15 different organic pollutants occurring at trace level in UWWE. The study included the optimization of working conditions at bench scale and the scaling up at pilot scale. Using a dose of 10 mg/L of hydrogen peroxide, these authors reported the UV-C/H_2O_2 system as a successful option to remove water PhACs. Between the most important remarks, they pointed out the important variations of •OH exposure as a consequence of the fluctuations on nitrite concentration and the scavenging effect of substances in water. A common feature in most of the studies is the significant differences in the removal efficiency depending on the specific substance, as a direct consequence of their chemical structure and properties. For instance, Miklos et al. [68] reported the total removal of diclofenac (DCF) but a null efficiency over tris(2-carboxyethyl)phosphine (TCEP).

On the other hand, Zhang et al. [70] reported the comparison between UV/H_2O_2, UV/free chlorine, and UV/chloramines in the removal of 0.5 log units of 1,4-dioxane

Table 2 Summary of recent pilot-scale UV/H_2O_2 studies for the removal of PhACs and other micropollutants in different water matrices

Reference	Location	Water matrix	[Oxidant]	pH	Operation mode	Reactor characteristics	UV dose	Flow rate (m^3/h)	Reactor volume (L)	Pollutants	Yield (%)
Sarathy et al. [80]	Canada	Surface water	H_2O_2 = 10 mg/L	N.S.	Continuous	Trojan UVSwift 4L12 Trojan UVPhox 8AL20	N.S.	MP reactor: 22–65 LP reactor: 11–22	N.S.	Natural organic matter	> 58%
Sichel et al. [81]	Germany	Tap water	H_2O_2 = 5 mg/L	7	Continuous	WTL 40, 80, and 200, Siemens AG	N.S.	0.25	N.S.	BTA, TTA, DEA, CBZ, SMX, DCF, IPM	50–100%
Lester et al. [82]	Boulder (USA)	Secondary effluent wastewater	H_2O_2 = 10 mg/L	N.S.	Continuous	Trojan 100 W LP Trojan 400 W MP	750 mJ/cm^2 1,500 mJ/cm^2	0.11–0.23	N.S.	SCL, LMT, GMB, CLAR, ATN, 2,4-D, CBZ, MTP, TMP, VEN, DPD	750 mJ/cm^2: 18–54% 1,500 mJ/cm^2: 37–71%
Chu et al. [83]	Singapore	Fresh water	H_2O_2 = 0.5–10 mg/L	6.3	Continuous	Trojan UVSwift 4 MP UV lamps; each of 3 kW	N.S.	40–60	N.S.	BPA, E2, PFOA, PFOS, NDMA, DCF	45–100%, except for PFOS and PFOA, 5–20%
Miralles-Cuevas et al. [84]	Almería (Spain)	Simulated fresh water	H_2O_2 = 10, 25, and 50 mg/L PS = 10, 25, and 50 mg/L	N.S.	Continuous	3 UV-C lamps, 230 W	115 W/m^2 (254 nm)	1.2	5.21	AP, CFN, CBZ, CPX, SMX	65–99%

(continued)

Table 2 (continued)

Reference	Location	Water matrix	[Oxidant]	pH	Operation mode	Reactor characteristics	UV dose	Flow rate (m^3/h)	Reactor volume (L)	Pollutants	Yield (%)
Sarasidis et al. [85]	Thessaloniki (Greece)	Tap water	H_2O_2 = 10–80 mg/L	7.8	Continuous	PURO 4S, 4 UV-C lamps	N.S.	0.027	15	DCF	100%
Krystynik et al. [86]	Czech Republic	Groundwater	H_2O_2 = 2 mM/h	7.62	Continuous	TUV UVC TL-D 36 W G13 Philips	N.S.	1.27	42.4	TCE, PCE	100%
Miklos et al. [68]	Munich (Germany)	Secondary effluent wastewater	H_2O_2 = 10 mg/L	N.S.	Continuous	WEDECO LBX 90e WEDECO LBX 10	800 mJ/cm^2 (254 nm)	LBX 90e: 35 LBX 10: 11	LBX 90e: 45 LBX 10: 13	DCF, IOP, SMX, CLI, TRA, BZN, VEN, SOT, PRI, CBZ, GBP, TCEP	44–99.6% 1–90% 1–81%
Baresel et al. [87]	Stockholm (Sweden)	Wastewater	H_2O_2 = 10–50 mg/L	6.8–7	Continuous	Van Remmen UV Technology	3,000–10,000 mJ/m^2 (254 nm)	1–2	50	Sum of pharmaceuticals	50–80%
Rodríguez-Chueca et al. [74]	Madrid (Spain)	Secondary effluent wastewater	H_2O_2; PS and PMS = 0.01–5 mM	7.23	Continuous	Philips TUV PL-L 95 W/4 P HO 1CT/25	5.7–57 J/L	0.06–0.6	0.47	DCF, IBP, SMX, CBZ, ATN, SMZ, ATZ, TCS, CFN, SCL	23–100%
Wünsch et al. [88]	Germany	Rhine river	H_2O_2 = 4 mg/L	8.1 ± 0.1	Continuous	N.S.	6,000 mJ/m^2 (254 nm)	N.S.	N.S.	EDTA, ACE, IPA, IME, MET, BTZ, IPR	10–95%
Zhang et al. [70]	Fountain Valley (USA)	Permeate reverse osmosis	H_2O_2 = 3.1–6.2 mg/L	5.7 and 7	Continuous	Trojan 257 W LPHO	≈1,000 mJ/cm^2 (254 nm)	1.33	9.7	1,4-Dioxane	65–78%

N.S., Not specified

Table 3 Summary of recent full-scale UV/H_2O_2 studies for the removal of PhACs and other micropollutants in different water matrices

Reference	Location	Water matrix	pH	Operation mode	Reactor characteristics	[Oxidant]	UV dose	Flow rate (m^3/h)	Reactor volume (L)	Pollutants	Yield (%)
Wang et al. [89]	Cornwall (Canada)	St. Lawrence River	6.5, 7.5, and 8.5	Continuous	Trojan UVSwift 8 L24	H_2O_2 = 1, 2.9, 4.8 mg/L	2,000 ± 150 mJ cm^{-2} (200–400 nm)	360	720	Geosmin, MIB, CFN	93% for geosmin; 83% for MIB; 77% for CFN
Rodríguez-Chueca et al. [69]	Toledo (Spain)	Secondary effluent wastewater	7.17 ± 0.04	Continuous	WEDECO LBX400e	H_2O_2; PS and PMS =0.05–0.5 mM	42–170 J/L	28, 75, 114	140	LDC, TRA, ODT, VLX, NVNF, CBZ, OCBZ, MIR, GBP, METO, SFX, BTR, TT, DEET, TBY, TCEP, TCPP, TPP, HHCB, HHCB-I, HHCB-II, AHTN, ATII, OTNE	Average removal of 48% with PMS; 55% with H_2O_2
Wetterau et al. [72]	California (USA)	Permeate reverse osmosis	5.6	Continuous	Trojan 30AL50 UVPhox Trojan 72AL75 UVPhox	H_2O_2 = 1–4 mg/L	N.S.	568–947	N.S.	1,4-Dioxane	1.2 log
Rodríguez-Chueca et al. [90]	Toledo (Spain)	Secondary effluent wastewater	7.17 ± 0.04	Continuous	WEDECO LBX 400e	H_2O_2; PS and PMS	42–170 J/L	28, 75, 114	140	Antibiotics (AB): AZT, CLAR,	AB: average removal

(continued)

Table 3 (continued)

Reference	Location	Water matrix	pH	Operation mode	Reactor characteristics	[Oxidant]	UV dose	Flow rate (m^3/h)	Reactor volume (L)	Pollutants	Yield (%)
						=0.05–0.5 mM				CPFX, OFX, SMX, SFD, SFP, TMP, MTZ, CLI Antibiotic resistance genes (ARG): intI1, sul1, sul2, blaOXT, blaTEM, qnrS	of 29% with PMS; 36% with H_2O_2 ARG: average removal of 0.31 log with PMS; 0.55 log with H_2O_2
Wang et al. [71]	Mississauga (Canada)	UF effluent	6.5 and 8	Continuous	Trojan UVSwift ECT16L30	H_2O_2 = 5–10 mg/L	16 medium-pressure lamps (12.3 kW). Three UV ballast power levels (50, 70, and 100%)	625	316	SCL, CFN, DBP (THMs, HAA, and AOX)	At pH 6.5, 0.3–1 log; at pH 8, 0.2–0.8 log

N.S., Not specified

(1,4-D) to regenerate wastewater, as well as the assessment of the generation of disinfection by-products (DBPs). Their main conclusion is that the UV/free chlorine system is the most efficient at pH 5.7, while at pH 7, because of the scavenging caused by the OCl^-, the most efficient treatment was UV/H_2O_2. In any case, the UV/free chlorine system achieved a cheaper cost than UV/H_2O_2 and UV/chloramines treatments, and the generation of DBPs was similar in all the cases but promoting the generation of different substances. Other authors, as Baresel et al. [87], also reported a good efficiency of UV/H_2O_2 in removing micropollutants but confirm that they are still not economically competitive regarding other technologies such as ozonation or activated carbon.

Miralles-Cuevas et al. [84] compared from the economical point of view the application of UV/H_2O_2 using UV-C and solar radiation. They concluded that the main contribution to the operating costs comes from reagents and electricity from lamps operation. However, according to their results, the main difference between using solar and UV systems on the removal of PhACs is the optimal concentration of the oxidant. Rodríguez-Chueca et al. [74] compared the efficiency of UV/H_2O_2 with modification treatments using PS and PMS as oxidants with very low residence times. In general, the use of PMS and PS as oxidants led to higher removal yield than using hydrogen peroxide. Both persulfate anions allowed to reach the total removal of some PhACs under the most demanding operating conditions. Just the removal of sucralose (SCL) was more effective by using H_2O_2.

Other authors, such as Sarasidis et al. [85], studied the efficiency of UV/H_2O_2 to regenerate powdered activated carbon (PAC) used on the removal of PhACs. With promising results on the removal of DCF, the authors proposed a further investigation on the mechanisms of regeneration of PAC because it involves adsorption, desorption, and oxidation. Some authors reported the use of UV/H_2O_2 on the treatment of fresh surface water, as it is the case of Wünsch et al. [88], studying the impact of a UV/H_2O_2 before soil aquifer treatment (SAT). They assessed the efficiency on the abatement of different PhACs and other micropollutants present in Rhine river water, using a pilot plant consisting of a UV-C reactor and two parallel soil columns. These coupled systems enhanced the removal of micropollutants by an additive effect of the unit processes. However, in almost all cases, the UV/H_2O_2 system is responsible for the abatement of the studied micropollutants, being only the metformin primarily abated by SAT.

The disadvantages reported by some authors in terms of efficiency and cost of UV/H_2O_2 in comparison with other AOPs have slowed down the implementation of technology on a full scale. Despite this, there are reports about the application of UV/H_2O_2 at full-scale, in some cases as research studies, but in others in real water facilities using this technology in the removal of PhACs and other micropollutants (Table 3). For instance, Wang et al. [71] compared the efficiency of UV/H_2O_2 and UV/Cl_2 on the removal of micropollutants in water adapting the tertiary treatment of the Lorne Park Water Treatment Plant in Canada. Both AOPs were applied over the permeate of the filtration step. The conclusions obtained are quite similar to the ones showed by Zhang et al. [70] at pilot scale. At pH 6.5, the generation of •OH was almost twice by UV/Cl_2 than UV/H_2O_2, but at pH 8.0, •OH formation was negligible during UV/Cl_2 system. Similar conclusions comparing UV/Cl_2 and UV/H_2O_2 were reached by other authors, such as Wang et al. [89] applying the technologies in the Cornwall Water Purification Plant (Canada) and Wetterau et al. [72] in WRD Leo J. Vander Lans Advanced Water Treatment Facility in California (USA). In this last case, Wetterau et al. [72] pointed that the same level of 1,4-dioxane reduction can be achieved with UV/Cl_2 at roughly one-third the dose as UV/H_2O_2, reducing the use of reagents or the size of the UV reactor.

Finally, Rodríguez-Chueca et al. [69] modified the tertiary treatment of municipal wastewater treatment plant in Toledo (Spain) to test different UV-driven AOPs in the existing UV-C equipment. Among them, UV/H_2O_2, UV/PMS, and UV/PS were tested using different flow rates, or in other words different UV-C doses and UV-C contact time. In general, photolysis of H_2O_2 and PMS led to similar results considering the average removal of the sum of the studied micropollutants, with a slightly higher performance using H_2O_2 (55% with H_2O_2 vs. 48% with PMS). These results, together with the lower cost of H_2O_2, made UV/H_2O_2 more attractive than the use of PMS for the removal of micropollutants. In a similar study, Rodríguez-Chueca et al. [90] analyzed the removal of antibiotics and antibiotic resistance genes from the effluents. In general, the addition of the oxidants improved significantly the efficiency of the removal of the antibiotics in comparison with the UV alone. However, the removal of antibiotic resistance genes was maximum when only UV is dosed, without oxidants. This fact suggested a compromising situation because the highest antibiotics removal corresponded to the lowest removal of genes. The reason is that the generation of hydroxyl and sulfate radicals, triggered by light absorption of the oxidants, actually reduces the availability of photons for the direct damages on the DNA. Consequently, although in theory both processes can take place simultaneously, to achieve a high efficiency in the removal on the PhACs and the antibiotic resistance genes, much higher UV doses have to be used.

5 Concluding Remarks

In this chapter the fundamentals of several advanced oxidation processes for the removal of PhACs in urban wastewater effluents have been summarized, remarking that the most suitable oxidation technique would depend mainly on the water matrix and the final use of the effluent.

Nowadays, O_3, O_3/H_2O_2, and UV/H_2O_2 treatments constitute good options to succeed in the removal of PhACs as tertiary treatments in municipal wastewaters. Modified processes such as catalytic ozonation or the use of alternative peroxides are under study.

Applying conventional Fenton at acidic pH to urban wastewater effluents would dramatically increase effluent salinity, environmental impact, and treatment costs disregarding it as an option for PhACs removal from this type of effluents. Many iron-chelating agents (carboxylic acids, humic substances, EDTA) to operate Fenton-like processes at circumneutral pH presented important drawbacks to be applied for PhACs removal in UWWE. EDDS appears as a possible iron chelate for UWWE treatment, and further reuse in agriculture seems to be a plausible option if the treatment would be completed in just a few minutes. Modifications of Fenton and Fenton-like processes at circumneutral pH are under study in many research labs, but they are still far away from their application.

Independently on the oxidation treatment used to remove PhACs, it should be considered the potential formation of by-products during the oxidative treatment.

Treating urban wastewater effluents for removal of PhACs should be mild, minimizing the use of additional chemicals, and economically competitive with best available technologies as ozonation and adsorption on active carbon, reported to range between 0.10 and 0.15 €/m^3.

References

1. Glaze WH, Kang JW, Chapin DH (1987) The chemistry of water treatment processes involving ozone, hydrogen peroxide and ultraviolet radiation. Ozone Sci Eng 9:335–352. https://doi.org/10.1080/01919518708552148
2. Andreozzi R, Caprio V, Insola A, Marotta R (1999) Advanced oxidation processes (AOP) for water purification and recovery. Catal Today 53:51–59. https://doi.org/10.1016/S0920-5861(99)00102-9
3. Parsons S (2004) Advanced oxidation processes for water and wastewater treatment. IWA Publishing, London. https://doi.org/10.2166/9781780403076
4. Stefan MI (2017) Advanced oxidation processes for water treatment: fundamentals and applications, vol 16. IWA Publishing, London, p 9781780407197. https://doi.org/10.2166/9781780407197
5. Lado Ribeiro AR, Moreira NFF, Li Puma G, Silva AMT (2019) Impact of water matrix on the removal of micropollutants by advanced oxidation technologies. Chem Eng J 363:155–173. https://doi.org/10.1016/j.cej.2019.01.080
6. Ortiz I, Mosquera A, Lema J, Esplugas S (2015) Advanced technologies for water treatment and reuse. AICHE J 61:3146–3158. https://doi.org/10.1002/aic.15013
7. Ahmed MB, Zhou JL, Ngo HH, Guoa W, Thomaidis NS, Xuca J (2017) Progress in the biological and chemical treatment technologies for emerging contaminant removal from wastewater: a critical review. J Hazard Mater 323:274–298. https://doi.org/10.1016/j.jhazmat.2016.04.045
8. Kanakaraju D, Glass BD, Oelgemoller M (2018) Advanced oxidation process-mediated removal of pharmaceuticals from water: a review. J Environ Manag 219:189–207. https://doi.org/10.1016/j.jenvman.2018.04.103
9. Wang J, Wang S (2016) Removal of pharmaceuticals and personal care products (PPCPs) from wastewater: a review. J Environ Manag 182:620–640. https://doi.org/10.1016/j.jenvman.2016.07.049
10. Yang Y, Ok YS, Kim K, Kwon EE, Tsang YF (2017) Occurrences and removal of pharmaceuticals and personal care products (PPCPs) in drinking water and water/sewage treatment plants: a review. Sci Total Environ 596–597:303–320. https://doi.org/10.1016/j.scitotenv.2017.04.102
11. Langlais B, Reckhow DA, Brink DR (1991) Ozone in water treatment. Application and engineering. Lewis Publishers, Chelsea
12. Tiwari G, Bose P (2007) Determination of ozone mass transfer coefficient in a tall continuous flow counter-current bubble contactor. Chem Eng J 132(1–3):215–225. https://doi.org/10.1016/j.cej.2006.12.025

13. Beltran FJ (2004) Ozone reaction kinetics for water and wastewater systems. CRC Press, Boca Raton. https://doi.org/10.1201/9780203509173
14. von Sonntag C, von Gunten U (2012) Chemistry of ozone in water and wastewater treatment: from basic principles to applications. IWA Publishing, London. https://doi.org/10.2166/9781780400839
15. Pocostales JP, Sein MM, Knolle W, von Sonntag C, Schmidt TC (2010) Degradation of ozone-refractory organic phosphates in wastewater by ozone and ozone/hydrogen peroxide (peroxone): the role of ozone consumption by dissolved organic matter. Environ Sci Technol 44:8248–8253. https://doi.org/10.1021/es1018288
16. Acero JL, von Gunten U (2000) Influence of carbonate on the ozone/hydrogen peroxide based advanced oxidation process for drinking water treatment. Ozone Sci Eng 22:305–328. https://doi.org/10.1080/01919510008547213
17. Lee Y, Gerrity D, Lee M, Bogeat AE, Salhi E, Gamage S, Trenholm RA, Wert EC, Snyder SA, von Gunten U (2013) Prediction of micropollutant elimination during ozonation of municipal wastewater effluents: use of kinetic and water specific information. Environ Sci Technol 47:5872–5881. https://doi.org/10.1021/es400781r
18. Merényi G, Lind J, Naumov S, von Sonntag C (2010) Reaction of ozone with hydrogen peroxide (peroxone process): a revision of current mechanistic concepts based on thermokinetic and quantum-chemical considerations. Environ Sci Technol 44:3505–3507. https://doi.org/10.1021/es100277d
19. von Gunten U, Oliveras Y (1997) Kinetics of the reaction between hydrogen peroxide and hypobromous acid: implication on water treatment and natural systems. Water Res 31:900–906. https://doi.org/10.1016/S0043-1354(96)00368-5
20. Pisarenko AN, Stanford BD, Yan D, Gerrity D, Snyder SA (2012) Effects of ozone and ozone/peroxide on trace organic contaminants and NDMA in drinking water and water reuse applications. Water Res 46:316–326. https://doi.org/10.1016/j.watres.2011.10.021
21. Malvestiti JA, Cruz-Alcalde A, López-Vinent N, Dantas RF, Sans C (2019) Catalytic ozonation by metal ions for municipal wastewater disinfection and simultaneous micropollutants removal. Appl Catal B Environ 259:118104. https://doi.org/10.1016/j.apcatb.2019.118104
22. Nawrocki J, Kasprzyk-Hordern B (2010) The efficiency and mechanisms of catalytic ozonation. Appl Catal B Environ 99:27–42. https://doi.org/10.1016/j.apcatb.2010.06.033
23. Bourgin M, Beck B, Boehler M, Borowska E, Fleiner J, Salhi E, Teichler R, von Gunten U, Siegrist H, Mcardell CS (2018) Evaluation of a full-scale wastewater treatment plant upgraded with ozonation and biological post-treatments: abatement of micropollutants, formation of transformation products and oxidation by-products. Water Res 129:486–498. https://doi.org/10.1016/j.watres.2017.10.036
24. Rizzo L, Malato S, Antakyali D, Beretsou VG, Maja BD, Gernjak W, Heath E, Ivancev-Tumbas I, Karaolia P, Lado AR, Mascolo G, Mcardell CS, Schaar H, Silva AMT, Fatta-Kassinos D (2019) Consolidated vs new advanced treatment methods for the removal of contaminants of emerging concern from urban wastewater. Sci Total Environ 655:986–1008. https://doi.org/10.1016/j.scitotenv.2018.11.265
25. Justo A, González O, Aceña J, Pérez S, Barceló D, Sans C, Esplugas S (2013) Pharmaceuticals and organic pollution mitigation in reclamation osmosis brines by UV/H_2O_2 and ozone. J Hazard Mat 263:268–274. https://doi.org/10.1016/j.jhazmat.2013.05.030
26. Cruz-Alcalde A, Esplugas S, Sans C (2019) Abatement of ozone-recalcitrant micropollutants during municipal wastewater ozonation: kinetic modelling and surrogate-based control strategies. Chem Eng J 360:1092–1100. https://doi.org/10.1016/j.cej.2018.10.206
27. Cruz-Alcalde A, Esplugas S, Sans C (2020) Continuous H_2O_2 addition in peroxone process: performance improvement and modelling in wastewater effluents. J Hazard Mater 387:121993. https://doi.org/10.1016/j.jhazmat.2019.121993
28. Lee M, Blum LC, Schmid E, Fenner K, von Gunten U (2017) A computer-based prediction platform for the reaction of ozone with organic compounds in aqueous solution: kinetics and mechanisms. Environ Sci Process Impacts 19:465–476. https://doi.org/10.1039/c6em00584e

29. Chon K, Salhi E, von Gunten U (2015) Combination of UV absorbance and electron donating capacity to assess degradation of micropollutants and formation of bromate during ozonation of wastewater effluents. Water Res 81:388–397. https://doi.org/10.1016/j.watres.2015.05.039
30. Cruz-Alcalde A, Esplugas S, Sans C (2020) New insights on the fate of EfOM during ozone application for effective abatement of micropollutants in wastewater effluents. Sep Pur Tech 237:116468. https://doi.org/10.1016/j.seppur.2019.116468
31. Federal Office for the Environment (FOEN), Switzerland. https://www.bafu.admin.ch/bafu/en/home.html
32. Fenton HJH (1894) Oxidation of tartaric acid in presence of iron. J Chem Soc 65:899–910. https://doi.org/10.1039/CT8946500899
33. Pignatello JJ, Oliveros E, MacKay A (2006) Advanced oxidation processes for organic contaminant destruction based on the Fenton reaction and related chemistry. Crit Rev Environ Sci Technol 36:1–84. https://doi.org/10.1080/10643380500326564
34. Sychev AY, Isaak VG (1995) Iron compounds and the mechanisms of the homogeneous catalysis of the activation of O_2 and H_2O_2 and of the oxidation of organic substrates. Russ Chem Rev 64:1105–1129. https://doi.org/10.1070/RC1995v064n12ABEH000195
35. Bishop DF, Stern G, Fleischmann M, Marshall LS (1968) Hydrogen peroxide catalytic oxidation of refractory organics in municipal wastewaters. Ind Eng Chem Process Des Dev 7:110–117. https://doi.org/10.1021/i260025a022
36. Stieber M, Putschew A, Jekel M (2011) Treatment of pharmaceuticals and diagnostic agents using zero-valent iron - kinetic studies and assessment of transformation products assay. Environ Sci Technol 45:4944–4950. https://doi.org/10.1021/es200034j
37. Wu J, Wang B, Cagnetta G, Huang J, Wang Y, Deng S, Yua G (2020) Nanoscale zero valent iron-activated persulfate coupled with Fenton oxidation process for typical pharmaceuticals and personal care products degradation. Sep Pur Technol 239:116534. https://doi.org/10.1016/j.seppur.2020.116534
38. Clarizia L, Russo D, Di Somma I, Marotta R, Andreozzi R (2017) Homogeneous photo-Fenton processes at near neutral pH: a review. Appl Catal B Environ 209:358–371. https://doi.org/10.1016/j.apcatb.2017.03.011
39. Rowe DR, Abdel-Magid IM (1995) Handbook of wastewater reclamation and reuse. CRC Press, p 550
40. FAO (2003) Users manual for irrigation with treated wastewater. FAO regional Office of the near East, Cairo, Egypt. http://www.fao.org/tempref/GI/Reserved/FTP_FaoRne/morelinks/Publications/English/Usersmanual-en.pdf
41. Drechsel P, Mara DD, Bartone C, Scheierling SM (2010) Improving wastewater use in agriculture: an emerging priority. World Bank policy research working paper no. 5412, 111 p. https://doi.org/10.1596/1813-9450-5412
42. Bigda RJ (1995) Consider Fenton's chemistry for wastewater treatment. Chem Eng Prog 91:62–66
43. Krýsová H, Jirkovský J, Krýsa J, Mailhot G, Bolte M (2003) Comparative kinetic study of atrazine photodegradation in aqueous $Fe(ClO_4)_3$ solutions and TiO_2 suspensions. Appl Catal B Environ 40:1–12. https://doi.org/10.1016/S0926-3373(01)00324-1
44. Sulzberger B, Laubscher H, Karametaxas G (1994) Photoredox reactions at the surface of iron (III)(hydr-)oxides. In: Helz GR, Zepp RG, Crosby DG (eds) Aquatic and surface photochemistry. Lewis Publishers, Boca Raton, pp 53–74. https://doi.org/10.1201/9781351069847-4
45. Carra I, Malato S, Jiménez M, Maldonado MI, Sánchez Pérez JA (2014) Microcontaminant removal by solar photo-Fenton at natural pH run with sequential and continuous iron additions. Chem Eng J 235:132–140. https://doi.org/10.1016/j.cej.2013.09.029
46. Estrada-Arriaga EB, Cortés-Muñoz JE, González-Herrera A, Calderón-Mólgora CG, Rivera-Huerta ML, Ramírez-Camperos E, Montellano-Palacios L, Gelover-Santiago SL, Pérez-Castrejón S, Cardoso-Vigueros L, Martín-Domínguez A, García-Sánchez L (2016) Assessment of full-scale biological nutrient removal systems upgraded with physico-chemical processes for

the removal of emerging pollutants present in wastewaters from Mexico. Sci Total Environ 571:1172–1182. https://doi.org/10.1016/j.scitotenv.2016.07.118
47. Soriano-Molina P, Miralles-Cuevas S, Esteban García B, Plaza-Bolaños P, Sánchez Pérez JA (2020) Two strategies of solar photo-Fenton at neutral pH for the simultaneous disinfection and removal of contaminants of emerging concern. Comparative assessment in raceway pond reactors. Catal Today. https://doi.org/10.1016/j.cattod.2019.11.028
48. FAO (1985) Ayers RS, Westcot DW (eds) Water quality for agriculture. Food and Agriculture Organization of the United Nations, Rome. http://www.fao.org/3/t0234e/t0234e00.htm
49. De la Cruz N, Gimenez J, Esplugas S, Grandjean D, de Alencastro LF, Pulgarin C (2012) Degradation of 32 emergent contaminants by UV and neutral photo-Fenton in domestic wastewater effluent previously treated by activated sludge. Water Res 46:1947–1957. https://doi.org/10.1016/j.watres.2012.01.014
50. Fukushima M, Tatsumi K (2001) Degradation pathways of pentachlorophenol by photo-Fenton systems in the presence of iron(III), humic acids and hydrogen peroxide. Environ Sci Technol 35:1771–1778. https://doi.org/10.1021/es001088j
51. Lipczynska-Kochany E, Kocjany J (2008) Effect of humic substances on the Fenton treatment of wastewater at acidic and neutral pH. Chemosphere 73:745–750. https://doi.org/10.1016/j.chemosphere.2008.06.028
52. Lian L, Yao B, Hou S, Fang J, Yan S, Song W (2017) Kinetic study of hydroxyl and sulfate radical-mediated oxidation of pharmaceuticals in wastewater effluents. Environ Sci Technol 51:2954–2962. https://doi.org/10.1021/acs.est.6b05536
53. Sirés I, Brillas E (2012) Remediation of water pollution caused by pharmaceutical residues based on electrochemical separation and degradation technologies: a review. Environ Int 40:212–229. https://doi.org/10.1016/j.envint.2011.07.012
54. Mackulak T, Nagyová K, Faberová M, Grabic R, Koba O, Gál M, Birosová L (2015) Utilization of Fenton-like reaction for antibiotics and resistant bacteria elimination in different parts of WWTP. Environ Toxicol Pharmacol 40:492–497. https://doi.org/10.1016/j.etap.2015.07.002
55. Segura Y, Martínez F, Melero JA (2013) Effective pharmaceutical wastewater degradation by Fenton oxidation with zero-valent iron. Appl Catal B Environ 136–137:64–69. https://doi.org/10.1016/j.apcatb.2013.01.036
56. Zhang Y, Zhou M (2019) A critical review of the application of chelating agents to enable Fenton and Fenton-like reactions at high pH values. J Hazard Mater 362:436–450. https://doi.org/10.1016/j.jhazmat.2018.09.035
57. Schmidt CK, Fleig M, Sacher F, Brauch HJ (2004) Occurrence of aminopolycarboxylates in the aquatic environment of Germany. Environ Pollut 13:107–124. https://doi.org/10.1016/j.envpol.2004.01.013
58. Egli T (2001) Biodegradation of metal-complexing aminopolycarboxylic acids. J Biosci Bioeng 92:89–97. https://doi.org/10.1016/S1389-1723(01)80207-3
59. Jaworskal JS, Schowanek D, Feijtel TCJ (1999) Environmental risk assessment for trisodium [s, s]-ethylene diamine disuccinate, a biodegradable chelator used in detergent applications. Chemosphere 38(15):3591–3625. https://doi.org/10.1016/S0045-6535(98)00573-6
60. Tandy S, Ammann A, Schulin R, Nowack B (2006) Biodegradation and speciation of residual SS-ethylenediaminedisuccinic acid (EDDS) in soil solution left after soil washing. Environ Pollut 142:191–199. https://doi.org/10.1016/j.envpol.2005.10.013
61. Vandevivere PC, Saveyn H, Verstraete W, Feijtel TCJ, Schowanek DR (2001) Biodegradation of metal-[S,S]-EDDS complexes. Environ Sci Technol 2001(35):1765–1770. https://doi.org/10.1021/es0001153
62. Hernández-Apaolaza L, Lucena JJ (2011) Influence of irradiation time and solution concentration on the photochemical degradation of EDDHA/Fe^{3+}: effect of its photodecomposition products on soybean growth. J Sci Food Agric 91:2024–2030. https://doi.org/10.1002/jsfa.4414
63. Huang W, Brigante M, Wu F, Mousty C, Hanna K, Mailhot G (2013) Assessment of the Fe (III)–EDDS complex in Fenton-like processes: from the radical formation to the degradation of bisphenol A. Environ Sci Technol 47:1952–1959. https://doi.org/10.1021/es304502y

64. Rodríguez-Chueca J, Ormad MP, Mosteo R, Ovelleiro JL (2015) Kinetic modeling of *Escherichia coli* and *Enterococcus* sp. inactivation in wastewater treatment by photo-Fenton and H_2O_2/UV–vis processes. Chem Eng Sci 138:730–740. https://doi.org/10.1016/j.ces.2015.08.051
65. Wang WL, Wu QY, Huang N, Xu ZB, Lee MY, Hu HY (2018) Potential risks from UV/H_2O_2 oxidation and UV photocatalysis: a review of toxic, assimilable, and sensory-unpleasant transformation products. Water Res 141:109–125. https://doi.org/10.1016/j.watres.2018.05.005
66. Baxendale JH, Wilson JA (1957) Photolysis of hydrogen peroxide at high light intensities. Trans Faraday Soc 53:344–356. https://doi.org/10.1039/TF9575300344
67. Pablos C, Marugán J, van Grieken R, Serrano E (2013) Emerging micropollutant oxidation during disinfection processes using UV-C, UV-C/H_2O_2, UV-A/TiO_2 and UV-A/TiO_2/H_2O_2. Water Res 47:1237–1245. https://doi.org/10.1016/j.watres.2012.11.041
68. Miklos DB, Hartl R, Michel P, Linden KG, Drewes JE, Hübner W (2018) UV/H_2O_2 process stability and pilot-scale validation for trace organic chemical removal from wastewater treatment plant effluents. Water Res 136:169–179. https://doi.org/10.1016/j.watres.2018.02.044
69. Rodríguez-Chueca J, Laski E, García-Cañibano C, Martín de Vidales MJ, Encinas Á, Kuch B, Marugán J (2018) Micropollutants removal by full-scale UV-C/sulfate radical based advanced oxidation processes. Sci Total Environ 630:1216–1225. https://doi.org/10.1016/j.scitotenv.2018.02.279
70. Zhang Z, Chuang YH, Szczuka A, Ishida KP, Roback S, Plumlee MH, Mitch WA (2019) Pilot-scale evaluation of oxidant speciation, 1,4-dioxane degradation and disinfection byproduct formation during UV/hydrogen peroxide, UV/free chlorine and UV/chloramines advanced oxidation process treatment for potable reuse. Water Res 164:114939. https://doi.org/10.1016/j.watres.2019.114939
71. Wang C, Moore N, Bircher K, Andrews S, Hofmann R (2019) Full-scale comparison of UV/H_2O_2 and UV/Cl_2 advanced oxidation: the degradation of micropollutant surrogates and the formation of disinfection byproducts. Water Res 161:448–458. https://doi.org/10.1016/j.watres.2019.06.033
72. Wetterau G, Bruce Chalmers R, Schulz C (2018) Full-scale testing of UV/chlorine for potable reuse. IUVA News 20(3):11–15. https://iuvanews.com/stories/pdf/IUVANews_2018_Quarter3_links-Wetterau.pdf
73. Matafonova G, Batoev V (2018) Recent advances in application of UV light-emitting diodes for degrading organic pollutants in water through advanced oxidation processes: a review. Water Res 132:177–189. https://doi.org/10.1016/j.watres.2017.12.079
74. Rodríguez-Chueca J, García-Cañibano C, Lepistö RJ, Encinas Á, Pellinen J, Marugán J (2019) Intensification of UV-C tertiary treatment: disinfection and removal of micropollutants by sulfate radical based advanced oxidation processes. J Hazard Mater 372:94–102. https://doi.org/10.1016/j.jhazmat.2018.04.044
75. Neta P, Huie RE (1985) Free-radical chemistry of sulfite. Environ Health Perspect 64:209–217. https://doi.org/10.1289/ehp.8564209
76. Herrmann H (2007) On the photolysis of simple anions and neutral molecules as sources of O^-/OH, SO_x^- and Cl in aqueous solution. Phys Chem Chem Phys 9:3935–3964. https://doi.org/10.1039/B618565G
77. Cai L, Li L, Yu S, Guo J, Kuppers S, Dong L (2019) Formation of odorous by-products during chlorination of major amino acids in east Taihu Lake: impacts of UV, UV/PS and UV/H_2O_2 pre-treatments. Water Res 162:427–436. https://doi.org/10.1016/j.watres.2019.07.010
78. Qiu W, Zheng M, Sun J, Tian Y, Fang M, Zheng Y, Zhang T, Zheng C (2019) Photolysis of enrofloxacin, pefloxacin and sulfaquinoxaline in aqueous solution by UV/H_2O_2, UV/Fe(II), and UV/H_2O_2/Fe(II) and the toxicity of the final reaction solutions on zebrafish embryos. Sci Total Environ 651(1):1457–1468. https://doi.org/10.1016/j.scitotenv.2018.09.315
79. Yoon Y, Chung HJ, Di Wen DY, Dodd MC, Hur HG, Lee Y (2017) Inactivation efficiency of plasmid-encoded antibiotic resistance genes during water treatment with chlorine, UV, and UV/H_2O_2. Water Res 123:783–793. https://doi.org/10.1016/j.watres.2017.06.056

80. Sarathy SR, Stefan MI, Royce A, Mohseni M (2011) Pilot-scale UV/H_2O_2 advanced oxidation process for surface water treatment and downstream biological treatment: effects on natural organic matter characteristics and DBP formation potential. Environ Technol 32 (15):1709–1718. https://doi.org/10.1080/09593330.2011.553843
81. Sichel C, Garcia C, Andre K (2011) Feasibility studies: UV/chlorine advanced oxidation treatment for the removal of emerging contaminants. Water Res 45(19):6371–6380. https://doi.org/10.1016/j.watres.2011.09.025
82. Lester Y, Ferrer I, Thurman EM, Linden KG (2014) Demonstrating sucralose as a monitor of full-scale UV/AOP treatment of trace organic compounds. J Hazard Mater 280:104–110. https://doi.org/10.1016/j.jhazmat.2014.07.009
83. Chu X, Xiao Y, Hu J, Quek E, Xie R, Pang T, Xing Y (2016) Pilot-scale UV/H_2O_2 study for emerging organic contaminants decomposition. Rev Environ Health 31(1):71–74. https://doi.org/10.1515/reveh-2016-0008
84. Miralles-Cuevas S, Darowna D, Wanag A, Mozia S, Malato S, Oller I (2017) Comparison of UV/H_2O_2, UV/$S_2O_8{}^{2-}$, solar/Fe(II)/H_2O_2 and solar/Fe(II)/$S_2O_8{}^{2-}$ at pilot plant scale for the elimination of micro-contaminants in natural water: an economic assessment. Chem Eng J 310:514–524. https://doi.org/10.1016/j.cej.2016.06.121
85. Sarasidis VC, Plakas KV, Karabelas AJ (2017) Novel water-purification hybrid processes involving in-situ regenerated activated carbon, membrane separation and advanced oxidation. Chem Eng J 328:1153–1163. https://doi.org/10.1016/j.cej.2017.07.084
86. Krystynik P, Masin P, Kluson P (2018) Pilot scale application of UV-C/H_2O_2 for removal of chlorinated ethenes from contaminated groundwater. J Water Supply Res T AQUA 67 (4):414–422. https://doi.org/10.2166/aqua.2018.144
87. Baresel C, Harding M, Junestedt C (2019) Removal of pharmaceutical residues from municipal wastewater using UV/H_2O_2. IVL Swedish Environmental Research Institute 2019. Report number B 2354. https://sjostad.ivl.se/download/18.2299af4c16c6c7485d05de/1567162616009/B2354%20UV-H2O2%20rapport.pdf
88. Wünsch R, Plattner J, Cayon D, Eugster F, Gebhardt J, Wülser R, von Guten U, Wintgens T (2019) Surface water treatment by UV/H_2O_2 with subsequent soil aquifer treatment: impact on micropollutants, dissolved organic matter and biological activity. Environ Sci Water Res Technol 5:1709–1722. https://doi.org/10.1039/C9EW00547A
89. Wang D, Bolton JR, Andrews SA, Hofmann R (2015) UV/chlorine control of drinking water taste and odour at pilot and full-scale. Chemosphere 136:239–244. https://doi.org/10.1016/j.chemosphere.2015.05.049
90. Rodríguez-Chueca J, Varella Della Giustina S, Rocha J, Fernandes T, Pablos C, Encinas A, Barceló D, Rodríguez-Mozaz S, Manaia CM, Marugán J (2019) Assessment of full-scale tertiary wastewater treatment by UV-C based-AOPs: removal or persistence of antibiotics and antibiotic resistance genes? Sci Total Environ 652:1051–1061. https://doi.org/10.1016/j.scitotenv.2018.10.223

Removal and Degradation of Pharmaceutically Active Compounds (PhACs) in Wastewaters by Solar Advanced Oxidation Processes

Sixto Malato, Jaime Giménez, Isabel Oller, Ana Agüera, and José Antonio Sánchez Pérez

Contents

1 Introduction 300
2 Solar Semiconductor Photocatalysis 303
3 Degradation of PhACs by Solar Photo-Fenton Process 306
4 Application of Advanced Analytical Techniques for Assessing the Efficiency of AOPs 310

S. Malato (✉)
Plataforma Solar de Almería (CIEMAT), Almería, Spain
e-mail: Sixto.malato@psa.es

J. Giménez
Departamento de Ingeniería Química, Facultad de Química, Universitat de Barcelona, Barcelona, Spain
e-mail: j.gimenez.fa@ub.edu

I. Oller
Plataforma Solar de Almería (CIEMAT), Almería, Spain
e-mail: Isabel.oller@psa.es

A. Agüera
Solar Energy Research Centre (CIESOL), Joint Centre University of Almeria-CIEMAT, Almeria, Spain
e-mail: aaguera@ual.es

J. A. Sánchez Pérez
Solar Energy Research Centre (CIESOL), Joint Centre University of Almeria-CIEMAT, Almeria, Spain

Department of Chemical Engineering, University of Almería, Almeria, Spain
e-mail: jsanchez@ual.es

Sara Rodriguez-Mozaz, Paqui Blánquez Cano, and Montserrat Sarrà Adroguer (eds.), *Removal and Degradation of Pharmaceutically Active Compounds in Wastewater Treatment*, Hdb Env Chem (2021) 108: 299–326, DOI 10.1007/698_2020_688,

5 Modelling of Solar Photoreactors 315
6 Concluding Remarks 319
References 320

Abstract Most advanced oxidation processes (AOPs) are based on combination of oxidants, catalysts and radiation. Main disadvantages of these processes are their high energy consumption and operation costs. The use of solar radiation could be a way to overcome these drawbacks, but only two processes fit with this approach: heterogeneous photocatalysis with semiconductors and homogeneous photocatalysis through photo-Fenton-like processes. Solar AOPs (SAOPs) present special interest for the degradation of pharmaceutically active compounds (PhACs) as they can be easily adapted to comply with fundamental principles of green chemistry. This chapter states the main concepts of (1) solar semiconductor photocatalysis including its management with specific solar photoreactors; (2) solar photo-Fenton processes application, with special emphasis in circumneutral pH operation, for PhAC elimination; (3) optimized treatment validation under actual conditions with sensitive and selective analytical approaches; and (4) finally photoreactor modelling as a key tool to integrate the effect of temperature, PhAC concentration and radiation absorption in kinetics for a proper prediction of treated volume of water per unit of photoreactor surface and time.

Keywords Kinetic modelling, Photocatalysis, Photo-Fenton, Photoreactor, Solar treatment

1 Introduction

Most advanced oxidation processes (AOPs) are based on combination of oxidants (O_3/H_2O_2), oxidant and catalysts (Fe^{2+}/H_2O_2), oxidant and radiation (UV/H_2O_2) or radiation and catalysts (UV/TiO_2), although there are AOPs that combine all of the above with other processes [1]. One of the main disadvantages of these processes is their high energy consumption, mainly electrical, which makes them economically demanding. One of the strategies to reduce costs is their combination with other technologies, or the use of renewable energies, as solar radiation. In addition, advances in photoreactors, modelling, modes of operation and control strategies can reduce costs [2]. SAOPs are "near ambient temperature and pressure water treatment processes driven by solar energy which involve the generation of hydroxyl radicals in sufficient quantity to effective water purification", using an adaption of the pioneer definition made by [3].

Nowadays, there are two processes that can be carried out with solar energy: heterogeneous photocatalysis with semiconductors and homogeneous photocatalysis through photo-Fenton-like processes. Both are catalytic, which represents a certain advantage by compensating the consumption of reagents as in other AOPs. In the case of heterogeneous photocatalysis, TiO_2 is, currently, the most effective catalyst, as is discussed in Sect. 2 of this chapter. In photo-Fenton-like processes, iron no longer acts based on its aquacomplexes, and H_2O_2 is even replaced by other oxidants [4], as is discussed in Sect. 3. In this sense, literature on SAOPs and their applications have experienced an exponential growth in recent years and today are remarkably active and productive research fields within the scope of environmental chemistry and engineering. Evidence of this productivity includes the large number of related research works published recently. For example, in between 2010 and 2019, a search of SAOPs topic on the Scopus database returned more than 13,500 entries. These numbers have been continuously increasing since 1996 (26 entries) up to >2,500 (2019). This demonstrates the worldwide interest in this research field with multiple applications including PhAC treatment in water.

SAOP technology and equipment have much in common with solar thermal applications. Consequently, reactors and other components of the photocatalytic plants initially followed the design of solar thermal collectors, such as parabolic trough collectors or non-concentrating collectors [5]. However, from the first developments, their non-feasibility was shown [6] due to the following issues: (1) the fluid must be exposed to solar UV radiation, so the absorber must be transparent, (2) high temperature (>50°C) must be avoided, and (3) SAOPs must capture only short wavelength radiation (UVA and part of visible radiation) to promote photoreactions. Note that the concentration factor, RC, is the ratio between the area of the collector and the absorber. Therefore, in SAOPs, RC = 1 is required. Finally, such collectors must be simple and inexpensive and work with little pressure drop. In this way, the use of tubular absorbers (photoreactors) has clear advantages, and they are available in different sizes and materials, being the most obvious choice in systems that work under pressure conditions.

SAOPs present special interest for degradation of PhACs in secondary effluents from municipal wastewater (WW) treatment plants (MWWTPs) as they can be easily adapted to comply with many of the 12 fundamental principles of green chemistry. Anastas and Eghbali [7] reviewed 10 years ago those principles in the work entitled "design of chemical products and processes to reduce or eliminate the use and generation of hazardous substances". This definition and the concept of green chemistry were first formulated at the beginning of the 1990s [8]. SAOPs cover 9 of 12 fundamental principles of green chemistry: (1) prevent waste, (2) generate substances with little or no toxicity to human health and the environment, (3) do not use auxiliary substances (e.g. solvents, separation agents, etc.), (4) minimize energy requirements, (5) do not use derivatization steps, (6) are catalytic, (7) use chemicals that are recovered at the end of their function (catalysts) and they do not persist in the environment (or break down into innocuous degradation products as H_2O_2), (8) could be controlled by real-time monitoring and, finally, (9) are safe (near ambient temperature and pressure, not hazardous chemicals are used).

Since the pioneering work on SAOP cost comparison published by Bauer et al. [9], the studies related to SAOP costs are normally based on data obtained at laboratory or pilot plant scale. Although the final goal is always the possible industrial implementation, little information is found about the cost involved at such a scale. There is a need for unification of cost assessment criteria that would help in the comparison of results obtained by different research groups as well as cost scaling-up [10]. Photo-Fenton system seems to be the one that provides the best reaction rates when treating WW containing organic contaminants, thus requiring photoreactors with lower volume and lower economic investments. Although the best performance of solar photo-Fenton process compared to solar heterogeneous photocatalysis might not be a universal rule, a huge body of published works clearly point towards the existence of faster reaction rates for the photo-Fenton system [2].

Muñoz et al. [11] compared solar photo-Fenton and heterogeneous photocatalytic processes by using Life Cycle Assessment (LCA) showing less environmental impact from the former. LCA is thus increasingly important to understand the environmental impact of implementing any SAOP and for comparing alternative SAOPs with different processes [12]. Indeed, LCA applied to SAOP in combination with other technologies could substantially help in taking decisions for the efficient removal of PhACs [13]. The first step in a LCA is the definition of the main goal (for instance, environmental impact related to each m^3 of treated secondary effluents from MWWTPs). The inventory analysis step involves data collection to quantify system inputs and outputs, that is, evaluation of raw materials, energy consumption and pollutant production. The proposed methodologies have been demonstrated to be valid tools for the analysis of cost and environmental consequences of SAOPs. This would also help to inform future policy development on PhACs and their elimination, which is currently in its infancy.

The efficiency of SAOPs for PhAC removal is deeply related to ambient conditions, as it depends on solar irradiance and operation temperature. Therefore, it is highly influenced by MWWTP location. In spite of their importance, the combined effect of these variables on SAOPs has not been explored yet though lately some interesting studies have been reported [14]. The feasibility of any SAOPs should be based on two different scaling-up strategies for each location: one based on the average ambient conditions and the other on the most unfavourable ambient conditions, always monthly based. Different photoreactor sizes, treatment costs and LCA would be obtained, being a not obvious decision.

Green chemistry constitutes a good framework for SAOPs, but the relative "greenness" cannot be evaluated solely based on compliance with these principles, due to their qualitative nature. LCA can be a better tool for environmental assessment. Results and conclusions obtained by means of streamlined LCA studies based exclusively on laboratory-derived data do not match correct conclusions. Detailed LCA studies applied to SAOPs require at least demonstration projects in order to supply enough quality data. Environmental impact of SAOPs increases in locations where less solar radiation is available, due to the progressive increase in solar collector area and water flow requirements. Therefore, the differences in environmental impact among the different SAOPs and with regard to non-solar AOP or conventional technology prevent from speaking about SAOPs as a group of "green" technologies for PhAC elimination, unless the discussion is on particular options for treating particular WWs.

2 Solar Semiconductor Photocatalysis

Photocatalysis is based on the use of semiconductors able to promote electrons from the valence band to the conduction band, by absorbing radiation of adequate wavelength and with the sufficient energy to overcome the semiconductor bandgap.

The beginning of modern photocatalysis could be the work of Fujishima and Honda [15], although titanium dioxide was used as an electrode and not as suspended particles. The objective of the work was to obtain hydrogen from water, which was the initial objective of photocatalysis with titanium dioxide during the 1970s and until the mid-1980s [16, 17]. Since then, the number of works on the subject has grown continuously. According to the Web of Science, introducing the keywords photocatalysis-TiO_2 and during the last 40 years (1980–2019), 27,011 works on the subject have been published. Using the keywords photocatalysis-TiO_2-drugs, the number of references is 1,265, and, delimiting with photocatalysis-TiO_2-solar-drugs, they are 301. All this indicates the impact of the topic focused on PhAC elimination. In addition to the papers, numerous books have been published, as an example the books of Ollis and Al-Ekabi [18] and Marcí and Palmisano [19] could be cited, both show the evolution of photocatalysis, and the last one is very recent, which indicates the topicality of the subject. Reviews also continue to appear allowing for an update and a retrospective view of the state of the art [20–22].

Titanium dioxide is one of the most used semiconductors in photocatalysis because it is cheap, stable and easy to obtain. Other semiconductors, such as zinc oxide, iron oxide, etc., were also tested, but they gave worse results and presented stability problems. A handicap of water photocatalysis for hydrogen production was the low yield. It was tried to solve the problem by doping the catalyst with different elements, mainly noble metals, favouring the charge separation and avoiding the electron/hole pair recombination.

At the end of the 1970s, photocatalysis began to be used for the treatment of inorganic contaminants [23], taking advantage of the electrons generated to reduce cations. The electrons in the conduction band can produce the photo-reduction (Eq. 1) of the compound (C) or the generation of radicals from oxygen (Eq. 2). The next step, almost immediately, was its use in the treatment of organic pollutants, taking advantage of the generation of hydroxyl radicals with a high oxidation potential [24, 25]. The holes generated in the valence band are responsible for the production of hydroxyl radicals (Eq. 3) or for the direct photo-oxidation (Eq. 4) of some compounds (C).

$$C + e^- \rightarrow C^{\cdot -} \tag{1}$$

$$O_2 + e^- \rightarrow O_2^{\cdot -} \tag{2}$$

$$H_2O + h^+ \rightarrow HO^{\bullet} + H^+ \tag{3}$$

$$C + h^+ \rightarrow C^{\bullet +} \quad (4)$$

Solar radiation was also early used in the elimination of contaminants by photocatalysis [26–29]. In the late 1990s and early this century, titanium dioxide was specifically applied for the abatement of different types of PhACs [30, 31], such as antibiotics (sulfamethoxazole, trimethoprim, etc.), β-blockers (propranolol, metoprolol, etc.), nonsteroidal anti-inflammatory PhACs-NSAIDs (diclofenac, naproxen, ibuprofen, etc.), analgesics (paracetamol, etc.), etc. The percentage of elimination depends on the type of contaminant and its functional groups. Likewise, attacks on the molecule occur primarily by hydroxyl radicals but also by other types of radicals and agents. In addition, it is interesting to determine and to eliminate the reaction intermediates formed since these may be more dangerous than the initial contaminant.

The elimination of PhACs has been studied by using different types of photoreactors with different geometries: annular, flat, cylindrical, tubular, etc. Different types of lamps have been employed, such as Xe lamps, used in solar simulators, since they have an emission spectrum very similar to solar in the ultraviolet range. BLB lamps, Hg UV lamps and other types of radiation have also been utilized. Natural sunlight as radiation source was also used in the photocatalytic degradation of PhACs with different types of reactors [32, 33]. Those that offered better yields are compound parabolic collectors (CPCs; see Figs. 1 and 2), since they present better sunlight capture taking profit of both direct and diffuse radiation. The percentage of pollutant removal depends a lot on the particular operating conditions: type of pollutant, initial concentration, TiO_2 concentration, reactor geometry, etc. In this regard, it should be noted that, as the catalyst concentration increases, the reaction rate increases but a value is reached (it can range between 0.2 and 1 g/L of TiO_2) from which the reaction rate does not increase any more since opacity and light-scattering phenomena occur.

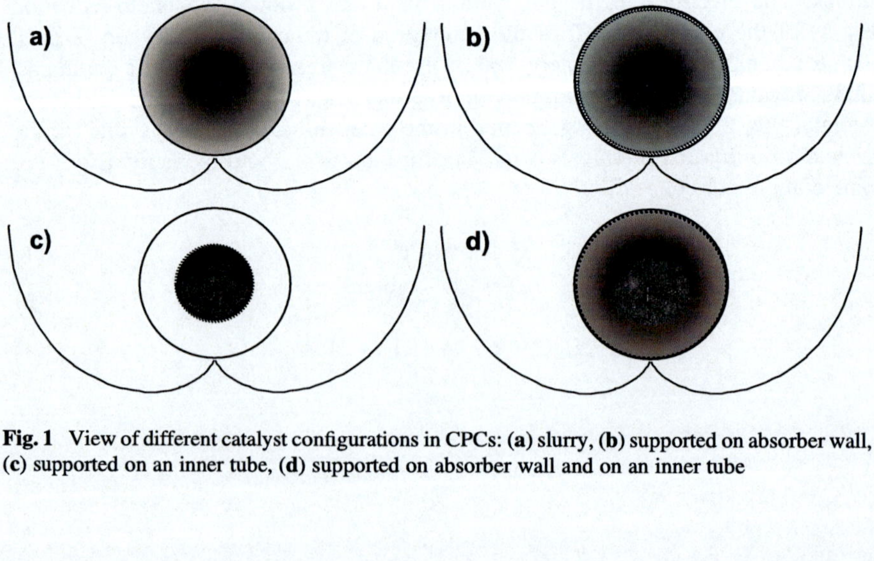

Fig. 1 View of different catalyst configurations in CPCs: (**a**) slurry, (**b**) supported on absorber wall, (**c**) supported on an inner tube, (**d**) supported on absorber wall and on an inner tube

Fig. 2 View of CPCs (**a**) and basic schematic concept of RPR (**b**)

The use of artificial light usually makes the process faster, but the electricity consumption makes treatment more expensive. From this point of view, solar reactors can make the process competitive since the cost of energy is zero, but obviously not the cost of fixed assets. Efficiency and costs can be calculated in different ways. One of them is the electric energy per order (E_{EO}, the electric energy in kWh required to degrade a contaminant C by one order of magnitude) or the electric energy per mass (E_{EM}, the electric energy in kWh required to bring about the degradation of a unit mass), both proposed by Prof. Bolton [34]. It is also possible to estimate kW/ppm, Euros/ppm, kW/L, Euro/L, etc. [12].

Titanium dioxide absorbs radiation in the UV band, which represents a very small percentage of the solar radiation that reaches the earth. In order to broaden the absorption spectrum, the catalyst has been doped with different elements (noble metals, B, N, etc.), although the improvement has not been very substantial in the majority of cases. On the other hand, the fact of working with doped catalysts makes the process more expensive and can generate problems of stability and loss of activity along treatment time.

The catalyst used as particles in suspension presents the problem of its separation from the reaction medium, although in some cases and depending on the particle size, it is possible to separate it by sedimentation. To avoid this problem, attempts have been made of working with the catalyst (TiO_2) supported on the reactor walls, deposited on balls with which the reactor was filled or on materials, such as graphene or other carbon products. This avoids the problem of sedimentation, but presents two additional problems: on the one hand, a worse use of radiation with respect to the catalyst in suspension and, on the other hand, the stability of the supported catalysts can be lower due to poor adhesion of TiO_2 on the support. Figure 1 shows different possible configurations of supported catalysts in CPCs (see also Fig. 2 for a general view) in comparison with a slurry, where radiation field is conceptually shown (as darker, less photons would be available).

The time required for the degradation of PhACs is dependent on the water characteristics, and it is often established through studies using the WW of interest.

The main causes why PhAC degradation rate diminishes through matrix effects are [35] (1) scavenging of hydroxyl radicals by bicarbonate, chloride, sulphate and other anions, producing lower oxidant potential radicals; (2) screening effect, when matrix components have light absorption at same wavelengths than catalyst; (3) turbidity, which avoids light transmission through the water; and (4) adsorption onto catalyst surface of some organic and inorganic species (e.g. NOM or phosphate).

A key issue for scaling-up is the modelling of photocatalytic processes with titania in suspension. It is a complex matter, since the reaction rate depends not only on conventional kinetic parameters but also on the radiation absorbed in the medium. Therefore, when modelling the system, the following parameters must be considered for semiconductor photocatalysis (more details could be found in Sect. 5, including modelling of homogeneous photocatalysis):

- Reaction medium: pH, concentration of reactants, adsorption of reactants on catalyst surface, intermediate concentration, final concentration of products, colour, temperature. In this way, it should be noted that actual WW treatment with TiO_2 could present problems because of the presence of organic matter, particles, etc., which represents a competition for light absorption and light-scattering problems, too.
- Catalyst: type of semiconductor-catalyst used, catalyst concentration, particles diameter, specific area, structure.
- Reactor: geometry (tubular, annular, cylindrical, etc.), dimensions (volume, diameter, height, etc.), radiation concentrators (geometry, location), location with respect to light and flow model.
- Fluid dynamics: feed flow rate, O_2 feed flow, mixing.
- Operation (batch, continuous and recirculation): volume, total experimental time, time and volume of samples, feed flow rate, air feed flow, recirculation flow.
- Radiation: power of radiation source, spectrum, radiation reaching the reactor, radiation entering the reactor, radiation absorbed, use and type of filters. Since light plays a fundamental role in the reaction rate, the influence of radiation on the kinetic equation must be introduced. Different models have been used to estimate the absorbed radiation, from analytical models to probabilistic models [36–38]. Radiation models will be widely explained in Sect. 5 of this chapter.

3 Degradation of PhACs by Solar Photo-Fenton Process

Among those AOPs successfully applied for the remediation of different types of WW, Fenton and photo-Fenton processes have shown to be highly efficient for the elimination and degradation of a wide range of contaminants and bio-recalcitrant materials [39]. Both have shown great potential to be commercialized and finally applied at industrial scale [40].

Fenton process is extensively covered in another chapter of this book, but briefly the mechanism is based in H_2O_2 decomposition by iron (II) at acidic pH [39]. In this mechanism, the regeneration of iron (II) from iron (III) is considered the limiting step in the catalytic cycle, even more when the iron is present at low concentrations (Eqs. 5 and 6). In this sense, Fenton's reactions are clearly accelerated in the presence of light, due to the catalytic regeneration of iron (II). This process is known as photo-Fenton, occurring at wavelengths lower than 580 nm and so opening the possibility to be powered by solar energy as renewable irradiation source [1]. Precisely, a large amount of studies have been published on the use of solar UV radiation in the photo-Fenton process bringing out the potential on the reduction of electric energy consumption and increasing its environmental sustainability [2]. Another advantage of using solar radiation is that certain complexes which intervene in Fenton's reaction absorb UV light and part of the visible light of the solar spectrum, increasing the generation of hydroxyl radicals and so the efficiency of the process [41]. Iron (III) complexes suffer a ligand to metal charge transfer, provoking the dissociation in iron (II) and the oxidized ligand (Eq. 7) [39].

$$Fe^{+2} + H_2O_2 \rightarrow Fe^{+3} + OH^- + HO^\bullet \ (K\tilde{7}0\ M/s) \tag{5}$$

$$Fe^{+3} + H_2O_2 \rightarrow Fe^{+2} + HO_2^\bullet + H^+ \ (K\tilde{1} - 2 \times 10^{-2}\ M/s) \tag{6}$$

$$Fe^{+3}(L)_n + h\upsilon \rightarrow Fe^{+2}(L)_{n-1} + L_{ox}^\bullet \tag{7}$$

In this context, it is important to stress that the optimum pH to operate photo-Fenton process is 2.8, as, under this condition, precipitation of iron is avoided and the maximum generation of active species of iron is also assured ($FeOH^+$ and Fe $(OH)^{2+}$). At pH higher than 3, the reactivity of the system starts to decrease, and at values higher than 4, dissolved iron starts to precipitate as ferric hydroxide [42]. This issue can be clearly considered as one of the main challenges investigated by the scientific community in the last years, with the main objective of successfully applying the highly powerful photo-Fenton process for the elimination of PhACs and other microcontaminants. PhACs and their metabolites, detergents, personal care products, flame retardants, industrial additives, steroids and hormones are present at extremely low concentrations in, for instance, superficial waters, non-treated urban WWs and effluents of MWWTPs (between 50 and 150 μg/L) [43–45]. Special awareness is placed on PhACs as they have been discovered in almost all environmental water matrices including surface water (lakes, rivers, streams, estuaries and seawater), groundwater, WWTP effluents and influents and sludge. Even, they have been broadly detected in polar regions [46].

This global challenge must face important modifications in existing WWTP infrastructures, in which the integration of photo-Fenton as a technical and economical feasible process has become an important research focus from the green, environmental and sustainable point of view [47–49]. One of the main requirements in such investigation is the importance of performing photo-Fenton tertiary treatment at neutral pH, as addition of reagents for adjusting this parameter in urban WWs would provoke different economic and environmental disadvantages: increase in effluent salinity, environmental impact and increasing treatment operating costs (including the generation of sludge) [50].

As it has been widely studied, the application of photo-Fenton process at neutral pH is considered a promising alternative, thanks to the use of iron complexing agents for maintaining iron in solution, as well as the possibility of being powered by solar irradiation as a renewable, abundant and pollution-free energy source [42, 51]. Nevertheless and from the operational point of view, it is always necessary to avoid the addition of high concentrations of such complexing agents, as an implicit increase in total organic carbon will occur, with the consequent reduction in the efficiency of the treatment as well as an increase in operating costs. In general, iron complexing agents commonly used are macromolecules with carboxylic and/or amino groups which absorb light in the UV-visible range suffering also self-photocatalytic degradation: oxalate [52], ferrioxalate [53], EDTA [50, 54], EDDS [55, 56], NTA [57], citrate [58–60] and humic acids [61]. Low biodegradability and toxicity of EDTA and the reduced pH range in which citric acid or oxalate could be used, as well as the high cost of EDDS, have pushed recent studies to evaluate the feasibility of using natural waters (containing polyphenolic compounds) [62] and agro-industrial residues [63] as complexing agents. Though their efficiency has been proved, it was demonstrated that the iron complexes generated are not stable enough showing activity only for a few minutes at the beginning of the process and so attaining low degradation efficiencies at circumneutral pH.

Nowadays, [S,S]-ethylenediamine-N,N′-disuccinic acid (EDDS) is one of the chelating agents most used in the application of solar photo-Fenton tertiary treatments to maintain iron in solution. In recent years, EDDS has gained the attention of researchers as a stable iron complexing agent, and Fe^{3+}-EDDS complex has been widely studied since the pioneering work of Gilles Mailhot group in France [54, 64]. In addition, EDDS is highly biodegradable [65], fulfilling the principle of environmental sustainability. As already mentioned, different parameters can influence the chelate-modified photo-Fenton process and the contaminants removing, including the type and load of chelating agents and reagents (iron, hydrogen peroxide, etc.), pH, applied energy, type of organic pollutant, water matrices, etc. [66, 67].

Looking for low-cost and more competitive commercial applications, the use of solar energy is being widely studied to perform photo-Fenton tertiary treatment for the elimination of PhACs. Thus, in locations with water scarcity problems and abundant solar radiation, the development of solar technologies for WW reusing looks as a plausible alternative water resource [14].

Solar photoreactors mostly tested and reported in literature for PhAC degradation in different water matrices are:

- Solar simulators at laboratory scale. In the last years, neutral pH is stated, thanks to chelating agents and different water matrices evaluated. However, many studies still deal with distilled water or natural water in which target PhACs are spiked and even low pHs applied (closer to optimal photo-Fenton pH). In addition, extremely high concentrations of PhACs are sometimes considered far from those actually found in WWs [4, 68, 69]. Traditionally, best-operating conditions for solar photo-Fenton process performance are evaluated at laboratory scale in solar simulators (Xe lamps). Trovó et al. [60] studied the elimination of the antibiotic amoxicillin by solar photo-Fenton by using this methodology. In the last years, the degradation of PhAC contained in simulated and actual effluents at more realistic concentrations (even near neutral pH) is also tackled under solar simulation devices at laboratory scale [70–72].

- Compound parabolic collectors (CPCs). CPCs are static collectors with surface following an involute around a cylindrical reactor tube (made of borosilicate glass) placed in the axis of an aluminium reflector. CPCs can make efficient use of both direct and diffuse solar radiation, without the need for solar tracking. Most of literature on solar photo-Fenton applications, at pilot plant scale, has reported the use of CPC for microcontaminant removal, including PhACs [56, 73]. When industrial WWs (with high organic content) have to be treated by photo-Fenton, high concentrations of oxidant species (hydroxyl radicals and others) are required. Besides, many WWs must be confined without contact with the environment for a safe operation. In this sense, CPC photoreactors (Fig. 2a) have been demonstrated to be highly efficient for solar irradiation capture with adequate absorber tube diameter and iron concentration ([74]) and so must be selected as the best option. Nevertheless, increasingly studies are focused on synthetic or actual WW spiked with PhACs or even monitoring actually present PhACs by advanced analytical tools such as liquid chromatography coupled to mass spectrometry [75, 76]. The advanced analytical techniques are a very powerful tool to follow the degradation of PhACs, allowing detection and monitoring of the reaction intermediates produced.
- Raceway pond reactors (RPR). One of the main weaknesses of CPC photoreactors lays on their manufacture and amortization costs. But, when the objective is the elimination of PhACs present at concentrations at least thousand times lower than other organics in industrial WW, the process needs less concentration of hydroxyl radicals and, consequently, less iron concentration and solar irradiance density. In such a case, the possibility of using open photoreactors, as RPRs, is promising, because they are less efficient in photon capture but much more simple to manufacture and with lower costs (per surface unit). RPRs have been widely applied for microalgal mass culture [77, 78]. They consist of extensive reactors with channels in a closed loop, provided with a paddle wheel connected to an engine to recirculate the water (Fig. 2b). They can be designed and operated at different light path lengths (liquid depth can be varied), higher than CPC, therefore permitting lower iron concentration. Worse solar irradiation capture could be compensated by higher surface due to lower costs. Treating non-toxic superficial waters, non-treated urban WWs or effluents of MWWTPs permits to operate open photoreactors. Few published works have already demonstrated the efficiency of solar photo-Fenton at neutral pH in RPRs for the elimination of PhACs contained in different actual effluents of MWWTPs [79].

Combining technologies have also been demonstrated to be promising, though further research is still lacking in this topic. For instance, combination of Fenton with reverse osmosis reported removal rate up to 99% of amoxicillin according to Zhang et al. [57]. Nanofiltration membrane system followed by a mild solar photo-Fenton tertiary treatment for the elimination of PhAC has been also successfully reported by Miralles-Cuevas et al. [58, 59, 80]. Such published works showed how the preconcentration of microcontaminants (including PhACs) as well as the reduction of the total volume of water to be treated resulted in a significant reduction of reagent consumption and surface area of solar photoreactor, compared to the application of solar photo-Fenton alone. From the work published by Miralles-Cuevas et al., it was also concluded that the combination of NF membranes with solar photo-Fenton process improved total costs, though operating costs at neutral pH when using EDDS were higher than conventional photo-Fenton at pH 3, mainly due to the high cost of this chelating agent [81]. In addition, it has been confirmed that several transformation products could be more resistant than their original parent compounds. Therefore, the existence of transformation products must be essentially investigated after the treatment in order to evaluate the effectiveness of the solar photo-Fenton process [82–84].

Another important challenge lies on the proper selection of the technology to be applied for the elimination of PhACs. In this regard, it is always crucial to assess how environmentally effective is the proposed treatment in reducing ecotoxicity impact of such microcontaminants and if, by doing so, potential ecotoxicity could increase elsewhere or even cause other environmental impacts. This aspect is tackled by performing Life Cycle and Risk Assessments [13, 85–87].

4 Application of Advanced Analytical Techniques for Assessing the Efficiency of AOPs

Nowadays, it is well known that the application of conventional analytical techniques is insufficient for assessing the efficiency of AOPs in removing PhACs in WW. Dedicated methodologies, focused on the identification and monitoring of organic microcontaminants (OMCs), are required to provide a comprehensive evaluation of the treatment efficiency.

In any case, evaluating the behaviour of "all contaminants" present in WW is a difficult or rather impossible task, given the complexity of the samples; the low concentration at which these compounds are present, compared to other components of the matrix; and the large number of potential contaminants. Thus, most reported degradation studies focus on a limited group of compounds.

Test protocols generally include adding the selected OMCs at high concentrations in simulated or actual WW and their monitoring by high-performance liquid chromatography (HPLC) with classical detectors (UV or DAD) during the treatments. These tests are very useful as preliminary studies and can be used for the initial optimization of treatment parameters (type and concentration of reagents, treatment times, etc.) or even for comparison purposes, to determine the most suitable treatment for a certain application. As an example, Klamerth et al. [88] selected 15 OMCs as model compounds to optimize a modified photo-Fenton treatment at neutral pH by using EDDS. Selected compounds included PhACs representing various therapeutic classes, such as antibiotics (flumequine, ofloxacin, sulfamethoxazole), analgesic and antipyretic drugs (acetaminophen, ibuprofen, antipyrine), nonsteroidal anti-inflammatory drugs (ketorolac, diclofenac), antiepileptic drugs (carbamazepine), hormones (progesterone), psychoactive drugs (caffeine) and also personal care products (triclosan) and pesticides (atrazine, isoproturon, hydroxybiphenyl). All of them provide a reliable response in a conventional UPLC-UV/DAD system, with limits of detection and quantification from 1.5 to 10 μg/L. MWWTP effluents were spiked at two concentration levels (100 and 15 μg/L), and the concentration profile of each compound was monitored by recording the UV signal for each compound at the wavelength of maximum absorption. Experiments with the highest concentration level were monitored by direct injection of samples into the UPLC-UV/DAD system, thus providing a simple, inexpensive and rapid evaluation of results. When the lowest concentration level was selected, preconcentration of the samples (50-fold) was required to obtain a useful OMC transformation profile along the treatment.

Despite the interest of this approach, validation of the optimized treatments under actual concentration levels and operation conditions is required, since, in general, degradation percentages depend on the type and concentration of compounds present in WW. To provide a comprehensive assessment of a treatment process in real WW, the use of more sensitive and selective analytical approaches is required. Gas chromatography (GC) and HPLC coupled to mass spectrometry (MS) represent the best choice. HPLC-MS is mostly applied in PhAC determination due to its versatility and wide scope of analysis, extended to thermolabile, highly polar or with limited volatility compounds. GC-MS is reserved only for volatile and non-polar compounds, in order to avoid the application of derivatization processes, required to improve detectability of more polar and less volatile compounds [89]. Derivatization is a time-consuming technique, prone to analyte losses and non-recommended in multiresidue analysis. Therefore, HPLC or, more recently, ultra-high-performance liquid chromatography (UHPLC), which provides higher resolution and shorter analysis times, has become the preferred advanced technique for multiclass PhAC analysis.

Another advantage associated to the use of LC-based methods is the possibility of using direct sample injection. Direct injection (DI) technique is becoming a popular option in the analysis of environmental matrices due to its inherent advantages, such as (1) reduction of solvent consumption, in line with fundamental principles of green chemistry, (2) analysis time reduction and (3) lower cost, compared to the use of conventional extraction procedures, such as liquid-liquid extraction (LLE) or solid-phase extraction (SPE). This technique also increases sample throughput, eliminates analyte losses associated with the extraction procedures and provides adequate reproducibility and low sample contamination. Campos-Mañas et al. [90] developed a DI-based analytical method to be applied in the analysis of 115 OMCs (including PhACs) in WW effluents. Sample pre-treatment was reduced to acetonitrile addition and filtration of the mixture previous to HPLC-MS analysis. 92% of the studied compounds showed limits of quantification ≤100 ng/L and mean recoveries in the range of 70–120%, with RSD values ≤20%. Up to 67 compounds could be determined in actual WW effluent samples at concentrations ranging from 10 ng/L to 26 μg/L, thus demonstrating the applicability of the method. This strategy was also successfully applied to validate, in realistic and variable conditions, a solar photo-Fenton process at neutral pH with Fe^{3+}-EDDS [79]. The treatment was applied to five WW effluents of diverse origin and composition. The removal of more than 80% of total load of PhACs and other OMCs (from 18 to 45 compounds) could be monitored, and the efficiency of the photo-Fenton process to treat effluents of very different physicochemical composition was demonstrated.

Although DI has indisputable advantages and allows a rapid and inexpensive monitoring of PhACs along the treatments, an enrichment step may still be required to evaluate the reaction kinetics at very low concentrations, mainly at the end of treatment, when most compounds have reached maximum degradation. In this case, SPE is the recommended sample preparation technique, due to its ease of use, low solvent consumption and great availability of sorbent materials, which facilitate the adaptation of the technique to multiple applications. In multiresidue analysis, polymeric sorbents, such as Isolute ENV+, Strata-X or more frequently Oasis HLB material, represent the best choice and provide high capacity and retention of a broad class of compounds in a wide range of polarities. Other alternatives, such as mixed-mode polymeric sorbents, which improve selectivity and retention of ionic compounds, or highly specific molecularly imprinted polymers (MIP), have also been applied to improve the selectivity of the methods, but their use is restricted to a limited number of selected compounds.

A typical multiclass SPE procedure involves a previous filtration of the sample by a 0.7 μm glass fibre filter, to remove suspended solids, and a pH adjustment, typically at pH 7–8 as a compromise solution, although some more acidic PhACs, such as some antibiotics, may require lower pH for proper recovery [91]. Final elution of retained compounds is usually accomplished with methanol.

After pre-treatment step, identification and quantification of pharmaceuticals in WW samples is usually performed by MS techniques. Modern MS systems using hybrid triple quadrupole (QqQ-MS) or quadrupole linear ion trap (QqLIT-MS) analysers allow very low detection limits to be reached when operating in MRM (multiple reaction monitoring) mode, and they represent, without any doubt, the best choice for target quantitative analysis. Despite the excellent performance characteristics of these systems, matrix effects involving ion suppression or enhancement are frequent, contributing to a significant degradation of the analytical accuracy and reproducibility. Electrospray ionization sources, which are routinely used in WW analysis, due to its suitability for polar and ionic compounds determination, are particularly prone to matrix effects. To avoid or minimize matrix effects, several strategies have been proposed, such as improvement on clean-up protocols or HPLC separations, matrix dilution, use of standard addition method, external matrix matched standards, analogue-structure internal standards or isotopically labeled standards [92]. The latter represent the more robust approach, but the cost and availability of the standards limit their extensive use.

Although most degradation studies focus on the quantification and monitoring of selected compounds originally present in raw WW, one aspect that is becoming increasingly relevant concerns the identification of transformation products (TPs), generated during the treatments. It is well known that the absence of parent contaminants does not guarantee the good quality of treated water, and WW treatments must consider the formation of TPs to minimize the environmental impact of the effluents. However, the identification and structure proposal of a priori unknown compounds entail great difficulty, and in many cases, only a tentative proposal can be accepted, as the analytical reference standards of such TPs are not always available for confirmation [93]. The analytical techniques proposed for quantitative analysis of target compounds are inefficient in TP identification, and the application of high-resolution mass spectrometry (HR-MS) is required due to its high confirmatory capabilities, derived from the high resolving power and the mass accuracy in MS and MS/MS modes. Time-of-flight (TOF) and Orbitrap analysers, or more frequently their hybrid configurations (QTOF, QOrbitrap), represent the best choice for this type of analysis. MS/MS fragmentation data is used for structure elucidation, as spectral databases are not useful in most cases. Previous knowledge of the most common transformations taking place in the different treatments, the structural similarity with the precursor and the information on the elemental composition of the TPs candidates, provided by the mass spectrometer, are essential in the proposal of their structures.

As a first approach, assays for the identification of TPs must be carried out individually for each compound, in distilled water and at relatively high concentrations of the selected contaminant, in order to obtain detectable concentrations of TPs without applying any preconcentration step, thus avoiding losses of the more polar TPs. Samples are then analysed by HR-MS. The use of suspect screening strategies has been recently proposed to facilitate the TP detection and identification [94]. Suspect analysis consists of the creation of searching lists containing the monoisotopic masses of tentative TPs, collected from literature or by using computational (in silico) prediction tools, such as PathPred [95] or EAWAG-BBD pathway prediction system [96]. The exact masses of the predicted TPs are extracted from the chromatograms, and a series of confirmation criteria, based on the plausibility of the retention time, isotopic pattern and observed fragmentation pattern, are applied to confirm the proposed structure. Jaén-Gil et al. [97] have successfully applied this strategy to the identification of the TPs of metoprolol and its metabolite metoprolol acid in UV/H_2O_2-treated WWs, identifying 24 TPs with potential ecotoxicological implications. Once the TPs are identified and the transformation pathway is proposed, the formation of the reported TP can be investigated under more realistic conditions.

As a final remark, it can be said that the monitoring of PhACs, other OMCs and their TPs during WW treatments is essential to evaluate their efficiency and guarantee the quality and safety of treated water. There are several analytical strategies available, and their selection will be based on the objective of the analysis. Figure 3 summarizes the most frequently used analytical procedures and techniques.

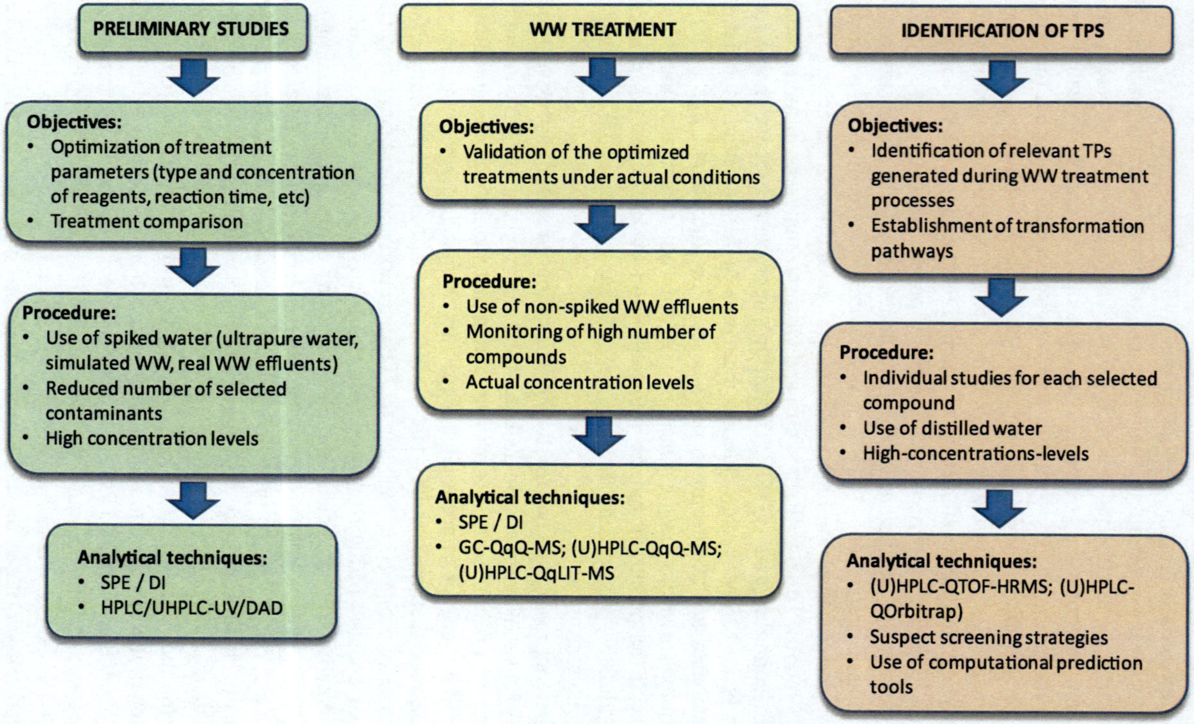

Fig. 3 Flow diagram of procedures and analytical strategies applied in the degradation studies of PACs in WW

5 Modelling of Solar Photoreactors

Reactor modelling starts with accurate kinetic models. In addition to temperature and reactant concentrations, light distribution inside the reactor and photon absorption phenomena play an essential role on the modelling solar photoreactors. Difficulty is higher when considering the myriad of reactions taking place simultaneously in radical chain mechanisms as happens in AOPs. In this regard, different approaches have been considered from empirical to mechanistic models. The former approach uses a mathematical function to fit the response of the main variables to the factors causing changes in the system. The latter uses rate law equations and kinetic parameters. Among SAOPs for PhAC removal in complex media, such as MWWTP secondary effluents, homogeneous photo-Fenton process has several advantages with regard to heterogeneous TiO_2 photocatalysis. Photo-Fenton process yields higher reaction rates, and fouling on catalyst surface is not a problem. Conversely, it is strongly dependent on pH. Most of literature about modelling of solar photoreactors deals with photo-Fenton, at both acidic and neutral pH.

As for the empirical approach, response surface methodology and neural network models provide valuable information on PhAC degradation in a selected experimental framework. Schenone et al. [98] applied a three-level factorial experimental design to model the photo-Fenton degradation of the herbicide 2,4-dichlorophenoxyacetic acid at circumneutral pH with the ferrioxalate complex. Rivas et al. [99] used a Langmuir-Hinshelwood-type model to predict apparent first-order rate constant for the removal of the neonicotinoid pesticide acetamiprid as well as H_2O_2 consumption, by solar photo-Fenton at acidic pH in a raceway pond reactor. Giannakis et al. [100] developed second-degree models for the removal of the PhAC venlafaxine by photo-Fenton by using the response surface methodology (3 pH-specific central composite designs). The models were used as input for the desirability functions to obtain the optimal regions to operate the photo-Fenton process.

These empirical models provide mathematical expressions useful for reactor design in the experimental range, but their main drawback is that they do not provide information about the reaction mechanism. In addition, they cannot be applied globally and consequently extrapolation is not possible.

To overcome these limitations, attention is paid to mechanistic models. To this end, the knowledge of photoreactors' solar field is crucial. To determine the radiation field, the radiative transfer equation (RTE) must be solved. It describes the radiation intensity at any position along a ray pathway through the reaction bulk. In homogeneous systems (without scattering), the photon emission by the absorbing species can be neglected, and the RTE is defined by Eq. (8) [101], where I_λ is the spectral specific intensity; Ω and s denote the direction and the distance of photon transport, respectively; and κ_λ is the volumetric absorption coefficient.

$$\frac{dI_\lambda(s,\Omega,t)}{ds} = -\kappa_\lambda(s,t)\, I_\lambda(s,\Omega,t) \tag{8}$$

In a homogeneous medium, the change in the radiation intensity along the ray trajectory is only due to the absorption process in the reaction bulk. This variation can be expressed by an equation that involves the volumetric absorption coefficient. This coefficient is function of the wavelength and variables such as temperature and composition. The absorption coefficient could be determined accurately by studying the photon absorption by atoms or molecules exposed to a radiation field. Macroscopically, the volumetric absorption coefficient is linearly dependent on the concentration of absorbing species. In Eq. (9) [101] i denotes each absorbing species, $\kappa_{T,\lambda}$ is the volumetric absorption coefficient, $\alpha_{\lambda,i}$ is the molar absorptivity, and C_i stands for the molar concentration of the absorbing species.

$$\kappa_{\lambda,i} = \alpha_{\lambda,i} C_i \tag{9}$$

The local volumetric rate of photon absorption (LVRPA) represents the amount of photons absorbed per unit of time and reaction volume. It depends on the photon source, the concentration of absorbing species, the optical properties of the system and the reactor geometry. Although it is defined for monochromatic radiation, it can be extended to polychromatic sources by integrating over the wavelength range. When working under solar UV radiation, LVRPA values corresponding to the direct and diffuse radiation are calculated in a different way. In homogeneous systems, in which radiation can be modelled with one spatial coordinate, the monochromatic LVRPA corresponding to the absorption of direct radiation can be calculated by Eq. (10) [102], where $q_{w,\lambda}$ is the discretized incident radiation, κ_λ is the volumetric absorption coefficient of the absorbing species and $\kappa_{T,\lambda}$ is the volumetric absorption coefficient of the medium.

$$e_\lambda^a(x,t) = q_{w,\lambda}\kappa_\lambda(t)\exp\left[-\kappa_{T,\lambda}(t)x\right] \tag{10}$$

The concept of diffuse radiation absorption is more complex. Brandi et al. [103] applied the hypothesis of azimuthal symmetry (rays propagate in one spatial coordinate (x) and one angular coordinate (θ)) to model the absorption of diffuse radiation in a simple lab-scale photoreactor. In this case, RTE for homogeneous systems can be defined by Eq. (11) [103]. Integrating Eq. (11) with the boundary conditions corresponding to photons moving both forward and backward, the diffuse LVRPA can be calculated by Eq. (12):

$$\cos\theta\frac{dI_{\lambda,\Omega}(x,t)}{dx} + \kappa_\lambda(x,t)I_{\lambda,\Omega}(x,t) = 0 \tag{11}$$

$$e_{\lambda}^{a}(x,t) = 2\Pi I_{\lambda}^{0}\kappa_{\lambda}(t)\int_{-\frac{\Pi}{2}}^{\frac{\Pi}{2}} \exp\left[-\frac{\kappa_{\lambda}(t)}{cos\theta}x\right] \text{sen}\theta d\theta \tag{12}$$

$$e^{a}(x,t) = \int_{\lambda \min}^{\lambda \max} e_{\lambda}^{a}(x,t)d\lambda \tag{13}$$

$$\langle e^{a}(x,t)\rangle_{V_R} = \frac{1}{L_R}\int_{0}^{L_R} e^{a}(x,t)dx \tag{14}$$

As mentioned above, LVRPA is defined for monochromatic radiation. For solar radiation (polychromatic radiation), LVRPA can be computed by Eq. (13). Furthermore, in perfectly mixed reactors with constant cross-sectional under photolimitation conditions (the reaction rate linearly depends on the photon absorption), LVRPA can be averaged across the reactor volume giving the VRPA, Eq. (14) [103], where L_R and V_R stand for reactor length and reactor volume, respectively.

For reported kinetic models for PhAC removal by photo-Fenton process at acidic pH taking into account the VRPA, Andreozzi et al. [104] studied the oxidation of an aromatic heterocyclic compound, benzothiazole, taking into account 21 reactions. The model successfully predicted the influence of H_2O_2, Fe^{3+} and sulphate concentrations on the system reactivity. Conte et al. [105] proposed a kinetic model of 15 reactions for the degradation of the herbicide 2,4-dichlorophenoxyacetic acid (2,4-D) including the effects of iron concentration, H_2O_2:2,4-D initial concentration ratio, temperature and radiation level. Cabrera Reina et al. [106] proposed a simplified kinetic model of nine reactions for the mineralization of the PhAC paracetamol in raceway pond reactors, and the effect of the VRPA and the liquid depth were included. With the same PhAC, Audino et al. [107] developed a kinetic model in an annular photoreactor computing the VRPA. These studies were carried out to remove organic contaminants in the concentration range of mg/L. Nonetheless, most of PhACs appear in WW at concentrations in the μg/L range. In this regard, a mechanistic model consisting of a set of only eight reactions was proposed for the removal of 100 μg/L of acetamiprid in synthetic MWWTP secondary effluent, taking into account the effect of temperature and photon absorption. The model was validated in raceway pond reactors of 5 and 15 cm of liquid depth (120 and 360 L capacity) under natural sunlight [108]. The model predicted that reaction rate linearly increases with the VRPA up to VRPA values of 866 μEinstein/m^3/s and at higher values the process became photosaturated.

Regarding photo-Fenton process operated at neutral pH, chelating agents are used to keep iron in solution as mentioned before. For instance, Conte et al. [109] studied the influence of different inorganic anions on ferrioxalate complex photo-activity and proposed a kinetic model for the removal of 2,4-dichlorophenoxyacetic acid at circumneutral pH, including the hydroxyl radical scavenging effect of anions. In the case of using EDDS as iron complexing agent, the higher absorptivity of Fe^{3+}-EDDS in comparison with Fe^{3+} makes photosaturation of the reaction rate to take place at VRPA values higher than 1,547 μEinstein/m^3/s pointing out the relevant role of light absorbing species in the photo-Fenton kinetics. In particular, a new mechanistic

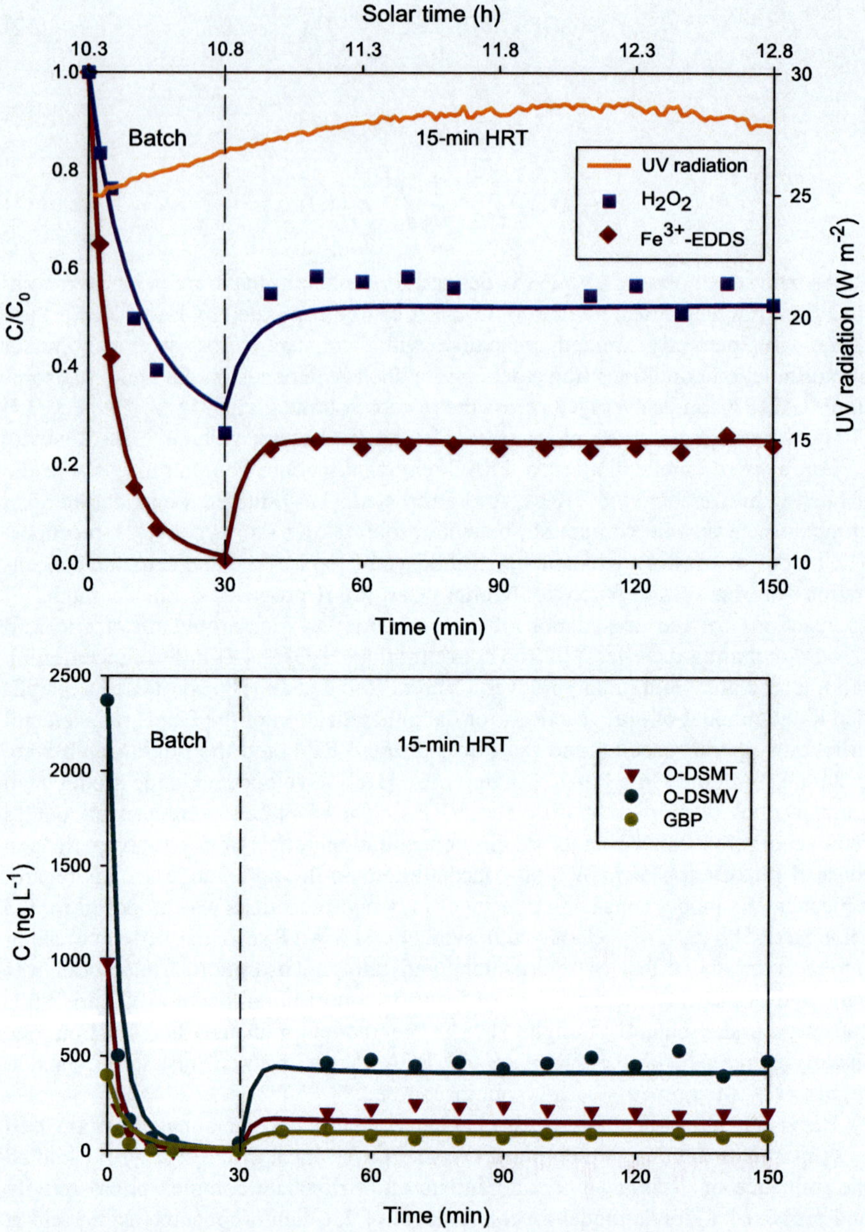

Fig. 4 Model predictions of the performance of a 5-cm-deep RPR operated in continuous flow mode at 15 min HRT to treat MWWTP secondary effluent containing PhACs (O-desmethyltramadol (O-DSMT), O-desmethylvenlafaxine (O-DSMV), gabapentin (GBP)) by solar photo-Fenton with 0.1 mM Fe^{3+}-EDDS and 0.88 mM H_2O_2. The reactor started in batch mode for 30 min and then continuous flow lasted for 120 min. Solid lines represent model predictions and points, experimental data

model of 18 reactions has been reported for Fe^{3+}-EDDS system, following the removal of the 3 most abundant PhACs detected in 5 different actual secondary MWWTP effluents (gabapentin, *O*-desmethyltramadol and *O*-desmethylvenlafaxine). The model constitutes a tool for the design of continuous raceway pond reactors (RPRs) for PhAC removal by solar photo-Fenton process [110]. The treatment capacity of the RPR in terms of volume of water treated per unit of reactor surface and time was successfully predicted as a function of the liquid depth and the hydraulic residence time (HRT), taking into account the availability of solar UV radiation. Figure 4 shows the model predictions compared with the experimental data during the treatment of an actual MWWTP secondary effluent in a 5-cm-deep RPR at pilot plant scale. For instance, the reactor's surface required to treat 100 m^3/day can be calculated as follows: at 15 min of HRT, the treatment capacity would be 600 L/m^2/day for 6 h of continuous flow operation per day. The ratio 100/0.6 gives a reactor surface of 167 m^2. The RPR volume for 5 cm of liquid depth is 8.35 m^3.

6 Concluding Remarks

Application of SAOPs to PhAC elimination in MWWTP secondary effluents must procure beneficial impact on the environment, public health and a greener economy. However, some barriers still need to be overcome, such as the design of new photoreactors specifically for the degradation of contaminants present in the range of micro-gramme per litre; optimization of operating conditions, in particular, solar radiation-catalyst interaction; and long-term reliability of solar operation. Industrial and market acceptance of SAOPs will increase considerably if some of these issues are correctly addressed in the near future. Total costs of best available technologies (BAT), already applied at actual scale in a few countries for PhAC elimination from treated MWWTP secondary effluents, have been informed to range from 0.10 to 0.15 €/m^3 for ~50,000 population equivalent. Capitalized investment-related and operation costs, for the given life cycle, are stated by Rizzo et al. [48], decreasing further with increasing plant size. Therefore, development of SAOPs for PhAC removal should consider these figures to gain a market niche.

Elimination of PhACs in large water volumes needs mild operating conditions (low catalyst and oxidant concentrations). Therefore, environmental impact evaluation (by Life Cycle Assessment) is always crucial for a proper selection of the technology to be applied. For that, validation of the optimized treatments at real conditions using advanced analytical approaches based on mass spectrometry and confident (reproducible) kinetic models are needed.

Acknowledgements I. Oller and S. Malato wish to thank the Spanish Ministry of Economy and Competitiveness for funding under the ECOSAFEFARMING Project (International Joint Programming Actions, reference: PCIN-2017-005) and 2016 Water and FACCE JPIs Joint Call. J. Giménez wish to thank the Spanish Ministry of Economy and Competitiveness (project CTQ2017-86466-R,

MINECO/FEDER, UE) and AGAUR-Generalitat de Catalunya (project 2017SGR-131) for funds received. A. Agüera and J.A. Sánchez Pérez wish to thank contribution of the LIFE ULISES project funded by the European Union under the LIFE Financial Programme Grant Agreement No. LIFE18 ENV/ES/000165.

References

1. Gogate PR, Pandit AB (2004) A review of imperative technologies for WW treatment I: oxidation technologies at ambient conditions. Adv Environ Res 8:501–551. https://doi.org/10.1016/S1093-0191(03)00032-7
2. Malato S, Fernández-Ibáñez P, Maldonado MI, Blanco J, Gernjak W (2009) Decontamination and disinfection of water by solar photocatalysis: recent overview and trends. Catal Today 147:1–59. https://doi.org/10.1016/j.cattod.2009.06.018
3. Glaze W, Kang JW, Chapin DH (1987) The chemistry of water treatment processes involving ozone, hydrogen peroxide and ultraviolet radiation. Ozone Sci Eng 9:335–352. https://doi.org/10.1080/01919518708552148
4. Ahmed MM, Chiron S (2014) Solar photo-Fenton like using persulphate for carbamazepine removal from domestic WW. Water Res 48:229–236. https://doi.org/10.1016/j.watres.2013.09.033
5. Dillert R, Cassano AE, Goslich R, Bahnemann D (1999) Large scale studies in solar catalytic WW treatment. Catal Today 54:267–282
6. Malato S, Blanco J, Vidal A, Richter C (2002) Photocatalysis with solar energy at a pilot-plant scale: an overview. Appl Catal B Environ 37:1–15
7. Anastas P, Eghbali N (2010) Green chemistry: principles and practice. Chem Soc Rev 39:301–312. https://doi.org/10.1039/b918763b
8. Collins TJ (1997) In green chemistry, Macmillan encyclopedia of chemistry, vol 2. Simon and Schuster, Macmillan, New York, pp 691–697
9. Bauer R, Waldner G, Fallmann H, Hager S, Klare M, Krutzler T (1999) The photo-Fenton reaction and the TiO2/UV process for waste water treatment-novel developments. Catal Today 53:131–144
10. Giménez J, Esplugas S, Malato S, Peral J (2019) Economic assessment and possible industrial application of a (photo) catalytic process: a case study. In: Marci G, Palmisano L (eds) Heterogeneous photocatalysis: relationships with heterogeneous catalysis and perspectives. Elsevier, Amsterdam, pp 235–267
11. Muñoz I, Peral J, Ayllón JA, Malato S, Passarinho P, Domènech X (2006) Life cycle assessment of a coupled solar photocatalytic-biological process for WW treatment. Water Res 40:3533–3540
12. Giménez J, Bayarri B, González O, Malato S, Peral J, Esplugas S (2015) Advanced oxidation processes at laboratory scale: environmental and economic impacts. ACS Sustain Chem Eng 3:3188–3196. https://doi.org/10.1021/acssuschemeng.5b00778
13. Gallego-Schmid A, Tarpani RRZ, Miralles-Cuevas S, Cabrera-Reina A, Malato S, Azapagic A (2020) Environmental assessment of solar photo-Fenton processes in combination with nanofiltration for the removal of micro-contaminants from real. Sci Total Environ 650:2210–2220. https://doi.org/10.1016/j.scitotenv.2018.09.361
14. Cabrera Reina A, Miralles-Cuevas S, Cornejo L, Pomares L, Polo J, Oller I, Malato S (2020) The influence of location on solar photo-Fenton: process performance, photoreactor scaling-up and treatment cost. Renew Energy 145:1890–1900
15. Fujishima A, Honda K (1972) Electrochemical photolysis of water at a semiconductor electrode. Nature 238:37–38

16. Borgarello E, Kiwi J, Pelizzetti E, Visca M, Grätzel M (1981) Photochemical cleavage of water by photocatalysis. Nature 289(5794):158–160
17. Herrmann JM, Disdier J, Mozzanega MN, Pichat P (1979) Heterogeneous photocatalysis: in situ photoconductivity study of TiO2 during oxidation of isobutane into acetone. J Catal 60 (3):369–377
18. Ollis DF, Al-Ekabi H (eds) (1993) Photocatalytic purification and treatment of water and air. Elsevier, New York. https://doi.org/10.1016/j.apcatb.2012.05.036
19. Marcí G, Palmisano L (eds) (2019) Heterogeneous photocatalysis. Relationships with homogeneous catalysis and perspectives. Elsevier, Amsterdam
20. Awfa D, Ateia M, Johnson MS, Yoshimura C (2018) Photodegradation of pharmaceuticals and personal care products in water treatment using carbonaceous-TiO2 composites: a critical review of recent literature. Water Res 142:26–45. https://doi.org/10.1016/j.watres.2018.05.036
21. Fagan R, McCormack DE, Dionysiou DD, Pillai SC (2016) A review of solar and visible light active TiO2 photocatalysis for treating bacteria, cyanotoxins and contaminants of emerging concern. Mater Sci Semicond Process 42:2–14. https://doi.org/10.1016/j.mssp.2015.07.052
22. Kumar A, Khan M, He J, Lo IMC (2020) Recent developments and challenges in practical application of visible-light-driven TiO2-based heterojunctions for PPCP degradation: a critical review. Water Res 170:115356. https://doi.org/10.1016/j.watres.2019.115356
23. Frank SN, Bard AJ (1977) Heterogeneous Photocatalytic oxidation of cyanide and sulfite in aqueous solutions at semiconductor powders. J Phys Chem 81(15):1484–1488
24. Fujihira M, Satoh Y, Osa T (1981) Heterogeneous photocatalytic oxidation of aromatic compounds on TiO2. Nature 293(5829):206–208
25. Ollis DF, Pelizzetti E, Serpone N (1991) Photocatalyzed destruction of water contaminants. Environ Sci Technol 25(9):1522–1529
26. Curcó D, Giménez J, Marco P (1996) Photocatalytic degradation of phenol: comparison between pilot plant scale and laboratory results. Sol Energy 56(5):387–400
27. Malato S, Blanco J, Richter C, Curcó D, Giménez J (1997) Low-concentrating CPC collectors for photocatalytic water detoxification. Comparison with a medium concentrating solar collector. Water Sci Technol 35(4):157–164
28. Matthews RW, McEvoy SR (1992) Destruction of phenol in water with sun, sand, and photocatalysis. Sol Energy 49(6):507–513
29. Pelizzetti E, Carlin V, Minero C, Pramauro E, Vincenti M (1992) Degradation pathways of atrazine under solar light and in the presence of TiO2 colloidal particles. Sci Total Environ 123–124:161–169
30. Calza P, Sakkas VA, Medana C, Baiocchi C, Dimou A, Pelizzetti E, Albanis T (2006) Photocatalytic degradation study of diclofenac over aqueous TiO2 suspensions. Appl Catal B Environ 67:197–205. https://doi.org/10.1016/j.apcatb.2006.04.021
31. Molinari R, Pirillo F, Loddo V, Palmisano L (2006) Heterogeneous photocatalytic degradation of pharmaceuticals in water by using polycrystalline TiO2 and a nanofiltration membrane reactor. Catal Today 118:205–213. https://doi.org/10.1016/j.cattod.2005.11.091
32. Méndez-Arriaga F, Maldonado MI, Giménez J, Esplugas S, Malato S (2009) Abatement of ibuprofen by solar photocatalysis process: enhancement and scale up. Catal Today 144:112–116. https://doi.org/10.1016/j.cattod.2009.01.028
33. Pérez-Estrada L, Malato S, Agüera A, Fernández-Alba AR (2007) Degradation of dipyrone and its main intermediates by solar AOPs identification of intermediate products and toxicity assessment. Catal Today 129:207–214. https://doi.org/10.1016/j.cattod.2007.08.008
34. Bolton JR, Bircher KG, Tumas W, Tolman CA (2001) Figures-of-merit for the technical development and application of advanced oxidation technologies for both electric- and solar-driven systems (IUPAC technical report). Pure Appl Chem 73(4):627–637
35. Lado Ribeiro AR, Moreira NFF, Li Puma G, Silva AMT (2019) Impact of water matrix on the removal of micropollutants by advanced oxidation technologies. Chem Eng J 363:155–173. https://doi.org/10.1016/j.cej.2019.01.080

36. Acosta-Herazo R, Monterroza-Romero J, Mueses MA, Machuca-Martinez F, Li Puma G (2016) Coupling the six flux absorption-scattering model to the Henyey-Greenstein scattering phase function: evaluation and optimization of radiation absorption in solar heterogeneous photoreactors. Chem Eng J 302:86–96. https://doi.org/10.1016/j.cej.2016.04.127
37. Cassano AE, Alfano OM (2000) Reaction engineering of suspended solid heterogeneous photocatalytic reactors. Catal Today 58(2–3):167–197. https://doi.org/10.1016/S0920-5861(00)00251-0
38. Curcó D, Giménez J, Addardak A, Cervera-March S, Esplugas S (2002) Effects of radiation absorption and catalyst concentration on the photocatalytic degradation of pollutants. Catal Today 76:177–188. https://doi.org/10.1016/S0920-5861(02)00217-1
39. Pignatello JJ, Oliveros E, MacKay A (2006) Advanced oxidation processes for organic contaminant destruction based on the Fenton reaction and related chemistry. Crit Rev Environ Sci Technol 36:1–84. https://doi.org/10.1080/10643380500326564
40. Gupta VK, Ali I, Saleh TA, Nayak A, Agarwal S (2012) Chemical treatment technologies for waste-water recycling-an overview. RSC Adv 2:6380–6388. https://doi.org/10.1039/C2RA20340E
41. Oppenländer T (2003) Photochemical purification of water and air: advanced oxidation processes (AOPs)-principles, reaction mechanisms, reactor concepts. Wiley, Weinheim. https://doi.org/10.1002/9783527610884
42. Clarizia L, Russo D, di Somma I, Marotta R, Andreozzi R (2017) Homogeneous photo-Fenton processes at near neutral pH: a review. Appl Catal B Environ 209:358–371. https://doi.org/10.1016/j.apcatb.2017.03.011
43. Bui XT, Vo TPT, Ngo HH, Guo WS, Nguyen TT (2016) Multicriteria assessment of advanced treatment technologies for micropollutants removal at large-scale applications. Sci Total Environ 563-564:1050–1067. https://doi.org/10.1016/j.scitotenv.2016.04.191
44. Luo Y, Guo W, Ngo HH, Nghiem LD, Hai FI, Zhang J, Liang S, Wang XC (2014) A review on the occurrence of micropollutants in the aquatic environment and their fate and removal during WW treatment. Sci Total Environ 473:619–641. https://doi.org/10.1016/j.scitotenv.2013.12.065
45. Pal A, Gin KY-H, Lin AY-C, Reinhard M (2010) Impacts of emerging organic contaminants on freshwater resources: review of recent occurrences, sources, fate and effects. Sci Total Environ 408:6062–6069. https://doi.org/10.1016/j.scitotenv.2010.09.026
46. Patel M, Kumar R, Kishor K, Mlsna T, Pittman CUJ, Mohan D (2019) Pharmaceuticals of Emerging Concern in aquatic systems: chemistry, occurrence, effects, and removal methods. Chem Rev 119:3510–3673. https://doi.org/10.1021/acs.chemrev.8b00299
47. Alharbi SK, Price WE (2017) Degradation and fate of pharmaceutically active contaminants by advanced oxidation processes. Curr Pollut Rep 3:268–280. https://doi.org/10.1007/s40726-017-0072-6
48. Rizzo L, Malato S, Antakyali D, Beretsou VG, Đolić MB, Gernjak W, Heath E, Ivancev-Tumbas I, Karaolia P, Lado Ribeiro AR, Mascolo G, McArdell CS, Schaar H, Silva AMT, Fatta-Kassinos D (2019) Consolidated vs new advanced treatment methods for the removal of contaminants of emerging concern from urban WW. Sci Total Environ 655:986–1008. https://doi.org/10.1016/j.scitotenv.2018.11.265
49. Rodríguez-Narváez O, Rodríguez OM, Peralta-Hernández JM, Goonetilleke A, Bandala E (2017) Treatment technologies for emerging contaminants in water: a review. Chem Eng J 323:361–380. https://doi.org/10.1016/j.cej.2017.04.106
50. De Luca A, Dantas RF, Esplugas S (2014) Assessment of iron chelates efficiency for photo-Fenton at neutral pH. Water Res 61:232–242. https://doi.org/10.1016/j.watres.2014.05.033
51. Zhan Y, Zhou M (2019) A critical review of the application of chelating agents to enable Fenton and Fenton-like reactions at high pH values. J Hazard Mater 362:436–450. https://doi.org/10.1016/j.jhazmat.2018.09.035

52. Klamerth N, Malato S, Maldonado MI, Agüera A, Fernández-Alba AR (2011) Modified photo-Fenton for degradation of emerging contaminants in municipal WW effluents. Catal Today 161:241–246. https://doi.org/10.1021/es903455p
53. Durán A, Montegudo JM, San Martín I (2018) Operation costs of the solar photo-catalytic degradation of pharmaceuticals in Water: a mini-review. Chemosphere 211:482–488. https://doi.org/10.3390/nano9091194
54. Huang W, Brigante M, Wu F, Hanna K, Mailhot G (2012) Development of a new homogenous photo-Fenton process using Fe(III)-EDDS complexes. J Photochem Photobiol A Chem 239:17–23. https://doi.org/10.1021/es304502y
55. De la Obra I, Ponce-Robles L, Miralles-Cuevas S, Oller I, Malato S, Sánchez Pérez JA (2017) Microcontaminant removal in secondary effluents by solar photo-Fenton at circumneutral pH in raceway pond reactors. Catal Today 287:10–14. https://doi.org/10.1016/j.cattod.2016.12.028
56. Klamerth N, Malato S, Agüera A, Fernández-Alba A (2013) Photo-Fenton and modified photo-Fenton at neutral pH for the treatment of emerging contaminants in WW treatment plant effluents: a comparison. Water Res 47:833–840. https://doi.org/10.1016/j.watres.2012.11.008
57. Zhang G, Ji S, Xi B (2006) Feasibility study of treatment of amoxicillin wastewater with a combination of extraction, Fenton oxidation and reverse osmosis. Desalination 196(1–3):32–42. https://doi.org/10.1016/j.desal.2005.11.018
58. Miralles-Cuevas S, Oller I, Sánchez-Pérez JA, Malato S (2014b) Removal of pharmaceuticals from MWTP effluent by nanofiltration and solar photo-Fenton using two different iron complexes at neutral pH. Water Res 64:23–31. https://doi.org/10.1016/j.watres.2014.06.032
59. Miralles-Cuevas S, Oller I, Ruiz Aguirre A, Sánchez Pérez JA, Malato Rodríguez S (2014a) Removal of pharmaceuticals at microg L-1 by combined nanofiltration and mild solar photo-Fenton. Chem Eng J 239:68–74. https://doi.org/10.1016/j.cej.2013.10.047
60. Trovó AG, Pupo Nogueira RF, Agüera A, Fernandez-Alba AR, Malato S (2011) Degradation of the antibiotic amoxicillin by photo-Fenton process – chemical and toxicological assessment. Water Res 45:1394–1402. https://doi.org/10.1016/j.watres.2010.10.029
61. Ballesteros SG, Costante M, Vicente R, Mora M, Amat AM, Arques A, Einschlag FSG (2017) Humic-like substances from urban waste as auxiliaries for photo-Fenton treatment: a fluorescence EEM-PARAFAC study. Photochem Photobiol Sci 16:38–45. https://doi.org/10.1039/C6PP00236F
62. Papoutsakis S, Pulgarin C, Oller I, Sánchez-Moreno R, Malato S (2016) Enhancement of the Fenton and photo-Fenton processes by components found in WW from the industrial processing of natural products: the possibilities of cork boiling WW reuse. Chem Eng J 304:890–896. https://doi.org/10.1016/j.cej.2016.07.021
63. Ruíz-Delgado A, Roccamante MA, Oller I, Agüera A, Malato S (2019) Natural chelating agents from olive mill WW to enable photo-Fenton-like reactions at natural pH. Catal Today 328:281–285. https://doi.org/10.1016/j.cattod.2018.10.051
64. Wu Y, Brigante M, Dong M, De Sainte-Claire P, Mailhot G (2014) Toward a better understanding of Fe(III)–EDDS photochemistry: theoretical stability calculation and experimental investigation of 4-tert-butylphenol degradation. J Phys Chem A 118:396–403. https://doi.org/10.1021/jp409043e
65. Vandevivere PC, Saveyn H, Verstraete W, Feijtel TCJ, Schowanek DR (2001) Biodegradation of metal-[S,S]-EDDS. Complexes. Environ Sci Technol 35:1765–1770. https://doi.org/10.1021/es0001153
66. Gutierrez-Mata AG, Velazquez-Martínez S, Álvarez-Gallegos A, Ahmadi M, Hernández-Pérez JA, Ghanbari F, Silva-Martínez S (2017) Recent overview of solar Photocatalysis and solar photo-Fenton processes for WW treatment. Hindawi Int J Photoenergy:27. https://doi.org/10.1155/2017/8528063
67. Souza BM, Marinho BA, Moreira FC, Dezotti MW, Boaventura RAR, Vilar VJP (2017) Photo-Fenton oxidation of 3-amino-5-methylisoxazole: a by-product from biological

breakdown of some pharmaceutical compounds. Environ Sci Pollut Res 24:6195–6204. https://doi.org/10.1007/s11356-015-5690-1

68. Ahmed MM, Brienza M, Goetz V, Chiron S (2014) Solar photo-Fenton using peroxymonosulfate for organic micropollutants removal from domestic WW: comparison with heterogeneous TiO2 photocatalysis. Chemosphere 117:252–256. https://doi.org/10.1016/j.chemosphere.2014.07.046
69. Mirzaei A, Chen Z, Haghighat F, Yerushalmi L (2017) Removal of pharmaceuticals from water by homo/heterogonous Fenton-type processes- a review. Chemosphere 174:665–688. https://doi.org/10.1016/j.chemosphere.2017.02.019
70. Costa EP, Roccamante M, Amorim CC, Oller I, Sánchez Pérez JA, Malato S (2020) New trend on open solar photoreactors to treat micropollutants by photo-Fenton at circumneutral pH: increasing optical pathway. Chem Eng J 385:123982. https://doi.org/10.1016/j.cej.2019.123982
71. Cuervo Lumbaque E, Araújo DS, Klein TM, Lopes-Tiburtius ER, Argüello J, Sirtori C (2019) Solar photo-Fenton-like process at neutral pH: Fe(III)-EDDS complex formation and optimization of experimental conditions for degradation of pharmaceuticals. Catal Today 328:259–266. https://doi.org/10.1016/j.cattod.2019.01.006
72. Koltsakidou A, Antonopoulou M, Sykiotou M, Evgenidou E, Konstantinou I, Lambropoulou D (2017) Photo-Fenton and Fenton-like processes for the treatment of the antineoplastic drug 5-fluorouracil under simulated solar radiation. Environ Sci Pollut Res 24:4791–4800. https://doi.org/10.1007/s11356-016-8138-3
73. Alalm MG, Tawfik A, Ookawara S (2015) Degradation of four pharmaceuticals by solar photo-Fenton process: kinetics and costs estimation. J Environ Chem Eng 3:46–51. https://doi.org/10.1016/j.jece.2014.12.009
74. Malato S, Blanco J, Maldonado MI, Fernandez-Ibañez P, Alarcon D, Collares M, Farinha J, Correia J (2004) Engineering of solar photocatalytic collectors. Sol Energy 77:513–524
75. Hinojosa Guerra MM, Oller Alberola I, Malato Rodriguez S, Agüera López A, Acevedo Merino A, Quiroga Alonso JM (2019) Oxidation mechanisms of amoxicillin and paracetamol in the photo-Fenton solar process. Water Res 156:232–240. https://doi.org/10.1016/j.watres.2019.02.055
76. Ioannou-Ttofa L, Raj S, Prakash H, Fatta-Kassinos D (2019) Solar photo-Fenton oxidation for the removal of ampicillin, total cultivable and resistant E. coli and ecotoxicity from secondary-treated WW effluents. Chem Eng J 355:91–102. https://doi.org/10.1016/j.cej.2018.08.057
77. Chisti Y (2007) Biodiesel from microalgae. Biotechnol Adv 25:294–306
78. Davis R, Aden A, Pienkos PT (2011) Techno-economic analysis of autotrophic microalgae for fuel production. Appl Energy 88:3524–3531
79. Soriano-Molina P, Plaza-Bolaños P, Lorenzo A, Agüera A, García-Sánchez JL, Malato S, Sánchez Pérez JA (2019) Assessment of solar raceway pond reactors for removal of contaminants of emerging concern by photo-Fenton at circumneutral pH from very different municipal WW effluents. Chem Eng J 366:141–149. https://doi.org/10.1016/j.cej.2019.02.074
80. Miralles-Cuevas S, Oller I, Agüera A, Ponce-Robles L, Sánchez Pérez JA, Malato S (2015) Removal of microcontaminants from MWTP effluents by combination of membrane technologies and solar photo-Fenton at neutral pHs. Catal Today 252:78–83. https://doi.org/10.1016/j.cattod.2014.11.015
81. Miralles-Cuevas S, Oller I, Agüera A, Sánchez Pérez JA, Malato S (2017) Strategies for reducing cost by using solar photo-Fenton treatment combined with nanofiltration to remove microcontaminants in real municipal effluents: Toxicity and economic assessment. Chem Eng J 318:161–170. https://doi.org/10.1016/j.cej.2016.06.031
82. Della-Flora A, Wilde ML, Pinto IDF, Lima ÉC, Sirtori C (2020) Degradation of the anticancer drug flutamide by solar photo-Fenton treatment at near-neutral pH: identification of transformation products and in silico (Q)SAR risk assessment. Environ Res 183:109223. https://doi.org/10.1016/j.envres.2020.109223
83. De Melo Santos MM, da Silva TD, de Lucena ALA, Napoleão DC, Duarte MMMB (2020) Degradation of Ketoprofen, Tenoxicam, and meloxicam drugs by photo-assisted peroxidation

and photo-Fenton processes: identification of intermediates and toxicity study. Water Air Soil Pollut 231:35. https://doi.org/10.1007/s11270-020-4401-9
84. Majumder A, Gupta B, Gupta AK (2019) Pharmaceutically active compounds in aqueous environment: a status, toxicity and insights of remediation. Environ Res 176:108542. https://doi.org/10.1016/j.envres.2019.108542
85. Ioannou-Ttofa L, Foteinis S, Chatzisymeon E, Michael-Kordatou I, Fatta-Kassinos D (2017) Life cycle assessment of solar-driven oxidation as a polishing step of secondary-treated urban effluents. J Chem Technol Biotechnol 92:1315–1327. https://doi.org/10.1002/jctb.5126
86. Salazar LM, Grisales CM, Garcia DP (2019) How does intensification influence the operational and environmental performance of photo-Fenton processes at acidic and circumneutral pH. Environ Sci Pollut Res 26:4367–4380. https://doi.org/10.1007/s11356-018-2388-1
87. Tarpani RRZ, Azapagic A (2018) Life cycle environmental impacts of advanced WW treatment techniques for removal of pharmaceuticals and personal care products (PPCPs). J Environ Manag 215:258–272. https://doi.org/10.1016/j.jenvman.2018.03.047
88. Klamerth N, Malato S, Agüera A, Fernández-Alba AR, Mailhot G (2012) Treatment of municipal WW treatment plant effluents with modified photo-Fenton as a tertiary treatment for the degradation of micro pollutants and disinfection. Environ Sci Technol 46:2885–2892. https://doi.org/10.1021/es204112d
89. Rosal R, Rodríguez A, Perdigón-Melón JA, Petre A, García-Calvo E, Gómez MJ, Agüera A, Fernández-Alba AR (2010) Occurrence of emerging pollutants in urban WW and their removal through biological treatment followed by ozonation. Water Res 44:578–588. https://doi.org/10.1016/j.watres.2009.07.004
90. Campos-Mañas MC, Plaza-Bolaños P, Sánchez-Pérez JA, Malato S, Agüera A (2017) Fast determination of pesticides and other contaminants of emerging concern in treated WW using direct injection coupled to highly sensitive ultra-high performance liquid chromatography-tandem mass spectrometry. J Chromatogr A 1507:84–94. https://doi.org/10.1016/j.chroma.2017.05.053
91. Maia AS, Paíga P, Delerue-Matos C, Castro PML, Tiritan ME (2020) Quantification of fluoroquinolones in WWs by liquid chromatography-tandem mass spectrometry. Environ Pollut 259:113927. https://doi.org/10.1016/j.envpol.2020.113927
92. González-Antuña A, Domínguez-Romero JC, García-Reyes JF, Rodríguez-González P, Centineoc G, García Alonso JI, Molina-Díaz A (2013) Overcoming matrix effects in electrospray: quantitation of β–agonists in complex matrices by isotope dilution liquid chromatography–mass spectrometry using singly 13C-labeled analogues. J Chromatogr A 1288:40–47. https://doi.org/10.1016/j.chroma.2013.02.074
93. Schymanski EL, Jeon J, Gulde R, Fenner K, Ruff M, Singer HP, Hollender J (2014) Identifying small molecules via high resolution mass spectrometry: communicating confidence. Environ Sci Technol 48:2097–2098. https://doi.org/10.1021/es5002105
94. Bletsou AA, Jeon J, Hollender J, Archontaki E, Thomaidis NS (2015) Targeted and non-targeted liquid chromatography-mass spectrometric workflows for identification of transformation products of emerging pollutants in the aquatic environment. TrAC Trends Anal Chem 66:32e44. https://doi.org/10.1016/j.trac.2014.11.009
95. Kyoto University (2018) Pathpred: pathway prediction server [WWW document]. https://www.genome.jp/tools/pathpred
96. EAWAG (2018) EAWAG-BBD pathway prediction system [WWW document]. http://eawag-bbd.ethz.ch/predict
97. Jaén-Gil A, Buttiglieri G, Benito A, Gonzalez-Olmos R, Barceló D, Rodríguez-Mozaz S (2019) Metoprolol and metoprolol acid degradation in UV/H2O2 treated WWs: an integrated screening approach for the identification of hazardous transformation products. J Hazard Mater 380:120851. https://doi.org/10.1016/j.jhazmat.2019.120851
98. Schenone AV, Conte LO, Botta MA, Alfano OM (2015) Modeling and optimization of photo-Fenton degradation of 2,4-D using ferrioxalate complex and response surface methodology (RSM). J Environ Manag 155:177–183. https://doi.org/10.1016/j.jenvman.2015.03.028
99. Rivas G, Carra I, García Sánchez JL, Casas López JL, Malato S, Sánchez Pérez JA (2015) Modelling of the operation of raceway pond reactors for micropollutant removal by solar

photo-Fenton as a function of photon absorption. Appl Catal B Environ 178:210–217. https://doi.org/10.1016/j.apcatb.2014.09.015

100. Giannakis S, Hendaoui I, Rtimi S, Fürbringer JM, Pulgarin C (2017) Modeling and treatment optimization of pharmaceutically active compounds by the photo-Fenton process: the case of the antidepressant venlafaxine. J Environ Chem Eng 5:818–828. https://doi.org/10.1016/j.jece.2016.12.050
101. Cassano AE, Martin CA, Brandi RJ, Alfano OM (1995) Photoreactor analysis and design: fundamentals and applications. Ind Eng Chem Res 34:2155–2201. https://doi.org/10.1021/ie00046a001
102. Conte LO, Schenone AV, Alfano OM (2016) Photo-Fenton degradation of the herbicide 2,4-D in aqueous medium at pH conditions close to neutrality. J Environ Manag 170:60–69. https://doi.org/10.1016/j.jenvman.2016.01.002
103. Brandi RJ, Citroni MA, Alfano OM, Cassano AE (2003) Absolute quantum yields in photocatalytic slurry reactors. Chem Eng Sci 58:979–985. https://doi.org/10.1016/S0009-2509(02)00638-3
104. Andreozzi R, D'Apuzzo A, Marotta R (2000) A kinetic model for the degradation of benzothiazole by Fe3+-photo-assisted Fenton process in a completely mixed batch reactor. J Hazard Mater 80:241–257. https://doi.org/10.1016/S0304-3894(00)00308-3
105. Conte LO, Farias J, Albizzati ED, Alfano OM (2012) Photo-Fenton degradation of the herbicide 2,4-dichlorophenoxyacetic acid in laboratory and solar pilot-plant reactors. Ind Eng Chem Res 51:4181–4191. https://doi.org/10.1021/ie2023228
106. Cabrera Reina A, Santos-Juanes L, García Sánchez JL, Casas López JL, Maldonado Rubio MI, Li Puma G, Sánchez Pérez JA (2015) Modelling the photo-Fenton oxidation of the pharmaceutical paracetamol in water including the effect of photon absorption (VRPA). Appl Catal B Environ 166–167:295–301. https://doi.org/10.1016/j.apcatb.2014.11.023
107. Audino F, Conte LO, Schenone AV, Pérez-Moya M, Graells M, Alfano OM (2018) A kinetic study for the Fenton and photo-Fenton paracetamol degradation in an annular photoreactor. Environ Sci Pollut Res 26(5):4312–4323. https://doi.org/10.1007/s11356-018-3098-4
108. Sánchez Pérez JA, Soriano-Molina P, Rivas G, García Sánchez JL, Casas López JL, Fernández SJM (2017) Effect of temperature and photon absorption on the kinetics of micropollutant removal by solar photo-Fenton in raceway pond reactors. Chem Eng J 310:464–472. https://doi.org/10.1016/j.cej.2016.06.055
109. Conte LO, Schenone AV, Giménez BN, Alfano OM (2019) Photo-Fenton degradation of a herbicide (2,4-D) in groundwater for conditions of natural pH and presence of inorganic anions. J Hazard Mater 372:113–120. https://doi.org/10.1016/j.jhazmat.2018.04.013
110. Soriano Molina P, García Sánchez JL, Malato S, Plaza-Bolaños P, Agüera A, Sánchez Pérez JA (2019) On the design and operation of solar photo-Fenton open reactors for the removal of contaminants of emerging concern from WWTP effluents at neutral pH. Appl Catal B Environ 256:117801. https://doi.org/10.1016/j.apcatb.2019.117801

Advanced Electrochemical Processes for the Elimination of Pharmaceutical Compounds in Contaminated Waters

Jelena Radjenovic and Luis Baptista-Pires

Contents

1 Introduction 328
2 Mechanisms of Electrochemical Oxidation of Pharmaceutical Residues in Water 330
3 Limitations of State of the Art in Electrochemical Water Treatment 332
4 Electrochemical Treatment for the Removal of Pharmaceutical Residues from Real Contaminated Water 333
5 Three-Dimensional, Flow-Through Electrochemical Systems and Development of Electrochemical Membranes 338
6 Conclusions and Outlook 342
References 343

Abstract Electrochemical systems have often been investigated for the removal and degradation of pharmaceuticals in water, mainly focusing on their electrooxidation kinetics and transformation pathways and sometimes relying on sophisticated high-resolution mass spectrometry (HRMS) analysis. Most of these studies are based on ideal experimental designs of high electrolyte conductivity and high contaminant

The title of the original version of this chapter was corrected from "Chemical Processes for the Elimination of Pharmaceutical Compounds in Contaminated Waters" to "Advanced Electrochemical Processes for the Elimination of Pharmaceutical Compounds in Contaminated Waters". A correction to this chapter can be found at https://doi.org/10.1007/698_2021_757

J. Radjenovic (✉)
Catalan Institute for Water Research (ICRA), Girona, Spain

Catalan Institution for Research and Advanced Studies (ICREA), Barcelona, Spain
e-mail: jradjenovic@icra.cat

L. Baptista-Pires
Catalan Institute for Water Research (ICRA), Girona, Spain

University of Girona, Girona, Spain
e-mail: l.baptistapires@gmail.com

Sara Rodriguez-Mozaz, Paqui Blánquez Cano, and Montserrat Sarrà Adroguer (eds.), *Removal and Degradation of Pharmaceutically Active Compounds in Wastewater Treatment*, Hdb Env Chem (2021) 108: 327–348. DOI 10.1007/698_2020_689,

Published online: 13 October 2020

concentration, resulting in outputs that can hardly be extrapolated to a real case scenario. This chapter discusses a limited number of studies in which the performance of electrochemical systems in removing and degrading pharmaceutical residues was evaluated using real waste streams. The main mechanisms of electrochemical oxidation of pharmaceuticals are explained, with a particular emphasis on the recently reported participation of strong oxidant species, sulfate radicals, in electrooxidation of organic contaminants. Finally, as electrode and reactor design evolves toward three-dimensional geometries with reduced mass transfer limitations, significant attention has been paid to the progress made using flow-through, porous electrode materials for the electrooxidation of pharmaceuticals. Novel, electrocatalytically active, stable membrane and foam/sponge-type electrodes are thoroughly discussed as such materials represent an essential enabler of a wider-scale implementation of electrochemical treatment systems and their integration into water and wastewater treatment infrastructure.

Keywords Boron-doped diamond (BDD), Dimensionally stable anode (DSA), Electrochemical water treatment, PhACs, Reactive electrochemical membrane (REM)

1 Introduction

The ability of electrolysis to remove persistent organic and inorganic pollutants from water has triggered an enormous amount of research toward the end of the twentieth century. This research was intensified by the invention of a new material, boron-doped diamond (BDD) anode in 1987 by Pleskov [1], which was later patented for organic pollutant removal from water by Carey et al. [2]. Compared with other commercial electrode materials based on mixed metal oxide (MMO) coatings, BDD has demonstrated superior performance in terms of its oxidizing power and capacity to form strong oxidizing species, hydroxyl radicals ($HO^{\bullet}$) (Table 1). BDD electrodes also display excellent electrochemical and corrosion stability, longer service life, and stability under polarity reversal, which is crucial for field applications of electrochemical treatment to prevent scaling and fouling at the cathode [5]. Nevertheless,

Table 1 Onset potentials for oxygen evolution reaction (E_{OER}) expressed versus standard hydrogen electrode (SHE) and characteristics of typical anode materials

Electrode	E_{OER}, V/SHE	Conditions	Ref.
Ti/RuO_2	1.5	0.5 M H_2SO_4	[3]
Ti/IrO_2	1.5	0.5 M H_2SO_4	[3]
Ti/SnO_2	1.9	0.05 M H_2SO_4	[3]
Ti/PbO_2	1.9	0.05 M H_2SO_4	[3]
BDD	2.3	0.5 M H_2SO_4	[3]
Ti_4O_7	2.7–3.0[a]	3.7 M H_2SO_4	[4]

[a]No generic OER as oxygen evolution strongly depends on the pore size distribution of Ti_4O_7, where decrease in pore size shifts the OER to more positive values

BDD suffer from two major limitations: (1) synthesis based on chemical vapor deposition (CVD) or high-pressure high-temperature (HPHT) process keeps the price of BDD electrodes at ~6,000 € m^{-2} [6] and makes it difficult to upscale the technology due to low production rates, and (2) anodic polarization of BDD electrode results in termination with various functional groups (e.g., hydroxyl, carbonyl, carboxyl) and leads to a dramatic increase in surface and charge transfer resistance, thus increasing the electrode potentials (i.e., cell voltage) and consequently the energy consumption. Furthermore, BDD electrodes can be produced only in plate (Si, Nb) and mesh (Nb) geometries. It is difficult to produce them in three-dimensional (3D) forms that would offer high specific surface area and thus minimized mass transfer limitations and lower energy consumption for the degradation of trace organic pollutants.

As an alternative to pricey BDD anodes, MMO-type electrodes based on the oxides of ruthenium (RuO_2), iridium (IrO_2), lead (PbO_2), and tin (SnO_2) coated on titanium have been intensively investigated for the removal of organic contaminants from water. In particular, SnO_2-based electrodes, often doped with antimony (Sb_2O_5) to increase the coating conductivity, have drawn significant attention due to their large overpotential for oxygen evolution reaction (OER) and thus capability to efficiently form $HO^{\bullet}$ (Table 1) [7]. The commercial fabrication of Ti/SnO_2-Sb_2O_5-coated titanium electrodes is mostly based on thermal decomposition process and has been used to produce mesh and plate-type geometries. However, Ti/SnO_2-Sb_2O_5 anodes are characterized by the short service life and low corrosion stability, which can be somewhat improved by the interlayer insertion (e.g., IrO_2), addition of polymers, and/or modification of Ti substrate, for example, by the insertion of a TiO_2 nanotube array (NTA) layer [8]. Mesh-type geometries still have limited specific surface area for the degradation of pollutants, although significant improvements in the electrical efficiency in low-conductivity electrolytes can be achieved by using solid polymer electrolytes (e.g., proton exchange membranes) compressed between the mesh anode and cathode [9].

More recently, Magnéli phase materials based on sub-stoichiometric titanium oxide, Ti_4O_7, have gained attention due to their high electric conductivity, comparable to that of graphite, high specific surface area and high stability toward corrosion [10]. The lab-scale fabrication costs were determined to be significantly lower compared with BDD, although the cost and feasibility of a large-scale production of Ti_4O_7 anodes is yet to be determined given that its manufacture involves combustion under H_2 atmosphere at temperatures above 1,000°C [11]. The onset potentials for OER at Ti_4O_7 anodes were reported to vary between 2.7 and 3.0 V/SHE depending on the anode porosity, with smaller pore size yielding higher onset for OER (Table 1) [4].

2 Mechanisms of Electrochemical Oxidation of Pharmaceutical Residues in Water

Electrochemical degradation of trace pollutants is rarely achieved by direct electron transfer (DET), except in the cases when contaminants are resistant to $HO^{\bullet}$ and other oxidant species produced at the anode, and/or when the sorption of contaminant onto the electrode surface is favored. For example, in electrochemical oxidation of reverse osmosis concentrate at $Ti/RuIrO_2$ anode, antidepressant drug sertraline was the only compound among 28 pharmaceuticals and pesticides analyzed that was removed by direct electrolysis at low applied current densities, likely due to its high $\log K_{OW}$ (5.3) and thus better adsorption onto the anode surface [12]. To achieve degradation of organic contaminants, electrochemical water treatment is normally conducted at high current densities at which the contribution of DET is significantly lowered. Under these conditions, electrochemical oxidation of pharmaceutical residues and other trace organic pollutants is mediated by the strong oxidant species formed during electrolysis of water such as $HO^{\bullet}$ (Eq. 1), H_2O_2 (Eq. 2) and other reactive oxygen species, as well as $Cl_2/HClO/OCl^-$ (Eqs. 3 and 4), sulfate radical ($SO_4^{\bullet-}$), and persulfate, in the presence of Cl^- and SO_4^{2-} (Eqs. 5–8) [13]:

$$H_2O \rightarrow HO^{\bullet} + H^+ + e^- \tag{1}$$

$$2H_2O \rightarrow H_2O_2 + 2H^+ + 2e^- \tag{2}$$

$$2Cl \rightarrow Cl_2 + 2e^- \tag{3}$$

$$Cl_2 + H_2O \rightarrow HOCl + H^+ + Cl^- \tag{4}$$

$$SO_4^{2-} \rightarrow SO_4^{\bullet -} + e^- \tag{5}$$

$$HSO_4^- + HO^{\bullet} \rightarrow SO_4^{\bullet -} + H_2O \tag{6}$$

$$H_2SO_4 + HO^{\bullet} \rightarrow SO_4^{\bullet -} + H_3O^+ \tag{7}$$

$$SO_4^{\bullet -} + SO_4^{\bullet -} \rightarrow S_2O_8^{2-} \tag{8}$$

Electrochemical generation of reactive species such as $HO^{\bullet}$ is affected by the properties of the electrode material, i.e., OER overpotential [13]. The amount of $HO^{\bullet}$ generated in the system will determine not only the oxidation efficiency of organic contaminants but also the impact of inorganic and organic matrix on the treatment outcome. For example, electrooxidation of highly saline reverse osmosis brine stream at MMO anodes results in the increase in toxicity to *Vibrio fischeri* due to the formation of chlorinated and brominated organic byproducts [12]. In the case of BDD anode, no such increase in toxicity was noted, in spite of the observed increase in concentrations of adsorbable organic halogen (AOX), suggesting that organochlorine and organobromine compounds formed were likely comprised of lower molecular weight compounds, as a consequence of more of $HO^{\bullet}$ produced at the anode and thus enhanced bond breakage [14]. Another interesting feature of BDD anodes is their ability to form sulfate radical ($SO_4^{\bullet-}$) directly from the electrolysis of sulfate [15]. Although sulfate-based solutions have long been used as inert electrolytes, several studies conducted by Farhat et al. [15] demonstrated the participation of anodically formed $SO_4^{\bullet-}$ in electrooxidation of persistent pharmaceuticals such as carbamazepine, iodinated contrast media (ICM) diatrizoate and iopromide, and lower production of chlorinated organic byproducts when SO_4^{2-} and Cl^- are present simultaneously [16, 17]. Furthermore, BDD/SO_4^{2-} system exhibits oxidative capacitance during an intermittent power supply operation, with prolonged oxidation of diatrizoate observed in the power-off mode [18]. Application of 100 A m^{-2} in cycles of 5 min on/15 min off enabled a saving in the electrical energy consumption by ~35% compared to constant current experiment. Nevertheless, the exact mechanism behind this prolonged oxidation could not be determined and was hypothesized to involve radical chain propagators of $SO_4^{\bullet-}$ formed during the oxidation [19, 20]. Since then, research on electrochemically formed sulfate-based oxidants has intensified, given that in the currently applied oxidation methods based on $SO_4^{\bullet-}$, their formation from persulfate or peroxymonosulfate needs to be induced by not so practical methods based on UV, transition metals, heat, or high pH. Recent studies assigned the formation of $SO_4^{\bullet-}$ to heat activation of anodically formed persulfate that occurs at higher cell potentials due to ohmic heating [21, 22]. Some controversial studies hypothesized a non-radical electrochemical activation of the $S_2O_8^{2-}$ formed by anodic oxidation of sulfate [23, 24]. For example, Song et al. [24] hypothesized the formation of a transition state structure between persulfate and Ti/Pt anode, which enhanced the removal of carbamazepine, sulfamethoxazole, propranolol, and benzoic acid. Similar non-radical persulfate activation at the BDD anode was suggested to enhance the removal of antibiotic ciprofloxacin [23]. However, these observations were based on the decreased but not completely halted oxidation of pollutants in the presence of typical $HO^\bullet$ and $SO_4^{\bullet-}$ radical scavengers, *tert*-butanol, and methanol. Although these saturated alcohols are efficient quenchers of dissolved oxidant species, they do not react readily with electrode surfaces and alternative probes, for surface-bound, anodically produced radical oxidants may be better suited, such as allyl alcohols [25]. In a recent study by Cai et al. [26], direct electrogeneration of $SO_4^{\bullet-}$ from SO_4^{2-} at BDD anode was demonstrated using in situ electron paramagnetic resonance spectrometry. The authors observed continued degradation of di-n-butyl phthalate (DnBP), a typical endocrine disruptor, in the power-off phase during an intermittent power supply, which was assigned to the accumulation of $S_2O_8^{2-}$ and its reactivation to $SO_4^{\bullet-}$ via specific degradation intermediates. Similar to the study of Farhat et al. [18], electrical energy consumption was reduced by over 30% compared with continuous power supply operation. Continued generation of $SO_4^{\bullet-}$ was demonstrated by the spin trapping in the absence of applied potential [26] and was observed to lead to continued pollutant oxidation both inside the electrochemical system and in the sample vial upon sampling in the absence of quenchers [18]. Yet, oxidative capacitance and sustained oxidation of organic pollutants in the power-off phase of the BDD/SO_4^{2-} system need to be further investigated to elucidate the underlying propagation mechanisms [26].

The impact of $SO_4^{\bullet-}$, Cl_2/HOCl, $Cl_2^{\bullet-}$ and other oxidants is highly dependent on the nature and reactivity of a given pharmaceutically active pollutant. For instance, reaction rate constants of aromatic compounds with $Cl_2^{\bullet-}$ can vary from $k < 10^5$–$10^7\ M^{-1}\ s^{-1}$ for compounds containing substituents with negative inductive effect to $k = 10^7$–$10^9\ M^{-1}\ s^{-1}$ for compounds containing substituents with positive inductive effect [27]. In addition, electrooxidation of trace organic contaminants will be impacted by the organic and inorganic matrix of the contaminated water. For example, elevated concentrations of HCO_3^- can scavenge reactive species like chlorine radicals and reduce the transformation rates of electron-poor contaminants such as β-blockers [25]. Nevertheless, Cl^- present in virtually any water and wastewater will undoubtedly pose limitations to electrochemical treatment due to the formation of chlorinated organic byproducts from the trace organic pollutants and organic matrix, leading to an increase in the toxicity of the treated water [12, 28, 29]. Electrochlorination of the bulk organic matter yields a range of toxic low-molecular-weight byproducts such as trihalomethanes (THMs) and haloacetic acids (HAAs) [30–32]. Also, at BDD-, Ti_4O_7-, and Ti/SnO_2-type anodes with higher oxidizing power, Cl^- is easily oxidized to toxic chlorate and perchlorate that pose significant challenges to the discharge of the treated effluent [33–35]. Application of an electrochemical system as a single treatment step may be compromised depending on the characteristics of the organic and inorganic matrix of the contaminated water. In this case, downstream treatment options should be considered to reduce the effluent toxicity, such as granular activated carbon and/or reductive posttreatment.

3 Limitations of State of the Art in Electrochemical Water Treatment

Research on electrochemical oxidation for the removal of pharmaceuticals and other pollutants has been mainly performed using clean supporting electrolytes and often unrealistic conditions of both high contaminant concentration (i.e., mg L^{-1} range) and high electrolyte conductivity. This makes it very difficult to extrapolate the obtained results to a real case scenario in terms of the required energy consumption and reported transformation kinetics and pathways of a given compound. More importantly, one of the main bottlenecks of electrochemical treatment systems – generation of toxic chlorinated byproducts, both organic and inorganic – has not been sufficiently addressed as most researchers either used sulfate-, nitrate-, and perchlorate-based electrolytes, or they did not consider that the observed disappearance of trace pharmaceuticals in NaCl supporting electrolyte is achieved by electrochlorination. Another limitation in the field of electrochemical water treatment is dependence of the figures of merit on the experimental design, as it is difficult to compare batch versus continuous mode or batch beaker test versus a batch recirculation reactor. There is a great disparity in the way the obtained results are reported and expressed, with most authors opting for using the observed pseudo-first-order removal rate constants (k_{OBS}, h^{-1}) of target pollutants, in some cases normalized to the electrode surface area, and current efficiencies (CE, in %) calculated using Eq. (9):

$$CE = 100 \times nFV \frac{(C_0 - C_t)}{It} \quad (9)$$

where n is the number of electrons required for complete oxidation of a specific pollutant (i.e., to CO_2, H_2O, etc.), F is the Faraday constant (96,487 C mol^{-1}), V the electrolyte volume (L), C_0 and C_t are concentrations of compound measured at time $t = 0$ and time t (in mol L^{-1}), I is the applied current (A), and t is the duration of electrolysis (s). Yet, for processes targeting removal of pharmaceuticals and other trace pollutants, electric energy per order (E_{EO}) kWh m^{-3}) is a more appropriate figure of merit that expresses energy required to reduce the concentration of a contaminant by one order of magnitude in a unit volume of water (Eqs. 10 and 11) [13]:

$$E_{EO} = \frac{Pt}{V \log \frac{C_0}{C_t}} \text{ for batch processes} \quad (10)$$

$$E_{EO} = \frac{P}{q \log \frac{C_0}{C_t}} \text{ for continuous processes} \quad (11)$$

where P is the rated power of the system (W), V is volume of water (L) treated in time t (h) in a batch system, q is the flow rate applied in a continuous system (L h^{-1}), and C_0 and C_t are initial and final concentration of the target pollutant where C_t is an order of magnitude lower than C_0 (i.e., $C_0 = 10_{\times} C_t$ and thus log $\frac{C_0}{C_t}$ =1).

4 Electrochemical Treatment for the Removal of Pharmaceutical Residues from Real Contaminated Water

The first study on electrochemical transformation of a pharmaceutical compound was published in 1964 by Merkle and Discher [36]. They investigated electrolysis of an antipsychotic drug chlorpromazine at Pt electrode as a means of quantitative analysis. Although the potential of this technology for the degradation of organic compounds was known for a long time, it was not until the 2000s when reporting on the occurrence of pharmaceuticals in sewage, surface, and groundwater started to intensify that electrochemical water treatment gained attention. Given that oxidative degradation typically offers more complete degradation of organic pollutants, electrochemical oxidation has been in the focus of the scientific community much more than the cathodic electrolysis of contaminants.

The number of studies where electrochemical removal of pharmaceuticals was evaluated using real contaminated waste streams is very limited. In electrooxidation using Ti/RuIrO_2 anode of reverse osmosis (RO) brine amended with pharmaceuticals at low μg L^{-1} concentrations, most of the target pharmaceuticals completely disappeared when applying 461.5 Ah m^{-3} (cell energy consumption of 1.87 kWh m^{-3}) in continuous mode and somewhat lower charge of 437.9 Ah m^{-3} in batch mode (1.77 kWh m^{-3}), due to the higher accumulation of bulk chlorine species and thus faster transformation of target pollutants [12]. Antibiotics trimethoprim, enrofloxacin, norfloxacin, sulfadiazine and roxithromycin, ß-blockers metoprolol and tramadol, anticonvulsant carbamazepine, nonsteroidal anti-inflammatory drugs (NSAIDs) gemfibrozil and diclofenac, and others were rapidly removed at currents $\geq$150 A m^{-2}. Pharmaceuticals more recalcitrant to oxidation were NSAID ibuprofen, anti-epileptic drug phenytoin, and ICM iopromide, for which continuous mode achieved less than 70% removal, whereas in batch mode their removal was more complete. In electrooxidation of RO brine using BDD anode, all target pharmaceuticals disappeared after 2 h of electrolysis at 100 A m^{-2} (i.e., applied charge of 700 Ah m^{-3}), except for ibuprofen which was removed up to 55% [32]. The energy consumption for the removal of pharmaceuticals was not reported but at lower current density (50 A m^{-2}) 6.4 kWh m^{-3} was needed for complete removal of chemical oxygen demand (COD). Nevertheless, disappearance of target pollutants does not imply that the treatment was successful. The observed effluent toxicity to *Vibrio fischeri* was increased up to 37-fold to 53-fold in both continuous and batch mode electrooxidation of RO brine at Ti/RuIrO_2 anode, suggesting electrochlorination as the dominant oxidation mechanism [12]. Electrochlorination of organic pollutants in the presence of high chloride concentration was demonstrated in a subsequent study focused on identification of transformation products of ß-blocker metoprolol formed using Ti/RuIrO_2 and Ti/SnO_2-Sb_2O_5 anode [29]. Metoprolol and its degradation products were transformed to chloro, chloro-bromo, and bromo derivatives at both anodes in the presence of 1.5 g L^{-1} of Cl^- and 1.5 mg L^{-1} of Br^- in RO concentrate (Fig. 1). Metoprolol and its derivatives were transformed faster at Ti/SnO_2-Sb_2O_5 anode, likely due to its capacity to form $HO^•$ that participate in the oxidation besides active chlorine species (Cl_2/HOCl/OCl^-).

In another study, lower residual chlorine concentration was measured in electrooxidation of wastewater effluent spiked with 50 μM of carbamazepine than in the same effluent without the added pharmaceutical suggesting its reaction via electrochlorination mechanism, which was also confirmed by the increase in toxicity [37]. Also, in electrooxidation of hospital wastewater effluent using BDD anode, the monitored pharmaceuticals were removed faster than expected from their $HO^•$ radical rate constants due to their reaction with chlorine species [38]. Hydroxy-ibuprofen and hydroxy-diclofenac were more persistent than their parent compounds, indicating the need for including major human metabolites of pharmaceuticals in the evaluation of the treatment technology.

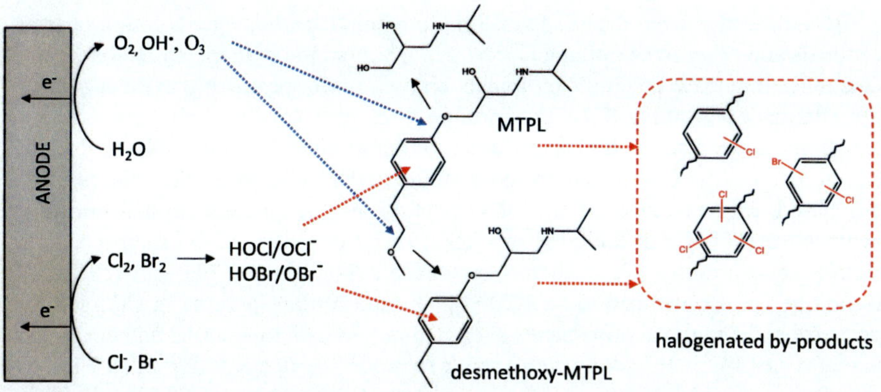

Fig. 1 Schematic representation of the competition between electrochlorination and electrobromination and oxidation by HO• and other reactive oxygen species formed at the anode, during electrooxidation of metoprolol in reverse osmosis brine using BDD anode. Adapted from [29]

Very similar results in terms of electrochlorination of trace organic contaminants were obtained in the treatment of secondary domestic sewage effluent that contained significantly lower Cl^- concentration than typically encountered in RO brines. In electrooxidation at BDD anode, most of the identified degradation products of pharmaceuticals carbamazepine, diclofenac, trimethoprim, and venlafaxine were chlorinated [39]. In electrooxidation of pharmaceuticals spiked in secondary municipal wastewater effluent at low μg L^{-1} concentrations, most of the pharmaceuticals were removed after 2 h of electrolysis at 196 A m^{-2} using BDD anode (energy consumption of anodic compartment of 0.64–0.85 kWh m^{-3}) at both acidic and neutral pH, with the exception of ibuprofen and ICM diatrizoate and iopromide [34]. The molecular structures of these compounds are characterized by the presence of electrophilic halogen groups with a negative inductive effect at the aromatic ring (e.g., ICM) or by the absence of nucleophilic substituents with an activating effect on the aromatic ring (e.g., ibuprofen), which diminishes the reactivity of these compounds toward HO•, chlorine, and other oxidant species. Ibuprofen was also reported to be the most recalcitrant pharmaceutical residue by Urtiaga et al., with 70% removed after 1 h of treatment at 100 A m^{-2} using BDD anode [40]. They applied ultrafiltration (UF) and reverse osmosis (RO) to treat secondary wastewater effluent, whereas electrooxidation of RO concentrate reduced the total load of pharmaceutical residues from 149 μg L^{-1} to less than 10 μg L^{-1}. The extent of electrochlorination can be decreased by applying lower current densities [12, 41]. In electrooxidation of hospital wastewater at BDD anode, application of 15.6 A m^{-2} yielded low amounts of residual chlorine (0.6 mg L^{-1}) and complete removal of bezafibrate, gemfibrozil, indomethacin, and sulfamethoxazole after 40 min of electrolysis, resulting in the energy consumption of 0.055 kWh m^{-3}. Electrochemically treated effluent was successfully subjected to biological posttreatment with activated sludge.

BDD has often been the anode material of choice for the degradation of pharmaceuticals and other trace pollutants in water. Yet, variations in the structure of BDD electrodes may have a significant impact on their performance and oxidizing capacity. Ultrananocrystalline (UNCD) diamond films with smaller grain sizes are more active and have a much higher surface to volume ratio than their microcrystalline counterparts [42]. Given that in most studies BDD electrodes are commercially purchased, characteristics of the BDD coating and parameters applied during its synthesis are often not included or considered. For example, optimum boron-to-carbon concentration ($C_{B\text{-}C}$) in the gas phase applied in hot filament (HF) CVD technique was determined to be 10,000 ppm, and further increase in $C_{B\text{-}C}$ led to a decrease in the removal efficiencies of carbamazepine, diclofenac, trimethoprim, and venlafaxine [39]. In this study, venlafaxine removal was decreased from 60% to 10% after 240 min of electrolysis at 500 A m^{-2}, when $C_{B\text{-}C}$ was increased from 10,000 to 15,000 ppm. Also, the use of BDD layer grown on structured Si substrate (i.e., Si substrate subjected to wet etching to increase its roughness) yielded nearly 100% of removal efficiency for all target pharmaceuticals, implying the determining impact of the surface morphology of the Si substrate on the performance of anodic oxidation process. Furthermore, the performance of BDD anodes is dependent on the nature of the surface termination groups [13, 43]. BDD anodes produced using CVD method are H-terminated with reduced surface functional groups (i.e., $=CH_2$, $\equiv$C–H), and anodic polarization, typically in concentrated sulfuric acid, generates oxygenated-functional groups such as $\equiv$C-OH, =C=O, and -COOH that are more hydrophilic, yield better anode wettability and act as electroactive sites for the generation of $HO^{\bullet}$ and other oxidants [42]. Electrooxidation of ICM diatrizoate at as-received Nb/BDD anode yielded k_{OBS} of 0.003 h^{-1} in $NaNO_3$ and Na_2SO_4 electrolyte of the same conductivity (i.e., 5.8 mS cm^{-1}), whereas conditioning of the anode increased the k_{OBS} to 0.16 h^{-1} (in $NaNO_3$) and 1.06 h^{-1} (in Na_2SO_4) electrolyte, due to the enabled formation of sulfate radicals ($SO_4^{\bullet-}$) [44].

To obtain a more complete oxidative degradation of pharmaceuticals, some studies used both anodic and cathodic half-cell of an electrochemical unit [45, 46]. For example, reductive deiodination of ICM iopromide at nano Pd-doped graphite felt cathode was combined with oxidative degradation at BDD anode to avoid the formation of persistent iodinated intermediates, typically formed due to the preferential attack of $HO^{\bullet}$ at the alkyl side chains [46]. Cathodic transformation of pharmaceuticals and other trace pollutants is a less attractive process due to the necessity to use noble metal catalyst such as Pd and Pt to achieve enough efficiency of reductive transformation of pollutants. The cathodic half-cell of an electrochemical system has typically been used to produce hydrogen peroxide from the reduction of oxygen, according to Eq. (12):

$$O_2 + 2H^+ + 2e^- \rightarrow H_2O_2 \quad (12)$$

H_2O_2 is a weak oxidant with $E°(H_2O_2/H_2O) = 1.763$ V/SHE (standard hydrogen electrode) in acidic solution and is activated to $HO^•$ by the addition of catalytic amounts of ferrous ion, Fe^{2+} (Eq. 13), in a process known as electro-Fenton [47]:

$$Fe^{2+} + H_2O_2 \rightarrow Fe^{3+} + HO^{\bullet} + OH^- \quad (13)$$

This enables a more complete oxidative degradation of trace organic pollutants that react with the $HO^•$ formed in the bulk liquid, in addition to the anodically formed $HO^•$ that remain in a proximity of the anode surface. Furthermore, electro-Fenton can be further enhanced by the addition of UV light in a so-called photoelectro-Fenton and solar photoelectro-Fenton processes. The performance of these systems in removing pharmaceutical residues from water has been extensively reviewed in literature [47, 48]. However, great majority of the research on electro-Fenton and photoelectro-Fenton processes has been performed using single component solutions, using high initial contaminant concentration (e.g., in mg L^{-1} range) and high conductivity electrolytes. Very few studies have investigated the process performance when dealing with real contaminated water. For example, electrochemical Fenton-based treatment was successfully applied for the removal of tetracaine, anesthetic drug, in synthetic and urban wastewater [49]. In another study, electro-Fenton was applied to remove NSAIDs diclofenac, naproxen, and ibuprofen from the sewage effluent [50]. In these and other studies, real wastewater had to be amended with 0.3–0.5 mM Fe^{2+} and acidified to pH 2.8–3 to enable the Fenton's reaction. The requirement for the addition of iron and acid and the need to neutralize the treated effluent represent major limitations for the upscale of Fenton-based electrochemical processes.

Due to their exceptional modularity, electrochemical systems can be easily combined with other physicochemical water treatment processes. For example, Zhi et al. [51] combined ozonation and Ti_4O_7 porous anode typically considered as a reactive electrochemical membrane (REM), to achieve a more complete removal of tetracycline, propranolol, ciprofloxacin, and sulfamethoxazole. The energy consumption of the O_3-REM-coupled process was 101.5 kWh $kg_{TOC}{}^{-1}$, significantly lower than those of O_3 (241.2 kWh $kg_{TOC}{}^{-1}$) and Ti_4O_7 REM (121.5 kWh $kg_{TOC}{}^{-1}$), for achieving similar TOC removal (around 77%). In another study, 3D electrochemical system with a packed bed of granular activated carbon (GAC) was combined with ozonation for the treatment of pharmaceutical wastewater [52]. Addition of ozone and electric field enabled in situ regeneration of GAC, enhanced TOC removal, and caused a significant decrease in toxicity to *Vibrio fischeri*.

5 Three-Dimensional, Flow-Through Electrochemical Systems and Development of Electrochemical Membranes

Poor current efficiency and consequently high energy consumption for the removal of pharmaceuticals and other trace organic pollutants have been a major impediment for the upscaling and application of electrochemical systems in water and wastewater treatment. Conventional 2D plate-type electrochemical reactors are characterized by high diffusion limitations and reactions with the electrogenerated $HO^{\bullet}$ limited to a thin boundary layer (~100 μm) at the anode surface. These diffusion limitations can be significantly reduced using flow-through, 3D systems, and porous electrodes that offer higher specific surface area and reduced mass transfer limitations for electrolytic degradation of trace contaminants [13, 43]. However, prerequisite for such systems is the existence of electrocatalytically active, stable porous 3D anode materials. Currently, BDD- and MMO-coated anodes cannot be developed in 3D geometries. Some researchers used carbon-based anodes coated with active catalysts, given that typical carbon-based materials have low electrocatalytic activity for both oxidative and reductive transformation of pollutants. For example, Zhai et al. used MnO catalyst confined in graphite felt for electro-assisted catalytic wet air oxidation of ibuprofen, tetracycline, and other pollutants [53]. Bubbling of oxygen at the bottom of the reactor and polarization of the anode at low potential of 1.0 V/SHE enabled chemisorption of O_2 on MnO, activation of the chemisorbed oxygen species via electrooxidation, and oxidation of organic pollutants by the activated chemisorbed oxygen species. The stability of Mn and other metal catalysts on carbon electrodes may be an issue when dealing with real waste streams due to their high reactivity and potential for leaching.

In an effort to reduce the energy consumption of electrochemical systems, researchers have proposed different porous electrode materials for advection-enhanced mass transfer of trace organic pollutants, based on carbon nanotubes (CNTs), Ti_4O_7, and graphene, among others (Table 2) [60]. Due to the ease and simplicity of producing porous thin films by filtration of CNTs-based ink, CNT-based conductive membranes have been widely investigated for filtration, adsorption, and electrochemical degradation of pharmaceuticals and other organic pollutants [61–63]. Liu et al. applied a multi-walled CNTs (MWCNTs)-based filter for adsorption and electrochemical degradation of tetracycline [54]. They obtained efficient removal of tetracycline with an oxidative flux of 0.025 mol h^{-1} m^{-2} at total cell potential of 2.5 V when using MWCNT anode and Ti cathode, whereas oxidative flux was calculated according to Eq. (14):

$$\text{Electrooxidation flux} = \frac{\left[Cin - Cout\left(\frac{\text{mol}}{\text{L}}\right)\right] * \text{flow rate}\left(\frac{L}{h}\right)}{\text{Effective filter area (m2)}} \tag{14}$$

Table 2 Overview of the studies on electrochemical removal of pharmaceuticals using porous, flow-through electrode materials

Reactor configuration		Operational conditions						
Anode	Cathode	Supporting electrolyte	Flux (L m^{-2} h^{-1})	E_{WE} (V/SHE)	Target pharmaceutical, initial concentration (µM)	% removal, oxidation flux (µmol h^{-1} m^{-2})	Energy consumption	Ref.
CNT (WE)	CNT	10 mM Na_2SO_4, –	52	0.7	Tetracycline, 200 µM	>99%, 3.80	0.7 kWh kg_{COD}^{-1}	[54]
Carboxilated CNTs (WE)	Ti	10 mM NaCl, pH 6	70.6	2.2	Ibuprofen, 96 µM	79–82%, 1.11	n.r.	[55]
MWCNTs PTFE (WE)	SS	20 mM NaCl, pH 3	176.5	3.2	Amoxicillin, 13 µM Sulfamethoxazole, 20 µM Ciprofloxacin, 15 µM	Amox. 98%; 0.07 Sulfameth. 90%; 0.10 Ciproflox. 75%, 0.06	n.r.	[56]
Graphene nanoplateles/CNT (WE)	Ti	10 mM Na_2SO_4, –	52.9	1.0	Tetracycline, 100 µM	88%, 1.66	n.r.	[57]
Pt	Graphene e-Fenton PTFE	100 mM Na_2SO_4, pH 5.9	20.8	−0.4	Florfenicol, 2.7 µM	90%, 0.11	111 ± 5.5 kWh kg^{-1}	[58]
BDD	Pd-Cu/Ti_4O_7 (WE)	100 mM KH_2PO_4, pH 4.7	300	−1.14	Sulfamethoxazole, 100 µM	96.1%, 0.32	0.33 ± 0.04 kWh m^{-3}	[59]

CNT carbon nanotube membrane electrode, *SS* stainless steel, *BDD* boron-doped diamond, *WE* working electrode

Oxidative flux was increased 2.3-fold when Ti cathode was replaced with MWCNT cathode, which also resulted in a decrease in total cell potential to 1 V due to a higher cathode surface area and thus larger distribution of current, resulting in lower cathodic current density. When the electrochemical filter was applied on drinking water and wastewater effluent samples spiked with tetracycline (0.2 mM), the target antibiotic was completely removed within short residence time (<2 s) and had an oxidative flux of 0.015 and 0.022 mol h^{-1} m^{-2}, respectively (energy consumption of 0.084 kWh m^{-3}). Similar MWCNT-based anode filter was employed for the removal of sulfamethoxazole, ciprofloxacin, and amoxicillin [56]. Although the target antibiotics were adsorbed initially to some extent, breakthrough and increase in effluent concentrations were observed already after the first 5–10 min of continuous operation. Nevertheless, application of anode potentials above 2 V vs Ag/AgCl yielded a steady 70–80% removal of target antibiotics, and the major degradation pathways involved hydroxylation, cleavage of sulfonamide bond and oxidation of aniline moiety (in sulfamethoxazole), oxidation of quinolone ring and the reaction at the piperazinyl ring (in ciprofloxacin), and fracture of the β-lactam ring and oxidation of sulfur atom (in amoxicillin). Besides pharmaceutical and other trace organic pollutant removal (e.g., pesticides, dyes), CNT-based porous electrochemical filters have also been explored for a variety of applications ranging from anti-fouling to viral and bacterial inactivation [64–66]. To enhance the CNT membrane performance and increase its electrocatalytic activity, several authors used atomic doping (boron and nitrogen), nanoparticle functionalization (addition of iron to induce Fenton reactions), and polymeric coatings (e.g., addition of polyaniline to increase conductivity and enhance anodic stability) [67, 68]. For example, electroactive membrane based on MWCNTs and MWCNTs doped with nano-zerovalent iron (nZVI) was applied for sorption and electrochemical oxidation of metoprolol at very low applied potentials, i.e., 0.75 and 1.2 V/SHE, achieving 55% and 97% removal, respectively [69]. nZVI was considered to enhance the electroactive membrane performance through Fenton-type reactions and liberation of Fe^{2+}, and main degradation products identified (e.g., α-hydroxy and α-keto metoprolol, O-desmethyl metoprolol, metoprolol-carboxylic acid) suggested $HO^{\bullet}$-induced transformation. Addition of iron functionalities was also employed to enable the electro-Fenton in a flow-through system using a porous cathodic membrane based on reduced graphene oxide (RGO) [58]. The authors reported 60% mineralization efficiency of antibiotic florfenicol RGO membrane functionalized by -$COOFe^{2+}$ groups and cathodically polarized to −0.6 V/SHE and no mineralization when using RGO membrane without the added iron functionalities. This was enabled by the simultaneous reduction of supplied O_2 to H_2O_2 and its activation to $HO^{\bullet}$ by the -$COOFe^{2+}$ groups introduced into the RGO layer. Despite the efforts made to stabilize CNT-based electrodes, their mechanical stability and electrode degradation remain a handicap for application of this material in electrochemical water treatment. In addition, CNTs are characterized by excessive cost, toxicity, as well as several intrinsic limitations, i.e., low $HO^{\bullet}$ production efficiency; low onset potential for OER (approximately 1 V vs SHE), and consequently low current efficiency for the degradation of pollutants; and continuous degradation of the anode membrane/electrode at anodic potentials higher than 4 V/SHE [70–72].

In the last 5 years, Magnéli phase Ti_4O_7 anodes have gained significant attention as a promising porous electrode material with high conductivity (1,000 mS cm^{-1}), high corrosion resistance (both to acids and bases), and higher stability under anodic polarization compared with CNTs electrodes [11, 72]. Porous Ti_4O_7 also have a large specific surface area (2.9×10^6 m^{-1}) relative to BDD electrodes (1.0×10^3 m^{-1}), although the electroactive surface area is typically significantly lower than the total surface area due to the poor accessibility of the electrolyte to a large portion of the electrode [6]. All these properties, along with the high capacity for generation of $HO^•$, have been exploited for electrooxidation of a range of trace organic pollutants, including antibiotics and other pharmaceuticals [59, 60]. The manufacturing procedure of Ti_4O_7 enables the production of flow-through materials with controlled porosity. The general fabrication method consists of sintering a porous TiO_2 material (tubes, plates, rods, or pellets), followed by a second thermal process where temperatures are increased to 1,000°C under hydrogen atmosphere [11]. Chaplin et al. first applied Magnéli phase-based reactive electrochemical membranes for the degradation of oxalic and terephthalic acid by the electrogenerated $HO^•$ [11]. Trellu et al. applied a cylindrical porous anode based on a mixture of Ti_4O_7 and Ti_5O_9 Magnéli phases in cross-flow filtration mode, with the stainless steel rod cathode placed in the center, for the degradation of refractory organic pollutants [73]. Acetaminophen (0.1–2.3 mM) in 50 mM Na_2SO_4 electrolyte was mineralized with 47% current efficiency at 150 A m^{-2}. The authors emphasized the role of both $HO^•$ and DET in degrading this pharmaceutical, given that some of the intermediates formed by $HO^•$ attack (e.g., 1,4-benzoquinone, oxalic acid) are susceptible only to direct electrolysis. In addition, the authors observed anode fouling at higher organic flux due to the polymerization of aromatic compounds. Magnéli phase materials are also a suitable substrate for the incorporation of catalytic materials or nanoparticles in order to further increase the HO• production or the reductive capacity of REM (e.g., by the addition of Pd). For example, bismuth-doped tin oxide has been incorporated in a Ti_4O_7 substrate by simple electrodeposition and subsequent mild thermal oxidation in water, resulting in higher HO• production compared to the undoped electrode [72]. Also, electrochemical reduction of sulfamethoxazole was improved from 3.8% for the Ti_4O_7 REM to 96.1% for the Pd-Cu-doped REM at -1.14 V/SHE (energy consumption 0.33 kWh m^{-3}) [59]. The reduction was found to proceed through N-O bond scission at the isoxazole ring. When Ti_4O_7 REM was applied as an anode, 95.7% of sulfamethoxazole was removed at 2.03 V/SHE, with an energy consumption of 0.27 kW hm^{-3}, but also enabled mineralization, on the contrary to the reductive transformation. Ti_4O_7 electrodes seem to be a very promising material for the removal of pharmaceuticals and other trace organic pollutants from water. Nevertheless, their performance is yet to be evaluated in environmentally relevant conditions of typical conductivity of wastewater (1.5–3 mS cm^{-1}), groundwater (0.5–2 mS cm^{-1}) and other contaminated water matrices, and also for low ng – μg L^{-1} concentrations of pharmaceuticals. Ti_4O_7 anodes suffer from similar limitations as other anode materials in terms of the formation of toxic chlorinated byproducts in the presence of Cl^-, i.e., organochlorine compounds and inorganics ClO_3^- and ClO_4^- [35]. Also, more research is needed about the stability of Ti_4O_7 REMs during long-term operation, given that some authors reported irreversible passivation during anodic polarization [74], while other observed restoration of the anode electrocatalytic activity after cathodic polarization [75].

6 Conclusions and Outlook

In order to decrease the energy consumption of electrochemical systems applied for the removal of pharmaceuticals and other persistent trace contaminants, research on novel porous, 3D anode materials with high electrocatalytic activity and stability is essential. Nevertheless, it is crucial to conduct the experiments under environmentally relevant conditions in terms of electrolyte conductivity and composition and contaminant concentration. If a new electrode material is developed, the authors should report its electrochemical stability, for example, by performing accelerated service life tests. Special attention should be paid to the potential of anode materials to generate chlorine and thus induce the formation of chlorinated byproducts, as this is one of the main bottlenecks of electrochemical water treatment. Wider-scale application of electrochemical water treatment systems can only be enabled by addressing their limitations and searching for solutions to minimize the energy consumption and formation of organic and inorganic halogenated byproducts.

Several porous electrode materials have been recently proposed as means to reduce the energy consumption of electrochemical process. Among these, CNT- and Magnéli-based electrochemical membranes are attractive materials due to their high electrical conductivity, specific surface area, and possibility of functionalization. These two materials open the way to pore control and subsequently production of flow-through electrodes with confined active areas in the nanometer to micrometer range. Their production is relatively simple and based on the use of ink solutions (i.e., CNTs) or combustion under reductive atmosphere and sintering (i.e., Magnéli). However, low E_{EO} values and high removal rates for the degradation of trace pollutants are typically achieved when surface functionalization is introduced, e.g., through doping of Sb, Sn, Bi, or Fe_2O_3. This introduces a risk of leaching of toxic metals that needs to be carefully evaluated during electrode operation, particularly at high anode potentials typically required for the degradation of more persistent pollutants. In addition, both CNT and Magnéli phase REMs are yet to be evaluated for low-conductivity electrolyte solutions and low initial contaminant concentration.

More recently, graphene-based membranes were proposed to precisely control the molecular size of elements passed to the permeate side by regulating the interlayer distance between graphene 2D sheets [76]. The ability of graphene membranes to be functionalized (by π-π stacking and oxygen functional groups) is similar to CNTs, although graphene-based materials have more standardized properties and homogeneity in their production in terms of final size, thickness, and electrical properties [61]. In their pioneering study, Zhou et al. demonstrated that it was possible to precisely control water permeation and flux of graphene-based membranes by applying current, from ultrafast permeation to complete blocking [77]. This was achieved by the application of current to the conductive filaments in the graphene oxide membrane that leads to ionization of the water molecules inside the graphene capillaries and impedes water transport. Graphene-based membranes have been rarely adopted for electrochemical removal of trace organic pollutants [57, 58, 78]. Nevertheless, RGO is a low-cost material that offers interesting opportunities for producing 3D macroscale foams with controlled pore structure, size, and degree of RGO reduction by simple methodologies such as hydrothermal or 3D printing [79, 80]. During the synthesis of RGO materials, their electrical, mechanical, and physicochemical properties (e.g., hydrophobicity) can be tailored for a specific water treatment application. In addition, graphene offers an excellent matrix for functionalization and introduction of other 2D materials (e.g., MXenes, hexagonal boron nitride – hBN) and thus further fine-tuning of their electrocatalytic performance. These and other novel materials offer emerging opportunities for electrochemical systems and different ways to architecture a water treatment system, by designing reactors with directional flows depending on the nanostructures and catalytic improvements made to enhanced surface exposure.

References

1. Pelskov YV, Sakharova AY, Krotova MD, Bouilov LL, Spitsyn BV (1987) Photoelectrochemical properties of semiconductor diamond. J Electroanal Chem Interfacial Electrochem 228:19–27
2. Carey JJ, Christ Jr CS, Lowery SN (1995) Electrolysis of liquid wastes using a doped diamond anode to oxidize solutes. US Patent 5, 247 (Ed.)
3. Panizza M, Cerisola G (2009) Direct and mediated anodic oxidation of organic pollutants. Chem Rev 109:6541–6569
4. Kolbrecka K, Przyluski J (1994) Sub-stoichiometric titanium oxides as ceramic electrodes for oxygen evolution—structural aspects of the voltammetric behaviour of TinO2n−1. Electrochim Acta 39:1591–1595
5. Vinokur N, Miller B, Avyigal Y, Kalish R (1996) Electrochemical behavior of boron-doped diamond electrodes. J Electrochem Soc 143:L238–L240
6. Chaplin BP (2019) The Prospect of electrochemical technologies advancing worldwide water treatment. Acc Chem Res 52:596–604
7. Comninellis C, Chen G (2010) Electrochemistry for the environment. Springer, Berlin
8. Yang Y, Kao LC, Liu Y, Sun K, Yu H, Guo J, Liou SY, Hoffmann MR (2018) Cobalt-doped black TiO_2 nanotube array as a stable anode for oxygen evolution and electrochemical wastewater treatment. ACS Catal 8(5):4278–4287
9. Oriol R, Clematis D, Brillas E, Cortina JL, Panizza M, Sirés I (2019) Groundwater treatment using a solid polymer electrolyte cell with mesh electrodes. ChemElectroChem 6(4):1235–1243
10. Bunce NJ, Bejan D (2014) Pollutants in water – electrochemical remediation using Ebonex electrodes. In: Kreysa G, Ota K-i, Savinell RF (eds) Encyclopedia of applied electrochemistry. Springer, New York, pp 1629–1633

11. Nayak S, Chaplin BP (2018) Fabrication and characterization of porous, conductive, monolithic Ti4O7 electrodes. Electrochim Acta 263:299–310
12. Radjenovic J, Bagastyo A, Rozendal RA, Mu Y, Keller J, Rabaey K (2011) Electrochemical oxidation of trace organic contaminants in reverse osmosis concentrate using RuO2/IrO2-coated titanium anodes. Water Res 45:1579–1586
13. Radjenovic J, Sedlak DL (2015) Challenges and opportunities for electrochemical processes as next-generation technologies for the treatment of contaminated water. Environ Sci Technol 49:11292–11302
14. Bagastyo AY, Batstone DJ, Kristiana I, Escher BI, Joll C, Radjenovic J (2014) Electrochemical treatment of reverse osmosis concentrate on boron-doped electrodes in undivided and divided cell configurations. J Hazard Mater 279:111–116
15. Farhat A, Keller J, Tait S, Radjenovic J (2015) Removal of persistent organic contaminants by electrochemically activated sulfate. Environ Sci Technol 49(24):14326–14333
16. Farhat A, Keller J, Tait S, Radjenovic J (2017) Assessment of the impact of chloride on the formation of chlorinated by-products in the presence and absence of electrochemically activated sulfate. Chem Eng J 330:1265–1271
17. Radjenovic J, Petrovic M (2017) Removal of sulfamethoxazole by electrochemically activated sulfate: implications of chloride addition. J Hazard Mater 333:242–249
18. Farhat A, Keller J, Tait S, Radjenovic J (2018) Oxidative capacitance of sulfate-based boron-doped diamond electrochemical system. Electrochem Commun 89:14–18
19. Rashid R, Mark F, Schuchmann HP, von Sonntag C (1991) SO·−4-induced oxidation of 1,3,6-trimethyluracil and 1,3,5-trimethyluracil (1,3-dimethylthymine) by potassium peroxodisulphate in aqueous solution: an interesting contrast. Int J Radiat Biol 59:1081–1100
20. Schuchmann H-P, Deeble DJ, Olbrich G, Von Sonntag C (1987) The SO− ·4-induced chain reaction of 1,3-dimethyluracil with Peroxodisulphate. Int J Radiation Biol Related Stud Phys Chem Med 51:441–453
21. Cai J, Zhou M, Liu Y, Savall A, Groenen Serrano K (2018) Indirect electrochemical oxidation of 2,4-dichlorophenoxyacetic acid using electrochemically-generated persulfate. Chemosphere 204:163–169
22. Shin Y-U, Yoo H-Y, Ahn Y-Y, Kim MS, Lee K, Yu S, Lee C, Cho K, Kim H-i, Lee J (2019) Electrochemical oxidation of organics in sulfate solutions on boron-doped diamond electrode: multiple pathways for sulfate radical generation. Appl Catal B Environ 254:156–165
23. Matzek LW, Tipton MJ, Farmer AT, Steen AD, Carter KE (2018) Understanding electrochemically activated Persulfate and its application to ciprofloxacin abatement. Environ Sci Technol 52:5875–5883
24. Song H, Yan L, Ma J, Jiang J, Cai G, Zhang W, Zhang Z, Zhang J, Yang T (2017) Nonradical oxidation from electrochemical activation of peroxydisulfate at Ti/Pt anode: efficiency, mechanism and influencing factors. Water Res 116:182–193
25. Barazesh JM, Prasse C, Sedlak DL (2016) Electrochemical transformation of trace organic contaminants in the presence of halide and carbonate ions. Environ Sci Technol 50:10143–10152
26. Cai J, Niu T, Shi P, Zhao G (2019) Boron-doped diamond for hydroxyl radical and sulfate radical anion Electrogeneration, transformation, and voltage-free sustainable oxidation. Small 15:1900153
27. Zhang K, Parker KM (2018) Halogen radical oxidants in natural and engineered aquatic systems. Environ Sci Technol 52:9579–9594
28. Jasper JT, Yang Y, Hoffmann MR (2017) Toxic byproduct formation during electrochemical treatment of latrine wastewater. Environ Sci Technol 51:7111–7119
29. Radjenovic J, Escher BI, Rabaey K (2011) Electrochemical degradation of the β-blocker metoprolol by Ti/Ru0.7Ir0.3O2 and Ti/SnO2-Sb electrodes. Water Res 45:3205–3214
30. Anglada Á, Urtiaga A, Ortiz I, Mantzavinos D, Diamadopoulos E (2011) Boron-doped diamond anodic treatment of landfill leachate: evaluation of operating variables and formation of oxidation by-products. Water Res 45:828–838

31. Bagastyo AY, Radjenovic J, Mu Y, Rozendal RA, Batstone DJ, Rabaey K (2011) Electrochemical oxidation of reverse osmosis concentrate on mixed metal oxide (MMO) titanium coated electrodes. Water Res 45:4951–4959
32. Pérez G, Fernández-Alba AR, Urtiaga AM, Ortiz I (2010) Electro-oxidation of reverse osmosis concentrates generated in tertiary water treatment. Water Res 44:2763–2772
33. Bagastyo AY, Batstone DJ, Kristiana I, Gernjak W, Joll C, Radjenovic J (2012) Electrochemical oxidation of reverse osmosis concentrate on boron-doped diamond anodes at circumneutral and acidic pH. Water Res 46:6104–6112
34. Garcia-Segura S, Keller J, Brillas E, Radjenovic J (2015) Removal of organic contaminants from secondary effluent by anodic oxidation with a boron-doped diamond anode as tertiary treatment. J Hazard Mater 283:551–557
35. Wang L, Lu J, Li L, Wang Y, Huang Q (2020) Effects of chloride on electrochemical degradation of perfluorooctanesulfonate by Magnéli phase Ti4O7 and boron doped diamond anodes. Water Res 170:115254
36. Merkle FH, Discher CA (1964) Electrochemical oxidation of chlorpromazine hydrochloride. J Pharm Sci 53:620–623
37. García-Espinoza JD, Mijaylova-Nacheva P, Avilés-Flores M (2018) Electrochemical carbamazepine degradation: effect of the generated active chlorine, transformation pathways and toxicity. Chemosphere 192:142–151
38. Ouarda Y, Bouchard F, Azaïs A, Vaudreuil M-A, Drogui P, Dayal Tyagi R, Sauvé S, Buelna G, Dubé R (2019) Electrochemical treatment of real hospital wastewaters and monitoring of pharmaceutical residues by using surrogate models. J Environ Chem Eng 7:103332
39. Mordačíková E, Vojs M, Grabicová K, Marton M, Michniak P, Řeháček V, Bořík A, Grabic R, Bruncko J, Mackuľak T, Vojs Staňová A (2020) Influence of boron doped diamond electrodes properties on the elimination of selected pharmaceuticals from wastewater. J Electroanal Chem 862:114007
40. Urtiaga AM, Pérez G, Ibáñez R, Ortiz I (2013) Removal of pharmaceuticals from a WWTP secondary effluent by ultrafiltration/reverse osmosis followed by electrochemical oxidation of the RO concentrate. Desalination 331:26–34
41. Rodríguez-Nava O, Ramírez-Saad H, Loera O, González I (2016) Evaluation of the simultaneous removal of recalcitrant drugs (bezafibrate, gemfibrozil, indomethacin and sulfamethoxazole) and biodegradable organic matter from synthetic wastewater by electro-oxidation coupled with a biological system. Environ Technol 37:2964–2974
42. Yang N, Yu S, Macpherson JV, Einaga Y, Zhao H, Zhao G, Swain GM, Jiang X (2019) Conductive diamond: synthesis, properties, and electrochemical applications. Chem Soc Rev 48:157–204
43. Chaplin BP (2014) Critical review of electrochemical advanced oxidation processes for water treatment applications. Environ Sci Process Impacts 16:1182–1203
44. Radjenovic J, Petrovic M (2016) Sulfate-mediated electrooxidation of X-ray contrast media on boron-doped diamond anode. Water Res 94:128–135
45. Mu Y, Radjenovic J, Shen J, Rozendal RA, Rabaey K, Keller J (2010) Dehalogenation of Iodinated X-ray contrast media in a bioelectrochemical system. Environ Sci Technol 45:782–788
46. Radjenovic J, Flexer V, Donose BC, Sedlak DL, Keller J (2013) Removal of the X-ray contrast media diatrizoate by electrochemical reduction and oxidation. Environ Sci Technol 47:13686–13694
47. Oturan N, Oturan MA (2018) Chapter 8 – electro-fenton process: background, new developments, and applications. In: Martínez-Huitle CA, Rodrigo MA, Scialdone O (eds) Electrochemical water and wastewater treatment. Butterworth-Heinemann, pp 193–221
48. Brillas E (2014) A review on the degradation of organic pollutants in waters by UV photoelectro-Fenton and solar photoelectro-Fenton. J Braz Chem Soc 25:393–417

49. Ridruejo C, Centellas F, Cabot PL, Sirés I, Brillas E (2018) Electrochemical Fenton-based treatment of tetracaine in synthetic and urban wastewater using active and non-active anodes. Water Res 128:71–81
50. Villanueva-Rodríguez M, Bello-Mendoza R, Hernández-Ramírez A, Ruiz-Ruiz EJ (2019) Degradation of anti-inflammatory drugs in municipal wastewater by heterogeneous photocatalysis and electro-Fenton process. Environ Technol 40:2436–2445
51. Zhi D, Wang J, Zhou Y, Luo Z, Sun Y, Wan Z, Luo L, Tsang DCW, Dionysiou DD (2020) Development of ozonation and reactive electrochemical membrane coupled process: enhanced tetracycline mineralization and toxicity reduction. Chem Eng J 383:123149
52. Zhan J, Li Z, Yu G, Pan X, Wang J, Zhu W, Han X, Wang Y (2019) Enhanced treatment of pharmaceutical wastewater by combining three-dimensional electrochemical process with ozonation to in situ regenerate granular activated carbon particle electrodes. Sep Purif Technol 208:12–18
53. Zhai L-F, Duan M-F, Qiao M-X, Sun M, Wang S (2019) Electro-assisted catalytic wet air oxidation of organic pollutants on a MnO@C/GF anode under room condition. Appl Catal B Environ 256:117822
54. Liu Y, Liu H, Zhou Z, Wang T, Ong CN, Vecitis CD (2015) Degradation of the common aqueous antibiotic tetracycline using a carbon nanotube electrochemical filter. Environ Sci Technol 49:7974–7980
55. Bakr AR, Rahaman MS (2016) Electrochemical efficacy of a carboxylated multiwalled carbon nanotube filter for the removal of ibuprofen from aqueous solutions under acidic conditions. Chemosphere 153:508–520
56. Tan T-Y, Zeng Z-T, Zeng G-M, Gong J-L, Xiao R, Zhang P, Song B, Tang W-W, Ren X-Y (2020) Electrochemically enhanced simultaneous degradation of sulfamethoxazole, ciprofloxacin and amoxicillin from aqueous solution by multi-walled carbon nanotube filter. Sep Purif Technol 235:116167
57. Liu Y, Dustin Lee JH, Xia Q, Ma Y, Yu Y, Lanry Yung LY, Xie J, Ong CN, Vecitis CD, Zhou Z (2014) A graphene-based electrochemical filter for water purification. J Mater Chem A 2:16554–16562
58. Jiang W-L, Xia X, Han J-L, Ding Y-C, Haider MR, Wang A-J (2018) Graphene modified electro-Fenton catalytic membrane for in situ degradation of antibiotic Florfenicol. Environ Sci Technol 52:9972–9982
59. Misal SN, Lin M-H, Mehraeen S, Chaplin BP (2020) Modeling electrochemical oxidation and reduction of sulfamethoxazole using electrocatalytic reactive electrochemical membranes. J Hazard Mater 384:121420
60. Trellu C, Chaplin BP, Coetsier C, Esmilaire R, Cerneaux S, Causserand C, Cretin M (2018) Electro-oxidation of organic pollutants by reactive electrochemical membranes. Chemosphere 208:159–175
61. Ando T (2009) The electronic properties of graphene and carbon nanotubes. NPG Asia Mater 1:17–21
62. Sadia AJ, Zhi Z (2016) Electrochemical carbon nanotube filters for water and wastewater treatment. Nanotechnol Rev 5:41–50
63. Vecitis CD, Gao G, Liu H (2011) Electrochemical carbon nanotube filter for adsorption, desorption, and oxidation of aqueous dyes and anions. J Phys Chem C 115:3621–3629
64. Ajmani GS, Goodwin D, Marsh K, Fairbrother DH, Schwab KJ, Jacangelo JG, Huang H (2012) Modification of low pressure membranes with carbon nanotube layers for fouling control. Water Res 46:5645–5654
65. de Lannoy C-F, Jassby D, Gloe K, Gordon AD, Wiesner MR (2013) Aquatic biofouling prevention by electrically charged Nanocomposite polymer thin film membranes. Environ Sci Technol 47:2760–2768
66. Vecitis CD, Schnoor MH, Rahaman MS, Schiffman JD, Elimelech M (2011) Electrochemical multiwalled carbon nanotube filter for viral and bacterial removal and inactivation. Environ Sci Technol 45:3672–3679

67. Gao G, Vecitis CD (2011) Electrochemical carbon nanotube filter oxidative performance as a function of surface chemistry. Environ Sci Technol 45:9726–9734
68. Gao G, Vecitis CD (2012) Doped carbon nanotube networks for electrochemical filtration of aqueous phenol: electrolyte precipitation and phenol polymerization. ACS Appl Mater Interfaces 4:1478–1489
69. Yanez H, Wang Z, Lege S, Obst M, Roehler S, Burkhardt CJ, Zwiener C (2017) Application and characterization of electroactive membranes based on carbon nanotubes and zerovalent iron nanoparticles. Water Res 108:78–85
70. Duan W, Ronen A, Walker S, Jassby D (2016) Polyaniline-coated carbon nanotube ultrafiltration membranes: enhanced anodic stability for in situ cleaning and electro-oxidation processes. ACS Appl Mater Interfaces 8:22574–22584
71. Gao G, Zhang Q, Hao Z, Vecitis CD (2015) Carbon nanotube membrane stack for flow-through sequential regenerative electro-Fenton. Environ Sci Technol 49:2375–2383
72. Gayen P, Chen C, Abiade JT, Chaplin BP (2018) Electrochemical oxidation of atrazine and Clothianidin on Bi-doped SnO2–TinO2n–1 electrocatalytic reactive electrochemical membranes. Environ Sci Technol 52:12675–12684
73. Trellu C, Coetsier C, Rouch J-C, Esmilaire R, Rivallin M, Cretin M, Causserand C (2018) Mineralization of organic pollutants by anodic oxidation using reactive electrochemical membrane synthesized from carbothermal reduction of TiO2. Water Res 131:310–319
74. Waraksa CC, Chen G, Macdonald DD, Mallouk TE (2003) EIS studies of porous oxygen electrodes with discrete particles II. Transmission line modeling. J Electrochem Soc 150:E429–E437
75. El-Sherif S, Bejan D, Bunce NJ (2010) Electrochemical oxidation of sulfide ion in synthetic sour brines using periodic polarity reversal at Ebonex® electrodes. Can J Chem 88:928–936
76. Mi B (2014) Graphene oxide membranes for ionic and molecular sieving. Science 343:740–742
77. Zhou KG, Vasu KS, Cherian CT, Neek-Amal M, Zhang JC, Ghorbanfekr-Kalashami H, Huang K, Marshall OP, Kravets VG, Abraham J, Su Y, Grigorenko AN, Pratt A, Geim AK, Peeters FM, Novoselov KS, Nair RR (2018) Electrically controlled water permeation through graphene oxide membranes. Nature 559:236–240
78. Nidheesh PV (2017) Graphene-based materials supported advanced oxidation processes for water and wastewater treatment: a review. Environ Sci Pollut Res 24:27047–27069
79. Guo H, Lv R, Bai S (2019) Recent advances on 3D printing graphene-based composites. Nano Mater Sci 1:101–115
80. Yousefi N, Lu X, Elimelech M, Tufenkji N (2019) Environmental performance of graphene-based 3D macrostructures. Nat Nanotechnol 14:107–119

Sonochemical Advanced Oxidation Processes for the Removal of Pharmaceuticals in Wastewater Effluents

Efraím A. Serna-Galvis, Judy Lee, Félix Hernández, Ana María Botero-Coy, and Ricardo A. Torres-Palma

Contents

1 Fundamental Aspects 350
1.1 Ultrasound as an Advanced Oxidation Process and Environmental Problem of Pharmaceuticals 350
1.2 Effect of Frequency and Power on Pharmaceuticals Degradation by Ultrasound 351
1.3 Effect of Pharmaceutical Nature on the Sonochemical Degradation 355
2 Chemical Transformations of Pharmaceuticals upon Sonochemical Action 359
2.1 Analytical Techniques to Follow Pharmaceuticals Degradation and Transformation During Sonochemical Treatment 359
2.2 Transformations of Pharmaceuticals Promoted by Sonochemical Action 363
3 Treatment of Wastewater Polluted with Pharmaceuticals by Ultrasound 364
3.1 Removal of Pharmaceuticals in Industrial, Municipal and Hospital Wastewaters Effluents 364
3.2 Ultrasound Generators and Engineering Aspects for the Wastewater Treatments by Ultrasound 369
3.3 Combinations of Ultrasound with Membrane Filtration Techniques as a Strategy for Overcoming the Issues Related to Large-Scale Application 373
4 Concluding Remarks 374
References 375

E. A. Serna-Galvis and R. A. Torres-Palma (✉)
Grupo de Investigación en Remediación Ambiental y Biocatálisis (GIRAB), Instituto de Química, Facultad de Ciencias Exactas y Naturales, Universidad de Antioquia UdeA, Medellín, Colombia
e-mail: efrain.serna@udea.edu.co; ricardo.torres@udea.edu.co

J. Lee
Department of Chemical and Process Engineering, University of Surrey, Guildford, Surrey, UK
e-mail: j.y.lee@surrey.ac.uk

F. Hernández and A. M. Botero-Coy
Research Institute for Pesticides and Water (IUPA), University Jaume I (UJI), Castellón, Spain
e-mail: hernandf@uji.es; botero@uji.es

Sara Rodriguez-Mozaz, Paqui Blánquez Cano, and Montserrat Sarrà Adroguer (eds.), *Removal and Degradation of Pharmaceutically Active Compounds in Wastewater Treatment*, Hdb Env Chem (2021) 108: 349–382, DOI 10.1007/698_2020_665,

Abstract Pharmaceuticals in wastewater effluents represent a current environmental concern, which demands application of effective processes. In such regard, sonochemical advanced oxidation processes emerge as an attractive alternative, as illustrated herein. This chapter begins presenting some fundamental aspects of sonochemical treatments (e.g. effects of frequency, power and nature of pharmaceutical pollutants). In second place, the transformations of pharmaceuticals are described, considering topics such as the used analytical techniques and structural modifications of pollutants by chemical effects of ultrasound. Then, treatment of diverse wastewater containing pharmaceuticals are shown, paying special attention to degradations in complex matrices, reactors configurations and combination of ultrasound with membrane filtration processes. Final part is dedicated to highlight the key points presented along the chapter.

Keywords Advanced oxidation process, Environmental remediation, Pharmaceuticals degradation, Ultrasound, Wastewater treatment

1 Fundamental Aspects

1.1 Ultrasound as an Advanced Oxidation Process and Environmental Problem of Pharmaceuticals

Ultrasound comprises mechanical waves having frequencies from 20 to 10,000 kHz (20 kHz to 10 MHz). These waves are utilized in areas such as biomedical diagnostics, material synthesis and environmental chemistry/engineering, among other fields [1]. Herein, we will focus on environmental applications, specifically on the use of ultrasonic waves for degrading organic pharmaceuticals.

Pharmaceuticals are considered contaminants of emerging concern (CECs) because these substances are continuously introduced into the environment at very low concentrations (e.g. ng L^{-1} to μg L^{-1}); they may affect water quality, natural water supplies and ecosystems equilibrium [2]. Additionally, the regulation by governments about pharmaceuticals input into environment is still incipient. Thus, the environmental discharge of pharmaceutical must be limited, and ultrasound-based techniques can be utilized for such purpose.

The generation of radicals in ultrasound systems involves transient acoustic cavitation phenomenon. This consists of formation of bubbles (also called cavities, which contain oxygen gas and/or water vapour) in the liquid; due to the pass of ultrasound waves (expansion and contraction cycles), the cavities grow up until a critical size, and then, they violently collapse [3, 4]. During the collapse temperature and pressure within the cavity is enormous (e.g. 5,000 K and 1,000 atm); these critical conditions lead to rupture of water vapour and oxygen to form radicals

(Eqs. 1–4). Additionally, cavitation can also produce the emission of light (called sonoluminescence) [5, 6].

$$H_2O+))) \rightarrow H\bullet + HO\bullet \quad (1)$$

$$O_2+))) \rightarrow 2\,O\bullet \quad (2)$$

$$H_2O + O\bullet \rightarrow 2\,HO\bullet \quad (3)$$

$$O_2 + H\bullet \rightarrow O\bullet + HO\bullet \quad (4)$$

The formation of radicals by ultrasound can be demonstrated be means of spin trapping and electron spin resonance (ESR) technique [4]. Besides, the formation/accumulation of hydrogen peroxide (coming from recombination of radicals, Eq. 5) is an indirect way to prove the radical generation in an ultrasonic system [7, 8].

$$HO\bullet \rightarrow H_2O_2 \quad (5)$$

Ultrasound is considered an advanced oxidation process (AOP) because this can be used for degrading organic pollutants in aqueous matrices by action of the generated radicals [9]. In the aqueous systems submitted to ultrasound, it has been figured out three degradation zones: (1) interior of cavitation bubble, (2) interface of bubble-liquid and (3) bulk of liquid. Volatile compounds are pyrolysed inside bubble, whereas hydrophobic non-volatile and hydrophilic substances are mainly destroyed at the interface and liquid-bulk zones, respectively [3].

It must be mentioned that the efficiency of sonochemical degradation is influenced by operational parameters, chemical nature of pollutant, matrix components and reactor configuration. All these aspects are analysed and discussed in the following sections.

1.2 *Effect of Frequency and Power on Pharmaceuticals Degradation by Ultrasound*

Physical effects dominate at lower frequencies (20–100 kHz), chemical effects are more significant at frequencies in the range 100–1,000 kHz (1,000 kHz = 1 MHz), and at higher frequencies (>1 MHz) acoustic streaming effect are dominant [10].

For degradation of organic pollutants, frequencies among 20 kHz and 1.2 MHz are commonly used (Table 1). Pharmaceuticals such as cephalexin, diclofenac, estrogenic hormones and ofloxacin have been treated at low frequencies (20–100 kHz) and at very high powers or power densities. Likewise, high frequencies (among 100 kHz and 1.2 MHz) have been widely utilized for degrading acetaminophen, ampicillin, atenolol, levodopa, losartan and triclosan; nevertheless, lower powers (or power densities) than the required at frequencies below 100 kHz can be used. Table 1 shows that if frequencies are varied in the range 100 kHz to

Table 1 Pharmaceuticals degradation at different frequencies

Compound	Frequency (kHz)	Removal	Time (min)	Power conditions	Reference
Cephalexin	24	70% COD	150	Electric 200 W (acoustic 17.3 W)	[13]
Diclofenac	20	~50% Degradation	60	Electric: 100 W L^{-1}	[14]
Oestrogens mixture	20	~80–90% Degradation	25	Electric: 320 W (2,100 W L^{-1})	[15]
Mix of amoxicillin, carbamazepine and diclofenac	20	~20–50% Degradation	60	Electric: 100 W L^{-1}	[16]
Ofloxacin	20	31% Degradation	240	Electric: 640 W L^{-1}	[17]
Atenolol	200	[a]k: 0.01649	–	Electric: 50 W	[18]
	350	[a]k: 0.04476	–		
	620	[a]k: 0.03717	–		
	1,000	[a]k: 0.03720	–		
Acetaminophen	574	95% Degradation	240	Acoustic: 32 W	[19]
	860	92%	240		
Triclosan	574	90%	15	Acoustic: 140 W L^{-1}	[20]
	856	75%	15		
	1,134	72% Degradation	15		
Ampicillin	375	39% Degradation	30	Acoustic: 24.4 W	[8]
	990	26% Degradation	30		
Losartan	375	70% Degradation	30	Acoustic: 88 W L^{-1}	[12]
	575	54% Degradation	30		
	990	15% Degradation	30		

[a]Pseudo-first-order degradation rate constant (k, min^{-1})

1.2 MHz, the highest degradations are observed between 200 and 400 kHz, due to in such range occurs the highest production of radical species [11], which favours the pharmaceuticals degradation [8, 12].

In general, at a fixed power (or power density), the increasing of frequencies results in higher nucleation and production of bubbles that are comparatively smaller in size, for example, at 20 kHz, ~20,000 implosions per second could occur, whereas at 300 kHz, ~100,000 cavitation events per second are possible. This enhances the production of radicals and leads to a higher degradation of pharmaceuticals

[1, 7]. However, it should be indicated that an upper limit exists for the applied frequency above which the active cavitation for pollutants degradation diminishes. Then, if the bubbles have not enough size, the extent of collapse is low and fewer radicals are formed [1].

On the other hand, in the sonochemical systems, a transduction phenomenon (i.e. the transformation of electrical power into acoustic power) is involved. During such transformation, not all electrical power is converted into acoustic power. Then, the actual acoustic power should be determined to establish the sonochemical efficiency of the transducers [21, 22]. The actual acoustic power determination is typically done by calorimetric methods. The sonochemical energy can be assimilated to the calorimetric energy due to the rising up of the temperature in a fixed volume of water. This is represented by the Eq. 6, where *Cp* is the specific heat capacity of the solvent (in J kg^{-1} K^{-1}), *m* is the mass of solvent (in kg) and $\Delta T/\Delta t$ means the temperature difference (in K) after a specific sonication time (in s) [21]. Therefore, the efficiency of the energy transduction is calculated as the ratio between the actual acoustic power and the electric power supplied to the system.

$$\text{Acoustic power} = m\, Cp\, (\Delta T/\Delta t) \tag{6}$$

It is well-known that the increasing in the electric power (and consequently in the acoustic power) augments the degradation of pharmaceuticals [1, 18, 19, 22, 23]. A higher power delivers an increasing in the number of cavitation bubbles and the magnitude of the collapse, which lead to a higher production of radicals (and consequently a higher accumulation of H_2O_2, Fig. 1a), favouring the pollutants degradation [24–26], as exemplified in Fig. 1b. However, it should be indicated that excessive powers induce a decrease in the number of bubbles per volume, which is attributed to an increase in bubble coalescence, degassing and liquid agitation. Also, it can be remarked that there is a lower limit or threshold of power. Below such lower limit, the sound field is very small to induce nucleation or bubble growth; consequently, bubbles readily succumb to surface tension effects and dissolve, resulting in a limited cavitation activity [1].

It is important to indicate that although power and frequency are the primary physical parameters that influence sonochemical process, the elimination of pharmaceutical by ultrasound can also be modified by other operational parameters such as aqueous sample temperature, dissolved gas and operation mode (i.e. pulsed vs. continuous waves). Such topics are not considered in this chapter; however valuable information about them is available for reader in the references [1, 23, 27–30].

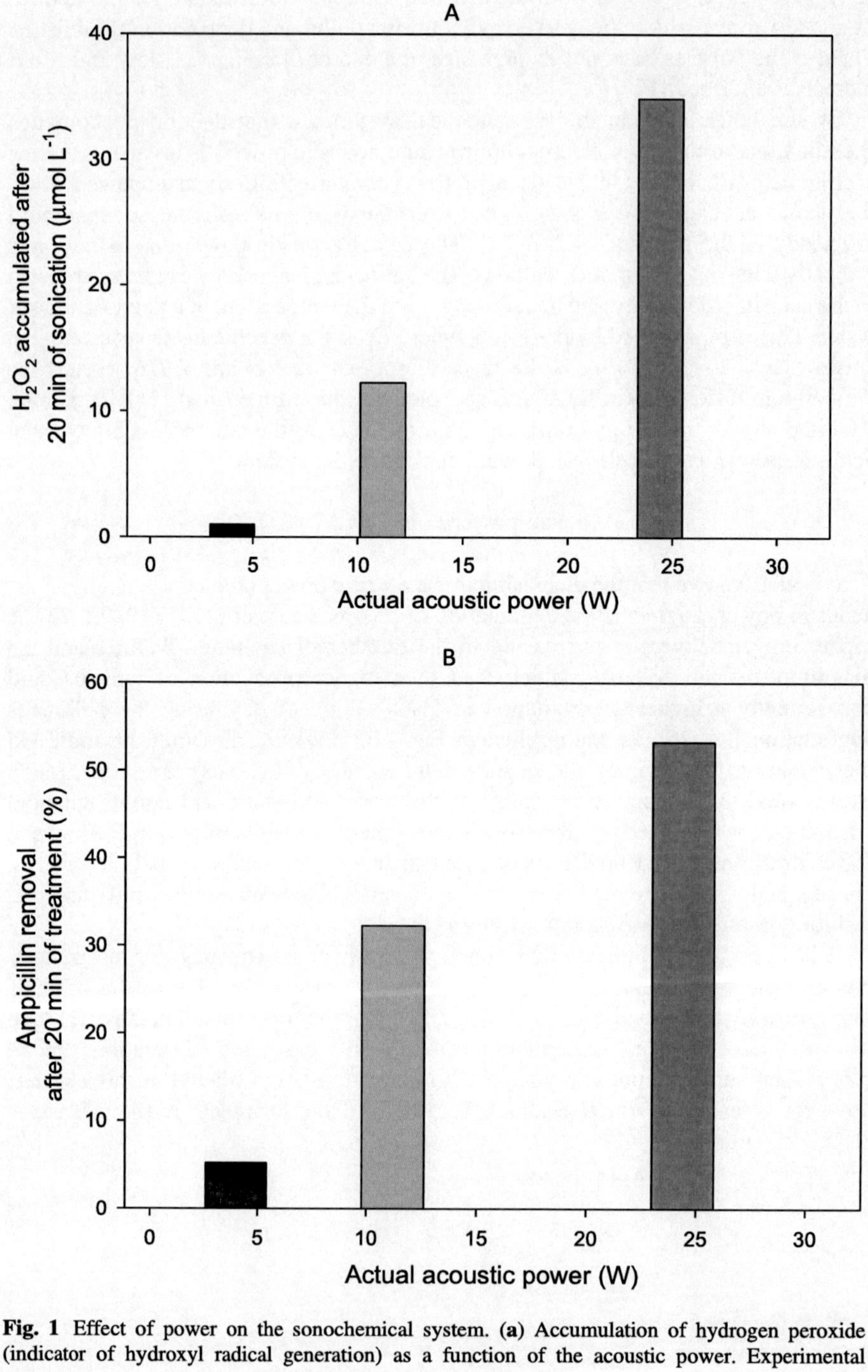

Fig. 1 Effect of power on the sonochemical system. (**a**) Accumulation of hydrogen peroxide (indicator of hydroxyl radical generation) as a function of the acoustic power. Experimental

1.3 *Effect of Pharmaceutical Nature on the Sonochemical Degradation*

Most of organic pharmaceuticals are non-volatile substances. Thereby, degradation of such pollutants in the sonochemical system mainly occurs at the interfacial zone and/or in the solution bulk. Some works have studied the correlation between sonochemical degradation rate for pollutants and physico-chemical parameters (e.g. molar mass, Henry's law constant, second-order reaction constant with hydroxyl radical), indicating that only one property cannot be utilized to predict the degradation of any compound by ultrasound. However, simple correlations may be useful when the considered compounds have similar/related structure [28, 31].

For example, a research about the sonochemical degradation of some antibiotics showed a good correlation between the initial degradation rate (Rd) and the octanol-water partition coefficient (Log *P*). The pharmaceuticals having higher Log *P* values were faster sonodegraded (Table 2). The Log *P* is a parameter related to hydrophobicity. The hydrophobic compounds tend to accumulate at the cavitation bubble interface, where they can have a high interaction with hydroxyl radicals and are prone to a faster degradation by ultrasound action [32].

The closeness of some pharmaceuticals to cavitation bubbles can be indirectly inferred by using interferent substances such as isopropanol (a volatile scavenger of radicals) and ferrous ion (a promoter of radicals in the solution bulk through Fenton reaction, Eq. 7) [12, 28]. Then, substances that due to their hydrophobic nature are placed close to the cavitation bubble will experiment a lower inhibition degradation degree in presence of the radical scavenger than the more hydrophilic ones. On the contrary, in the presence of the radical promoter, the hydrophobic pharmaceuticals are less accelerated than the hydrophilic compounds [12, 32, 33].

$$Fe^{2+} + H_2O_2 \rightarrow Fe^{3+} + HO^- + HO\bullet \tag{7}$$

On the other hand, many pharmaceuticals have acid or basic functional groups on their structure. Hence, structural variations can be promoted by pH changes. This markedly influences the sonochemical degradation, due to modification of the hydrophilic/hydrophobic nature of the pollutants [15, 16, 23, 26]. In the case of molecules having acid moieties such as carboxylic acids and phenols (e.g. acetaminophen, dicloxacillin and naproxen), their degradations are more favoured at pH values below their pK_a of such functional groups, where predominates their non-charged form (Table 3). Meanwhile, pollutants with basic moieties such as amines (e.g. fluoxetine) have higher degradations at pH values above their

Fig. 1 (continued) conditions, 250 mL of distilled water; $pH_{initial}$, 5.6, 20°C, sonication at 375 kHz by using a Meinhardt reactor. (**b**) Removal of ampicillin at diverse acoustic powers. Experimental conditions, 250 mL of aqueous solution of ampicillin at 30 μmol L^{-1}; $pH_{initial}$, 6.5, 20°C, sonication at 375 kHz by using a Meinhardt reactor. Figure was built with own data of the authors

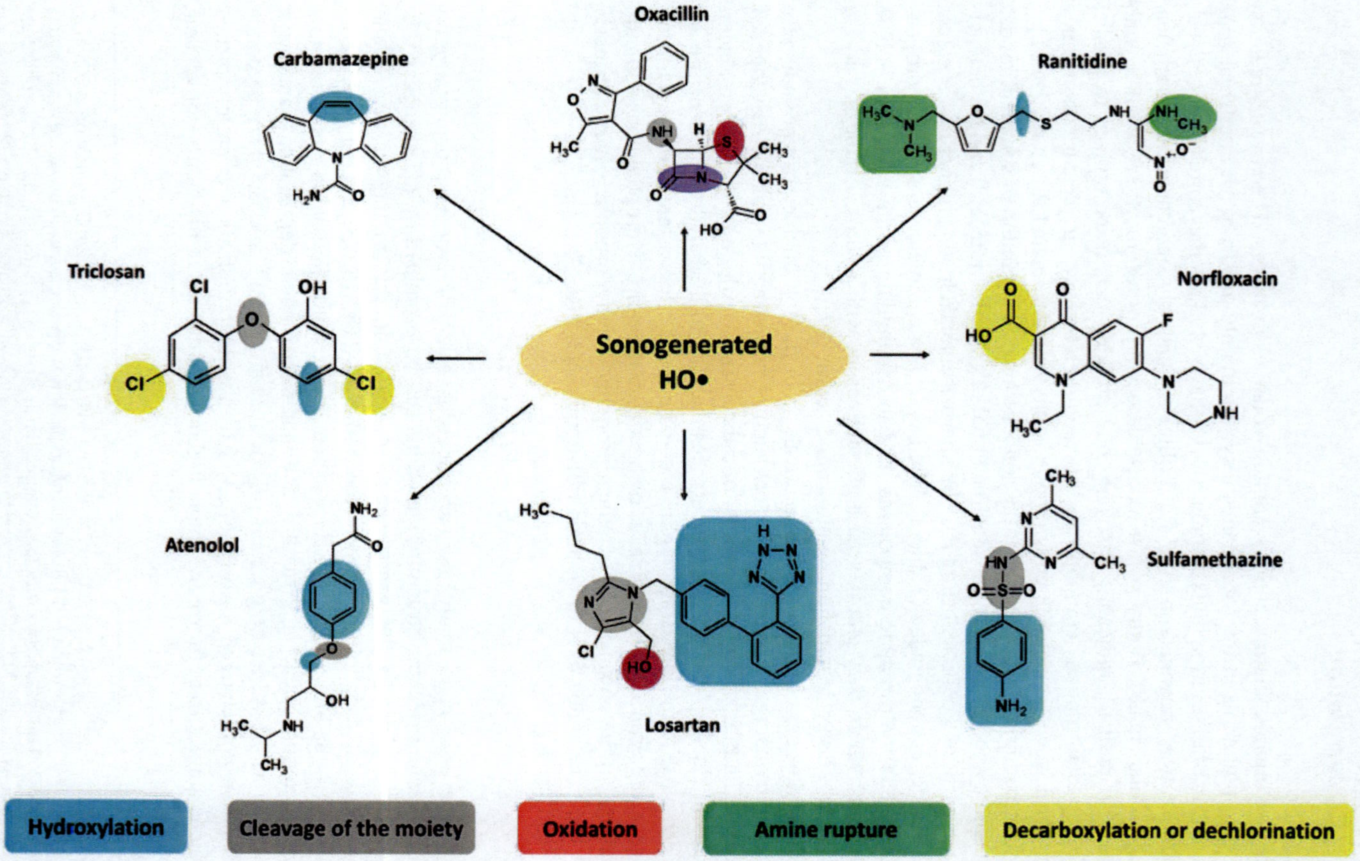

Fig. 2 Some examples of moieties on pharmaceuticals initially attacked by sonogenerated hydroxyl radical during degradation in pure water. Figure was made by the authors. In the case of sulfamethazine (antibiotic), the primary transformations lead to products of the aniline moiety hydroxylation, plus a breakdown of

the sulphonamide group (i.e. the S–N bond) [24]. Besides, the treatment of triclosan (antifungal) by sonochemistry generates compounds coming from hydroxylation, dechlorination and ether cleavage [20]. It can be highlighted that the hydroxyl radical (which has an electron-deficient nature) initially attacks the electron-rich regions (e.g. C=C double bonds, aromatic rings, thioethers, amines and ethers) or unstable moieties (e.g. highly strained rings as β-lactams) on the pharmaceuticals (Fig. 2). Moreover, the hydroxyl radical can also act through a hydrogen abstraction mechanism, which promotes decarboxylation and/or fragmentations

Table 2 Relationship between sonodegradation rate (Rd) of some pharmaceuticals and octanol-water partition coefficient (Log *P*)[a]

Pharmaceutical	Rd (μmol L^{-1})	Log *P*
Cefadroxil	0.4150	−0.40
Ciprofloxacin	0.5600	0.28
Cephalexin	0.7300	0.65
Cloxacillin	0.8600	2.48

[a]Table made by the authors based on Serna-Galvis et al. [32]

Table 3 Effect of pH on sonodegradation of pharmaceuticals

Pharmaceuticals	Acid/base groups	pH values tested	Best degradation at pH	References
Acetaminophen	Phenol (pK_a: 9.5)	3.0, 5.6, 9.5, 11.0 and 12.0	3.0 and 5.6	[26]
Dicloxacillin	Carboxylic acid (pK_a: 2.8)	3.0, 5.7 and 9.0	3.0	[25]
Oestrogens mixture	Phenols ($pK_a > 10$)	3.0, 7.0 and 9.0	3.0	[15]
Ibuprofen	Carboxylic acid (pK_a: 4.9)	3.0, 5.0 and 11.0	3.0	[34]
Naproxen	Carboxylic acid (pK_a: 4.2)	3.0, 6.0 and 10.5	3.0	[35]
Fluoxetine	Amine (pK_a: 10.05)	6.2 and 11.0	11.0	[36]
Cephalexin	Carboxylic acid (pK_a: 3.3–5.2) and amine (pK_a: 7.3)	4.5, 6.5, 7.5, 8.5 and 10	6.5	[13]
Sulfadiazine	Sulphonamide (pK_a: 2.1) and aniline (pK_a: 6.28)	3.0, 5.5, 7.0, 9.0 and 11.0	5.5	[37]

pK_a, where predominates their non-charged species. In the case of pharmaceuticals having both acid and basic moieties, which are able to form zwitterions (e.g. cephalexin and sulfadiazine), their degradations are more favoured at pH values where the zwitterion predominates (i.e. a pH between the pK_a of both groups). This is because such species has an equilibrium with its corresponding neutral non-charged structure (Eq. 8). Overall, the non-charged structures of molecules are more hydrophobic than their corresponding ionic forms (i.e. cation, anion or zwitterion). Therefore, the non-charged forms are closer to the cavitation bubble, which favours their degradation by action of the sonogenerated radicals.

$$^{+}H_3N - R - COO^{-} \leftrightarrow H_2N - R - COOH \quad (8)$$

2 Chemical Transformations of Pharmaceuticals upon Sonochemical Action

2.1 Analytical Techniques to Follow Pharmaceuticals Degradation and Transformation During Sonochemical Treatment

Studies on sonochemical treatments require the measurement of the concentrations of the target compounds under different treatment conditions and along the time. To this aim, it is necessary to apply reliable and sensitive analytical techniques in order to follow the concentrations changes of the parent compound during the treatment, as well as to evaluate the transformations suffered along the process. Ideally, the transformation products (TPs) formed should be elucidated and, if possible, the changes in their concentrations monitored. This implies the use of quantitative techniques for detection and identification of the compounds, particularly when experiments are performed in complex matrix samples, and for the determination of low analyte concentrations. In addition, powerful techniques able to give information about the chemical structures of the compounds and about potential modifications produced in the process, i.e., elucidation techniques, are also needed.

In this field, the importance of chromatographic techniques hyphenated to mass spectrometry is unquestionable. Liquid chromatography coupled to tandem mass spectrometry (LC-MS/MS) with triple quadrupole (QqQ), or related analysers, such as QqQ-linear ion trap, e.g., QTRAP, is the technique of choice for quantitative analysis, due to the medium-high polarity of most pharmaceuticals and TPs, while liquid chromatography coupled to high-resolution mass spectrometry (LC-HRMS) is typically used to identify/elucidate the TPs formed during the treatments [23, 38–42].

2.1.1 Quantitative Analysis

Quantitative analysis of pharmaceuticals in aqueous media, both in environmental studies and in laboratory degradation experiments, is typically performed by LC-MS/MS. This is the technique of choice by several reasons: excellent sensitivity and selectivity, robustness, wide linear dynamic range and fully compatible with aqueous samples without the need to make solvent exchange. There are hundreds of papers reporting the use of LC-MS/MS in the field of pharmaceutical analysis, and a detailed review is out of the scope of this chapter. Instead, some references can be given for those readers interested in this topic [12, 38, 50, 41, 43–49].

Although the advantages of LC-MS/MS are obvious in comparison with LC-based methods using conventional detectors, the latter are still applied in many laboratory experiments that are performed under controlled conditions using clean water (e.g. distilled or Milli-Q) or synthetic water of known composition. In these experiments, "clean" samples are spiked at known concentrations with the pharmaceuticals under study, and spiking levels use to be at nearly, or above, the ppm levels. Under this “ideal” situation, conventional techniques, such as LC-UV, LC-FD or LC-DAD, can be used to follow the evolution of the concentrations, because there are not sensitivity constrains and/or the need of confirmation of the identity of the compound, as occurs in real-world complex water samples. Thus, a number of publications dealing with pharmaceutical analysis and/or sonochemical treatments report data produced by techniques other than LC-MS/MS [12, 49, 51]. However, when the experiments are performed with real-world (waste) water samples, it is necessary to apply more sensitive techniques, able to determine concentration levels at the ppb (μg L^{-1}) and ppt (ng L^{-1}) levels similar to those found in the samples. The situation can be in part alleviated if the samples are spiked at sufficiently high levels to facilitate monitoring the evolution of the concentrations along the treatment. However, the most critical issue is the correct identification and interferences removal when dealing with complex samples, e.g. wastewater. It is in these cases when the potential of LC-MS/MS is actually exploited due to its excellent performance for quantification. Besides, the acquisition of, at least, two selected reaction monitoring (SRM) transitions and the evaluation of ion ratios and chromatographic retention time allow the reliable identification of the compounds [43, 47]. An illustrative example is the elimination of 15 relevant pharmaceuticals in hospital wastewater after combination of a biological system and a sono-photo-chemical process, which was monitored by LC-MS/MS QqQ [52]. Similarly, LC-MS/MS analysis was used to assess the degradation by sonochemical advanced oxidation processes of 17 contaminants of emerging concern, selected on the basis of their occurrence in urban wastewaters [53].

Although LC-MS/MS is widely accepted for pharmaceutical analysis, there is an issue of major concern when applied to complex-matrix samples. This is matrix effects, which may dramatically affect the ionization of the analytes and consequently the quantification. Matrix effects, if no properly corrected or eliminated, may lead to notable errors in quantification, by default (ionization suppression) or by excess (ionization enhancement). The use of isotope-labelled internal standards (ILIS) is among the best solutions for matrix effects correction, although the own analyte-ILIS should be used for an efficient correction. This implies the acquisition of a notable number of ILIS in multi-residue multiclass methods where many pharmaceuticals are determined. The high cost and low availability of ILIS for some compounds is a limitation to apply this approach, which however is the recommended when only a few pharmaceuticals are under study [43].

Another interesting topic that has arisen in the last few years is the possibility of direct injection of samples in LC-MS/MS methods. Traditionally, a pre-concentration step has been applied for pharmaceuticals determination, typically based on solid-phase extraction (SPE) using different cartridges [38, 44, 46, 49]. However, recent advances in MS instrumentation, particularly in terms of sensitivity improvement, have made possible the direct injection of samples, reaching concentrations at the sub-ppb levels without any type of sample treatment. This is highly convenient, as it saves time and costs, and reduces the possibility of analytical errors associated with sample manipulation. With the LC-MS/MS new-generation instrumentation, it is possible to avoid the SPE step [43, 54]. We will see in the near future an increase in applications of LC-MS/MS in the pharmaceuticals field making use of direct injection of samples.

We cannot end this section without emphasizing the importance of the quality control of the analyses. There is not discussion about the need to validate the analytical methods applied. However, even if a previous validation is made, it is necessary to apply strict rules for quality control in every batch of samples analysed. The fact that a method works well in a given sample (e.g. analysed by quintuplicate, with satisfactory recovery and precision) does not mean that it will give satisfactory results when applied to other samples, even if they are from the same type than the sample used for validation. Environmental water and wastewater samples, among others, are very different in nature and may have quite different composition and very distinct matrix effects. Therefore, it is necessary to ensure the quality of analysis in real-world samples. One of the best ways to guarantee the reliability of an analytical method is to analyse a notable number of quality control (QC) samples together with the samples under study. QCs are prepared by spiking "real-world blank" samples at known analyte concentrations. QC recoveries are used to assess the quality of data and must be typically between 60 and 140% to be acceptable. QC analysis is not only used to evaluate the recoveries, but the ion intensity ratios used for identification of the compounds [43]. On some occasions, it is hard to find true blank samples for preparation of QCs, due to the presence of the studied compound in all the samples analysed (e.g. some pharmaceuticals in wastewater). This fact makes problematic the recoveries calculation for QCs at low analyte concentration.

2.1.2 Techniques Used for Identification/Elucidation of Transformation Products (TPs)

In the most comprehensive studies, not only the evolution of the pharmaceutical concentration is pursued but also the potential transformation/degradation of the compound. While LC-MS/MS is highly appropriate for quantitative analysis, it is however limited for elucidation purposes. Here, HRMS is the preferred technique due to the useful information contained in accurate-mass full-spectrum acquisition data provided [55–57]. Unfortunately, other common elucidation techniques, such as NMR, IR or X-ray, have little application in degradation experiments because of their limited sensitivity and the low amount of the compound to be elucidated, which is present at very low concentrations in the aqueous media used in the degradation experiments. Gas chromatography-based techniques are rarely used because most of the species are in aqueous phase and contain polar groups [23]. So, LC-HRMS (with analysers such as (Q)Orbitrap or (Q)TOF) is the most applied. In general, the data obtained by this technique is the only information available [39, 40, 42, 58, 59]. This fact, added to the unknown nature of the TPs, makes the elucidation process very complicated [41, 60].

In most cases, only a tentative identification is possible. However, if such identification is well founded and supported by accurate-mass data, with a deep knowledge of fragmentation rules, the information obtained and the chemical structure proposed for the TPs can be sufficient for the purpose of the work. Thus, the primary transformations of six antibiotics during the sonochemical treatment were studied through establishing the initial degradation products by LC-QTOF MS [32]. The final acquisition of the reference standard is the last step to confirm the identity of the TP, which is a hard task when many TPs are involved due to the low availability of reference standards. Under this situation, the only possibility for full confirmation of the identity is the synthesis of the candidate compound.

HRMS allows the application of different workflows for efficient TPs identification: target analysis for already known TPs with standards available; suspect screening for potential TPs reported in literature and/or predicted by in silico models; and non-target analysis for discovering unknown, non-reported nor previously identified TPs. Searching in environmental samples or in laboratory experiments for target TPs, with reference standards available, is not a big problem, as the accurate-mass information together with the chromatographic retention time and experimental fragments resulting from the injection of the reference standard communicates high confidence in the identification. Suspect screening approaches make use of lists of tentative compounds. The use of a purpose-built database containing TPs reported in the literature (e.g. including elemental composition as well as information about the product ions reported) is a good approach that facilitates the identification of TPs. This strategy allowed the tentative identification by LC-QTOF MS of 12 TPs from eight pharmaceuticals (gemfibrozil, nimesulide, furosemide, paracetamol, propranolol, dipyrone, fluoxetine and diazepam) after application of the Fenton process [61].

The information to create a list of suspect compounds can be collected from software prediction tools and/or databases containing a broad number of compounds to be likely detected [62]. In some cases, several lists (from literature, in-house library, and created by the software used for compound prediction) can be combined for a more comprehensive searching [59]. The prediction of TPs is usually performed in silico using different tools, such as the Eawag-Pathway (enviPath) prediction system (formerly UM-PPS; http://eawag-bbd.ethz.ch/). After TP prediction, suspect screening of expected exact masses of TPs is conducted in HRMS data, usually comparing versus a control sample. The plausibility of retention time, isotopic pattern and ionization mode can be used to reduce the candidate peaks. Finally, MS/MS acquisition can also be performed, which allows matching the chemical structure of the suspected TP to the observed fragmentation.

Another interesting approach to investigate the presence of TPs relies on the assumption that many of them maintain similarity in their chemical structure with the parent compound and therefore present several common fragment ions [55, 60]. Searching for common fragments is a powerful strategy, able to detect and tentatively identify TPs/metabolites in the samples, although it is time-consuming and requires a notable knowledge of mass fragmentation rules. This approach has been applied to the identification of valsartan and losartan TPs produced by application of sonochemical systems or other advanced oxidation processes [12, 52, 63].

2.2 *Transformations of Pharmaceuticals Promoted by Sonochemical Action*

The use of the above detailed techniques for identification of transformation products has shown that the sonochemical treatment initially induce modifications such as hydroxylations, oxidations, fragmentations, decarboxylations, dealkylations or dehalogenations of the pharmaceuticals. For example, oxacillin (antibiotic) has as primary transformation pathways: oxidation of the thioether group, opening of its β-lactam and fragmentation of the central amide, all by attacks of the sonogenerated hydroxyl radical [64]. In the case of norfloxacin (antibiotic), a decarboxylation has been reported, while for cephalexin (antibiotic), the opening of its β-lactam ring is an initial degradation pathway [32]. Acetaminophen (analgesic) experiments hydroxylations initially [65]. Also, atenolol (beta-blocker) exhibits hydroxylation of its aromatic ring and propoxyl group combined to a rupture of the central ether [18].

It is reported that losartan (antihypertensive) experiments cleavage of its imidazole ring, hydroxylations of the biphenyl-tetrazole nucleus and alkyl moiety, plus oxidation of its alcohol group [12]. In turn, carbamazepine (psychiatric pharmaceutical) is initially transformed through hydroxylations and epoxidation at the double bond in the middle of its structure, followed by fragmentations of the primary products [22]. Meanwhile, the ranitidine (antiulcer) is degraded by the HO• attacks to its N,N-dimethylamine, N-methylamine and α-carbons to the sulphur [66].

In general, the pharmaceuticals degradation by action of ultrasound can be schematized as a sequence of three phases (Fig. 3): phase (1) formation of primary intermediates from parent pollutants; phase (2) rupture of the primary intermediates to generate shorter-chain or ring-opened structures and phase (3) transformation of substances from phase 2 into small molecules such as aliphatic carboxylic acids [23]. Although the sonochemical process is able to induce severe transformations on

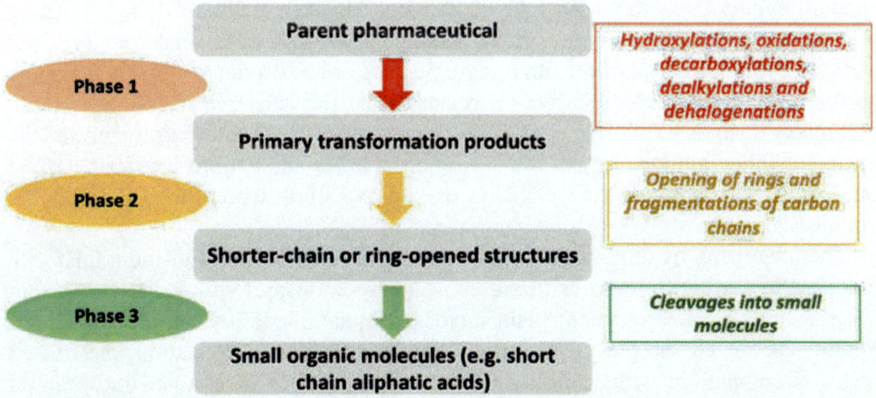

Fig. 3 Schematic sequence of phases during pharmaceuticals degradation by sonochemistry. Figure was made by the authors

pharmaceuticals, it is recognized that this system has a moderate mineralizing ability [7, 8, 23, 64]. This is because most products formed in the phase 3 (e.g. short chain carboxylic acids) have a very hydrophilic character, which limits their interaction with the sonogenerated hydroxyl radical. However, the mineralization degree can be improved by adding to the sonochemical system some reagent (e.g. ferrous ions, persulfate anion, ozone or hydrogen peroxide) that promotes generation of extra radical able to reach the hydrophilic products in the solution bulk [23].

Despite the moderate mineralizing ability of sonochemistry acting alone, during the treatment of antibiotics, this process is able to generate by-products without antimicrobial activity due to the sonogenerated hydroxyl radical attacks the moieties responsible for such activity [32, 64]. Furthermore, in many cases the sonodegradation of pharmaceuticals produces water having smaller molecules more biodegradable and/or less toxic than the parent pollutants [23]. These aspects indicate positive changes exerted by the sonochemical treatment of water polluted with pharmaceuticals.

3 Treatment of Wastewater Polluted with Pharmaceuticals by Ultrasound

3.1 Removal of Pharmaceuticals in Industrial, Municipal and Hospital Wastewaters Effluents

Ultrasound is a versatile advanced oxidation process, which has been applied to treat pharmaceuticals within diverse industrial, municipal and hospital wastewater (Table 4). For instance, degradation of oxacillin or dicloxacillin (antibiotics) in simulated effluents of pharmaceutical industries (the pharmaceuticals plus excipients such as mannitol, calcium carbonate or glucose) has been evaluated [25, 64]. These works show that the elimination of the pollutants or its antimicrobial activity is not affected by the presence of the other ingredients, even when the excipients are much more concentrated than the target pharmaceutical. The selective treatment of these pharmaceuticals is explained considering that due to the hydrophilic nature of the excipients, they are far away of the cavitation bubbles and had no interfering effect on the degradation of the oxacillin or dicloxacillin, which are hydrophobic antibiotics.

The treatment by ultrasound of a mixture of diclofenac (anti-inflammatory), amoxicillin (antibiotic) and carbamazepine (anticonvulsant) spiked on an effluent of urban wastewater treatment plant increased biodegradability and decreased the toxicity of treated samples. Thus, after sonication process application, the treated water is compatible with a biological process or can be discharged into natural aqueous media [16]. Also, the efficiency of ultrasound to reduce hormones oestrogen (E1) and 17-beta-oestradiol (E2) from municipal wastewater has been tested, reporting that after 45 min the process removed 85 and 96% of E1 and E2, respectively [67].

Table 4 Treatment of pharmaceuticals in industrial, municipal and hospital wastewaters by ultrasound

Pharmaceuticals	Wastewater[a]	Ultrasound conditions	References
Dicloxacillin	Simulated PWW (dicloxacillin and glucose)	Frequency: 600 kHz Electric power: 60 W Volume: 300 mL	[25]
Oxacillin	Simulated PWW (oxacillin, mannitol and calcium carbonate)	Frequency: 275 kHz Electric power: 60 W Volume: 250 mL	[64]
Mixture of diclofenac, amoxicillin and carbamazepine	Pharmaceuticals were spiked to an EWWTP	Frequency: 20 kHz Electric power density: 100 W L^{-1} Volume: 200 mL	[16]
Oestrogen (E1) and 17-beta-oestradiol (E2)	Municipal WW	Frequency: 30–60 kHz Electric power density: 70–110 W Volume: 1.5 L	[67]
5-Fluorouracil, ibuprofen, clonidine, estriol, nifedipine and lovastatin	The pharmaceuticals were individually added to an EWWTP	Frequency: 205 kHz Acoustic power density: 45 W L^{-1} Volume: 300 mL	[27]
Seventeen pharmaceuticals Bogotá	Actual EWWTP	Frequency: 375 kHz Acoustic power density: 88 W L^{-1} Volume: 300 mL	[53]
Losartan and valsartan	Pharmaceuticals individually spiked to an EWWTP	Frequency: 375 kHz, Acoustic power density: 88 W L^{-1}, Volume: 300 mL	[12]
Trimethoprim	EWWTP	Frequency: 20 kHz Acoustic power density: 36 W L^{-1} Volume: 200 mL	[68]
Cloxacillin and cefadroxil	Simulated HWW	Frequency: 375 kHz Acoustic power density: 88 W L^{-1} Volume: 300 mL	[32]
Losartan and valsartan	Simulated hospital wastewater	Frequency: 375 kHz Acoustic power density: 88 W L^{-1} Volume: 300 mL	[12]
Fifteen pharmaceuticals	HWW prior treated by an aerobic biological process	Frequency: 375 kHz Acoustic power density: 88 W L^{-1} Volume: 350 mL	[52]

(continued)

Table 4 (continued)

Pharmaceuticals	Wastewater[a]	Ultrasound conditions	References
Amantadine	PWW	*Sono-Fenton:* Frequency: 25 kHz Electric power: 200 W Volume: 50 mL Fe (II): 0.1 mol L^{-1} and H_2O_2: 2.0 mol L^{-1}	[69]
Ciprofloxacin	Pharmaceutical spiked to an EWWTP	*Sono-Fenton:* Frequency: 580 kHz Actual power: 30.6 W Volume: 350 mL Ratio H_2O_2/Fe (II): 6 and H_2O_2: 14.2–31.8 mmol L^{-1}	[70]
Ampicillin and nafcillin	Simulated EWWTP	*Sono-photo-Fenton:* Frequency: 375 kHz Acoustic power density: 24.4 W L^{-1} Volume: 250 mL Fe (II): 90 μmol L^{-1} UVC lamp: 4 W	[33]
Tinidazole	Simulated PWW and actual PWW	*Sono-peroxide:* Frequency: 120 kHz Electric power: 750 W Volume: 100 mL H_2O_2: 333 mmol L^{-2}	[71]
Mixture of paracetamol, chloramphenicol and diclofenac	Simulated PWW	*Sono-photolysis:* Frequency: 20 kHz Electric power: 80 W Volume: 7.0 L *(Pilot reactor)* H_2O_2: 900 mg L^{-1} UVC lamp: 13 W	[72]
Diclofenac	Pharmaceutical spiked on an actual EWWTP	*Sono-ozonolysis:* Frequency: 20 kHz Electric power density: 400 W L^{-1} Volume: 150 mL Ozone: 31 g h^{-1}	[73]
Diclofenac, sulfamethoxazole and carbamazepine	Mixture of pharmaceutical	*Sono-ozonolysis:* Frequency: 20 kHz Electric power density: 370 W L^{-1} Volume: 200 mL Ozone: 1.3–3.3 g h^{-1}	[74]

[a]*WW* raw wastewater, *EWWTP* effluent of municipal wastewater treatment plant and *PWW* pharmaceutical wastewater

Xiao et al. investigated the individual degradation by sonochemistry of fluorouracil (5-FU), ibuprofen (IBU), clonidine (CLND), estriol (ESTO), nifedipine (NIFE) and lovastatin (LOVS) in a wastewater effluent. Because of the higher concentration of the effluent components, the matrix slowed degradation of the pharmaceuticals compared to elimination in deionized water. Degradations of 5-FU and LOVS were inhibited by the wastewater components in a greater extent than the other pharmaceuticals. The elimination trends indicated that the pharmaceuticals reacting primarily on the bubble surface were slightly affected by the matrix components, whereas substances reacting mainly in the bulk were strongly affected by the effluent. Indeed, some organic matter of the effluent are able to migrate to the cavitation bubble surfaces and compete for the sonogenerated hydroxyl radicals [27].

Losartan and valsartan are other pharmaceuticals treated by ultrasound in municipal wastewater plant effluents. For these compounds, degradations in the effluents were very similar to the obtained in distilled water, indicating that the matrix components had a very low effect on the removal of losartan and valsartan by ultrasound. This was associated with the high hydrophobic character of such pharmaceuticals [12].

In addition to matrices spiked with the pollutants, treatment of actual wastewater, containing pharmaceuticals at realistic $\mu g\ L^{-1}$ levels, has been considered. In fact, the degradation of 17 pharmaceuticals (diclofenac, carbamazepine, venlafaxine, ciprofloxacin, norfloxacin, valsartan, losartan, irbesartan, sulfamethoxazole, clarithromycin, azithromycin, erythromycin, metronidazole, trimethoprim and clindamycin) in an effluent of municipal wastewater plant (which had only primary processes) has been reported. The ultrasound action increased the concentrations of four pollutants (diclofenac, sulfamethoxazole, ciprofloxacin and norfloxacin), whereas for the rest of pharmaceuticals, the concentration decreased with the treatment. Such results are rationalized based on both physical and chemical effects of ultrasound. Owing to the effluent contained suspended solids, some pollutants sorbed on such solids increased their concentrations by the physical action of ultrasound (i.e. turbulence and high shearing in the liquid medium by cavitation phenomena), which induces a size reduction of solids and releasing of adsorbed pharmaceuticals. Meanwhile, the removal of the pollutants was associated with the chemical effects of the process (i.e. degrading action of hydroxyl radical, [53]).

As well as effluents from pharmaceutical industry and municipal treatment plants, hospital wastewaters loaded with pharmaceuticals have been also processed by sonication. Indeed, synthetic hospital wastewater contaminated with cloxacillin, cefadroxil, losartan or valsartan has been submitted to ultrasound action. The elimination of cloxacillin, losartan or valsartan in the hospital effluent is very efficient. In contrast, cephalexin (which is very hydrophilic) removal is affected by the matrix components [12, 32].

Other example of elimination of pharmaceutical in wastewater is the sonication of a real effluent of hospital prior treated by an aerobic biological process. In this case the removal of 15 pollutants (acetaminophen, diclofenac, carbamazepine, venlafaxine, loratadine, ciprofloxacin, norfloxacin, valsartan, irbesartan, sulfamethoxazole, trimethoprim, clarithromycin, azithromycin, erythromycin and clindamycin) was considered. Ultrasound application led to a decreasing of the concentrations of pharmaceuticals except for acetaminophen and valsartan, suggesting that such compounds were liberated from active sludge flocs proceeding from the bio-treatment. When the adsorbed pharmaceuticals are released, they also are susceptible to the chemical effects of ultrasound. The sonogenerated hydroxyl radicals are the main responsible of the elimination of pollutants, leading to ~59% of pondered removal of pharmaceuticals after 90 min of treatment [52].

Although sonication alone has shown a high ability to degrade pharmaceuticals in complex matrices, as an alternative to improve the pollutants degradation and/or reduce the treatment times, the ultrasound action has been complemented by addition of reagents to generate hybrid advanced oxidation processes. Such is the situation of amantadine (antiviral and antiparkinsonian) in a wastewater of the pharmaceutical industry, which was degraded by sono-Fenton (by adding Fe (II) and H_2O_2 at the water sample). The hybrid process lead to a synergistic pharmaceuticals degradation and a significant reduction of wastewater toxicity, thanks the extra formation of radicals by means of the Fenton reaction (Eq. 7) [69]. On the other hand, the feasibility of ciprofloxacin (antibiotic) degradation in a real matrix from a municipal wastewater treatment plant by sono-Fenton was informed. The addition of H_2O_2 and Fe (II) to the sonication system improved both the pharmaceutical elimination and mineralization [70].

Sono-photo-Fenton, which is formed by addition of Fe (II) and UV light to the sonochemical system, has been applied for degrading ampicillin and nafcillin (antibiotics) in a simulated effluent of wastewater treatment plant. This system induced removals of the antibiotics ≥90% at 60 min of treatment [33]. Likewise, the simultaneous degradation of several pharmaceuticals in hospital wastewater previously bio-treated or in actual effluents of wastewater treatment plant is significantly increased by the utilization of sono-photo-Fenton. Such system may produce additional radical species at the solution bulk through interaction among iron, hydrogen peroxide and UV light (Eqs. 7 and 9) [52, 53]. Also, it should be mentioned that ultrasound plus hydrogen peroxide, persulfate anion and iron swarf is another hybrid process able to generate extra amounts of radicals. This system has successfully applied to reduce the chemical oxygen demand of pharmaceutical effluents [75].

$$Fe^{3+} + H_2O + \text{UV light} \rightarrow Fe^{2+} + HO\bullet + H^+ \quad (9)$$

The addition of hydrogen peroxide solely is another way to intensify sonochemical degrading action (i.e. sono-peroxide process, Eqs. 10–11), specially at low frequencies. This procedure has been reported for tinidazole (antiparasitic), which was treated in both synthetic and actual pharmaceutical industrial effluents, achieving at 150 min of treatment, 75 and 68%, respectively [71]. Furthermore, low frequency ultrasound (30 kHz) intensified by H_2O_2 and activated carbon (AC) has been applied to pharmaceutical wastewater, decreasing ~79% of chemical oxygen demand of such matrix [76].

$$H_2O_2 +))) \rightarrow 2HO\bullet \tag{10}$$

$$H\bullet + H_2O_2 \rightarrow HO\bullet + H_2O \tag{11}$$

The merging of low frequency ultrasound (20 kHz), hydrogen peroxide and UVC light generates a sono-photolytic system. This kind of process has been utilized for treating paracetamol, chloramphenicol and diclofenac in a simulated pharmaceutical wastewater. The sono-photolytic system exhibited a synergistic effect for the mineralization of this wastewater, achieving 91% of total organic carbon removal after 120 min of treatment, due to the high ability of such system to produce hydroxyl radicals (Eqs. 10–12) [72].

$$H_2O_2 + UVC \rightarrow 2HO\bullet \tag{12}$$

Aside from addition of hydrogen peroxide, activated carbon, iron and light, another possibility is the simultaneous action of low frequency ultrasound and ozone (sono-ozonolysis), which has also shown to be useful for degrading pharmaceuticals in complex matrices. This system is able to improve the mass transfer of ozone to the aqueous medium and the collapsing bubbles yield additional free radicals. For example, the treatment of a mixture of diclofenac, sulfamethoxazole and carbamazepine (simulating a pharmaceutical wastewater) by sono-ozonolysis resulted in an enhanced removal of these pharmaceuticals [74]. Similarly, the removal of organic carbon content from a wastewater treatment plant effluent polluted with diclofenac was improved by the join action of ultrasound and ozone [73].

It must be mentioned that other cases related with the degradation of pharmaceuticals in wastewater by ultrasound alone, intensified or combined with other process, can be found in the following references [77–85]. Besides, from a critical point of view, it should be remarked that the most of works on elimination of pharmaceuticals pollutants in wastewaters lack toxicity or biodegradability analyses to evidence the positive impact of treatments. Moreover, few works on hybrid/intensified processes report or discuss the synergy; thus it is not clear the actual effects of addition of reagents or light to ultrasound. Finally, despite varied application of the ultrasound-based processes to very complex water matrices, the volumes of treatment are typically low (<1.0 L). Therefore, information about performance of these processes at large-scale is still scarce.

3.2 Ultrasound Generators and Engineering Aspects for the Wastewater Treatments by Ultrasound

Presented in this section is information on different types of ultrasound generators available and the engineering aspects associated with treating wastewater by ultrasound from lab scale to pilot scale.

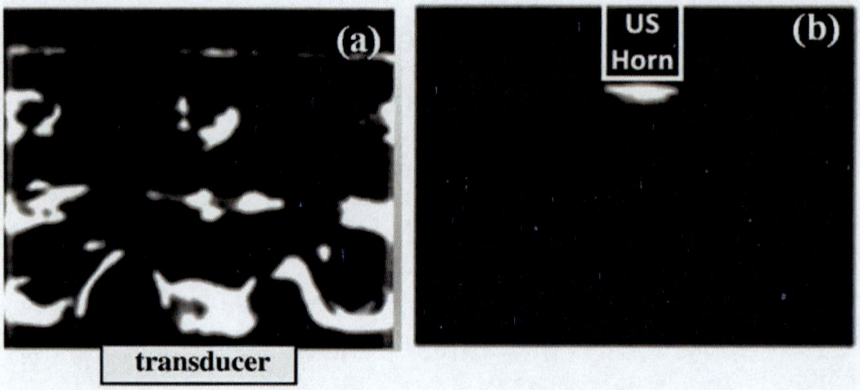

Fig. 4 Sonoluminescence activity (white part) generated by (**a**) 98 kHz plate transducer and (**b**) 20 kHz ultrasonic horn. Reprinted with permission from reference [86]

3.2.1 Type of Ultrasound Generators

Ultrasound is usually generated by a piezoelectric ceramic transducer with a specific resonance frequency that can range from 20 kHz up to 10 MHz and is powered by an amplifier. The transducers are usually glue to a vibrating plate or bottom of an ultrasonic bath with localized cavitation activity (revealed through sonoluminescence) at the pressure antinodal planes perpendicular to the direction of the ultrasound waves (Fig. 4a). Also, power ultrasonic horns have been designed to allow the transduction at a tip, which can vary in diameter from 2 mm up to 45 mm. This leads to high intense cavitation and fluid mixing mainly localized near the horn tip (Fig. 4b) [86]. These ultrasonic horn and bath are usually at the lower end of the ultrasound frequencies (20–100 kHz). For frequencies between 100 kHz and 2 MHz, a specialized reactor setup is usually required coupled with a high-power amplifier. For laboratory scale, one ultrasonic horn or a single plate transducer are generally used for volumes ranging from 1 mL to 1 L and in a batch mode. Meanwhile, for large-scale processing, multiple horns or plate transducers are employed and regularly involve a continuous flow system that requires several passes.

3.2.2 Laboratory-Scale Sonication

As above presented, ultrasound has shown to be effective for degrading pharmaceuticals in diverse water matrices. In this sense, to determine the engineering aspects in terms of the applied frequency, power density and geometry is critical. What is even more important is the spatial distribution of the cavitation activity within the reactor,

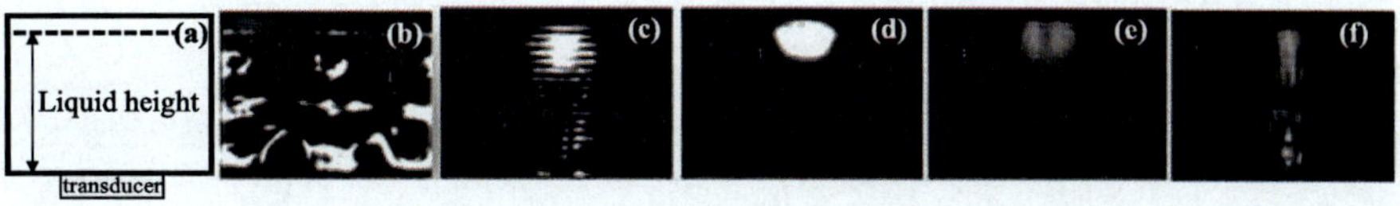

Fig. 5 Spatial distribution of sonoluminescence (SL) intensity for different frequencies: (**a**) configuration of the sonication cell, (**b**) 37 kHz, (**c**) 168 kHz, (**d**) 448 kHz, (**e**) 726 and (**f**) 448 kHz with a surface stabilizer. All images were taken with an exposure time of 30 s and amplifier of 20 W. Reprinted with permission from references [86, 88]

which determines the efficiency of the system and is affected by frequency, power and solution conditions [87].

Figure 5 shows the variation in the spatial distribution of sonoluminescence (SL) from cavitating bubbles generated by different ultrasound frequencies. An optimum frequency for high sonochemical yield has been reported to be related to bubbles characteristics (e.g. population and collapse intensity) with increasing frequency [88]. However, attenuation and spatial distribution of ultrasound within the reactor could also contribute to the optimum frequency for sonochemical yield, and it could be altered by reactor design as illustrated by Fig. 5f, where the spatial distribution is enhanced when the reflectivity of the waves is increased by stabilizing the liquid surface. This would explain the different optimum frequencies being reported between 200 kHz and 800 kHz [1, 89, 90]. Therefore, in addition to the sonochemical yield, it is important to evaluate the spatial distribution of cavitation activity for a given setup.

Laboratory-scale systems are limited from 10 to 1,000 mL, with the cavitation mainly located above the transducer. Often, different sonochemical reactor geometries are used, which includes immersing small vials containing the wastewater solution. For a cylindrical cell with a diameter matching that of the transducer diameters, the entire liquid above the transducer are filled with cavitation bubbles. However, for the rectangular reactor where the width of the reactor extends beyond the transducer diameter, there are regions near the wall that do not have active cavitation taking place. This would thus decrease the overall efficiency of the system [91]. When all the cavitation is concentrated within the vial, the power densities for the different systems would therefore vary and consequently modify the overall pollutants degradation efficiencies [92].

3.2.3 Industrial-Scale Sonication

Full-scale development of ultrasound reactors for wastewater treatment in industry is still incipient. There are a few reports on in-house built pilot-scale studies involving multiple transducers attached to the side of a reactor with volumes ranging from 6 L up to 91 L [93–95]. Figure 6 shows some large-scale ultrasound reactors reported in the literature. The highest ultrasonic reactor was 91 L reported by Gole et al. [95] for the degradation of organic pollutants (per- and poly-fluorinated alkyl sulfonate

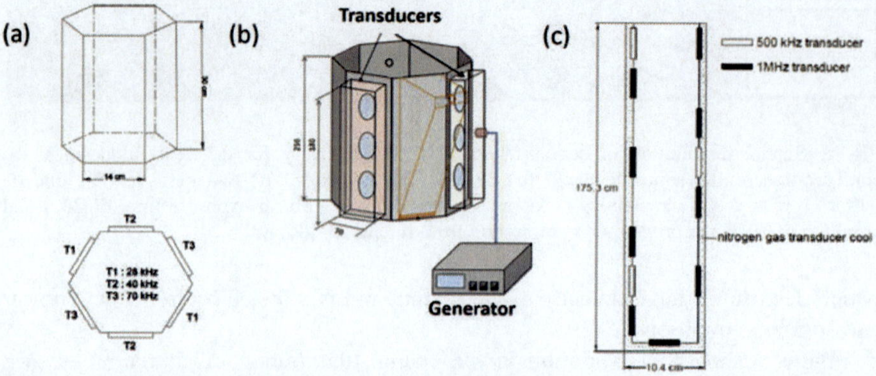

Fig. 6 Large-scale ultrasound reactors reported in the literature. (**a**) 15 L triple frequency reactor (adapted with permission from [94]), (**b**) 6 L reactor with four 332 kHz (adapted with permission from [93]) and (**c**) 91 L multifrequency reactor (adapted with permission from [95])

substances (PFAS)). The reactor was fabricated by PCT System Inc. (Fremont, CA, USA), and it has dimensions of 20.9 × 26.0 × 175.3 cm with 12 transducers, three of these transducers were 500 kHz, and nine were 1 MHz transducers, positioned as indicated in Fig. 6c. It was concluded that the ultrasound treatment was effective at reducing the fluorocarbon substances and it led to a moderate removal of total organic carbon with both single and dual frequency modes [95]. In turn, it has been reported the use of a 140 L ultrasonic reactor equipped with 24 transducers at 350 kHz, each capable of generating free radicals at a rate of 5.2 μmol L^{-1} min^{-1} [96]. However, results on the performance of sonication treatment for specific organic pollutants were no reported (only global parameters).

In the large-scale ultrasound reactor studies, no justification for the chosen design, frequencies and powers were given, and there was a lack of focus on optimizing the conditions controlling cavitation activity and distribution. Upscaling of reactors can potentially be influenced by many parameters that affect the efficiency of sonolysis. This is because changes in liquid volume, liquid height, vessel geometry and solution properties will have significant effect on the ultrasound wave field inside the vessel [97]. As a result, the cavitation activity, which is depended on ultrasound wave field, is affected, and this in turn will affect the efficiency of the sonochemistry. To scale-up ultrasound reactors, there are many optimization factors that can be explored such as sound field distribution, bubbling gas in-line, pH changes and changing the external temperature and pressure. Therefore, there is a need for much more studies on the incremental upscaling ultrasound reactors, especially at high frequencies.

On the other hand, as outlined above, large-scale ultrasound reactors usually require multiple transducers that would need multiple amplifiers to drive each transducer and often cooling is required due to the heating of the transducers. These additional costs associated with the upscaling further hinder the development of pilot-scale ultrasound reactors. An alternative that seems promising to face such concerns associated with large-scale ultrasound applications is the combination with membrane filtration techniques, allowing to concentrate pollutants and significantly reduce the volumes to be treated, which could be processed by using small ultrasound reactors. This topic is developed in the next section.

3.3 *Combinations of Ultrasound with Membrane Filtration Techniques as a Strategy for Overcoming the Issues Related to Large-Scale Application*

Membrane filtration is a separation technique that is based on either size or charge, or both depending on the type of membrane material and the compound to be separated. Nanofiltration (NF) and reverse osmosis (RO) have shown high rejection (>85%) of pharmaceuticals from wastewater [98–100]. However, as a separation technology, the process concentrates the pharmaceutical wastes and needs to be combined with other treatment that is capable of destroying the pollutant.

Although ultrasound has shown to be effective for degrading pharmaceutical pollutants, most of the literature on the combination of ultrasound and membrane filtration focuses on the physical impact of ultrasound on enhancing the permeate flux, reduce fouling and membrane surface cleaning [101–103]. A unique hybrid ultrasound-activated carbon-ultrafiltration membrane (USAMe) process to treat three active pharmaceutical ingredients (API), diclofenac, carbamazepine and amoxicillin has been reported [104]. The USAMe composed of glass tubing, encasing a single hollow fibre membrane with powdered activated carbon deposited within the fibre, that is immersed in an ultrasonic bath with a frequency of either 35 kHz or 135 kHz. It was reported that the presence of ultrasound increased the percentage removal of the pharmaceuticals during the treatment. Additionally, 35 kHz was found to be more effective than 135 kHz at sustaining a high removal rate over the duration of the experiment, which led to the conclusion that the main mechanism of ultrasound is to enhance the adsorption of the API onto the activated carbon rather than sonochemically degrading the API [104].

Caretti et al. had used a pilot plant-scale ultrasound reactor to treat a wastewater effluent prior to ultrafiltration (UF) treatment compared to the use of ultrasound to treat the permeate from the UF [96]. It was found that pre-treatment of the effluent prior to UF was more effective (the efficiency of the pilot-scale ultrasound reactor was not reported). However, the configurations proposed in this last work consider an ultrasound reactor at large-scale, which has the limitations previously indicated. Therefore, a new strategy should be considered. In the future, different membrane filtration technologies may be used to concentrate the waste stream (target organic pollutants as pharmaceuticals) and reduce to a more manageable volume to be treated by ultrasonication at small-scale, enhancing the efficiency of pollution elimination and the extraction of clean water permeate that can be recycled.

4 Concluding Remarks

From information presented in this chapter, it can be highlighted the following aspects:

- Elimination of pharmaceuticals by ultrasound is strongly dependent of operational parameters such as frequency and power (or density power).
- The hydrophobic/hydrophilic nature of pharmaceutical pollutants determines their degradation rates by sonochemistry. Furthermore, when the molecules have acid/base functional groups, the hydrophobic/hydrophilic character may be modified through variation of the solution pH.
- The sonogenerated hydroxyl radicals are able to induce primary modifications such as hydroxylations, oxidations, fragmentations, decarboxylations, dealkylations, or dehalogenations to the pharmaceuticals. Meanwhile, the intensive action of radicals leads to transformation of these pollutants into small non-toxic or biocompatible substance; even some degree of pharmaceuticals mineralization is achieved. However, it should be indicated that the most studies about the pharmaceuticals transformations by ultrasound are carried out in very simple water matrices (distilled water, deionized water or milli-Q water). Hence, some future researches on the influence on matrix components on transformations could be developed.
- Ultrasound is able to degrade pharmaceuticals in diverse wastewater; even selective elimination is possible in some cases. The sonochemical action can be enhanced by adding reagents (e.g. ferrous ions) or light. Nevertheless, further investigations should include more information on toxicity and biodegradability of treated wastewater, in addition to synergy calculations to evidence the actual effects of addition of reagents or light to ultrasound.
- Applications at pilot and full scales of ultrasound for wastewater treatment are still incipient. Additionally, it is recognized that the main drawbacks of sonochemical processes for application at large-scale are the high energy costs and the requirements of multiple of amplifiers and cooling systems to avoid the transducers heating. An option to overcome such concerns could be the combination of ultrasound with membrane filtration techniques to concentrate pollutants and significantly reduce the wastewater to volumes which can be treated by using the small available ultrasound reactors.

Acknowledgements The authors thank the financial support provided by the Royal Society (UK) through the project "Sound" methods of remediating emerging contaminants in hospital wastewater (RA4056, ICA\R1\191053) and the financing from COLCIENCIAS through the project No. 111577757323 (Convocatoria 777 de 2017). R.A. Torres-Palma and E.A. Serna-Galvis thank Universidad de Antioquia UdeA for the support provided to their research group through "Programa de Sostenibilidad".

References

1. Wood RJ, Lee J, Bussemaker MJ (2017) A parametric review of sonochemistry: control and augmentation of sonochemical activity in aqueous solutions. Ultrason Sonochem 38:351–370. https://doi.org/10.1016/j.ultsonch.2017.03.030
2. Rivera-Utrilla J, Sánchez-Polo M, Ferro-García MÁ et al (2013) Pharmaceuticals as emerging contaminants and their removal from water. A review. Chemosphere 93:1268–1287. https://doi.org/10.1016/j.chemosphere.2013.07.059
3. Adewuyi YG (2001) Sonochemistry: environmental science and engineering applications. Ind Eng Chem Res 40:4681–4715. https://doi.org/10.1021/ie010096l
4. Riesz P, Berdahl D, Christman CL (1985) Free radical generation by ultrasound in aqueous and nonaqueous solutions. Environ Health Perspect 64:233–252. https://doi.org/10.2307/3430013
5. McNamara WB, Didenko YT, Suslick KS (1999) Sonoluminescence temperatures during multi-bubble cavitation. Nature 401:772–775. https://doi.org/10.1038/44536
6. Beckett MA, Hua I (2001) Impact of ultrasonic frequency on aqueous sonoluminescence and sonochemistry. J Phys Chem A 105:3796–3802. https://doi.org/10.1021/jp003226x
7. Tran N, Drogui P, Brar SK (2015) Sonochemical techniques to degrade pharmaceutical organic pollutants. Environ Chem Lett 13:251–268. https://doi.org/10.1007/s10311-015-0512-8
8. Montoya-Rodríguez DM, Ferraro F, Serna-Galvis EA, Torres-Palma RA (2020) Degradation of the emerging concern pollutant ampicillin in aqueous media by sonochemical advanced oxidation processes – parameters effect, removal of antimicrobial activity and pollutant treatment in hydrolyzed urine. J Environ Manage 261:110224. https://doi.org/10.1016/j.jenvman.2020.110224
9. Mahamuni NN, Adewuyi YG (2010) Advanced oxidation processes (AOPs) involving ultrasound for waste water treatment: a review with emphasis on cost estimation. Ultrason Sonochem 17:990–1003. https://doi.org/10.1016/j.ultsonch.2009.09.005
10. Castelo-Grande T, Augusto PA, Barbosa D (2020) Chapter 14 ultrasound-assisted remediation methods. In: The handbook of environmental remediation: classic and modern techniques. The Royal Society of Chemistry, Cambridge, pp 379–409
11. Kang JW, Hung HM, Lin A, Hoffmann MR (1999) Sonolytic destruction of methyl tert-butyl ether by ultrasonic irradiation: the role of O3, H2O2, frequency, and power density. Environ Sci Technol 33:3199–3205. https://doi.org/10.1021/es9810383
12. Serna-Galvis EA, Isaza-Pineda L, Moncayo-Lasso A et al (2019) Comparative degradation of two highly consumed antihypertensives in water by sonochemical process. Determination of the reaction zone, primary degradation products and theoretical calculations on the oxidative process. Ultrason Sonochem 58:104635. https://doi.org/10.1016/j.ultsonch.2019.104635
13. Guo W, Wang H, Shi Y, Zhang G (2010) Sonochemical degradation of the antibiotic cephalexin in aqueous solution. Water SA 36:651–654. https://doi.org/10.4314/wsa.v36i5.61998
14. Naddeo V, Belgiorno V, Kassinos D et al (2010) Ultrasonic degradation, mineralization and detoxification of diclofenac in water: optimization of operating parameters. Ultrason Sonochem 17:179–185. https://doi.org/10.1016/j.ultsonch.2009.04.003
15. Fu H, Suri RPS, Chimchirian RF et al (2007) Ultrasound-induced destruction of low levels of estrogen hormones in aqueous solutions. Environ Sci Technol 41:5869–5874. https://doi.org/10.1021/es0703372
16. Naddeo V, Meriç S, Kassinos D et al (2009) Fate of pharmaceuticals in contaminated urban wastewater effluent under ultrasonic irradiation. Water Res 43:4019–4027. https://doi.org/10.1016/j.watres.2009.05.027
17. Hapeshi E, Achilleos A, Papaioannou A et al (2010) Sonochemical degradation of ofloxacin in aqueous solutions. Water Sci Technol 61:3141–3146. https://doi.org/10.2166/wst.2010.921

18. Nejumal KK, Manoj PR, Aravind UK, Aravindakumar CT (2014) Sonochemical degradation of a pharmaceutical waste, atenolol, in aqueous medium. Environ Sci Pollut Res 21:4297–4308. https://doi.org/10.1007/s11356-013-2301-x
19. Quesada-Pañate I, Julcour Lebigue C, Jáuregui-Haza U-J et al (2009) Sonolysis of levodopa and paracetamol in aqueous solutions. Ultrason Sonochem 16:610–616. https://doi.org/10.1016/j.ultsonch.2008.11.008
20. Vega LP, Soltan J, Peñuela GA (2019) Sonochemical degradation of triclosan in water in a multifrequency reactor. Environ Sci Pollut Res 26:4450–4461. https://doi.org/10.1007/s11356-018-1281-2
21. Gogate PR, Shirgaonkar IZ, Sivakumar M et al (2001) Cavitation reactors: efficiency assessment using a model reaction. AICHE J 47:2526–2538. https://doi.org/10.1002/aic.690471115
22. Tran N, Drogui P, Zaviska F, Brar SK (2013) Sonochemical degradation of the persistent pharmaceutical carbamazepine. J Environ Manage 131:25–32. https://doi.org/10.1016/j.jenvman.2013.09.027
23. Rayaroth MP, Aravind UK, Aravindakumar CT (2016) Degradation of pharmaceuticals by ultrasound-based advanced oxidation process. Environ Chem Lett 14:259–290. https://doi.org/10.1007/s10311-016-0568-0
24. Gao YQ, Gao NY, Deng Y et al (2013) Factors affecting sonolytic degradation of sulfamethazine in water. Ultrason Sonochem 20:1401–1407. https://doi.org/10.1016/j.ultsonch.2013.04.007
25. Villegas-Guzman P, Silva-Agredo J, Giraldo-Aguirre ALAL et al (2015) Enhancement and inhibition effects of water matrices during the sonochemical degradation of the antibiotic dicloxacillin. Ultrason Sonochem 22:211–219. https://doi.org/10.1016/j.ultsonch.2014.07.006
26. Villaroel E, Silva-Agredo J, Petrier C et al (2014) Ultrasonic degradation of acetaminophen in water: effect of sonochemical parameters and water matrix. Ultrason Sonochem 21:1763–1769
27. Xiao R, Wei Z, Chen D, Weavers LK (2014) Kinetics and mechanism of sonochemical degradation of pharmaceuticals in municipal wastewater. Environ Sci Technol 48:9675–9683. https://doi.org/10.1021/es5016197
28. Xiao R, Diaz-Rivera D, Weavers LK (2013) Factors influencing pharmaceutical and personal care product degradation in aqueous solution using pulsed wave ultrasound. Ind Eng Chem Res 52:2824–2831. https://doi.org/10.1021/ie303052a
29. Xiao R, Diaz-Rivera D, He Z, Weavers LK (2013) Using pulsed wave ultrasound to evaluate the suitability of hydroxyl radical scavengers in sonochemical systems. Ultrason Sonochem 20:990–996. https://doi.org/10.1016/j.ultsonch.2012.11.012
30. Torres-Palma RA, Serna-Galvis EA (2018) Chapter 7 – sonolysis. In: Ameta SC, Ameta RBT (eds) Advanced oxidation processes for waste water treatment. Academic Press, Cambridge, pp 177–213
31. Suri RPS, Singh TS, Abburi S (2010) Influence of alkalinity and salinity on the sonochemical degradation of estrogen hormones in aqueous solution. Environ Sci Technol 44:1373–1379. https://doi.org/10.1021/es9024595
32. Serna-Galvis EA, Montoya-Rodríguez DM, Isaza-Pineda L et al (2018) Sonochemical degradation of antibiotics from representative classes-considerations on structural effects, initial transformation products, antimicrobial activity and matrix. Ultrason Sonochem 50:157–165. https://doi.org/10.1016/j.ultsonch.2018.09.012
33. Montoya-Rodríguez DM, Ávila-Torres Y, Serna-Galvis EA, Torres-Palma RA (2020) Data on treatment of nafcillin and ampicillin antibiotics in water by sonochemistry. Data Br 29:105361. https://doi.org/10.1016/j.dib.2020.105361
34. Méndez-Arriaga F, Torres-Palma R, Pétrier C et al (2008) Ultrasonic treatment of water contaminated with ibuprofen. Water Res 42:4243–4248
35. Im J-K, Heo J, Boateng LK et al (2013) Ultrasonic degradation of acetaminophen and naproxen in the presence of single-walled carbon nanotubes. J Hazard Mater 254–255:284–292. https://doi.org/10.1016/j.jhazmat.2013.04.001

36. Serna-Galvis EA, Silva-Agredo J, Giraldo-Aguirre AL, Torres-Palma RA (2015) Sonochemical degradation of the pharmaceutical fluoxetine: Effect of parameters, organic and inorganic additives and combination with a biological system. Sci Total Environ 524–525:354–360. https://doi.org/10.1016/j.scitotenv.2015.04.053
37. Lastre-Acosta AM, Cruz-González G, Nuevas-Paz L et al (2015) Ultrasonic degradation of sulfadiazine in aqueous solutions. Environ Sci Pollut Res 22:918–925. https://doi.org/10.1007/s11356-014-2766-2
38. Carmona E, Picó Y (2018) The use of chromatographic methods coupled to mass spectrometry for the study of emerging pollutants in the environment. Crit Rev Anal Chem 48:305–316. https://doi.org/10.1080/10408347.2018.1430555
39. Hernández F, Ibáñez M, Bade R et al (2014) Investigation of pharmaceuticals and illicit drugs in waters by liquid chromatography-high-resolution mass spectrometry. TrAC Trends Anal Chem 63:140–157. https://doi.org/10.1016/j.trac.2014.08.003
40. Kosjek T, Heath E (2008) Applications of mass spectrometry to identifying pharmaceutical transformation products in water treatment. TrAC Trends Anal Chem 27:807–820. https://doi.org/10.1016/j.trac.2008.08.014
41. Hernández F, Bakker J, Bijlsma L et al (2019) The role of analytical chemistry in exposure science: focus on the aquatic environment. Chemosphere 222:564–583. https://doi.org/10.1016/j.chemosphere.2019.01.118
42. Ferrando-Climent L, Gonzalez-Olmos R, Anfruns A et al (2017) Elimination study of the chemotherapy drug tamoxifen by different advanced oxidation processes: transformation products and toxicity assessment. Chemosphere 168:284–292. https://doi.org/10.1016/j.chemosphere.2016.10.057
43. Botero-Coy AM, Martínez-Pachón D, Boix C et al (2018) An investigation into the occurrence and removal of pharmaceuticals in Colombian wastewater. Sci Total Environ 642:842–853. https://doi.org/10.1016/j.scitotenv.2018.06.088
44. Bussy U, Li K, Li W (2016) Application of liquid chromatography-tandem mass spectrometry in quantitative bioanalyses of organic molecules in aquatic environment and organisms. Environ Sci Pollut Res 23:9459–9479. https://doi.org/10.1007/s11356-016-6433-7
45. Gracia-Lor E, Sancho JV, Hernández F (2011) Multi-class determination of around 50 pharmaceuticals, including 26 antibiotics, in environmental and wastewater samples by ultra-high performance liquid chromatography–tandem mass spectrometry. J Chromatogr A 1218:2264–2275. https://doi.org/10.1016/j.chroma.2011.02.026
46. Gusmaroli L, Insa S, Petrovic M (2018) Development of an online SPE-UHPLC-MS/MS method for the multiresidue analysis of the 17 compounds from the EU "Watch list". Anal Bioanal Chem 410:4165–4176. https://doi.org/10.1007/s00216-018-1069-8
47. Hernández F, Castiglioni S, Covaci A et al (2018) Mass spectrometric strategies for the investigation of biomarkers of illicit drug use in wastewater. Mass Spectrom Rev 37:258–280. https://doi.org/10.1002/mas.21525
48. Ibáñez M, Gracia-Lor E, Bijlsma L et al (2013) Removal of emerging contaminants in sewage water subjected to advanced oxidation with ozone. J Hazard Mater 260:389–398. https://doi.org/10.1016/j.jhazmat.2013.05.023
49. Kim C, Ryu H-D, Chung EG et al (2018) A review of analytical procedures for the simultaneous determination of medically important veterinary antibiotics in environmental water: sample preparation, liquid chromatography, and mass spectrometry. J Environ Manage 217:629–645. https://doi.org/10.1016/j.jenvman.2018.04.006
50. Rodriguez-Mozaz S, Vaz-Moreira I, Varela Della Giustina S et al (2020) Antibiotic residues in final effluents of European wastewater treatment plants and their impact on the aquatic environment. Environ Int 140:105733. https://doi.org/10.1016/j.envint.2020.105733
51. Michael SG, Michael-Kordatou I, Nahim-Granados S et al (2020) Investigating the impact of UV-C/H2O2 and sunlight/H2O2 on the removal of antibiotics, antibiotic resistance determinants and toxicity present in urban wastewater. Chem Eng J 388:124383. https://doi.org/10.1016/j.cej.2020.124383

52. Serna-Galvis EA, Silva-Agredo J, Botero-Coy AM et al (2019) Effective elimination of fifteen relevant pharmaceuticals in hospital wastewater from Colombia by combination of a biological system with a sonochemical process. Sci Total Environ 670:623–632. https://doi.org/10.1016/j.scitotenv.2019.03.153
53. Serna-Galvis EA, Botero-Coy AM, Martínez-Pachón D et al (2019) Degradation of seventeen contaminants of emerging concern in municipal wastewater effluents by sonochemical advanced oxidation processes. Water Res 154:349–360. https://doi.org/10.1016/j.watres.2019.01.045
54. Boix C, Ibáñez M, Sancho JV et al (2015) Fast determination of 40 drugs in water using large volume direct injection liquid chromatography–tandem mass spectrometry. Talanta 131:719–727. https://doi.org/10.1016/j.talanta.2014.08.005
55. Kern S, Fenner K, Singer HP et al (2009) Identification of transformation products of organic contaminants in natural waters by computer-aided prediction and high-resolution mass spectrometry. Environ Sci Technol 43:7039–7046. https://doi.org/10.1021/es901979h
56. Nurmi J, Pellinen J, Rantalainen A-L (2012) Critical evaluation of screening techniques for emerging environmental contaminants based on accurate mass measurements with time-of-flight mass spectrometry. J Mass Spectrom 47:303–312. https://doi.org/10.1002/jms.2964
57. Schollée JE, Schymanski EL, Avak SE et al (2015) Prioritizing unknown transformation products from biologically-treated wastewater using high-resolution mass spectrometry, multivariate statistics, and metabolic logic. Anal Chem 87:12121–12129. https://doi.org/10.1021/acs.analchem.5b02905
58. Boix C, Ibáñez M, Bagnati R et al (2016) High resolution mass spectrometry to investigate omeprazole and venlafaxine metabolites in wastewater. J Hazard Mater 302:332–340. https://doi.org/10.1016/j.jhazmat.2015.09.059
59. Jaén-Gil A, Buttiglieri G, Benito A et al (2019) Metoprolol and metoprolol acid degradation in UV/H2O2 treated wastewaters: an integrated screening approach for the identification of hazardous transformation products. J Hazard Mater 380:120851. https://doi.org/10.1016/j.jhazmat.2019.120851
60. Ibáñez M, Borova V, Boix C et al (2017) UHPLC-QTOF MS screening of pharmaceuticals and their metabolites in treated wastewater samples from Athens. J Hazard Mater 323:26–35. https://doi.org/10.1016/j.jhazmat.2016.03.078
61. Cuervo Lumbaque E, Cardoso RM, Dallegrave A et al (2018) Pharmaceutical removal from different water matrixes by Fenton process at near-neutral pH: Doehlert design and transformation products identification by UHPLC-QTOF MS using a purpose-built database. J Environ Chem Eng 6:3951–3961. https://doi.org/10.1016/j.jece.2018.05.051
62. Bletsou AA, Jeon J, Hollender J et al (2015) Targeted and non-targeted liquid chromatography-mass spectrometric workflows for identification of transformation products of emerging pollutants in the aquatic environment. TrAC Trends Anal Chem 66:32–44. https://doi.org/10.1016/j.trac.2014.11.009
63. Martínez-Pachón D, Ibáñez M, Hernández F et al (2018) Photo-electro-Fenton process applied to the degradation of valsartan: effect of parameters, identification of degradation routes and mineralization in combination with a biological system. J Environ Chem Eng 6:7302–7311. https://doi.org/10.1016/j.jece.2018.11.015
64. Serna-Galvis EA, Silva-Agredo J, Giraldo-Aguirre AL et al (2016) High frequency ultrasound as a selective advanced oxidation process to remove penicillinic antibiotics and eliminate its antimicrobial activity from water. Ultrason Sonochem 31:276–283. https://doi.org/10.1016/j.ultsonch.2016.01.007
65. Jagannathan M, Grieser F, Ashokkumar M (2013) Sonophotocatalytic degradation of paracetamol using TiO2 and Fe3+. Sep Purif Technol 103:114–118. https://doi.org/10.1016/j.seppur.2012.10.003
66. Elias MT, Chandran J, Aravind UK, Aravindakumar CT (2019) Oxidative degradation of ranitidine by UV and ultrasound: identification of transformation products using LC-Q-ToF-MS. Environ Chem 16:41–54

67. Roudbari A, Rezakazemi M (2018) Hormones removal from municipal wastewater using ultrasound. AMB Express 8:1–8. https://doi.org/10.1186/s13568-018-0621-4
68. Arvaniti OS, Frontistis Z, Nika MC et al (2020) Sonochemical degradation of trimethoprim in water matrices: effect of operating conditions, identification of transformation products and toxicity assessment. Ultrason Sonochem 67:105139. https://doi.org/10.1016/j.ultsonch.2020.105139
69. Zeng P, Du J, Song Y et al (2015) Efficiency comparison for treatment of amantadine pharmaceutical wastewater by Fenton, ultrasonic, and Fenton/ultrasonic processes. Environ Earth Sci 73:4979–4987. https://doi.org/10.1007/s12665-015-4204-2
70. González Labrada K, Alcorta Cuello DR, Saborit Sánchez I et al (2019) Optimization of ciprofloxacin degradation in wastewater by homogeneous sono-Fenton process at high frequency. J Environ Sci Heal A Tox Hazard Subst Environ Eng 53:1139–1148. https://doi.org/10.1080/10934529.2018.1530177
71. Rahmani H, Gholami M, Mahvi AH et al (2014) Tinidazole removal from aqueous solution by sonolysis in the presence of hydrogen peroxide. Bull Environ Contam Toxicol 92:341–346. https://doi.org/10.1007/s00128-013-1193-2
72. Ghafoori S, Mowla A, Jahani R et al (2015) Sonophotolytic degradation of synthetic pharmaceutical wastewater: statistical experimental design and modeling. J Environ Manage 150:128–137. https://doi.org/10.1016/j.jenvman.2014.11.011
73. Naddeo V, Ricco D, Scannapieco D, Belgiorno V (2012) Degradation of antibiotics in wastewater during sonolysis, ozonation, and their simultaneous application: operating conditions effects and processes evaluation. Int J Photoenegy 2012. 624270, 7 p. https://doi.org/10.1155/2012/624270
74. Naddeo V, Uyguner-Demirel CS, Prado M et al (2015) Enhanced ozonation of selected pharmaceutical compounds by sonolysis. Environ Technol 36:1876–1883. https://doi.org/10.1080/09593330.2015.1014864
75. Nachiappan S, Muthukumar K (2013) Treatment of pharmaceutical effluent by ultrasound coupled with dual oxidant system. Environ Technol 34:209–217. https://doi.org/10.1080/09593330.2012.689367
76. Sivakumar R, Muthukumar K (2011) Sonochemical degradation of pharmaceutical wastewater. Clean (Weinh) 39:136–141. https://doi.org/10.1002/clen.200900289
77. Adityosulindro S, Barthe L, González-Labrada K et al (2017) Sonolysis and sono-Fenton oxidation for removal of ibuprofen in (waste)water. Ultrason Sonochem 39:889–896
78. Mowla A, Mehrvar M, Dhib R (2014) Combination of sonophotolysis and aerobic activated sludge processes for treatment of synthetic pharmaceutical wastewater. Chem Eng J 255:411–423. https://doi.org/10.1016/j.cej.2014.06.064
79. Brüggemann H, Köser H, Meyer E, Nguyen T-H (2003) Sonolytic debromination of ambroxol process wastewater. Water Res 37:674–680. https://doi.org/10.1016/S0043-1354(02)00363-9
80. Pulicharla R, Brar SK, Rouissi T et al (2017) Degradation of chlortetracycline in wastewater sludge by ultrasonication, Fenton oxidation, and ferro-sonication. Ultrason Sonochem 34:332–342. https://doi.org/10.1016/j.ultsonch.2016.05.042
81. Monteagudo JM, Durán A, San Martín I (2014) Mineralization of wastewater from the pharmaceutical industry containing chloride ions by UV photolysis of H2O2/Fe(II) and ultrasonic irradiation. J Environ Manage 141:61–69. https://doi.org/10.1016/j.jenvman.2014.03.020
82. Naddeo V, Landi M, Scannapieco D, Belgiorno V (2013) Sonochemical degradation of twenty-three emerging contaminants in urban wastewater. Desalin Water Treat 51:6601–6608. https://doi.org/10.1080/19443994.2013.769696
83. Yang B, Zuo J, Li P et al (2016) Effective ultrasound electrochemical degradation of biological toxicity and refractory cephalosporin pharmaceutical wastewater. Chem Eng J 287:30–37. https://doi.org/10.1016/j.cej.2015.11.033

84. Chandak S, Ghosh PK, Gogate PR (2020) Treatment of real pharmaceutical wastewater using different processes based on ultrasound in combination with oxidants. Process Saf Environ Prot 137:149–157. https://doi.org/10.1016/j.psep.2020.02.025
85. Papoutsakis S, Afshari Z, Malato S, Pulgarin C (2015) Elimination of the iodinated contrast agent iohexol in water, wastewater and urine matrices by application of photo-Fenton and ultrasound advanced oxidation processes. J Environ Chem Eng 3:2002–2009. https://doi.org/10.1016/j.jece.2015.07.002
86. Ashokkumar M, Lee J, Iida Y et al (2009) The detection and control of stable and transient acoustic cavitation bubbles. Phys Chem Chem Phys 11:10118–10121. https://doi.org/10.1039/B915715H
87. Lee J, Yasui K, Tuziuti T et al (2008) Spatial distribution enhancement of sonoluminescence activity by altering sonication and solution conditions. J Phys Chem B 112:15333–15341. https://doi.org/10.1021/jp8060224
88. Lee J, Ashokkumar M, Yasui K et al (2011) Development and optimization of acoustic bubble structures at high frequencies. Ultrason Sonochem 18:92–98. https://doi.org/10.1016/j.ultsonch.2010.03.004
89. Kanthale P, Ashokkumar M, Grieser F (2008) Sonoluminescence, sonochemistry (H2O2 yield) and bubble dynamics: frequency and power effects. Ultrason Sonochem 15:143–150. https://doi.org/10.1016/j.ultsonch.2007.03.003
90. Lesko T, Colussi AJ, Hoffmann MR (2006) Sonochemical decomposition of phenol: evidence for a synergistic effect of ozone and ultrasound for the elimination of total organic carbon from water. Environ Sci Technol 40:6818–6823. https://doi.org/10.1021/es052558i
91. Ashokkumar M, Lee J, Iida Y et al (2010) Spatial distribution of acoustic cavitation bubbles at different ultrasound frequencies. ChemPhysChem 11:1680–1684. https://doi.org/10.1002/cphc.200901037
92. Nalesso S (2019) Antisolvent and reactive crystallisation under variations of ultrasound parameters: effects and comparisons. University of Surrey, Guildford
93. Cui M, Jang M, Ibrahim S et al (2014) Arsenite removal using a pilot system of ultrasound and ultraviolet followed by microfiltration. Ultrason Sonochem 21:1527–1534. https://doi.org/10.1016/j.ultsonch.2014.01.001
94. Manickam S, Zainal Abidin N, Parthasarathy S et al (2014) Role of H2O2 in the fluctuating patterns of COD (chemical oxygen demand) during the treatment of palm oil mill effluent (POME) using pilot scale triple frequency ultrasound cavitation reactor. Ultrason Sonochem 21:1519–1526. https://doi.org/10.1016/j.ultsonch.2014.01.002
95. Gole VL, Fishgold A, Sierra-Alvarez R et al (2018) Treatment of perfluorooctane sulfonic acid (PFOS) using a large-scale sonochemical reactor. Sep Purif Technol 194:104–110. https://doi.org/10.1016/j.seppur.2017.11.009
96. Caretti C, Coppini E, Fatarella E, Lubello C (2011) Membrane filtration and sonication for industrial wastewater reuse. Water Sci Technol 64:2500–2507. https://doi.org/10.2166/wst.2011.800
97. Lee J (2016) Importance of sonication and solution conditions on the acoustic cavitation activity BT – handbook of ultrasonics and sonochemistry. Springer, Singapore, pp 137–175
98. Kimura K, Amy G, Drewes JE et al (2003) Rejection of organic micropollutants (disinfection by-products, endocrine disrupting compounds, and pharmaceutically active compounds) by NF/RO membranes. J Membr Sci 227:113–121. https://doi.org/10.1016/j.memsci.2003.09.005
99. Amy G, Kim T-U, Yoon J et al (2005) Removal of micropollutants by NF/RO membranes. Water Supply 5:25–33. https://doi.org/10.2166/ws.2005.0035
100. Xu P, Drewes JE, Bellona C et al (2005) Rejection of emerging organic micropollutants in nanofiltration–reverse osmosis membrane applications. Water Environ Res 77:40–48. https://doi.org/10.2175/106143005X41609

101. Muthukumaran S, Kentish SE, Stevens GW, Ashokkumar M (2006) Application of ultrasound in membrane separation processes: a review. Rev Chem Eng 22:155–194. https://doi.org/10.1515/REVCE.2006.22.3.155
102. Naddeo V, Borea L, Belgiorno V (2015) Sonochemical control of fouling formation in membrane ultrafiltration of wastewater: effect of ultrasonic frequency. J Water Process Eng 8:e92–e97. https://doi.org/10.1016/j.jwpe.2014.12.005
103. Reuter F, Lauterborn S, Mettin R, Lauterborn W (2017) Membrane cleaning with ultrasonically driven bubbles. Ultrason Sonochem 37:542–560. https://doi.org/10.1016/j.ultsonch.2016.12.012
104. Secondes MFN, Naddeo V, Belgiorno V, Ballesteros F (2014) Removal of emerging contaminants by simultaneous application of membrane ultrafiltration, activated carbon adsorption, and ultrasound irradiation. J Hazard Mater 264:342–349. https://doi.org/10.1016/j.jhazmat.2013.11.039

Correction to: Advanced Electrochemical Processes for the Elimination of Pharmaceutical Compounds in Contaminated Waters

Jelena Radjenovic and Luis Baptista-Pires

Correction to: Chapter "Advanced Electrochemical Processes for the Elimination of Pharmaceutical Compounds in Contaminated Waters" in: Jelena Radjenovic and Luis Baptista-Pires, Hdb Env Chem, https://doi.org/10.1007/698_2020_689

This chapter was inadvertently published with an incorrect title "Chemical Processes for the Elimination of Pharmaceutical Compounds in Contaminated Waters". The title has now been updated correctly to read as "Advanced Electrochemical Processes for the Elimination of Pharmaceutical Compounds in Contaminated Waters".

The updated online version of this chapter can be found at
https://doi.org/10.1007/698_2020_689

Sara Rodriguez-Mozaz, Paqui Blánquez Cano, and Montserrat Sarrà Adroguer (eds.), *Removal and Degradation of Pharmaceutically Active Compounds in Wastewater Treatment*, Hdb Env Chem (2021) 108: 383–384, https://doi.org/10.1007/698_2021_757,

Understanding Physics Using Mathematical Reasoning

Andrzej Sokolowski

Understanding Physics Using Mathematical Reasoning

A Modeling Approach for Practitioners and Researchers

Springer

Andrzej Sokolowski
Division of Mathematics and Science
Lone Star College
Houston, TX, USA

ISBN 978-3-030-80204-2   ISBN 978-3-030-80205-9 (eBook)
https://doi.org/10.1007/978-3-030-80205-9

This Springer imprint is published by the registered company Springer Nature Switzerland AG
The registered company address is: Gewerbestrasse 11, 6330 Cham, Switzerland

Preface

The book is addressed to physics practitioners, researchers, and students who strive to present physics beyond the boundaries of traditional curricula. It aspires to propose instructional approaches that make the physics content exciting to teach, learn, and explore. Situated in contemporary physics education research findings, it encourages the readers to consider mathematical reasoning as a convenient tool to predict, extend, and verify scientific theories. While mathematical reasoning can take various layers of complexity, this book focuses on developing covariational sense and graphing techniques supported often by limiting case analysis. A synthesis of research supported a design of an empirical–mathematical modeling scheme and ready-to-use classroom instructional materials guided by the design. Emphasized correlation to mathematics methods makes the book an appealing resource also for mathematics educators and researchers who are interested in mathematical modeling.

The Book Structure

The book consists of three parts composed of 13 chapters whose scope and sequence will lead the reader to discover many opportunities for extending inquiry in physics by taking advantage of mathematical reasoning tools.

Part I consists of two chapters; it brings forth physics building blocks such as laws, principles, theories, and theorems and reflects on these constructs through the prism of the structural math domain. While these constructs are frequently used in teaching practice, their algebraic entanglements are often overshadowed by their scientific meanings. Research in scientific modeling as an environment nurturing the link of scientific and mathematical reasoning skills is also discussed. Chapter 1 of Part I examines the potential of the laws and principles to be expressed as covariate algebraic structures. While mathematics and physics are different subjects, they share similar philosophies that when realized make students' math knowledge transitioning to physics comfortable. Chapter 2 delves deeper into the methodologies of physics and mathematics, revealing common teaching goals that help to establish pathways to allow the structural domain of math knowledge to transition to the physics classroom.

Part II, consisting of three chapters, is dedicated to establishing links between current recommendations in physics education and how the research in mathematics education can respond to these recommendations. Chapter 3 of Part II summaries recent findings on using mathematical modeling in physics practice. Although the literature provides several modeling cycles, mathematical reasoning is not explicitly underlined; therefore, a necessity to develop a new model emerged. In Chap. 4, empirical–mathematical modeling scheme is proposed, and its phases are described. The scheme provides a general framework of labs and lecture structure designs that extends students' scientific reasoning by considering natural phenomena as covariate relations. Mathematical reasoning has its established place in physics education; however, its more powerful version, covariate reasoning, is still being researched. Chapter 5 provides a more pragmatic view on using covariate reasoning, and in addition, it discusses the main underpinnings of limiting case analysis that serves as a tool to quantify a hypothetical system's behavior. An attempt to arrange physics formulas as classes of various types of covariate entities is also made.

Part III is considered the core of the book. It consists of eight case studies where the theoretical underpinnings outlined in Parts I and II are placed into practice and evaluated. While in Chap. 6 of Part II, limiting case analysis extends the investigation of the motion of a system of two objects, Chap. 7 proposes a reconstruction of Newton's law of universal gravity to include gravitational fields as a mediating parameter nurturing covariate reasoning. In Chap. 8, applications of parametric equations to enhance the understanding of projectile motion as a two-dimensional motion are examined. Chapter 9 investigates the opportunity to generalize image characteristics stemming from the lens equation considered an algebraic function. In the traditional science curriculum, the concept of the mole is associated with the content of the chemistry curriculum; however, it is also a physics concept. An effort to induce math reasoning to understand the unique measure of mole and its conversions is included in Chap. 10. An attempt to reconstruct the Einstein formula for the photoelectric effect that allows reason mathematically is presented in Chap. 11. Chapters 12 and 13 aim differently and examine students' perspectives of applying math structural domain to understand physics. Each case study concludes with suggestions for further explorations.

Houston, TX, USA Andrzej Sokolowski

Contents

Part I Conceptual Background

1 Physics Constructs Viewed Through the Prism of Mathematics 3
1.1 Mathematics as an Indispensable Part of Physics Inquiry 3
1.2 Laws of Physics and Their Mathematical Embodiments 4
1.3 Principles and Their Relations to Laws 10
1.4 Theories and Laws 12
1.5 Theories and Theorems 13
References 14

2 The Interface Between the Contents of Physics and Mathematics ... 15
2.1 Mathematics as a Language in Physics Classroom 15
2.2 Philosophy and the Substance of the Knowledge of Mathematics . 16
2.3 Procedural and Conceptual Mathematical Knowledge 17
2.4 Unifying Classification of Math Knowledge Used in Physics Education 18
2.5 Arrays of Applying Mathematics in Physics 19
2.6 Search for Tools and Methods 21
2.7 Mathematical and Scientific Reasoning; Are These Mental Actions Equivalent? 22
2.8 Synthesis of Students' Challenges with Math Knowledge Transfer 22
References 24

Part II Designing Learning Environments to Promote Math Reasoning in Physics

3 Modeling as an Environment Nurturing Knowledge Transfer 29
3.1 Scientific Modeling and Models 29
3.2 Modeling Cycles in Physics Education 30
3.3 Merging Mathematics and Physics Representations 32
References 33

4 Proposed Empirical-Mathematical Learning Model ... 35
4.1 Didactical Underpinnings of the Design ... 35
4.2 Description of the Learning Phases ... 35
4.3 Hypotheses as Learners' Proposed Theories ... 37
4.4 Mainstream of the Inquiry and Its Confirmation ... 38
4.5 Methods of Enacting Mathematical Structures ... 38
4.6 Concluding Phases of the Learning Process ... 39
References ... 39

5 Covariational Reasoning – Theoretical Background ... 41
5.1 Quantities, Parameters, and Variables ... 41
5.2 Formulas in Science and Mathematics ... 43
5.3 Covariational Reasoning in Mathematics Education ... 45
5.4 Covariational Reasoning in Physics Education ... 48
5.4.1 Viewing Phenomena as Covariations of Their Parameters ... 49
5.4.2 Proposed Categories of Covariations Embedded in Physics Formulas ... 51
5.4.3 Discussing Covariations of Parameters in Experiments ... 55
5.5 Limiting Case Analysis ... 57
5.5.1 Evaluating Limits when the Variable Parameter Is Getting Very Large; $x \to \infty$ ... 58
5.5.2 Evaluating Limits when the Variable Parameter Is Close to a Specific Value; $x \to a$ ... 60
5.5.3 Is Limiting Case Analysis Really "Limiting"? ... 62
References ... 63

Part III From Research to Practice

6 Extending the Inquiry of Newton's Second Law by Using Limiting Case Analysis ... 67
6.1 Limits - Tools for Extending Scientific Inquiry ... 67
6.2 Research Methods ... 68
6.2.1 Research Questions, Logistics, and Participants ... 68
6.2.2 Criteria for the Study Content Selection ... 68
6.2.3 Discussion of the Applied Algebraic Tools ... 70
6.3 Description of the Instructional Unit ... 71
6.3.1 Analyzing Acceleration of the System in the Function of Mass m_2 ... 71
6.3.2 Analyzing Acceleration of the System in the Function of Mass m_1 ... 75
6.4 Data Analysis ... 76
6.4.1 Analysis of the Pretest Results ... 76
6.4.2 Analysis of the Posttest Results ... 77
6.5 Conclusions ... 78
References ... 80

7 Reconstructing Newton's Law of Universal Gravity as a Covariate Relation 81
7.1 Prior Research Findings 81
7.2 Theoretical Framework 82
7.2.1 Historical Perspective 83
7.2.2 Contemporary Presentations of the Law of Universal Gravity 84
7.3 Methods 85
7.4 Didactical Underpinnings of the Instructional Unit 85
7.5 The Lecture Component 86
7.5.1 Gravitational Field Intensity and the Effects of Covariate Quantities 87
7.5.2 Reconstructing the Formula to Calculate Mutual Gravitational Force 90
7.6 Analysis of Pretest - Posttest Results 93
7.6.1 Analysis of the Pretest Results 93
7.6.2 Analysis of the Posttest Results 96
7.7 Conclusions and Suggestions for Further Research 98
References 100

8 Parametrization of Projectile Motion 101
8.1 Prior Research Findings 101
8.2 Theoretical Framework 103
8.2.1 Categories of Motion Studied in High School and Undergraduate Physics Courses 103
8.2.2 Why Parametric Equations? 104
8.2.3 Foundations of Constructivist Learning Theory 105
8.3 Methods 106
8.3.1 Study Description and the Research Question 106
8.3.2 The Participants 107
8.3.3 Lecture Component Sequencing 107
8.3.4 Topics Embedded within the Curriculum to Enhance the Treatment 108
8.4 General Lab Description 114
8.4.1 Lab Logistics 115
8.4.2 Gathering Data to Construct Positions Functions for a Projected Object 115
8.4.3 Constructing Representations of the Position Functions 117
8.4.4 Finding Velocities and Acceleration Functions 119
8.4.5 Verification Process 120
8.5 Treatment Evaluation 121
8.6 Summary and Conclusions 124
References 125

9 Reimaging Lens Equation as a Dynamic Representation 127
9.1 Introduction 127
9.2 Prompts Used for the Instructional Unit Design 128
9.2.1 Mathematical Background 128
9.2.2 Lab Equipment 129
9.2.3 Conversion of Lens Equation into a Covariational Relation 130
9.2.4 Sketching and Scientifically Interpreting the Graph of the Lens Function 131
9.2.5 Formulating Magnification Function 134
9.2.6 Merging Mathematical and Experimental Representations into One Inquiry 136
9.3 Suggested Independent Student Work 140
9.4 Summary 142
References 143

10 Embracing the Mole Understanding in a Covariate Relation 145
10.1 Introduction and Prior Research Findings 145
10.2 Theoretical Framework 146
10.2.1 Weaknesses of the Mole Understanding 146
10.2.2 Proportional Reasoning, Rates, and Ratios 147
10.3 Methods 150
10.4 The Lecture Component 151
10.4.1 The Mole as a Fundamental Unit of the Substance Amount 151
10.4.2 Converting the Number of Atoms to the Units of Moles 152
10.4.3 Converting Mass of Substance to Moles 154
10.4.4 Converting Mass of a Substance to the Number of Atoms 156
10.5 Pretest Posttest Analysis 157
10.5.1 Analysis of the Pretest Results 157
10.5.2 Comparisons of the Pretest and Posttest Results 158
10.6 Summary and Conclusions 160
References 161

11 Enabling Covariational Reasoning in Einstein's Formula for Photoelectric Effect 163
11.1 Prior Research 163
11.2 Theoretical Background 164
11.3 Embracing the PE into the Framework of Covariational Representation 164
11.3.1 Weaknesses of the Graph of K_{MAX} Versus Photons' Frequency Presented in Physics Resources 165
11.3.2 Covariation of Photon's Energy and Frequency as a Linear Function 167

11.3.3 Electrons' Binding Energy as a Function of Photons Threshold Frequency 168
11.3.4 Maintaining a Minimum Number of Covariational Parameters During the Inquiry. 169
11.4 Reassembling the PE Formula to Assure a Coherence of Representations 169
11.4.1 Graph Constructing 170
11.4.2 Finding Algebraic Representation of the Graph 171
11.4.3 Linking the Photons Threshold Frequency and the Work Function $hf_o = W_o$. 173
11.5 Summary and Conclusions 173
References. 174

12 Are Physics Formulas Aiding Covariational Reasoning? *Students' Perspective* 177
12.1 Introduction and Prior Research Findings 177
12.2 Theoretical Background and Methods. 179
12.2.1 Foundations of Covariation Reasoning 179
12.2.2 Study Description, Participants, Research Questions, and Evaluation Instrument. 179
12.3 Data Analysis 181
12.4 Summary and Conclusions 183
12.4.1 Traditional Formula Notation Does Not Aid Covariational Reasoning in Physics. 184
12.4.2 Physics Depends on the Mathematical Rules and Notation. 184
References. 185

13 Adaptivity of Mathematics Representations to Reason Scientifically *Students' Perspective* 187
13.1 Prior Research Findings. 187
13.2 Theoretical Framework, Research Questions, and Study Logistics 188
13.3 Study Instrument 189
13.3.1 General Characteristics of the Treatment: How Did Covariational Reasoning Emerge?. 189
13.3.2 Actions Taken to Exercise Covariation Model Using Laboratory. 194
13.4 Data Analysis 196
13.5 Summary and Conclusions 199
References. 201

Teaching Physics Using Mathematical Reasoning 203

Index 205

Part I
Conceptual Background

Chapter 1
Physics Constructs Viewed Through the Prism of Mathematics

1.1 Mathematics as an Indispensable Part of Physics Inquiry

Among academic disciplines, physics is one of the oldest (Krupp, 2003), and it is defined as a general analysis of nature conducted to understand how nature behaves (Young & Freedman, 2004). Physicists observe phenomena in the pursuit to identify patterns, explain these patterns following scientific inquiry, and apply mathematical apparatus, if plausible, to express the patterns concisely. These patterns can be supported by theories and advance to formulas called laws, principles, or theorems. Physics constructs are often expressed using algebraic symbols and the rules of mathematics. Some physicists claimed that the elegance and logic of physical laws are only apparent when expressed in the appropriate mathematical framework (Schwartz, 2012). According to Feynman (2005), mathematics provides the tools to express the phenomena' dynamics using a universal language. Mathematics appears as the perfect tool in which one can narrate a natural phenomenon (Agnew et al., 2009) because nature is the domain of measure and order that responds only to questions expressed in a mathematical language (Koyré, 1957). While mathematics rules provide a framework for phenomena quantification, phenomena' representations cannot be reduced to constructing mathematical formulas and manipulating their parameters using mathematical algorithms. Mathematics helps model data, formulate the relations of interests in symbolic forms and extend the inquiry into a hypothetical domain. However, what distinguishes these representations in physics from how they are used in a context-free mathematical sense is that the scientific equations are built of quantities with definite physical interpretations reflecting their derivation from nature. Once embraced in algebraic structures, they provide opportunities to apply algorithms to find unique values and reason beyond the boundaries of their measurable magnitudes.

The content of physics has been formulated through investigations conducted by experimental and theoretical physicists. Experimental physicists apply scientific

A. Sokolowski, *Understanding Physics Using Mathematical Reasoning*, https://doi.org/10.1007/978-3-030-80205-9_1

methods often followed by inductive reasoning to formulate theories, laws, principles, and theorems. They observe the behaviors, identify parameters, select parameters of interest, gather data, and formulate general mathematical models, if plausible. Theoretical physicists apply different scientific methods, but the aim is similar; they make observations, construct theories, formulate mathematical models, but apply a deductive inquiry to determine if their models support the theories. Theoretical physicists also use mathematics as a reasoning tool while applying deductive inquiry. The success of their investigations depends on how well the theoretical results agree with observations of nature. In the case of dissonance, the mathematical representations, not the facts observed, are being reexamined.

The products of investigations by experimental and theoretical physicists are similar. The verbal explanations of these constructs are called theories. The following sections provide more details about each of these fundamental physics constructs.

1.2 Laws of Physics and Their Mathematical Embodiments

Phenomena behavior is typically described by establishing relationships between selected quantities, leading to formulating a law. Physics laws reflect casual relationships within phenomena and are formulated through data taking and mathematical modeling (Wigner, 1990). Laws do not explain the phenomena. Laws describe the phenomena. To be called a law, the report must be conducted and validated over a wide range of observed experiments of a similar nature and aim. Bhushan and Rosenfeld (2000) claimed that the derived laws are the best approximation of nature's more profound behavior that is still waiting to be fully discovered. While the laws are established and widely accepted, according to Bhushan, a room for discoveries or modifications of the existing ones still exists, which can be an inspiring message for the students. Laws as descriptive principles of nature must hold in all circumstances covered by the laws (Oxford University Press, 2015, p.212). Scientists give the title *law* to certain concise but general statements about how nature behaves.

In most cases, laws are formulated mathematically, and they can be applied to support other discoveries, for example, in biochemistry or biophysics. Laws are generally discovered by observations; therefore, they stem from empirical evidence. Physics educators' task is to reflect on these methods while delivering these constructs to students as closely as possible. Developing empirical-mathematical models of the laws of physics emerges as the most profound method. While mathematical laws are considered to have absolute certainty, scientific laws can be questioned, contradicted, restricted, or modified by future observations and investigations (Fischbein, 1987).

Laws of physics, when expressed using algebraic symbols, can take various forms. The complexity of the structures depends on the mutual relationships of the quantities that constitute the law. A vast majority of the laws in physics are expressed in the forms of multivariable formulas. A science formula is defined as a concise way of expressing information symbolically or a general construct of a relationship between given quantities (Rautenberg, 2010). Formulas are context-related and thus are restricted to describe a specific phenomenon or its parts. However, they can be classified according to how their parameters' behavior is expressed mathematically. For example, if a direct proportionality is observed $f(x) = mx$ then structures of linear functions are applied. Formulas like $v(\omega) = R\omega$, $P(h) = \rho gh$, or $p(v) = mv$ can be interpreted scientifically following the meanings of the constant rates of change called also slopes when graphing is concerned. If an indirect proportionality is observed, then rational functions can be considered. For example, $R(A) = \frac{\rho l}{A}$ or $\alpha(I) = \frac{\tau_{Net}}{I}$. A more detailed classification of formulas concerning the complexity of the covariation is included in Chap. 5.

Laws expressed as algebraic functions can be regarded as dynamic algebraic structures that can reveal more details about phenomena. How students perceive formulas versus algebraic functions will be discussed in Part III of the volume and especially in Chaps. 12 and 13. Laws contain quantities that represent measurable entities. Quantities are called parameters when embraced in algebraic entanglements. Until the middle of the nineteenth century, algebra was almost identical to solving algebraic equations, particularly polynomials (Koch, 2000). The concepts of algebraic functions were not extensively applied in mathematics, and perhaps, they were not applied extensively in physics. For instance, Newton's law of universal gravity, constructed as a mathematical structure, was called a formula $F = \frac{Gm_1m_2}{d^2}$, and it has been used similarly nowadays to calculate one of the quantities labeled as either m_1, m_2, d, or F. Such application of this fundamental law of gravity is exercised widely and can be found in traditional physics textbooks. However, this formula can be perceived as an excellent example of a rational function of one variable if the parameter representing the distance between the masses' centers, d, is considered a variable and the remaining parameters are constant. Following this condition, the formula can be expressed as $F(d) = \frac{Gm_1m_2}{d^2}$, and its functional embodiment is illustrated in Fig. 1.1. While in mathematics, a graph position and appearance in XY coordinates are usually determined by the allowable function domain, position of $F(d)$ in F vs. d coordinates is determined by the restrictions that stem from its scientific image.

For example, to graph $F(d) = \frac{Gm_1m_2}{d^2}$, each mass can be considered 1 kg, and the distance d between the centers of the masses can be chosen as $d \geq 1$ m. Using this condition, the force of the mutual attraction is expressed in terms of the universal gravitational constant G; $F(d) = \frac{G(1\,\text{kg})(1\,\text{kg})}{d^2}$.

The graph illustrates that the force per unit of distance decreases, and the rate of the decrease can be computed by taking the derivative with respect to the distance: $F'(d) = \frac{-2G(1\,\text{kg})(1\,\text{kg})}{d^3}$. The graph also illustrates that the force decreases more

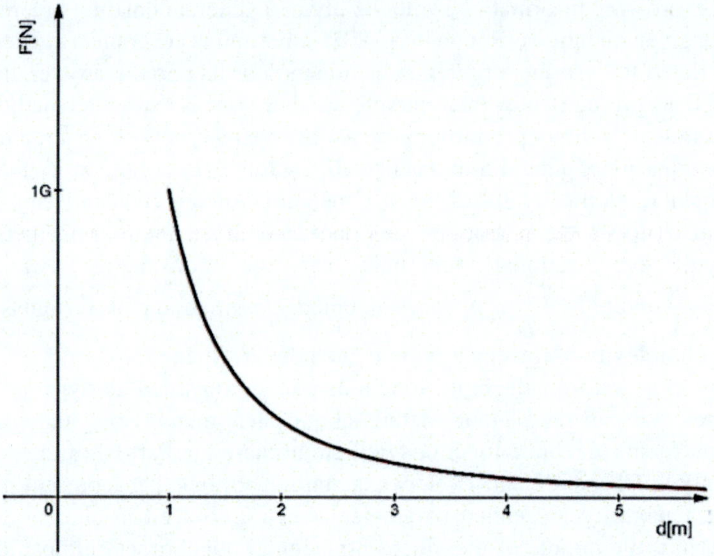

Fig. 1.1 Force of gravitational attraction between two masses being moved away; $d \geq 1$ m

rapidly for 1 m $\leq d \leq$ 2 m, and the rate of change and the force values are getting near zero when $d \to \infty$. For a considerable significant distance, the magnitude of the force can be approximated by employing limits that prove that the force is near the value of zero $F(d \to \infty) = \lim_{d \to \infty} \frac{Gm_1 m_2}{d^2} = 0$.

The law can also be investigated considering the distance between the objects as a constant parameter and the masses considered varying parameters. For instance, establishing the mass $m_1 = m_2 = m$, results in $F(d) = \frac{Gmm}{d^2} = \frac{Gm^2}{d^2}$. The function can be presented as $F(m) = \frac{G}{d^2}(m^2)$, and by considering e.g., $d = 1$ m, the general algebraic form can be simplified to $F(m) = Gm^2$. The formula is now a polynomial function of the second degree (Fig. 1.2).

The force is increasing due to an increase of masses, and the rate of increase of the force per unit of mass change is represented by a linear function $F'(m) = 2Gm$.

Both functions $F(d) = \frac{Gm_1 m_2}{d^2}$ and $F(m) = \frac{Gm^2}{d^2}$, when sketched in respective coordinates, evoke different inquiries about how the forces of mutual attraction change based on the functions' structural attributes. By considering different parameters as variables, different covariation is formulated, and its graphical representations take different forms, yet they are still about investigating the same phenomenon of gravitational attraction. By explicitly labeling the parameter that changes, its effects on the quantity of interest called the dependent parameter can be easily identified and its scientific behavior examined.

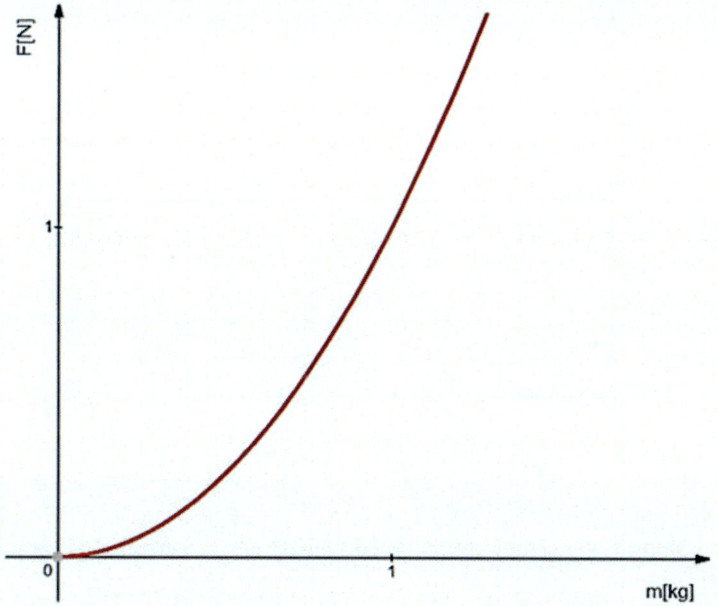

Fig. 1.2 Graph representing gravitational attraction between variable masses separated by 1 m

Another example of a physics law that can be converted to a one-variable function is the law of conservation of momentum (LCM) discovered and formulated by Isaak Newton. When expressed as equity of the initial and final momentum, the LCM takes the general form of $\sum_{n}^{i=1} \overrightarrow{p_i} = \sum_{n}^{i=1} \overrightarrow{p_f}$, and this form of the law dominates physics resources. A handful of problems can be found in literature where one of the unknown quantities (either mass, velocity, or momentum) is considered an unknown and all remaining parameters constant. While the earlier discussed form of the law of universal gravity resembled a function at first, the mathematical structure of the LCM does not resemble a function. Thus, the covariate relation between its parameters cannot be explicitly determined except that the left and the right sides must be numerically and vectorially identical. Analyzed by algebra rules, the law is merely a statement of equity of two sides of sums of algebraic expressions. In the absence of external forces, the total amount of kg·m/s remains constant before, during, and after the collision; thus, the graph of the system's total momentum versus time must represent a horizontal line (Fig. 1.3).

Can the law be converted to a one-variable covariate function? Such modification is possible. Once converted to one variable function, its behavior can be analyzed using either its graph or algebraic representation. For example, when considering that the two moving objects of masses m_1 and m_2 collide and got entangled after the collision, one parameter established as a variable can be used to theoretically determine the direction of motion of both masses after the impact. The direction of the

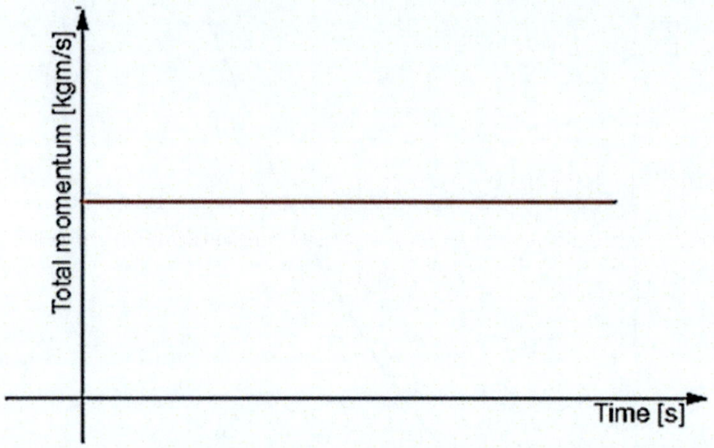

Fig. 1.3 Graph of the total momentum of two objects before, during, and after the collision

objects' motion when they are coupled can be derived from a general LCM and represented by:

$$\vec{v_o} = \frac{\vec{v_1}m_1 + \vec{v_2}m_2}{m_1 + m_2} \tag{1.1}$$

Suppose that one is interested in exploring the direction of the joint velocity of both objects after a perfectly inelastic collision when the mass m_1 can take various magnitudes and the velocities remain constant in each trial before the collision. To convert the problem to a one-variable relation, restrictions on its parameters must be made; thus the velocities of both objects remain constant; e.g., each is of a magnitude of 2 m/s as well as the mass m_2 must be constant, suppose $m_2 = 1$ kg. By these restrictions, the only variable parameter is the mass m_1. With that condition, the expression (1.1) takes the following covariate form

$$\vec{v_o}(m_1) = \frac{2m_1 - 2}{m_1 + 1}. \tag{1.2}$$

This rational function can be used to predict the direction of motion of the coupled objects after the collision and then to verify it using a simulated experimental setup (e.g., Fig. 1.4). The formulated covariate expression (Eq. 1.2) explicates the condition related to the total system mass when the coupled objects stop after the collision and when they move to the right or left after they are coupled. Students realize that to predict these behaviors, it is sufficient to analyze the numerator of the rational function because its denominator takes only positive values. While the simulated experiment allows only a definite range of magnitudes of the velocities and masses, the rational function form allows employing limiting case analysis and

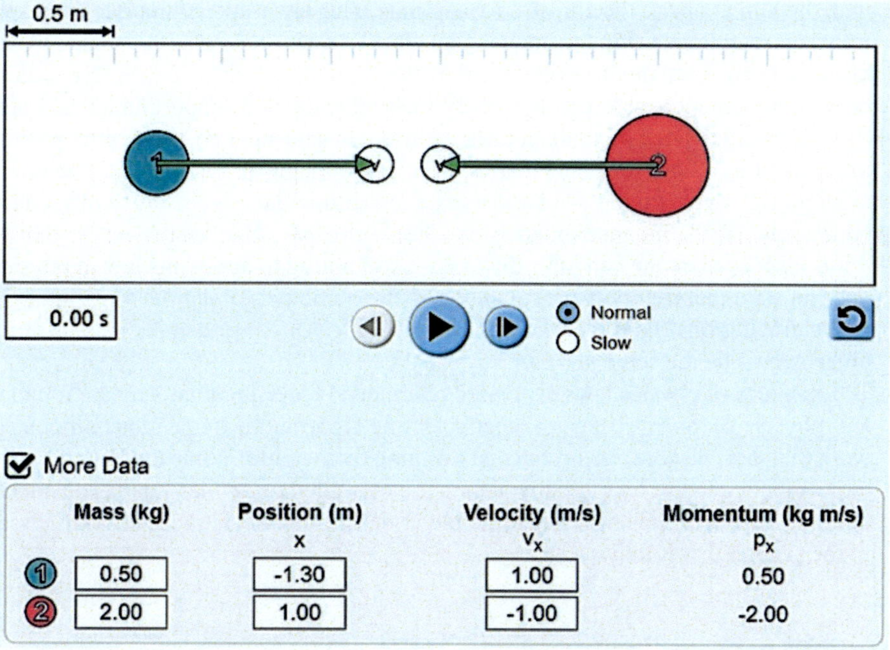

Fig. 1.4 Snapshot of an experiment investigating the law of conservation of momentum. (Source: Phet Interactive Simulations (n.d.))

hypothetically predict the systems' speed if the values of m_1 are considerably high. Such a task can be accomplished by taking the limit of the function (1.2) with respect to m_1 when $m_1 \to \infty$. Thus $\overrightarrow{v_o}(m_1 \to \infty) = \lim_{m_1 \to \infty} \frac{2m_1 - 2}{m_1 + 1} = \frac{2m}{s}$. The simulated experiment provides ways to verify the predictions using other parameters arrangements; the mathematical reasoning extends the analysis beyond the boundary of the lab setup.

When converted to covariate representations, a high diversity of algebraic forms of laws can shed light on different parts or phases of the phenomena and enrich its conceptual understanding. Upon reclassifying some quantities using the language of mathematics as independent, dependent, or constant according to their restricted behavior, such expressions are transferred to richer forms whose attributes can serve as a basis for employing deeper hypothetical thinking. These modifications add dynamics to the analysis and help zoom into how one quantity affects the other beyond the phenomenon classroom boundaries.

While most physics laws are expressed using the language of mathematics, some have been expressed qualitatively. For instance, Lentz's law that describes the direction of induced electric current in a solenoid due to a magnet's movement is an example of such a form that is often supported graphically. Another example can be Kepler's law for a planetary motion that states that the orbit of a planet is an ellipse with the Sun at one of the ellipse's focal points. This law is also supported by graphical representations using the image of an ellipse and its geometry. As some physics laws are formulated qualitatively and the others quantitatively, these representations often complement each other to amplify their meaning and support understanding.

Can abstract laws of mathematics be brought and applied to support physics understanding? Laws of mathematics are being applied for algorithmic operations at large. However, attempts are and can be made to include others. One of the laws commonly used in mathematics is the *transitive law* which says that if $a \times b = c \times b$, then one can infer that $a = c$. Attempting to apply this law in physics, one can state that if gravity acting on two different objects under the same intensity of the gravitational field is the same, thus if $m_1 \times g = m_2 \times g$, then the masses of the objects must be the same $m_1 = m_2$.

Examples of physics laws expressed qualitatively and quantitatively are found in any physics section. All laws reflect scientific rigor and a tremendous amount of work that their discoverers invested in bringing them to life. While the laws describe more significant spectra of reality, their subparts called *principles* can be considered building blocks of the laws. The principle as describing the nature will be discussed in the section that follows.

1.3 Principles and Their Relations to Laws

Principles are like laws, but they are usually less general than laws and refer to fundamental relationships between more than one physical quantity within a given phenomenon (Giancoli, 2008). Principles constitute a part of physics knowledge like axioms in mathematics. Principles are also defined as inherently clear statements used to guarantee the truth of the theorems from which they were derived (Easwaran, 2008). Since principles support a more comprehensive description of laws, they are often a part of the law, illustrated in Fig. 1.5.

Physics provides many examples of supportive roles of principles. For instance, the principle of wave superposition based on the algebraic property of function additivity supports a quantitative description of interference. If periodic functions are used to model the motion of the particles of the wave, then when two waves interact, the superposition principle informs that the resulting wave function is the sum of the two individual wave functions. The principle of wave superposition is often described graphically, and the resulting patterns are classified according to the

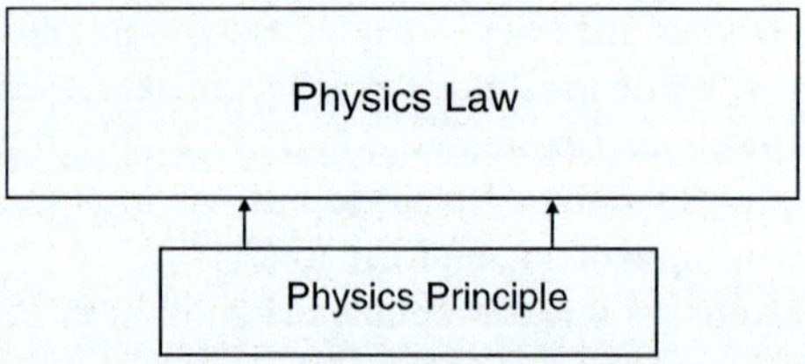

Fig. 1.5 Relation between principles and laws in physics

law of interference. While in high school practice, this principle is often introduced conceptually and embedded in graphical representations, it can also be supported by combining periodic functions expressed algebraically, presenting concurrently a gateway to more advanced Fourier analysis. The amplitude of interference is interpreted using a general law of interference, and consequently, the resulting patterns are classified as constructive, destructive, and totally destructive. The principle supports the quantification process, and the results are classified according to the law of interference. In its different yet congruent manner, the principle of superposing is also used in mechanics to compute the net force when two or more forces act on a particle simultaneously or when computing a net torque when various torques are exerted on an object simultaneously. The principle of superposition, in these contexts, enhances the process and advises that to find a net force or torque, a vectorial sum of all forces or torques acting on the object must be computed. Ultimately, using these principles can lead to formulating the operational form of Newton's laws of motion. It can be further inferred that the result of applying the principle of superposition can be called the action on the object and constitute one of the sides of Newton's law when it is expressed algebraically $\sum_{k}^{i=1} F_k = ma$. The right side expressed as a product of the object's mass, and acceleration can be called the result of the action. Interchanging these sides will still produce the same final answers, and students are usually given freedom in this regard. Literature does not set preference over the sides' order. However, from a cause-and-effect viewpoint, each representation $\sum_{k}^{i=1} F_k = ma$ and $ma = \sum_{k}^{i=1} F_k$ generates, to some extent, different scientific interpretations, even though the forms of the law are congruent.

Archimedes principle is yet another commonly taught physics principle. While it is used to quantify the buoyant force, it supports the law of floating, and it also aids the algebraic formulation of Newton's law to learn about an object's state of motion in the fluid. The Archimedes principle subsidized by the buoyant force's computations emerges as the building block to constructing the net force and, consequently, finding the object's acceleration. Thus, in describing an object's motion in fluids, two principles, Archimedes' principle and superposition of forces, can be considered supporting principles of Newton's second law when used to describe an object's behavior in that fluid. Principles are not parts of the laws' description; they support the law's formulations and enhance their applications, especially when quantification is concerned. Although not examined in physics research, making students aware of the contributions of these building blocks to the content of physics may have a profound effect on enhancing the pathways for their understanding.

The algebraic form of a law or principle is one side of the coin; interpreting the relation qualitatively and providing a supporting theory for its truthfulness is the other side that can evidence conceptual understanding. Theories are sometimes mistakenly considered theorems. The following section attempts to shed light on the differences between these two constructs and their relations to laws and principles.

1.4 Theories and Laws

The theory is described as an invented explanation of natural phenomena formulated using scientific methods and accepted fundamental principles or hypothetical deductive testing (Haig, 2018). The theory is also described as a qualitatively expressed testable prediction or hypothesis often accompanied by a great precision (Giancoli, 2008). A theory is not a random thought or an unproven concept. Theory needs to be experimentally verified to be accurate to speak the behavior of future similar events. To develop a scientific theory, a scientist must ask appropriate questions, design experiments, and draw appropriate conclusions from the results. While laws and principles describe the phenomena, theories are usually formulated verbally, and they explain the phenomena rather than describe them. To have a comprehensive understanding of a phenomenon, one needs to understand the law's formulation and its surrounding theory. Due to possessing an explanatory nature, theories are not formally proven; however, theory and data must be interwoven in a relationship, and the relationships must be rational and logical (Simon, 2012).

The relationships between theory and data must lead to correct conclusions about the extent to which those data support the theory. The physicist's task is to make explicit that the intuitively employed principles in justifying corresponding theories are correct (Giere, 2010). In this sight, taking data and analyzing their patterns can be used to verify the theory, but it is also applied to formulate a law quantitatively or qualitatively. Systematically taken data with clearly defined variable parameters can constitute the simplest form of an algebraic function, which can lead to enacting a formula called a law.

Laws and theories are both the products and tools of science, but each has a distinct heritage and role. One does not become the other (McComas, 2013). How law and theory can complement each other during class discussion illustrates the following example. Suppose that a system of objects possesses kinetic and potential energy and that the system satisfies the condition of having the total mechanical energy conserved. The law of conservation of mechanical energy states that at any time instant, the sum of the kinetic and potential energies of the system remains unchanged, which symbolically can be expressed as $K_{i\ sys} + U_{i\ sys} = K_{f\ sys} + U_{f\ sys}$, where K and U represent kinetic and potential energies, respectively. The law can be supported by the theory that energy cannot be created nor destroyed. Theory similarly to theorem can also be defined as a collection of axioms or definitions used to support a phenomenon's behavior. In general, an axiom is primarily used in mathematics, and it is a statement that is true to serve as a premise for reasoning (Duran-Guerrier, 2008).

Theory can elicit a model, primarily when theory is concerned with a simplified and idealized physical system or phenomenon (e.g., gas laws). A physical model usually mediates between scientific theory and reality. In contrast, laws are well established in physics education; theories that explain the laws perhaps not deliberately play a supportive role in physics teaching. Such relation of laws and theories might diminish the potential to develop students' deeper scientific reasoning and the phenomena understanding. When used in their mathematical forms, physics laws provide means for quantifications; they do not explain the phenomenon, nor can they justify students' conceptual understanding adequately. For instance, when asked to explain Newton's second law, students often claim that the law informs that force is equal to mass times acceleration. Such a statement does not explain conditions for accelerated motion, but it is merely a verbal description of the law. Explanation of the law demands an understanding of the actions and object's mass on its acceleration. The law is constructed as equity of action determined by the amount of net external force F acting on the object and its observable state of speeding up or slowing down, known as acceleration.

1.5 Theories and Theorems

There is a difference between theory and theorem; while theorems are not often used in physics, it is worthwhile to discuss these entities as well. Theorems are predominantly used in mathematics and defined as general statements not self-evident but proved by a chain of reasoning from a set of axioms established by mathematical logic (Loveland, 2016). A fundamental theorem of calculus (FTC) that states that change of function values is equal to accumulation under its derivative is one example of a mathematical theorem that is symbolically expressed as $\int_b^a f(x)dx = F(b) - F(a)$. This theorem, while not explicitly highlighted, is often applied in physics. For instance, in algebra-based physics courses, students calculate the area under the object's velocity-time graph and conclude its displacement. The fundamental theorem of calculus can also lead to deriving an integral function to compute, for example, a specific object's position when the initial position along with the velocity function is provided: $x(t) = x(0) + \int_t^0 v(t)dt$. It can also support finding the object's momentum at a specific time instant, t, when a variable impulse of force, $F(t)$ acts on it; $p(t) = p(0) + \int_t^0 F(t)dt$. Rich physics contexts can also be used to model the mathematical underpinnings of the FTC in calculus courses (Sokolowski, 2021).

Another example of a theorem is *the parallel-axis theorem* used to find the moment of inertia, I, of an object about an axis, parallel to the axis passing through the center of mass; $I = I_o + Mh^2$. The Mean Value Theorem is another example of a mathematical theorem that remains silent in physics that can enrich analyzing systems in physics. This theorem allows finding, for example, time instants when an object's instantaneous velocity is equal to its average velocity on the interval where the position function is differentiable.

Mathematical theorems support physics quantification and increase the set of tools to deploy to solve physics problems or enrich the inquiry. While theorems are statements often expressed symbolically, theories explain phenomena qualitatively.

The preliminary synthesis of the leading physics constructs illustrates that being derived empirically or hypothetically each contributes to physics knowledge in its own and distinctive right. While principles often support laws, theories initiate the formulation of the laws. It is beneficial to make students aware of each of these constructs' qualitative and quantitative roles to support holistic comprehension of physics discoveries. These constructs will arise while attempting to integrate mathematical reasoning into the context of physics that will be further deployed in this book.

References

Agnew, A. F., Bobe, A., Boskoff, W. G., & Suceavă, B. D. (2009). Gheorghe Ţiţeica and the origins of affine differential geometry. *Historia Mathematica, 36*(2), 161–170.

Bhushan, N., & Rosenfeld, S. (Eds.). (2000). *Of minds and molecules: New philosophical perspectives on chemistry*. Oxford University Press.

Duran-Guerrier, V. (2008). Truth versus validity in mathematical proof. *ZDM, 40*(3), 373–384.

Easwaran, K. (2008). The role of axioms in mathematics. *Erkenntnis, 68*(3), 381–391.

Feynman, R. P. (2005). *The pleasure of finding things out: The best short works of Richard P. Feynman*. Basic Books.

Fischbein, H. (1987). *Intuition in science and mathematics: An educational approach* (Vol. 5). Springer Science & Business Media.

Giancoli, D. C. (2008). *Physics for scientists and engineers with modern physics*. Pearson Education.

Giere, R. N. (2010). *Explaining science: A cognitive approach*. University of Chicago Press.

Haig, B. D. (2018). An abductive theory of scientific method. In *Method matters in psychology* (pp. 35–64). Springer.

Koch, H. (2000). *Number theory: Algebraic numbers and functions*. (No. 24). American Mathematical Society.

Koyré, A. (1957). *From the closed world to the infinite universe* (Vol. 1). Library of Alexandria.

Krupp, E. C. (2003). *Echoes of the ancient skies: The astronomy of lost civilizations*. Dover Publications.

Loveland, D. W. (2016). *Automated theorem proving: A logical basis*. Elsevier.

McComas, W. F. (Ed.). (2013). *The language of science education: An expanded glossary of key terms and concepts in science teaching and learning*. Springer Science & Business Media.

Oxford University Press. (2015). Conservation laws. In *Oxford dictionary of science* (7th ed., p. 212). Oxford University Press.

PhET Interactive Simulations. (n.d.). *The University of Colorado at Boulder*. Retrieved from http://phet.colorado.edu. September 2020.

Rautenberg, W. (2010). *A concise introduction to mathematical logic*. Springer.

Schwartz, M. (2012). *Principles of electrodynamics*. Courier Corporation.

Simon, H. A. (2012). *Models of discovery: And other topics in the methods of science* (Vol. 54). Springer Science & Business Media.

Sokolowski, A. (2021). Modelling the fundamental theorem of calculus using scientific inquiry. In F. K. S. Leung, G. A. Stillman, G. Kaiser, & K. L. Wong (Eds.), *Mathematical modelling education in east and west. International perspectives on the teaching and learning of mathematical modelling*. Springer. https://doi.org/10.1007/978-3-030-66996-6_36

Wigner, E. P. (1990). The unreasonable effectiveness of mathematics in the natural sciences. In *Mathematics and science* (pp. 291–306). World Scientific.

Young, H. D., & Freedman, R. A. (2004). *University physics with modern physics* (11th ed.). Addison Wesley.

Chapter 2
The Interface Between the Contents of Physics and Mathematics

2.1 Mathematics as a Language in Physics Classroom

The following sections shed light on how mathematics' structural domain can enrich physics understanding from research standpoints. While widely used, mathematics with its inaccessibility surpasses all the other scientific disciplines (Sfard, 1991). It is expected that students use mathematics to quantify scientific phenomena and apply mathematical reasoning to hypothetically predict phenomena behaviors. Quantifying the inputs/outputs of scientific experiments can be accomplished by applying simple algorithmic operations. However, activating students' hypothetical thinking to enrich the quality of inferences requires time and effort on the learners' and the instructors' sides.

According to research, scientists describe mathematics as a language of nature, which in practice is inadvertently reduced to routine calculations, especially when problems do not require a deeper mathematical insight. If the range of mathematics tools is not supported by its structural domain, it does not fully reflect what the phrase intends to mean practically. Such a view results in physics concepts delivered to students without their more profound and more thought-provoking scientific interpretations. Narrowing mathematics influx of using only its technical domain disables investigations from zooming into details of the scientific underpinnings and proving a new hypothesis or verifying laws. Thus, a different, more comprehensive view on introducing and applying physics concepts is sought. When defined, this task will require underlying these connections daily in the physics classroom while new concepts are introduced. Providing students with opportunities to explore such connections will position the students on the trajectory to endure the decisive role that mathematics plays in supporting their learning and, ultimately, on their desire to continue studying physics.

The effects of mathematics on discoveries in physics are unquestionable. As a source of models and abstractions, mathematics enables to obtain new insights into

A. Sokolowski, *Understanding Physics Using Mathematical Reasoning*,
https://doi.org/10.1007/978-3-030-80205-9_2

how natural phenomena behave and how nature operates (Justi & Gilbert, 2002). Thus, bringing it to the physics classroom appears as a task worthy of pursuing. Mathematics allows for symbolically expressing laws of nature and prompting redesigning already formulated laws in physics. For example, Dirac's formulated a wave function (Dirac's equation) that allowed him to discover a new particle, called positron (Farmelo, 2009). Rather than using purely scientific methods, including mathematics, with its powerful hypothetical–deductive tools, often empowered scientific discoveries and allowed for presenting them from different viewpoints. Enriching the teaching of physics concepts using the math structural domain can also initiate formulations of new representations of already known laws which can have profound effects on encouraging physics students to initiate their scientific inventions (e.g., Sokolowski, 2021). While appreciation of mathematics tools requires perhaps some level of maturity and experience, addressing the effects of mathematics on physics may also encourage students to delve deeper into the structures of mathematics.

2.2 Philosophy and the Substance of the Knowledge of Mathematics

The discussion of mathematical structures and their effects on physics reasoning depends on the complexity of these structures. However, it is worthy of discussing the nature and categories of mathematical knowledge as shaped by its development.

There exist two traditional assumptions concerning the nature of mathematics that (a) mathematical knowledge is a product of human knowledge and rationality that is secure and that (b) mathematical constructs, e.g., numbers, sets, and geometric objects, exist in some objective realm (Ernest, 1994). Consequently, there are two general paradigms of the philosophy of knowledge of mathematical (a) *epistemology* that is a study of beliefs about the acquisition of knowledge, and (b) *ontology* that is a study of beliefs about the nature of reality (Schraw & Olafson, 2008). Thom (1973, p. 204) claimed that "All mathematical pedagogy, even if scarcely coherent, rests on a philosophy of mathematics." Because the philosophy of mathematics has its classroom consequences (Steiner, 1987), the consequences must affect the content of mathematics and the teaching methods. The history of scientific discoveries shows that most physics laws and principles were discovered when mathematics was embraced in *epistemological* philosophy, which likely influenced its applications in science. This might perhaps explain why many physics laws are provided as static formulas that do not explicitly enhance exploratory and hypothetical didactical reasoning. Situating a framework for teaching mathematics following either *epistemology* or *ontology* affects many facets of mathematics pedagogy. These facets can include selecting the concepts for teaching, the objectives, teaching methodologies, didactical principles, learning theorems, models, and the theorems used in research (Steiner, 1987).

Since the middle of the 1980, mathematics philosophy took a different aim and moved away from complete epistemological systems toward ontology (Shapiro, 1997).

This shift influenced the content of mathematics curricula and the way mathematics has been taught. Consequently, it also polarized differently the way students use and perceive mathematics in science classes. As opposed to superficially doing mathematical operations, relational understanding of mathematics was a priority in which concept understanding and ideas came forth before the rules (Skemp, 1976). Such shift generated further needs for reclassifying the categories of mathematical knowledge taught to students and perhaps encouraged venues for exploring mathematical concepts in other disciplines such as physics.

2.3 Procedural and Conceptual Mathematical Knowledge

The shift in the philosophy of mathematics initiated the categorization of mathematical knowledge as *procedural* and *conceptual* (Hiebert & Wearne, 1986). These categories are the building blocks of competencies in mathematics education when developing the knowledge of concepts and procedures is the priority (Rittle-Johnson & Schneider, 2015).

Procedural knowledge encompasses familiarities with isolated facts or pieces of information, and it is applied to perform mathematical operations. Procedural knowledge does not explicitly develop students' conceptual thinking (Watson & Mason, 2006). An example of procedural knowledge can be familiarity with the symbols used in formulas and with tasks that lead to the symbols' desired configurations to solve problems. Mastering and retaining procedural knowledge is acquired by practice. For instance by solving equations of either algebraic, transcendental, or trigonometric forms.

Conceptual knowledge encompasses the understanding of structures of mathematics concepts (Hiebert & Wearne, 1986). It is gained by either (a) integrating elements of various knowledge within a discipline, (b) creating the relationships between prior and new knowledge, or (c) by linking knowledge of more than one discipline. Conceptual knowledge can also be thought of as an interconnected web of knowledge or a network in which linking relationships is as prominent as the discrete pieces of information (Hiebert & Lefevre, 1986). As learners progress with mathematics education, understanding conceptual math knowledge and abstract mathematical theorems and laws become the priority.

Procedural and conceptual math knowledge, see a diagram in Fig. 2.1, contribute to developing students' mathematical and scientific literacy and general STEM disposition. However, being able to possess and apply conceptual knowledge is inferior because it develops students' mathematical reasoning that reflects the needs of STEM education (Honey et al., 2014; Sokolowski, 2018) and the demands for twenty-first-century technological advances.

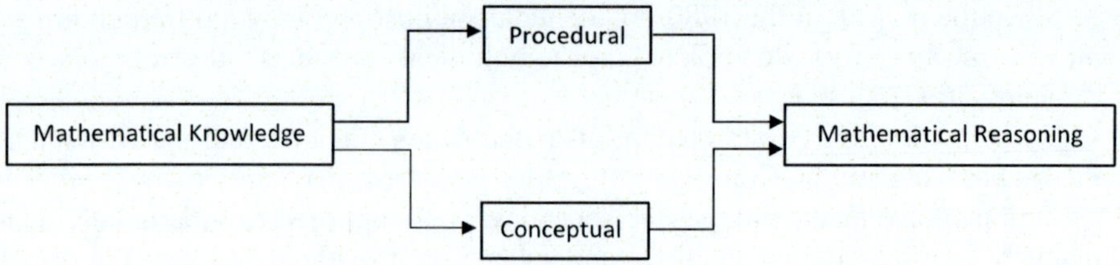

Fig. 2.1 Mathematical knowledge as categorized by mathematics education research

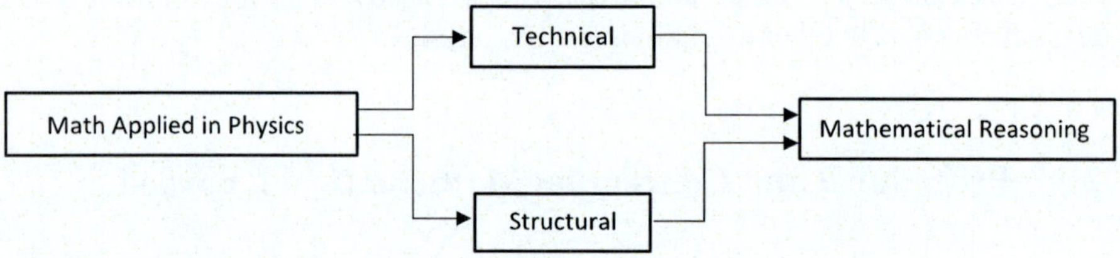

Fig. 2.2 Mathematical knowledge categorized by physics education research

The ability to reason mathematically is founded on managing both conceptual and procedural math knowledge with a superiority given to the conceptual counterpart.

2.4 Unifying Classification of Math Knowledge Used in Physics Education

While the terminologies used in math and physics to describe math knowledge do not change their interpretations, it is worthy of attempting to find parallelism. Physics research (Karam & Pietrocola, 2010) identifies two domains by which mathematics supports the meanings of physics concepts; (a) a technical domain that includes algorithmic operations and (b) a structural domain that includes these mathematical tools that help analyze system behavior in deeper theoretical analyses (Fig. 2.2).

While the technical domain requires employing students' technical skills, the structural domain requires activating students' algebraic reasoning and applying it to learn more about physical systems beyond evaluating formulas. Students' conceptual mathematics skills do not automatically convert into mathematics structural physics skills (Angell et al., 2008). This finding is perceived in physics education as one of the obstacles preventing students' physics knowledge from advancing to a higher level. It appeared beneficial to seek parallelism and unification of terminology used to also communicate such parallelism with the mathematics research community. Since the math knowledge used in physics courses correlates with that students learn in their math courses, mathematical knowledge used in physics must

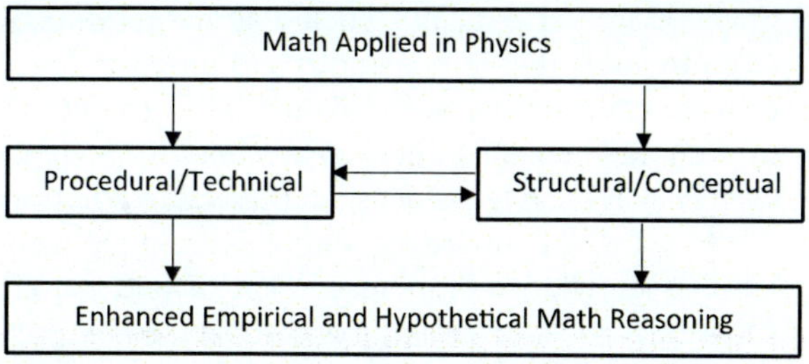

Fig. 2.3 Dual categorization of mathematics knowledge used in physics

have equivalences in mathematics education. As a result, a synthesis (see Fig. 2.3) has emerged.

Following the mapping, a general conclusion emerges that concepts and ideas encompassing *procedural math knowledge* in mathematics are considered *technical math knowledge* in physics. Concurrently, *conceptual knowledge of mathematics* can be categorized as *structural* in physics. Such mapping attempted to unify math knowledge categorization, and it is meant to assist math and physics educators in realizing the terminological similarities when addressing them to students.

In sum, efforts are made to enhance mathematical reasoning in physics by emphasizing *conceptual* – as defined in math, and *structural* – as defined in physics knowledge of mathematics. All four descriptors are being adapted in mathematics and physics education; therefore, to allow the transfer of terminology, all four descriptors will be used interchangeably yet with priority established by physics education communities.

2.5 Arrays of Applying Mathematics in Physics

The role of mathematics in physics has multiple dimensions. It (a) serves as a tool that constitutes pragmatic aspects, (b) acts as a language that represents communicative functions, (c) provides a logical framework for describing, ordering, and classifying physical processes and theories that encompass structural aspects (Krey, 2019). Delving deeper, mathematics as a tool provides technical/procedural algorithms, and mathematics as a language and framework provider supplies conceptual/structural aspects.

Considering the nature of physics students' difficulties to reason mathematically during physics inquiries, it seems that mathematics is underrepresented in communicative and structural aspects that stem from the domain of conceptual math knowledge. Possessing the ability to apply these structures can be acquired through familiarity with using various mathematical representations, mainly functions that can illustrate system dynamism, and interpreting these structures in the context of physics laws and principles. In this book, these aspects will be developed through identifying covariate relationships between selected phenomenon parameters and viewing that behavior by attending to the resulting covariate functions and their graphs. More specifically, such a structural domain of mathematics will be exercised by tools that help identify and interpret function attributes, such as function change, rates, limits, accumulation, or intercepts. Students learn and use functions in their mathematics courses as *domain-specific knowledge*. By being *domian-specific* that knowledge does not transfer to other subjects such as physics. Enabling this transfer of knowledge from one discipline within another requires reconstructing that knowledge in the new contexts and settings (Pugh et al., 2015). Research shows that such undertaking is complex, and it proves to be more challenging than many physics teachers anticipate (Orton & Roper, 2000). Following the research suggestions, such reconstruction requires teacher guidance in pinpointing critical phases or integration, especially in extracting scientific knowledge from the behavior and outputs of algebraic structures.

The research documented that procedural knowledge encompasses congruent structures and procedures in mathematics and physics. However, conceptual mathematics knowledge encompassing the structural domain does not have its equivalent counterpart in physics at hand. This means that to support the transfer of conceptual math knowledge to physics, math concepts must be reconstructed to be readily available to physics students. Such reconstruction is not about reteaching these concepts or letting students review these concepts on their own. To succeed, such reconstructions must provide a web of impulses that the students will be attracted to retain and apply to support their theories during analyses of natural phenomena. In contemporary research on physics education, such attempts are rare. While a wealth of theoretical theses about a need for such didactical enterprises can be found, a lack of projects is seen as the main reason for fragile progress. To effectively stipulate students' mathematical reasoning in physics, conceptual knowledge of mathematics and the structures that these concepts are built upon are vital in the physics classroom and central in formulating conceptual and algebraic models. Organizing physics learning experiences that gravitate toward such an aim is to transcend students' reasoning skills to a level that will enable them to make mental connections between phenomena behaviors and their corresponding mathematical embodiments. One of the fundamental methods the scientific community uses to uprise new ideas, theories, and laws is evaluating whether and how they cohere with existing knowledge (Thagard, 2007). Therefore, mathematical reasoning that can serve as a determinant to support predictions, hypotheses, and reconstructing or deriving physics knowledge will also need to adhere to existing norms. Some scientists (e.g., Branchetti et al., 2019) are more explicit with such description claiming that mathematics is much more than a language for dealing with the physical world; it is a source of models and abstractions which enable to obtain new insights into how nature operates. Since this book's goal was to propose didactical methods to enrich students' scientific inquiry by mathematical reasoning, an attempt to review mathematics concepts that can engage physics students in deeper mathematical reasoning will also be made.

2.6 Search for Tools and Methods

Mathematical reasoning or thinking has multiple definitions. Harel and Tall (1991) defined it as a process that enables learners to enrich and expand their ideas. Several mental processes such as conjecturing, generalizing, convincing, comparing, reversing, generalizing, explaining, justifying, and verifying can be induced (Mason, 1989). Mathematical reasoning can also include auxiliary processes such as calculating, solving, drawing and measuring. All these mental and physical activities can involve either procedural or conceptual knowledge or their blend, and as a result, new knowledge can be inferred. While there is specific variability in how these two constructs are defined, there is consensus that the relations between conceptual and procedural knowledge are often bi-directional and iterative. Thus, applying these attributes, conceptual knowledge can be derived from procedural, and procedures can be altered due to what conceptual knowledge is sought. Such bidirectional relation has its place in physics; algebraic computations can lead to unique inferences about the system behavior; formulating a theory can trigger specific algebraic procedures that will lead to a law formulation. Iteration also has its roots in scientific methods, which are entrenched in the procedures to reach a decision or a desired result by repeating rounds of analysis or a cycle of operations. A consensus has been reached that the knowledge of both categories, procedural and conceptual, affects mathematical reasoning development. The literature suggested two critical practices to enhance mathematical reasoning: (a) justifying and generalizing and (b) symbolizing, representing, and communicating. For some researchers (e.g., Kilpatrick et al., 2001), justifying as to "provide sufficient reason for" (p. 130) is a critical element of being fluent to reason mathematically. These mental processes can be exercised from analyses of covariational representations and many of such metal actions like generalizing, symbolizing, representing, and communicating, are included in STEM modeling frameworks designed for enhancing scientific inquiry in mathematics (Marginson et al., 2013; Sokolowski, 2018). The prospect of using algebraic representations in physics courses derived from natural phenomena can enrich students' physics and math skills. In an earlier study, Sokolowski (2015) found out that modeling activities that use scientific contexts in mathematics courses present math ideas in a way that makes the students view these ideas in a different more pragmatic perspective. Many mathematical representations: symbols, numbers, tables, diagrams, graphs, and algebraic expressions are being used to represent physical constructs. Beyond this representational role, mathematics allows for conceptual insight that might produce richer inferences (Quale, 2011). Therefore, to develop scientific literacy, physics should be taught to rely not only on observing experiments, but these experiments should provide bases for empirical-mathematical reasoning concluded with symbolically expressed behavior of selected experiment's parameters.

2.7 Mathematical and Scientific Reasoning; Are These Mental Actions Equivalent?

Enhancing physics inquiry by mathematical reasoning can take diverse paths depending on several factors such as the layer of embedded reasoning, the grade level, content taught, and the alignment of mathematics and physics curricula. Though, there are certain general factors and strategies that can help flesh out these paths.

This paragraph is to delve deeper into identifying conceptual and procedural knowledge and their effects on developing students' mathematical reasoning skills in physics. In concluding, a general link between mathematics knowledge learners acquire in mathematics and that needed in physics will emerge. These connections will be established through a preliminary analysis of the possible reasoning approaches developed and used in mathematics and physics.

Theorems, laws, and axioms in mathematics differ from these studied in sciences. They are introduced as context-free because they need to be universal to any subject application. Because of the lack of context, mathematics laws are presented very abstractly. Physics constitutes a natural framework for testing, applying, and elaborating mathematical theorems, methods, and concepts, or even motivating, stimulating, instigating, and creating all kinds of mathematical innovations (Tzanakis & Thomaidis, 2000). However, these methods are not widely exercised in mathematics; thus, students entering a physics classroom typically possess the tools, but they do not possess the skills how to use these tools to support scientific analyses. To initiate the transfer of knowledge, mathematical symbols, concepts, and laws that students learn in their math classes should be refreshed and augmented in physics considering their purpose of being used in physics. For this to sustain, the semantics that informs how the mathematical symbols are interpreted in math classes should also be transferrable to physics classrooms and interpreted congruently with their mathematical meanings. The final products of such actions should lean on scientific interpretations.

2.8 Synthesis of Students' Challenges with Math Knowledge Transfer

Research shows that students can easily activate their procedural mathematics knowledge in physics classes, but they face difficulties activating more sophisticated conceptual mathematical knowledge (Wilcox et al., 2013). This can be explained in the way how mathematics knowledge is delivered to students. Procedural math skills such as calculating, solving, drawing, or measuring, can be easily transferred to physics due to the nature of these activities classified as domain-specific developed and used in a similar nature in mathematics as in physics. Learners do not encounter difficulties in transferring procedural *domain-specific* math knowledge because they construct the understanding of these concepts within similar disciplinary norms in both subjects. Calculating or solving for a variable are elements of procedural math knowledge and such knowledge is transferred effortlessly from mathematics to other subjects and also to physics. The only element of the new knowledge encountered in physics is the syntactic (not conceptual) understanding of physics symbols if calculating or solving for a parameter is involved. While working on such physics problems, intuitive mathematics knowledge, symbolic forms, and interpretive devices are not put into play (Tuminaro & Redish, 2007). Transferring conceptual math knowledge is different. The barriers that arise when transferring conceptual math knowledge to physics classes are researched, and these difficulties are well documented by studies encompassing multiple countries, curricula, and classroom settings (Pospiech, 2019). A handful of these studies report these difficulties as shortfalls of students' abilities to activate conceptual math knowledge to support physics reasoning (e.g., Fraser et al., 2014). Planinic et al. (2013) proposed three categories of possible sources of these difficulties (a) either that the required resource does not exist (students did not learn required math concepts in their math classes), or (b) the resource exists but is not activated due to the incorrect framing of the problem (this could result from a misalignment of the terminology used) or (c) the resource is activated, but its mapping to the problem is not appropriate (this could result from a misalignment of symbols or structures). Research provides recommendations on eliminating these barriers; Redish (2005) suggested augmenting mathematics tools so that their syntaxes fit into these used in physics. Others found out that transfer is more likely to occur when students realize a given idea in at least two diverse contexts or receive metacognitive scaffolding (see Hammer et al., 2005). Branchetti et al. (2019) used a model developed by Uhden et al. (2012) to design a teaching tutorial to help mathematics and physics university students with knowledge transfer. Wilcox et al. (2013) developed an analytical framework to assist instructors and researchers in aligning students' difficulties with specific mathematical tools when solving problems of the upper physics division. Kuo et al. (2020) found out that students sought coherence between formal mathematics and conceptual understanding and suggested that mathematical sense making – the practice of seeking coherence between formal mathematics and conceptual understanding – is a critical element of physics problem-solving. diSessa (2008) suggested that (a) inquiry is a more successful way to develop ontological meanings than other methods and (b) that students can become much wiser with this kind of process and enhance future learning on that basis.

This synthesis illustrates that students' difficulties can be accounted for by a lack of parallel activities in physics courses that would activate their conceptual mathematical knowledge departing from the way they learned these concepts in math classes. It can be inferred that expecting that students activate that knowledge automatically without the teacher's guidance can be deceptive. Organizing physics learning experiences that gravitate toward knowledge transfer is seen as equipping students with reasoning skills to help them make mental connections between phenomena behaviors and interpretations of corresponding mathematical embodiments.

In sum, it is reasonable to say that obstacles with transferring math knowledge stem from a lack of parallelism between methods, approaches, and perhaps syntaxes that students learned in math and what the expectations in physics are. Establishing a conceptual framework that will attempt to elevate the challenges and allow this transition will be discussed and proposed in Part II of the book.

References

Angell, C., Kind, P. M., Henriksen, E. K., & Guttersrud, Ø. (2008). An empirical-mathematical modeling approach to upper secondary physics. *Physics Education, 43*(3), 256.

Branchetti, L., Cattabriga, A., & Levrini, O. (2019). Interplay between mathematics and physics to catch the nature of a scientific breakthrough: The case of the blackbody. *Physical Review Physics Education Research, 15*(2), 020130.

diSessa, A. A. (2008). A "theory bite" on the meaning of scientific inquiry: A companion to Kuhn and Pease. *Cognition and Instruction, 26*(4), 560–566.

Ernest, P. (1994). The philosophy of mathematics and the didactics of mathematics. In *Didactics of mathematics as a scientific discipline* (pp. 335–350). Kluwer Academic Publishers.

Farmelo, G. (2009). *The strangest man: The hidden life of Paul Dirac, quantum genius*. Faber & Faber.

Fraser, J. M., Timan, A. L., Miller, K., Dowd, J. E., Tucker, L., & Mazur, E. (2014). Teaching and physics education research: Bridging the gap. *Reports on Progress in Physics, 77*(3), 032401.

Hammer, D., Elby, A., Scherr, R. E., & Redish, E. F. (2005). Resources, framing, and transfer. In *Transfer of learning from a modern multidisciplinary perspective* (pp. 89–119). IAP.

Harel, G., & Tall, D. (1991). The general, the abstract, and the generic in advanced mathematics. *For the Learning of Mathematics, 11*(1), 38–42.

Hiebert, J., & Lefevre, P. (1986). Conceptual and procedural knowledge in mathematics: An introductory analysis. In J. Hiebert (Ed.), *Conceptual and procedural knowledge: The case of mathematics* (pp. 1–27). Erlbaum.

Hiebert, J., & Wearne, D. (1986). Procedures over concepts: The acquisition of decimal number knowledge. In J. Hiebert (Ed.), *Conceptual and procedural knowledge: The case of mathematics*. Erlbaum.

Honey, M., Pearson, G., & Schweingruber, H. A. (Eds.). (2014). *STEM integration in K-12 education: Status, prospects, and an agenda for research* (Vol. 500). National Academies Press.

Justi, R. S., & Gilbert, J. K. (2002). Modelling, teachers' views on the nature of modelling, and implications for the education of modellers. *International Journal of Science Education, 24*(4), 369–387.

Karam, R., & Pietrocola, M. (2010). Recognizing the structural role of mathematics in physical thought. In *Contemporary science education research: International perspectives* (pp. 65–76). Pegem Akademi.

Kilpatrick, J., Swafford, J., & Findell, B. (2001). *Adding it up: Helping children learn mathematics* (Vol. 2101). National Research Council (Ed.)). National Academy Press.

Krey, O. (2019). What is learned about the roles of mathematics in physics while learning physics concepts? A mathematics sensitive look at physics teaching and learning. In *Mathematics in physics education* (pp. 103–123). Springer.

Kuo, E., Hull, M. M., Elby, A., & Gupta, A. (2020). Assessing mathematical sensemaking in physics through calculation-concept crossover. *Physical Review Physics Education Research, 16*(2), 020109.

Marginson, S., Tytler, R., Freeman, B., & Roberts, K. (2013). *STEM: Country comparisons: International comparisons of science, technology, engineering, and mathematics (STEM) education*. Final report.

Mason, J. (1989). Mathematical abstraction as the result of a delicate shift of attention. *For the Learning of Mathematics, 9*(2), 2–8.

Orton, T., & Roper, T. (2000). Science and mathematics: A relationship in need of counselling? *Studies in Science Education, 35*, 123–154.

Planinic, M., Ivanjek, L., Susac, A., & Milin-Sipus, Z. (2013). Comparison of university students' understanding of graphs in different contexts. *Physical Review Special Topics – Physics Education Research, 9*(2), 020103.

Pospiech, G. (2019). Framework of mathematization in physics from a teaching perspective. In *Mathematics in physics education* (pp. 1–33). Springer, Cham.

Pugh, K. J., Linnenbrink-Garcia, L. I. S. A., Phillips, M. M., & Perez, T. O. N. Y. (2015). *Supporting the development of transformative experience and interest* (pp. 369–383). AERA.

Quale, A. (2011). On the role of mathematics in physics. *Science & Education, 20*(3–4), 359–372.

Redish, E. F. (2005). Changing student ways of knowing: What should our students learn in a physics class. *Proceedings of World View on Physics Education*, 1–13.

Rittle-Johnson, B., & Schneider, M. (2015). Developing conceptual and procedural knowledge of mathematics. In *Oxford handbook of numerical cognition* (pp. 1118–1134). Oxford University Press.

Schraw, G. J., & Olafson, L. J. (2008). Assessing teachers' epistemological and ontological worldviews. In *Knowing, knowledge, and beliefs* (pp. 25–44). Springer.

Sfard, A. (1991). On the dual nature of mathematical conceptions: Reflections on processes and objects as different sides of the same coin. *Educational Studies in Mathematics, 22*(1), 1–36.

Shapiro, S. (1997). *Philosophy of mathematics: Structure and ontology*. Oxford University Press on Demand.

Skemp, R. R. (1976). Relational understanding and instrumental understanding. *Mathematics Teaching, 77*(1), 20–26.

Sokolowski, A. (2015). The effects of mathematical modelling on students' achievement-meta-analysis of research. *IAFOR Journal of Education, 3*(1), 93–114.

Sokolowski, A. (2018). Formulating conceptual framework for multidisciplinary STEM modeling. In *Scientific inquiry in mathematics – Theory and practice* (pp. 53–62). Springer.

Sokolowski, A. (2021). Enabling covariational reasoning in Einstein's formula for photoelectric effect. *Physics Education, 56*(3), 035029.

Steiner, H. G. (1987). Philosophical and epistemological aspects of mathematics and their interaction with theory and practice in mathematics education. *For the Learning of Mathematics, 7*(1), 7–13.

Thagard, P. (2007). Coherence, truth, and the development of scientific knowledge. *Philosophy of Science, 74*(1), 28–47.

Thom, R. (1973). Modern mathematics: Does it exist? In A. G. Howson (Ed.), *Developments in mathematical education* (pp. 194–209). Cambridge University Press.

Tuminaro, J., & Redish, E. F. (2007). Elements of a cognitive model of physics problem solving: Epistemic games. *Physical Review Special Topics – Physics Education Research, 3*(2), 020101.

Tzanakis, C., & Thomaidis, Y. (2000). Integrating the close historical development of mathematics and physics in mathematics education: Some methodological and epistemological remarks. *For the Learning of Mathematics, 20*(1), 44–55. Retrieved January 18, 2021, from https://eric.ed.gov/?id=ej607175

Uhden, O., Karam, R., Pietrocola, M., & Pospiech, G. (2012). Modelling mathematical reasoning in physics education. *Science & Education, 21*(4), 485–506.

Watson, A., & Mason, J. (2006). Seeing an exercise as a single mathematical object: Using variation to structure sense-making. *Mathematical Thinking and Learning, 8*(2), 91–111.

Wilcox, B. R., Caballero, M. D., Rehn, D. A., & Pollock, S. J. (2013). Analytic framework for students' use of mathematics in upper-division physics. *Physical Review Special Topics – Physics Education Research, 9*(2), 020119.

Part II
Designing Learning Environments to Promote Math Reasoning in Physics

Chapter 3
Modeling as an Environment Nurturing Knowledge Transfer

3.1 Scientific Modeling and Models

The case studies proposed in this book (Part III) conclude with models, often expressed as covariate algebraic expressions with modeling considered a process that will lead students to formulate these algebraic models. This section delves deeper into the learning advantages that stem from immersing students in modeling activities to pursue math knowledge transfer. A model in science has many definitions. It is perceived as a surrogate object or a conceptual representation of a part of reality that enables its understanding and theory formulation. As theory explains a phenomenon behaviour, the model aids in understanding specific aspects of the phenomenon. The primary purpose of formulating models is enabling an epistemological link between realities, theory, and law (see Fig. 3.1).

A model is formulated by applying a suitable strategy, and it is to reflect on the relations of system parameters in the search for a deeper understanding. Model is also called a kind of analogy or mental image of the phenomena expressed in other structures that we are already familiar with (Giancoli, 2014). To serve its purpose, the model needs to be simple and provide a structural similarity to the phenomena under investigation. While the theory is based on certain assumptions, the model is typically free from assumptions and is restrained in its representation.

Modeling is considered the most popular and effective way of developing algebraic representations to interpret system behavior (Selden & Selden, 1992). While modeling is the action, the model is the product of that action. Modeling activities bring students closer to science's epistemic view and warrant more insightful ideas about the scientific inquiry that they can further deploy when working on other scientific investigations or solving real problems. Physics is a modeling enterprise, and as such, it should train students to become competent modelers and interpreters of these models (Angell et al., 2008).

A. Sokolowski, *Understanding Physics Using Mathematical Reasoning*,
https://doi.org/10.1007/978-3-030-80205-9_3

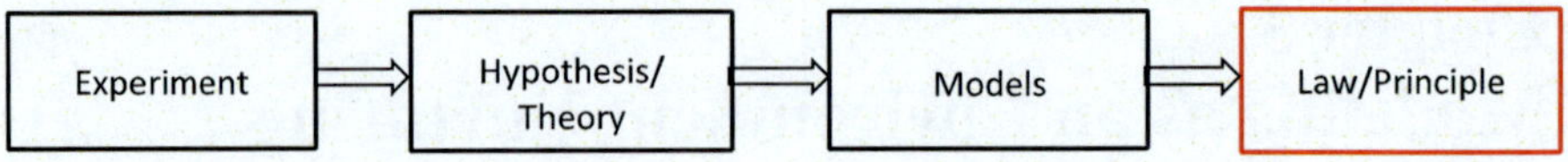

Fig. 3.1 Models as mediators between experiment, theory, and law

Modeling allows blending representations naturally, and it is accepted as an essential tool in science and mathematics education at all levels (Lesh et al., 2010). There are more descriptions of modeling found in research. Odenbaugh (2005) defines modeling as a collection of cognitive strategies formulated to pursue scientific inquiry aims. Wilkinson (2011) perceived modeling as an attempt to develop an understanding of elements of a system, considered a part of reality, their states, and interactions with other elements.

Models of natural phenomena similar to representations can take different forms; conceptual, mental, verbal, physical, statistical, logical, graphical, etc. Research identifies models created using mathematical structures as the most prominent and most frequently deployed to modeling physics activities (Redish & Kuo, 2015). While being formulated, physics constructs mediate between more than one type of model before reaching their final representational forms. Creating mathematical models that allow both predictions and theory is perceived as a primary physics education goal (Hestenes, 1987). When mathematical representations are the final products, they must accord with formulating empirical-mathematical models suggested by research in mathematical modeling. Such alignment will nurture math knowledge transfer.

One of the essential advantages of using models is that they reduce a need to transfer domain-specific knowledge by providing meaningful ways for students to construct, explain, describe, manipulate, or predict patterns and regularities associated with the situation under investigation (Michelsen, 2005). The development of the competency of using mathematics reasoning in science is not just about applying mathematics algorithms. The competency is developed by merging interpretations of the phenomena behavior with attributes of algebraic functions. Physics as an experimental science provides multiple opportunities for organizing such learning venues.

3.2 Modeling Cycles in Physics Education

Traditional development of the competency of using mathematics in science is about applying mathematics algorithms. The book is posited to change this perception. Using mathematics can be enriched by merging interpretations of the phenomena behavior with mathematical representations where modeling is a mediator during these enterprises. While the modeling products can take diverse forms, modeling that concludes with the formulation of an algebraic function representing a formula will be the priority in this book. Being such, modeling applied in experiments is to generate the basis for developing algebraic representations of the part of reality. Applying scientific methods that delve deeper in formulating and testing hypotheses and merge math and physics reasoning appears a rich learning experience. Such tasks enhance the students' scientific inquiry skills as they have the prospective to make the concepts of science and mathematics more tangible and more attractive to explore and understand.

There exist several modeling cycles in the physics education literature that assist in organizing modeling activities. Hestenes (1995) proposed to initiate the modeling process by isolating a part of a situation called the phenomenon. More detailed phases of these enterprises, such as purpose and validity, were included in that process, and all supported the model formulation. Analysis and justification of the enacted model concluded the modeling cycle. Students' depth of reasoning that emerges from this modeling depends on the specific nature of the investigations at hand. While it was assumed that models, as the products of such investigations, could take various forms, algebraic representations were assumed to be one of them.

Redish and Gupta (2009) developed a modeling cycle that explicitly linked algebraic representations with a physical system by mental actions such as processing, interpreting, and evaluating. Modeling appeared as the catalyst of transferring properties of the physical system into a mathematical representation. Such formulated representations were to be interpreted and further appraised for their accuracy to reflect on the physical system's behavior under investigation.

More recently, Uhden et al. (2012) developed a scheme how mathematics is used in physics and how it should enhance mathematical reasoning. The model was intended to provide a guiding framework when aspects related to mathematical reasoning in physics were concerned. It exemplified various levels of knowledge of mathematics to constructing mathematical-physical representations. The sophistication of using the tools of mathematics depended on the degree of mathematization endured with technical mathematical operations regarded as pure mathematics tools and structural skills regarded as students' capacity of employing mathematical knowledge for structuring physical situations. The interpretation process included a mathematical-physical model and was to enhance the scientific part of the analysis.

The modeling cycles discussed herein were to support instructors in designing lessons and laboratory activities. As such, they serve the general purpose of establishing a didactical framework. By being broadly defined, they can be applied to any grade level/physics section and help set up the modeling processes. According to research, there is a need for more detailed designs of lab activities that will illustrate classroom interactions while exemplifying mathematical reasoning. Thus, a general framework and specific content-related actions that will support these frameworks are compulsory. Reflecting on these actions and identifying the prompts that will activate students' math knowledge appeared a priority that will be discussed in the following sections.

3.3 Merging Mathematics and Physics Representations

Prain and Waldrip (2006) called for establishing a prominent role of representations in physics education to enhance scientific inquiry because they fit into students' learning styles and engage students to learn science. To reflect on this recommendation, in mathematics courses students are presented, for instance with various function forms followed by the *rule of five*, which suggests using five function representations: verbal description, symbolic representation, graphs, table of values, and mapping (Yerushalmy, 1997). Students are to use these representations and be able to transfer between them. Representations that resonate with students' realities are easily converted into mental impulses and stored in students' long-term memory. Being able to produce new representations enables deriving new theories. Prominent scientists made their discoveries by carefully selecting and analyzing existing representations and then inventing new (Cheng, 1999). Such prospect of tasks seems reasonable to apply in physics courses. More sophisticated enterprises can include applying the concept of limits, called also *limiting case analyses* that will be used more extensively in Chaps. 6 and 9. These analyses will enhance scientific inquiries in situations where direct evaluations of derived covariation relationships will not be plausible.

Mathematics plays an essential role in quantifying science and embracing it in concise mathematical language. Thus, being an expert in understanding and formulating scientific laws and theories and applying these constructs to solve physics problems requires recognizing essential elements between various scientific representations and their mathematical counterparts. Upon understanding these entanglements, learners are set to become versed in applying these entities to construct new laws or reconstruct laws that had already been discovered. Merging mathematical reasoning with its scientific counterpart means assuring coherence between interpretations and enabling transitions not only between various mathematics representations but also between their physics equivalent counterparts. Acting upon enabling such transitioning appears to strengthen understanding of empirical evidence.

Using the mathematical way of reasoning to enhance scientific inquiry is also the key to success in engineering programs where expert problem-solvers need to exploit opportunities to use blended conceptual reasoning (Fauconnier & Turner, 2003). Working on lab activities or solving physics problems often concludes with formulating algebraic models of the phenomena behavior. All these mental processes inherent in identifying prompts and reconstructing them in various representations merge into one coherent thought process. Diverse physics curricula comprising laws, principles, and theorems expressed qualitatively or quantitatively provide a wealth of opportunities for enacting these lines of inquiry. Research showed that reasoning, and especially this supported by empirical evidence, has proven difficult for physics students (Hammer, 1996); therefore, developing learning environments that will produce a web of stimuli that students can use to engage in these processes to reach fluency in this domain is needed. In Part II, Chap. 4, such a modeling scheme is proposed. The scheme provides a general design for lab activities, and lesson conducts that prioritize the phase of merging the methods of physics and mathematics.

References

Angell, C., Kind, P. M., Henriksen, E. K., & Guttersrud, Ø. (2008). An empirical-mathematical modeling approach to upper secondary physics. *Physics Education, 43*(3), 256.

Cheng, P. C. H. (1999). Unlocking conceptual learning in mathematics and science with effective representational systems. *Computers & Education, 33*(2), 109–130.

Fauconnier, G., & Turner, M. (2003). *The way we think: Conceptual blending and the mind's hidden complexities*. Basic Books. ISBN: 9780465087860.

Giancoli, E. (2014). *Physics: Principles with applications* (AP ed.). Pearson.

Hammer, D. (1996). More than misconceptions: Multiple perspectives on student knowledge and reasoning, and an appropriate role for education research. *American Journal of Physics, 64*(10), 1316–1325.

Hestenes, D. (1987). Toward a modeling theory of physics instruction. *American Journal of Physics, 55*(5), 440–454.

Hestenes, D. (1995). *Modeling software for learning and doing physics*. In Thinking physics for teaching (pp. 25–65). New York: Springer US.

Lesh, R., Galbraith, P. L., Haines, C. R., & Hurford, A. (2010). *Modeling students' mathematical modeling competencies*. Boston, MA: Springer US. doi, 10, 978-1.

Michelsen, C. (2015). Mathematical modeling is also physics—interdisciplinary teaching between mathematics and physics in Danish upper secondary education. Physics *Education, 50*(4), 489.

Odenbaugh, J. (2005). Idealized, inaccurate but successful: A pragmatic approach to evaluating models in theoretical ecology. *Biology and Philosophy, 20*(2–3), 231–255.

Prain, V., & Waldrip, B. (2006). An exploratory study of teachers' and students' use of multi-modal representations of concepts in primary science. *International Journal of Science Education, 28*(15), 1843–1866.

Redish, E. F., & Gupta, A. (2009). Making meaning with math in physics: A semantic analysis. *GIREP-EPEC & PHEC 2009*, 244.

Redish, E. F., & Kuo, E. (2015). Language of physics, the language of math: Disciplinary culture and dynamic epistemology. *Science & Education, 24*(5), 561–590.

Selden, A., & Selden, J. (1992). Research perspectives on conceptions of function: Summary and overview. In *The concept of function: Aspects of epistemology and pedagogy* (Vol. 25, pp. 1–16). Mathematical Association of America.

Uhden, O., Karam, R., Pietrocola, M., & Pospiech, G. (2012). Modelling mathematical reasoning in physics education. *Science & Education, 21*(4), 485–506.

Wilkinson, D. J. (2011). *Stochastic modelling for systems biology*. Boca Raton, FL: CRC Press.

Yerushalmy, M. (1997). Designing representations: Reasoning about functions of two variables. *Journal for Research in Mathematics Education, 28*(4), 431–466.

Chapter 4
Proposed Empirical-Mathematical Learning Model

4.1 Didactical Underpinnings of the Design

While the general design of the scheme aims at merging the math structural domain with physics concepts, other objectives such as sparkling students' interest in studying physics and encouraging their investigations were also considered. The empirical-mathematical modeling scheme was inspired by an earlier designed modeling cycle where physics aided the link between natural phenomena and their engineering and technology applications (Sokolowski, 2018). As the primary inquiry process remains scientific in both designs, the one presented in Fig. 4.1 highlights physics's understanding as the primary learning objective. Instructional suggestions on how to use it in practice follow.

The process is designed for organizing lab activities, however it can also serve as a backbone to developing instructional units that intend to immerse students in mathematical reasoning while exploring the physics concepts. The scheme consists of mental or physical actions, represented by the cells in the middle column of the diagram. These actions will reflect on the specific lab objective and goals. The cells on the right and left sides of the stem actions indicate phases when integrating mathematics and physics takes priority. When completed, the formulated mathematical model can be deployed to problem solving or it can constitute new knowledge derived from the investigations.

4.2 Description of the Learning Phases

The labelings in the first two cells as real or simulated contexts pertain to preparation and formulating the lab purpose. These tasks are to be prepared by the instructor. The remaining tasks in the stem actions reflect on the methods of scientific

A. Sokolowski, *Understanding Physics Using Mathematical Reasoning*,
https://doi.org/10.1007/978-3-030-80205-9_4

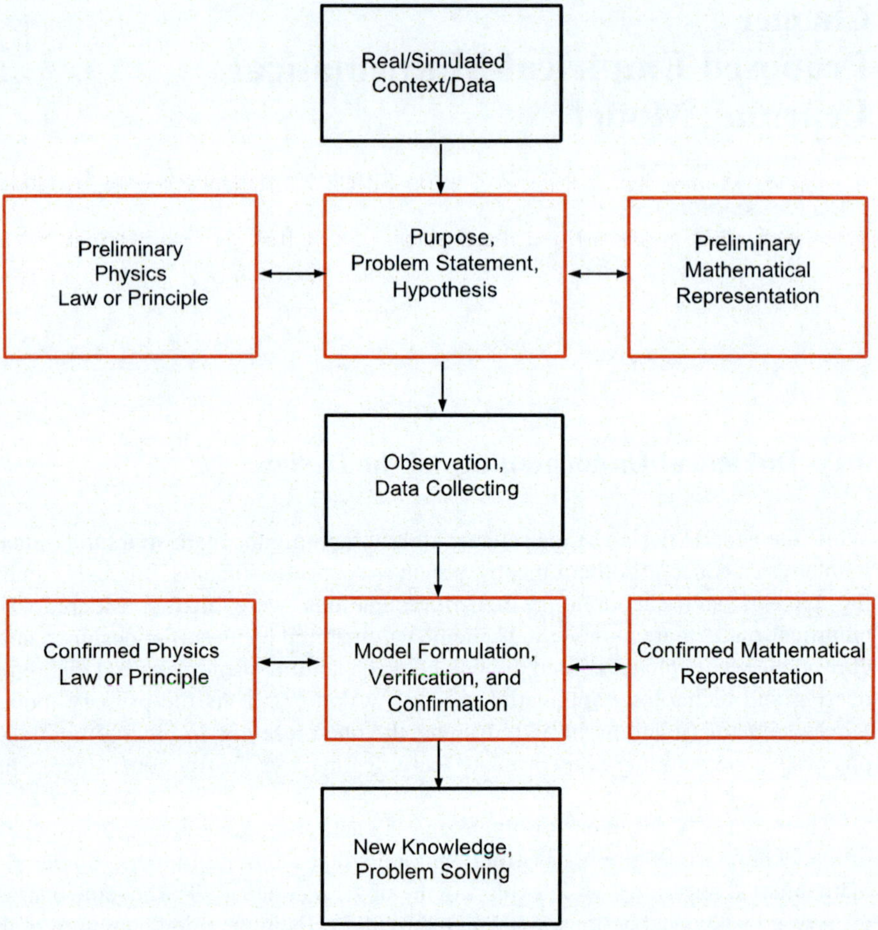

Fig. 4.1 Blended empirical-mathematical modeling scheme enhancing covariational reasoning

inquiry and will be assigned to be performed by the students. Students should be provided with instructional support that will guide them through these actions. The degree of support depends on the complexity of the phenomena and covariate relations that will emerge from the lab analyses.

The stem of the lab intertwines with two parallel chambers positioned on the left and right sides of the main actions that suggest merging math and physics constructs as they fit lab constraints. These constructs will fuse during investigations to produce an integrated inquiry. Such alignment is purposeful. Research showed that physics students have difficulties applying conceptual math knowledge to physics; thus, by explicitly highlighting this parallelism and mutual coherence, such merging will be more likely to occur.

The forms of representations that students will develop during the lab conducts can range from data tables and graphs to covariate structure formulation. A constant interplay between abstract mathematical symbolism and physics realities is to engage the learners in generating new knowledge. While the depth of the structural domain of mathematics is not elaborated explicitly, its baseline is drawn on the entanglements between the lab parameters embedded in the problem statement, hypothesis, and math knowledge that students acquired. The complexity of the algebraic structure brought to life during specific investigations will parallel with the traditional high school and undergraduate physics curricula. Due to its general purpose, the scheme can serve as a reference for using students' mathematical reasoning skills in other science branches as well.

The modeling process can be exercised using actual equipment, virtual simulations, or data provided by a table of values or other forms. It is suggested that if virtual simulations are used, they possess an exploratory character that allows for controlling variable parameters, data taking, and graphs plotting. It is also essential that the virtual context provides opportunities for verifications of the derived covariate relations. Simulations that meet all these conditions are often classified as computational (Davis et al., 2007). Students must be able to use their knowledge of physics and the language of mathematics to identify parameters of interest and classify them as dependent, independent. They might be encouraged to further categorize these variables as covariate parameters using more detailed scientific classification that is proposed in Sect. 5.2.

It is suggested that the instructor explain the lab's purpose and provide suggestions about the law or principle to be discovered or verified during the lab. The extend of the explanation will depend on the lab context and the grade level of the students. The instructor might introduce the experiment conduct and explain how to utilize the measuring devices to quantify parameters whether they are real or virtual. The lab introduction should be done, if possible, in a manner that will present it as a discovery adventure. The students always welcome a general lab presentation and how the experiment's parameters are set up.

4.3 Hypotheses as Learners' Proposed Theories

A problem statement typically formulated by the instructor is a form of unknown law and theory to discover upon lab completion. A catalyst of the lab conduct is formulating the hypothesis that reflects on the problem statement. The hypothesis is defined as the investigators' proposed theory supported by their prior knowledge explaining why the phenomenon under investigation behaves the way it does. Thus, depending on the nature of the investigation, hypotheses have the potential to be called law or theory (Simon, 2012). A reward for correctly hypothesized outcomes should be optimal, yet students undoubtedly welcome their hypotheses to be correct. Empowering the learners to formulate the hypotheses can be considered as passing the ownership to discover the underpinning of the experiment behavior to the students. Such action can inspire the students and increase their motivation to complete the lab. By formulating the hypotheses, students are placed in the position of being experts in understanding and merging physics and mathematics' laws and principles. Learners need time and *space* to develop, verify and revise their hypotheses if needed. While most typical lab activities are to verify already discovered scientific laws, there are plenty of opportunities for designing activities that enable new law formulation.

To ensure that the learners reflect on their mathematics knowledge and skills, they need to be unambiguously prompted to do so during the lab. This parallelism will be maintained throughout the lab by inserting questions that will require the knowledge merging. It is to assure that the switch from a verbally stated theory to a computational form of law is vital during the lab conduct.

4.4 Mainstream of the Inquiry and Its Confirmation

Conceptual questions precipitated during the lab conduct will enrich data taking and nurture the reasoning development that will merge into algebraic relationships. Research on problem-solving in physics stated that physics experts begin solving problems from a conceptual analysis of the physical situation then move into a more sophisticated mathematical analysis. In contrast, novices tend to start by selecting and manipulating equations (Larkin et al., 1980). To assure physics experts' inquiry order, proactive tasks will be embedded in the instructional support to direct the learners to take such path. In the proposed lab conduct design (Fig. 4.1), structural (conceptual) mathematics knowledge will constitute the main pilar of verifying algebraic structures that symbolically express the system's behavior and extend its hypothetical predictions.

4.5 Methods of Enacting Mathematical Structures

How will the algebraic relations be derived? The significant phases include identifying the parameters of interest, observing their mutual relationship, data gathering, graph plotting, and then zooming in on the graph possible algebraic representation. Depending on the system's complexity level, students might be offered supportive questions to select a correct algebraic expression for the data gathered. For example, to differentiate between selecting a linear or quadratic model, students might be prompted to discuss if the quantity of interest attains a maximum value or a constant rate of change. Questions of such prompts will help activate students' mathematical reasoning and identify such relations in the lab and problem-solving. Formulated mathematical models should correspond with what students predicted and observed in the experiment. Following the collections of such inferences, students will formulate an algebraic representation for the data. This phase of the inquiry might require revision of their hypothesized outcome and theory. There can be multiple supportive questions embeded in the verification phase to ensure that students reach the correct conclusions and revise and support them if needed.

4.6 Concluding Phases of the Learning Process

Students must be aware that revising, refining, and modifying their theories are necessary actions taken during scientific conduct. These actions enhance the quality of the investigations rather than diminish them. Revisions do not devaluate the investigator's knowledge and efforts, but they improve the scientific enterprise's validity. The revision process requires verification and establishing coherence between both scientific and mathematics reasoning outcomes. Students will be required to adhere to the scientific principles and their mathematical correctness when verifying formulated algebraic structures. While the reasoning process will develop an algebraic model of the phenomena, efforts will be made to extend the lab's inferences, bridge, and deploy these findings to problem solving that the students will encounter in their daily school practice. This extension is to have the students realize that the discovered and derived algebraic models are meaningful and helpful to succeed on assessment items. It is hoped that the inferences derived from the investigations will solidify in students' long-term memory as valuable impulses of knowledge that are ready to be retrieved when needed.

The proposed modeling scheme does not aspire to represent a universal theorem of modeling. It is rather to serve as a general guide to design labs in physics that aim at integrating physics inquiry with the structural domain of mathematics.

References

Davis, J. P., Eisenhardt, K. M., & Bingham, C. B. (2007). Developing theory through simulation methods. *Academy of Management Review, 32*, 480–499.

Larkin, J., McDermott, J., Simon, D. P., & Simon, H. A. (1980). Expert and novice performance in solving physics problems. *Science, 208*(4450), 1335–1342.

Simon, H. A. (2012). *Models of discovery: And other topics in the methods of science* (Vol. 54). Springer Science & Business Media.

Sokolowski, A. (2018). Formulating conceptual framework for multidisciplinary STEM modeling. In *Scientific inquiry in mathematics-theory and practice* (pp. 53–62). Springer.

Chapter 5
Covariational Reasoning – Theoretical Background

5.1 Quantities, Parameters, and Variables

Covariate reasoning can be exercised by observing selected parameters behavior and considering algebraic functions or formulas as structures depicting the behavior. Formulas are built up of varying and constant parameters that stem from a specific algebraic arrangement of quantities that warrant the experiment behavior. To realize their positions within the formula structures, the elementary elements need to be defined.

Quantities are the building blocks of physics constructs, laws, principles, and theorems. Quantities are inherent attributes of a specific object or parts of the system that support its numerical descriptions (Gilliard, 2020). For example, if the mass of the electron is concerned, the mass is an attribute of an electron. If the diameter of the Earth is concerned, the diameter represents an attribute of the Earth. These attributes are measurable, and their values possess certain magnitudes and units. Quantity can also be defined as someone's conceptualization of an object if the object has a measurable attribute (Fey, 1990). A variable represented by a symbol can be any one of a class of elements that may or may not be quantized. Following these definitions, quantities represented by parameters appear to be subsets to variables. How to differentiate between variables and parameters? A parameter in mathematics represents a constant quantity in the case under consideration but varies in other cases (Oxford University Press, 2015). Parameters in mathematics are usually denoted by alphabetical letters, a, b, c,etc., and the example (Eq. 5.1) of a quadratic function signifies these meanings.

$$g(x) = ax^2 + bx + c \tag{5.1}$$

The domain of this function is unrestricted and it can take any real value $x \in R$; however, there is a restriction on the parameter a in $g(x)$ to depict a quadratic

A. Sokolowski, *Understanding Physics Using Mathematical Reasoning*,
https://doi.org/10.1007/978-3-030-80205-9_5

function. The parameter cannot take a value of zero, $a \neq 0$, because the quadratic function would be reduced to a linear one $g(x) = bx + c$.

Following this analysis, a bridge to the quadratic formula (Eq. 5.2) that allows computing the x-intercepts of $g(x)$; can be made.

$$x = \frac{-b \pm \sqrt{b^2 - 4ac}}{2a} \tag{5.2}$$

The quadratic formula is dependent only on the values of its parameters, not on the independent variable x. In its covariate form, it can be expressed as $x(a,b,c) = \frac{-b \pm \sqrt{b^2 - 4ac}}{2a}$.

While this is apparent for teachers, for the students, it might not be. Is there a case in physics to parallel an application of this algebraic structure? Consider an object projected upward with an initial velocity v_1 from a height of y_1. The formula (Eq. 5.3) to find the vertical position of the object considered a function is

$$y(t) = y_1 + v_1 t - \frac{gt^2}{2} \tag{5.3}$$

A common task associated with function analysis is finding their horizontal intercepts, also called function zeros. The position function $y(t)$ can be used to find the time when the object lands on the ground by letting the vertical position of the object be 0, thus $0 = y(t)$. By rearranging the terms of Eq. (5.3), a quadratic equation in its standard form can be formulated $0 = gt^2 - 2v_1 t - 2y_1$, which mirrors the case of finding the solutions of $ax^2 + bx + c = 0$ in mathematics by applying the formula (Eq. 5.2). Such mapping results in

$$t = \frac{v_1 \pm \sqrt{v_1^2 + 2y_1 g}}{g} \tag{5.4}$$

Since the formula produces positive and negative time values, one would consider only $t \geq 0$ as the acceptable solution(s). There is a specific condition for the parameters to produce real-time values; the value of the expression under the square root, also called the discriminant in mathematics must be greater than or equal to zero; $v_1^2 + 2y_1 g \geq 0$ otherwise, it would produce complex solutions. This condition does not surface physics resources, yet it presents itself as a compelling case for investigations. Furthermore, the formula to find the quadratic function x and y position of the vertex is:

$$x_v = \frac{-b}{2a} \text{and} \, y_v = f(x_v) \tag{5.5}$$

Referring to Eqs. (5.3) and (5.5) the time when the object reaches the maximum height can be calculated by $t_v = \frac{v_1}{g}$ and consequently, the maximum height by evaluating $y_{\max}(t_v) = y_1 + v_1(t_v) - \frac{g(t_v)^2}{2}$. The evaluation will result in $y_{\max}\left(\frac{v_1}{g}\right) = y_1 + v_1\left(\frac{v_1}{g}\right) - \frac{v_1^2}{2g} = y_1 + \frac{v_1^2}{2g}$. One will notice that $t = \frac{v_1}{g}$ can also result from formulating the velocity function and finding the time when the velocity of the object reaches a zero value. All these answers can be generated using kinematics formulas. However, offering students an alternative way can be an exciting experience of proving that interpretation of parameters in physics and mathematics leads to congruent conclusions. Using the properties of quadratic functions does not diminish the physics behind it; it instead equips students with more tools to solve physics problems and adds restrictions on the parameters to produce real output quantities.

When symbols of parameters are present in the function equation, the arrangement of variables defines a general rule for the family of functions. The values of parameters map up the structure to its particular case or graph. The idea can, to some degree, reflect on finding a general antiderivative and then generating a family of solutions by provided coordinate values and calculating the constant parameter c in the antiderivative. For instance, if $v(t) = \int(3t^2 + 2)dt$ is concerned, then the velocity functions are typically expressed as $v(t) = t^3 + 2t + c$. Thus, if an object accelerates with $a(t) = 3t^2 + 2$, its velocities are given by a family of functions $v(t) = t^3 + 2t + c$ that differ by the initial velocity denoted by c. The symbol c is called a constant in mathematics and physics resources; however, more precisely, it should perhaps be called a parameter whose value can be calculated due to the specificity of a particular case of motion. The symbol c represents a constant value when referred to a specific case motion, but it can take other values within the boundaries of the experiment setup. Parameters do not alter the general function form and the shapes of their graphs even though they constitute separate entities within function or equation. Parameters do alter the values of the function outputs or the values of the quantity of interest by the algebraic rules represented by the function structure. For example, $f(x) = x^2$ and $f(x) = 2x^2$ will produce different outputs for the same x values even though both equations will depict two quadratic functions whose general positions in the XY coordinates are similar. In a congruent fashion, the position function reflecting the formula $y(t) = y_1 + v_1 t - \frac{gt^2}{2}$ will generate similar graphs even though initial motion parameters like initial position and velocity might differ.

5.2 Formulas in Science and Mathematics

The term formula is widely used in sciences and mathematics as it manifests a mathematical model of real-world phenomena. The formula is defined in mathematics as an established rule of law expressed in algebraic symbols (Stewart, 2016). The building blocks of formulas are often called parameters, and they refer to specific quantities. Following this reasoning, one can conclude that quantities are being called parameters when embraced in algebraic formulas. The distinctions between variables and parameters are not often discussed in mathematics education. Research (Sokolowski, 2020) shows that even precalculus students have difficulties in differentiating between variables and parameters when a function or formulas are concerned. Exposing physics students to differentiating between these entities and developing their skills in this regard emerged as a task benefitting their general scientific and mathematical dispositions.

In science, a formula is a concise way of expressing a relation between quantities of a system using algebraic symbols. For example, $pV = nRT$ or $E = mc^2$. As such, a formula can entangle more than two quantities, and its complexity depends on the number of parameters that contribute to the phenomena behavior or investigators' research aim. Formulas are considered the most frequently used algebraic entities to solve or execute application problems in physics and other branches of science when quantification is concerned. While in physics, formulas help quantify specific instants of a phenomena behavior, in mathematics and especially in geometry, formulas are associated with an object's general properties and are used to calculate static assets such as volume, perimeter, area, or length, height, width, etc. At their vast applications, formulas in science and mathematics or other subjects resemble structures of algebraic functions. For instance, the formula for an area of a rectangle $A = ab$ can be expressed as $A(a) = ab$ or $A(b) = ab$ depending on what side is changing. Under that assumption, the formula can be called a linear function. While the right sides of the structures are unchanged, the left sides are different. Placing a parameter in parentheses directly, informs which parameter changes and which is constant.

Formulas are generally not considered as entities that can be sketched or analyzed from a function standpoint. Their applications are not embraced or restricted to specific values, although such restrictions are often needed. Consider, for example, the formula for the apparent frequency in the Doppler effect when the source is moving, $f'(v_s) = f_o\left(\frac{v \pm v_o}{v \pm v_s}\right)$. In this formula $f'(v_s)$ represents apparent frequency due to source movement, f_o the frequency produced by the source, v_s the speed of the source, v_o the speed of the observer and v the speed of the sound energy propagation in the given medium. When $v \pm v_s = 0$, the apparent frequency $f'(v_s)$ is undefined that creates a vertical asymptote on the graph of $f'(v_s)$, consequently, in the scientific sense, this situation produces a shock wave. The scientific interpretation of this case can be supported by taking sided limits of $f'(v_s)$ in the proximity of v_s which generates an exciting hypothetical analysis. Such analysis unpacks the scientific behavior of the phenomena not directly concludable from the formula notation. The hypothetical reasoning has its scientific support when the formula is interpreted as an algebraic function. Formulas are specific to a particular aspect of reality or *snapshot* of reality, but they can still be mapped up into their functional algebraic forms.

There are various ways of denoting algebraic functions and specifically underlying symbols of these entities that change. Such differentiating is not commonly applied when formulas are used, which when graphing or taking derivative or antiderivative is concerned might inevitably hide the variable nature of some parameters.

The notation that explicitly highlights the variable or variable parameter is called function notation and is placed in the parenthesis of the right side of the expression, e.g., *f*(*x*) or *d*(*t*). Function notation is typically not being applied in formulas to highlight the quantity that changes when presented in the general case. For example, in$U_G = -\frac{Gm_1m_2}{r}$, all the building blocks, objects' masses, and the separation between their center can change. The formula provides a general view of what parameters contribute to the change of the quantity of interest. However, in this regard, the formulas' scientific interpretations are reduced, perhaps therefore they are considered static entities. Considering formulas as static entities might disassociate these algebraic structures from the dynamism of the scientific processes they expose. By narrowing the usage of formulas to such applications, students miss the opportunities of perceiving formulas as covariate relations that can be sketched and whose behavior can be generalized. Consequently, students are unable to extract more general covariational relationships embedded within these structures. In the end, students lose the opportunity to develop conceptual skills for modeling function relationships in which the function output variable changes continuously in tandem with continuous changes in the input variable. Such reasoning abilities are essential for representing and interpreting the changing nature of a wide array of natural phenomena. The lack of such notation transcends to using these formulas in physics classrooms even when these entities are to be sketched. Moreover, the lack of highlighting the variable parameter by using function notation can dilute the process of finding the derivative or antiderivative. In calculus classes, the task of taking the derivative is embraced in the strict notation of the dependent and independent variable; $\frac{dv}{dt}$ or $v'(t)$ making these operations very explicit. By not emphasizing the notation that students learned in their math classes, the transitioning of math skills to physics might not be enabled.

5.3 Covariational Reasoning in Mathematics Education

Mathematical reasoning is not explicitly taught the same way mathematical concepts are being taught. It develops when students experience using numbers, quantities, variables, and relations between these quantities discussed earlier in Part II. Covariation reasoning (CR) appears as a more complex mental action that in this book will be considered a special category of mathematical reasoning. The relation of CR to mathematical reasoning is depicted in Fig. 5.1.

Mathematical reasoning generates a big picture of mathematical thinking methods. Covariational reasoning will be considered its more sophisticated derivative that includes attending to dynamic relations between selected parameters. The definitions and subsequent tasks of covariational reasoning as seen through the prisms

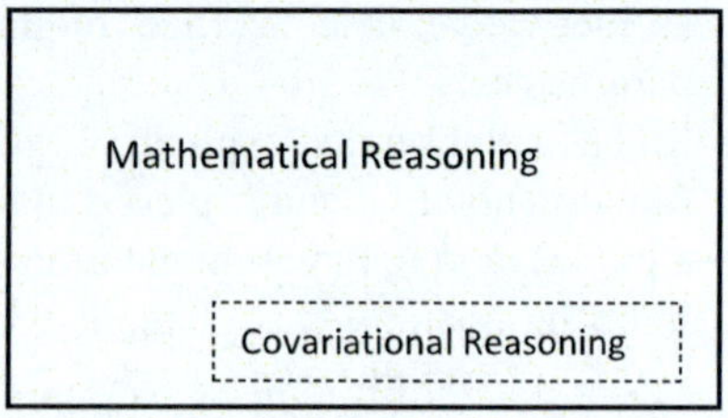

Fig. 5.1 Relation of covariational and mathematical reasoning

of physics and mathematics will be discussed, and an attempt to extract these meanings that will enhance the book's objective will be made.

While functions provide general frames for representing specific behavior concisely, formulas dressed up with contexts narrow applications of the representation to a specific part of reality. The mathematical mechanism of the formulas can be unearthed by analyzing covariation between the formula parameters. The meaning of the term covariation is closely related to variation. While *variation* refers to the diversity of one type of variable, covariation concerns the association of at least two different variables. Such association is also called a correspondence of the variations (Watson & Moritz, 2000). Covariational relations and reasoning are universally used in many branches of mathematics, and recently it caught the attention of physics education researchers. The definitions of a covariate parameter or variable are often augmented for the subject where they are being applied; thus, the definitions differ in mathematics, statistics, and science. At the same time, the definitions reflect on the subject domains and all exhibit certain commonalities. This section provides contemporary interpretations of covariation and suggestions for inducing such reasoning in physics teaching. A general definition of covariation that encompasses the meaning of this reasoning type found in mathematics research also emerged.

Covariation is a relation of mutual engagement of two quantities that becomes a gate to a broader cognitive tool of mathematical reasoning. Studies show that the function's perception as a process that accepts specific inputs and produces outputs is essential for a mature image of algebraic function (Johnson, 2015). This entanglement is essentially advanced in mathematics courses by embracing it in covariational reasoning. While mathematical reasoning and covariation are traditionally not taught, they can be developed through practicing and applying mathematical tools. Covariational reasoning as a theoretical construct appeared in mathematics in the late 1980s and early 1990s (Thompson & Carlson, 2017). Covariational relations strictly reflect on the dynamic behavior of a part of reality and can lead to formulating a symbolic representation of that reality.

There are several definitions of covariates and covariation that literature in mathematics and statistics education suggests. Carlson et al. (2002) defined covariation as a cognitive activity involving coordinating two varying quantities while attending to how they change with the relation to each other and advocated for emphasizing more covariate relations between functions and their derivatives to enhance students reasoning skills about function change. More explicitly, covariation is defined in statistics; Ott and Longnecker (2010) described a covariate as a variable related to the response variable (or parameter of interest) used to reduce the variability of the response variable. A covariate is also called a secondary variable (Moore, 2000). Following this definition, the independent variable is called primary. In contrast, covariate relationships would, according to this definition, be secondary to direct *cause and effect* relationships. These definitions place covariate relation as secondary to direct independent-dependent relation which does not correspond with the definitions of covariates by Thompson and Carlson (2017). Covariational variables as *secondary* variables are further classified as (a) *mediators* that support and help explain the relationship between the dependent and independent variables and (b) a *confounding* or *extraneous variable* whose effects are traditionally not welcomed in the experiments (Everitt, 2002). According to the definition, when the scientific experiment is concerned, confounding variables as extraneous variables that affect both the dependent and independent variables, which perhaps could transcend to factors that produce all kinds of errors during investigations related to accuracy and precision of data taking. To precisely estimate X's effect on Y, a researcher must prevent the effects of *cofounding* variables (Pearl, 2009). Cheng (1997) combined causal and covariation relations, stating that both must be induced from observable events. A causal relation between two parameters exists if the changes of the independent variable cause changes in the dependent variable. In this regard, the independent parameter is called the cause and the other the effect. Following Cheng's definition, the existence of casual and secondary covariational relationships must be observable and measurable. Johnson (2012) formulated three categories of covariation called *perspectives* which highlight the difference the investigator can perceive the relationships in a particular situation: (a) static covariation, (b) continuous dynamic, and (c) discrete dynamic. Research in mathematics education (Thompson et al., 1994) suggests that curriculum and instruction increase emphasis on advancing math students from a coordinated image of two variables to a coordinated image of their instantaneous rates of change. In Thompson's theory of quantitative reasoning, a person develops covariational reasoning when she/he envisions two quantities' values varying simultaneously. Following these descriptions covariation in conceptualizing individual quantities' values as varying and then conceptualizing two or more quantities as varying simultaneously.

In developing students' skills of coordinating changes of two quantities and describing the changes qualitatively and quantitatively, covariational appears as a precursor to understanding the concept of algebraic functions that is central to graphing and mathematizing phenomena using the rules and symbols of mathematics. Developing students' covariation reasoning abilities also concerns physics research communities; therefore, a brief perspective on how it is being used in physics courses will follow. Mixing all these perspectives will emerge as a proposal on how covariate relations will be considered and applied in the case studies in Part III of the volume.

5.4 Covariational Reasoning in Physics Education

Covariational reasoning and its effects on students' physics understanding is gaining momentum. However, research on covariational reasoning in physics research is still limiting (Panorkou & Germia, 2020). At the current stage, it is being developed conceptually and is similar in aim to its mathematical counterpart. For instance, the learners are provided with a bird's eye view or side view of a path of an object's undergoing periodic motion, and using these contexts; they are asked to sketch a graph of a distance of the object versus the total distance the object makes concerning a selected reference point (see Zimmerman et al., 2019). This relation can be described as the coordination of parameters viewed from different frames of reference. In other examples, students are asked to draw graphs of the object's speed versus position along know path of motion (see Monk, 1992). The cognitive value of such enterprises is unquestionable. However, a substantial focus on the conceptual development of covariation reasoning might unintentionally disjoin the accumulated students' skills from converting these prompts into concise algebraic representations. Constructing algebraic equations seems to be a task that students often encounter in their physics or other science courses; thus, not moving forward with formulating such for emerged covariance is making the efforts inadvertently missing their aim. Developing a conceptual understanding of covariates is one side of the coin; merging these covariations with algebraic structures and graphical representation of these functions is the other. Converting information to its graphical representation as a function requires identifying parameters that change (variables) and their classification as independent and dependent. Currently, these phrases are silent in physics education research. Can neglecting precise mathematical notations and parameter classifications in physics be one of the reasons for students' difficulties to effectively apply their math reasoning skills to graphing dynamic relations in physics? Findings of case studies described in Chaps. 12 and 13 provide evidence that a lack of precision of parameters labeling in physics humpers not only the interpretation of physics formulas but also jeopardizes graph sketching. Reforms in mathematics education (Hughes-Hallett, 2006) emphasize presenting functions in various forms; graphical, symbolic, verbal, and numerical to meet students' diverse learning styles and train them to transition between these representations. This didactical enterprise also inspired the design of the case studies clustered in Part III.

The term covariation already exists in physics education, and it broadly describes the casual relation between selected variables of the experiment. Covariation reasoning is applied to describe natural phenomena; thus, it involves actual quantities that are reimaged to parameters and assembled together in various algebraic relationships. These inquiries begin with identifying the parameters of interest and coordinating their intermediate relationships to their final product consisting of mental and symbolic images of their coordination. Following these processes, images of covariation are considered developmental phases that conclude with the formulation of the entire event. Applications of covariational reasoning are vast, ranging from their standard function relations and moving into the coordination of their rates of change or coordination of the functions of their components (e.g., two-dimensional motion).

In sum, while there are various definitions of covariations in mathematics, perceiving these special relations between two or more parameters will be used in this book and this definition will be a departing phase to propose a more detailed categorization of variables and parameters in physics teaching.

5.4.1 Viewing Phenomena as Covariations of Their Parameters

While there are several classifications of covariations in mathematics, literature does not provide equivalent classifications for sciences and, more specifically, physics, where such classification of the parameters is the priority. Thus, an attempt to synchronize these definitions is made. The classification was inspired by the synthesis of research and it is to generalize the meanings of various terms used and help understand the scientific underpinnings of the effects of parameters on the output of the investigation.

Physics is considered an experimental subject. Thus, investigations in physics are mainly experimental as opposed to just observational. The investigator actively manipulates certain variables associated with lab design, observes the parameters, takes measurements, and analyzes the data for potential covariate relationships. The quantities that the investigator manipulates can be called explanatory or independent covariates. Explanatory quantities generate specific changes in the quantity of interest, called also a response quantity, determined by the experiment design. There are usually other parameters involved in the experiment that affect the response quantity like confounding that generate measurement errors and mediating that link independent and dependent parameters. It is suggested, they these parameters are called secondary covariate quantities. Thus, secondary parameters will be further classified as confounding and mediators. The effects of the confounding quantities that fall into the basket of these factors that produce errors during data taking could be minimalized by, for example, employing more precise measuring devices or increasing the accuracy of data taking. Should such classifications be included during labs? Such classification does not seem to be necessary to assure that students focus on primary parameters; however, making students aware of all effects that contribute to the dependent parameter would enhance and broaden the perspectives on the experiment conducts. More detailed classification of parameters can be applied to formulas considered dynamic representations that can enrich students' perspectives on the techniques to solve problems.

This classification includes the terms, parameters, and variables as possessing similar meanings, although the term *variable parameter* appears to be more suitable. A *variable* in a mathematical sense carries the notion of a value (number) without a unit, and it commonly denotes an entity that can take various yet restricted values. Variables are classified as independent and dependent, thus using the term *variable parameter* will link the terminology used in both subjects.

Furthermore, the term parameter embraces scientific interpretation on the variables, thus both adjectives; variable independent or dependent parameters seem to be suitable. In practice quantities that have the potential to change could be called e.g., variable force, variable mass, or variable resistance, and depending on the specifics of the experiment these would be further reclassified as dependent and independent. Thus to link the meanings, both terms will be used, and this is purposeful. Research (Sokolowski, 2020) shows that students do not have a clear image of how to differentiate between parameters and variables. Thus making these terms explicitly defined would help use them adequately.

Certain variations of each parameter are often present during experiments and each of the parameters (Fig. 5.2) will fit into that classification. This classification can be applied to enhance experiment descriptions and formula's understanding. Both venues are to help students master *cause-effect* reasoning and develop more profound insight into how phenomena behave also from their covariances viewpoints. For simplicity, the secondary covariates could also be perceived as indirectly affecting the experiment response variable (dependent parameter). The proposed classification does not include covariations of the rates of either the primary or secondary parameters. Such covariation can be derived using techniques of integration or differentiation and perhaps does not require an explicit categorization. Many physics students take a statistics course concurrently; thus, using the terms *mediating* and *confounding* in the same sense should help transition knowledge and enhance meanings. It is, though, not mandatory.

The section (Sect. 5.4.2) proposes how to use the classifications (Fig. 5.2) to support understanding of formulas' empirical-mathematical entanglements. Preliminary work on this section called for an attempt to categorize physics formulas due to how their covariates are executed which is also included. Section 5.4.3 attempts to provide an example of how to use classification (Fig. 5.2) while designing experiments.

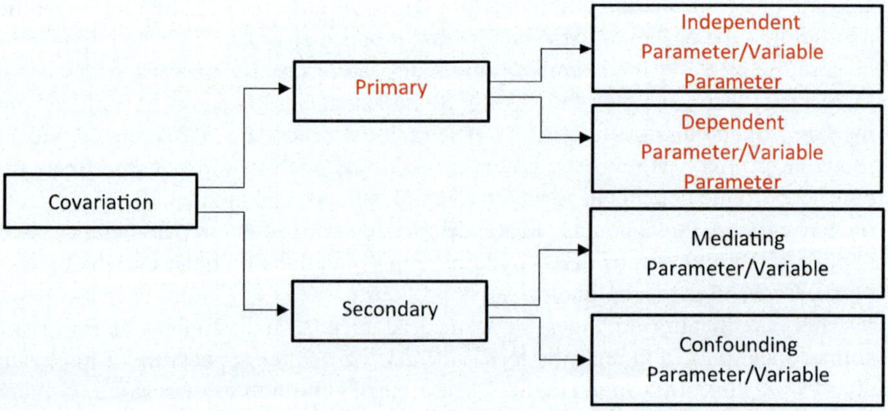

Fig. 5.2 Proposed classification of parameters in physics education

5.4.2 *Proposed Categories of Covariations Embedded in Physics Formulas*

Research shows that covariation can take various layers of complexity. Identifying primary and secondary covariates during experiments is one way of applying mathematical reasoning in sciences. Another could be examining the structures of various physics formulas and categorizing their structures considering the algebraic complexity of their covariations. The purpose of such an enterprise would be to offer students a more comprehensive view of describing natural phenomena algebraically. It is hypothesized that such discussions will support students' judgments in selecting correct algebraic representations to describe certain phenomena and let them search for their inventions. Physics formulas are built based on algebraic rules. The proposed classification of covariations reflects primarily on the underlying algebraic structures coupled with scientific interpretations. There are five categories identified and all are subject to possible revisions based on more exhaustive research. These categories do not include formulas presented as derivatives or antiderivatives for an earlier stated reason.

Type (1): Covariation involving a linear dependence

Examples of such variation can be formulas to calculate, for instance; the force of gravity acting on an object $F = mg$, wave equation $v = f\lambda$, or force exerted by stretched spring $F = -kx$. When embraced in typical functional notations, the formulas can take the following forms $F(m) = mg$, $v(f) = f\lambda$, or $F(x) = -kx$. Thus, when regarded as specifically formulated covariates, parameters of interest can be embraced in functional notation and positioned on the equations' left sides. The independent parameters are placed on the right side of these equations, however, their variable natures are denoted by placing their symbols in parenthesis. A rather stimulating discussion emerges when describing the effects of other parameters embedded within each of these formulas. From the mathematical standpoint in $F(m) = mg$, the parameter g representing gravitational field constant will be called a proportionality constant or the slope of the $F(m)$ function; from the proposed classification point of view, g can be considered a mediating parameter. As such, mediating parameter, g, also generates particular scientific insight. The gravitational field constant is, in this case, independent of the object's mass. It does, though, affect the object's weight or the force of gravity acting on it. Object's weight depends on g, not vice versa. As viewed through the prism of the proposed classification, all these constant quantities in; $F(m) = mg$, $v(f) = f\lambda$, and $F(x) = -kx$, thus g, λ, and k can be classified as *mediating* parameters. While g describes an external gravitational field, the wavelength depicts the medium that when coupled with vibration supplied by an external generator, affects the energy (wave) propagation speed. As the same mass has different weights depending on the magnitudes of the gravitational fields, the same energy vibration will move at different speeds depending on the medium that will alter the wave wavelengths. Referring to $F(x) = -kx$ spring constant k defines the property of the medium that is the spring under the extension labeled x. The same extensions will require different forces depending on the mediating spring constant. While in algebra, a general form of a linear function is usually expressed as $f(x) = mx + b$, thus the slope listed as the first factor in the product mx, there seems to be no such rule in physics. For instance, in $F = mg$, the slope shows as the second factor, in $F(x) = -kx$, the slope k shows as the first factor. While this is not an obstacle in interpreting the formulas, pointing this lack of consistency to students will make them aware of the flexibility they need to exhibit while interpreting the physics formulas.

Formulas can consist of more than one parameter, e.g., $F_B = \sigma Vg$. A general interpretation of such formulas would gear toward considering one parameter as the mediating, in this case, g, and the density of the fluid σ along with fluid displaced volume V as constituting independent parameters. Such classification would perhaps depend solely on the specific experimental setup or problem to analyze.

For example, one of these parameters σ or V can also be considered a mediating one, and the other labeled as a variable one. Certainly, referring to specific objectives of the experiments will make such classifications more explicit. All these quantities of interest increase in magnitude when the independent parameter increases, and the changes are proportional to mediating parameters. Such discussions seem to bring new insight into these formulas and their graphical interpretations. While the slopes of each of these formulas visualize mediating parameters, the mediating parameters do all exist independently from the external actions. These actions can be used to visualize and quantify all these mediators mathematically. There are certainly more physics formulas of Type 1 structures.

Type (2): Covariation involving a rational dependence

Examples of such covariations can be $a = \frac{Gm}{d^2}$ or $R = \frac{\sigma A}{l}$. When converted to functional dependencies, they can take the following forms $a(d) = \frac{Gm}{d^2}$ or $R(l) = \frac{\sigma A}{l}$. The effects of the independent parameters on the parameter of interest depend on a selected parameter's algebraic position within the symbolic formula representation. If the independent parameter takes numerical values, its variation is described by sequencing it in increasing order, thus from lower to larger magnitudes, and respectively the effects are justified due to covariation. For example, the dependence of the gravitational field intensity from the object's mass or the distance from its center. This covariation typically takes two different forms as $a(m) = \frac{Gm}{d^2}$ which is linear or $a(d) = \frac{Gm}{d^2}$ which is rational. In special cases, the magnitude of the intensity can be affected by a varying mass m and varying distance d which could be denoted as $a(m,d)$. All examples describe the properties of gravitational field produced by the mass m. Can G, the universal gravitational constant, be considered a mediator of the phenomenon? This is an open question. G undoubtedly supports quantification of the phenomena, but it is a coefficient in this formula that makes the formula universal. The other formula, $R = \frac{\sigma A}{l}$, describes the quantification of a conductor resistance considering its geometrical parameters and resistivity.

When considered $R(l)=\frac{\sigma A}{l}$, the formula resembles a rational function which tells that the resistance decreases as the length of the wire increases. When considering the cross-sectional area as the independent parameter, $R(A)=\frac{\sigma A}{l}$, the resistance R depends linearly on the area. From students' points of view, such flexibility in interpretations might not be apparent. While standard covariations of two variable parameters cannot be sketched in the traditional XY plane e.g. $a(m,d)$ or $R(l,A)$, physics provides many examples when students must find an expression for a new value when two independent parameters change. While such algebraic manipulations seem to be straightforward, the mathematics curriculum does not provide many opportunities for practicing such skills; therefore, to enhance knowledge transfer, physics teachers' guidance is welcomed and appreciated by the students.

Type (3): Linking two or more dependent parameters

Linking the variation of two dependent parameters can also lead to learning new information about the phenomena. A good example to discuss can be two-dimensional motion. While the causes of motion in the vertical direction do not affect the horizontal motion, coordinating, for example, vertical and horizontal positions, thus $x(t)$ and $y(t)$ can emerge as the path of the object's motion. This is possible because the independent parameter, t, is shared for both functions, and evaluation of both of these functions $x(t)$ and $y(t)$ generates position coordinates (x,y) and the table of values for $y(x)$. It is to note that in the final form, the *path of motion* does not contain that parameter. Type 3 covariation can also include finding component velocity and acceleration functions; $v_x(t)$ and $v_y(t)$ along with $a_x(t)$ and $a_y(t)$ that in a similar manner to position functions can be used to finding resultant velocity and acceleration functions.

Another example of using such covariation could be analyzing the temperature in two different scales $C(t)$ and $K(t)$. For example, a scientist measures the temperature of a substance at selected time instants using two thermometers with two different scales Celsius and Kelvin, and creates $C(t)$ and $K(t)$ functions. If time is considered the independent parameter, the temperature will be called the dependent. The thermometers will measure the temperature independently, producing different values because of the different scales. Finding an algebraic function that shows a covariation of the temperatures can be used to derive conversion equations, either $C(K)$ or $K(C)$. When mediating parameters are sought in this enterprise, one could pinpoint the specific heat capacity of the substance used in the experiment, affecting the temperature increase rate. However, the heat capacity will not affect the general conversion formulas $C(K)$ or $K(C)$ because measurements on both will be altered on both thermometers at the same rates dictated by the substance heat capacity.

Type (4): Covariation of multiple parameters within a system

This category would involve systems of objects or one object simultaneously examined using more than one physical concept. These structures are often called laws, e.g., the law of conservation of momentum, law of conservation of mechanical energy, gas law. The parameters constituting these structures can also be embraced in analyzing their dependent, independent, or mediating effects. Since laws encompass usually more objects and parameters, there is greater flexibility in classifying the parameters. Let us consider the law of conservation of mechanical energy with kinetic and potential energies involved. For instance, if an object of mass 2 kg that moves in two dimensions possesses only kinetic and gravitational energies and no frictional effects are present, then at any two selected time instants or positions, the total mechanical energy remains the same, and thus $2gh_1 + \frac{2v_1^2}{2} = 2gh_2 + \frac{2v_2^2}{2}$ holds to be true. The law can be converted to a more direct covariate expression upon assigning the dependent and independent parameters and its analysis enriched by considering covariations of the system parameters. Suppose that the total initial mechanical energy of the object is 200 J, and one is interested in attending to the velocity function expressed in terms of the variable height to which the object can rise. Following this condition, $200 = 2gh2 + v_2^2$ and solving for $v_2(h_2) = \sqrt{200 - 2gh_2}$ provides means of calculating the magnitude of the velocity at any height h_2 along the motion path and sketch it in respective coordinates. The independent variable in this expression is the vertical position of the object above a reference level labeled h_2, and g is a mediator. Since a new algebraic representation was derived, there can be a new analysis imposed on it. The radical sign generates certain restrictions on the on variable heigh h_2 such that $200 - 2gh_2 \geq 0$, which leads to $h_2 \; {''} \; \frac{100}{g}$. This restriction tells that with the conditions provided, the maximum height reached by the object is about 10 m. Students can extend their hypothetical reasoning and discuss the effects of the mediating parameter; the gravitational intensity on the object's maximum height. The analysis would take a similar aim, resulting in a different final product if the dependent parameter is the object height h_2 instead of speed labelled v_2. Laws involving systems of objects provide more opportunities to generate algebraic structures because of the diversity and richness of the parameters from which they are built.

Type (5): Scaffolding covariation

This covariation is not surfacing physics formulas, but it is useful when composites of two functions are concerned or when the independent parameter depends yet on another formula. Algebraically, such covariations are expressed as $y = f(g(x))$ where $g(x)$ is considered an inner function of $f(x)$. While problem-solving typically does not require creating composite formulas, such algebraic operations can support graphing and formulas deriving from different, perhaps more insightful perspectives. For example, the gravitational potential energy of an object is calculated using $U(y) = mgy$, where y represents the position of the object above or below the established frame of reference. If the object's position changes according to $h(t) = 2t + 4$, then the object's gravitational potential energy can be expressed in terms of time t thus $U(h(t)) = U(t) = mg(2t + 4)$. The units stemming of the new formula must still be *joules*, but the independent variable is no longer the object's position but it is the time, t. Such representations of the gravitational potential energy add dynamism to the formulas and allow to predict and verify the energy for a range of time values.

Another example could be $K = \frac{1}{2} I\omega^2$. If the variable parameter is the angular velocity ω, then the formula can be expressed as a function of that parameter, thus $K(\omega) = \frac{1}{2} I\omega^2$. If the behavior of the angular velocity function is given, e.g., $\omega(t) = \omega_0 + \alpha t$, then $K(t) = \frac{1}{2} I(\omega_0 + \alpha t)^2$.

Students study the techniques of finding composite functions in a typical high school math curriculum, thus, transferring these skills to physics should be possible. Another example of this category can be the formula for pendulum period $T = 2\pi\sqrt{\frac{l}{g}}$, which in school practice is considered as $T(l) = 2\pi\sqrt{\frac{l}{g}}$. This formula already embeds an inner function $\frac{l}{g}$ which likely has no special physical meaning.

Discussions about finding commonalities in physics formulas provide a different perspective on these entities and open up a gate to broaden student's perspectives on the universality of these structures and their fit to describing scientific phenomena. Consequently demystifying the algebraic rules that the formulas are build from should also help students set up equations to solve problems in physics.

The categories suggested in this section are not exhaustive and they can take more aims. It seems that as long as they enrich students' views on physics methods and help with understanding, they serve their purposes.

5.4.3 Discussing Covariations of Parameters in Experiments

This section is an attempt to provide examples of classifying parameters when experiments or laboratory activities are concerned. Experiments provide an opportunity to include confounding parameters, thus involved parameters can be classified as dependent, independent, mediating, and confounding. Any experiments, especially at their designing stage, can involve several parameters, yet not all are necessarily used to achieve the lab goal. Selecting and classifying variables will be followed by selecting respective measuring devices needed for the lab completion; for example, a sensor to measure the object's initial speed, tape to measure length, etc. The discussion that follows is preliminary, and indeed, other venues can emerge due to range of available measuring devices, equipment, or creativity of the lab designer.

Example 1: Suppose that an object is projected straight up with different velocities, and an algebraic relation between the height and the initial velocity is sought.

The response variable is the height, and the input variable is the object's initial velocity. The height reached by the object depends on the gravitational field and the time. These two parameters will be called *mediators*. They affect the height, but they will not be included in the analysis. There is also a force of air resistance that decreases the object's height because of its effect on the object's net force. The air resistance can be called a confounding variable. This experiment could aim differently; it could investigate the time of motion in the object's initial velocity function. Such lab profile requires reclassifying the parameters. The response variable will be now the time, and the independent parameter the initial velocity. The intensity of the gravitational field would retain to be called a mediator and the air resistance as the confounding parameter.

Example 2: Investigating the coefficient of kinetic friction between the wooden block and the floor by pushing an object along the floor so that it moves a definite distance and it stops.

Suppose that the available measuring devices are tape to measure the length and a stopwatch. The process of Identifying and classifying the parameters involved in computations of the coefficient of friction is not that straightforward due to how the parameter of interest (response variable) relates to the distance covered and time of motion. While $F_f = \mu F_N$, the parameter of interest μ is not directly related to kinematics of the motion that could be measured using available devices. Thus, Newton's law $F_{Net} = -F_f$ must be used to find the value of the left side of the equation. Considering $F_{Net} = ma$ and $x = \frac{at^2}{2}$ mediating formulas, the coefficient of friction can be found. While unfolding the covariates, more parameters are being included, and all enrich the process. Is it necessary to classify all the parameters? Such a task is interesting and it would enhance the connections between all the parameters. By realizing the covariate links between the quantities, students would have an opportunity to envision a holistic view of the process, understand its building blocks that in return will help them remember it.

Example 3: Measuring the effects of differentiating the potential difference on the current flowing a resistor.

This experiment would resemble Type 1 covariation, with the independent parameter the potential difference and the current the dependent variable. The internal properties of the resistors will emerge as the mediating parameter representing the electrical resistance. An extension of such lab could be having two such resistors and investigating their equivalent resistance when connected in series and parallel. The connection types resulting in different equivalent resistance in this case will mediate with the value of the electric current. Thus it can be categorized as yet another mediating parameter that graphically is represented as a slope in potential difference versus the current graph.

Classifying parameters as dependent, independent, or mediating requires a more profound analysis when experiments are concerned. Bringing forth a dynamic nature of the formulas should enable students to perceive these mathematical entities as functions that depict the experiment behavior. Being provided with various lab contexts will let the students' exhibit flexibility in assigning quantities as variables that should also influence their problem-solving skills. It is hoped that these preliminary classifications provide departing points for a more detailed analysis that would extend generalizations of the phenomenon behavior stemming from either experiments or formulas.

5.5 Limiting Case Analysis

Limiting case analysis has already been surfaced in this book content, and it will be used more extensively in Chaps. 6 and 9. This section provides more algebraic details about limits evaluation in the context of physics. Using the idea of limits allows for extending both predictions and explanations of physical phenomena. According to Redish (2017), these types of analyses should be essential goals of teaching physics because they allow for generalizing inferences by evaluating or estimating values of algebraic formulas considered covariate functions. In practice, such a transition from formulas to functions provides opportunities for deriving valid conclusions for cases when direct laboratory measurements or formula evaluations are not possible. While using limits is common for scientists, the idea of applying limits in school practice is not visible, and testing students' ability in this area is also rare (Sokolowski, 2018). Historically, the idea of limits was developed by Archimedes of Syracuse in the third century BC. The theory of limits was independently developed by Newton and Leibnitz in the seventeenth century (Burton, 2007), and it sets the foundation of modern calculus. The concept of limits has been used to derive the concept of derivative and accumulation as well as other fundamental ideas in calculus. Due to its diverse applications and being described in various, often very abstract ways, its understanding by the students of mathematics causes challenges (Vinner, 2002), therefore presenting it to students from the application point of view should help with understanding. To use limits in physics, its definition will be augmented so that students see its relevance, usefulness, and help to understand physics ideas. The following definition of limits is suggested:

> *Limits are mathematical tools that help evaluate formulas expressed as covariate algebraic representations in situations when a direct evaluation is not possible.*

As such, limits are induced to covariate dynamic representations and help explicate the nature of these covariates. Being able to pinpoint the independent parameter and its algebraic relation with the dependent one appears the first step toward understating applications of limits in science. Consequently, to understand why limits of formulas can be taken, students have to perceive formulas as dynamic algebraic relations that have their equivalent representations in typical XY representations. Limits are closely related to functions diverse expressions, symbolic, graphical, or tables of values with explicitly defined variables. To help students link techniques of taking limits that they learn in mathematics with inducing these techniques to physics formulas, physics formulas need to be reimagined, and the parameters need to be reclassified as constant or variable. The pathway to structuralize the induction of the limiting case analysis to analyze natural phenomena is illustrated in Fig. 5.3.

There will be generally two types of limits applied in this book (a) when the values of the independent parameter are getting very large and (b) when the values of the independent parameter are minimal and close to a zero magnitude. Both methods result in producing approximated values of the dependent parameter that can be considered by the definition of limits its best approximation. The discussion

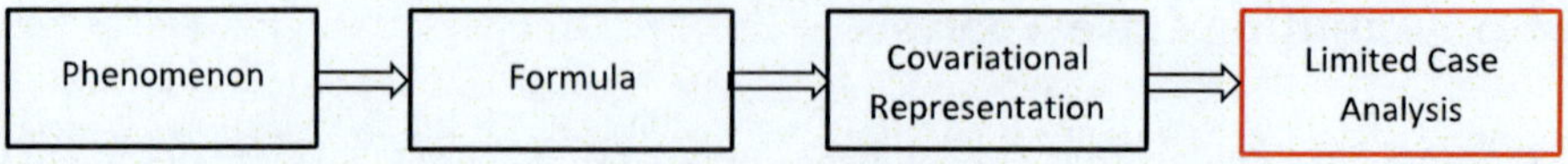

Fig. 5.3 Schematic representation of the theoretical framework when limiting case analysis is used

of the techniques of taking limits for very large and very small values of the independent variable will be conducted using an electric circuit with two resistors connected in parallel as a model. Presenting these ideas in a context should enhance the understanding of the result. A complete discussion of this topic is presented in The Physics Teacher (Sokolowski, 2019a).

5.5.1 Evaluating Limits when the Variable Parameter Is Getting Very Large; x→∞

Suppose that one analyzes a parallel connection of two resistors such that one resistor has a constant resistance, e.g., $R_1 = 10\ \Omega$ and the other has a variable resistance. Note that one of the resistors needs to have a constant value to represent a one-variable function. Alternatively, the value of one resistor could be expressed in terms of the other.

The specific task is to determine the equivalent or total resistance when the resistance of the variable resistor is getting a substantial and infinite value. First, we need to identify a correct formula related to finding equivalent resistance of two resistors connected in parallel that is:

$$\frac{1}{R_T} = \frac{1}{R_1} + \frac{1}{R_2} \tag{5.6}$$

The formula (Eq. 5.6) does not explicitly display a covariate representation. Thus, applying the techniques of taking limits using students' math knowledge is not recommended, although possible using Eq. (5.6). Furthermore, Eq. (5.6) does not allow for visualizing the idea by sketching the function and its approximated value at infinity; thus, relying on the rules of mathematics and rearranging the formula is needed. To apply the techniques of taking limits that students study in mathematics and visualize the result, the formula needs to be reimaged to a rational functional form. It is convenient to label the variable resistor as x because it will resonate with forms that students analyzed in their mathematics courses.

$$R_T(x) = \frac{10x}{x+10} \tag{5.7}$$

While both forms (Eqs. 5.6 and 5.7) can lead the students to reach similar final answers, students can easily recognize the function equation in Eq. (5.7) as a rational function and the appropriate technique to taking the limit. The limit must be induced because substituting infinity for x results in an undetermined form $\frac{\circ}{\circ}$ whose value can not be approximated. While the phrase *take limit* dictates specific tasks in mathematics, in physics populated by contexts, the phrase might be modified to say, for example, determine the value of the equivalent resistance of the circuit for a considerable significant value of the resistor labeled R_2. How students denote the algorithm of taking function limits in mathematics? Suppose that $f(x) = \frac{10x}{x+10}$, then the process is typically initiated by embracing the function expression in the limit notation $\lim_{x \to \infty} \frac{10x}{x+10}$. Realizing that the function is rational, one would consider only its dominant terms as the terms contributing to the function value for very large x values; $\lim_{x \to \infty} \frac{10x}{x+10} = \lim_{x \to \infty} \frac{10x}{x} = 10$. One can notice that the notation does not explicate the critical information that the value of the limit represents the approximation of the function $f(x)$ for substantial x values. While it is expected that students realize this result, and perhaps they do, omitting the notation might cause some uncertainties. To make the association of the limit output to function (formula) value more apparent this process can be explicitly labeled. Referring back to our example; to enhance this understanding, the following notation is suggested:

$$R_T(x \to \infty) = \lim_{x \to \infty} \frac{10x}{x+10} = \lim_{x \to \infty} \frac{10x}{x} = 10\Omega \tag{5.8}$$

Including the function notation on the left side with the symbol of the independent parameter explicates the results making it clear to interpret. By using $x \to \infty$, not $R(\infty)$, the rules of mathematics are also respected. One would read Eq. (5.8) following; the value of the equivalent resistor of the circuit as the resistance of R_2 is very large is 10 Ω. There is another nuance worthy of discussing; the interpretation of the limit value in the context of the function. While the equal sign is being used to denote the limit value, the value of 10 Ω approximates the equivalent resistance in these circumstances. It is not the accurate value of the total resistance. This interpretation can be visualized (see Fig. 5.4) by sketching the limit value (the dotted red line) and the Eq. (5.7). Thus the result of taking limit of Eq. (5.8) represents the horizontal asymptote of $R_T(x) = \frac{10x}{x+10}$ and the value of the limit. Horizontal asymptotes are described as the boundaries of function values for very large (positive or negative) values of independent variables. Asymptotes are not considered parts of functions, but they well approximate their values.

Graphical representations are easier accessible by students than symbolic; thus, illustrating the idea not only conveniently shows the limiting value of the equivalent resistance, but also allows for tracing the dynamism of the total resistance when the resistance of one resistor increases.

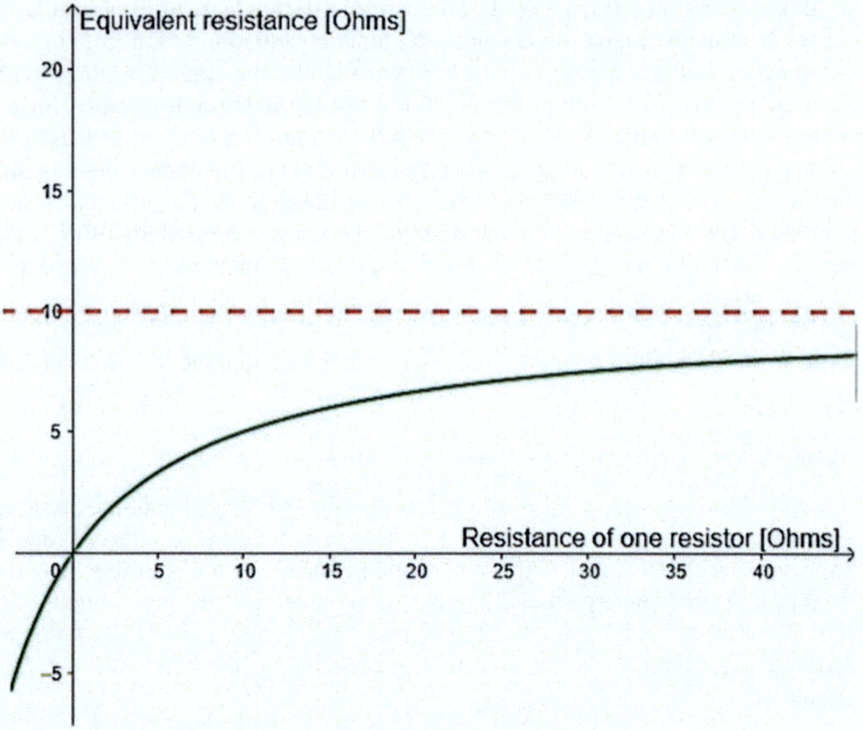

Fig. 5.4 Graph of $R_T(x) = \dfrac{10x}{x+10}$ restricted to $x > 0$ showing the value of the total resistance when $x \to \infty$

5.5.2 *Evaluating Limits when the Variable Parameter Is Close to a Specific Value;* x→a

Evaluating function limits for a very small value of the independent variable is rare in physics education as well, yet its output often provides even more interesting facts about the phenomena behavior than limits taken for large value of the independent parameter. Limit when $x \to a$ is often rewritten as sided limits, and it carries two independent operations as $x \to a^+$ and $x \to a^-$ that tell the direction of function evaluation; respectively from the right or the left side of of a. The technique of evaluating such limits requires a more detailed analysis. In contrast, applying it to evaluate the phenomena behavior can be very rewarding.

Let us consider again the parallel connection discussed in Sect. 5.5.1 and focus this time on analyzing the value of the total electric current drawn from the battery when the resistor R_2 is reaching a value close to zero. Let us assume that the potential difference in the battery is 9 V. The electric current depends on the total resistance of the circuit by $I = \dfrac{V}{R_T} = \dfrac{9V}{R_T}$. Since the expression for R_T has been formulated, then $I(R_T)$ provides means for formulating the electric current function.

$$I(R_T(x)) = I(x) = \frac{9}{\frac{10x}{x+10}} = \frac{9x+90}{10x} \tag{5.9}$$

This composite function (scaffolding covariate) represents another rational function whose graph is illustrated in Fig. 5.5.

When $x = 0\ \Omega$, the function is undefined, consequently one can say the electric current has an undefined value, but the idea of a sided limit can be induced to approximate the value by taking the right-sided limit of the function $I(x \to 0^+) = \lim_{x \to 0^+} I(x) = \lim_{x \to 0^+} \frac{9(0.001)+90}{10(0.001)} = \infty$. This result proves that the electric current could take an infinite value when the resistance R_2 is close to zero ($x \to 0\ \Omega$). The infinite right-sided limit of the *current* function suggests a boundary for the value that is illustrated in Fig. 5.5 as a vertical asymptote $x = 0\ \Omega$. The vertical asymptote is depicted by the red vertical dashed line positioned along the y-axis. To expand on

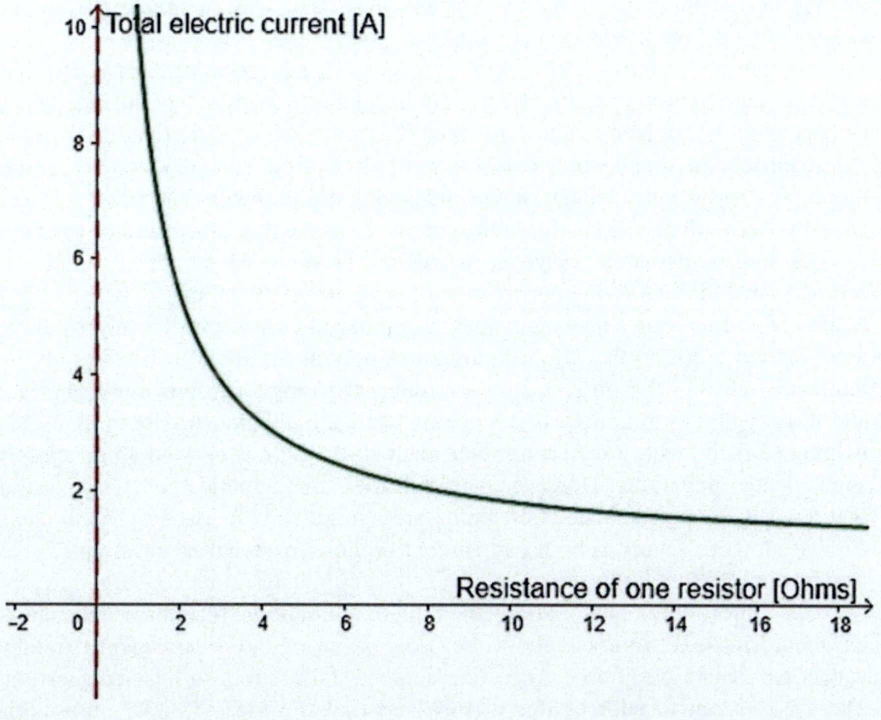

Fig. 5.5 Electric current expressed in the function of the variable resistance $I(x) = \frac{9x+90}{10x}$

the underlying scientific behavior, one can conclude that when the resistance of one resistor in a parallel connection takes the value of zero, then almost all electrons take the path with the zeroth resistance to flow creating a subsential electric current. It is to note that the left-sided limit $x \to 0^-$ was not be evaluated because the analysis is valid only when $x > 0$.

The physics curriculum provides multiple possibilities for inducing limits and extend the scientific inquiry of natural phenomena beyond classroom experiments and textbook problems. Students find it exciting to learn that complex calculus concepts have applications, and these applications extend their knowledge about nature.

5.5.3 Is Limiting Case Analysis Really "Limiting"?

In commonly used calculus textbooks, problems requiring students to take limits are usually phrased like: *state the value of each quantity*, *state the value of the limit* or *find the limit* (Stewart, 2016). Thus, a phrase *applying limiting analysis* is not being used in mathematics education as a standard task dictating limit application; instead, students are prompted to use function evaluation techniques and take function limits. Taking the limit of a function appears as another way of function evaluation rather than a different way of analyzing function.

While termed *limiting case analysis* in physics education such task command does not seem to be necessary. The term *limiting* might suggest that the independent or dependent parameter's values are small or limited to specific fixed magnitudes. While this can be true in some cases, in most cases, limits are applied to extend the inquiry for substantial values of the independent parameter. The phrase *limiting case analysis* will be used in this book to allow transitioning of terminology between physics and mathematics, however, it will not be emphasized, and its alternative prompt: *evaluation of the covariational expression* by taking a limit, will be employed. Generating a new mathematics language to use in physics might inadvertently create a notion that the technique of applying limits is restricted only to a domain of physics that in fact does not reflect the range of limits applicability. In short, using the terminology that students use in mathematics is seen as a factor helping students with transferring their math skills to physics even when applying sophisticated apparatus. However, while in their mathematics courses, questions that ask students to take function limits are explicitly verbalized or symbolically expressed, there seems to be no agreement on how to verbalize questions to have students realize that limits are to be taken when applications are concerned in physics. Will a direct command of taking the limit of a functional formula be reasonable? Or should instead, a phrase about the progression of the independent parameter values be clearly described suggesting students take limits to answer questions? There is a room to address this nuance and find the most optimal phrases that will satisfy both mathematics rigor and scientific interpretation.

References

Burton, D. (2007). *The history of mathematics: An introduction*. McGraw-Hill.

Carlson, M., Jacobs, S., Coe, E., Larsen, S., & Hsu, E. (2002). Applying covariational reasoning while modeling dynamic events: A framework and a study. *Journal for Research in Mathematics Education, 33*, 352–378.

Cheng, P. (1997). From covariation to causation: A causal power theory. *Psychological Review, 104*(2), 367–405.

Everitt, B. S. (2002). *The Cambridge dictionary of statistics* (2nd ed.). Cambridge University Press.

Fey, J. T. (1990). Quantity. In *On the shoulders of giants: New approaches to numeracy* (pp. 61–94). National Academy Press.

Gilliard, R. P. (2020). Fundamental units and constants of physics. *Advanced Studies in Theoretical Physics, 14*(4), 209–217.

Hughes-Hallett, D. (2006). What have we learned from calculus reform? The road to conceptual understanding. In N. Hastings (Ed.), *Rethinking the courses below calculus*. Mathematics Association of America.

Johnson, H. L. (2012). Reasoning about variation in the intensity of change in covarying quantities involved in the rate of change. *The Journal of Mathematical Behavior, 31*(3), 313–330.

Johnson, H. L. (2015). Secondary students' quantification of ratio and rate: A framework for reasoning about changes in covarying quantities. *Mathematical Thinking and Learning, 17*(1), 64–90.

Monk, S. (1992). Students' understanding of a function given by a physical model. In G. Harel & E. Dubinsky (Eds.), *The concept of function: Aspects of epistemology and pedagogy* (MAA Notes) (Vol. 25, pp. 175–193).

Moore, T. (2000). *Teaching statistics: Resources for undergraduate instructors*. Mathematical Association of America.

Ott, R. L., & Longnecker, M. T. (2010). *An introduction to statistical methods and data analysis* (6th ed.). Brooks/Cole.

Oxford University Press. (2015). Parameter. In *Oxford dictionary of science* (7th ed., p. 586). Oxford University Press.

Panorkou, N., & Germia, E. F. (2020). Integrating math and science content through covariational reasoning: The case of gravity. *Mathematical Thinking and Learning*, 1–26. https://doi.org/10.1080/10986065.2020.1814977

Pearl, J. (2009). Simpson's paradox, confounding, and collapsibility. In *Causality: Models, reasoning and inference* (2nd ed.). Cambridge University Press.

Redish, E. F. (2017). Analyzing the competency of mathematical modelling in physics. In *Key competencies in physics teaching and learning* (pp. 25–40). Springer.

Sokolowski, A. (2018). Modeling acceleration of a system of two objects using the concept of limits. *The Physics Teacher, 56*(1), 40–42.

Sokolowski, A. (2019a). Applying structural mathematics in physics: Case of parallel connection. *The Physics Teacher, 57*(9), 627–629.

Sokolowski, A. (2020). Like terms in algebra-is the current definition adequate? Proposal for an instructional unit for high school students. *Australian Mathematics Education Journal, 2*(2), 13.

Stewart, J. (2016). *Calculus: Single variable calculus* (8th ed.). Brooks/Cole.

Thompson, P. W., & Carlson, M. P. (2017). Variation, covariation, and functions: Foundational ways of thinking mathematically. *Compendium for Research in Mathematics Education*, 421–456.

Thompson, P. W., Hatfield, N. J., Yoon, H., Joshua, S., & Byerley, C. (2017). Covariational reasoning among US and South Korean secondary mathematics teachers. *The Journal of Mathematical Behavior, 48*, 95–111.

Thompson, P. W., Philipp, R., & Boyd, B. (1994). Calculational and conceptual orientations in teaching mathematics. In *1994 Yearbook of the NCTM* (pp. 79–92). NCTM.

Vinner, S. (2002). The role of definitions in the teaching and learning of mathematics. In *Advanced mathematical thinking* (pp. 65–81). Springer.

Watson, J. M., & Moritz, J. B. (2000). The longitudinal development of understanding of average. *Mathematical Thinking and Learning, 2*(1-2), 11–50.

Zimmerman, C., Olsho, A., Loverude, M., Boudreaux, A., Smith, T., & Brahmia, S. W. (2019). Towards understanding and characterizing expert covariational reasoning in physics. *arXiv preprint arXiv:1911.01598.*

Part III
From Research to Practice

Chapter 6
Extending the Inquiry of Newton's Second Law by Using Limiting Case Analysis

6.1 Limits - Tools for Extending Scientific Inquiry

This research aims to present an example of applying limits to extend the inquiry in physics. While most current developments in this area take the stance that students' mathematical reasoning can be enriched by improving students' conceptual understanding of physics formulas, this study aims differently. It aims to extract phenomena behavior from formulas considered functional covariation relations. What prompted such aim? Research (Sokolowski 2020) has shown that physics students do not consider formulas as mathematical entities that describe phenomena dynamic nature. Therefore, the traditional aim of using formulas in their current representations was not considered. The main algebraic structures in mathematics are covariational relations called functions. Thus instead of using formulas, their function representations will be applied, because limits are traditionally used to evaluate functions. The phrase *limiting case analysis* will be used to allow transitioning of terminology between physics and mathematics. However, it will not be emphasized, and its alternative task command *evaluation of the covariational expression by taking limit* will be employed. Such undertaking is to enhance familiarity with situations when limits can be applied. While this study was published in its initial version earlier (Sokolowski, 2019), this one offers more consistent terminology and limit interpretation.

A. Sokolowski, *Understanding Physics Using Mathematical Reasoning*,
https://doi.org/10.1007/978-3-030-80205-9_6

6.2 Research Methods

6.2.1 Research Questions, Logistics, and Participants

This undertaking can be classified as a pretest/posttest one-group experimental case study (Shadish et al., 2002). The participants of the study consisted of a group of twenty-five (N = 25) college-level algebra-based physics students. Within this group, nine (N = 9) were female students and 16 (N = 16) male students. All participants possessed a prior calculus background, and they were familiar with sketching rational functions and evaluating the limits of these types of functions for the independent variable approaching an infinitely large value. They also possessed a physics background in analyzing the motion of systems of objects from high school. The pretest was assigned within the first week of the course. It was to assess students' awareness that limits are considered to quantify the formula's outputs. The results were not disclosed to students. The instructional unit was delivered within the section of dynamics about a month later. Students took posttest after 2 weeks from participating in the instructional unit. The posttests results were used to assess their gain in handling algebraic structures (functions) and tasks to learn about the system behavior. The pretest/posttest questions were similar except that on the posttest students were given a choice to use algebraic tools to answer the questions. The following emerged as the research questions for this study:

(a) *Are physics students comfortable with converting formulas to a designated covariational representation?*
(b) *Can physics students reassign the variables in formulas to a covariational relationship?*
(c) *Can physics students interpret the limit results in the scientific contexts?*

The first question was assessed using the pretest results. The posttest results were used to determine if guiding students by converting formulas to covariate relationships and introducing limits as a tool for evaluating the covariate relationships enrich their techniques of solving problems.

6.2.2 Criteria for the Study Content Selection

Analyzing the motion of a system of two objects, with or without friction, is a shared context in high school and undergraduate physics curricula rooted in a standard formula to analyze the system's motion under a constant net force.

$$\sum \vec{F} = m\vec{a} \tag{6.1}$$

When considering the mass of the system as a constant parameter, which would describe the most common scenarios, the action described symbolically as • $\vec{F}$

directly affects the system's acceleration, $\vec{a}$ with a proportionality constant m that is the system's total mass. If mapping the formula into a typical *input-output* functional relation, its covariational scientific underpinnings could be better highlighted when the formula is presented as $m\vec{a} = \sum \vec{F}$. The form $m\vec{a} = \sum \vec{F}$ informs that the independent parameter of the covariation is the action represented by the sum of acting forces or the net force denoted by $\Sigma \vec{F}$. The result of the action is the object's acceleration with the mass m considered the system's constant parameter. Zooming more into the algebraic entanglements the acceleration of the system can be expressed as $\vec{a}(F) = \frac{\sum \vec{F}}{m} = \frac{1}{m} \sum \vec{F}$. This representation better aligns with students'daily experiences in math courses, and thus it should be easier interpreted; the higher the force, the higher the system acceleration. The traditional school laboratory exercises on exploring such motion involve deriving expressions for the system acceleration, also known as Newton's second law's operational form:

$$\vec{a} = \frac{\sum \vec{F}}{m} \tag{6.2}$$

Upon being introduced to this formula, students conclude that the acceleration depends directly upon the magnitude of the net force computed using the superposition principle $\Sigma \vec{F}$ and inversely proportional to the system's total mass m. These conclusions can be visualized and confirmed by generating two separate graphs, acceleration versus the net force and acceleration versus the total system's mass. Such pedagogy, albeit commonly used in physics education, embraces two different types of algebraic covariations, direct (linear) and indirect (rational) within one inquiry, which might inadvertently diminish the nature of the phenomenon's cause and effect. The underlying notion is that when investigating the acceleration, changing simultaneously the force and the mass is not convenient. For the clarity of the inquiry, two lines of investigations seem to emerge:

(a) $\vec{a}(m) = \frac{\sum \vec{F}}{m}$, which investigates the effects on the system acceleration of the varying total mass assuming a constant acting force.

(b) $\vec{a}(F) = \frac{\sum \vec{F}}{m}$, which investigates the effects of varying acting force with aconstant total system mass.

Such analyses would be recommended before students can embrace a more complex analysis of the modified Atwood machine illustrated in Fig. 6.1 that is commonly being used to investigate both covariates (a) and (b) presented above.

While for an experienced teacher, finding the system acceleration is mainly driven by setting up equations for each mass and solving the system, the scientific underpinnings and the parameters covariate entanglements have much more to offer regarding their effects on the systems' acceleration.

The problem with using a half-Atwood machine is that the hanging object m_2 (see Fig. 6.1) contributes to the system acceleration in two different natures; as an

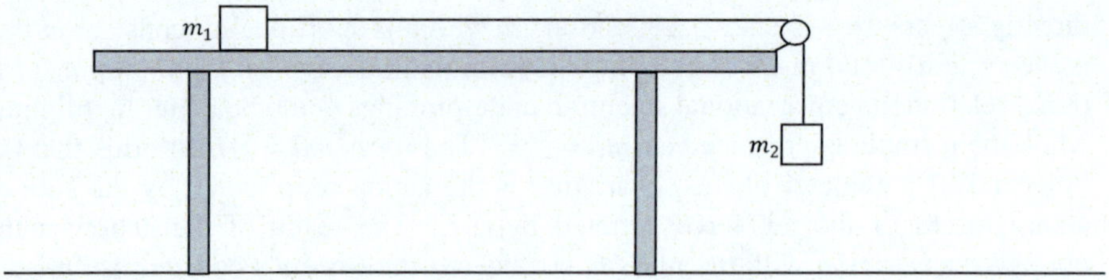

Fig. 6.1 The system of two objects connected by a massless string

inertial mass and as a gravitational mass, thus it seems that the corresponding graphs and covariates cannot be explicitly classified as direct or indirect. It is reasonable to say that its effects cannot be fully apprehended without producing a graph of acceleration when a hypothetical range of higher mass values can be considered. The effects of one of mass varying can be *observed* using the function behavior and hypothetical deductive reasoning in the form of *limiting case analyses* can be applied. Using the system to conclude that the acceleration depends directly on the magnitude of the gravitational force acting on m_2 might not precisely reflect the system's principles. While for small values of the mass, the graph appears to be linear, high range values, which enrich the analysis and extend the graph show that the graph is not linear when the force of gravity generated by m_2 is getting substantial values. Students' confusions related to interpreting experiments about Newton's laws were reported (Chief Reader Report, 2017). For instance, students faced difficulties identifying reasonable forces that affected the acceleration or could not explicitly explain their effect on the acceleration. It is seen that the bi-variational effects of the mass producing the action (gravitational and inertial) on the system acceleration can be a cause of this confusion. While this study does not aim to prove that using limits eliminates these difficulties, it is hoped that students will describe more precisely the system's motion in terms of the cause and effect when one variable parameter is considered.

6.2.3 Discussion of the Applied Algebraic Tools

While scientists and engineers commonly apply limits, using limits in school practice and examining students' ability to apply limits to enhance scientific inquiry is not often undertaken in science education research. In this study, I suggest conducting such an analysis using the context of two objects connected by a massless string (see Fig. 6.1). While the study was conducted with college students, its context is also suitable for advanced high school physics courses. In practice, such transition provides opportunities for deriving valid conclusions, especially in extreme or very small values for which direct laboratory measurements are not plausible. Although the idea seemed to illustrate applications of limits in science at first, rather than enhancing students' scientific inquiry, developed instructional units and students' responses have advanced the topic to a promising endeavor. The necessary students' math background was their familiarity with the techniques of evaluating limits of rational functions at infinity, also known in mathematics as evaluating *function end behavior*. The operational form of Newton's second law of motion when the modified Atwood machine is concerned, $\vec{a}(m) = \frac{\sum \vec{F}}{m}$ represents a rational expression, and the idea of taking limits should produce results that can be interpreted as the system acceleration. In general, using the idea of limits should extend both predictions and explanations of this type of motion that, according to Redish (2006), are fundamental goals of teaching physics. The limiting case analyzes typically do not embrace the formal processes of evaluating limits considering formulas as functions but assumes these structures (Savage & Williams, 1990). This study is posited to bring forth precise mathematics tools and apply the techniques of taking function limits in a fashion consistent with what students learn in their mathematics courses. Such an approach allows also for sketching these functions and view the limits outputs referring to these functions graphs (and asymptotes). It is hypothesized that such an approach helps students transfer their skills and understand the outputs of the processes applied in physics more clearly.

6.3 Description of the Instructional Unit

While initially, this study's intended to assess students' readiness in applying and interpreting limits in physics, the pretest results prompted to extend the mathematical part and propose an instructional unit during which the underpinnings of applying limits would be discussed in more detail. In their prior physics education, the participants studied how to construct a formula to find the acceleration using the Atwood machine model. However, it appeared that constructing it from *scratch* and guiding students on how Newton's laws can transition to more sophisticated algebraic covariation appeared as an exciting learning experience.

6.3.1 Analyzing Acceleration of the System in the Function of Mass m_2

The instructional unit was initiated by posting a problem about predicting the magnitude of the system's acceleration when either of the masses m_1 or m_2 (see Fig. 6.1) had a zero or infinite magnitude. The instructor then presented Part 1 (Sect. 6.3.1) and discussed the tasks that lead to embracing the process of concluding the answers using limits.

Problem 1 Consider a system of two blocks (Fig. 6.1). The block m_1 of 2 kg is placed on a horizontal frictionless surface. The block of m_2, with an initial mass of

2 kg, is hanging over the side of the table. However, more masses are continually being added to m_2, eventually increasing its amount to an infinite value.

(a) Predict the value of the system acceleration when m_2 reaches an infinite value.
(b) Formulate an expression for the system's acceleration using Newton's 2nd law and find its value when the mass m_2 reaches an infinite amount. Use the technique of limits to derive your answer.

Students recorded their predictions, and the instructor guided them through the solution process. The inquiry was initiated by labeling primary forces acting on and within the system. The pulley was frictionless, so the system's motion was driven only by the force of gravity acting on the mass, m_2, as shown in Fig. 6.2.

By considering the forward and downward directions of the forces acting on the masses as positive and acceleration a shared parameter for all masses, the net force acting on each object was formulated.

$$\begin{cases} m_1 a = T \\ m_2 a = -T + m_2 g \end{cases} \tag{6.3}$$

Adding the equations *by sides* and factoring out the symbol of acceleration resulted in $(m_1 + m_2)a = m_2 g$. Solving the equation for the acceleration results in:

$$a = \frac{m_2 g}{m_1 + m_2} \tag{6.4}$$

Equation (6.4) represents an operational form of Newton's 2nd law. It allows for computing system acceleration by substituting known mass values for m_1 and m_2. However, converting the formula to a covariate representation and approximating the function value for a large mass m_2 will answer the problem's questions more explicitly. The pretest disclosed that the students were unsure how to handle the formula's evaluation for a considerable value of m_2; therefore, more attention to this phase of the reasoning was dedicated. It seems that classifying the parameters as dependent, constant, and independent can be conveniently processed using Eq. (6.4). Since the mass m_2 is changing, therefore this quantity will constitute the

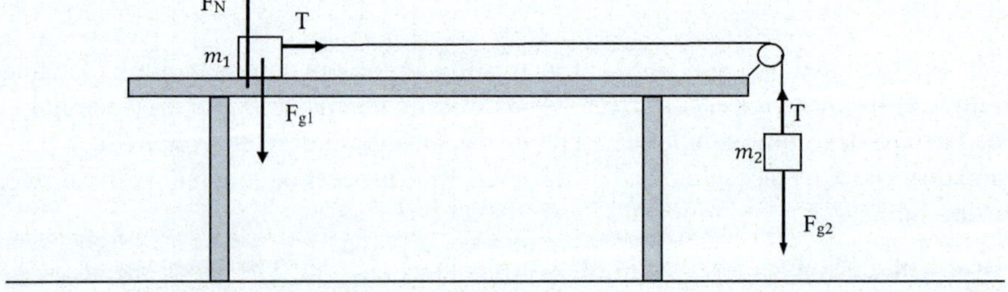

Fig. 6.2 Primary forces acting on the system

variable independent parameter of the function. To highlight the variable nature of the acceleration and dependence on the mass m_2, parameter a was embraced in a functional notation; $a(m_2)$. The mass $m_1 = 2$ kg, thus, it was replaced by the magnitude of 2. The equation then took the form of a rational covariation:

$$a(m_2) = \frac{m_2 g}{2 + m_2} \tag{6.5}$$

It is to note that the mass m_2 contributes to the system acceleration as gravitational because it is coupled with g (the numerator of the expression) and inertial as m_2 (the denominator of the expression) thus from a physics point of view, its effect on the system acceleration has a dual contribution. The effect of each depends on its relative value with m_1. Once the formula is embraced in its covariational nature, its graph (see Fig. 6.3) can be sketched, and the limits evaluated using the graph or by applying formal conduct. The limit value will be used to justify the parameter of interest; the system acceleration. Both representations should lead to the same conclusions. Referring to the graph, one will notice that when the mass m_2 increases significantly, the acceleration is getting close to 9.8 m/s^2. The value of the limit represents also the horizontal asymptote or the boundary of the rational function, depicted by the dashed horizontal line in Fig. 6.3. It is to note that the system's acceleration will never reach the value of 9.8 m/s^2 because the graph cannot cross

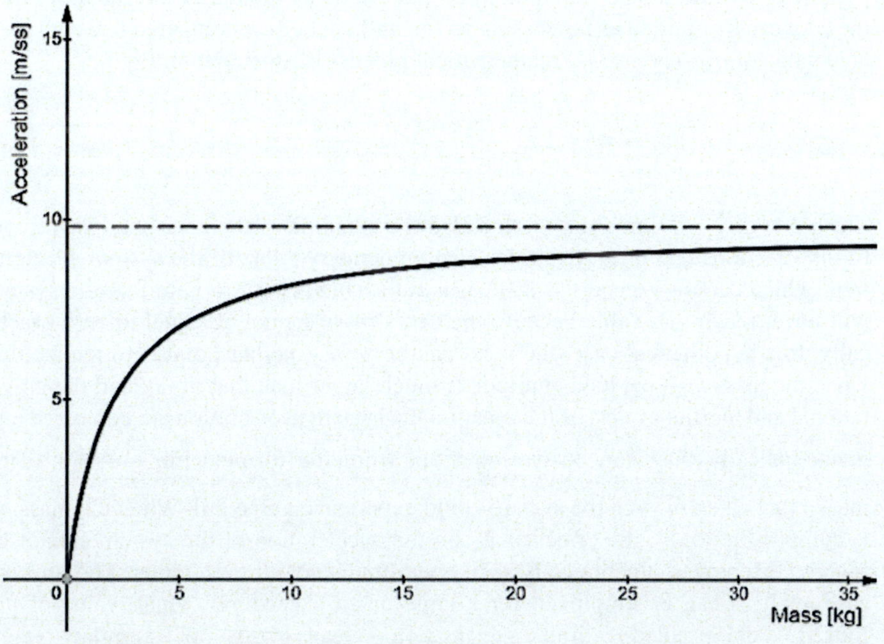

Fig. 6.3 Graph of the acceleration of the system when the hanging mass m_2 increases

the horizontal asymptote at infinity (assuming that the experiment is conducted on the Earth), and this conclusion corresponds with the system behavior.

The graph in Fig. 6.3 is generated arbitrarily for $0\ kg \leq m_2 \leq 35\ kg$. The value of 0 kg was chosen to highlight the physical domain of the function; $m_2 \geq 0\ kg$. While this graph appears to illustrate an exponential function due to applied restriction on its domain, it represents a rational function due to the properties of the algebraic structure. Evaluating the function for $m_2 = 0$ produces zero acceleration of the system that is represented by the vertical intercept of the graph. This situation transcends to no mass hanging, thus no system motion. It is to note that the graph provides an effective means to approximate the acceleration for various values of the hanging mass. For instance, when the mass is 5 kg, the acceleration is about 7 m/s^2. It is to note that in classroom experiments, when the values of the hanging mass are not large, perhaps when $0\ kg < m_2 < 2\ kg$, the graph appears to be linear, suggesting a direct relationship. However, this is not entirely true when the analysis extends to a substantial value of the mass. Thus, considering the mass m_2 as directly affecting the system's acceleration does not correctly reflect the system's scientific underpinnings. This is primarily attributed to the fact that at the same time, the mass contributes (increases) the total system's inertia, which transcends that the acceleration is inversely proportional to the sum of the masses m_1 and m_2. Such duality of mediating effects of the mass on the system acceleration does not emerge if the corresponding graph of a rational function is not generated.

To completely support the scientific inferences by mathematical reasoning a formal way of evaluating function limits is needed. If m_2 increases to infinity, this parameter will dominate the numerator and the denominator of the function, thus the 2 kg can be omitted as possessing an insignificant effect compared to a considerable value of m_2. The formal mathematical process is illustrated in Eq. 6.6.

$$a\left(m_2 \to \infty\right) = \lim_{m_2 \to \infty} \frac{m_2 g}{m_2 + 2} = \lim_{m_2 \to \infty} \frac{m_2 g}{m_2} = g = 9.8\, m/s^2 \tag{6.6}$$

The value of $g = 9.8$ m/s^2 represents also the equation of the horizontal asymptote of the function. The limit illustrates also a boundary value of the system acceleration, which cannot exceed 9.8 m/s^2. It is rather interesting to note that the system will never reach that value because the cart's mass cannot be equal to zero practically. In a hypothetical case, the mass can be zero. Another point is to realize that when the mass $m_2 \to \infty$, its magnitude is much larger than that of m_1, and the gravitational and inertial effects of the mass on the system acceleration are equalized and mathematically they can be canceled out from the formula $\lim_{m_2 \to \infty} \frac{m_2 g}{m_2} = g$. This means that when $m_2 \to \infty$ the system could experience a free fall. When the mass m_2 is comparable to m_1, the effect of m_2 on the acceleration of the system cannot be ignored. Moreover, algebraically such operation is not allowed either. The limiting case analysis can be employed for a large or an exceedingly small value of m_2. Indeed the sided limit evaluation leads to a similar conclusion $a\left(m_2 \to 0^+\right) = \lim_{m_2 \to 0^+} \frac{m_2 g}{m_2 + 2} = \lim_{m_2 \to \infty} \frac{0g}{2+0} = 0\frac{m}{s^2}$.

6.3.2 *Analyzing Acceleration of the System in the Function of Mass* m_1

Problem 2 Suppose that the mass m_2 remains constant (2 kg) and that the value of the mass m_1 increases to infinity.

(a) Predict the magnitude of the system acceleration
(b) Formulate an expression for the system's acceleration using Newton's 2nd law and determine its value when the mass m_1 reaches a substantial value. Use the concept of limits to support your answer.

The phases of the solution process are like in Sect. 6.3.1. Since the system's general setup does not change, its acceleration formula will also not change completely. Thus, considering m_1 a variable parameter and m_2 possessing the value of 2 kg, the formula $a = \frac{m_2 g}{m_1 + m_2}$ takes the form of $a(m_1) = \frac{2g}{m_1 + 2}$. To answer part *a* of the problem, the limit of the function for $m_1 \to \infty$ needs to be taken. Following the techniques of evaluating limits of rational functions, the mass of 2 kg can be removed as insignificant compared to a substantial value of m_1. This reasoning results in evaluating the quotient of $2g/m_1$ for a large m_1 that produces an acceleration of zero. A formal limit evaluation ilustrates Eq. 6.7.

$$a(m_1 \to \infty) = \lim_{m_1 \to \infty} \frac{2g}{m_1 + 2} = \lim_{m_1 \to \infty} \frac{2g}{m_1} = 0 \tag{6.7}$$

A similar conclusion can also be reached using graph $a(m_1)$ that is illustrated in Fig. 6.4.

Graph (Fig. 6.4) also provides a valid physical interpretation for its vertical intercept. It proves that when mass $m_1 = 0$, the system's acceleration equals in magnitude to the acceleration due to gravity 9.8 m/s^2. Likewise, Fig. 6.3 graph Fig. 6.4 can also be used to find the system acceleration for various masses of the cart and help identify and interpret its horizontal asymptote that has an equation of the limit when $m_1 \to \infty$, $a\ (m_1) = 0$. The asymptote is denoted with a horizontal dotted line on the graph. The horizontal asymptote can be used to conclude that the value of the acceleration will never reach the value of zero, but it will be remarkably close to zero when the mass $m_1 \to \infty$. The graph interpretations and conclusions reached using the function equation were to convince the students about its precision in depicting the phenomenon's nature. This conclusion has its scientific support in the system behavior; since the surface is frictionless, even when the mass m_1 is exceptionally large, the constant pulling force produced by the force of gravity acting on the 2 kg mass will still accelerate the mass.

Furthermore, increasing m_1 increases the system's resistance to more, decreasing the system acceleration. An actual demonstration can be used as a medium of partial experimental verification and conceptualization of this case. The graph generated in the classroom most likely will not exhibit its asymptotic behavior. Realizing this limitation, the students appreciated mathematical reasoning that extended the analysis beyond classroom experimentation.

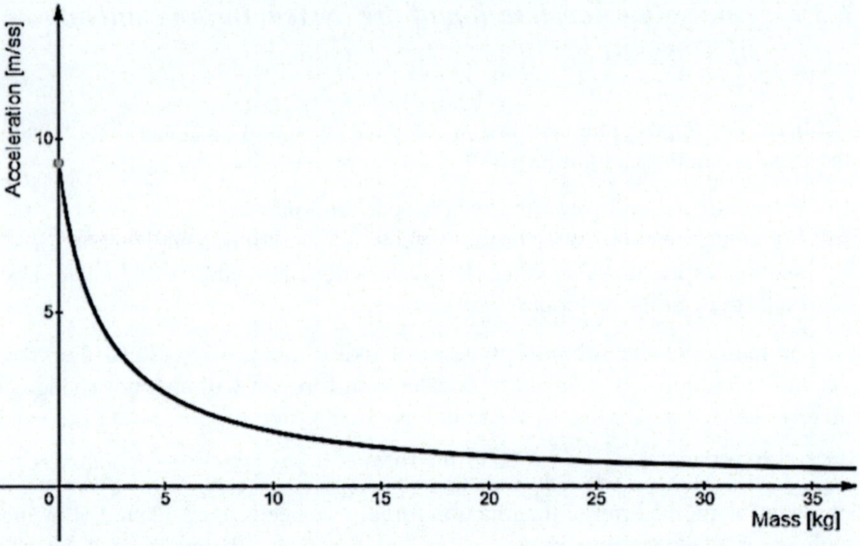

Fig. 6.4 Graph of the acceleration function if the mass m_1 varies

6.4 Data Analysis

The pretest and posttest questions were similar to the ones presented in Sects. 6.3.1 and 6.3.2. While the pretest was administered within the first week of the course, the posttests question was assigned as one of the exam questions on the cumulative chapter exam.

6.4.1 Analysis of the Pretest Results

In problem #1 about analyzing acceleration when $m_2 \rightarrow \infty$, the students derived the formula for the system acceleration; however, they could not find adequate algebraic tools to evaluate the formula when the mass m_2 increased. Even though the pretest provided clues on what mathematical processes to apply to answer the questions, the students' responses and the percent of correct answers were low. In problem 1, 20% of the students (N = 5) concluded that the acceleration would be near 9.8 m/s^2. The majority of the students, 80% (N = 19), concluded that the system's acceleration would increase, yet they did not explicate on the boundary of acceleration magnitude. Some verbatim samples of student responses are as follows: *The acceleration will increase to infinity; the acceleration will be tremendous because there is more and more mass hanging, thereby making the acceleration larger; acceleration will increase because higher mass will increase the force of gravity; acceleration will approach infinity because of the heavier weight pulled;*

acceleration is infinite; the acceleration would significantly increase; the acceleration will increase quickly. None of the students zoomed into inertial and gravitational effects that contributed to the system's acceleration. While the physics reasoning was acceptable, the students most likely recalled from their introductory physics courses that acceleration is directly proportional to the acting force; they did not attempt to explicitly transfer the formula into a covariate expression and formally take the limit. Without such association, they did not notice that the pulling mass has a dual contribution to the system's acceleration, inertial and gravitational. Some students substituted large values for m_2 and evaluated the formula, but they did not foresee the use of the idea of limits as a valid quantification tool. While all of these students did possess a background on the limit evaluation either from precalculus or calculus courses, the background was insufficient to realize that this problem called for applying the limiting analysis.

In problem 2, whose virtue was rooted in $m_1 \rightarrow \infty$, many students concluded that the acceleration would be zero (N = 18, 75%), which was accepted as a correct answer. However, the most accurate answer stating that the acceleration will be near the value zero but will never reach that value was concluded only by six (N = 6, 25%) students. Verbatim samples of student responses are as follows: *The acceleration will be positive but never close to zero; the acceleration will be very small because m_1 would require movement at the same acceleration; the object will have acceleration because of the gravitational field; the acceleration will be zero; the acceleration will decrease.*

In both parts of the problems, students faced difficulty supporting their reasoning using the outputs by consistently applying algebraic operations. While their physics reasoning was fine, their written work showed a lack of consistency and clarity of thought process, which illustrated their lack of algebraic tools to confirm their physics reasoning. This was explicated from their algebraic manipulations. A lack of attention to the limit interpretation could be accounted for a rare students' exposure to applying and interpret limits in their prior physics courses and perhaps also in their mathematics.

6.4.2 *Analysis of the Posttest Results*

On the posttest, the students were given an option to conclude the magnitude of acceleration using any method. Such formulated task was to activate their other forms of reasoning.

In problem 1, the results showed an improvement in handling the limit evaluation. The students' work showed more consistency in identifying constant and variable quantities, which guided them to be more successful. This problem was correctly answered by 54% (N = 13) of the students. Some students still faced difficulties with identifying the independent and constant parameters in the mathematical sense. Interestingly, most students who renamed the independent variable m to x computed the limit and the acceleration correctly. These students who did not

Table 6.1 Summary of Pre-test/Post-test Results

	Pretest	Post-test
Problem 1	20%	54%
Problem 2	25%	90%

rename the variables usually stated that the limit has the value of one instead of *g*. In supporting their answers, the majority (N = 20, 80%) analyzed the function's algebraic structure, and the remaining (N = 5, 20%) used simple scientific reasoning. Some verbatim samples of student responses for Problem 1 follow: *The acceleration of the system will be g because the two infinity values will cancel out; acceleration goes to infinity because increasing m_2 gives a high force of gravity and the system has the same inertia, if m_2 increases to infinity, then the acceleration would essentially just beg; the acceleration is infinity, and it will continue to increase when m_2 increases; the object cannot accelerate faster than g; therefore the answer g from the limit makes sense.*

In problem #2, the percent of correct answers was high, 90% (N = 22). Almost all students used the algebraic form and applied the limiting analysis to conclude and comment on their answers. This case was easier to analyze because the variable parameters were placed in the expression denominator. Verbatim samples of student responses are as follows: *As m_1 approaches* ∞, *the acceleration would be close to zero due to increased denominator; as* $m_1 \rightarrow \infty$, $a \rightarrow \infty$, *the larger the mass of* m_1, *the lower the rate of change of acceleration become; the acceleration would be very slow, but it would be there due to the presence of* m_2; *the acceleration of the system is 0 m/s*^{2} *because as in the limit, the denominator would be infinite where the numerator would become relatively insignificant, and the value would be approaching zero; inertia increases to infinity then the acceleration gets close to zero;* $\frac{2g}{\circ}$ *is going to be* 0; *as m_1 increases, the acceleration will be smaller and smaller.* It was noted that when asked to support their answers, the students explained how they used the techniques of limits evaluation instead of scientific reasoning. This way was not marked as a weakness against the students but it suggested including more analyses to improve on the math-physics interface. A summary of pretest/posttest results is shown in Table 6.1.

6.5 Conclusions

The posttest results indicated that the students improved their skills in applying the technique of taking limits to calculate physical quantities. The way they derived the solutions illustrated that they felt more comfortable following through the process of reclassifying parameters as constant and varying and taking limits. While on the pretest, most of these students did not realize that limits must be applied to answer these questions; upon being instructed, the students recalled the techniques and attempted to use these techniques and succeed. Their verbal supports also showed more consistency and clarity. Did such a lesson enhance their general disposition of including limits as an algebraic tool to solve physics problems? While questions targeting this objective were not developed, the students were very eager to learn how to quantify physics problems using limits. It seemed that such curriculum enhancement sparked their curiosity to explore more phenomena using that exciting math tool.

Observing students' genuine interest in knowing how limits can support physics understanding was encouraging to consider creating more contexts with such learning opportunities that are included in this book. Upon being instructed on applying limits, their solutions' processes were more consistent with what they had learned in their calculus classes. They seemed to appreciate the opportunity to integrate that knowledge because that knowledge helped them with solving physics problems. While solving physics formulas with parameters is common in physics, converting these entities to functions and learn more about system behavior is still rare that surfaced on the pretest.

Another remark voiced earlier in this chapter is that the system of two masses as set up in the problem can help learn to eliminate the misconception that the acceleration is proportional to the force of gravity acting on the hanging mass. After analyzing the acceleration graph (Fig. 6.2) and its algebraic function, claiming a direct proportionality of the acceleration was not accurate. The graph appeared linear only for limited values of the hanging mass. Still, the relation cannot be labeled as linear. Examining the graph behavior on the entire domain determined by a broader set of the hanging masses provided the evidence.

A nuance about interpreting function limits also emerged from analyzing the phenomenon. Computed function limit is typically denoted in mathematics with an equal sign (=) rather than with a sign of approaching that value (→). For example, one writes that $\lim_{m_2 \to \infty} \frac{m_2 g}{2+m_2} = g$, not that $\lim_{m_2 \to \infty} \frac{m_2 g}{2+m_2} \to g$ which changes the verbal interpretation. The notation of approaching that value (→) was not used during the instructional unit to assure its consistency with calculus notation. Nevertheless, an emphasis was given to limit interpretation as a value approaching the output. Indeed, if, for example, $m \to \infty$, then the output should also represent an approaching result at its end, e.g., $a(m) \to 0$. The students realized that using an arrow instead of an equal sign would more accurately describe the meaning of the limit value in the phenomenon's context. However, the notation was not augmented because the limit value in such cases shows the boundary of the quantity of interest that transcends to an equation of the function horizontal asymptotes, not to actual function value.

While this paper presented an analysis of two motion cases, the experimental setup provided opportunities for creating more variations, where the idea of inducing algebraic functions could be further explored. For example, formulating a system consisting of three different masses with two of them hanging below the top of the table would diversify the application of limits and extend the idea of modeling the properties of inertial mass. Furthermore, by applying differential calculus and taking the first derivative of the acceleration function concerning established variable mass, students could learn that the acceleration rate of change concerning the hanging mass reaches zero for a substantial magnitude of the mass placed on the table. It is prudent that formulas converted to algebraic functions have the potential to return many more exciting scientific facts about the physical system's behavior than their formula counterparts do.

References

Chief Reader Report. Accessed from https://secure. Accessed 17 Oct 2017.

Redish, E. F. (2006). Problem-solving and the use of math in physics courses. *arXiv preprint physics/0608268*.

Savage, M., & Williams, J. (1990). *Mechanics in action: Modeling and practical investigations*. Cambridge University Press.

Shadish, W. R., Cook, T. D., & Campbell, D. T. (2002). *Experimental and quasi-experimental designs for generalized causal inference/William R. Swedish, Thomas D. Cook, Donald T. Campbell*. Houghton Mifflin.

Sokolowski, A. (2019). Enhancing scientific inquiry by mathematical reasoning: Case of applying limits to model motion of a system of objects. *Journal of Physics: Conference Series, 1287*(1), 012051. IOP Publishing.

Sokolowski, A. (2020). Unpacking structural domain of mathematics to aid inquiry in physics: a pilot study. *Physics Education, 56*(1), 015009.

Chapter 7
Reconstructing Newton's Law of Universal Gravity as a Covariate Relation

7.1 Prior Research Findings

Research on teaching and learning Newton's law of universal gravitation is not extensive, perhaps because the law is mainly used to calculate the magnitudes of forces of the mutual attraction between two objects. University library search engines such as ERIC, ProQuest Educational Journals, Science Direct, and Google Scholar returned several research papers in this domain that at large related to electrostatic and electromagnetic fields and forces. Temiz and Yavuz (2014) found out that students perceive that Newton's second law did not apply in space and suggested providing more scientific evidence to change this perception. Galili (1995) stated that the core concepts of mechanics and cosmology are built strictly on the force concept and that the effects of gravitational fields are rarely mentioned. He further contended that misunderstandings of fields and forces in electricity and magnetism (E&M) stem from a "mismatch of the methodology of teaching fields in mechanics" p.380. Saeli and MacIsaac (2007) reached a similar conclusion. Sokolowski (2008) noticed that such lack of parallelism in teaching fields and gravitational forces impedes analogy as a learning method to support the introduction of fields and forces in the sections of E&M.

As exploring the concept of gravity focuses primarily on the computations of the mutual gravitational force, so do the assessment items targeting the law understanding. For example, the force concept inventory items (Hestenes et al., 1992) concentrate on testing students' ability to quantify the effects of the gravitational field on objects placed near Earth's surface. In some of the items, the gravitational field generated by the falling mass is neglected, suggesting that the only mass-producing gravitational field within this system is the larger one, the Earth, whose effect is observed. Focusing on the object on which *action-at-a-distance* is exerted without understanding the force source is also visible in electrostatics. Maloney (1985) found out that students associate the size of a moving electric charge with the

A. Sokolowski, *Understanding Physics Using Mathematical Reasoning*,
https://doi.org/10.1007/978-3-030-80205-9_7

magnitude of the force exerted on the charge ignoring the effects on the external electrostatic field. Others Cao and Brizuela (2016) and Pocovi and Finley (2002) noted that while solving problems on electromagnetic forces, students associated electric interaction as an *action-at-a-distance* due to the presence of changes rather than as an action initiated by the presence of the fields generated by the charges. Asghar and Libarkin (2010) found out that college students frequently confound gravity with magnetism and atmospheric pressure.

While students' difficulties with understanding the nature of forces associated with fields might be rooted in interpreting the phrase *action-at-a-distance,* some support the theory that the learners do not consider the forces mediated by fields, whether gravitational, electrostatic, or electromagnetic. Referring to students' difficulties in E&M, Galili (1995) contended that without proper instruction in mechanics regarding fields and forces, students would carry out their personal beliefs about the physical world in E&M. Such notion can be difficult to correct in E&M when the complexities of these phenomena do not provide much room for revisiting gravitational effects and building up the theories on these concepts. A similar conclusion was reached by Hestenes (2006). He argued that the differences underlying principles of Newtonian concepts and the students' personal beliefs are significant and that "conventional introductory physics courses are not effective at reducing the gap" p.3.

The preliminary synthesis of research points out a missing chain of realizing that an object must be placed in an external gravitational field produced by another object to experience gravitational interaction. An object produces a gravitational field, but it cannot, and it will not mediate with it. Furthermore, a lack of parallelism in teaching methodology between forces and fields in mechanics and E&M can perhaps be partially accounted for the fact that Newton's law of universal gravitation was published much earlier than when the concept of fields was introduced to physics by Faraday. Since this historical development affects how the law of universal gravitation is taught today, a brief historical perspective of how fields and forces were developed in physics follows.

7.2 Theoretical Framework

The theoretical framework is followed by the historical development of fields and forces to enrich their methodological entanglements and support the proposed didactical enterprise of introducing the law of universal gravity to students. To highlight a coherence of the methodology, it also reflects on current didactical presentations of these ideas found in physics teaching resources.

7.2.1 Historical Perspective

Gravity is traditionally perceived as the phenomenon by which objects are attracted and by which they possess weight. The methods to quantify thc force of gravitational interaction between two objects were formulated by Sir Isaac Newton in 1687 and published in *The Principia* (1999). Newton's inverse-square law deduced from Kepler's laws (Hamming, 1980) supported the algebraic structure of the formula called the law of universal gravitation. Newton proved that if the orbits under a central force are conics, then the magnitude of that force varies inversely as the square of the distance. He also established the converse; if the force obeys the inverse-square law, then the orbits are conics (Hamming, 1980).

Furthermore, Newton (1999) postulated that "All bodies gravitate toward each of the planets, and at any given distance from the center of any planet the weight of anybody whatever toward that planet is proportional to the quantity of matter which the body contains" (p. 503). It is to note that the concept of fields remained silent in that theory. Without considering fields, the concept of force as *action-at-a-distance* prevailed even when Coulomb generalized the law of electrostatic interactions in the late 1700s (Smith, 2002). While discussing gravity, a need for considering fields gained its importance after Faraday and Maxwell constructed the field theory of electromagnetism in the 1800s, and it was Faraday who introduced the field concept to physics called physical *lines of force*. Maxwell agreed with Newton that the gravitational force varies as some power of the distance. "If the force were any function of the distance except a power of the distance, the ratio of the forces at any two different distances would not be a function of the ratio of the distances" (see Hamming, 1980, p.86). During most of the eighteenth century, Newton's theory of gravity remained under dispute, primarily because of the absence of a machine producing gravitational forces (Maxwell, 1873). In 1915 Einstein attempted to provide a theory for *forces- acting- at- a -distance* and described gravity not as a force but as a consequence of the curvature of space-time caused by the uneven distribution of mass (Newton, 1999). This assumption was, in fact, the critical idea of Einstein's theory of general relativity. While the theory of gravity mainly concerns the fields outside of the masses, it enables the compute the intensity of gravitational field below the surface of a uniformly dense sphere that varies linearly with the distance from its center.

The law of universal gravity derived and formulated by Newton is also presented nowadays in physics textbooks and taught in a similar way. Its algebraic structure and the theory beyond it are unquestionable as they represented the phenomena well when it was formulated. As the summary of the history of the development of fields and interactions show, new theories and discoveries after Newton implemented it to physics did not affect much the form, and the interpretation of how Newton formulated it. New developments in physics education recommending induction of covariation reasoning are not yet applied to teaching the law. This study is posited to reflect on these recommendations.

7.2.2 Contemporary Presentations of the Law of Universal Gravity

The laws of physics are built following developed models that reflect proposed theories. Newton's law of universal gravity (Eq. 7.1) is crucial because it allows to compute and predict the behavior of specific physical systems. It is seen that the conceptualization of the law, as presented in physics education resources, does not emphasize enough the effects of field interaction that Faraday has discovered. It is hypothesized that developing a solid conceptual understanding of gravitational fields will support understanding gravitational interactions and contribute to students' better understanding of electrostatic and magnetic fields.

A traditional sequence of introducing Newton's law of gravity (e.g., Giancoli, 2014) is initiated by presenting two masses, m_1 and m_2, separated by a distance r and introducing the inverse square law $F_G \propto \frac{1}{r^2}$. This statement is followed by an explanation of the inverse square law, and Newton's law is presented as Eq. (7.1) along with a conceptual description: "Every particle in the universe attracts every other particle with a force that is proportional to the product of their masses and inversely proportional to the square of the distance between the masses" (Giancoli, 2014 p.357).

$$F_G = \frac{Gm_1m_2}{d^2}. \tag{7.1}$$

Some introductory textbooks on mechanics introduce fields at the concluding phases of the course as a refinement of gravitation (see Serway, 1996 p.391). In other textbooks, the refinement points to the curvature of space rather than that of the fields. As research demonstrates, these methods do not explicitly develop such interpretation of the force -fields' interactions that would help students quantify the interaction based on the fields' magnitudes. By using direct and inverse proportional reasoning, the description of the law merely verbalizes the mutual entanglement of the parameters that constitute its algebraic structure as presented. The law supports the computation of the gravitational forces and emphasizes the masses' magnitudes as the primary sources of gravitational attraction. As such, the law formulation perhaps unintentionally hinders the effects of the gravitational fields produced by the masses on the gravitational force existence. Furthermore, since force acts on an object, the algebraic form, as informed to the learners, does not explicate what object the computed force acts and how to establish the direction of the computed force. Consequently, the law's verbal description does not explicitly help understand the theory beyond it. If the effects of fields are not discussed, the students might associate the forces with the magnitudes of the masses, claiming mistakenly that if the masses are of different magnitudes, the forces acting on each mass are also of different magnitudes (Kavanagh & Sneider, 2007). Informing students that the Eq. 7.1 computes the force of gravitational attraction does not nurture the conceptual understanding of the phenomena, which was disclosed in the pretest for this study. Following this analysis, it is hypothesized that the current presentation of the law might inadvertently imply that the primary source of the gravitational force between two objects is the existence of the masses. This study is posited that such reasoning does not reflect on the scientific underpinnings of the phenomena. Narrowing the phenomenon to its algebraic representation overshadows the effects of the gravitational fields and reduces the opportunity for linking the strength of the fields with the magnitudes of joint forces and consequently quantification of the masses' accelerations. One of the reasons for the delay in introducing these phenomena to physics can be accounted for the historical development; the interactions of fields with objects were introduced to physics almost a century after Newton formulated the law of universal gravitation. Newton did not include the effects of the fields, which might account for students' difficulties understanding the cause-effect of the phenomena. While understanding the effects of gravitational fields is not necessary to compute the forces using Eq. 7.1, it is suggested that the inquiry should move beyond these relatively straightforward algorithms. Considering the historical development, prior research findings, and pretest results, an instructional intervention (Sect. 7.5) emerged.

7.3 Methods

This enterprise can be classified as one group mixed-method study (Shadish et al., 2002). While a control group would allow for applying inferential statistical measures with effect size as the primary evaluation determinant, this was not plausible. Thus, the evaluation of the proposed method was primarily conducted by employing descriptive statistics and qualitative analysis. The instructional unit was presented to a group of 25 freshman college physics students who possessed a prior high school physics background about Newton's law of universal gravitation. Twenty of these students (N = 20, 80%) had a precalculus background, and five (N = 5, 20%) were concurrently enrolled in a calculus course. Many of these students majored in science and engineering fields. The students took a pretest during the first week of the course and a posttest after two weeks.

7.4 Didactical Underpinnings of the Instructional Unit

The purpose of designing the lesson was to emphasize that Newton's law of universal gravity is rooted in the interaction of mass m_1 and an external gravitational field produced by another mass m_2. Departing from this statement, a covariate effect of the other the mass m_2 was induced, and the formal Newton's Law of universal gravity was assembled. There was an emphasis on differentiating between the concept of gravity, gravitational forces, and gravitational fields as often these three different yet mutually related concepts, when not explicitly differentiated, might confuse interpretation. Considering the prior research and the students' pretests' level of knowledge, it was evident that a more robust conceptual underpinning and thus a modeling-type approach was needed. The modeling type approach is usually associated with an inductive inquiry initiated by specific observations that lead to reaching a general conclusion (Temiz & Yavuz, 2014). A deficiency of students' statements that interaction of the gravitational field with an external mass generates gravitational force suggested placing more emphasis on the properties of the field before introducing the traditional form of the law. To enrich the process, a reference was given to students' prior knowledge, e.g., Newton's second and third laws of motion. It was hypothesized that extending the range of tools would broaden the conceptual foundation for a deeper understanding of the law. The idea of mathematical modeling applied during the unit was to help formulate a dynamic representation of the phenomena. The following are the operational definitions used during the modeling process:

- The system consisted of one object (see Sect. 7.5.1) and then two objects (see Sect. 7.5.2) defined by their masses and spherical structures.
- The interaction among the system objects (Tipler, 2008) was mediated by gravitational fields produced by the masses and forces of gravitational attraction produced between the masses.
- The interpretation of medium of interaction; *the gravitational field* was defined as a natural phenomenon that alters properties of space around matter by giving it the ability to attract other objects placed in its proximity.

The definitions suported the development of the lecture exploratory process.

7.5 The Lecture Component

Objectives

- Develop a conceptual understanding of the gravitational field.
- Develop an understanding of gravitational force as an action generated by the gravitational field on an external mass.
- Use mathematical reasoning to understand the algebraic structure of the law of universal gravitation.

The treatment emerged as one lecture unit consisting of two parts developed in Sects. 7.5.1 and 7.5.2. In Sect. 7.5.1, a formulation of an algebraic expression to compute the magnitude of the *field intensity at a distance* produced by a spherical mass m was developed. In Sect. 7.5.2, Newton's law of universal gravity was reassembled by starting from an expression for the derived intensity of the gravitational field. The inquiry was centered on three short problems presented to the students before exploring the phenomenon underpinnings. This exploratory-type lesson allowed for highlighting to students the covariational reasoning while attending to verifying predictions. Students hypothesized the solutions/answers, and then the instructor guided the learners through an integrated reasoning process.

7.5.1 *Gravitational Field Intensity and the Effects of Covariate Quantities*

As the Earth considered a uniform solid sphere produces a gravitational field around its body, and inside it, so does any mass, even a 1-kg object. Gravitational fields produced by objects of other shapes, e.g., cubes, cylinders, are modeled using methods that include the process called in calculus the integration. The following analysis refers to the field produced outside of the sphere, thus for distances greater or equal to the sphere radius.

Problem 1 How to graphically depict and calculate the field produced by the mass?

The field must have a similar nature to the Earth's field. Thus, it must be directed toward the center of the mass, as illustrated in Fig. 7.1. The subscript "1" corresponds to the mass that produces the field. The vector on the right side was labeled $\vec{a}(d)$ since, in this location, an external mass will be placed to develop a general formula to quantify the magnitude and the direction of the force. While the symbol g denoting the intensity of the gravitational field is typically used, the symbol a was chosen to bridge the interpretation of the intensity of the field with the object's acceleration.

To formulate an algebraic representation to compute the field strength, one must determine the quantities that affect its value: the mass, m_1 of the object producing the field, and the distance, d from the center of the mass to where the field is to be computed. While the mass is the parameter describing the object producing the field, the distance is the space's parameter. The distance parameter can be considered a mediating one. It must be included in the intensity computations if different

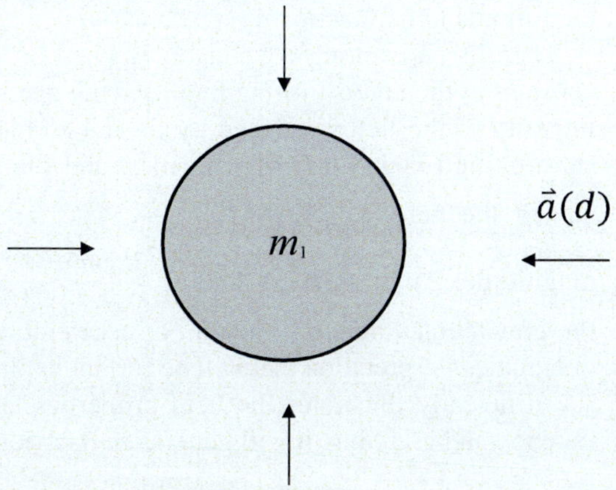

Fig. 7.1 Vectors depicting the intensity of gravitational field produced by the mass m_1

distances are concerned. In sum, direct and inverse proportional reasoning led the students' thinking toward realizing that the larger the mass m_1 is, the more intense the field is, thus using algebraic notation $a_1 \sim m_1$. Respectively the larger the distance, the lesser the intensity is. The instructor explained that the second premise was called the inverse square law and that Newton adopted it using Kepler's laws. The law can be seen as a distribution of the intensity into a spherical surface area with radius d, which supports the algebraic form of the inverse square law as $a_1 \sim \frac{1}{d^2}$. Due to a direct proportionality of the mass, the interaction between the mass and the surrounding space takes the following form:

$$a_1 \sim \frac{m_1}{d^2} \tag{7.2}$$

To convert this statement to an equation, a proportionality constant was needed. The value of the constant called the universal gravitational constant $G = 6.67 \times 10^{-11} \frac{Nm^2}{kg^2}$ was experimentally determined by Lord Henry Cavendish nearly a century later (1798) by using a torsion balance (Galili, 1995). The final form of the formula for computing the field produced by the mass m_1 is:

$$a_1 = \frac{Gm_1}{d^2} \tag{7.3}$$

Where m_1 is the mass producing the field that is the independent parameter, and d is the distance from the center of the mass to the location where the intensity is to be computed called another independent parameter that deems its value when calculating a specific value of the intensity, and G the gravitational constant that can be called a mediating parameter.

The instructor included restrictions on the function domain by stating that the function is valid only for spherical uniformly distributed masses provided the distance d be outside the radius of the sphere thus for $d \geq r$. This element was to link the formula to its functional representation.

There is an exciting conclusion reached by including the interpretation of the constant in the context of the inquiry. One can interpret it as the magnitude of the intensity of the field produced by the 1-kg object at 1 m from its center. Indeed, substituting 1 kg in Eq. 7.3 for the mass and 1 m for the distance results in the intensity of the field $a_1 = 6.67 \times 10^{-11} \frac{N}{kg}$. The intensity proves to have nearly zero magnitude. By expressing 1 N as $\frac{1kgm}{s^2}$, $a_1 = 6.67 \times 10^{-11} \frac{m}{s^2}$. The intensity of the gravitational field has the unit of acceleration m/s^2 and as such it takes a more pragmatic interpretation that will be further explored in this study. While this analysis sufficiently illustrates the field properties, expressing Eq. 7.3 as an algebraic function and graphing, it will generalize its meaning.

Problem 2 How to graphically representant the formula $a_1(d) = \frac{Gm_1}{d^2}$?

The formula consists of two independent variable parameters m_1 and d. To graphically represent it in the traditional XY coordinate system, only one independent parameter is allowed, and traditionally it is the distance d from the center of the mass, thus $a_1(d) = \frac{Gm_1}{d^2}$. A corresponding algebraic structure that can serve as a reference for sketching the graph is $f(x) = \frac{k}{x^2}$, where k is a positive constant. Considering the product Gm_1 a constant value, $a_1(d) = \frac{Gm_1}{d^2}$ resembles a rational function whose graph is a hyperbola $f(x) = \frac{k}{x^2}$ illustrated in Fig. 7.2. The graph is hypothetical; it does not include magnitudes on its axes.

Students realize that the graph is built on two different algebraic rules, linear and rational, due to possessing two different shapes; thus, a piecewise function is suitable to describe it symbolically.

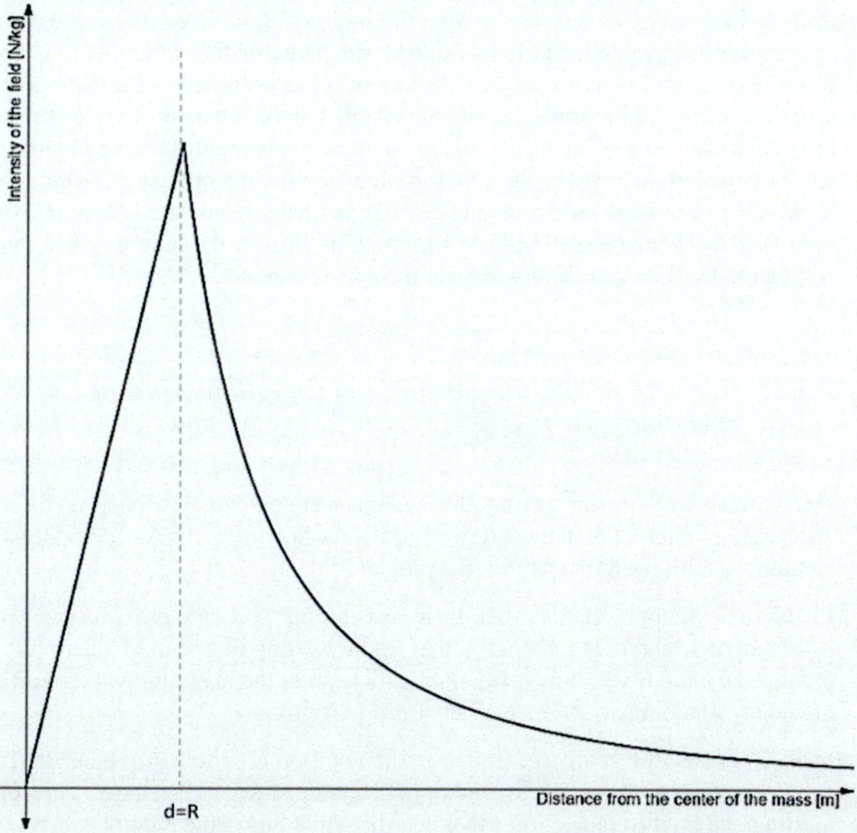

Fig. 7.2 Gravitational intensity in the function of the distance from the center for a spherical solid mass

$$a(d) = \begin{cases} \dfrac{Gm_1}{R^3} d, & \text{for } 0 \le d < R \\ \dfrac{Gm_1}{d^2}, & \text{for } d \ge R \end{cases} \tag{7.4}$$

Providing students with an expression for the intensity inside of the uniform sphere is to complete the phenomena algebraic representation. The function changes its rule at $d = R$, and this was labeled with a vertical segment on Fig. 7.2. To enrich the inquiry by mathematical reasoning, limits were used. For example, the intensity of the field at the infinite distance can be approximated by $a_1(d \to \infty) = \lim_{d \to \infty} \frac{Gm_1}{d^2} = 0$. Using a more precise interpretation of the limit value, one can claim the intensity of the field is close to zero at an infinite distance from the source producing the field. The intensity will never reach that value of zero because the expression $\frac{Gm_1}{d^2} = 0$ has no solutions when d is considered a variable parameter. This conclusion further suggests that the intensity can only be approximated to a zeroth value at an infinite distance. The analysis focused on the field interpretation when $d > R$ and a discussion of linear dependence of the field from the planet radius when $0 \le d \le R$ completed the picture. It was seen that without providing students with that part of the graph, they could mistakenly get an impression that the intensity at the sphere center has an infinite value. Such a conclusion could be inadvertently reached by the rational function behavior that is getting infinite when the denominator is getting close to zero. The scientific model to calculate the intensity inside the sphere tells otherwise; as the distance is getting close to zero, thus close to the sphere center, the field is linearly decreasing, reaching zero at the sphere's center.

7.5.2 Reconstructing the Formula to Calculate Mutual Gravitational Force

During this part of the discussion, the students were guided to reconstruct the law of gravitational interaction between two masses. Following a similar method, the discussion was followed by task-related problems.

Problem 3 Suppose that another mass m_2 is brought into the proximity of m_1, and it is being held there. Is there a force of gravity acting on the mass due to the field produced by m_1? If yes, how to algebraically express the magnitude of the force acting on m_2 assuming an absence of any third-party force?

F_{NET} denoted the net force. However, the notation F_{12} could also be used. To add a more tangible analysis, the instructor held another 1-kg mass located 1 m from the existing one (in the plane of the blackboard). While this demo did not generate data, it was to illustrate its potential applicability to do so. The instructor suggested finding an expression for the net force acting on the mass at the location as indicated in Fig. 7.3.

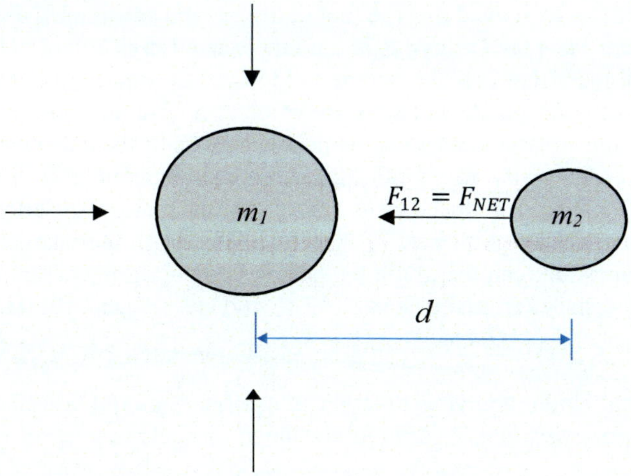

Fig. 7.3 The vector of force showing the action of the field produced by m_1 on the mass m_2

Recalling Newton's law that the net force acting on any object can be computed using $Fnet = ma$, students realize that the *force* acting on m_2, while explicitly not known at this stage of inquiry, can be found using the product of the gravitational field intensity produced by m_1 and the magnitude of the external mass m_2. The instructor emphasized that the gravitational field intensity supplied the medium for enacting the force. Nevertheless, the *strength of the field* was not the action. The gravitation field produced by m_1 coupled with the mass m_2 generated the force of action. Thus, mathematical rules paired with scientific reasoning led to a pivotal phase in the modeling process. It was hypothesized that a lack of establishing this connection earlier in the students' knowledge diverted their attention from selecting a correct response to pretest Problem 1 and 2. Omitting this reasoning essential phase might also account for students' difficulties with problem-solving when the formulation of F_{NET} acting on an object in *outer space* is considered; students do not realize the importance of the gravitational field when formulating such equations. Continuing this line of inquiry, since the intensity of the field produced by m_1 at that specific location is $a_1(d) = \frac{Gm_1}{d^2}$, and $F_{NET} = m_2a_1(d)$, then the net force acting on $m_2; F_{NET}(d) = m_2\left(\frac{Gm_1}{d^2}\right) = \frac{Gm_1m_2}{d^2}$. The students have emphasized the direction of the net force following the direction of the gravitational field intensity at that location. To parallel the notations with the graph, the instructor expressed the net force using a bi-numerical notation:

$$F_{12}(d) = \frac{Gm_1m_2}{d^2} \tag{7.5}$$

It was further explained that the first digit, $_1$, of the subscript $_{12}$ denotes the *mass that produces the gravitational field,* and the second digit refers to the labeling of the external mass that is placed in the field of the first mass. While using cascade covariation and composite functions, the students realize how one parameter affects the other. Such covariational entanglement led the analysis to reconstruct Newton's law of universal gravity. Its significance is rooted in the mutual interaction between the gravitational field produced by m_1 and an external mass, m_2 inserted into the field.

To highlight the significance of the graph (Fig. 7.2), the instructor posited the following question: can the graph be used to compute force acting on any known mass at any distance from the mass m_1? The students realized that this was possible. A graph of $F_{12}(d) = \frac{Gm_1m_2}{d^2}$ could be easily generated by referring to students' mathematical skills, applying function transformation, and considering Eq. (7.3) so-called *parent function*. This part of the inquiry revealed an opportunity to develop a bi-numerical notation further and denote the force as $F_{12} = F_{NET}(d) = \frac{Gm_1m_2}{d^2}$.

To parallel the exploration, the vectors of the field produced by m_2 were labeled, and an algebraic representation for the gravitational field was formulated as $a_2(d) = \frac{Gm_2}{d^2}$. Applying the process of finding the net force on m_1; the formula took the following shape $F_{21}(d) = m_1a_2 = m_1\left(\frac{Gm_2}{d^2}\right) = \frac{Gm_1m_2}{d^2}$. This statement assured that the magnitudes of the forces of mutual attraction are equal, disregarding the sizes of the masses. This was a surprising conclusion as the pretest has shown that most students thought that the larger the mass, the larger the force is, which perhaps was accounted for the misinterpreting of the idea of *force acting at a distance*. The equity of the forces $F_{21}(d) = F_{12}(d)$, despite different masses, could also be supported by Newton's 3rd law of motion. A pictorial representation of the conclusion is illustrated in Fig. 7.4. To further visualize the differentiation between the forces and their origins, different colors were used. The red color denotes field and forces due to m_1, and the black color represents fields and forces due to m_2.

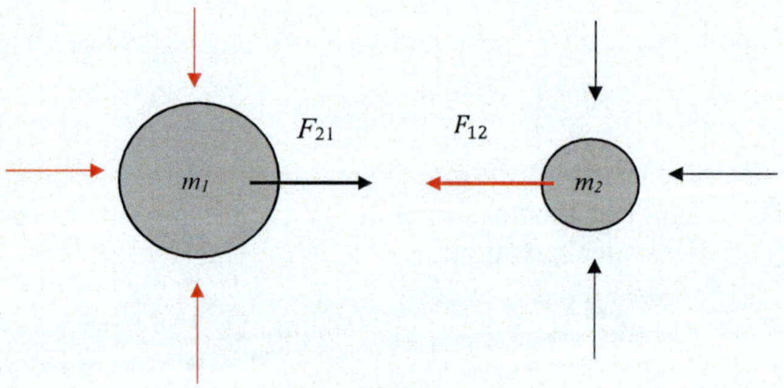

Fig. 7.4 Vectors of forces of attraction between two masses and fields

Students realized that the field is to be computed first at a distance, *d*, from m_2, before computing the net force acting on the external mass, m_1. While such scaffolding is not necessary for calculating forces between the objects, emphasizing this reasoning was to serve as a reference for applying the principle of superposition of fields and undertaking more advanced computations of net field and force when the system consisted of more than two masses. Extracting this operation also emphasized that the coexistence of fields' intensities plays a pivotal role in inducing gravitational force and describing objects' motion. Foremost, in the gravitational field, gravitational force can be described as an action produced by the gravitational field on another mass. It is apparent that to find the mutual force of attraction, the magnitudes of both masses and the distance between their centers are sufficient, though the modeling process was to amplify the effects of gravitational fields and to have students realize that this dominated the scientific underpinning of the phenomenon. If the mass is not present at the given location in space, that space does not produce the necessary field to attract other objects. Thus, the primary condition for gravitational force is the mass and the gravitational field it produces. Another conclusion that emerged from the analysis was that mass cannot interact with its gravitational field — only mediating between the field and an external mass placed in the field generates the force of gravitational attraction.

7.6 Analysis of Pretest - Posttest Results

While the pretest results were used to guide the instructional unit, pretests and posttest results were clustered in one section to allow for better comparisons of the results, thus allowing for a better assessment of the instructional unit.

7.6.1 Analysis of the Pretest Results

There were three pretest problems that the students were to answer. Problem #1 and problem 2 required a qualitative response, and Problem #3 was a multiple-choice question that required a selection of a correct answer. The author of this study designed these problems and consulted with colleagues who teach similar courses to target conceptual understanding of the differences between gravitational force, gravitational field, and the covariational interpretation of Newton's law of universal gravitation.

Question 1 The force of gravitational attraction between two masses m_1 and m_2 can be calculated using $F = \frac{Gm_1m_2}{d^2}$.

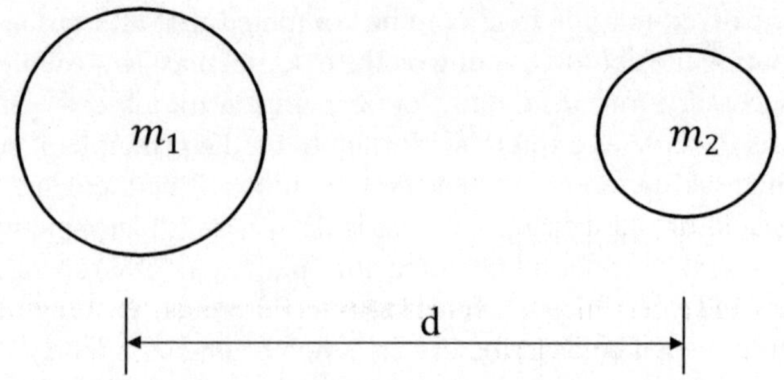

Why are the forces of the same magnitude even though one of the mass is larger than the other? Support your answer. In this diagram, $m_1 > m_2$.

The purpose of the questions was to determine if the students realize that the forces are products of two parameters mass and external fields. More specifically, the smaller mass and the larger gravitational field will produce the same force as a product of the more significant force and the smaller gravitational field.

Question 2 A mass M is held stationary, as shown in the diagram below.

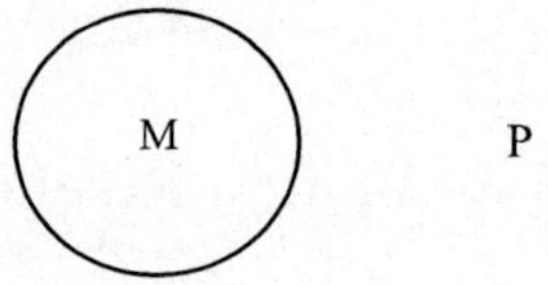

Does the mass exert a gravitational force at the location labeled P?

Students confuse gravitational force and gravitational field. The purpose of this question was to find out if there is a need to address this deficiency in the instructional unit explicitly. To avoid confusion about a presence of a mass at that location, no dot was marked.

Question 3 Two masses m_1 and m_2 are separated by a distance d and held stationary. What force is being computed by using $F = \frac{Gm_1m_2}{d^2}$?

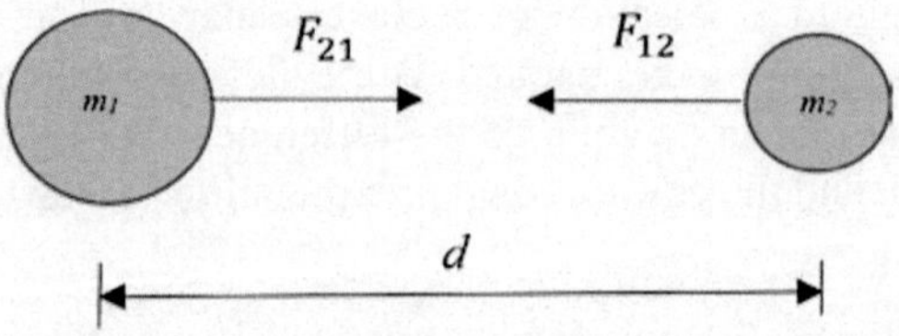

A. The net force acting on the masses.
B. Only the force that the larger mass exerts on the small mass.
C. The magnitude of the force F_{21} and F_{12}
D. None of the above because the subscripts in the general formula are not provided.

When asked for the gravitational force between two objects, students often blindly use the formula Eq. 7.1 and compute the force without realizing that they calculate the magnitudes of either F_{21} or F_{12} or each simultaneously. The purpose of this question was to find out if the students realize this nuance. The answer to this question determined the degree of details presented while deriving the equity of these forces.

Question 1 was correctly answered by three students (N = 3, 12%), and one student supported the answer by the equity of the forces. These students used Newton's 3rd law, which was accepted as correct. The underlying interaction between fields on external masses and the structure of Newton's law of universal gravity was silent in these responses. Selected students' answers are provided in Table 7.1. The correct answers were labeled with an asterisk (*).

Question 2 was answered incorectly by most students (N = 18, 72%) who suggested a gravitational force was acting at the point P, and the rest of the group (N = 7, 28%) claimed otherwise. While the question was asked to support the answer, only six students (24%) did so. Their responses are verbatim listed in Table 7.2. The first four responses (1–4) support the incorrect claim about the gravitational force's presence at that location, and the responses (5–6) support a claim of a lack of such force.

From these responses, one could infer that some of the students associate the gravitational field with gravitational force even though the question addressed the force explicitly at the location d distant from the mass-producing the field. It was inferred that the mistake could be attributed to either students' assumption that there must be a mass at the location P or due to a weak understanding of the underlying algebraic structure of the formula representing the law. None of these students attempted to support the answer by using the formula $F = \frac{Gm_1m_2}{d^2}$ and showing that it would produce a zero force when there is no other (e.g., mass $m_2 = 0$) in the proximity of m_1.

Highlighting and realizing the differences between field and force properties is fundamental for problem-solving in physics. It is concluded that the traditional textbook presentation of the law hinders the fields' effects and does not sufficiently address its fundamental effects on the force existence due to an external mass.

The correct answer for this Question 3 was choice C. Only one of the students (N = 1, 4%) answered the question correctly. Most of the students (N = 20, 80%)

Table 7.1 Students' pretest responses to Question 1

Student	Response
1	The forces are the same because the net force is zero between these objects.
2*	Due to Newton's 3rd law: action is equal to reaction.
3	The forces are the same because the value of G is constant. Each object is affected, but the force is the same. The force of gravitational pull connects these objects.
4	Because they have the same distance and gravitational pull.
5	The forces are the same because the two masses exchange forces making it so that even though one is bigger than the other.
6	There are the same forces. Just as the force of tension is the same on connected objects.

Table 7.2 Samples of students' pretest responses to question 2

Student	Response
1	Yes, there will be a small force.
2	Yes, just the downward force of gravity.
3	Yes, there is a gravitational pull on everything in space.
4	Technically, there will be a force, but very small.
5*	No, because there is no mass for the gravity to act on.
6*	No, because P does not have a mass and the gravitational force exists between two masses

claimed that the law of universal gravitation is used to compute the net gravitational force and selected the choice A. This answer was not marked as correct because net force is calculated as a sum of forces acting on a specific object, not between the objects. A small subgroup ($N = 4$, 16%) selected choice B. While partially correct, this choice did not include the possibility of simultaneously computing the force that the smaller mass exerted through its field on the larger one.

These results further supported the hypothesis that the problem with formulating expressions for the net force when dealing with forces generated by gravitational fields might be rooted in a lack of understanding of the cause of the force *as action-at-a-distance.* There can be other factors causing students' diversion from selecting choice C. For instance, a lack of understanding of the meaning of the net force or a lack of interpretation of the subscript notations; F_{21} and F_{12}. Nevertheless, the pretest questions' analysis suggested that the traditional method of presenting Newton's law of universal gravitation does not support students' natural intuitions about that phenomena. Hestenes (2010) argued that students' intuitions are essential cognitive resources developed through years of real-world experience and that "The chief problem in learning physics is not to replace intuitions but to *tune* the mapping to produce a veridical image of the world in the imagination" p. 27. The pretest results served as a basis for analyzing students' prior intuitions and prompts for formulating interventions. The pretest was not handed back to the students, and the correct answers were not revealed until the instructional unit was delivered and the posttest administered.

7.6.2 Analysis of the Posttest Results

The posttest questions were assigned on a summative assessment and other problems about *Rotational Motion* and the *Law of Universal Gravity*. The problems were similar in structure to assess a change in students' thinking. To reduce the effect of *fact recalling*, modifications of the question contents were made. For example, in question #2, the magnitude of the mass and the distance were provided quantitatively. In question #2, the subscripts were changed to A and B, and the symbol for the distance has been replaced by *l*.

Question 1 was correctly answered by fifteen (N = 15, 60%) of these students. While the emphasis during instructional intervention was the interaction of fields with masses, students often used Newton's III law of motion to support their answers. This law was also mentioned during the lecture but was labeled as an additional law that supports the equity of the masses. It seemed that presenting other covariate relations, e.g. if $m_1a_1 = m_2a_2$, then $\frac{m_1}{m_2} = \frac{a_2}{a_1}$ could better highlight the covariate.

Question 2 was answered correctly by (N = 18, 72%) of the students; most of them (N = 10, 56%) provided verbal supports for their answers. Out of 25 students, seven (N = 7, 28%) did not provide a correct answer. Some of the responses are verbatim listed in Table 7.3. The responses from (1 to 4 marked with an asterisks) were considered as displaying acceptable reasoning, whereas responses from 5 to 10 were marked as incorrect (Table 7.3).

Twenty-one students selected the correct answer for Question 3 (N = 21, 84%), and four students selected either choice A or C, which showed a visible improvement compared to the pretest.

A general question emerged what factors could still detour some students from differentiating between gravitational force and field (Question 2)? To find a possible explanation, these students' verbal responses were further scrutinized. Some of them (see response #6 and #10) associated the formula for gravitational field $g = \frac{Gm}{d^2}$ with the formula for gravitational force. Two inferences could be drawn from this misinterpretation; (a) students, and perhaps the instructors, often use the

Table 7.3 Posttest responses to question 2

Student	Response
1*	No, there is only a gravitational field. Until a second mass is introduced, there will be no force of gravity produced. To have a force of gravity produced, there must be two objects.
2*	There is no force acting at d = 1 *m*. There is a gravitational field from the mass, but the field needs another mass to act as a force on that object.
3*	There is no gravitational field, but there is no second object for the field to exert a force.
4*	The gravitational field is present everywhere and draws objects together, but the F_G will only be acting if there are two objects.
5	Yes, because at any distance, there is a force of gravity according to $F = \frac{Gm_1m_2}{d^2}$.
6	Yes, $g = \frac{Gm}{d^2}$ calculates the force of gravity (g) at a distance d from the mass.
7	Yes, because the gravitational field from 3 kg extends that far.
8	Yes, there is always going to be a force of gravity from that mass.
9	Yes, $\frac{Gm_1m_2}{d^2} > 0$ given *m* and d $\neq 0$.
10	Yes, because every object has a gravitational field. The only thing is that the force of gravity at this distance so weak (used $\frac{G3}{1^2}$ to support the magnitude).

term *gravity* when they mean either intensity of the field or the magnitude of gravitational force. The lack of verbal precision might transfer to students' inability to differentiate between the force of gravity and gravitational field, (b) in the standard curriculum; students first learn the formula $F_G = mg$ to find the force of gravity, which unintentionally might suggest that when mass is used in the formula, like in $g = \frac{Gm}{d^2}$ gravitational force is also computed. A suggestion for a more profound differentiation on the mediating parameters in each equation, in $F_G = mg$; g is the external field and in $g = \frac{Gm}{d^2}$; g is produced by the mass m emerged. The effects of the mass m, are unparallel, and highlighting more the differences might help.

Several students (e.g., response #8, Table 7.3) still associated predictable action on a mass without realizing that in the problem, only one mass was given. Indeed, a qualitative study allowing for more detailed coding of students' verbal responses would allow for a more precise location of the misconception.

A high percentage of correct answers for question #3 (84%) on posttest was promising. Students realized that using the formula $F = \frac{Gm_A m_B}{l^2}$ generates the magnitudes of both forces, the gravitational force acting on the mass m_A and the force acting on m_B simultaneously. The physics curriculum does not provide many examples of such formula outputs interpretations, therefore addressing this nuance explicitly turned to be a move in the right direction.

7.7 Conclusions and Suggestions for Further Research

This study presented a blended hypothetical mathematical modeling approach of introducing Newton's law of universal gravity. Rather than the masses only, the gravitational fields coupled with the external masses were exemplified. The unit's far-reaching goal was to have students realize the differences between gravitational forces, fields, and the concept of gravity as a phenomenon. While it is true that the advantage of the field treatment might be genuinely appreciated in classical physics on account of many masses/charges, this study focused on analyzing only two masses. This was purposeful because using fewer objects reduced the cognitive load of applying advanced vectorial algebra, thus allowed focusing on the conceptual understanding. Adding more masses to the system will include a superposition principle for fields and forces but will not change the analysis's line reasoning.

The prior research highlighted students' misconceptions about gravity as one of the obstacles in teaching this physics section. It appeared that using precise language to address either gravitational field, force, or the phenomenon of gravity might, to a certain degree, support eliminating these misconceptions. The precision of indicating the quantity of interest evokes its property, units, and contribution to its understanding. the quantity described.

Linking the effects of the gravitational fields and their algebraic embodiments to calculate the force of gravity was to provide background for more advanced explorations of fields with subsequent inclusions of kinematics analysis. While the physics curriculum does not suggest more insightful analyses in these regards in introductory physics courses, students appreciated the opportunities of being introduced to methods that they will use in E&M. It was hoped that the unit equips students with methods to undertake more complex parallel computations of fields and forces in E&M.

There are certain limitations of the study. One of them is a small sample size of the experimental group and a lack of a controlled group that would allow for computing effect sizes of the treatment. It is seen that the students' posttest responses provided evidence that using this study treatment establishes the basis for further unit development and that this enterprise is worthy of considering.

This study also generated other possible research questions. For example, (a) will introducing the formula $g = \frac{Gm}{d^2}$ before $F = mg$ better highlight their scientific underpinnings and help realize the differences between the quantities each calculates and, consequently, help distinguish between the gravitational fields and forces? (b) do students mistakenly consider the product of Gm as an expression for gravitational force? (c) to what extent would the proposed introduction of Newton's law of gravity help students understand the superposition of electrostatic or electromagnetic fields?

Another line of inquiry worthy of further examining is the students' interpretation of the phrases' *forces-at-a distance* or *action-at-a-distance* that is frequently used in physics textbooks. Does the phrase *action-at-a-distance*, if not explicitly addressed and explained by the teacher, suggest to a novice learner either presence of force just due to the field without an additional mass or force produced just by mass without a field. Should such phrase be instead augmented to *forces by action originated from the presence of the field?* Such nuance, when examined, can magnify the students' misconceptions that might not necessarily occur due to their naïve perceptions but perhaps due to how these concepts are verbalized.

Although not discussed during the instructional unit, the proposed reasoning pathway also provided opportunities to differentiate between the terms *inertial* and *gravitational masses*. Can a mass, broadly defined as a property of a physical object, exhibit simultaneously inertial and gravitational nature? Indeed, there are not many real scenarios for a mass to exhibit such duality, which in short can depend on the specific nature of the mass interaction due to its position with reference to the direction of motion. If a mass causes an object's resistance to change its state of motion, it will represent its inertial properties. For example, if a net force F acts on a mass m, then the mass m represents the object's inertial mass. If the mass, m, is considered a source of the gravitational field, and thus gravitational force acting on another mass M, both these masses will experience their gravitational nature. However, the true notion of this differentiation perhaps lies in what mass is considered to be in motion. For exmaple, if the gravitational force acting on mass M is considered, M will exhibit its gravitational effects because it contributes to the magnitude of gravitational force $F = \frac{Gm_1M}{d^2}$. If the acceleration of the mass M is concerned, the M will also exhibit its inertial effects; $Ma_1 = F$. Since $Ma_1 = \frac{Gm_1M}{d^2}$, one will notice that the acceleration of the mass M is independent of its inertial mass because M is a common factor on both sides of the equation, and it can be canceled out, resulting in $a_1 = \frac{Gm_1}{d^2}$. This is instead a unique case that perhaps does not have an equivalent when the object moves along a surface.

It is seen that the differentiation between inertial and gravitational mass does not affect the understanding of the law of universal gravity but instead enhances it, and this inquiry provides another opportunity for the study extension.

References

Asghar, A., & Libarkin, J. C. (2010). Gravity, magnetism, and. *Science Educator, 19*(1), 42–55.

Cao, Y., & Brizuela, B. M. (2016). High school students' representations and understandings of electric fields. *Physical Review Physics Education Research, 12*(2), 020102.

Galili, I. (1995). Mechanics background influences students' conceptions in electromagnetism. *International Journal of Science Education, 17*(3), 371–387.

Giancoli, E. (2014). *Physics: Principles with applications* (AP ed., p. 357). Pearson.

Hamming, R. W. (1980). The unreasonable effectiveness of mathematics. *The American Mathematical Monthly, 87*(2), 81–90.

Hestenes, D., Wells, M., & Swackhamer, G. (1992). Force concept inventory. *The Physics Teacher, 30*(3), 141–158.

Hestenes, D. (2006). Notes for a modeling theory. In *Proceedings of the 2006 GIREP conference: Modeling in physics and physics education* (Vol. 31, p. 27). University of Amsterdam.

Hestenes, D. (2010). Modeling theory for math and science education. In *Modeling students' mathematical modeling competencies* (pp. 13–41). Springer.

Kavanagh, C., & Sneider, C. (2007). Learning about gravity I. Free fall: A guide for teachers and curriculum developers. *Astronomy Education Review, 5*(2), 21–52.

Maxwell, J. C. (1873). *A treatise on electricity and magnetism* (Vol. 1). Clarendon Press.

Maloney, D. P. (1985). Charged poles? *Physics Education, 20*(6), 310.

Newton, I. (1999). *The principia: Mathematical principles of natural philosophy*. Univ of California Press.

Pocovi, M. C., & Finley, F. (2002). Lines of force: Faraday's and students' views. *Science & Education, 11*(5), 459–474.

Saeli, S., & MacIsaac, D. (2007). Using gravitational analogies to introduce elementary electrical field theory concepts. *The Physics Teacher, 45*(2), 104–108.

Serway, R. A. (1996). *Physics for scientists and engineers* (4th ed., p. 391). Saunders.

Shadish, W. R., Cook, T. D., & Campbell, D. T. (2002). *Experimental and quasi-experimental designs for generalized causal inference*. Houghton Mifflin.

Smith, G. E. (2002). From the phenomenon of the ellipse to an inverse-square force: Why not? In *Reading natural philosophy: Essays in the history and philosophy of science and mathematics* (pp. 31–70).

Sokolowski, A. (2008). Using gravitational analogies to introduce electric field theory concepts—A response. *The Physics Teacher, 46*(3), 132–133.

Temiz, B. K., & Yavuz, A. (2014). Students' misconceptions about Newton's second law in outer space. *European Journal of Physics, 35*(4), 045004.

Tipler, P. A. (2008). *Physics for scientists and engineers* (5th ed., p. 367). W.H. Freeman.

Chapter 8
Parametrization of Projectile Motion

8.1 Prior Research Findings

Search engines such as Ebsco, JStar Google Scholar, or ERIC returned a substantial amount of literature documented students' misconceptions about the causes of parabolic curves of objects undergoing projectile motion. The most common misconception was that objects need some *impetus* that keeps them moving in a parabolic curve. Subsequent theories of motion provide qualitative rather than quantitative supports focusing students' attention on realizing that object's paths of motion in the gravitational field are parabolic (Jörges & López-Moliner, 2019), which might be a reason for such perception. Two-dimensional motion typically encompasses circular motion or its part (e.g., the motion of a pendulum), parabolic motion (e.g., projectile motion with varying velocity components), or an elliptical motion (e.g., planetary motion). Mihas and Gemousakakis (2007) formulated several recommendations that are to help students understand two-dimensional analysis. For instance, they highlighted the property that an object's vector of velocity is always tangential to the path of motion or that centripetal acceleration is perpendicular to tangential velocity if the tangential speed is constant. Springuel et al. (2007) also focused on students' understanding of objects' accelerations in curved paths. They applied cluster analysis to group students' answers by common traits to learn how they conceptualize vectors to depict acceleration. While objects undergoing a two-dimensional motion are typically projected from a stationary launchpad and analyzed from a stationary frame of reference, several studies conceptualized projectile motion by situating the launchpad on a moving platform (see Millar & Kragh, 1994; Klein et al., 2014). A moving launchpad was to have students realize that an object released from a moving carrier possessed an initial (usually horizontal) velocity equal to the carrier's velocity. Such situating while contextually rich added a concept of a relative motion. Many students believe (Prescott & Mitchelmore, 2005) that these projectiles do not possess forward velocity when released, and

A. Sokolowski, *Understanding Physics Using Mathematical Reasoning*, https://doi.org/10.1007/978-3-030-80205-9_8

consequently, they think that dropped objects move backward or fall straight down. By using moving horizontal platform projectile motion displays a great deal of exploratory nature. Still, it also depends on understanding another motion; the idea of relative motion perhaps unintentionally adds another conceptual challenge to understanding projectile motion.

Several studies suggest advanced technology to have students realize parabolic shapes of the trajectories of the projected objects. For instance, Wee et al. (2012) suggested tracing the motion path by using computer software. Jimoyiannis and Komis (2001) used computer simulations and vector analysis to enhance students' conceptual understanding of the motion in the gravitational field to overcome their cognitive constraints. In a similar accord, Klein et al. (2014) used tablet computers to record and analyze a ball's motion thrown vertically from a moving skateboard. The software allowed to generate data for the object's position, plot the data to generate graphs for motion components and find their function equations. Some studies included a *damping factor* in a form of a force proportional to the square of the object's velocity with projectile motion (see Das & Roy, 2014). Espinoza (2004) sought to enhance understanding of projectile motion by linking it with an understanding of the momentum of a projected object and found out that students who studied momentum before forces were more successful in demonstrating a correct direction of the momentum, by referring to the tangential nature of the object's velocity. This idea followed Gamble (1989), who found out that reconstructing physics knowledge and making it more relevant to mechanics concepts benefits physics understanding.

Students' understanding of projectile motion is often tested on advanced nationally administrated physics exams in the USA. Reviewers of such exams voiced students' lack of clarity on understanding the different modes of movement in the horizontal and vertical direction that a projected object possesses. For instance (College Board, 2015), the reviewers pointed out that students mix kinematics horizontal and vertical components; they confuse velocities or apply acceleration due to gravity to the motion's horizontal components which reflects on their poor understanding of the motion properties.

The survey of the literature on teaching and learning projectile motion displays vast arrays of didactical enterprises ranging from eliminating misconceptions and using technology to generate mathematical representations of the motion. Nevertheless, little attention is given to have students construct algebraic representations of the motion using data and apply mathematical reasoning to uncover the motion properties. Students' abilities to describe projectile motion using mathematical reasoning appeared to be overshadowed by the conceptual description of the motion. It is seen that understanding the techniques of resolving the position velocity and acceleration functions into components is paramount to handling problem solving and graphing projectile motion component functions. It is further hypothesized that setting up the development of students' perception of two-dimensional motion using parametric equations (PE) will come across these recommendations.

8.2 Theoretical Framework

The theoretical framework is posited to take a more comprehensive view. It will begin from situating the motion in specific categories and then it will embrace it in corresponding algebraic representations. Such a view is to broaden the perspective on multidimensional motion and simultaneously generalize the patterns.

8.2.1 Categories of Motion Studied in High School and Undergraduate Physics Courses

Kinematics as a study of motion is usually initiated from one-dimensional and progresses into two-dimensional motion with projectile motion representing a two-dimensional motion. While studying circular, harmonic, or wave motion, physics resources do not necessarily inform students that these motion types can also be classified as two-dimensional and that similar mathematical tools can be used to analyze these motions. Explicating the motion type does not change the general teaching approaches, yet it would help present these phenomena as other examples of applying parametric equations and help with the generalization of the mathematical methods. Figures 8.1 and 8.2 are examples of such classifications, and they were presented to the students within the first week of study. Figure 8.1 summarizes types of motions considering the degree of freedom of object's movement and the complexity of the motion path. These categories are not exhausted, and indeed, both can be enriched. They are to help students realize a progressive complexity of the types of motion and, followed more advanced mathematical apparatus to quantify the motion.

Projectile motion belongs to a class of two-dimensional motion like parabolic motion, circular and harmonic motion. All three types of motion require applications of PE to analyze these motions' properties. One-dimensional motion called rectilinear is the most fundamental category of motion from which kinematics is usually initiated. Three-dimensional motion can include all two-dimensional motions when the third dimension is added. Another category that emerged is the possible modes of object movement (see Fig. 8.2).

When the mode of movement is concerned, projectile motion can be described as uniform in the horizontal direction and non-uniform in the vertical direction represented by parametric equations also called vector functions. Realizing these categorizations should support mathematical descriptions of the object position, velocity, and acceleration and enhance subsequent graph sketching.

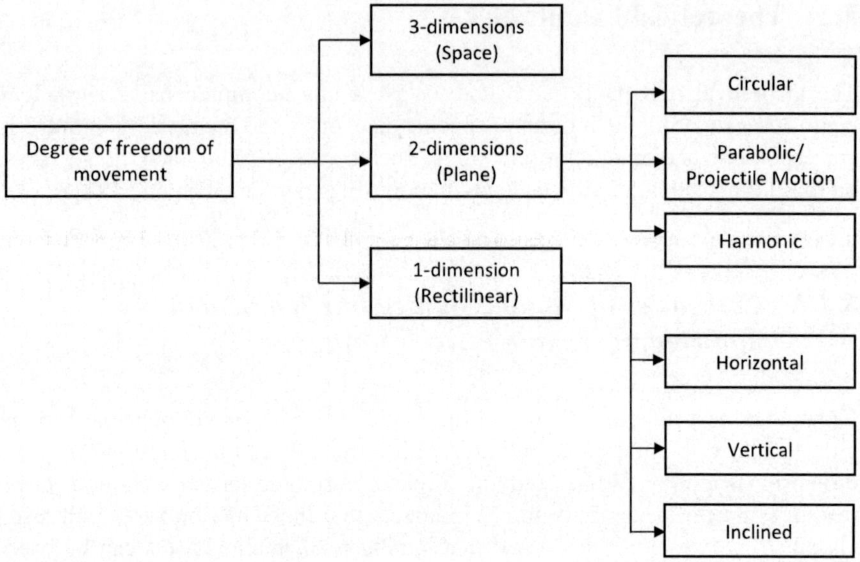

Fig. 8.1 Motion categorized by degree of movement freedom

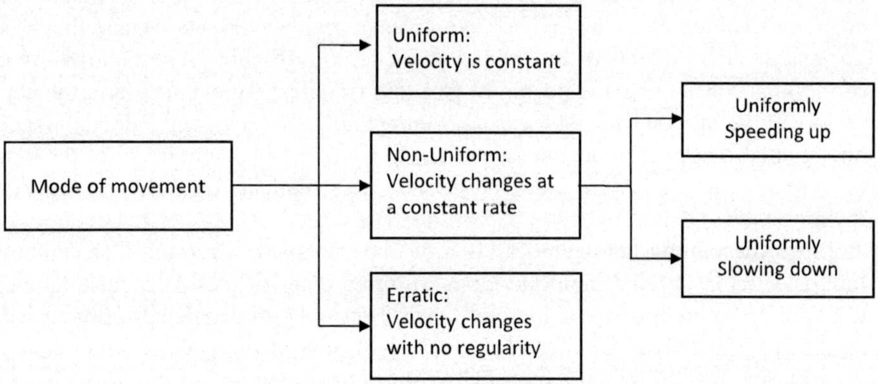

Fig. 8.2 Classifications of motion by the mode of movement

8.2.2 Why Parametric Equations?

Projectile motion is easily observable. It can be demonstrated in any classroom setting, yet the motion's scientific underpinnings, coupled with resolving it into vertical and horizontal components, are not readily observable in such demonstrations. Projectile motion is associated with motion in the gravitational field. A typical trajectory's parabolic shape might suggest that the gravitational field is the only covariate affecting the motion. It is to note that in typical problems on projectile motion, students are not explicitly directed to constructing component position or velocity functions and using these entities to answer questions. It is seen that emphasis on constructing such algebraic entities should take priority to develop in students' realities a prompt for such actions and succeed in solving projectile motion problems.

Developing this essential feature of the motion and placing more emphasis on students' skills of resolving the motion into components is not being highlighted in the current research. Therefore, to have students realize that projectile motion can be considered as being composed of two covariate modes of movement; horizontal and vertical appeared a priority. Research shows that that students mixed the components of velocities, e.g., they used horizontal components of the velocity to find the objects' vertical position. Such mistakes evidence a poor scientific understanding of the motion and perhaps a lack of foundation of familiarity with the methods of resolving the motion in components. Although taught in mathematics courses, PE properties, and their structures are not explicitly highlighted in physics textbooks and curricula, and thus, research on using PE to understanding motion is limited. Resolving the motion into components is typically rooted in finding the initial velocity components and enacting subsequent formulas. Mathematizing is reduced to formulating equations without deeper elaborations while it is possible to do that. While the component position functions $x(t)$ and $y(t)$ present an excellent opportunity to use covariational reasoning to support the phenomena' interpretation, this opportunity is lost.

This study emphasizes constructing a mathematical representation of projectile motion considering the motion's parametrization as a determinant at hand. Its main flow is designed using the scheme developed in Chap. 4. However, the learning theory chosen to guide the treatment design and the students' knowledge acquisition is the *constructivist learning theory,* whose foundations are summarized in Sect. 8.2.3.

8.2.3 Foundations of Constructivist Learning Theory

The cognitive effects of constructivism are highly appraised and advocated by von Glasserfield (1991). Constructivism is a theory in education that prioritizes gaining knowledge by constructing new understandings and integrating it with what learners already know (Jonassen, 1991). Knowledge acquired by constructing is more accessible, absorbed, stored, and retrieved from learners' long-term memory. Following this theory, Clark and Mayer (2016) suggested that effective knowledge acquisition is characterized by active processing. Learning occurs when people engage in appropriate cognitive tasks that require organizing provided information into coherent structures. While this learning theory is not visible in physics education, it is often applied in mathematics, helping students make sense of abstract mathematical structures. Since the structure of PE will be brought up during the treatment, using this theory to help students understand the suitability of these algebraic structures emerged as a prompt in the right direction. The students will actively construct PE for a given path of a projected object generated by a simulated experiment. Thus, students will derive algebraic equations for the motion using data generated from the observable experiment. Such task structuring creates appealing conditions for merging scientific and mathematical reasoning, thus nurturing new knowledge.

The advantage of the simulated dynamic representations is their potential also highlight the time parameter, t, and the combinations of x and y components of the vectorial kinematics quantities to form the object's path of motion. Observing how a change of time affects the object's position in both the x and y directions should highlight the mutual effect of time on both these motion components.

While learning or constructing mathematical structures involves manipulating mathematical symbols, graphical representations, especially their dynamic embodiments, have the highest potential to help students realize their scientific meanings. Being exposed to graphical representations, the learners acquire a set of tools that expand their capacity to model and interpret physical, social, and mathematical phenomena (NRC, 2012). Formulating internal or mental representations through understanding their external embodiments and retrieving their mental pictures plays an essential role in communicating messages in science and mathematics. Von Glasersfeld (1991) posited that the environment is construed by how one's internal representations are formulated. Enabling these experiences by engaging and intellectually stimulating learners through carefully designed learning environments deems nurturing effective learning.

8.3 Methods

8.3.1 Study Description and the Research Question

This study can be classified as one group experimental with posttest evaluation (Shadish et al., 2002). While a pretest would allow for a more accurate treatment evaluation, some of the study participants did not possess a physics background. Thus, the pretest would not objectively reflect on their prior physics experiences and provide reliable data for the treatment evaluation. The following question guided the study:

- *Will situating the learning of projectile motion as a two-dimensional motion representation help students realize its mathematical and scientific underpinning?*

The evaluation of the research question was assessed by analysing students' responses to the following qualitative questions:

- ***Question 1:*** *Projectile motion can be analyzed using its components. Evidence of what experiment/demonstration from the listed below convinced you that the motion has different modes of movement in the vertical and horizontal direction:*

    (a) *Observing the object move along its trajectory.*

(b) *Working on the virtual lab and using the path of motion to resolve the path into component function.*
(c) *Other experience from an introductory physics course.*

- ***Question 2:*** *Explain the process of analyzing mathematically projectile motion.* It was concluded from the prior research findings that confusion with using x and y components of the motion resulted from a poor understanding or a complete lack of possessing skills of constructing and using parametric functions to represent projectile motion. Therefore, mathematical reasoning and property of parametric equations came forth in this study. Students answered both questions after completing the treatment.

8.3.2 The Participants

Twenty (N = 22) first-year university physics students participated in this study. Seventeen of these students (N = 17) had a prior precalculus background and studied PE in their math classes, ten of these students (N = 10) were enrolled concurrently in a calculus course. Twelve of them (N = 12) were males, and eight (N = 8) females. Most of these students possessed a prior high school or college physics background.

8.3.3 Lecture Component Sequencing

Concepts in mathematics are usually introduced as context-free; thus, presenting these ideas as supporting physics understanding is always welcomed by physics students. While these students did have a precalculus background where these ideas are taught, PE and their applications were introduced during the first days of studying kinematics in this physics course. Upon introducing PE, these structures were used periodically to enhance one-dimensional motion by including the missing motion dimension; thus, $x(t)$ or $y(t)$ even though they were settled as either $x(t) = 0$ or $y(t) = 0$. Introducing the concept of parametric equations before studying projectile motion was to reduce its cognitive load and to better prepare the students to apply these mathematical structures in more advanced scientific contexts. The idea was introduced gradually, and its phases were embedded as they fitted the traditional scope and sequence of kinematics. The learning objectives embedded to enrich the traditional kinematics curriculum by using parametric equations were the following:

1. Enriching description of one-dimensional motion by using two-dimensional representation.
2. Resolving into components a uniform motion of a car moving in two-dimensions.

3. Constructing position functions for projectile motion given by its path of motion.
4. Constructing position functions for various types of two-dimensional motion.

8.3.4 Topics Embedded within the Curriculum to Enhance the Treatment

Objectives (1), (2), and (3) were considered prerequisites to Objective (4), which constituted the core lab activity of the study. It was hypothesized that inserting simple two-dimensional motion cases within the existing curriculum will provide students with the necessary mathematical tools to understand the algebraic underpinnings of the projectile and, eventually, circular, and harmonic motion. The subsections that follow describe how each of the objectives was introduced.

Objective 1 Enriching description of one-dimensional motion by using a two-dimensional representation.

PE is one of the objectives covered in a traditional Precalculus course. Thus, a reference to these concepts can be made. Could this be considered a prerequisite to apply these math tools in physics? Not necessarily. While vector components can depict instantaneous position, velocity, or acceleration, PE in physics courses can be considered a dynamic extension of vector components representation that allows for computing instantaneous values of these vectors at any time instant during the object motion, thus they can be called parametric functions (PF). The dynamic representations were further mapped into algebraic functions, with time t, considered the independent variable. The terms *component functions* and *parametric equations* were used interchangeably. PF cannot be narrowed to motion analysis only; in general, they can be considered mathematical tools that help describe a phenomenon from different perspectives. For example, a temperature change of water in a beaker can be measured simultaneously by two different thermometers, one in Celsius one in Fahrenheit scale, leading to concluding temperature conversion equations. Object's position undergoing two-dimensional motion can also be viewed using its component position functions, typically in the horizontal and vertical directions. In short, once the correlation between math and physics was established, the terms *parametric equations* were not used but instead, the instructor called these entities functions depicting components of an object's path of motion or components representing velocities or accelerations in both directions x and y.

More precise lines of thoughts on how PF were introduced to the physics curriculum are presented in the examples that follow. In short the learners were first presented with scenarios of various paths of object motion. A discussion of how to algebraically describe the motion led the students' thinking toward realizing that a need for more precision in motion descriptions is compulsory. While such motion descriptions are not being practiced, these didactical enhancements were purposeful on the premise that understanding the main idea of using PF will be more accessible when applied to simple motion cases. After such motion descriptions were understood, students were introduced to cases where position functions, in x and y directions, took non-zeroth forms. In any of these examples, it was essential to highlight the direction of the view used to formulate respective functions. These elements were considered the basis for enabling students to envision components of the motion and construct internal mental images of the functions and formulate symbolic representations. The view is traditionally established from the sides or /and from the top. Respective arrows labeling corresponds to the axis on which the object's position is projected. In all cases, the students were also supposed to classify the motion according to provided categorizations using Figs. 8.3 and 8.4.

Example 1 Starting from 4 *m* left, a student walks for 2 minutes along a horizontal line with a constant velocity of 2 m/s in the right direction. Formulate functions to describe the student's position in the horizontal and vertical directions.

Establishing a reference frame is necessary because that is how students initiate formulating functions in mathematics (Sokolowski & Capraro, 2013). The direction of view (projection) was also indicated to have students realize that this element helps build the mental model of the anticipated type of motion and, consequently, the function's algebraic form.

The position functions of interest were $x(t)$ and $y(t)$. The labeling in Fig. 8.3 was used to find $x(t)$ and Fig. 8.4 supported the finding of $y(t)$.

How to formulate $x(t)$? A constant rate of change of these parameters expressed as the student's speed implies a linear function $x(t) - x_i = v(t - t_i)$, where (t_i and x_i) are the coordinates of the initial point of the function. Substituting the given coordinates, the function takes the following form $x(t) = 2t - 4$. How to find $y(t)$? First students needed to formulate a mental view of the student's movement along the y-direction. Thus, determine if it changes and if it does, how it changes? It is apparent that the student's position as viewed from the left side of the picture does not change and can be described as $y = 0$.

To assure a consistent notation, the vertical position of the student was denoted as $y(t) = 0$.

In sum, both functions precisely describe the student's position while he is walking. While in math classes, the domain for PE is required to fully parametrize the motion, the time of motion $0 \le t \le 120s$ was also included. This helped realize that components of the motion share the same domain or that the time interval for both motion components is the same. This element will be more significant when the projectile motion will be discussed. A complete description of the student's position as he walks is:

$$\begin{aligned} x(t) &= 2t - 4 \\ y(t) &= 0 \qquad 0 \le t \le 120s \end{aligned} \tag{8.1}$$

An element visible in the diagram but not typically included in physics textbooks is the XY coordinates. The Cartesian plane does not often accompany physics

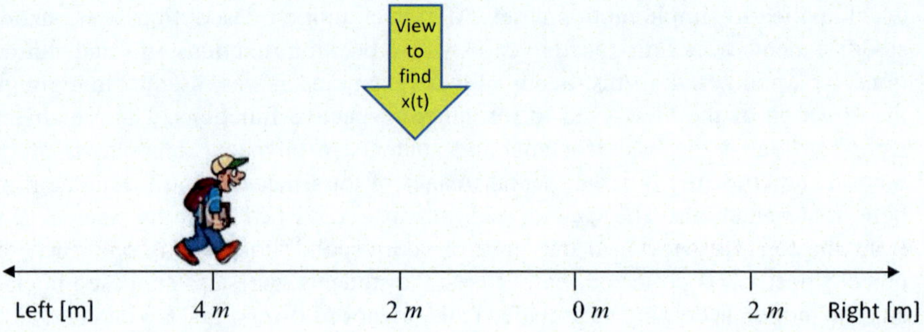

Fig. 8.3 One-dimensional frame of reference to determine the student's horizontal position

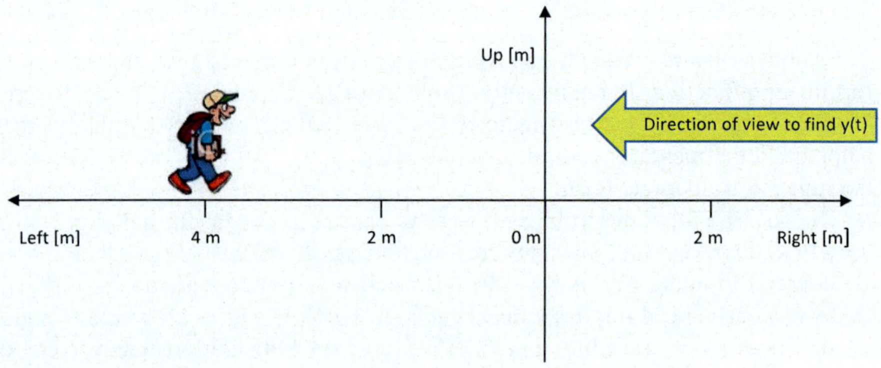

Fig. 8.4 Two-dimensional frame of reference to determine the student's position

formulas because physics formulas as static representations do not require such graphical representation. To describe projectile motion, the inclusion of the axes of projection will help construct a mental picture of the motion mode and conclude the type of algebraic functions to mathematize them.

Example 2 A baseball is projected from the ground with an initial velocity of 20 m/s. Referring to the coordinates provided, write PE for the object's position.

This experiment was demonstrated to students using the simulation (Fig. 8.5). It displayed a *free fall* vertical motion with horizontal position unchanged yet not positioned at $x = 0m$ but at $x = 2m$. Using a formula for uniformly accelerated motion in the vertical direction and a constant function in the horizontal, the students learned that:

$$\begin{aligned} y(t) &= 10t - \frac{9.8t^2}{2} \\ x(t) &= 2 \qquad\qquad 0 \le t \le 2.0s \end{aligned} \tag{8.2}$$

These functions mathematized the position of the projected object. To complete the representation the total time of the motion was calculated and included in the math description. Since some of the students were concurrently taken a calculus class, this example provided an excellent opportunity to apply their knowledge. Taking the derivative of each function, they derived the velocity functions:

$$\begin{aligned} v_y(t) &= 10 - 9.8t \\ v_x(t) &= 0 \qquad\qquad 0 \le t \le 2.0s \end{aligned} \tag{8.3}$$

Differentiating the equations again resulted in a set of acceleration functions:

$$\begin{aligned} a_y(t) &= -9.8 \\ a_x(t) &= 0 \qquad\qquad 0 \le t \le 2.0s \end{aligned} \tag{8.4}$$

These examples supplemented traditional textbook questions when one-dimensional motion was concerned. While finding both component functions was not necessarily, these examples were to expand students' view on motion and train them in considering two-dimensional analysis as richer algebraic representations to depict objects' position, velocity, or acceleration.

Objective 2 Resolving into components a uniform motion of a car moving in two-dimensions.

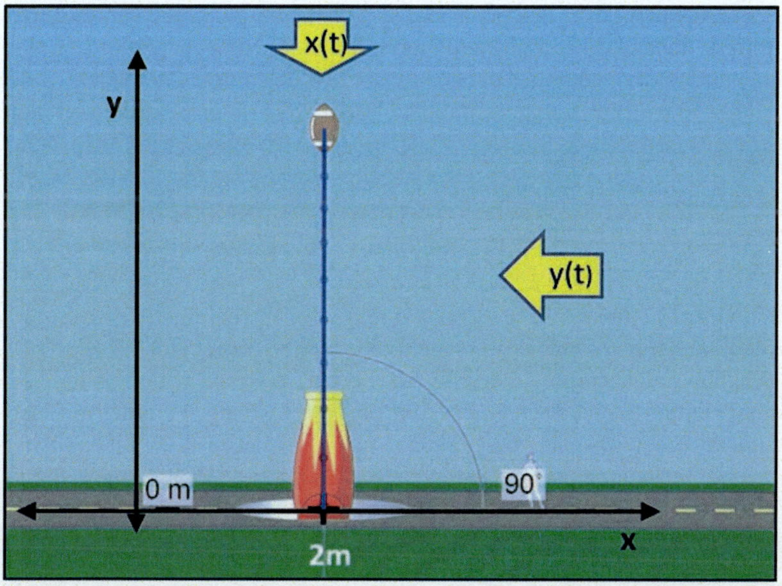

Fig. 8.5 Parametrization of vertical motion. (Source: PheET Interactive Simulations (n.d.))

The following example was discussed with students in the concluding lesson on one-dimensional motion. Its purpose was to present another technique of analyzing two-dimensional motion that involved an object's motion with a constant object yet at a certain angle concerning the horizontal axis. This case provided an excellent opportunity for extending vectors' meaning and linking these representations with their corresponding parametric functions.

Example 3 Starting from 20 km, North, a car moves in the direction 30° North of East at 60 km/h for three hours. Construct functions for the car's position along the East and North directions.

The bird's eye view of the motion is provided in Fig. 8.6. The axes of projection will be sketched *on the ground*; however, they will retain their orthogonal character.

The Cartesian plane's labeling is represented by a geographical reference in which the motion path was embraced. In this example, finding components of all vectorial quantities involved in describing the motion was necessary. The vectorial quantities in this scenario are the car's velocity and position. Thus, one needed to find the car's velocity components in the East and North directions and the components of the car's initial position in these directions.

The velocity components were: in the east direction $v_x = (60km)\cos 30^o = 52\dfrac{km}{h}$ and in the north direction $v_y = (60km/h)\sin 30^o = 30\dfrac{km}{h}$. The components of the initial position were: in the east direction $x_i = 0$ and $y_i = 20\ km$ in the north direction. The components of position and velocity were used to construct respective component functions. The vector nature of these quantities was changed to their dynamic functional representation that would, in return, allow for computing their instantaneous values at any time instant. The instructor highlighted the phase of transitioning from static vectorial values to dynamic forms of PF.

Since the car moved with a constant velocity, the component velocities were also constant. Thus, a linear algebraic model was used to represent the components of the car's position:

$$\begin{aligned} x(t) &= 0 + 52t \\ y(t) &= 20 + 30t \qquad 0 \le t \le 2h \end{aligned} \tag{8.5}$$

In a typical mathematical setting, a domain for the parameter t is or needs to be provided, which in this scenario is the total time the object is in motion $0 \le t \le 2h$. The PF allows an accurate description of the car's position within its time of movement and supports operations related to problem-solving. Parametric functions in this context can be considered *vectorial functions* because once formulated; they can be used to find the motion instantaneous position, velocity, or acceleration that possesses due to definite direction vectorial natures.

Objective 3 Parametrization of motion of an object sliding down on an inclined plane.

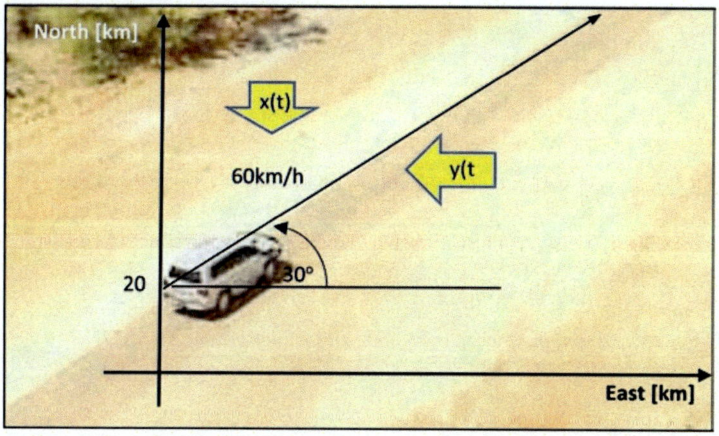

Fig. 8.6 Parametrization of inclined motion

Example 4 An Object is sliding down on a frictionless incline plane of height h and an angle of inclination θ. Establish a frame of references and write parametric equations for its position, velocity, and acceleration when it moves along the inclined plane.

This example provided students with an experience of finding components of the motion when the object's velocity was not explicitly provided and merged three representations: experimental, diagrammatic, and algebraic. This case was demonstrated to students by the instructor, and the students were guided to parametrize the motion. The available measuring devices were a ruler and a stopwatch. The instructor explained the lab's purpose and initiated the discussion on the tasks that must be taken to construct the motion position functions in x and y directions. Typically, the x axis is positioned along the incline, however standard positions of X and Y coordinates were selected to provide a bridge to the parametrization of projectile motion. Figure 8.7 depicts a diagram of the scenario's side view; the corresponding views and projections' axes are also labeled.

This example reinforced the idea of formulating all: position, velocity, and acceleration functions in x and y directions. Thus, vector components of all, acceleration, velocity, and initial position, were to be found beforehand. The process of parametrization consisted of two stages, first using general formulas for acceleration and constructing two functions that depicted the motion and then finding necessary numerical values. The students observed the motion and realized that the block accelerates uniformly along the incline. They realized that the block's acceleration could be computed by knowing the distance (the length of the ramp) and motion time. Computed acceleration was positioned along the incline. Therefore, it needed to be resolved along the x and y axes. Similarly, the velocity and the initial position needed to be embraced. The object accelerated uniformly thus, functions for uniform accelerated motion needed to be employed.

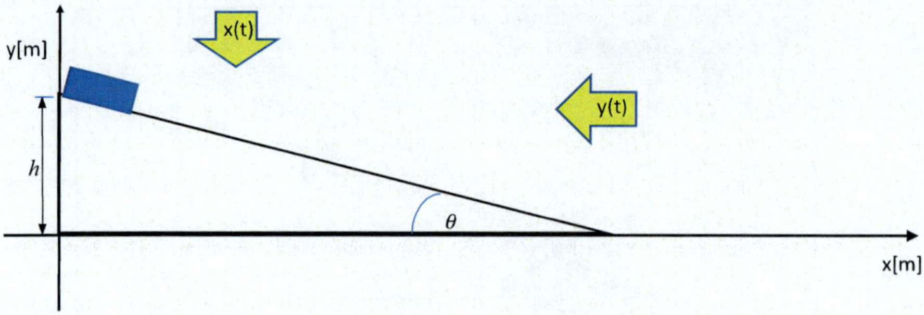

Fig. 8.7 Parametrization of object's motion on an inclined plane

The object started from rest, thus $v_x = 0$ and $v_y = 0$. By finding the angle of inclination needed to find the components of the acceleration a set of parametric position functions emerged:

$$x(t) = \frac{a(cos\theta)t^2}{2}$$
$$y(t) = h - \frac{a(sin\theta)t^2}{2} \qquad 0 \le t \le t_f \tag{8.6}$$

Students then developed velocity and acceleration component functions. Calculus students taking this physics course were asked to find the velocity and acceleration functions using the derived position functions. The parameter t_f represented the total time of the block's motion on the inclined plane. When the functions were formulated, an exciting discussion emerged about the effect of the angle of inclination on the shapes of the PF and their graphs. The angle affected the acceleration's value in both directions; thus, it also affected each parabola's vertical stretch or compression when the computed functions were considered the parent functions.

Objective 4 Using PF to construct functions for projectile motion given by its path of motion. This objective constituted a separate lab activity that is described in Sect. 8.4.

8.4 General Lab Description

The students conducted this activity in the physics lab. They took data by being provided with a snapshot of a simulated motion of a projected object demonstrated to them. They worked in groups of four to enable dialogs.

8.4.1 Lab Logistics

The idea of learning the properties of projectile motion considering its components is reminiscent of Galileo's approach to solving projectile motion problems.

However, by constructing covariate relations and realizing these functions' algebraic properties, students were to discover the unique properties of the motion when components are concerned. Linear position function represents motion with a constant velocity, and a quadratic position function represents motion with a constant acceleration. Attributes of functions such as *x* – and *y*-intercepts, slope, or maximum value coordinates were also considered. The continuity of reinforcing these properties was to establish a general strategy to mathematize multidimensional motion. The emphasis was placed on discovering the algebraic functions' attributes and using these attributes to determine the motion properties. According to the proposed modeling scheme (Part 2, Chap. 4), the general stem of instructional support was formulated focusing on merging algebraic structures and their scientific interpretations in the context of projectile motion. In its general form, the activity could also be suitable for other college-level physics courses. The differentiation would stem from the complexity of the motion under investigation. While projectile motion is often considered a phenomenon that helps students understand motion in two dimensions (see e.g., Jordan et al. 2018), in this activity, students used the techniques of deriving position functions for a two-dimensional motion to learn about specific properties of the projectile motion. Thus, a projectile motion was considered one of the multidimensional motion types rather than one supporting an understanding of two-dimensional motion.

8.4.2 Gathering Data to Construct Positions Functions for a Projected Object

The snapshot of the simulation (Fig. 8.8) was displayed in the physics lab using the classroom screen. Displaying the simulation on a classroom screen provided opportunities for directing students' attention to specific motion attributes and enhancing discussions. The simulated lab was used to generate a path of motion of an object projected at different angles and realize its different shapes. The selected path of motion was used to derive the object position in both the vertical and horizontal directions.

To link this motion with the types of multidimensional trajectories, the instructor posited a question about the motion classification considering the degree and mode of movement. All students considered this motion as two-dimensional, however realizing the mode of movement in each direction was not that apparent for all students. The instructor then directed the discussion towards finding out how to algebraically formulate the projectile motion's position functions in the horizontal and vertical directions. Some students were inclined to claim that the object accelerates in the vertical direction due to gravitational field and does not in the horizontal; however, supporting these claims by mathematical reasoning and algebra tools

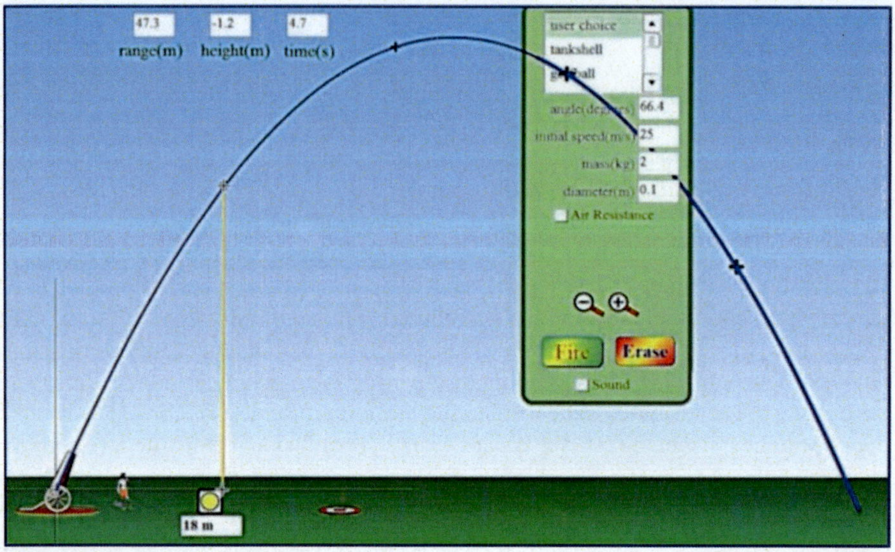

Fig. 8.8 Technique of data taking to construct parametric functions for projectile motion. (Source: PhET. Interactive Simulations (n.d.))

was not that apparent. Some considered finding components of the initial velocities but could not foresee a need to check the values of these velocities over the entire span of the motion. The question remained open about proving that the object does not accelerate in the horizontal direction. The discussion concluded with generating the table of values for the positions in x and y directions, finding these functions algebraic forms, and using these functions' attributes to determine movements' mode in each direction.

How to generate the table of values for these functions not being provided with these functions algebraic representations? To conclude the method, the learners were directed to positions of the projected object marked by (+) along the trajectory (see Fig. 8.8). A helpful element was the embedded timer that marketed the objects' positions after each second of the motion. Students realized that this attribute could depict the functions as a table of coordinates and then find their algebraic equations. The directions of view, while not labeled, remained the same as in the previous motion diagrams. Thus the top and side view were used. Figure 8.9 provides more details on how the snapshot was used to generate the data.

The following sequence of tasks was provided to students to generate the data:

(a) On the parabolic trajectory, label the position of the object after each second of motion as A, B, C, and D.
(b) Measure the positions OA_x and OA_y and formulate tables of positions in each direction separately for the remaining highlighted points.

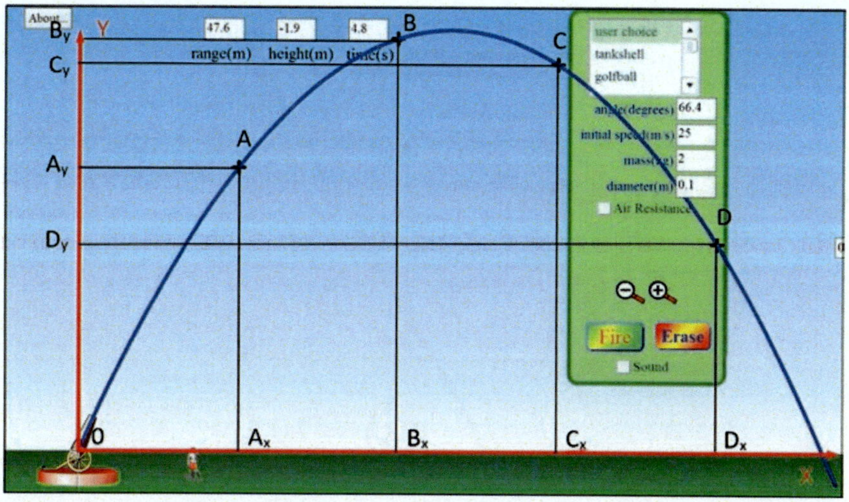

Fig. 8.9 Measuring the vertical and horizontal position of the object after each second of motion. (Source: PheET Interactive Simulations (n.d.))

Table 8.1 X and Y coordinates of the projected object

Time (s)	0	1	2	3	4	4.8
Coordinate	(0,0)	A	B	C	D	E
Position X(m)	0	10.3	21.1	30.1	40.4	48.5
Position Y(m)	0	18.6	26.1	24.5	13.3	0

Students realized that as on the *x*-axis, the position progresses with a constant rate of increase; on the vertical axis, the position increases and then decreases (Table 8.1).

The students also realized that the time variable linked the horizontal and vertical positions; they noticed that both motion components shared this variable. Following the data's progression, they further confirmed that the horizontal position changes by a constant rate, whereas the vertical position increases and then decreases, indicating an acceleration. These predictions were verified after the algebraic representations of these data tables were found.

8.4.3 *Constructing Representations of the Position Functions*

Functions' visual representations are the most convincing for the students (Gates, 2018). Thus, the next step of the inquiry was to graph $x(t)$ and $y(t)$ independently, realize the functions' attributes and thus motion properties in the designated directions. Students were prompted to draw the best-fit curves for each data set and find the function equations by the methods studied in their mathematic courses.

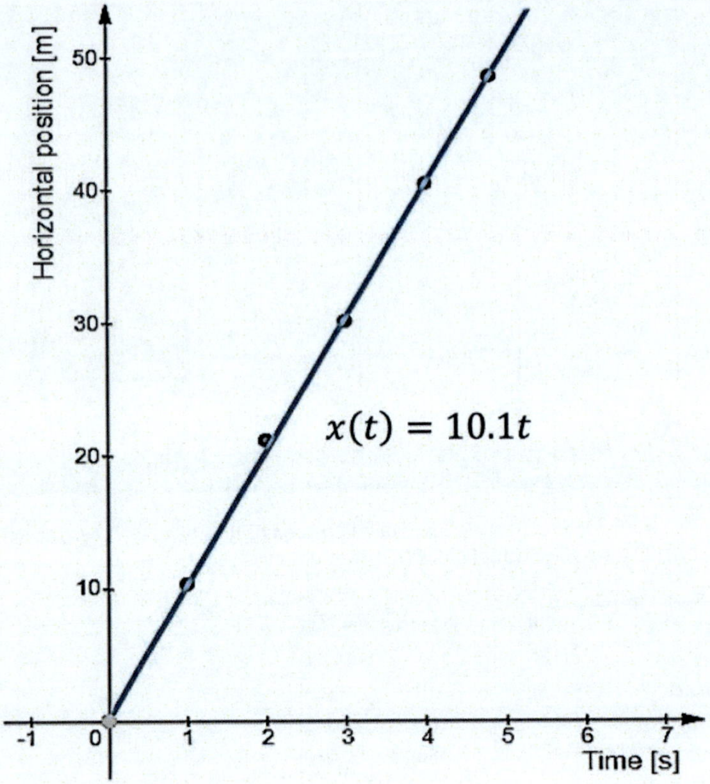

Fig. 8.10 The best-fit line and its equation for the horizontal position of the object

Figure 8.10 illustrates the graph of the object's horizontal position versus time, and Fig. 8.11 shows the graph of the vertical position of the object versus time.

The graph (Fig. 8.10) provided ample evidence that due to a constant slope of the horizontal position function, the rate at which the horizontal component of the object position changes is constant. Furthermore, the best-fit line returned the slope of 10 that represented the horizontal component of the velocity $v = 10\frac{m}{s}$. Thus, $x(t) = 10t$ showed the position function in the horizontal direction.

The best-fit curve uncovered a quadratic form for the vertical position, proving its accelerated mode in that direction (Fig. 8.11).

The students realized that this graph does not represent motion with a constant velocity because it did not resemble a linear model. The projected object was under the influence of the constant gravitational field, decreasing the object speed at a constant rate of 9.8 m/s^2 in the upward direction and increasing the speed at a constant rate of 9.8 m/s^2 in the downward direction. The students realized that the coordinates of both functions $x(t)$ and $y(t)$ generated coordinates for the motion path shown in Fig. 8.9. The complete algebraic representation of the motion is given in Eq. 8.7. It does include the time interval for the motion to follow up on the general

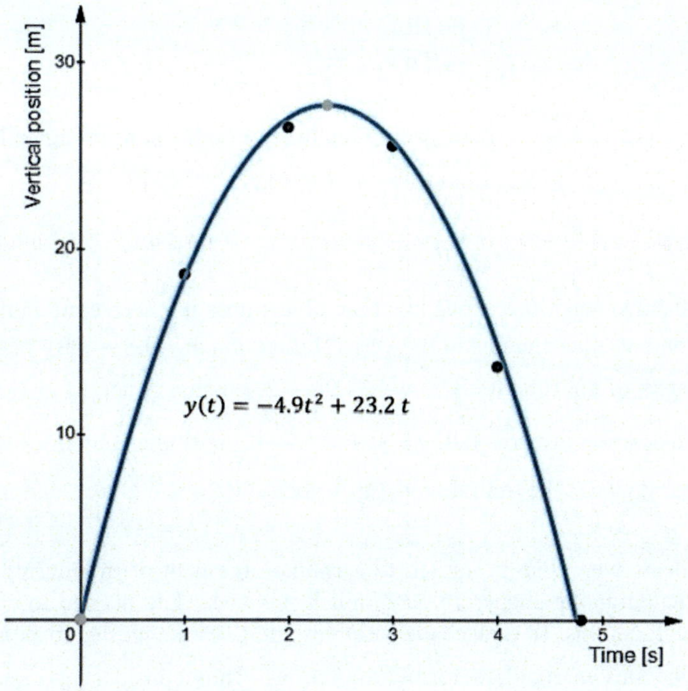

Fig. 8.11 The vertical position of the projected object in the function of time

algebraic description of parametric equations and assure the students that this parameter is shared by both component functions.

$$\begin{aligned} x(t) &= 10.1t \\ y(t) &= -4.9t^2 + 23.2t \qquad 0 \le t \le 4.8 \end{aligned} \tag{8.7}$$

As assessment items often ask students for graphs of velocities and acceleration of a projected object, the students were prompted to derive these functions as well which follows.

8.4.4 Finding Velocities and Acceleration Functions

Students who were concurrently taking a calculus class applied the concept of the derivate to formulate velocities and acceleration functions. The rest of the class was prompted to use standard kinematics formulas.

$$\begin{aligned} v_x(t) &= 10.1 \\ v_y(t) &= -9.8t + 23.2 \qquad 0 \le t \le 4.8 \end{aligned} \tag{8.8}$$

Taking the derivative with respect to the time of both equations again lead to

$$\begin{aligned} a_x(t) &= 0 \\ a_y(t) &= -9.8 \qquad 0 \le t \le 4.8 \end{aligned} \tag{8.9}$$

The above equations proved that the object does not accelerate in the horizontal direction but it accelerates in the vertical direction at $-9.8\frac{m}{s^2}$. By mapping up the parameters of the function $y(t) = -4.9t^2 + 23t$ with a general functional form for object's position in a free fall $y = y_o + vt + \frac{at^2}{2}$, the students realized that the initial position $y_o = 0$, the initial vertical velocity $v_o = 23.2$ m/s, and the acceleration is $a = -9.8\frac{m}{s^2}$.

Students were also prompted to formulate the path of motion algebraically to prove its parabolic shape. To accomplish this task, they needed to attend to both position functions. To replace the variables, they were suggested to express the position functions in standard formula notations. Thus considering $x = 10t$, $t = \frac{x}{10}$. Substituting $\frac{x}{10}$ for t in $y = -4.9t^2 + 23t$ resulted in $y(x) = -0.049x^2 + 0.23x$. The function was quadratic, which supported the shape of the path provided earlier in Fig. 8.9.

8.4.5 *Verification Process*

The students were also invited to verify algebraically the trajectory, $y(x)$, for its adherence to its graphical representation. They calculated the maximum position of the projected object using the formula for the vertex of the parabola and the parabola's zeros to find the object's maximum horiozntal range. The verification process also included derived parametric functions to assure that the object's position calculated using these entities correlated with that on the trajectory picture. They were also to find the resultant velocity at selected instants of time, the maximum height, and the maximum projectile range.

When problem-solving is concerned, other quantities are typically given, yet these examples provided a very comprehensive framework to organize the solution processes. On the next instructional day (not researched in this study), the students used trigonometric ratios to find necessary kinematics parameters and practiced writing parametric equations for provided motion scenarios.

8.5 Treatment Evaluation

The effects of the treatment were to answer the following research problem:

- *Will situating the learning of projectile motion as a two-dimensional motion representation help students realize its mathematical underpinning?*

The treatment's goal was to reflect on the research question of whether highlighting the construction of parametric equations helps students realize distinctive properties of the components of the projectile motion. Thus, supporting students' understanding of projectile motion by considering its path as a combination of two independent position functions was the priority of the enterprise. Stemming from this understanding, another essential property of projectile was expected to emerge; realizing two modes of movements of the projected object, a constant velocity in the horizontal direction and accelerated motion in the vertical one. During the lab activity, the students did not apply trigonometry to resolve motion vectorial parameters into components. This objective was realized on the following instructional units when the students found the components of either initial position, velocity, or acceleration using trigonometric ratios and used their knowledge about the properties of projectile motion developed ealier during the treatment to formulate the parametric functions.

There were two main qualitative-type questions that the students were invited to answer to reflect on the research problem (see Sect. 8.2.3). Students' responses were first categorized into classes based on their common themes and using descriptive statistics the percentages of each category were calculated. The students responded to these questions individually after a week from the date of the treatment. By analyzing the quality of the students' responses, their ability to apply parametric equations was also assessed.

Students' answers were not marked as right or wrong, but these answers allowed inferring how the treatment (labelled as choice (b)) affected their perception of projectile motion when compared to other means they experienced. Table 8.2 shows the percentage of students in each class, and Tables 8.3, 8.4, and 8.5 show verbatim selected responses from each class.

Students (9%, N = 2) who recalled projectile motion observation, choice (a) described the motion qualitatively. However, they could not extract specific quantities to provide legitimate support reflecting on the motion components. The scientific fact that the object does not accelerate in the horizontal direction, which is one of the central kinematics attributes to quantify the motion correctly, did not surface. Reading the students' responses clustered in choice (b) (64%, N = 14), it is reasonable to claim that this constructivist learning environment helped them realize the distinctive motion features of a projected object. The students especially highlighted the linearity of the function in the horizontal direction that proved a constant velocity. Some students mentioned a proven object's acceleration to be -9.8 m/s^2 which is not easily visualized in any experiment.

Choice (c) was selected by six students (27%, N = 6) who referred to the introductory physics courses and mentioned other experiences or labs they performed on

Table 8.2 Students' preference of representation to identify attributes to find the motion components

Response Type	Choice (a) Favored Observation	Choice (b) Favored the Lab Conduct	Choice (c) Favored Other Experiences
Percentage	9% (N = 2)	64% (N = 14)	27% (N = 6)

Table 8.3 Samples students' responses favoring observation: Choice (a)

Student	Response
1	By throwing a projectile. It goes until a max point where it freezes, then starts speeding up down. It goes in the ex-direction constantly (minus a little air friction.
2	There is no acceleration in the x-direction based on the ball's usual trajectory when it is parabolically changing its direction toward the y-axis.

Table 8.4 Sample students' responses favoring the lab conduct: Choice (b)

Student	Response
1	When measuring the horizontal distance at each second, the distances were multiples of each other, but the measurements were not proportional; therefore, there was accelerating.
2	We could visibly observe the vertical and horizontal components of the movement in projectile motion. Placing each table for x and y showed a linear graph showing no acceleration in the x-direction.
3	Finding PE enabled me to see the relationship between the motion in both frames of reference. The PE reinforced the idea that the x and y directions each had their functions but with the same parameter time.
4	The most significant piece of evidence about the object's having $-9.8\ m/s^2$ in the vertical direction, and no acceleration in x would be from the parametric lab. The distances and time would change at the same rate showing constant velocity in the x-direction. When the velocity is constant, there cannot be acceleration.
5	The parametric lab showed that while the range increased linearly, the max height could not surpass its quadratic form.

Table 8.5 Sample students' responses favoring other experiments: Choice (c)

Student	Response
1	Pitching a baseball requires you to throw the ball before crossing the horizontal line plane not to hit the dirt because a greater vertical angle yields more velocity.
2	I did a lab when marble rolled up and down in an arched vertical track and then was projected in the air and landed outside of the track (diagram was provided). We had to predict horizontal distance (x) traveled.
3	The monkey is falling from a tree simulator. If a cannon is shot at the monkey as the monkey lets go of the three, the monkey will catch it because it accelerates downward while the projectile takes its path. Therefore, both acceleration and the bullet's acceleration in the y-direction must be the same as the monkey.
4	A lesson I once learned showed the slope of the object shot from a cannon. Vertical velocity was not constant, but horizontal seemed constant.
5	A project about building a trebuchet would fire a pumpkin, and we had to plot the velocity, position, and acceleration for each second in the air.

projectile motion. The labs' contextual richness is unquestionable; however, these students did not recall these labs' specific features that would exemplify the motion's distinctive features when components are concerned.

Question 2 asked to explain the process of analyzing mathematically projectile motion and intended to inform if the students solidified a general mathematical strategy to get ready to solve projectile motion problems.

Debriefing analyses have revealed two categories of answers depending on their scientific validity. Category 1 was marked as showing sufficient understanding that included resolving the motion into components, and Category 2 was marked as showing insufficient understanding that did not provide such prompts. According to the classification, sixteen students (N = 16, 73%) acquired sufficient mathematical background, and six (N = 6, 27%) were still at the developmental stages. Tables 8.6 and 8.7 contain representative responses in each category.

Students in Category 1 provided evidence that they realized the significance of resolving the motion in the x and y components and constructing position and velocity functions, respectively. Resolving the position into components appeared as a necessary step toward analyzing the motion.

Students who did not show sufficient understanding (Category 2) did not explicate the specific motion parameters associated with the kinematics of projectile motion. Their responses missed the precision of describing the motion according to the axis of projection. One can infer from these rather loosely stated descriptions that these students will face challenges to work on projectile motion problems when finding components is necessary.

Table 8.6 Sample responses representing Category 1

Student	Response
1	When projected, an object will reach y-max and x-max. Without a y-component, there is no positive y-movement. Without an x-component, there is no distance in the x-direction
2	This is true because max in the vertical and horizontal planes of the motion acts independently. When the two are combined, they form the trajectory.
3	Visually it is easy to see vertical and horizontal movements, and they are independent. In the lab last week, hitting the tennis ball from the table at different speeds, we proved that motion does not depend on the horizontal velocity.
4	In the labs, finding a separate equation and proof for the horizontal distance and vertical distance traveled. This shows that it is composed of two equations.
5	Projectile motion is the movement of an object through space. The object moves in horizontal and vertical directions.
6	The projectile motion has two completely independent components. The horizontal component has a constant horizontal velocity as it were on a horizontal surface. Moreover, the vertical component accelerates due to gravity as it was falling straight down.
7	Projectile motion is when an object is sent at an initial speed and changes its position vertically and horizontally. These objects have negative acceleration in the y-direction and no acceleration in the x-direction.

Table 8.7 Sample responses representing Category 2

Student	Response
2	The path a projectile takes when fired or thrown
3	The whole motion of one thing plus the velocity, time, and acceleration
4	Projectile motion is when an opposing force acts upon an object with its initial velocity.
5	Projectile motion is a path when an object is projected according to vectors like initial velocity or gravity

Students' abilities to describe projectile motion using mathematical reasoning were considered the critical phase toward handling problem solving that often requires finding components. The majority of these students (N = 16, 73%) seem to acquire a sufficient understanding of the processes that supported the study's hypothesis that highlighting mathematical apparatus helped them build a rich scientific image of this phenomenon. By consistently writing the domain of both functions, the students realized that the time of the motion for both components must be the same. The shared parameter of time, t, for both component functions was to link both motion components into one that constituted the object motion path.

8.6 Summary and Conclusions

While focusing on enriching the analysis of projectile motion by parametric equations, this study reflected on the current recommendations of physics research communities to use the structural domain of mathematics to support the development of science concept's understanding. This study aimed to help students realize that projectile motion is an example of two-dimensional motion that can be resolved into components according to its vectorial representations with a constant velocity in the horizontal direction and varying velocity in the vertical. This understanding was supposed to help students link these ideas to mathematics and formulate parametric functions for any projectile motion. The instrument's evaluation results supported the study objective that the treatment can be considered a promising didactical approach that can help students with two-dimensional motion analyses. It was assumed that the continuity of enhancing traditional kinematics by including parametric equations starting from the first days of studying kinematics generated a better understanding of the distinctive properties of projectile motion. Can curriculum modification suggested in this study result that students will consider the formulation of the parametric equations an integrated part of the two-dimensional analysis? Such a statement is premature. However, this study seems to provide a green light in a direction to accepting this statement.

Physics textbooks (e.g., Giambattista et al. 2007 p.122) provide a stroboscopic photograph of two objects simultaneously falling alongside an edge: one with zeroth horizontal velocity and the other with a non-zero initial horizontal velocity. The stroboscopic pictures provide evidence that the two objects experience the same vertical acceleration and that the time of falling is the same for both objects. While the photographs provide observable evidence for both facts, the pictures do not help realize that each motion's vertical components are identical and they do not provide an insight into the distinctive properties of the motions' components. In the light of developing covariational thinking, this demonstration might not be as beneficial as it is intended to be; the lab includes two objects that in the students' realms generate two separate movements. The question might emerge on what basis the motion parameters of these objects can be compared? What is the mathematical proof for the vertical components of the motion to be the same? Such experiment provides an excellent opportunity for using parametric equations to prove algebraically that the total times of the motion for both objects are the same. This example was brought up to the students. They considered proving the experiment observable outputs as additional evidence of the power of parametric equations to support physics inquiries and make them explicit.

References

College Board. (2015). *Student performance Q&A 2015 AP® Physiscl Free Response questions.*. Retrieved from https://secure-media.collegeboard.org/digitalServices/pdf/ap/ap15_physics1_student_performance_qa.pdf

Clark, R. C., & Mayer, R. E. (2016). *E-learning and the science of instruction: Proven guidelines for consumers and designers of multimedia learning*. Wiley.

Das, C., & Roy, D. (2014). Projectile motion with quadratic damping in a constant gravitational field. *Resonance, 19*(5), 446–465.

Espinoza, F. (2004). Enhancing mechanics learning through cognitively appropriate instruction. *Physics Education, 39*(2), 181.

Gates, P. (2018). The importance of diagrams, graphics, and other visual representations in STEM teaching. In *STEM education in the junior secondary* (pp. 169–196). Springer.

Gamble, R. (1989). Force. *Physics Education, 24*(2), 79–82.

Giambattista, B., Richardson, McCarthy, & Richardson, R. (2007). *College Physics* (2nd ed., p. 122). Mc-Graw Hill Higher Education, Boston. MA.

Jimoyiannis, A., & Komis, V. (2001). Computer simulations in physics teaching and learning: A case study on students' understanding of trajectory motion. *Computers & Education, 36*(2), 183–204.

Jonassen, D. H. (1991). Objectivism versus constructivism: Do we need a new philosophical paradigm? *Educational Technology Research and Development, 39*(3), 5–14.

Jordan, C., Dunn, A., Armstrong, Z., & Adams, W. K. (2018). Projectile Motion Hoop Challenge. *The Physics Teacher, 56*(4), 200–202.

Jörges, B., & López-Moliner, J. (2019). Earth-gravity congruent motion benefits visual gain for parabolic trajectories. *bioRxiv*, 547497.

Klein, P., Gröber, S., Kuhn, J., & Müller, A. (2014). Video analysis of projectile motion using tablet computers as experimental tools. *Physics Education, 49*(1), 37.

Mihas, P., & Gemousakakis, T. (2007). Difficulties that students face with two-dimensional motion. *Physics Education, 42*(2), 163.

Millar, R., & Kragh, W. (1994). Alternative frameworks or context-specific reasoning? Children's ideas about the motion of projectiles. *School Science Review, 75*(272), 27–34.

National Research Council. (2012). *A framework for K-12 science education: Practices, crosscutting concepts, and core ideas*. National Academies Press.

PhET Interactive Simulations. (n.d.). The University of Colorado at Boulder. Retrieved from http://phet.colorado.edu, July 2020.

Prescott, A. E., & Mitchelmore, M. (2005). Teaching projectile motion to eliminate misconceptions. *PME*, 103–109.

Shadish, W. R., Cook, T. D., & Campbell, D. T. (2002). *Experimental and quasi-experimental designs for generalized causal inference*. Houghton Mifflin.

Sokolowski, A., & Capraro, M. M. (2013). Parametrization of motion. *Mediterranean Journal for Research in Mathematics Education, 12*(1–2), 121–133.

Springuel, R. P., Wittmann, M. C., & Thompson, J. R. (2007). Applying clustering to statistical analysis of student reasoning about two-dimensional kinematics. *Physical Review Special Topics-Physics Education Research, 3*(2), 020107.

Wee, L. K., Chew, C., Goh, G. H., Tan, S., & Lee, T. L. (2012). Using tracker as a pedagogical tool for understanding projectile motion. *Physics Education, 47*(4), 448.

Von Glasersfeld, E. (1991). Abstraction, re-presentation, and reflection: An interpretation of experience and Piaget's approach. In *Epistemological foundations of mathematical experience* (pp. 45–67). Springer, New York, NY.

Chapter 9
Reimaging Lens Equation as a Dynamic Representation

9.1 Introduction

Mathematics provides a computational system, reflects a physical idea, or conveniently encodes a quantification rule (Bing & Redish, 2008). A study conducted by Bossé et al. (2010) revealed that the standards and processes in the National Council of Teachers of Mathematics (NCTM's) *Principles and Standards for School Mathematics* (NCTM, 2000) and the five E's—engagement, exploration, explanation, elaboration, and evaluation—developed by the National Research Council (NRC, 2000) display similarities. Thus, a need for more teaching resources emerges. A substantial amount of instruction in mathematics courses is devoted to developing students' skills of graphing and analyzing functions (NCTM, 2000). Students evaluate function limits, determine domain and range, and describe the function behavior using the derivative concepts. Unfortunately, this wealth of techniques of graph analyses is not often applied in sciences. By working on formulas, students' ownerships are reduced to plugging in given quantities and finding a value that makes the resulting equation a true statement. There exist many areas in physics where structural mathematical apparatus can conveniently enrich scientific inquiries. Determining image characteristics using the graph of lens function emerged as a great opportunity of applying the structural domain of mathematics to physics. A more general version of the instructional unit was earlier published (see Sokolowski, 2012). This version is enriched by consistent terminology and a uniform approach to limit interpretation.

A. Sokolowski, *Understanding Physics Using Mathematical Reasoning*,
https://doi.org/10.1007/978-3-030-80205-9_9

9.2 Prompts Used for the Instructional Unit Design

Image characteristics are usually formulated using the *thin lens equation*: $\frac{1}{f}=\frac{1}{d_i}+\frac{1}{d_o}$ and evaluating it for given quantities. Due to a high multiplicity of possible image characteristics—*real* versus *virtual, upright* versus *inverted,* and *diminished* versus *enlarged*—a generalization of these cases is not readily achievable. It is indisputable that the lens equation and magnification formula provide a sound method for computing the characteristics, and a corresponding ray diagram further supports these inferences. However, these processes are applied to conclude isolated cases that do not provide direct access to summarizing image characteristics at its entire spectrum. To help students generalize the phenomenon of image production, a need for using a more general mathematical apparatus emerged. It seemed that converting the lens equation and magnification formula into algebraic functions and investigating the characteristics of the images using corresponding graphs provided a convenient means for enriching lessons about image characteristics. The process revealed that by using the graphs and applying limits, image characteristics were concluded more concisely. Generated graphs opened up opportunities for applying mathematical reasoning normally unreachable by using the traditional lens equation.

The inquiry has two general sections: Sect. 9.2.3 that is about using lens function, and a more detailed Sect. 9.2.4 that is about using magnification function. Considering the complexity of math knowledge needed, in lower level physics courses showing students how to investigate only the lens function would be sufficient. Students of advanced physics courses would also welcome the analysis of the magnification function, which would complete the entire spectrum of image analysis.

In sum, the goal of constructing the instructional unit was twofold (a) to help students with image classification and generalization, and (b) to provide students with opportunities to merge mathematics tools in a pursuit to delve deeper into how natural phenomena behave.

9.2.1 Mathematical Background

It is advised that participants be familiar with sketching rational functions and interpreting asymptotes to blend mathematical and scientific reasoning during this activity. According to *Principles and Standards for School Mathematics* (NCTM, 2000), these cognitive math concepts are primarily studied in precalculus and calculus courses. If learners have an insufficient background, a graphing calculator can assist. While general image characteristics can be concluded from the lens function without applying the concept of limits, it is suggested that students do use limits, e.g., left and right-sided and limits at infinity, and interpret the results in the contexts of the function asymptotes. Limiting case analysis will disclose more details about the image characteristics and show its powerful nature to enrich scientific inquiry.

9.2.2 Lab Equipment

This unit can be classified as a guided discovery lesson utilizing an optical bench or a virtual lab. I suggest using a virtual lab because it provides opportunities to verify students' hypotheses conveniently. The virtual lab utilized during the instructional unit is a physics simulation called *Geometric Optics* designed by the Physics Educational Team (PhET) at Colorado University and available for free at http://PhET.colorado.edu/sims/geometric-optics/geometric-optics_en.html. Figure 9.1 illustrates a general setup of the simulation.

Navigation of the simulation is simple, and it allows for exploring various paths of the image production. For this unit, the curvature of the lens is set to be equal to 30 cm. The corresponding focal length of the lens is $f = 15$ cm. The index of refraction is set at $n = 1.4$, and the diameter of the curvature of the lens is 100 cm. Other values of the index of refraction and the focal length are also possible to select. For specific values, the lens might produce images that will be too large to be observed on the computer screen. Thus, a domain for its values, in its mathematical sense, needs to be verified. One parameter calls for more elaboration, the distance of the object. As assigned in the simulation, numerical values of this distance are unparalleled with the functional values in the mathematical sense. The object's distance from the lens is typically considered positive if the object is positioned on the opposite side of the primary focal length. When the functions are graphed in XY coordinates, by convention, positive values are labelled on the right and negative on the

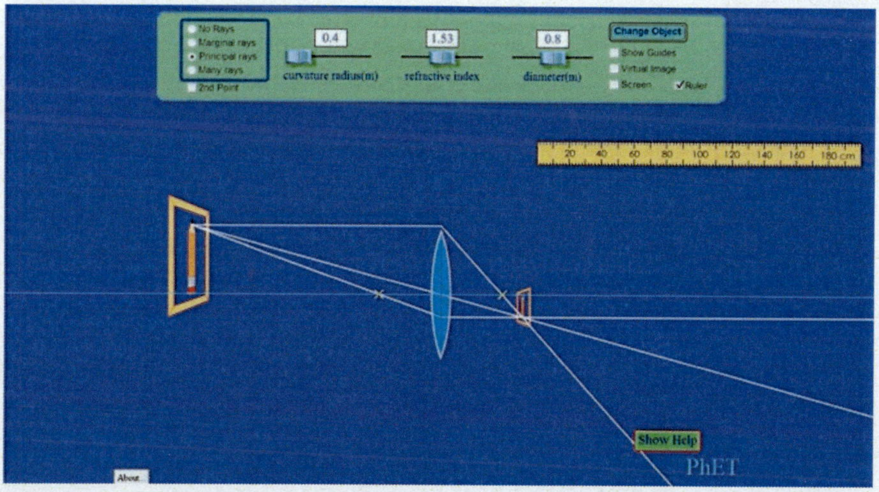

Fig. 9.1 Snapshot of the simulation *Geometric Optics*. (Source: PhET Interactive Simulations (n.d.))

left side of the horizontal axis. The simulation allows placing the object only on the left side of the lens, which might inadvertently suggest negative object's distances. This is not an obstacle, but the instructor might address this nuance to ensure correct scientific interpretations.

9.2.3 Conversion of Lens Equation into a Covariational Relation

This section provides teaching suggestions on converting the lens Eq. (9.1) into a covariational function. An algebraic function, for example, $y = f(x)$, is defined as a relation of two variables—independent x and dependent y—such that for each x value, there is only one corresponding y value (Stewart, 2006). In physics textbooks (e.g., Giancoli, 2005; Tipler and Mosca 2004), the lens formula is called the lens equation and is presented as follows:

$$\frac{1}{f} = \frac{1}{d_i} + \frac{1}{d_o} \tag{9.1}$$

The equation intertwines three different parameters: f, which represents the lens focal length, usually given as a constant; d_i, which represents the distance of the image from the plane of the lens; and d_o, which represents the distance of the object from the plane of the lens. Algebraic interpretation of this formula along with possible explorations are limited. Being derived from the similarity of triangles (Giancoli, 2005), it merely depicts an algebraic entanglement between these quantities that are warranted on the condition of algebraic equity of both its sides. Since it is not expressed as a mathematical function (e.g., using a function notation), its graphing representation is not directly generated, nor can its outcomes be generalized using its functional behavior. The formula needs to be converted into a covariation relation to providing opportunities for a more insightful interpretation that will, in return, offer means for generalizing the physics behind it. To accomplish this task, one needs to determine what scientific covariance is sought. Each constituting parameter can potentially be considered a constant, dependent, or independent. In a typical introductory lab on image analysis (McDermott, 1996), the focal length f is considered a constant parameter, d_i is the parameter of interest of the investigation, and d_o is the independent parameter. Because image characteristics vary as the object's position concerning the plane of the lens varies, it seems convenient to assign the distance of the object as varying independent parameters, the distance of the image as the output parameter, and the focal length as a constant parameter. Such association of the parameters will guide formulating the image characteristics using lens and magnification functions. The equation's parameters will be reimaged to better link and amplify the algebraic entanglement of the quantities with the formality of mathematical notation. This phase will help students to track the covariance between the parameters in their mathematical sense. By replacing the distance of the image, d_i with y, and the distance of the object, d_o with x, Eq. (9.1) takes the following form:

$$\frac{1}{f}=\frac{1}{y}+\frac{1}{x} \tag{9.2}$$

It is to note that the primary parameters of attention are the distances of the object and its images, not their heights. The function notation $f(x)$ was not used due to a possible confusion with the lensfocal length, f. To graph the function and evaluate its limits, the expression needs to be further rearranged to a functional form and illustrate y as a dependent parameter of x that shows Eq. (9.3). Function notation was not used to make a better association with the physical parameters.

$$y=\frac{fx}{x-f} \tag{9.3}$$

Considering the focal length of the lens 15 cm, the algebraic form (9.3) of the rational function that will be used for further analysis takes the following shape

$$y=\frac{15x}{x-15} \tag{9.4}$$

The instructor highlights the scientific interpretations of the variables; x represents the distance of the object from the lens, and y represents the distance of the image. For coherence of the interpretations, all dimensions will be expressed in centimeters.

The parameter considered a constant is a focal length, f, which can be assigned positive or negative values depending on the type of lens used in the experiment. According to the established convention, a converging lens with a real focal length is denoted by a positive value. A diverging lens having a virtual focal length is characterized by a negative value. In the proposed instructional unit, a converging lens will be used. Teaching suggestions on how to guide the inquiry follows in Sect. 9.2.4.

9.2.4 *Sketching and Scientifically Interpreting the Graph of the Lens Function*

The purpose of constructing the function (9.4) is to learn the *kind* of image, thus *real* or *virtual,* by analyzing a corresponding graph values as positive or negative. While the algebraic function is a concise representation, interpretation of its graph conveniently helps determine the image characteristics based on the object distance from the lens. To activate students' mathematical knowledge, a short review of the

Table 9.1 Summary of image characteristics for a converging lens

Position of the object	Image characteristics
$0 < x < f$	Kind (real/virtual) ________________ Orientation (upright/inverted) ________________ Magnification (diminished/enlarged/the same size) ________
$x = f$	Kind (real/virtual) ________________ Orientation (upright/inverted) ________________ Magnification (diminished/enlarged/the same size) ________
$x = 2f$	Kind (real/virtual) ________________ Orientation (upright/inverted) ________________ Magnification (diminished/enlarged/of the same size) ________
$x \to \infty$	Kind (real/virtual) ________________ Orientation (upright/inverted) ________________ Magnification (diminished/enlarged/the same size) ________

graphing techniques follows. The students always welcome such a review. As the analysis progressed, students were to complete Table 9.1 that summarized the findings of the investigation.

In mathematics, the process of sketching is supported by identifying the so-called parent function along with its domain and its possible asymptotes that determine the boundaries of the graph. The lens function, $f(x) = \frac{15x}{x-15}$, resembles $f(x) = \frac{g(x)}{k(x)}$, which classifies it as a rational function (Stewart, 2006) and its corresponding graph as a hyperbola. The function domain describes a set of function inputs that produce real outputs; for $y = \frac{15x}{x-15}, x \neq 15, \text{and}\, x \in R$.

While identifying the domain is sufficient for the mathematical purpose, further restrictions might be imposed due to scientific constraints on the formula from which the lens function was derived. Since x represents the object distance from the plane of the lens and the distance can take only positive values (in the experiment setup), the lens function contains an additional restriction on x that is $x \geq 0$, which results in $x \geq 0$, $x \neq 15$, and $x \in R$. The vertical or horizontal asymptotes will be used to bound the rational function. Subsequent scrutiny to determine these asymptotes for the lens function follows. Function (9.4) possesses two asymptotes, vertical at $x = 15$, due to mathematical restriction on its domain and a horizontal at $y = 15$, due to $y(x \to \infty) = \lim_{x \to \infty} \frac{15x}{x-15} = 15$. The graph of the lens function (Fig. 9.2) is presented to students with its restrictions.

To generalize *the kind* of image characteristics using the graph, it is recommended that students realize that all graphs of lens function with a converging lens (and concave mirrors) will resemble the one illustrated in Fig. 9.2. The difference in appearance will result from choosing focal lengths of different values that affect the position of the function vertical and horizontal asymptotes. Although the positions of vertical asymptotes will exhibit specific variation, all will be located on the positive side of the x-axis. Similarly, all horizontal asymptotes will be located above the

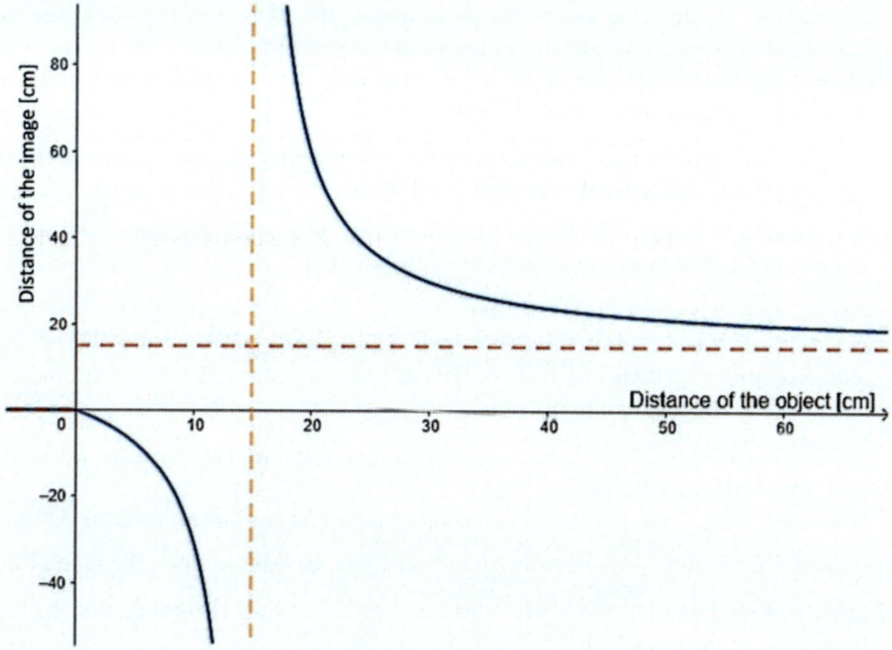

Fig. 9.2 Graph of the lens function $f(x) = \dfrac{15x}{x-15}$

x-axis. The effects of varying positions of the asymptotes could be supported by graphing technology. Lens function allows determining the kind of the image for specific distances of the object from the lens.

The discussion that follows merges three representations, graphical, algebraic, and scientific. The simulated lab serves as a tool for verification. However, attending to all representations might be overwhelming for the students; therefore, using the virtual lab for verification is suggested while analyzing the magnification function (Sect. 9.2.5). The analysis that follows is organized progressively considering the object's distance from the lens as established in the function domain.

Case 1. The object is placed at x = 0 cm

When the $x = 0,\ f(x) = \dfrac{(15)(0)}{x-15} = 0$. Thus no image is produced. This conclusion can also be reached by referring to graph Fig. 9.2.

Case 2. The object placed within 0 < x < 15

The y-coordinates of thegraph of the image distance function are negative denoting negative image distances. This graph behavior, according to the general convention of image classification proves that the image is virtual. Realizing the negative values of the distance function comes naturally, referring to the graph. Lens equation produces negative values; however, it does not further support why the distance shows as negative. To align the interpretation to reality, the term *position* would perhaps reflect better on the interpretation instead of distance. Students observe that as the object is being moved closer to the focal point of the lens, the distance of the image increases to infinity, but it remains negative.

Case 3. The object is being moved near the left proximity of x = 15 (x→15−)
To evaluate the function, in this case, the technique of evaluating sided limit is helpful. The notation $y(x \to 15^-)$ will denote such condition; $y\left(x \to 15^-\right) = \lim_{x \to 15^-} \frac{15x}{x-15} = -\infty$. This algebraic statement supports the graph behavior.

Case 4. Object is placed at x = 15 cm
At $x = 15$, the function is undefined, which shows on the graph as a vertical asymptote. Scientifically it means that there is no image produced when the object is positioned at the focal point of the lens.

Case 5. The object is being moved past the focal point, and it is located at the right proximity of x→15+
When the object passes the focal length, but it is still near it, a sided limit $f\left(x \to 15^+\right) = \lim_{x \to 15^+} \frac{15x}{x-15} = \infty$ helps with the image classification. The result illustrates that the image distance has large and positive values. This function attribute transcends to the discovery that the image is real.

Case 6. The object is being moved away to an infinite distance
This condition also calls for using the limiting case analysis; $f\left(x \to \infty\right) = \lim_{x \to \infty} \frac{15x}{x-15} = 15$. The limit value suggest that $f(x \to \infty) = 15$ *cm* which informs that the image is generated at the lens's focal length. A horizontal asymptote $y = 15cm$ of the function supports this conclusion.

Students realize that the distance of the object from the lens and the lens's focal length have a profound effect on the distance of the image and its *kind*. When the object's distance is greater than the focal length, the image is considered real because the function values are positive. The image is virtual when the object distance remains within the lens's focal length because the function values at that domain are negative. A more detailed analysis of the image function will be included in the section that follows. While the *kind* of image is important, its orientation and magnification will complete its optical characteristics. The discussion of the magnification function is more complex; thus, it is recommended for more advanced physics courses.

9.2.5 Formulating Magnification Function

This section advances the analysis, and it will support magnification and orientation of the image. Once the magnification function is derived, the verification process will also take a more sophisticated form; it will incorporate the outcomes of the virtual lab. The starting point of the analysis is the magnification formula

$M = \frac{-d_i}{d_o} = \frac{-y}{x}$. This formula is built of two covariate parameters, the distance of the image, y, and the distance of the object, x. From this form extracting the scientific sense could be a challenge. The magnification function can though be expressed as one variable function, and limiting case analysis can be conviniently applied to learn more about the image.

$$M(x) = \frac{-y}{x} = \frac{-\frac{fx}{x-f}}{x} = \frac{-f}{x-f} = \frac{f}{f-x} \tag{9.5}$$

This analysis will parallel with the one from Sect. 9.2.4. By letting $f = 15$ cm, the magnification function takes the following form:

$$M(x) = \frac{15}{15-x} \tag{9.6}$$

In this equation, $M(x)$ can take positive or negative values, yet it is dimensionless. The positive or negative value of the magnification will help determine the object's orientation, thus *upright* for $M > 0$ or inverted for $M < 0$. The absolute value of the magnification $|M|$ describes the actual size of the magnification. The magnification function $M(x)$ depends only on x, the object's distance from the lens and its graph are presented in Fig. 9.3.

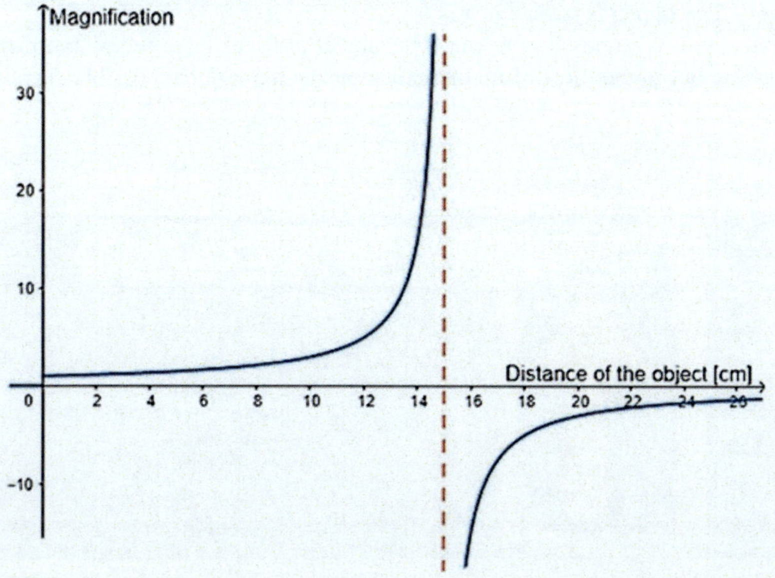

Fig. 9.3 Graph of magnification function. $M(x) = \frac{15}{15-x}$

The graph of *M(x)* also resembles a rational function. Both graphs' forms are similar, yet the scientific interpretations of these functions' behavior is different because each illustrates different aspects of the image production.

9.2.6 Merging Mathematical and Experimental Representations into One Inquiry

Since the students become acquainted with the interpretation of the lens function, the analysis of the magnification function includes one more representation; the experimental outputs supported by the virtual lab. The instructor might project the simulated lab (see Fig. 9.4) on the classroom screen to verify the conclusions reached from applying mathematical reasoning.

Expressing magnification as one variable function encourages implementing limiting case analysis similarly as it was applied to lens function. The independent parameter of the function represents the object's position concerning the plane of the lens as measured along the optical axis. The analysis will progress analogically to the prior one to highlight similarities of extracting scientific sense from $M(x)$ outputs that will be interpreted as *image orientation* and *magnification*. The interpretation will be consistent with convention found in physics textbooks; positive values of $M(x)$ will imply that the image is *upright*, negative values of $M(x)$ will indicate the image is *inverted*. The magnitudes of $M(x)$ with reference to the value of 1 will classify the image as (a) *diminished* when $0 < |M| < 1$ or (b) *enlarged* when $|M| > 1$. The discussion will begin from the object distance at $x = 0$ cm.

Case 1. The object is placed at x = 0

Physically, it is impossible to place an object at 0 cm from the plane of the lens. Reaching out to the tools of mathematics makes the inquiry possible. The value of

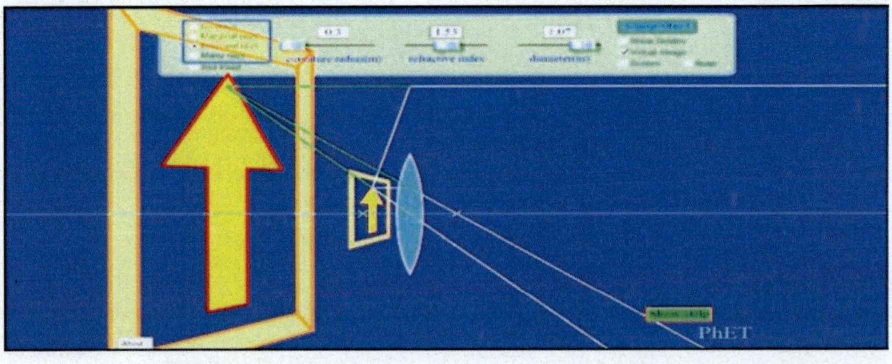

Fig. 9.4 Production of enlarged and upright image. (Source: PhET Interactive Simulations (n.d.))

$M(0) = 1$ appears as the vertical intercept of the magnification graph function on the graph of $M(x)$ (see Fig. 9.3). Magnification of 1 shows that the object and the image have the same heights. The y-intercept of the magnification function illustrates this situation.

This case is interesting; the function $M(x)$ can be evaluated when $x=0$; however, that experimental proof is not possible.

Case 2. The object is being moved from x = 0 to the proximity of x = 15 (0 < x < 15)

This distance interval does not require limiting case analysis because the magnification function can attain real values on that interval, and mathematical reasoning can be induced directly. $M(x)$ attains positive and greater than one value within this interval, which supports the claim that the image is *upright*. According to Fig. 9.3, the function outputs increase from 1 at $x = 0$ to infinitely large values, indicating that the image is *enlarged*. The lens resembles a magnifying glass in this domain. The observation of the image produced by the simulation (see Fig. 9.4) supports the inferences reached from mathematical reasoning. By placing the object close to the lens on its left side and then moving it slowly further to the left, the image appears to be upright and enlarged. By attending to the earlier developed image function, one can add one more characteristic of the image by describing it as *virtual*. The simulation supports this claim because the image is produced on the same side of the lens where the object is located.

Case 3. The object is near the proximity of the focal point on its left side (x→15−)

Evaluation of the sided limits for this case is also consistent with the graph of $M(x)$. When the distance of the image is getting very close to the focal point, thus when $x \to 15^-$ from the left side of the focal point (as seen on the graph function), the image is getting large and the magnification has positive values. This inference can also be concluded from the graph Fig. 9.3, and it can be calculated by evaluating the left-sided limit of the lens function using $M(x) = \lim_{x \to 15^-} \frac{15}{15 - x} = +\infty$. The image is upright because the magnification function takes positive values, and the magnification is very large.

Case 4. The object is placed at the focal point, thus at x = 15 cm

The graph of $M(x)$ displays a lack of corresponding y-values for $x = 15$, which is illustrated in the mathematical sense as a function vertical asymptote for that x value (see Fig. 9.3). Scientifically, this is interpreted as an absence of an image for this specific object's position. The simulation also supports this finding. Due to refracted rays being parallel, no image is produced when the object position is 15 cm (Fig. 9.5).

It is interesting to observe that the corresponding image magnification is also undefined when the object is placed at the focal point, and its algebraic proof

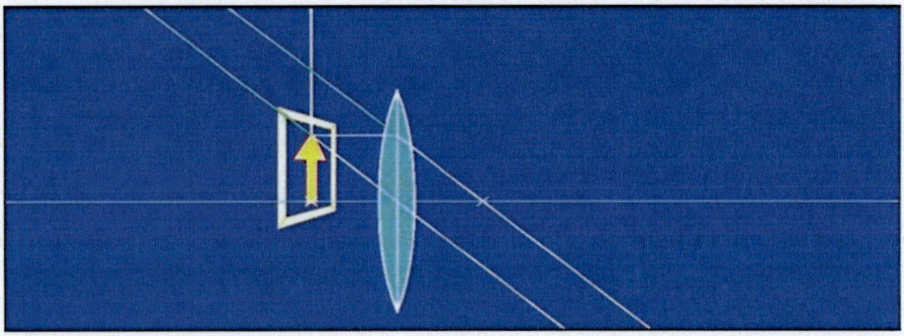

Fig. 9.5 The object is placed at the focal point of the lens, which results in no image being produced. (Source: PhET Interactive Simulations (n.d.))

follows. Since $M(x) = \frac{15}{15-x}$, and $x = 15$, then $M(15) = \frac{15}{15-15}$ is undefined. At $x = 15$, $M(x)$ has also a vertical asymptote; the image is not produced for that x-value and thus the magnification is undefined.

Case 5. The object is moved past the focal point, but it remains at its proximity $x \to 15+$ cm

At $x = 15$, the magnification function, similar to the lens function, is undefined; however, a right-sided limit can be used to describe the object's magnification and orientation when it in the proximity of 15 cm. $M(x \to 15^+) = \lim_{x \to 15^+} \frac{15}{15-x} = -\infty$ which informs that the magnification as an absolute value of the function values increases to infinity but that the image is inverted. Both the graph and simulation support the mathematical reasoning extracted from the magnification function.

The size of the image using directly the magnification relation, $M = -y/x$, is also very large. The simulation conveniently supports this interpretation (Fig. 9.6).

Case 6. The object is being moved further away from the focal point; $x > 15$ cm

When the object passes the focal length, but it is still near it, $M(x \to 15^+) = \lim_{x \to 15^+} \frac{15}{15-x} = -\infty$. This result illustrates that the object is inverted and is exceptionally large. As the object is being moved farther away, the image diminishes. This can be concluded from the graph in Fig. 9.3 and the simulation in Fig. 9.7. The image is also inverted due to the negative value of the magnification function. Limiting case analysis can be used to evaluate the magnification function; $M(x \to \infty) = \lim_{x \to \infty} \frac{15}{15-x} = 0$. The limit result suggests that when the

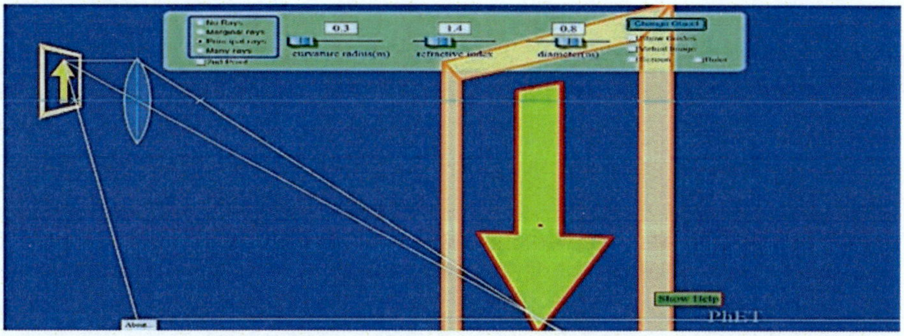

Fig. 9.6 Production of a real image. (Source: PhET Interactive Simulations (n.d.))

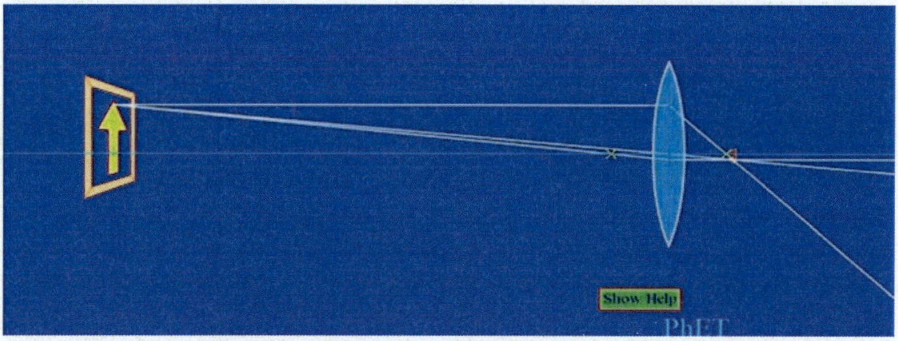

Fig. 9.7 The object is located very far away from the lens. The image is located near the primary focal point shown on the right side of the lens. (Source: PhET Interactive Simulations (n.d.))

distance is considerable, the height of the image is getting exceedingly small. The limit output though does not inform that the height will ever be zero, instead the horizontal asymptote of $y = 0$ places such boundary on its height. Mathematical reasoning can further upraise this result; there is no x-value for which the magnification function would ever be zero. Thus, the image will not disappear; it will exist even though its height will be near the value of zero centimeters (Fig. 9.7).

Analyzing the magnification function concludes the investigations. The structural domain of mathematical and reasoning that stemmed from its application supported the process of generalizing image characteristics that corresponded with images generated by the virtual lab. As the *lens function* helped determine the *kind* of image, *real* versus *virtual*, the magnification function helped delve deeper and describe the image *magnification* and its *orientation*. Coupled interpretation of both functions completed description of the image characteristics.

9.3 Suggested Independent Student Work

The independent students assignment is centered on several objectives, yet similar to the proposed instructional unit discussed herein. Problem 9.1 can also be considered a lab that provides students with opportunities to verify their graph inferences with real lab outputs.

Problem 9.1: Given is a convex lens with a focal length of 20 *cm* and an object initially located remarkably close to the lens, thus $x \to 0$. The object is being moved away from the lens.

(a) Formulate the lens and magnification functions, sketch their graphs and determine the characteristics of the image when the object is located at 10 *cm* and 50 *cm* from the plane of the lens.
(b) Formulate and sketch a lens function and magnification functions if instead of convex lens a concave lens is used and determine characteristics of an image when an object is located at 10 *cm* and 50 *cm*. Did you notice any similarities between the characteritics?

In Problem 9.2 a snapshot of a simulation needs to be provided with a special case of the image produced.

Problem 9.2: The focal length of the lens to generate the images is 40 cm. Write the lens and the magnification function for the image case provided below (Fig. 9.8) and prove algebraically its non-existance.

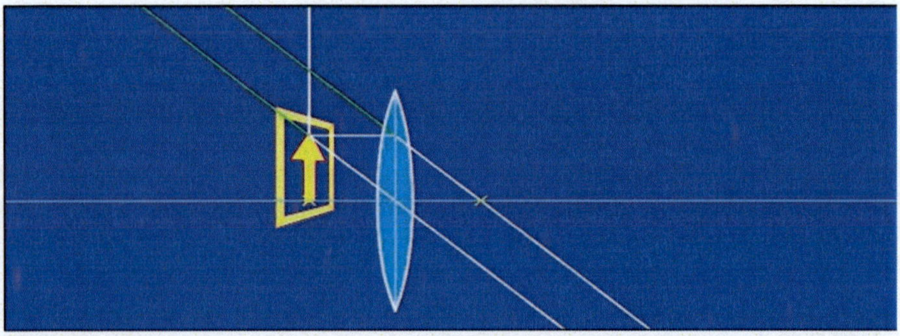

Fig. 9.8 The object is placed at the focal point of the lens. (Source: PhET Interactive Simulations (n.d.))

The context for Problem 9.3 was provided by a graph of a lens function.

Problem 9.3: An image of an arrow is created using a lens. The graph below shows the position of the image in the function of the object's position generated using that lens. All dimensions are given in centimeters.

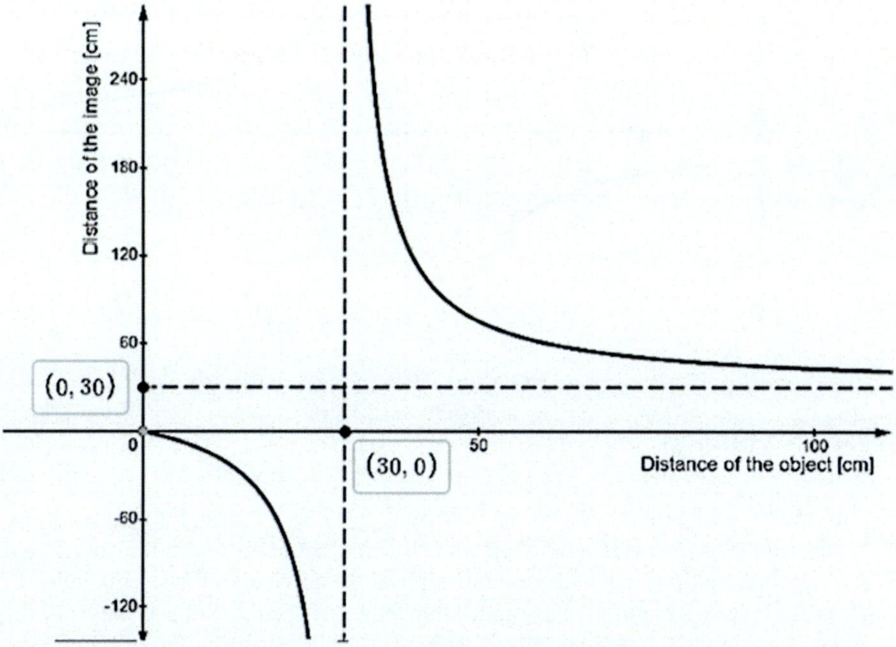

Using the graph, answer the following questions:

(a) What type of lens is used in the experiment?
(b) What is the approximated focal length of the lens?
(c) For what object distances is the image real?
(d) For what object distances is the image upright?
(e) Approximate the image position when the object distance is 10 cm.
(f) What is the magnification of the image when the object is placed at the position of *2f*?
(g) Find the radius of the curvature of the lens used in the experiment.

A graph of magnification function constituted a context for Problem 9.4.

Problem 9.4: The graph below illustrates a magnification function for the experiment described using this graph and the graph from Problem 9.3. (a) Write a lens function for this experiemnt, (b) Determine the characteristics of the image at the indicated object's distances: x=10 *cm*, x=30 *cm*, x=50 *cm*, and for the distance of the object considerably large. Use Table 9.1 to fill in your answers.

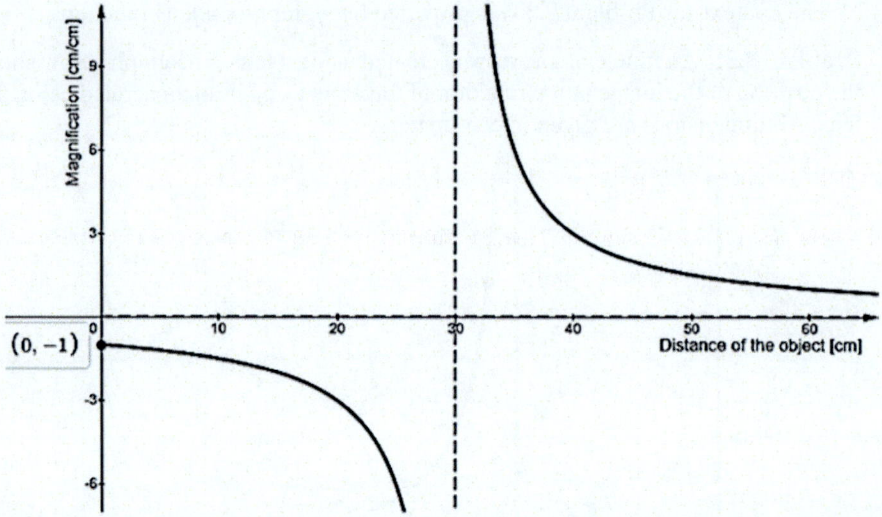

9.4 Summary

This study was not embraced in a formal evaluation. The lens function was though introduced to high school physics students. Students found it exciting to see how algebraic graphs can help generalize physics knowledge. It was concluded that a certain level of maturity and math knowledge is needed to have students fully appreciate such integrated lessons.

While the level of merging mathematical and scientific inquiries in the proposed instructional unit is sufficient to experience supporting role of mathematical reasoning to learn more about natural phenomena, opportunities for extending this inquiry, still exist. The focal length can be interpreted as yet another variable parameter, called perhaps mediating parameter. Such exploration could also constitute an independent student assignment and open up gates for students' to fulfill yet another journey of explorations. The focal length also depends on the lens index of refraction and the lens curvatures, classified as mediating parameters from a physics standpoint. The dependence encompasses these parameters is expressed as $\frac{1}{f} = (n-1)\left(\frac{1}{R_1} + \frac{1}{R_2}\right)$ and is often called the *lens maker's equation* (Giancoli, 2005). More mediating parameters can be identified to gain more insight into the characteristics of the image production and their effects on the image. This stage would depend on the students' physics and mathematical grade level and background. As lens function helps determine the kind of image, *real* versus *virtual*, and magnification function allows describing the size of the magnification and object orientation. Realizing this functional parallelism helps students remember tarditional textbooks rules on determining image characteristics.

References

Bing, T. J., & Redish, E. F. (2008). Symbolic manipulators affect mathematical mindsets. *American Journal of Physics, 76*, 418–424.

Bossé, M. J., Lee, T. D., Swinson, M., & Faulconer, J. (2010). The NCTM process standards and the five *Es* of science: Connecting math and science. *School Science and Mathematics, 110*, 262–276.

Giancoli, D. C. (2005). *Physics: Principles with applications* (6th ed.). Prentice-Hall.

McDermott, L. (1996). *Physics by inquiry: An introduction to physics and the physical sciences* (Vol. 2). Wiley & Sons.

National Council of Teachers of Mathematics (NCTM). (2000). *Principles and standards for school mathematics*. Retrieved from http://www.nctm.org/standards

National Research Council (NRC). (2000). *Inquiry and the national science education: A guide for teaching and learning*. National Research Press.

PhET Interactive Simulations (n.d.) accessed on July 2020.

Sokolowski, A. (2012). Enhancing interpretation of natural phenomena through math apparatus. *International Journal for Mathematics Teaching and Learning, 3*(399), 1–20.

Stewart, J. (2006). *Precalculus: Mathematics for calculus* (5th ed.). Brooks/Cole.

Tipler, P. A., & Mosca, G. P. (2004). *Physics for scientists and engineers* (5th ed.). New York, NY: W.H., Freeman.

Chapter 10
Embracing the Mole Understanding in a Covariate Relation

10.1 Introduction and Prior Research Findings

A diversity of research ranging from a perspective of psychology learning through historical and philosophical views (Furio et al., 2002; Siswaningsih et al., 2017, Feb.) had identified several deficiencies associated with an understanding of the mole; students (a) often identify the mole with mass expressed in kilograms and Avogadro's number of elementary entities; (b) avoid inserting the abbreviation of the amount in the units of moles that results in misinterpretations of the unit, (c) confuse *molar* mass with *molecular* mass. Other studies (Dahsah & Coll, 2007; Musa, 2009) showed that students' weak skills of handling conversion problems involving the mole stem from misinterpreting the meaning of *amount of substance* regarded as a number expressed in terms of moles. Johnstone (1971) found out that students' difficulties with the mole's concept are widespread and are rooted in the complexity of the mole's SI definition. Fang et al. (2014) suggested that "the research of the past 40 years is that the way the mole is conceptualized in educational settings is inconsistent with the meaning of the mole expressed in the SI definition" p. 351.

Attempts to help students understand the concept of the amount of substance as a mole has been made. Uce (2009) used a conceptual change method to develop the mole understanding. Staver and Lumpe (1995) pointed out that the difficulties originated from an insufficient understanding of the conversions' algorithmic routes. Tullberg et al. (1994) claimed that educators teach the mole's concept using extensive reasoning methods instead of direct algebraic equations. Fang et al. (2014) proposed using large rectangular shapes to depict schematically the mole and smaller rectangles inserted inside the larger to represent sub-concepts such as molar mass or atomic mass. Indriyanti and Barke (2017, August) suggested introducing the mole concept by having students count the number of items using weighing, which was employed to realize that the mole's concept was introduced because it

A. Sokolowski, *Understanding Physics Using Mathematical Reasoning*,
https://doi.org/10.1007/978-3-030-80205-9_10

was impossible to count the vast number of particles. Milton (2013) suggested introducing a different, more intuitive, definition of the amount of substance that measured the size of an ensemble of entities (atoms, molecules, ions, electrons, or other specified groups of particles). Erceg et al. (2016) identified several students' misconceptions about the kinetic molecular theory of gases that could be accounted for a weak understanding of units describing the mass of gas.

All the findings targeted various weaknesses of the mole understanding. Each pinpointed a specific aspect of the teaching process rather than suggested a more comprehensive mole introduction method. This study attempts to encompass all these suggestions and especially the one proposed by Scott (2010), who advocated for more integration between algebraic operations and scientific interpretations during the conversion processes. The instructional unit's central theme is to consider proportional reasoning as a backbone for setting up conversion processes and understanding the interpretation of the quantities computed.

10.2 Theoretical Framework

A position has been taken to design instructional support that would explicitly address students' weaknesses. To adequately reflect on the weaknesses, details about the nature of students' misinterpretations of the mole's understanding were summarized. These weaknesses along with suggestions on their elimination are tabularized in Table 10.1.

10.2.1 Weaknesses of the Mole Understanding

Prior research findings and challenges helped to formulate the framework for designing the treatment. The challenges are listed in the left column of Table 10.1, and the actions taken to overcome these challenges constitute the right column of the table.

Recommendations that emerged from the analysis called for unifying algebraic structures to enhance covariational relation between the mass expressed in the mass unit (e.g., kg) and mass expressed in terms of the moles. The review of the algebraic foundations of proportions, rates, and ratios emerged as the priority. Another recommendation targeted the terminology used to verbalize the quantities sought by applying the algebraic structures. For example, instead of *molar mass*, the phrase the mass of one mole of a substance *the mass of one mole of a substance* will be used, at least at the introductory stage of the unit development. This suggestion might seem to have a minor effect; however, when using the phrase *mass of one mole* or *mass of one atom*, the emphasis is placed on the quantity of mass, which students associate with grams or kilograms. It was reasonable to assume that the

Table 10.1 Prompt for formulating the theoretical framework

Challenges with the mole understanding	Suggested interventions
1. Overemphasized conceptual approaches during conversions. 2. High diversity of units used to describe the amount of quantities at a microscopic level. 3. Lack of systematization of the different units of measuring matter.	Use the structure of a proportion throughout various types of conversions systematically. Introduce the mole as a basic unit of the amount of substance in the form; $1mol = 6.02 \times 10^{23}\ atoms$ Introduce *molar mass* before introducing *atomic mass*.
4. Overemphasizing the mole's formal definition as regarded to Avogadro's number and the number of particles in carbon-10. 5. Lack of contrasting traditional units of mass (kg) with the amount of mass expressed in moles.	Introduce the formal definition of the mole after learners are fluent in basic conversion techniques. Provide students with opportunities for contrasting various units of mass and embracing it with conversions.
6. Lack of parallelism to algebraic operations while setting up converting the units of mass, e.g., expressed in kg to moles or a number of particles.	Using the idea of constructing *proportion* and the periodic table of elements to find the reference for molar and atomic masses.
7. Difficulties with interpretations of phrases like *molar mass* or *atomic mass* because the equivalent descriptions (e.g., *kilogram mass* or *pound-mass*) are not used in sciences.	Using the phrase *mass of 1 mole* instead of *molar mass* and *mass of 1 atom* instead of *atomic mass* and consequently apply these statements to build proportions.
8. Diminished effect of algebraic manipulations when applying a detailed dimensional analysis. 9. Lack of connection between atomic mass representing the mass of one mole of a substance.	Replace dimensional analysis by simplifying the units using algebraic algorithms for terms' reduction. Emphasize *mass number* (in the periodic table) as representing the mass, in grams, of 1 mole of a substance.

phrase *molar mass* might unintentionally emphasize the mole as the unit of measurements of the substance expressed in moles.

10.2.2 Proportional Reasoning, Rates, and Ratios

The structure of a proportion and its interpretations will serve as the primary algebraic structure to develop students' algebraic reasoning to help them understand the mole's underpinnings and subsequent conversions. Consequently, an outline of constructing proportions as viewed through the prism of covariation yet enriched to model scientific quantities emerged. While a similar concept called *the laws of proportions* (Mansoor & Rodrigues, 2001) can be found in the literature, this study introduces the learners to the modeling approach of formulating and solving proportions at the start. Departing from the theory of equation, fundamental scientific constants along with their interpretations will follow. Concise structures of mathematics will support quantification processes, and the corresponding scientific embodiments will stem from the outputs of the computations.

Proportional reasoning is a type of covariational reasoning referred to as the capacity to compare the values of two quantities within the same system (Confrey, 2008). The conceptual underpinning of a proportion is usually developed in junior high school (e.g., see Fielding-Wells et al., 2013). Coffield (2000) suggested its more formal introduction at the earlier grade levels to assure its solid conceptual and structural understanding. The applications of proportional reasoning extend to high school and higher-level mathematics. For example, in calculus courses, students formulate differential equations given by a specific entanglement between two quantities or algebraically model dependence on two or several quantities in related rates scenarios. Research has shown that while proportional reasoning is one of the most often applied covariational reasoning, its algebraic underpinnings are better understood with learners' maturity and experience (Clark & Kamii, 1996).

Proportional reasoning denotes reasoning with two variable quantities, y and x, between which there must exist a linear functional relationship of the form $y = mx$, where m represents the proportionality constant often expressed in the form of a constant rate or ratio. The proportionality constant can be picturized as a slope of the linear relationship when sketched. It can also be called a mediating parameter. The functional representation of proportionality is often regarded in mathematics as a *dynamic proportionality* (Miyakawa & Winsløw, 2009). The function can be used to find many values of the dependent or independent variables if a constancy of their rates or ratios within a particular domain is established. The other form of proportionality, more often applied in problem-solving, also used in this study, is called *static proportionality,* which is defined as a specific setup used for computing unique values. Static proportionality intertwines two different phases within a system of continuing the same behavior. Static proportionality or proportion compares two *rates* or two *ratios,* usually with one unknown quantity. What is the difference between ratio and rate?

(a) The ratio is a quotient of two similar or homogeneous quantities with typically the same units, for example, $\frac{10m^2}{6m^2}$, $\frac{1kg}{2kg}$, or $\frac{3L}{8L}$ that when simplified produce dimensionless values.

(b) The rate is a quotient of two different or heterogeneous quantities with different units—for example, $\frac{7m^2}{6s}$, $\frac{20kg}{5m^3}$, or $\frac{3mol}{6L}$. The units of rates cannot be canceled out.

When simplified by applying the division algorithm, ratios and rates represent a *unit* ratio or rate called the *proportionality constant.* For example, $\frac{10m^2}{6m^2} = 2$, $\frac{15kg}{5m^2} = 3\,kg/m^2$. In these processes, ratios might become dimensionless; however, rates retain the units. Furthermore, if the data set satisfies the condition that $y = 0$ if $x = 0$; formulating rates and ratios can establish direct dynamic proportionalities called linear functions. How to convert a static proportionality into a dynamic one or vice versa? Consider a linear functional relationship $y = mx$ and a direct proportionality constant expressed as a slope $m = \frac{y_1 - 0}{x_1 - 0} = \frac{y_1}{x_1}$. Substituting $\frac{y_1}{x_1}$ for m in $y = mx$ results in

$$y = \frac{y_1}{x_1} x \quad (10.1)$$

Equation (10.1) represents a dynamic proportionality allowing to compute y or x when either of these quantities value is provided and be sketched as a linear graph (Fig. 10.1). Suppose that the mass of one mole of magnesium is concerned. This value can be represented as a slope $m = 24\frac{g}{mol}$ and embraced in a dynamic proportional relation, $y = 24\frac{g}{mol}x$ that allows computing either the mass, y for a given number of moles of magnesium, x, or vice versa. This direct proportionality relation can be displayed in XY coordinates, with labeling corresponding to the quantities that it models.

The function inclination to the horizontal axis is the slope that can also be computed using $m = slope = \frac{rise}{run}$ with a value of m = 24. Sketching more graphs for different substances in the same XY axes would conveniently provide means for discussing their relative molar masses and chemical properties.

How to convert a dynamic proportionality into a static one? Static proportionality does not retain covariate relationships; instead, it refers to its selected values.

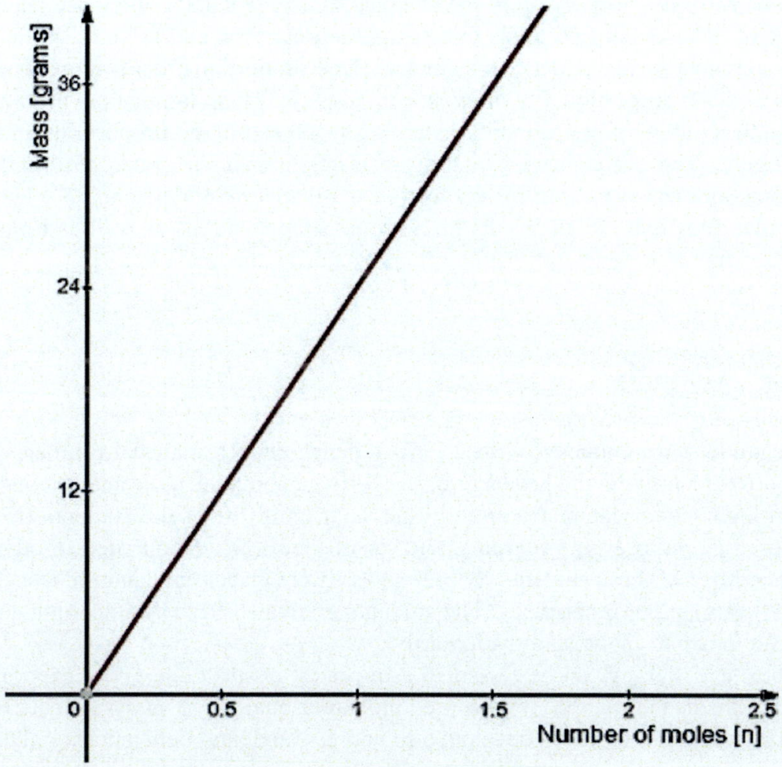

Fig. 10.1 Graph of mass versus the number of moles for magnesium

Replacing y by y_2 and x by x_2 in Eq. 10.1 results in $y_2 = \frac{y_1}{x_1} x_2$. Expressing the terms as ratios produces a static proportionality:

$$\frac{y_2}{x_2} = \frac{y_1}{x_1} \tag{10.2}$$

By applying a cross multiplication and division, this proportion can be converted to a proportion of two ratios, respectively

$$\frac{y_2}{y_1} = \frac{x_2}{x_1} \tag{10.3}$$

A static proportion, either containing two rates or ratios, can be formulated only for quantities satisfying the condition of a *direct proportionality*. For example, the proportions; $\frac{20kg}{5m^3} = \frac{x}{4m^3}$, or $\frac{4m^3}{5m^3} = \frac{x}{20kg}$ represent direct proportionality because, from an algebra point of view, a corresponding dynamic representation will lead to a linear function with the initial value of (0,0); $y = mx$. From a scientific point of view, the condition informs that e.g., if the object's volume is zero, its mass must also be zero. It is usually up to the students to decide whether they use an equivalence of two rates or two ratios to construct a proportion and solve it. While literature does not provide evidence on which of these proportionalities is more accessible to students, a proportionality of two ratios (Eq. 10.3) will dominate in this study.

Understanding proportionality extends beyond setting up proportions and computing the value of the missing quantity. It is linked with one of the most fundamental functions in algebra, a linear function that supports understanding other more complex functions. Realizing its high scope of applications as well as limitations will intertwine with scientific underpinnings during the treatment.

10.3 Methods

This study was conducted with 25 (10 females and 15 males) freshman college physics students who did have a prior background about the mole and its conversion from high school chemistry and physics courses. Most of the students (N = 18) majored in engineering programs, and six of them (N = 6) majored in other programs. Most of these students (N = 20, 80%) were Caucasian, and the rest (N = 5, 20%) consisted of other races. The students graduated from various high schools, and all took part in the study voluntarily.

This undertaking can be classified as a pretest-posttest one-group experimental study (Shadish et al., 2002) with predominantly qualitative analysis of its results. The instructional unit was developed to address students' deficiencies in handling the mole understanding listed in Table 10.1 and its subsequent conversions. The treatment design also took into account the results of the pretest taken by this study's participants. The purpose of the pretest was bifocal; it served (a) as an additional source of prompts to design the pedagogy of the instructional unit (b) to assess students' ability to adapt the proportional reasoning techniques to make conversions. The following question guided this study:

Will proportional reasoning help students conceptualized the meaning of one mole and subsequent mass conversions to the amount of substance and vice versa?

This study was also to determine if the students could adopt different computational techniques for conversion processes.

10.4 The Lecture Component

The instructional unit was designed for one class period (55 min), and it consisted of several parts organized deductively. Thus, it began with more general conversions and zoomed gradually into more detailed ones. Each part was initiated by conceptual questions whose purpose was to activate students' proportional reasoning before using formal algebraic conducts. The instructional unit took the form of a discovery-type lesson. The instructor posited inquiries, let students discuss the answers, and then the corrected conversion process was applied to prove or disprove students' claims. While the unit could be narrowed to conceptualizing the mole's idea, a more pragmatic approach was taken to also include the mole's conceptualization as an amount of substance and merge it with developing conversion techniques. It was hypothesized that providing students with opportunities to illustrate a higher diversity of mass conversions and offer techniques for handling them will provide tangible learning experiences that students could apply in their other science courses.

10.4.1 The Mole as a Fundamental Unit of the Substance Amount

From a scientific point of view, the mole's idea might be challenging because it does not contain the traditional mass units such as kilograms or grams that students are trained to applying, thus addressing this nuance beforehand can help. The lecture was initiated by stating that on a macrolevel, a sample of a known substance can be characterized by its mass (in kg). In contrast, the sample can be described on a microscopic level by the number of its entities. These entities are atoms whose samples of milligrams order, or less, are needed for microscopic experiments. How to handle counting the atoms? Suppose one needs 10^{30} atoms. Counting atoms as single entities is impossible, and a unit representing a more substantial amount was needed. The unit, convenient for calculations and laboratory practice, consisting of 6.02×10^{23} atoms was called *the mole*, and the base conceptualization of the quantity is called the *amount of substance,* not the amount of mass. If there are 6.02×10^{23} atoms in a container, then one can claim one mole of the atoms in the container. The students were assured that a more formal definition of the unit of one mole would be provided later. The teacher also highlighted that while the *amount of mass (not substance)* must be expressed in the units of mass (e.g., grams or kilograms), the phrase *amount of substance* does not contain the term *mass*. Thus, it does not induce the units of grams or kilograms. In sum, the mole's unit does not explicitly illustrate the amount of mass traditionally expressed in kilograms. Consequently, it does not account for the mass of one atom. However, subsequent algebraic operations allow various conversion processes. While the mass of a substance is usually denoted by a lower-case *m,* e.g., $m = 4\ kg$, the amount of substance is represented by *n*, e.g., $n = 10\ moles$ that is abbreviated to $n = 10\ mol$.

To further conceptualize the mole's idea, the teacher informed that as most fundamental physical quantities are measured using devices, e.g., temperature by a thermometer or electric current by an ammeter, no instrument measures the amount of substance (in moles) directly. However, by applying algebraic operations, a substance mass measured by a balance scale expressed in kilograms can be converted to moles. The ultimate question emerged how to convert the mass of a substance to a corresponding number of moles? Before immersing in mass-mole conversions, the instructor introduced a techniquc of converting a number of particles to moles to support conceptual understanding of the concept of one mole and to highlight the mechanism of applying proportional reasoning. These conversions are intended to have students realize that the mole's unit can link the micro - and macro - worlds and *vice* versa.

10.4.2 Converting the Number of Atoms to the Units of Moles

The teacher posited the following question: Does the number of moles of a substance depend on the number of atoms? After a short discussion confirming that it does not, he posited the following problem.

Example 10.1: Convert 10.04×10^{23} atoms of silver to moles.

To convert the number of atoms to moles, the teacher assigned the variable of interest as *n* represented by 10.04×10^{23} *atoms* and demonstrated the process of formulating two parallel statements that will lead to making up a proportion; first containing fundamental constant called in this study *a known (conversion) statement*; $1 mol = 6.02 \times 10^{23}$ *atoms* and the other containing the variable of interest;

$$\textit{Known statement}: 1 mol = 6.02 \times 10^{23}\ atoms$$

$$\textit{The statement with a variable}: n\ moles = 10.04 \times 10^{23}\ atoms$$

While to solve for n, the above setup of equations involved two algebraic operations instead of one if the *statement with a variable* were placed in the first line, this arrangement was purposeful and it was to highlight the mechanism of the proportional reasoning. Indeed the construction of all the proportions during this lecture was initiated from writing down the *known statements*. From an algebra point of view, the statements' order does not affect the final answer, and either is fine. The parallelism in formulating these statements is worthy of highlighting though; both contain moles on their left sides and atoms on their right sides, and the equal sign is vertically aligned. Such alignment is easier for students to remember. By embracing the statements in the algorithm of division, a proportion of two ratios emerged.

$$\frac{1}{mol} = \frac{6.02 \times 10^{23}\ atoms}{10.04 \times 10^{23}\ atoms} \tag{10.4}$$

By cross-multiplying, canceling the units of atoms, and solving the proportion for n, the students learned that

$$n = \frac{(1mol)(10.04 \times 10^{23}\ \cancel{atoms})}{6.02\ \times 10^{23}\ \cancel{atoms}} = 2\ mol \qquad 10.5$$

Some students had figured out the answer before solving the proportion. Nevertheless, the purpose of writing the statements was to have them understand the mechanism of setting up the equations whose general structures will be used for other types of conversions. Initiating the proportion using fundamental statements rather than ratios was to have students realize the underlying foundation of formulating the equity of the sides of the proportions. To enhance the understanding of the mole as a quantity linking the macro-world with the micro-world, the teacher posited a further question; can the amount of any substance expressed in moles be converted to atoms? After a short discussion, he wrote Example 10.2 on the board.

Example 10.2: How many atoms of monoatomic hydrogen are there in 4.2 moles of the substance?

The instructor pointed out that to solve the problem it is advisable to use a similar structure for formulating a proportion. Let x represent the number of atoms in 4.2 moles. The teacher did not use n to label the number of atoms to avoid confusion with n denoting the number of moles. The consistency happened to be unnecessary as almost all students who applied proportions on the posttest used x to denote any variable of interest which was fine.

$$Known\ statement : 1mol = 6.02 \times 10^{23}\ atoms$$

$$The\ statement\ with\ a\ variable : 4.2\ moles = x\ atoms$$

These statements lead to the following proportion:

$$\frac{1\,mol}{4.2\,mol}=\frac{6.02\times 10^{23}\ atoms}{x\ atoms} \tag{10.6}$$

When solved, $x = 25.28 \times 10^{23}$ atoms.

After solving a couple of more examples of a similar conceptual type, the teacher invited the students to determine if the kind of substance expressed in a number of atoms affected the corresponding amount expressed in moles. By observing the patterns of formulating the proportions, students realized independence of the properties of the unit of the mole from the atomic mass of the substance.

Students noted that mole-atoms conversions could be embraced in a dynamic proportion; $\frac{1\,mole}{n}=\frac{6.02\times 10^{23}\ atoms}{x}$ that lead to its more general version applied in the dimensional analysis; $x=\left(6.02\times 10^{23}\ \frac{atoms}{mole}\right)n$, where x represented the number of atoms and n, the number of moles. The shortcut stemmed from a static proportion, it was not given as a statement. Frequent use of 6.02×10^{23} *atoms* prompted an introduction of Avogadro's number. After assuring that the students realized the mole's unit as a designated number of atoms and were comfortable with mole-atoms conversion, the instructor proceeded to the second phase that dealt with the mass-mole conversion.

10.4.3 Converting Mass of Substance to Moles

This type of operation dominates in science, and it involves a reference to the periodic table of elements. Due to a low percentage of correct answers on the pretest (32%), (see Sect. 10.5.1 and suggestions of the prior research), this conversion path required a more detailed approach. To alert the students that if a substance mass in kilograms is given, then the number of moles will depend on the substance's atomic mass, the teacher posited the following conceptual problem. The problem will be solved after discussing with the students all involved lines of linking the quantities interpretations.

Example 10.3: Does 2 kg of mercury ${}^{200.59}_{80}Hg$ contain the same number of moles as 2 kg of potassium, ${}^{39.102}_{19}K$? Is it necessary to use Avogadro's number while converting these masses to the number of moles?

Most of the students claimed that the number of moles would not be the same and suggested that Avogadro's number be included in the proportion along with the mass of one atom of each of these substances. The students realized that when the mass in kilograms is given, a covariate relation of one atom's mass must be used first to find the number of atoms. Following this line of thought, the number of moles could be seen by setting up a proportion like in Example 10.1. Overusing Avogadro's number was coded by the prior research as students' *misinterpretations* leading them to incorrect conclusions regarding conversion. Thus, while this approach had its merit, the students realized that the Avogadro's number would have to be used twice in two inverse algebraic operations, multiplication and division, which would yield an identity operation. The teacher redirected the students' thinking toward interpreting the *mass number* as a mass, expressed in kg, of one mole of a substance. He reviewed the interpretations of numbers associated with atoms as described in the periodic table of elements; ${}^{A}_{M}X$ where A is called *the mass number*, and M (sometimes labeled as Z) is called the *atomic number*. As given in the periodic table of elements, the mass number is expressed in terms of u, the atomic mass unit that equals 1.66×10^{-27} *kg*. In conversion problems, the *mass number* can take a dual interpretation (see Fig. 10.2). It can be interpreted as a *mass of one mole of atoms* of a substance (also called *molar mass* and expressed in grams) or a *mass of one atom* of the substance (also called *atomic mass)* and expressed in kilograms if multiplied by the atomic mass unit u).

Both performances are mutually dependent, and one can be converted into the other or used as indicated by a specific task. Understanding the differences between these two different yet mutually coherent interpretations will be exemplified in the forthcoming conversion problems. The teacher informed the students that they would use the mass number as representing the mass in grams of one mole of a substance during this phase of inquiry. Realizing the differences in formulating proportion when finding the number of moles of a substance instead of finding the mass of a certain amount of substance moles was highlighted by the instructor as a critical determinant of setting up the proportion statements.

While in mathematics, setting up proportion is determined often by the symbols of variables, in science, correct interpretations of the tasks also contribute to correctly formulating algebraic statements. After that inclusion, the teacher referred to the problem stated and pointed out the elements' positions in the periodic table while asking the students to predict the answer before applying the formal process. The predicting phase was to have students use their mental image of the idea of 1mole, which, when coupled with the interpretation of notations used in the periodic table, provided a bridge to traditional interpretation of mass units. Students were ready now to construct proportions and convert both substances to moles. Example 10.3 is now converted to a task to prove.

Example 10.4: Prove that 2 kg of mercury, ${}^{200.59}_{80}\mathrm{Hg}$, contains a different amount of moles than 2 kg of potassium, ${}^{39.102}_{19}\mathrm{K}$.

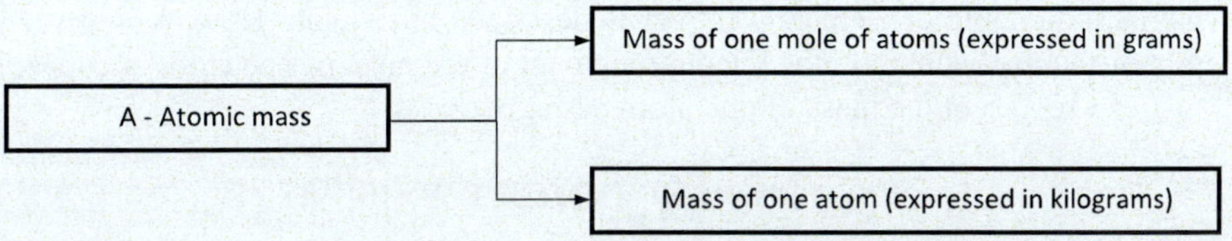

Fig. 10.2 Dual interpretation of the atomic mass number suggested in conversions

This problem requires setting up two independent proportions, finding the number of moles in 2 kg of mercury and finding the number of moles in 2 kg of potassium. Referring to the established in Example 10.1 structure for enacting a proportion, the teacher labeled n as representing the number of moles in 2 kg of mercury.

$$\textit{Known statement}: 1mol = 200.59g$$

$$\textit{The statement with a variable}: n\,mol = 2000g$$

$$\frac{1\,mol}{n\,mol} = \frac{200.59g}{2000g} \tag{10.7}$$

Thus $n = 9.97$ *moles*. By labeling the variable and setting up a similar proportion for potassium, the students learned that for potassium, ${}^{30.102}_{19}K$, $n = 66.44$ *mol*. Students confirmed that their predictions were correct that potassium with a lower mass of one mole of the substance generated more moles. When given by the number of atoms, the number of moles is independent of the substance type. However, the number of moles depends on the kind of substance when the mass of one atom characterizes the substance.

10.4.4 Converting Mass of a Substance to the Number of Atoms

This type of operation was designed to find conceptual connections between Examples 10.1 and 10.2 and practice using the mass number as representing the mass of one atom.

Example 10.5: How to set up a proportion to find the number of atoms in 4*kg* of copper?

Following the structures of previous conversions, the students suggested applying two steps to answer the question; first, to convert the mass to moles (as in Example 10.2) and then by using the property that $1mol = 6.02 \times 10^{23}$ *atoms* (as in Example 10.1) find the number of atoms. This path of thinking illustrated their conceptual understanding of the fundamental conversion techniques. After confirming the answer, the teacher intended to expand the type of conversions and suggested using the mass number as a mass of one atom of the substance expressed in kilograms. By the periodic table of elements, copper is described as ${}^{63.54}_{29}Cu$. How to construct a proportion that would use the known statement of the mass of one atom of copper?

Let x represent the mass of one atom of copper.

$$\textit{Known statement}: 1\,AMU = 1.67 \times 10^{-27}\,\text{kg}$$

$$\textit{The statement with a variable}: 63.54\,AMU = x\,kg$$

Creating a proportion and solving it for the variable results in $x = 1.06 \times 10^{-25}\ kg$ that represented the mass of one atom of copper. This process did not provide the answer for the initial question yet, and another proportion was needed. Let x represent the number of atoms in 4 kg of copper.

$$\textit{Known statement}: 1\,atom = 1.06 \times 10^{-25}\ \text{kg}$$

$$\textit{The statement with a variable}: x\,atoms = 4\,\text{kg}$$

By solving the proportion, one learned that $x = 3.77 \times 10^{25}$ *atoms* represent the number of atoms in 4 kg of copper. While the process depicted all the details, most students simplified it by formulating the proportion $\frac{1}{x} = \frac{1.06 \times 10^{-25}\ kg}{4kg}$ or just writing the final step $x = \frac{4kg}{1.06 \times 10^{-25}\ kg}$, which was also accepted.

After solving these examples, the students were invited to solve textbook problems and design a theoretical experiment that would formulate a regression line whose slope would represent the mole's unit. This task aimed to link static and dynamic representations of proportions and have the students become more fluent in translating between these representations. During the following lecture that is not discussed in this study, the instructor introduced a formal definition of one mole. He also made further connections to Avogadro's number and carbon -10. To bring forth applications, he used the conversion techniques to solve problems regarding ideal gas law; $pV = nRT$ and $pV = Nk_BT$, where p represented pressure inside the gas expressed in Pascal's, V volume of the gas expressed in meters cubed, n number of moles, $R = 8.31 \frac{J}{mol} \bullet K$ gas constant, T gas temperature expressed in Kelvin scale, N number of atoms, and $k_B = 1.381 \times 10^{-23} J/K$ Boltzmann's constant. While working on converting, students were given options of using any methods of their preference.

10.5 Pretest Posttest Analysis

The study used qualitative analysis of students' responses, which reflected on its aim. Statistical descriptive analysis was used to assess True/False multiple types of questions.

10.5.1 Analysis of the Pretest Results

Three questions, designed by the author and consulted with science professionals, were used to assess the students' understanding of the mole's unit and its conversion. Item #1 was a True/False type, item #2 required the students to convert the given mass to moles, and item #3 assessed students' level of understanding of the

Table 10.2 Students' pretest responses (Item #1, True/False)

Multiple choices	Answer/percent correct
(a) The mass of two moles of oxygen is the same as the mass of two moles of nitrogen.	False, 88% (N = 22)
(b) Two moles of iron contain the same number of atoms as two moles of zinc.	True, 56% (N = 14)
(c) Molar mass can be expressed in kilograms or grams.	True, 32% (N = 8)
(d) Both atomic mass and molar mass can be used to find the mass of a certain number of particles.	True, 72% (N = 18)

definitions of: 1 mole, molar, and atomic mass. Table 10.2 illustrates the descriptive analysis of item #1 and the evaluation of the remaining items follows.

The percent of correct answers on Item #1, part #a, was relatively high (88%), which showed that the students perceived the mole unit as the quantity measuring the amount of substance not mass. The percent of correct answers on the remaining parts of this assessment item varied. A relatively low percentage of the correct answers (32%) was reported on the molar mass interpretation (c). This deficiency was addressed while formulating the theoretical framework.

Item #2 asked the students to convert 2 kg of sodium ${}^{22.99}_{11}Na$ to moles. It was correctly solved by 20% (N = 5) of the students. About 24% (N = 6) of the students did not attempt to solve it, 15% (N = 4) got lost in setting up the conversion process (while using the dimensional analysis). The remaining 39% (N = 10) inserted the Avogadro's number in the dimensional analysis and could not carry out the units' cancellation. None of these students attempted to apply proportional reasoning to solve the problem.

Item #3 asked students to define the mole's unit, molar mass, and atomic mass using their own words. The analysis showed that the students understood the mole's conceptual definition; most of them (80%, N = 20) stated that the mole represents the number of particles/atoms/entities a substance has or an object is made of. The remaining 20% (N = 5) used general descriptions. Verbatim, for example: "*It is used to describe matter and can be converted to other things*," "*a unit of measure to express mass, volume, and particles*," "*a basic unit used for mass.*" The students could not conceptualize the phrases; *molar mass* and *atomic mass* on which the percentages of correct answers were respectively (25%, N = 6) and (30%, N = 8). The pretest questions were not returned to students, nor were they discussed until the study was completed.

10.5.2 Comparisons of the Pretest and Posttest Results

The students took the posttest after about 6 weeks from the date they took the pretest. The format of the posttest questions was like the one given on the pretest. A summary of the comparison of pretest/posttest responses to Question 1 is included in Table 10.3. The percentages show the correct answers.

Table 10.3 Students' pretests and posttest responses to Question #1

Question: classify as true/false	Pretest	Posttest
(a) The mass of two moles of oxygen is the same (b) As the mass of two moles of nitrogen	False, 88% (N = 22)	90% (N = 23)
(c) Two moles of iron contain the same number of atoms as two moles of zinc.	True, 56% (N = 14%)	85% (N = 21)
(d) Molar mass can be expressed in kilograms or grams.	True, 32% (N = 8)	78% (N = 20)
(e) Both atomic mass and molar mass can be used to find the mass of a certain number of particles.	True, 72% (N = 18%)	96% (N = 24)

The posttest results indicate an improvement in all areas tested, especially in interpreting *molar mass* (c). This improvement might be accounted for highlighting established verbal definition as *the mass of one mole* and suggest that the commonly used phrase *molar mass* does not explicitly convey the quantity's physical unit and its correct interpretation to students. At a similar rate (80%, N = 20), the students succeeded in converting 2 kg of sodium, $^{22.99}_{11}Na$, to moles. The results on both questions supported the hypothesis that using the structure of proportion and embracing the process with consistent and systematic reasoning helped students understand the interpretation of the mole's unit, learn, and apply the conversion techniques. Percentages of students who claim that both *molar mass* (86%, N = 22) and *atomic mass* (90%, N = 23) can be used to find a mass of particles were also higher on the posttest.

Several interesting inferences emerged from a more detailed descriptive analysis. While initially, I assumed that knowing the percent of correct/incorrect answers would assess the intervention, I decided to zoom more in-depth into how students solved the problems and evaluate their methods. Out of the 20 students who correctly converted 2 *kg* of sodium to the number to moles, (20%, N = 4) applied dimensional analysis, and 80% (N = 16) constructed proportions. On the pretest, no student used proportions in the conversion process. It is to mention that these students were free to use any method of their choice to answer these questions. While on the pretest, two of these students attempted to apply the dimensional analysis; these students adopted the method of proportions on the posttest.

Zooming deeper, it was interesting to learn what caused the remaining five students not to succeed in conversion problems. Two of the students did not attempt to solve it while having the proportion set up correctly; two included Avogadro's number in the dimensional analysis and got lost during the conversion process. One student incorrectly converted kilograms to grams, which resulted in incorrect mole computations. A strong association of the mole's unit to Avogadro's number still prevailed in these students' minds even though Avogadro's number was not emphasized during the instructional unit. Misuse of the Avogadro's number was also reported by other researchers (e.g., Furio et al., 2002).

The percentages of students who correctly interpreted (a) the unit of the mole were (N = 24, 96%), (b) the molar mass (86%, N = 22), and (c) atomic mass (90%,

N = 23). The percentages were higher on the posttest when compared to the pretest. It is believed that bringing forth students' mathematical reasoning in a manner consistent with what they had studied in their mathematics courses allowed the transitioning of these skills and supported the conversion processes.

10.6 Summary and Conclusions

The purpose of this study was to propose alternative ways of teaching the conversions of micromacro worlds by using proportional reasoning, fundamental constants, and the core idea of substance amount expressed in the units of moles. While the study's focus was to develop conversion techniques, understanding the idea of molar mass and atomic mass was also considered. I infer that augmenting molar and atomic mass verbal descriptions helped students understand the units' meaning and set up correctly the algorithms for conversions. Furthermore, consistent integration of mathematical structures with the scientific interpretation recommended by (Niaz, 1989; Tullberg et al., 1994; Scott, 2010) also had its high contributions. While the prior research suggested such integration, this study attempted to provide more in-depth didactical approaches readily available to be duplicated. It is somewhat misleading to assume that students will figure out how to integrate algebra rules during scientific quantification processes by themselves. The integration of both subjects' contents must be developed using involved contexts and delivered with attention to all details, algebraic and scientific so that students can realize their underlying mutual unity and adopt these methods in school practice.

It is reasonable to suggest that providing students with methods whose algebraic structures are familiar to students can improve the mole's understanding and other computations not necessarily related to conversions.

While using more pretest/posttest items and assessing the intervention from a wider angle would provide means for a more comprehensive evaluation of the intervention instrument, it is believed that the quantitative/qualitative nature of the items provided sound evidence of its didactical potential. One of the limitations of the study was the relatively low sample size of the group under treatment. However, it is believed that curriculum policymakers and other stakeholders can consider the encouraging posttest results to adopt the didactical merit of the unit in physics, chemistry, and lower levels of science courses.

I hope that this study will initiate further research to assess students' understanding of other scientific concepts where proportional reasoning can be applied. Furthermore, unifying the methods of teaching the mole across physics and chemistry and other science courses would not only help solidify the approaches but also encourage instructors and students to use covariate reasoning to unify other inquiries in science.

References

Clark, F., & Kamii, C. (1996). Identification of multiplicative thinking in children in grades 1–5. *Journal for Research in Mathematics Education, 27*(1), 41–51.

Coffield, F. (Ed.). (2000). *The necessity of informal learning*. The Policy Press.

Confrey, J. (2008, July). A synthesis of the research on rational number reasoning: a learning progressions approach to synthesis. *Paper presented at the 11th International Congress of Mathematics Instruction*, Monterrey, Mexico.

Dahsah, C., & Coll, R. K. (2007). Thai grade 10 and 11 students' conceptual understanding and problem-solving ability in stoichiometry. *Research in Science and Technology Education, 25*, 227–241.

Erceg, N., Aviani, I., Mešić, V., Glunčić, M., & Žauhar, G. (2016). Development of the kinetic molecular theory of gases concept inventory: Preliminary results on university students' misconceptions. *Physical Review Physics Education Research, 10*(2), 020139.

Fang, S. C., Hart, C., & Clarke, D. (2014). Unpacking the meaning of the mole concept for secondary school teachers and students. *Journal of Chemical Education, 91*(3), 351–356.

Fielding-Wells, J. (2013). Inquiry-Based Argumentation in Primary Mathematics: Reflecting on Evidence. *Mathematics Education Research Group of Australasia*.

Furio, C., Azcona, R., & Guisasola, J. (2002). The learning and teaching of the concepts 'amount of substance'and 'mole': A review of the literature. *Chemistry Education Research and Practice, 3*(3), 277–292.

Indriyanti, N. Y., & Barke, H. D. (2017, August). Teaching the mole concept with sub-micro level: Do the students perform better?. In *AIP Conference Proceedings* (Vol. 1868, No. 1, p. 030002). AIP Publishing.

Johnstone, A. H. (1971). Topic difficulties in chemistry. *Education in Chemistry, 8*(6), 210–213.

Mansoor, N. I. A. Z., & Rodrigues, M. A. (2001). Do we have to introduce history and philosophy of science or is it already 'inside'chemistry?. *Chemistry Education Research and Practice, 2*(2), 159–164.

Milton, M. J. (2013). The mole, amount of substance, and primary methods. *Metrologia, 50*(2), 158.

Miyakawa, T., & Winsløw, C. (2009). Didactical designs for students' proportional reasoning: An "open approach" lesson and a "fundamental situation". *Educational Studies in Mathematics, 72*(2), 199–218.

Musa, U. (2009). Teaching the mole concept using a conceptual change method at the college level. *Education, 109*(4), 683–691.

Niaz, M. (1989). The role of cognitive style and its influence on proportional reasoning. *Journal of Research in Science Teaching, 26*(3), 221–235.

Scott, F. J. (2010). Is mathematics to blame? An investigation into high school students' difficulty in performing calculations in chemistry. *Chemistry Education Research and Practice, 13*(3), 330–336.

Shadish, W. R., Cook, T. D., & Campbell, D. T. (2002). *Experimental and quasi-experimental designs for generalized causal inference*. Houghton Mifflin.

Siswaningsih, W., Firman, H., & Khoirunnisa, A. (2017, February). Development of two-tier diagnostic test pictorial-based for identifying high school students misconceptions on the mole concept. In *Journal of Physics: Conference Series* (Vol. 810, No. 1, p. 010117). IOP Publishing.

Sokolowski, A. (2019). Modeling the mole understanding with mathematical reasoning. *International Journal of Physics & Chemistry Education, 11*(4), 85–92.

Staver, J. R., & Lumpe, A. T. (1995). Two investigations of students' understanding of the mole concept and its use in problem-solving. *Journal of Research in Science Teaching, 32*(2), 177–193.

Tullberg, A., Strömdahl, H., & Lybeck, L. (1994). Students' conceptions of 1 mol and educators' conceptions of how they teach 'the mole'. *International Journal of Science Education, 16*(2), 145–156.

Uce, M. (2009). Teaching the mole concept using a conceptual change method at the college level. *Education, 109*(4), 683–692.

Chapter 11
Enabling Covariational Reasoning in Einstein's Formula for Photoelectric Effect

11.1 Prior Research

The research identified several students' misconceptions regarding the photoelectric effect's (PE) principles and difficulties with graphs' interpretations. McKagan et al. (2009) found that 42% of students "claimed (mistakenly) that a voltage is necessary and sufficient for current flow or to overcome the work function of the metal" (p. 92). Klassen (2008, July) found out that the main difficulty and, thus, the source of misconceptions originates from incorrect interpretation of the work function. Steinberg et al. (1996) reported that students faced problems in making predictions of relations embedded in the photoelectric effect. A series of experiments, computer-based tutorials, and simulations have been developed to help students understand the photoelectric effect. Earle et al. (2003) designed an experiment to help students understand that a photocurrent is dependent on *work function* when photons were incident on one side of a thin gold film, and electrons were ejected. Calvin (2004) created a module on quantum physics that had students reproduce some of the reasoning that won Einstein his Nobel Prize. Kovačević and Djordjevich (2006) developed a mechanical model of the photoelectric effect that enabled instructors to apply an analogy to support its understanding. Steinberg et al. (1996) created a computer-based tutorial on the photoelectric effect that addressed physics students' conceptual and reasoning difficulties. Sokolowski (2013) proposed an inductively designed instructional unit highlighting the PE's main phases and their entanglements. Research conducted by McKagan et al. (2009) proved that the virtual lab that emphasized the underlying principles of the photoelectric effect "provided significant improvement over traditional instruction," even though as they claimed, "there is still room for improvement in developing students' skills in reasoning from observations to inferences" (p. 94). While all these efforts aim to develop a conceptual understanding of the PE, the algebraic representation of the law of conservation of energy is not being analyzed for its adherence to algebraic function rules. Research

A. Sokolowski, *Understanding Physics Using Mathematical Reasoning*,
https://doi.org/10.1007/978-3-030-80205-9_11

shows that students consider functions as conceptually richer mathematical structures to learn about phenomena behavior than formulas (Sokolowski, 2020a); therefore, attempts to exemplify formulas' covariational attributes will be a priority in this study. While a linear function is widely taught and applied in the high math school curriculum, its interpretation in PE confuses the students. Considering the prior research findings, it was hypothesized that students' difficulties with applying mathematical reasoning to analyze PE might be rooted in how the central equation that models the law of conservation of energy of the PE, $K_{Max} = hf - W_o$, is presented and graphed. The following sections attempt to uncover the potential weaknesses of the equation and suggest its reimaging.

11.2 Theoretical Background

Covariational reasoning was discussed in more detail in Chap. 5. In this section, it is being used as a cognitive activity that involves the coordination of two varying parameters while trying to determine how one parameter affects the other in the context of the photoelectric effect. Covariational reasoning is primarily developed in mathematics classes; however, mathematics as a subject where students learn concepts and ideas without contexts does not offer students many opportunities to apply this type of synthesis in realistic settings. A preliminary survey conducted using academic research engines such as Google Scholar revealed that research in physics on covariational reasoning is staged on the conceptual level. Converting information to its graphical representation as a function requires identifying parameters that change (variables) and their classification as independent and dependent. Physics is a study of continuing dynamic phenomena; thus, it is reasonable to claim that covariational reasoning, as a way of inquiry, should earn its prominent role in physics education. This study is an example of such an undertaking.

11.3 Embracing the PE into the Framework of Covariational Representation

Einstein's formula to quantify the energy in the photoelectric effect includes–directly and indirectly–multiple concepts and principles; the Compton effect, Millikan's experiment, duality of light, photon, quantum, electric current, the intensity of the electric current, work function, threshold frequency, binding energy, potential difference, stopping potential and kinetic energy. All these principles and parameters aid the analysis as they all create a detailed picture of the phenomena' behavior. Following this complexity, a diversity of graphical representations is developed to support its quantitative and qualitative understanding. In this paper, a scrutiny of the main formula to quantify the energy in the photoelectric effect, $K_{Max} = hf - W_o$, will result in its reasembling to a covariate expression.

While the students' difficulties have a wide range, a substantial focus is given to identifying students' confusion with interpreting the graph of $K_{Max} = hf - W_o$. Following this finding, a discussion will be initiated to address this challenge by analyzing the graph of K_{Max}. This analysis will reflect on the effects of addends in the PE formula in exemplifying a linear covariation between external photons' frequency and the ejected electrons' kinetic energy. As a result, a piecewise linear function will emerge and be proposed in Sect. 13.4.

11.3.1 Weaknesses of the Graph of K_{MAX} Versus Photons' Frequency Presented in Physics Resources

The paper's purpose was to provide alternative graphical and algebraic representations of the PE. It was hypothesized that students are confused with the graph interpretation because the currently used algebraic and graphic models of PE do not foster a mutual coherence with its scientific understanding. The following sections provide support for that claim. The scrutiny will begin from analysing the graphical representation of the PE formula, shown in Fig. 11.1, and it will proceed toward discussing its algebraic equivalent. As the discussion unfolds, suggestions for modifications will emerge.

Figure 11.1 illustrates a graphical representation of $K_{Max} = hf - W_o$ that can be found in physics textbooks ranging from high school to college levels, e.g., Servay and Faughn (2002, p. 835), Allum and Talbot (2018, p. 509), Cutnell and Johnson (2007, p.119), and Giancoli (2005, p.760). The graph (Fig. 11.1) is meant to represent a linear function $f(x) = mx - b$. Some graph elements might challenge the students' mathematical reasoning if the general slope-intercept function is mapped into the PE formula. At first, the dotted part of the graph suggesting the electrons' zeroth kinetic energy can be misleading. In mathematics, a dotted line usually depicts attributes like function vertical or horizontal asymptotes. None of such graph attributes is meant to be labeled on this graph. Thus, to denote the electrons' zeroth kinetic energy, a different graphical representation should have been used. There are no frequency restrictions on the energy function $K_{Max} = hf - W_o$; thus, from a mathematical standpoint, there should be a way to evaluate the electrons' kinetic energy for any photons' frequency. This statement challenges the observation of the PE; the kinetic energy of the ejected electrons has a value of zero for $0 \leq f \leq f_o$. Does the graph intend to inform that the electrons' have negative kinetic energy or negative work function on its dotted part? Can the electrons' kinetic energy be concluded when its graph lays below the frequency axis? These are sample questions that students might have while trying to find a physical sense of such a graph using their math knowledge. Problematic can also be the interpretations of the graph's horizontal and vertical intercepts, which can stem from labeling two different quantities on

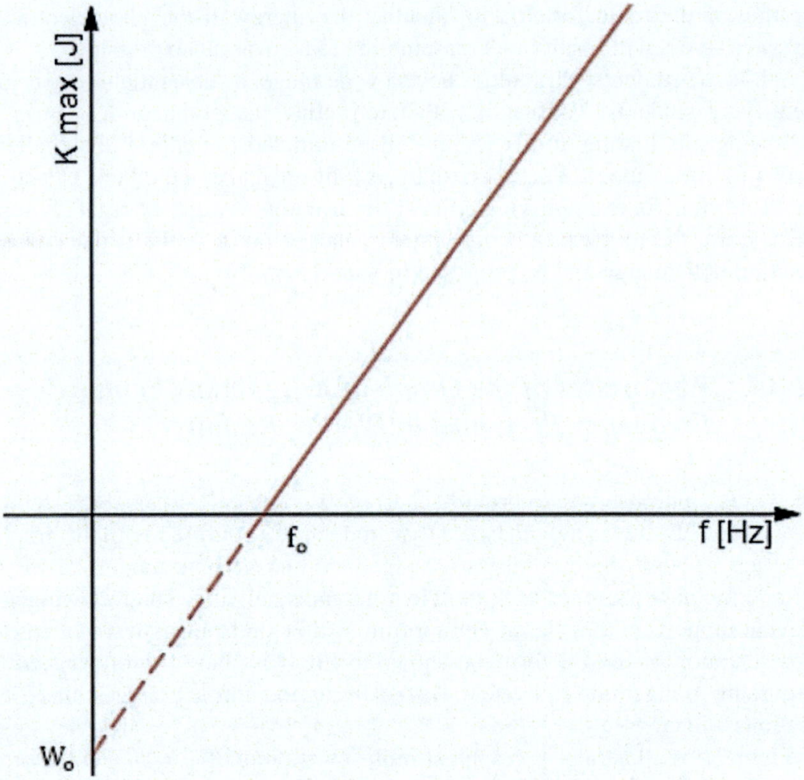

Fig. 11.1 A typical graph of the law of conservation of energy for PE found in physics textbooks

the graph's vertical axis: kinetic energy and the work function. In mathematics, the vertical intercept is a particular value of *f*(*x*), often called the initial value, whose units are similar to that of *f*(*x*). Work function represents in the equation $K_{Max} = hf - W_o$, *per see* the vertical intercepts of the linear function. However, this interpretation does not entirely reflect the scientific underpinnings of the PE formula. When the photons' frequency $f = 0$, there is no current observed, which is depicted by $K_{Max} = 0$. Thus, the vertical intercept of K_{Max} should be equal to zero. According to the PE formula, if $f = 0$, $K_{Max} = -W_o$. While the scientific interpretation can be (and typically is) augmented to satisfy such graphs, the algebraic reasoning does not fully support the scientific underpinnings if students attempt to interpret it directly using their mathematics knowledge. This element might cause some uncertainties; thus, a need emerged to re-evaluate and assemble the PE formula in a way that generates sound scientific conclusions coherent with observations. Students realize that kinetic energy is non-negative and that the external energy must be supplied to the system to free the electrons. This scientific fact has its right in the theory of algebraic equation; however, its covariational representation calls for revisiting. Suppose that one attempts to find the horizontal intercept of the graph. Students realize that in the search for the x-intercept and its scientific interpretation, they let the dependent parameter, thus $K_{Max} = 0$. Using the formula $hf = W_o$, which corresponds to the phase of freeing the electrons from the metal plate. In this case, $f_o = \frac{Wo}{h}$ is the threshold frequency of the metal plate or the incoming photons' minimum frequency that ejects the atoms' electrons. The kinetic energy of the electrons is still zero at this phase of analysis. Can the threshold frequency be called the x-intercept of the graph of K_{Max}?This is doubtful. According to the rules of algebra, x-intercepts are the x-values for which $f(x) = 0$. The function $f(x)$ that K_{Max} represents should have a value of zero on the entire interval $0 \leq f \leq f_o$. This behaviour is supported by the fact that if $0 \leq f < f_o$ no electrons are ejected or they are ejected with a zero kinetic energy when $f = f_o$. The electrons are ejected when the frequency of photons reaches the value of the threshold frequency f_o. However, they still have no kinetic energy to reach the collector. This scientific fact is not explicitly conveyed to students using $K_{Max} = hf - W_o$. The zeroth kinetic energy of the electrons on $0 \leq f \leq f_o$ will not emerge from using the algebraic form $K_{Max} = hf - W_o$. On the graph, this inquiry leads to the controversial dashed line (Fig. 11.1). If one attends to identifying a relationship between the photons' frequency and the ejected electrons' behavior, then the dashed part calls for a revision. It is hypothesized that students' confusion regarding the differences between the interpretation of the work function and electrons' kinetic energy might stem from trying to find a mathematical-physical sense of this part of the graph. The section that follows discusses the algebraic structure of the law of conservation of energy and attempts to align it with the phenomenon's behaviour.

11.3.2 Covariation of Photon's Energy and Frequency as a Linear Function

Photon's energy is commonly presented as $E = hf$. The scientific interpretation of the formula is straightforward. Photon's energy depends on the light source's frequency, with Planck's constant considered a proportionality constant or the relation's slope. A study (Sokolowski, 2020b) revealed that physics students have difficulties classifying parameters embedded in traditionally expressed physics formulas as dependent and independent, which creates uncertainty with graph interpretations. Thus, when covariational reasoning is concerned, function notation that signifies parameters' unique behaviour within the system is employed, which results in the formula expressed as $E(f) = hf$. This expression has a form of a linear function, and its slope must have a definite interpretation. Slope interpretations that link mathematical and scientific underpinnings also require attention to be correctly introduced to science students (Kwon et al., 2000). The unit of the slope in the PE is Js. The product, though, does not explicitly foster the slope interpretation as a rate. The slope in mathematics is considered as $m = \frac{\Delta y}{\Delta x}$, thus a rate or ratio of two quantities. If applications are concerned, the scientific units resulting from multiplying the slope and the independent variable must yield the unit of the variable of interest or the dependent variable. Does the formula $E(f) = hf$ convey the unit? In $E(f) = hf$, the Planck's constant $h = 6.626 \cdot 10^{-34}$ Js. Planck calculated the value from experimental data on black-body radiation and expressed the constant as a product of two quantities. While the unit's interpretation as a product is not an obstacle for physics instructors, it might not benefit the students. It is suggested that the unit be reconstructed to reflect on the slope interpretation more explicitly yet retain its original scientific sense. Since the unit of *hertz* is labelled on the horizontal axis, Planck's constant might be expressed in the units $\frac{J}{Hz}$, which is mathematically equivalent to *Js*. Expressing the Planck's constant as a rate of the energy change per unit of frequency allows enacting its precise interpretation using mathematical reasoning (Sokolowski, 2018); the energy of a photon changes by $6.626 \cdot 10^{-34} J$ per 1 Hz of photons' frequency change. Such interpretation also supports the quantized nature of the photon's energy. It is suggested that the traditional unit of the constant as *Js* is introduced to students after the PE, as a linear relation, is fully understood.

11.3.3 Electrons' Binding Energy as a Function of Photons Threshold Frequency

In the traditional algebraic setting, the independent variable of $f(x)$ affects the dependent. The vertical intercept of $f(x)$ represents a particular value of the dependent variable when the independent variable takes a value of zero. As earlier discussed, the scientific underpinnings of the formula $K_{Max} = hf - W_o$ do not entirely fit into this mathematical model. The binding energy of the electrons is an external parameter built in the phenomenon as an additional covariate. Its inclusion has its right, which is warranted in the work–energy principle stating that photons do work on electrons. Therefore, electrons are ejected. The work function is considered a metal plate parameter, which can be regarded as an additional object, thus a covariate quantity in the PE. From the covariational reasoning standpoint, its interpretation might be challenging to follow the PE dynamics. Therefore, instead of W_o, its equivalent expression hf_o will emerge and be used in the central Eq. (11.2) to quantify the PE. The symbols hf_o will correlate with the symbols of the function's established parameters. Thereby they also will enhance the math–physics interphase of the phenomena representation. The work function W_o will be included in Eq. (11.4) to support PE explorations and problem-solving.

11.3.4 Maintaining a Minimum Number of Covariational Parameters During the Inquiry

The number of varying parameters was initially reduced to assure clarity of the line of investigation. Physics formulas are traditionally built up as entanglements of two variable quantities. The photoelectric effect is more complicated if viewed from this angle. A high spectrum of contributing principles and quantities in the PE demands a high cognitive load; thus, a high capacity of the learners' working memory. Consequently, the multiple arrays of covariations that stem from the photocurrent production, while not requiring an advanced mathematical apparatus, can unintentionally dilute the PE's core idea. To maintain the number of the covariational parameters to its necessary minimum and yet attend to a full spectrum of the PT understanding, the following restrictions were imposed (a) the plate ejecting electrons; thus, hf_o remained unchanged, and (b) the external battery was removed initially from the consideration. Varying the plate's type certainly enriches the exploratory nature of the phenomena; however, this element does not directly contribute to the phenomena' algebraic representation. Analogically, the idea of the stopping potential was not included at first. While having its right by supporting the PE quantification when introduced at the initial stage, it might inadvertently create an impression that the battery enhances the electrons' movements. These steps were taken to eliminate possible distractions and clear the inquiry path to help students realize that the only condition for initiating the photocurrent was ejections of the electrons from the emitter and having them reach the collector. Lloyd (2020) provided further details supporting the removal of the stopping potential from PE teaching. As a result, the plate remained unchanged throughout the experiment demonstration, and the external battery was set up at zero potential. Thus, the analysis was restricted to Einstein's equation. Both these parameters remained silent and labelled as constants throughout the instructional unit conduct. These steps were taken to eliminate possible effects of other covariate parameters and thus clear the inquiry path to help students realize that the only condition for initiating the photocurrent was ejections of the electrons from the emitter and having them reach the collector.

11.4 Reassembling the PE Formula to Assure a Coherence of Representations

Research showed that physics students need to be guided to reimage formulas into algebraic functions (Sokolowski, 2019); more specifically, they can better identify and analyze casual relationships if the physics formulas resemble mathematical functions denoted using function notation. Function notation explicitly highlights the independent and dependent parameters (variables); thus, the path to focus investigator attention on their mutual entanglement is cleared. Following this finding, instead K_{Max}, $K(f)$ will be used to denote the kinetic energy of the ejected electrons. Using K_{Max} has its scientific validity, and it is suggested as an additional fact that enriches the scientific nature of the PE. The proposed notation $K(f)$ could retain the subscript, $_{Max}$. However, in students' realms, such subscript could evoke the maximum value of the kinetic energy, which in its mathematical sense triggers specific behavior and algorithms for its computations; these actions are not intended.

11.4.1 Graph Constructing

Visual representations are to decrease the demand for working memory to allow focusing on understanding the phenomena behavior. However, the graph is to be free from abstractions and clear to read its scientific underpinnings to serve its purpose. Graphs must also be coherent with algebraic representations of the phenomena' processes. Formulating the algebraic expression of the PE will be based on developing its graphical counterpart first. The virtual lab whose snapshot is presented in Fig. 11.2 enriches the visual representation and serves as a source of scientific validation of the enacted algebraic expression.

The experiment provided evidence that no electrons were ejected when $0 \leq f < f_o$. This scientific fact demanded that the kinetic energy graph should have zero value for that frequency range. Consequently, this observation suggested a piece-wise linear function and that the first piece of the $E(f)$ graph should have a zero slope and be positioned along the frequency axis. When the frequency of the photon reached the threshold value, f_o, electrons were ejected. A further increase in the photons'

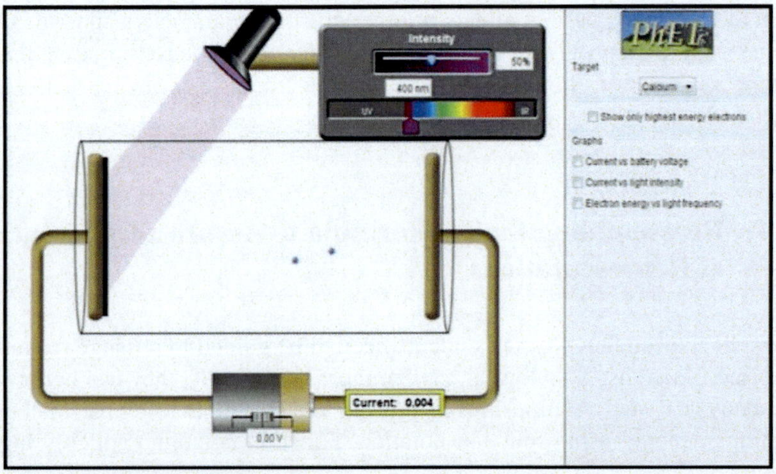

Fig. 11.2 Demonstration of photons kinetic energy. (Source: PhET Interactive Simulations (n.d.))

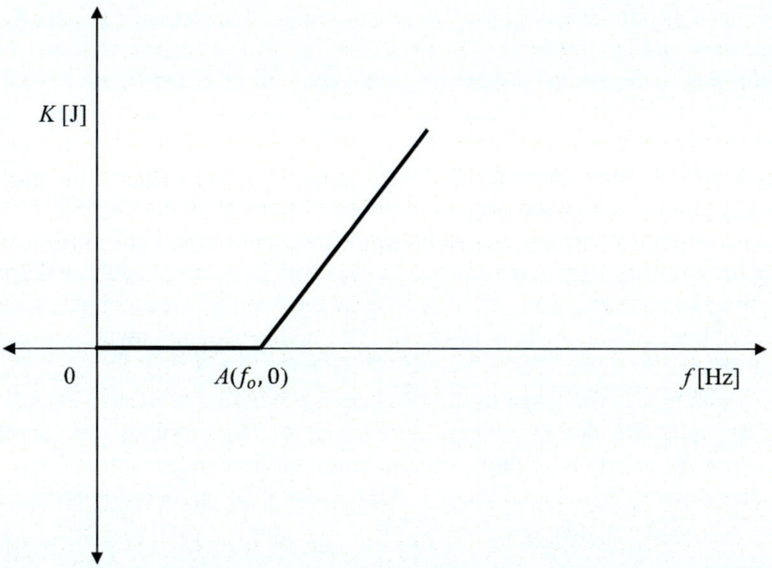

Fig. 11.3 A piecewise linear function representing kinetic energy of the ejected electrons

frequency caused a rise in the electrons' speeds, thus their kinetic energy. The electrons' kinetic energy depended linearly on the incoming photon's frequency. While justifying the linear covariance is difficult by observing the electrons' speeds, this can be followed by activating *Electrosn energy versus light frequency graph* made available on the right side on the simulation menu (Fig. 11.2). Therefore, the second part of the graph must also be linear but with a slope different than zero. A graph representing such a relation is illustrated in Fig. 11.3.

The graph does not cross the horizontal axis because kinetic energy cannot be negative. The function satisfies the vertical line test, following its math conditions (Stewart, 2000). It also meets the principle of continuity in its domain. Thus, mathematically, the graph depicts a linear covariate function that corresponds with the phenomena' behavior. In the earlier study (Sokolowski, 2021), the left and right-sided notations were denoted on the graph to indicate the experiment's possible directions to attain the threshold frequency. However, it seems that such insertion is not necessary.

11.4.2 Finding Algebraic Representation of the Graph

The horizontal and the inclined parts of the graph signify different algebraic rules within the function, yet two mutually consistent scientific behaviors. The vertical axis of the graph represents the kinetic energy of the ejected electrons on its entire range. Since the kinetic energy can be zero or greater than zero, the graph lays along and above the horizontal axis.

With such scientific and algebraic background, the students are ready to find the graph's algebraic representation. If the external frequency of the photons is not sufficient to overcome the electrons' binding energy, no electrons will be ejected, which is algebraically represented by $K(f) = 0$. Thus, according to the graph, the following can be concluded: $K(f) = 0$, when $0 \le f \le f_o$. If the external frequency reached f_o, the electrons are ejected. However, they have no capability of moving toward the collector. This scientific fact could be included as an additional third part of the function. Nevertheless, it was seen as an element of overusing the mathematical description and diminishing its clarity. To find the equation for the second piece of the graph, one can refer to the slope-point form for a linear function $y - y_A = m(x - x_A)$. The slope of the function is the earlier discussed Planck's constant. By assigning the experiment parameters to the variables, one learns that $K(f) - y_A = h(f - x_A)$. To complete the function, at least one additional coordinate from the graph of $E(f)$ is necessary. If the frequency of incoming photons reached the threshold frequency, f_o, $K(f_o) = \lim_{f \to f_o^-} (hf - hf_o) = hf_o - hf_o = 0$. The electrons are ejected, but they do not possess kinetic energy. The additional coordinate necessary to formulate $K(f)$ is then $A(f_o, 0)$. Substituting this coordinate for (x_A, y_A) in the slope-point linear model results in $K(f) - 0 = h(f - f_o)$, which leads to $K(f) = hf - hf_o$ when $f > f_o$. By combining both pieces of the function, one learns that:

$$K(f) = 0, \quad \text{when} \quad 0 \le f \le f_o \qquad (11.1)$$
$$K(f) = hf - hf_{o,} \ \text{when}\ f > f_o$$

The function $K(f)$ presented in (11.1) is a piecewise linear function, with two different rules on its domain. The parameter f_o represents the metal plate's threshold frequency and a specific value of the frequency of the incoming photons that ejects the electrons. It is to note that $f_{o,}$ called *cut off frequency*, has its specific value on the domain of $K(f)$, not on its range. Considering this condition, the kinetic energy function of the ejected electrons, $K(f)$, can be written in a more concise fasion using a piecewise function representation:

$$K(f) = \begin{cases} 0, & \text{when}\, 0 \le f \le f_o \\ hf - hf_{o,} & \text{when}\, f > f_o \end{cases} \qquad (11.2)$$

The electrons are not ejected at all if $f < f_o$. Students are to realize that when $f = f_o$, electrons are ejected but with a zero speed. This fact can also be inferred from the first piece of the function. If the frequency of the incoming photons is greater than that of the threshold frequency, thus if $f > f_o$, then electrons are ejected. According to the law of conservation of energy for the PE, electrons possess kinetic energy that can be computed from $K(f) = hf - hf_o$. The function illustrates the dynamic process of the phenomena, and the mathematical reasoning is aligned with it. Such function also satisfies the principle of continuity on its entire domain, $f \ge 0$, which transcends to the fact that the kinetic energy of the electrons can be calculated for any frequency within that domain. There can be a restriction on frequency maximum value due to the range of the light frequencies available for a specific experiment.

11.4.3 *Linking the Photons Threshold Frequency and the Work Function* $hf_o = W_o$

While the derived piecewise function satisfies the rules of algebra, students welcome its relevance to problem-solving. Using the law of conservation of energy, the threshold energy of the incoming photons is equal to the work function of the emitter, $hf_o = W_o$.

$$\left\{ \begin{array}{l} K(f) = \left\{ \begin{array}{ll} 0, & \text{when } 0 \leq f \leq f_o \\ hf - hf_{o,} & \text{when } f > f_o \end{array} \right. \\ hf_o = W_o \end{array} \right. \tag{11.3}$$

Creating such an algebraic system of equations allows linking W_o with $K(f)$, which is accomplished by replacing hf_o in $K(f)$ by W_o. By such algebraic manipulations, $K(f)$ takes the following form:

$$K(f) = \left\{ \begin{array}{ll} 0, & \text{when } 0 \leq f \leq f_o \\ hf - W_{o,} & \text{when } f > f_o \end{array} \right. \tag{11.4}$$

Once the students realize that the system of equations allows including W_o in the function equation, the confusion about the dashed line (Fig. 11.1) in the general graphical representation of the PE can be eliminated. Using W_o relates all involved entities in the photoelectric effect and supports problem-solving. However, it does not directly support the understanding of the virtue of the photoelectric effect and its graph. Certainly, including W_o, the work function of the emitter extends mathematical reasoning, thus enriches the inquiry. Using various work functions, thus different objects (emitters) will generate different threshold frequencies and strech or compress the graphs horizontally. However, the slopes of all graphs will remain identical and equal to Planck's constant.

11.5 Summary and Conclusions

The form of the law of conservation of energy for PE found in the textbooks works fine to support problem−solving; however, its graphical representation as constructed directly using its algebraic formula is problematic. The paper's purpose was to indicate its weaknesses as seen from the students' perspective and propose its alternative algebraic representation coherent with the phenomena' scientific underpinnings and a more rigorous algebraic notation.

While a formal evaluation of the proposed algebraic representation of the photoelectric effect was not possible to furnish at the time of the study development, an instructional unit with its context was delivered to a group of twenty (N = 20) high school physics students. The unit was conducted over two class periods. On the first day, the students investigated the scientific underpinnings of photons' energy and interactions with different metals, $W(f_o) = hf_o$ that generated a linear function with the slope of h. During the second day, PE's virtual setup was used, and the mathematical modeling of the PE proceeded. All students were familiar with formulating and interpreting piecewise functions. Thus, the process of enacting the graphical and algebraic representations moved smoothly. Graph (Fig. 11.1) was not discussed during that time. It was discussed on another occasion when the students learned the purpose of inserting the external battery into the circuit. While quantum physics is considered a complex physics section, the students appreciated the lesson's simplicity and the dynamic simulated processes that provided them with model-like scenarios and a better understanding. The lab helped the students construct the mental images of PE that enhanced the functional covariations of the parameters of interest. The students were excited to see applications of piecewise functions in science. While they have explored these functions' visual representations in kinematics and energy sections, piecewise formulas are rare in quantum physics. As a follow-up task, they were assigned to construct piecewise functions for the given threshold frequency of selected metal plates, e.g., sodium and gold. It was interesting to realize that knowing this parameter was sufficient to set up boundaries to precisely represent the PE's behavior graphically and algebraically.

Students' genuine involvement and participation illustrated that mixing mathematical structural knowledge with scientific inquiry generates productive learning environments worthy of further explorations. Emphasizing the coherence of scientific evidence and consistency of algebraic representations was also to send a message that although physics is considered a mature academic discipline, precision of algebraic embodiments of its laws can support its better understanding. Thus, an opportunity to formulate new models and representations of natural phenomena exists.

References

Allum, J., & Talbot, C. (2018). *Physics for the IB diploma, hodder education* (p. 509).

Calvin, S. (2004). Following in Einstein's footsteps: Teaching the photoelectric effect. *The Physics Teacher, 42*(6), 340–341.

Cutnell, J., & Johnson, K. W. (2007). *Physics* (Vol. 2, 7th ed., p. 919). New Wiley & Sons, Inc..

Earle, G. D., Copp, B. L., Klenzing, J. H., & Bishop, R. L. (2003). A novel empirical study of the photoelectric effect in thin gold films. *American Journal of Physics, 71*(8), 766–769.

Giancoli, D. C. (2005). *Physics: Principles with applications* (6th ed., p. 760). Prentice-Hall.

Klassen, S. (2008, July). The photoelectric effect: Rehabilitating the story for the physics classroom. In *Proceedings of the Second International Conference on Story in Science Teaching*, pp. 1–17.

Kovačević, M. S., & Djordjevich, A. (2006). A mechanical analogy for the photoelectric effect. *Physics Education, 41*(6), 551.

Kwon, Y. J., Lawson, A. E., Chung, W. H., & Kim, Y. S. (2000). Effect on development of proportional reasoning skill of physical experience and cognitive abilities associated with prefrontal lobe activity. *Journal of Research in Science Teaching, 37*(10), 1171–1181.

Lloyd, D. R. (2020). Should the concept of stopping potential be removed from teaching for A-level? *Physics Education, 56*(1), 015501.

McKagan, S. B., Handley, W., Perkins, K. K., & Wieman, C. E. (2009). A research-based curriculum for teaching the photoelectric effect. *American Journal of Physics, 77*(1), 87–94.

PhET Interactive Simulations. (n.d.). The University of Colorado at Boulder. Retrieved from http://phet.colorado.edu, March 2021.

Servay, R., & Faughn, J. (2002). Holt Physics, Holt, Rinehart and Winston, Austin, USA, p. 835.

Sokolowski, A. (2013). Teaching the photoelectric effect inductively. *Physics Education, 48*(1), 35.

Sokolowski, A. (2019, August). Enhancing scientific inquiry by mathematical reasoning: the case of applying limits to model motion of a system of objects. In *Journal of Physics: Conference Series* (Vol. 1287, No. 1, p. 012051). IOP Publishing.

Sokolowski, A. (2020a). Unpacking structural domain of mathematics to aid inquiry in physics: A pilot study. *Physics Education, 56*(1), 015009.

Sokolowski, A. (2020b). Developing Covariational reasoning among students using contexts of formulas: (are current formula notations in physics aiding graph sketching?). *The Physics Educator, 2*(04), 2050016.

Sokolowski, A. (2021). Enabling covariational reasoning in Einstein's formula for photoelectric effect. *Physics Education, 56*(3), 035029.

Steinberg, R. N., Oberem, G. E., & McDermott, L. C. (1996). Development of a computer-based tutorial on the photoelectric effect. *American Journal of Physics, 64*(11), 1370–1379.

Stewart, J. (2000). *Calculus concepts and contexts* (2nd ed., p. 45). Brooks/Cole.

Chapter 12
Are Physics Formulas Aiding Covariational Reasoning? *Students' Perspective*

12.1 Introduction and Prior Research Findings

Students acquire graphing skills in mathematics by developing covariational reasoning (Carlson et al., 2002). This competence is rooted in identifying independent and dependent variables in the given context and converting it into graphs satisfying the context's attributes and conditions for being functions. Physics students use quantities, and at large, formulas in their physics courses to solve problems and to graph relationships between quantities. Contemporary research on understanding physics formulas is often motivated by how physics students interpret these expressions in the physical sense (e.g., Domert et al., 2012). Little research is dedicated to how students perceive physics formulas in the mathematical sense. Focusing on formulas' physical sense has its own right, and it develops the notion that physicists are focusing on interpretations of symbols in formulas using different semiotics (Redish, 2006). Such research generates findings that support that aim; Sherin (2001) found out that students attempt to understand physics formulas by the vocabulary of formulas' component symbols. University physics students display their understanding of physics formulas by interpreting the symbols that formulas are built from (Domert et al., 2012). Dunn and Barbanel (2000) and McDermott et al. (1987) concluded that students' weak graphing skills might not necessarily originate from their inadequate math skills. In a similar accord, Hammer and Elby (2003) and Ceuppens et al. (2019) argued that students possess the needed math skills and knowledge but fail to activate these assets in physics. Sokolowski (2017, 2019) found out that physics students are not sensitive to verifying graphs for adherence to principles of functions, especially when continuity and differentiability are concerned. This finding ultimately led to the conclusion that physics graphs do not necessarily stem from covariational relationships from students' standpoints, even though they are supposed to depict functions in their mathematical sense.

A. Sokolowski, *Understanding Physics Using Mathematical Reasoning*,
https://doi.org/10.1007/978-3-030-80205-9_12

Physics contexts often surface in mathematics assessment items and research. A handful of these findings revealed that students have difficulties reading and interpreting graphs of motion in the context of reality (Nemirovsky & Monk, 2000). Students face problems distinguishing between a position-time graph and the actual path of the object's motion (Leinhardt et al., 1990). Some students confuse the object's velocity with its position (Monk, 1992). Other studies disclosed difficulties with sketching velocity and acceleration–time graphs provided with a position-time graph, which, according to Harel and Confrey (1994), is due to a weak understanding of the concept of rate. Bonotto (2013) showed that even when the problems and methods encountered in the classroom were similar to the real situations, students tended to disassociate the analysis from the problem-solving techniques learned in the school. Dufresne et al. (1997) found that problem-solving in introductory mathematics and science courses is often algorithmic or schematic that turns students to mimicking the teachers or textbook procedures. In an attempt to find a reasonable explanation for such difficulties, Planinic et al. (2013) stated that these difficulties are due to the nature of the contextual problem that requires interpretation and translation of meanings into mathematical language.

Identifying the sources of potential weaknesses in students' understanding is one step toward improvement; suggesting instructional interventions for that improvement is a way to make things better. Domert et al. (2012) formulated the epistemological *mindsets* of themes to make students aware of the importance of formulas in physics learning. These mindsets offered students practice recognizing symbols in formulas and opportunities to learn how and when to use these formulas. Cui et al. (2006 February) turned to mathematics and advocated for including more physics applications in mathematics courses. Sokolowski and Capraro (2013) focused on the differentiation between the position graph and the object's path of motion. They developed a modeling activity that offered students opportunities to contrast both these representations in considering their functional attributes. Uhden et al. (2012) proposed a scheme where mathematization and interpretation of physics formulas were grounded in their conceptual predicaments. More recently, Zimmerman et al. (2019) found out that students in introductory physics courses are not usually exposed to physics-specific reasoning skills in their math classes and suggested developing teaching strategies reflecting that reasoning.

Despite physics and mathematics research communities' efforts, students struggle to graph physics contextual problems in mathematics and apply the functions' attributes to display graphically such relationships in physics. The preliminary synthesis of research illustrates a strong emphasis on physical interpretations of symbols used in formulas (e.g., Domert et al., 2012; Uhden et al., 2012), which seem to overshadow the potential of unfolding covariational entanglement stored in the arrangements of the symbols. Focusing on the interpretation of a specific formula might not generalize these structures' properties as a means of cognitive prompts to bring forth their potential for sketching. It is prudent that formulas in physics are to support the quantification of phenomena or their contextual aspects. Though they are still founded on algebra rules, their covariational relationships seem to be missing from being conveyed to students. This paper is to shed more light on algebraic interpretations of formulas as sketchable entities. Departing from this perspective, an attempt will be made to develop a broader reasoning spectrum on formulas' interpretation in their sketching. In its initial form, this study was published in *The Physics Educator*, (Sokolowski 2020b).

12.2 Theoretical Background and Methods

12.2.1 Foundations of Covariation Reasoning

Detailed underpinnings of covariational reasoning were presented in Chap. 5; therefore, the following section is a general introduction to that subject.

Studies show that the function's perception as a process that accepts specific inputs and produces outputs is essential for a mature image of function (Johnson, 2015). This entanglement is advanced in mathematics courses by covariational reasoning. The process begins with identifying the parameters and coordinating their intermediate relationships. Following this process, images of covariation are considered developmental phases that conclude with the formulation of the entire event under consideration. Research in mathematics education (Thompson et al., 2017) suggests that curriculum and instruction should increase emphasis on advancing math students from a coordinated image of two variables to a coordinated image of their instantaneous rates of change. Such analyses would allow for zooming into the dynamics of the mathematized phenomena using designated quantities. Covariational reasoning is being researched in physics education as well, yet the impact on teaching physics is not satisfactory. One of the essential mental actions to acquire fluency in covariational reasoning is coordinating the value of one variable with changes in the other (Carlson et al., 2002). Can omitting strict mathematical notations and parameter classifications in physics lead students to effectively using their math reasoning skills about graphing in physics? The goal of this research is to assess students' abilities in this domain. Reforms in mathematics education (Hughes-Hallett, 2006) emphasize presenting functions in various forms; graphical, symbolic, verbal, and numerical to meet the demands of students' multiple learning styles and train students to make transitioning between these representations. Such didactical enterprises also inspired this study.

12.2.2 Study Description, Participants, Research Questions, and Evaluation Instrument

This study can be classified as diagnostic research following the description by Ketterlin-Geller and Yovanoff (2009). Diagnosis is an integral part of instructional decision-making in education. As the bridge between identifying students' weaknesses and delivering designed interventions, diagnosis also provides valuable information about students' misconceptions in the targeted domain (Fuchs et al., 2003). Diagnostic research is mainly applied for: cognitive assessments, skills analysis, or error analysis (Leighton & Gierl, 2007). In this study, the diagnostic test was used to assess students' skills. More specifically, a diagnosis of students' interpretation of selected mathematics structures used in physics will be scrutinized. Students' responses will be debriefed to shed light on possible interventions to strengthen the flow of students' covariational reasoning from mathematics to physics.

The study attempted to address the following questions:

1. *To what extent can the algebraic structure of a formula affect activating its covariational features related to sketching?*
2. *To what extent can physics students correctly classify mathematical symbols in formulas based on provided real context?*

The study participants consisted of 30 first-year college physics students (18 males and 12 females). Twenty-five of these students (N = 25, 83%) took one or two high school physics courses, and five (N = 5, 17%) took this course physics course for the first time. All students possessed a calculus background. The assessment instrument consisted of a kinematics context followed by two questions.

Starting from a position denoted as x_1, *an object moves with a constant velocity,v, along a horizontal line. Time t is expressed in seconds and the position in meters. Two symbolic representations are depicting the object's position:* $x(t) = vt + x_1$ *and* $x = vt + x_1$.

(a) *Classify each symbol in each equation as a parameter, independent variable, dependent variable, or none. Support your judgment.*
(b) *Can each of these equations* $x(t) = vt + x_1$ *and* $x = vt + x_1$ *depict the motion in x versus time axes graphically?*

Students took the assessment test during the second day of instruction. Thus, one can assume that most answered these questions, referring to their prior high school experiences. The first question was to diagnose students' ability to transform the physics equations into functions and categorize their symbols according to their algebraic interpretations. The algebraic forms of both formulas were identical, the difference laid in denoting their right sides; the first; $x(t) = vt + x_1$ was expressed in the function notation, the other $x = vt + x_1$ in the formula notation. Presenting these equations *side by side* was arranged on purpose. It was assumed that students who would be biased by anticipating differences would not apply covariational reasoning to the extent that would eliminate the bias. Though, students were provided with a physical context that was modeled by these equations.

12.3 Data Analysis

Data analysis consisted of two parts; a descriptive evaluation of the responses presented in Table 12.1 and debriefing analyses of responses whose samples are clustered in Tables 12.2 and 12.3.

Descriptive analysis revealed that the responses could be classified using four categories depending on the individual perception. These categories are described in Table 12.1. The number of students belonging to each category is expressed in a percent form.

Classifying the symbols for position expressed in the function notation appeared more understandable and accessible to extract the physics than that shown in the formula notation. Students are acquainted with *f(x)* interpretation as *f* depending on *x*, or as *x* being the independent variable of *f*. This familiar notation steered them to classify *t*, time as the independent variable and, *x*(*t*) the dependent one represented by the object's position. By identifying *x*(*t*) as a function, 93% (N = 28) of these students concluded that this formula could be sketched. Thus, one can conclude that they realized a functional covariation of *t* and *x*(*t*). The remaining symbols *v* and x_1 were classified as parameters by 90% for these students. The classification turned in a different direction if the formula notation $x = vt + x_1$ of the same expression - and representing the same context - was considered. Without highlighting *t* in the function notation as the independent variable, 13% (N = 4) considered *v* and x_1 the parameters and the same low percentage of students claimed that the equation could be sketched.

Similarly, a small percentage of students classified *t* and *x* as independent and dependent variables. Most of these students (N = 26, 85%) considered the symbols x_1, *v*, and *t* as all parameters or independent variables when analyzing the formula notation. These students were asked to support their answers verbally, and sixteen (N = 16, 53%) did so. Table 12.2 provides samples of students' responses solicited verbatim that showed the level of their ability to apply functional notation in physics.

Table 12.1 Descriptive analysis

Equation	Category 1 *v* and x_1 considered as the parameters	Category 2 *t* considered the independent variable	Category 3 *x*(*t*) or *x* as the dependent variable	Category 4 The equation can be graphed as given
Functional notation $x(t) = vt + x_1$	85% (N = 26)	90% (N = 27)	90% (N = 27)	93% (N = 28)
Formula notation $x = vt + x_1$	13% (N = 4)	20% (N = 6)	27% (N = 8)	13% (N = 4)

Table 12.2 Students' comments on decomposing $x(t) = vt + x_1$

Student	Response
1	X(t) is a dependent variable, dependent on time, t, which is an independent variable because it is unchanged by other values in the equation but does change on its own rule. V and x_1 are parameters that limit the outcome of the dependent variable, yet they are fixed values.
2	X(t) implies that t is the variable, and the parameter is being evaluated at time t (assuming that t is referring to time). The x1 value is referring to the position of the object, which is independent of how much time has passed, making x_1 a constant value.
3	x_1 furthermore, v are parameters because they are observable values, t is independent because x(t) will vary as t changes.
4	X is a function of t; thus, it is the dependent variable, and t is the independent variable; it is an independent change. V is the parameter as it affects the relationship between x and t. X_1 is constant.
5	X_1 depends on the initial position, meaning it depends on the application, and it is constant within the application. Since it is x(t), this means that x(t) depends on the value of t, not v, so t changes.

Table 12.3 Students' comments on decomposing $x = vt + x_1$

Student	Response
1	X is not a function of t, so each has set values that do not change, so they are all parameters.
2	X is dependent on the three parameters, v, t, and x.
3	V, t, x_1 are all observed values; thus, they all are parameters.
4	X is not defined to be a function of anything so that you could be given the velocity and the starting position. With that, you could find the time that has passed to reach that velocity or any other combination of these variables.
5	All of the letters v, t, and x_1 are not dependent on each other. Thus, they all are independent variables.

Table 12.3 provides verbatim responses that disclose students' attempts to make mathematical sense of the standard formula notation. The purpose of classifying these responses into two tables (categories) was to contrast the students' reasoning that motivated their choices and look for possible factors that deterred them from being correct.

Evaluation of these responses was problematic to a certain degree. These students concluded their answers referring to the interpretation of the functional notation; $f(x)$ or $y(x)$ learned in their mathematics courses. Indeed, without an explicit background, a statement like $z = k + y$ could be classified as built from variables or parameters. Considering such an assumption, some of the answers from Table 12.3 would be marked as correct. However, due to a provided context to be used to support the classification, these answers were not considered correct.

It seems that the absence of function notation in symbolically denoting the object's position confused the students. The majority of these students considered the task associated with $x = vt + x_1$ an example of a plugin and evaluate the formula. It seems that the provided prompts in the context were not attractive enough to suggest a covariational relation. This supports a notion also voiced in other research (e.g., Sherin, 2001; Uhden et al., 2012) that students consider formulas to calculate a specific quantity. A corollary conclusion can be drawn that these science students did not have an opportunity to experience situations when they would be required to envision covariational relations based on the provided context. These conclusions suggest that embracing formulas in functional notation makes the transfer of students' covariational reasoning to physics plausible. An absence of a variable in the function notation confused the identification of the independent variable. It generated - in students' realities - an algebraic structure that is not a function, and thus it is not sketchable. It is to note that the formula selected for the diagnosis has traditionally been used in physics and mathematics courses for problem-solving or sketching purposes. One can claim that if the formula is not expressed in the function notation, students doubt how to find a covariational sense in its algebraic entanglement. They also seem to be lost in assigning specific meanings to the symbols based on their math knowledge, thereby classifying the entire expression as a representation that can be sketched.

There was a visible disconnect between the provided context and the interpretations of the equations while analyzing $x = vt + x_1$. When asked for algebraic interpretations of the symbols, students did not refer to a provided real scenario but referred only to the expression structure. Strong relation to symbolic notations used in mathematics cannot be denied and blamed for students being misled. It should instead be appreciated and taken advantage of to benefit the learning of physics. The reason for that can be a lack of such experiences and heavy emphasis on formulas. This possible mediator could be verified by adding strictly qualitative research to this study and interviewing them, which was unfortunately impossible to arrange. The absence of linking the context with its algebraic embodiment could also be assigned to the responses that did not identify v as the equation's parameter.

Analyzing the structure of formulas and identifying parameters and variables is one element that seemingly calls for more attention; however, analyzing the structures in the context of the given problem for its covariational reasoning is another.

12.4 Summary and Conclusions

This study suggests possible arrays of assisting students to become versed and comfortable in graphing context-driven functional relations in physics. It is also hoped that this small facet has the potential to attract more research and encourage considering more often functional attributes of formulas during physics instruction. This study has certain limitations, one of which is a relatively low sample size. I hope the smaller sample size does not diminish the study's findings as diagnostic research. The larger sample size would undoubtedly increase the validity and allow employing more sophisticated inferential statistics. For example, pairing graphing choices with students' prior math background would shed light on possible mediators. The survey can be easily duplicated in any class setting, conclusions can be drawn, and treatment applied. Another fact that limited the study findings is the uniform mathematical background of the study participants manifested by similar math contents learned following the similar math curricula. Extending the study would shed more light on the effects of different math curricula.

12.4.1 Traditional Formula Notation Does Not Aid Covariational Reasoning in Physics

Formulas are essential to solving problems in science, business, and other academic branches that use quantitative measures. It is understood that function notation is not being used explicitly in these structures. In multivariable expressions, the variable of interest can be represented by any of the component quantities. Moreover, some of the well-known formulas in physics are typically expressed in more than one standard form—for example, $F = ma$ and $a = \frac{F}{m}$. Students might find these algebraic structures confusing when sketching is concerned. Therefore, it is the teacher's task to explain the purpose of these variations following the specificity of the investigation in the plan. It is believed that teaching students how to perceive formulas as covariational entities based on the provided context is essential. This skill can enable them to consider formulas as dynamic functions that can be sketched, differentiated, or integrated. Students need to transfer the reality of an experiment into algebraic symbols and formula equations and analyze the equations as a means of solving for a parameter and deploy further analysis. Envisioning formulas as covariational structures could be enhanced by developing labs where students make connections between what they observe and the mathematical structures they produce. Implementing, suggested in Chap. 5 classification of the parameters, would certainly generalize lab design and its analysis. Emphasizing verification of the coherence between these representations would extend the notion to problem-solving. It is anticipated that such merging would make the students realize that the way mathematical symbols are used in physics must be consistent with mathematics rules. Furthermore, the symbols' interpretations must be compatible with how they were assigned in the lab conducts.

12.4.2 Physics Depends on the Mathematical Rules and Notation

Attempts made in physics courses to improve the flow of math knowledge usually originate from physics points of view, and supportive suitability to algebra rules is established. This flow of knowledge seems to enhance the message that mathematics depends on physics. While, in many cases, this might be true, the literature suggests otherwise. Dirac (1940) once said, "The mathematician plays a game in which he invents the rules while the physicist plays a game in which Nature provides the rules, but as time goes on it becomes increasingly evident that the rules which the mathematician finds interesting are the same as those which Nature has chosen p.124". Understanding physics does depend on understanding and interpreting the rules of mathematics because to numerically express quantity, the rules of mathematics are needed. While physicists use algebraic structures differently in a traditional view because they are interested in analyzing the structures, paying attention to the effects of the precision of mathematics embedded in scientific inquiry design can be equally rewarding.

Referenced earlier, Hughes-Hallett (2006) emphasized presenting functions in various representations to meet the demands of multiple learning styles and train students to transition between these representations. It seems that exercising such transitioning in physics would benefit the learning and understanding of physics as well.

References

Bonotto, C. (2013). Artifacts as sources for problem-posing activities. *Educational Studies in Mathematics, 83*(1), 37–55.

Carlson, M., Jacobs, S., Coe, E., Larsen, S., & Hsu, E. (2002). Applying covariational reasoning while modeling dynamic events: A framework and a study. *Journal for Research in Mathematics Education, 33*(5), 352–378.

Ceuppens, S., Bollen, L., Deprez, J., Dehaene, W., & De Cock, M. (2019). 9th-grade students' understanding and strategies when solving $x(t)$ problems in 1D kinematics and $y(x)$ problems in mathematics. *Physical Review Physics Education Research, 15*(1), 010101.

Cui, L., Rebello, N. S., & Bennett, A. G. (2006, February). College students transfer from calculus to physics. In *AIP Conference Proceedings* (Vol. 818, No. 1, pp. 37–40). AIP.

Dirac, P. A. (1940). XI.—The relation between mathematics and physics. *Proceedings of the Royal Society of Edinburgh, 59*, 122–129.

Domert, D., Airey, J., Linder, C., & Kung, R. L. (2012). An exploration of university physics students' epistemological mindsets towards the understanding of physics equations. *Nordic Studies in Science Education, 3*(1), 15–28.

Dufresne, R. J., Gerace, W. J., & Leonard, W. J. (1997). Solving physics problems with multiple representations. *The Physics Teacher, 35*(5), 270–275.

Dunn, J. W., & Barbanel, J. (2000). One model for an integrated math/physics course focusing on electricity and magnetism and related calculus topics. *American Journal of Physics, 68*(8), 749–757.

Fuchs, L. S., Fuchs, D., Hosp, M. K., & Hamlett, C. L. (2003). The potential for diagnostic analysis within curriculum-based measurement. *Assessment for Effective Intervention, 28*(3&4), 13–22.

Hammer, D., & Elby, A. (2003). Tapping epistemological resources for learning physics. *The Journal of the Learning Sciences, 12*(1), 53–90.

Harel, G., & Confrey, J. (Eds.). (1994). *Development of multiplicative reasoning in the learning of mathematics*. The Sunny Press.

Hughes-Hallett, D. (2006). What have we learned from calculus reform? The road to conceptual understanding. In N. Hastings (Ed.), *Rethinking the courses below Calculus*. Mathematics Association of America.
Johnson, H. L. (2015). Secondary students' quantification of ratio and rate: A framework for reasoning about change in covarying quantities. *Mathematical Thinking and Learning, 17*(1), 64–90.
Ketterlin-Geller, L. R., & Yovanoff, P. (2009). Diagnostic assessments in mathematics to support instructional decision making. *Practical Assessment, Research, and Evaluation, 14*(1), 16.
Leighton, J. P., & Gierl, M. J. (2007). Why cognitive diagnostic assessment? In J. P. Leighton & M. J. Gierl (Eds.), *Cognitive diagnostic assessment for education: Theory and applications* (pp. 3–18). Cambridge University Press.
Leinhardt, G., Zaslavsky, O., & Stein, M. K. (1990). Functions, graphs, and graphing: Tasks, learning, and teaching. *Review of Educational Research, 60*(1), 1–64.
McDermott, L. C., Rosenquist, M. L., & Van Zee, E. H. (1987). Student difficulties in connecting graphs and physics: Examples from kinematics. *American Journal of Physics, 55*(6), 503–513.
Monk, S. (1992). Students' understanding of a function given by a physical model. In G. Harel & E. Dubinsky (Eds.), *The concept of function: Aspects of epistemology and pedagogy*, MAA Notes, 25, 175–193.
Nemirovsky, R., & Monk, S. (2000). 'If you look at it the other way..': An exploration into the nature of symbolizing. In *Symbolizing and communicating in mathematics classrooms: Perspectives on discourse, tools, and instrumental design* (pp. 177–221). Lawrence Erlbaum Associates.
Planinic, M., Ivanjek, L., Susac, A., & Milin-Sipus, Z. (2013). Comparison of university students' understanding of graphs in different contexts. *Physical Review Special Topics-Physics Education Research, 9*(2), 020103.
Redish, E. F. (2006). Problem-solving and the use of math in physics courses. *arXiv preprint physics/0608268*.
Sherin, B. L. (2001). How students understand physics equations. *Cognition and Instruction, 19*(4), 479–541.
Sokolowski, A. (2017). Graphs in kinematics—A need for adherence to principles of algebraic functions. *Physics Education, 52*(6), 065017.
Sokolowski, A. (2019). Graphs in kinematics—A need for adherence to function continuity and differentiability. *Physics Education, 54*(5).
Sokolowski, A. (2020b). Developing Covariational reasoning among students using contexts of formulas: Are the current formula notations in physics aiding graph sketching? *The Physics Educator, 2*(04), 2050016.
Sokolowski, A., & Capraro, M. M. (2013). A constructivist approach to embodying motion problems in mathematics. *Mediterranean Journal for Research in Mathematics Education, 12*(1–2), 121–133.
Thompson, P. W., Hatfield, N. J., Yoon, H., Joshua, S., & Byerley, C. (2017). Covariational reasoning among US and South Korean secondary mathematics teachers. *The Journal of Mathematical Behavior, 48*, 95–111.
Uhden, O., Karam, R., Pietrocola, M., & Pospiech, G. (2012). Modelling mathematical reasoning in physics education. *Science & Education, 21*(4), 485–506.
Zimmerman, C., Olsho, A., Loverude, M., Boudreaux, A., Smith, T., & Brahmia, S. W. (2019). Towards understanding and characterizing expert covariational reasoning in physics. *arXiv preprint arXiv:1911.01598*.

Chapter 13
Adaptivity of Mathematics Representations to Reason Scientifically
Students' Perspective

13.1 Prior Research Findings

Quantitative embodiments of physics laws are rooted in their extensive support by mathematics that provides pathways for setting up the inquiry and serves as a tool to express relationships between investigated quantities in symbolic or graphical forms. While mathematics is used extensively in physics, the knowledge about the supportive use of mathematics in physics education is still being developed (Uhden et al., 2012). The preliminary synthesis of research presented herein pertains to physics students' perception of the tools of mathematics used in physics and improvements suggested by the research. Sherin (2001) found out that physics students learn to understand physics formulas by the vocabulary of component elements rather than their algebraic structures. Domert et al. (2007) found out that even university students' majoring in physics typically do not highlight functional structures of physics formulas and instead focus on interpreting algorithms of operations by rote. Chi (2005) declared that using mathematics tools reduced to operations on formulas reflects students' lack of mathematical and problem-solving skills and a shallow understanding of the underlying mathematics concepts.

Attempts to improve students' transfer of the structural domain of mathematics knowledge to physics beyond the skills of manipulating formulas are undertaken. Domert et al. (2007) proposed the epistemological *mindsets* of themes to make students aware of the significance of formulas in physics. These mindsets included recognizing symbols and algebraic structures in formulas and how and when to use these formulas in everyday life. Consequently, a discussion of the meanings of the formulas was encouraged. Hestenes (2003, p. 104) called for more research targeting the design and use of mathematics in physics education. Cui et al. (2006, February) reflected on calculus and advocated for including more physics applications to prepare students to apply these tools in physics. Adams et al. (2006) suggested using multimedia to enhance a conceptual understanding of the phenomena

A. Sokolowski, *Understanding Physics Using Mathematical Reasoning*,
https://doi.org/10.1007/978-3-030-80205-9_13

before its mathematizing. Bagno et al. (2008) encouraged discussions of the applicability of the formulas to recognize their broader scientific relationship.

This synthesis illustrates that physics students face difficulties in considering physics quantities building blocks of formulas as covariate parameters. A large scale of attempts to improve that reasoning illustrates that it concerns physics communities. It seems that attempts proposed by research do not aid in seeking more direct links between what students have learned in their math courses but rather at analyzing physics formulas within the context of physics only. This study assesses physics students' awareness of formulas as covariate representations by reflecting on students' mathematical reasoning to uncover that awareness and draws on these findings to propose improvements.

13.2 Theoretical Framework, Research Questions, and Study Logistics

This study, in its initial version, was published earlier (Sokolowski, 2020).

The deduction of inferences in science is possible due to applying covariational reasoning at a larger scale, not just quantifying a physical argument. Mathematical representations that develop students' conceptual knowledge and reasoning skills are algebraic functions in all their forms: symbolic, tabularized, verbal, and graphical (NCTM, 2019). Thus, functions emerge as the foundations of symbolically expressing covariational reasoning. This conclusion guided the formulation of the research questions:

1. *To bring forth the covariational reasoning to the physics classroom, should more parallelism between algebraic functions and physics formulas be enabled to streamline the flow of mathematical reasoning to physics teaching and learning?*
2. *How do students perceive the usefulness of each type of these mathematical constructs: formulas and functions as dynamic functional counterparts to learn physics?*

Thus, rather than further explicating on differences in mathematics and physics perceptions of these two fundamental representations, the pursuit for linking the methodologies of algebra used in mathematics and physics emerged as the study's primary objective.

The study was conducted with a sample of 25 high school students in the US. These students (13 males and 12 females) were enrolled in an advanced physics course after taking an introductory level physics. Eighteen of these students were concurrently taking a precalculus course, and seven were enrolled in an advanced calculus course. While attending the same high school, these students took their prior math and physics courses taught by teachers of various backgrounds and professional preparation before enrolling in the advanced physics course.

The study was conducted over a two-month period that started at the beginning of the semester. Along with traditional applications of formulas to support problem-solving, the students were introduced to physics laws and principles perceived as functional relationships derived either by using an empirical-mathematical model supported by experimentation or by transferring formulas to algebraic functions due to constraints of the scientific context under investigation. At the concluding stage of the study, the students were invited to contrast the effects of using formulas and formulas considered functional relationship on physics understanding.

13.3 Study Instrument

This section summarizes the embodiments of covariational reasoning, its proposed interpretations along with the practices and didactical tasks introduced to the physics curriculum to enhance such reasoning. These methods did not change the scope of the chapters covered, thus kinematics, dynamics, and energy, but they enriched their contents and delivery methods.

The tasks of covariations were clustered into two general domains; (a) applied during experiments and (b) enacted by hypothetical – mathematical covariational models that stemmed from formulas. These structures were integrated within the course of study as they fitted to support the depth of the inquiry of the content under investigation.

13.3.1 General Characteristics of the Treatment: How Did Covariational Reasoning Emerge?

While a broader presentation of covariational reasoning was included in Chap. 5, this part adds specific ideas that were inserted in the traditional physics curriculum and delivered to the students. This section elaborates on the didactical routes taken to develop students' awareness about the opportunities stored in formulas when the formulas are considered as functions. Such formula reimaging was also used to support graphs' sketching and their scientific interpretations. The initial phase of such transitions was brought up by referring to functional representations that students learned in their mathematics courses. Contrasting *side by side* these functions to corresponding formulas was the initial step of the transitions. There were no categories of covariation presented at this time. Such generalization was considered to be deployed when students got familiar with more techniques and have more opportunities to explore the nuances between these categories.

Students study functions of one variable in a typical high school math curriculum, expressed in two different forms: in a coordinate notation, $y = kx$, and the function notation $f(x) = kx$. Functions as means of learning in physics are not emphasized;

instead, physics is populated by formulas perhaps because formulas are concept-specific entities. The formulas are often presented as multivariable expressions, e.g., $F = ma$, which can be converted to a one-variable covariational relationship by considering $F(a) = ma$ or $F(m) = ma$, which is also suggested by Karam et al. (2019). Students realize that the rate of change of each form generates different quantities and graphs in the Cartesian plane, and each, after all, will reflect on a different physical scenario. The nuance of modifying the left side of these formulas has a tremendous effect on the experiment analysis and information learned about its behavior. It was seen that being able to make that recognition was important not only to reflect on the phenomena line of inquiry but also to identify specific covariation embedded in the highlighted quantities. For example, when determining a correct graph of *kinetic energy vs. object's speed* using the formula $K = \frac{mv^2}{2}$, the students first needed to realize that this structure could be sketched considering, for example, the mass, m a constant parameter or a mediating parameter and the object's speed-v a variable parameter, thus $K\left(v\right) = \frac{mv^2}{2}$ would reflect the situation and the graph would emerge as a typical quadratic function. Such thinking would lead to realizing that this structure could be considered a quadratic function of the form $f(x) = ax^2$ and sketched as a parabola centered at (0,0) representing the *speed versus kinetic energy graph*. While Karam's suggested deploying this technique to primary functions, this study extended this idea and presented students with techniques of using composite or combinations of functions to support graphs' recognition, thus is targeted scaffolded covariation (as defined in Chap. 5). The following process illustrates the phases of transitioning from primary covariate relations to a scaffolded one. Suppose that an object of mass 10 kg moves with a variable velocity given by $v(t) = 3 + 2t$. Sketch the velocity function for the object for $t \geq 0$. Formulating a kinetic energy function one learns that $K(v(t)) = K(t) = \frac{10\left(2+2t\right)^2}{2} = 5\left(2+2t\right)^2$. One nuance emerged while analyzing this scaffolded covariation: if $K(v) = K(v(t)) = K(t)$ what independent parameter to label on the horizontal axis; the time, or the velocity? Will the graphs of $K(v)$ and $K(t)$ be identical when either the velocity v or the time t is labeled? While the general shape of kinetic energy function must remain the same, the independent parameter will affect the domain of the kinetic energy function. Two cases emerged:

(a) If time is labeled on the horizontal axes, then the domain of the functions is $t \geq 0$, and the graph will be initiated from a y-intercept labeled in Fig. 13.1 as $K = 20J$. The y-intercept represents the initial kinetic energy of the object with a mass of 10 kg.

(b) If the velocity is labeled on the horizontal axis, then the initial point of the graph will still occur when $t \geq 0$, yet the graph will show up for $v\left(t\right) \geq 2\frac{m}{s}$. This graph position in coordinate axes can be supported algebraically, if $t = 0$, $v(t) = 2m/s$, then $K(0) = \frac{10\left(2\right)^2}{2} = 20J$, which represents the initial kinetic energy of the object. The graphs are similar in shape but they differ by horizontal transformation (Fig. 13.2).

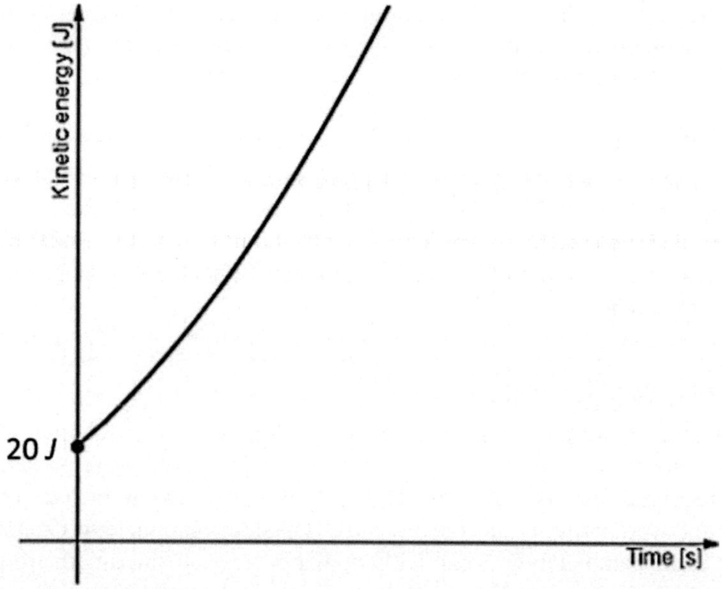

Fig. 13.1 Graph of the kinetic energy versus time of an object moving with a variable velocity $v(t) = 2 + 2t$, when $t \geq 0s$

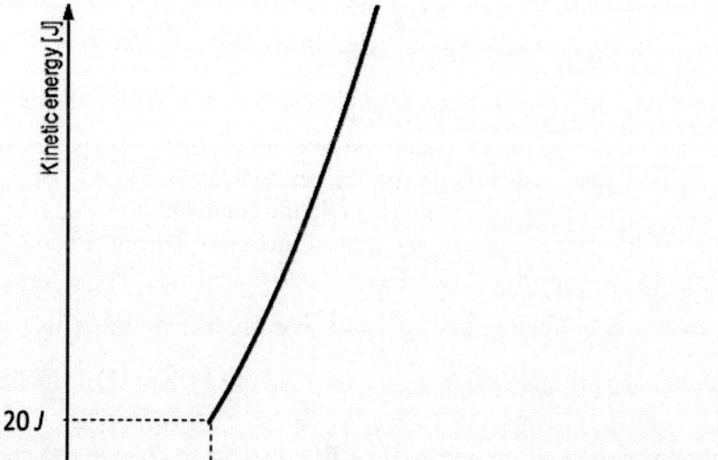

Fig. 13.2 Graph of kinetic energy versus velocity for $v(t) = 2 + 2t$ when $v \geq 2$

Departing from this line of thought, the students were further introduced to sketching composite functions, e.g., $K(v(t))$, where the velocity $v(t)$ was given, for example, as a periodic function $v(t) = Acos(\omega t)$. After creating such composite covariational relation $K(t) = \frac{m\left(Acos(\omega t)\right)^2}{2}$, the students found it easy to conclude the correct shape for the graphical representation of $K(v(t))$, which was almost impossible without applying the process of formulating explixitly a composite function of the kinetic energy and velocity. Formulas, considered as one covariational relationship, and their composite or scaffold relations helped students conclude the shape of the graph.

Another example of inducing covariational reasoning was the analysis of the formula $T = 2\pi\sqrt{\frac{m}{k}}$. While the formula helps calculate the period of motion of a mass attached to a spring, it comprises of two different objects; m, the mass of the hanging object, and k, the spring constant of the spring, representing two separate independent parameters. In contrast, many physics formulas include parameters related to the properties of one object whose dependent parameter is examined. This duality of the parameters was explicitly realized when considering the formula as a covariational relationship. When examining how each parameter contributed to the period magnitude, it was convenient to split up the analysis in two cases; (a) when one parameter was constant, for example, k and the mass changed, thus examine $T(m) = 2\pi\sqrt{\frac{m}{k}}$, and (b) when the mass was unchanged, and the spring constant was modified, thus examine $T(k) = 2\pi\sqrt{\frac{m}{k}}$. Using $T(m) = 2\pi\sqrt{\frac{m}{k}}$, the students realized that the period increased as the mass increased because as mass increases, the values of the resulting *radical function*, $2\neq\sqrt{\frac{m}{k}}$ also increase. The students suggested making a preliminary graph to support the prediction which was a convenient visual way of verifying the prediction. In (b), the period decreases as the spring value of the constant of the spring increases. This relation could be supported by the behavior of the combination of rational and radical functions. The inner part of the function $\frac{m}{k}$ decreases; thus, $T(k)$ also decreases as k increases. These formulas were further embraced in limiting case analyses. For instance, to prove that the period increases as the mass increases using a limit shows as $T(m \to \infty) = \lim_{m\to\infty} 2\pi\sqrt{\frac{m}{k}} = \infty$ and respectively if (k) is concerned then $T(k \to \infty) = \lim_{k\to\infty} 2\pi\sqrt{\frac{m}{k}} = 0$ that is illustrated in Fig. 13.3.

Both conclusions could be reached using scientific reasoning; however, limiting case analysis conveniently provided an elegant proof. Some students suggested investigating the effects of change of both the mass and the constant spring simultaneously. This case triggered a two-variable covariate that would require a

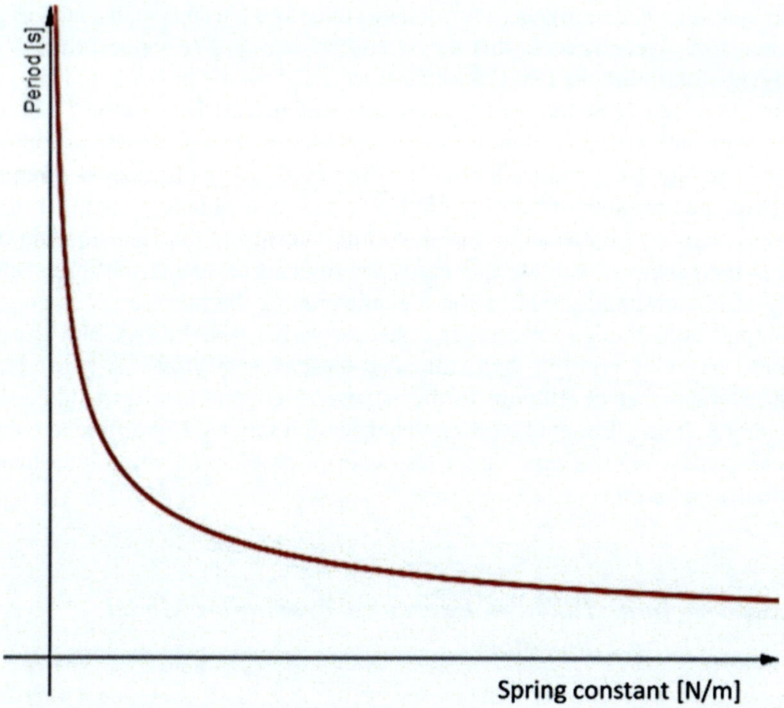

Fig. 13.3 Graph representing the period of a mass-spring system when spring parameter increases

differentiation process. Or such analysis could be supported by experimental data to learn how both the mass and the spring constant affect the period, which perhaps would not be easy to furnish in the realm, but the theoretical analysis could be employed. The graph Fig. 13.3 could also be used to employ the technique of taking sided limit and prove that the period becomes infinite when the spring constant is getting close to zero. This limit analysis created an interesting case to be explained back using scientific terms. Another thought emerged from this analysis; labeling the spring characteristic property as *constant* can, to a certain degree, be problematic. The parameter describing the spring can also be changed by using a different spring or its combinations (series or parallel arrangements), thus, perhaps calling it a mediating parameter or spring parameter, would better describe its property.

Attempts to verify the hypothetical analyzes were made by observing an actual motion of different masses attached to the same spring or the same mass attached to various springs using simulated experiments.

The students realized the limiting case analyzes extended the inquiry far beyond the capacity of the school physics lab. Limiting case analysis was also applied to examine the motion of two blocks on a frictional surface. The methodological details and subsequent evaluation of this method are discussed in Chap. 6. Followed by such inquiries, students' perception of formulas was evolving; the formulas were not considered static entities but as dynamic, continuously changing entities that enrich subsequent inquiries. Students appreciated the opportunities to learn the techniques of considering these formulas as functions because such a view expanded the range of algebraic tools that they could be deployed to support their scientific analyses. What were the areas of weaknesses, and how did physics students perform on questions requiring the induction of more sophisticated math tools? It seems that to answer such questions, more studies are needed. Physics students, perhaps by habits, did not pay much attention to the details of mathematical notation; for instance, the formality of carrying out the process of evaluating function limit. Such perception of mathematical tools did, at times, devaluate the precision and correctness of their answers, although generally, the students developed the correct answers. They also occasionally missed the classification of the parameters as constant or changing, even though experimental setups, verbal descriptions, and general lab conduct provided prompts that explicated these classifications. A general conclusion emerged that developing further students' covariational reasoning skills in physics requires a design of a different curriculum followed by different assessment items emphasising the exploratory character of problem-solving rather than static formula evaluation.

13.3.2 Actions Taken to Exercise Covariation Model Using Laboratory

Mathematical modeling supported by empirical processes was the primary method to encourage covariational reasoning in this study as its effects on developing students' mathematical reasoning in physics are highly recommended. For example, the students observed a bouncing ball and were supposed to derive an exponential decay function representing the amount of mechanical energy dissipated into the thermal energy presented in detail in an earlier study (Sokolowski, 2018). They had an opportunity to apply the limiting case analyses to approximate the total amount of energy lost during the experiment considering the law of conservation of energy as a covariational algebraic entity to initiate the derivation process. Such transitioning, though required the teacher guidance, perhaps due to several covariate parameters that contributed to the formulation of the final function; the students were unsure what parameters to use to build the function. Some included the kinetic energy of the bouncing ball in the general statement, which confused the algebraic process; some did not realize that $Q(n)$; the energy lost by the bouncing ball must be explicitly inserted into the law of conservation of energy during the experiment analysis. This covariate would be classified as Type 4 according to the established scale in Chap. 5. Thus, this experiment is worthy of being discussed in more detail. The initial potential energy of the ball was $U_g = mgh$. The gravitational potential energy retained after each bounce could be denoted as $U(n) = U_g(R)^n$, where $R = \frac{h_n}{h_{n-1}} < 1$ was the ratio of two consecutive bounces. The students concluded that the ratio R was constant by taking data and calculating the ratio; thus,

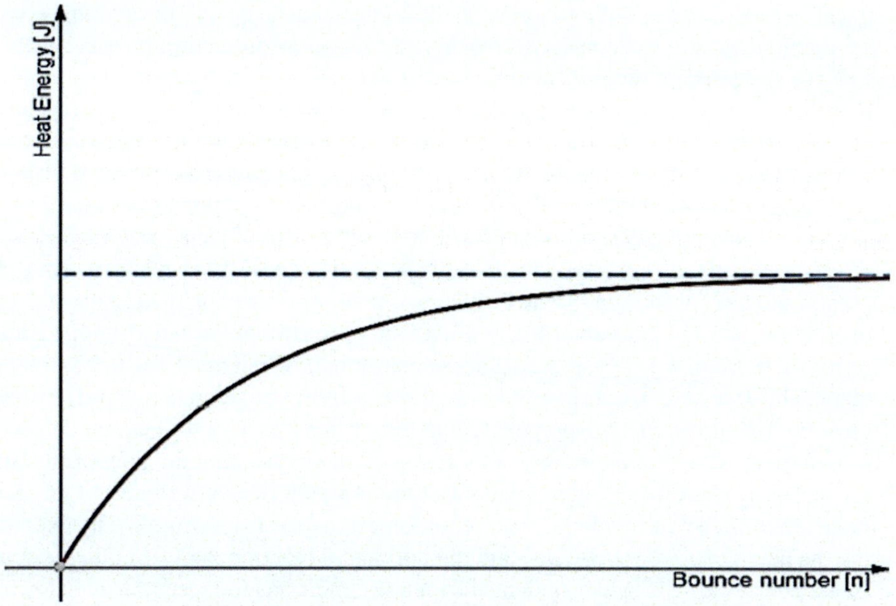

Fig. 13.4 Graph representing dissipation of energy of a bouncing ball

an exponential model could be used. Using the law of conservation of energy resulted in $U_g = U_g(R)^n + Q(n)$. Solving for $Q(n)$ resulted in $Q(n) = U_g - U_g(R)^n = U_G(1 - R^n)$. Since the ratio $R < 1$, the limit $Q(n \to \infty) = \lim_{n\to\infty} U_G(1 - R^n) = U_G$ represents the horizontal asymptote of $Q(n)$ and its graph is lustrated in Fig. 13.4.

The students performed another lab with a congruent context; they analyzed the loss of mechanical energy of a marble oscillating in a vertical parabolic track. After being guided on the *Bouncing Ball* lab, their handling of the lab with an oscilating marble improved.

Conceptual understanding of the cause and effect in experiments is essential; however, identifying the cause and effect and expressing these factors using independent, dependent and mediating parameters is one step toward improving the learning effects from lab conducts. Formulating a covariate expression is the concluding step that prepares the students for more advanced scientific endeavors. Acquiring such skills indeed requires practice, yet the rewards for their possessing are high.

The mathematical models enacted during lab processes are regarded as an axiomatic system that gains significance only after being interpreted through the physical model (Hudson & McIntire, 1977). A meta-analysis (Sokolowski 2015) revealed that modeling generates the largest learning effect sizes compared to traditional teaching methods. Laboratory activities in physics provide excellent opportunities for generating data and expressing the relation between the quantities of interest in a concise language of mathematics. Research shows that most of the graphing tasks in physics focus students' attention on determining the relationship between the quantities of interest without moving the inquiry further and formulating corresponding covariate relations. Without hypothesizing or predicting possible behaviors due to varied conditions, an opportunity to develop mathematical covariational reasoning is lost. Graphing is a crucial aspect of developing a broader perspective on phenomena behavior. The students performed several more lab activities about formulations of algebraic representations of the selected covariational relationships. One of such lab activities used the idea of parametric equations (see Chap. 8) to support the algebraic description of position functions for objects' moving in a two-dimensional plane, simultaneously exercising awareness between position-time graphs and path of motion. Some elements of that lab were explained in more detail in Sokolowski and Capraro (2013). Students were further provided with various types of motion, e.g., projectile, simple harmonic, accelerated on a frictionless plane, and free fall. The task was to establish a frame of reference and determine position functions in two perpendicular dimensions that described the object's motion path. The challenge faced was algebraically expressing the path of motion, for instance, for a free-falling object. The motion path does not represent an algebraic function, yet its symbolic representation was possible to formulate by expressing the height as $0 \leq y(t) \leq y_{MAX}$, and the horizontal location $x = 0$. Another inquiry (see Sokolowski, 2014) had students formulate conditions for speeding and slowing down using a virtual context. Students often associate the state of speeding up with an object's positive acceleration, which is not entirely true when an object possesses a negative velocity. While these states of motion are not signified in the physics curriculum, expressing such modes of movement using a correct sign of acceleration is crucial to formulating Newton's II law of motion. Nevertheless, another purpose of inserting such a lab into the physics curriculum was to eliminate the misconception that an object is speeding up if its acceleration is positive. The students exercised covariational reasoning by aligning the mode of movement of a virtual person with the behavior of acceleration and velocity functions generated by technology. By extracting specific attributes of these covariations, students identified the conditions for speeding up and slowing down and learned how to handle the sign of acceleration when forces and dynamics were concerned.

The prior research documented well students' struggle with graphing. A founding by Planinic et al. (2012) disclosed that physics students do not consider graphs in physics as a means of representing relationships among variables stands for an additional argument to further reimage formulas to functions under favorable conditions. Understanding the scientific underpinnings between quantities of interest was the initial step of formulating a general covariational representation and identifying the attributes of the algebraic expression that fitted that underpinnings concluded unpacking more sophisticated scientific knowledge behind it.

13.4 Data Analysis

After two months of being under treatment described herein, the students were invited to elaborate on the following problem statement:

There are various representations used in physics. For instance, formulas and formulas that can be considered functions.

(a) *Contrast formulas and functions as a means of learning physics.*
(b) *Which of these representations provided you with more opportunities to understand physics concepts?*

Data analysis was initiated by a descriptive assessment of the students' responses clustered in Table 13.1.

Forty percent of the students (N = 10) favored formulas over functions as means supporting the physics understanding. Following Karam et al.'s (2019) classification, these students preferred the technical domain of mathematics. Interestingly, these students (see Table 13.2) highlighted the convenience of applying formulas to solve textbook problems, which is undeniable. Nine students (36%) favored functions, and the remaining six students balanced their responses between both

Table 13.1 Students' preference of algebraic representations in physics

Response Type	Favors Functions Structural Domain	Favors Formulas Technical Domain	Favors Both Formulas and Functions Technical and Structural Domain
Percentage	36% (N = 9)	40% (N = 10)	24% (N = 6)

Table 13.2 Responses favoring formulas

Student	Response
1	In a realistic situation, there will be more than a single variable to plug in to produce an output variable. In physics, there are often many variables. Therefore one must use formulas.
2	Formulas are more valuable to solve problems; functions are used for sketching.
3	Formulas are used to find specific values; functions can change depending on the situation.
4	Formulas are more useful because they have something to do with numbers, while functions deal more with concepts.
5	Functions are essential for creating graphs and visualizing concepts, but formulas create physics.
6	The function is more of a changing thing with many x-values; I like using formulas because formulas are used to plug in specific values to find a realistic result (like $C = 2\pi r$ for the circumference of a circle).
7	I prefer formulas. A function is a set to mathematize a process to produce a value that changes depending on the situation; a formula is the same but useful to find a specific value.
8	Formulas are more straightforward; however, functions allow for more possibilities. Formulas allow changing the value of the variables, whereas functions are often more specific to the data set being used.
9	Formulas are simplified means of finding and explaining an answer in problem-solving. Formulas tend to be my focus, as math is something that more quickly can be researched.
10	Formulas are more frequently used in physics than in mathematics; they are a plug and solve approach without a need for deep thinking.

Table 13.3 Responses favoring functions

Student	Response
1	To solve physics problems, I find functions more useful because functions can be further graphed, and their limits or asymptotes are computed, which supports general comprehension.
2	Functions establish certain constancy when they are graphed. This cannot always be true for formulas.
3	The formula can convey much information regarding the situation, but it cannot display the whole picture and thus only shows instances for the most part. This leads to the strength of a function.
4	Functions can be used to explain concepts in physics on a more intensive level. Learning only formulas can lead to a robotic approach to physics. It can become merely a plug and solve approach without fully understanding what is happening.
5	Functions impart and require more conceptual understanding in their use than formulas do and thus are more useful for genuinely understanding physical concepts.
6	Functions can be much more helpful for certain types of problems than formulas because they allow greater mathematical manipulations such as finding limits, derivatives, and graphing.
7	Functions allow us to visually understand the theory by observing graphs, whereas formulas can be manipulated in various ways to solve for the unknown mathematically.
8	Functions seem to better fit for learning new information.
9	I feel functions are essential resources to utilize when learning concepts in physics because functions can be used in terms of time that allow us to visualize relationships between variables.

functions and formulas. The majority of these students (76%, N = 19) explicitly stated their preference and supported their choices by listing the *pros* and *cons* of applying either function or formula. Samples of students' verbal responses verbatim are clustered in Tables 13.2 and 13.3. While referring to specific answers, T# denotes the table from which the response is cited and S# a specifc student order number from that list. For instance, (T13.2; S4) represents the response of the fourth student from Table 13.2.

A preliminary analysis showed that students consider formulas as algebraic representations that provide a fast way of computing the final answer (T13.2; S6–9). Students highlighted the simplicity of using formulas to solve problems as the primary support for their preferences, which corresponds with conclusions reached in prior research (see Domert et al. 2007). Plugging in given quantities and solving for the unknown is a straightforward algebraic process that dominates current physics assessment items (Uhden et al., 2012). Such an approach leads the students to find solutions and ultimately succeed in problem-solving. This technique is very often exercised, especially in introductory physics courses. The question that arose was if this is the perception of physics, we intend the students to acquire? Even though some students favored formulas, they noted that formulas do not provide means for engaging in a more profound mathematical reasoning to analyze the system's change (T13.2; S4). It was interesting to note that none of these students considered formulas in their original forms as entities that enable graphing but mainly tools to finding solutions to problems. Table 13.3 consists of responses of these students who favored functions.

These students considered functions as entities possessing a high potential of carrying new information about phenomena (e.g., T13.3; S4). A graphing aspect of functions as helpful algebraic tools to understand physics was also prevalent in these responses. There is a visible appreciation of using calculus techniques to derive relations between physics quantities, which originated from students who were concurrently taking a calculus course (T13.3; S6). Students consider formulas converted to functions as a means to support understanding of laws and theorems. Students foresee mathematical reasoning as a gateway to strengthen knowledge of natural phenomena. Formulas are considered as facts or rules constructed using mathematical laws or as physical quantities that are linked with an equal sign. In this regard, formulas appear to students as sets of instructions for creating the desired results. Formulas are viewed as tools to solve the problem, not as tools to explore the system's behavior or learn about a deeper relationship between the parameters of interest. Even when favoring formulas, students perceive functions as superior representations providing more opportunities for exploring relations between quantities. More importantly, they consider functions as covariational relationships generating bases for visualizing these relationships graphically. The students realize that understanding functions require more effort and higher-level thinking skills to encode scientific knowledge efficiently, but such actions' cognitive benefits are high.

13.5 Summary and Conclusions

A traditional perception is that students associate graphing with procedures, not with formulas, perhaps because physics graphs in students' realities do not represent functions at a glance. It was concluded that as long as the students are not explicitly guided on how to view formulas through the prism of covariational relationships, the problem will persist. Graphs in mathematics are generated and classified by their algebraic structures, considering their strict rules that result in curves possessing unique variations. For example, sketching polynomial functions is methodologically different from sketching rational functions. A substantial emphasis on formulas in physics courses makes it almost impossible for students to identify covariational properties and apply them to sketch and analyze graphs for scientific inquiry purposes. This might have another unfavorable impact on physics understanding. If formulas are not converted to functional notations, students might not necessarily perceive sketching curves in physics as sketching equivalent covariate algebraic functions. Considering derived formulas as one variable functions and discussing with students adherence of the resulting graphs to the behavior of the physical phenomena was a technique that seemed to enhance the power of the structural domain of mathematics. Many physics laws and principles have been formulated before a formal definition of function was introduced in mathematics, which perhaps explains the fact that formulas, not functions, dominate the content of physics. Understanding the entanglement of the quantities behind the formulas is an essential component of physics learning because it aids in quantitative descriptions and their underlying theories. Formulas - have been and most likely will - constitute the primary mathematical tool to teach and learn science. However, to follow the recommendations of contemporary research in physics education thus to induce deeper mathematical reasoning to physics education, a shift is needed. Formulas considered functional/covariational algebraic structures to explicate quantities ease the sketching process by making these tasks parallel to what students encounter in their mathematics classes. The techniques of converting formulas to considering these entities as one- variable functions are not often explored in mathematics; therefore, students do not possess the flexibility to switch their perceptions between functions and formulas.

Although the sample size of the participants was small, the general themes that emerged from the data debriefing seem to produce exciting insights on learning physics. One can imply that developing mathematical reasoning in physics can be initiated by enabling covariation reasoning on formulas, which does not require drastic accommodations. The students under treatment realized that many opportunities for deriving new physics knowledge are stored in the functional representation of physics formulas, which was implied from discussions when such conversions were undertaken. It can be inferred from the study outcomes that the problems with knowledge transfer from math to physics lie not necessarily in students' inability to transfer that knowledge but in mathematical representations (formulas) used in physics that in students' realities are not designated to endure such transfer. Formulas are static representations that, in students' perceptions, are used to aid problem-solving rather than to explore system behavior that can, in return, allow for formulating more in-depth scientific theories.

Based on the research findings, it is claimed that teaching students the techniques of converting formulas to one-variable functions smoothens the activations of structural math knowledge, which opens up the gate for using the ideas of rates, limits, concavity, or monotonicity, and graphing. It is claimed that students cannot activate their mathematical knowledge in physics courses because they do not necessarily realize that formulas can be considered as dynamic algebraic representations. It seems that the didactical gap between math knowledge expected from physics students and their ability to handle it can be removed by augmenting and expanding the perceptions on the math tools used in physics. Allowing more functional analysis during physics explorations appears as a task worthy of further research. Undoubtedly, these attempts can not be made at the expense of experimentations that are the core of learning science. However, the study revealed that such attempts enhance the sophistication of students' perception of scientific inquiry moving it to a higher cognitive level.

References

Adams, W. K., Perkins, K. K., Podolefsky, N. S., Dubson, M., Finkelstein, N. D., & Wieman, C. E. (2006). A new instrument for measuring student beliefs about physics and earning physics: The Colorado Learning Attitudes about Science Survey. *Physics Review Special Topics, Physics Education Research, 2*, 010101.

Bagno, E., Berger, H., & Eylon, B. S. (2008). Meeting the challenge of students' understanding of formulae in high-school physics: A learning tool. *Physics Education, 43*(1), 713.

Chi, M. T. (2005). Commonsense conceptions of emergent processes: Why some misconceptions are robust. *The journal of the learning sciences, 14*(2), 161–199.

Cui, L., Rebello, N. S., & Bennett, A. G. (2006, February). College students transfer from calculus to physics. In *AIP Conference Proceedings* (Vol. 818, No. 1, pp. 37–40). AIP.

Domert, D., Airey, J., Linder, C., & Kung, R. L. (2007). An exploration of university physics students' epistemological mindsets towards the understanding of physics equations. *Nordic Studies in Science Education, 3*(1), 15–28.

Hestenes, D. (2003). Oersted medal lecture 2002: Reforming the mathematical language of physics. *American Journal of Physics, 71*(2), 104–121.

Hudson, H. T., & McIntire, W. R. (1977). Correlation between mathematical skills and success in physics. *American Journal of Physics, 413*(13), 470–471.

Karam, R., Uhden, O., & Höttecke, D. (2019). The "math as prerequisite" illusion: Historical considerations and implications for physics teaching. In G. Pospiech, M. Michelini, & B. S. Eylon (Eds.), *Mathematics in physics education*. Springer.

National Council of Teachers of Mathematics (NCTM). An Agenda for Action: Recommendations for School Mathematics of the 1980s. Reston, Va.: NCTM, 1980. Accessed November 13, 2019. http://www.nctm.org/flipbooks/standards/agendaforaction/index.html

Planinic, M., Milin-Sipus, Z., Katic, H., Susac, A., & Ivanjek, L. (2012). Comparison of student understanding of line graph slope in physics and mathematics. *International Journal of Science and Mathematics Education, 10*(6), 1393–1414.

Sherin, B. L. (2001). How students understand physics equations. *Cognition and Instruction, 19*(4), 479–1341.

Sokolowski, A. (2014). Modelling rate for change of speed in calculus proposal of inductive inquiry. *International Journal of Mathematical Education in Science and Technology, 45*(2), 174–189.

Sokolowski, A. (2015). The effects of mathematical modeling on students' achievement; a meta-analysis of research. *Journal of Education, 3*(1), 77–93.

Sokolowski, A. (2018). Modeling with exponential decay function. In *Scientific inquiry in mathematics-theory and practice* (pp. 65–82). Springer.

Sokolowski, A. (2020). Unpacking structural domain of mathematics to aid inquiry in physics: A pilot study. *Physics Education, 56*(1), 015009.

Sokolowski, A., & Capraro, M. M. (2013). A constructivist approach to embodying motion problems in mathematics. *Mediterranean Journal for Research in Mathematics Education, 12*(1–2), 121–133.

Uhden, O., Karam, R., Pietrocola, M., & Pospiech, G. (2012). Modelling mathematical reasoning in physics education. *Science & Education, 21*(4), 485–506.

Teaching Physics Using Mathematical Reasoning

Research and Practice

This book offers novice teaching strategies on applying mathematical reasoning to support the understanding of physics concepts. It proposes an empirical-mathematical scheme to design lectures and lab activities that invite the learners' reason mathematically while attending to physics understanding. The book provides didactical introductions of mathematical ideas that allow physics students to transfer their mathematical knowledge and embrace it in physics inquiries. While there are various levels of mathematical reasoning, this book prioritizes covariational reasoning crafted by limiting case analyses. The volume can interest a broad audience of scholars, practitioners, and curriculum policymakers. It is built on three pillars; (1) synthesis of contemporary research, (2) proposed conceptual framework, and (3) case studies that include readily available lecture ideas that highlight covariational reasoning to support physics understanding. Some of the ideas include; reconstructing Newton's law of universal gravity, parametrization of projectile motion, or enabling covariational reasoning in Einstein's formula for the photoelectric effect.

A. Sokolowski, *Understanding Physics Using Mathematical Reasoning*,
https://doi.org/10.1007/978-3-030-80205-9

Index

A
Algebraic form, 11
Algebraic functions, 5
Archimedes principle, 11
Avogadro's number, 159

C
Classifying parameters, 55
  kinetic friction, 56
  mediators, 55
  potential difference and current dependent variable, 56
Constructivist learning theory, 105, 106
Convert lens equation into covariational function, 130, 131
Covariate parameters, 188
Covariational reasoning, 41, 177, 179, 189
  algebraic interpretations, 183
  broader presentation, 189
  descriptive analysis, 181
  diagnostic research, 180
  experiment analysis, 194
  formula notation, 181, 184
  formula reimaging, 189
  formulas in sciences and mathematics, 43–45
  function notation, 45, 182, 189
  function zeros, 42
  hypothetical analyzes, 193
  independent and dependent variables, 181
  limiting case analysis, 57
    definition of limits, 57
    evaluating function limits in large value, 58, 59
    evaluating function limits in small value, 60, 62
    in mathematics education, 62
    in physics education, 62
    schematic representation, 58
mass-spring system, 193
mathematical modeling, 194
in mathematics education, 45
  cause and effect relationships, 47
  confounding/extraneous variable, 47
  relation of, 46
  in Thompson's theory, 47
  variation, 46
parameters, 41, 43
physics contexts, 178
in physics education, 179
  casual relation between selected variables, 48
  conceptual development, 48
  dependent parameter, 50
  gaining momentum, 48
  kinetic friction, 56
  linear dependence, 51, 52
  linking two/more dependent parameters, 53
  mediating and confounding, 50
  mediators, 55
  multiple parameters within system, 53, 54
  parameter classifications, 48–50, 55
  potential difference and current dependent variable, 56
  rational dependence, 52, 53
  scaffolding covariation, 54, 55
  variable parameter, 49

A. Sokolowski, *Understanding Physics Using Mathematical Reasoning*,
https://doi.org/10.1007/978-3-030-80205-9

Covariational reasoning (*cont.*)
physics formulas, 177, 178
quantities, 41
radical function, 192
students response, 182
study description, 179
using law of conservation of energy, 195
variables, 41
vertical parabolic track, 195

D
Doppler effect, 44

E
Electrosn energy *vs* light frequency graph, 171
Empirical-mathematical learning model, 36
algebraic relations, 38
conceptual questions, 38
concluding phases, 39
investigators' proposed theory, 37
learning phases, 35
mental or physical actions, 35
scientific classification, 37

F
Force of gravitational attraction, 6
Fundamental theorem of calculus (FTC), 13

G
Geometric optics, 129
Graphing, 196

H
Hypothetical deductive testing, 12
Hypothetical reasoning, 44

I
Image orientation, 136

K
Kepler's law, 9
Kinetic energy function, 190
Kinetic energy *vs*. object's speed, 190

L
Law of conservation of momentum (LCM), 7, 9
inelastic collision, 8
mathematical structure, 7
total momentum, 8
Laws and theories, 12
Laws of physics, 4
Lentz's law, 9
Limiting case analyses, 32
Limiting case analysis, 57, 193
algebraic tools, 70
behavior and hypothetical deductive reasoning, 70
bi-variational effects, 70
data analysis
posttest results, 77
pretest results, 76
definition of limits, 57
evaluating function limits in large value, 58, 59
evaluating function limits in small value, 60, 62
extend scientific inquiry inquiry, 67
half-Atwood machine, 69
input-output functional relation, 69
magnification function, 138
in mathematics education, 62
Newton's second law of motion, 71
Newton's second law's, 69
participants, 68
in physics education, 62
research questions, 68
schematic representation, 58
system acceleration
formal mathematical process, 74
function of mass, 75, 76
hanging mass, 73
massless string, 70
primary forces, 72
zero/infinite magnitude, 71

M
Mathematical knowledge, 16
conceptual knowledge, 17, 20
conceptual math knowledge, 23
domain-specific knowledge, 20
epistemology, 16
in physics classes, 22
in physics classroom, 15
mathematical reasoning, 20–22
mental processes, 21

nature of mathematics, 16
ontology, 16
parallel activities in physics, 23
physics education research, 18
procedural and conceptual math knowledge, 17
procedural knowledge, 17
procedural math skills, 22
role of mathematics in physics, 19
STEM education, 17
technical/procedural algorithms, 19
used in physics, 19
Mathematical language, 3
experimental physicists, 3
theoretical physicists, 4
Mathematical reasoning, 21, 187
scientific inquiry, 32
Mathematical rules and notation, 184
Mean Value Theorem, 13
Merging mathematics and physics, 32, 33
Mindsets, 187
Modeling
advantages, 30
algebraic representations, 29
blending representations, 30
in physics education, 30, 31
mediators, 30
purpose of, 29
Molar mass, 146
Mole understanding
concept of, 145
lecture component, 151–157
amount of mass, 152
amount of substance, 152
convert atoms to moles, 152–154
convert mass substance to atoms, 156, 157
convert mass substance to moles, 154–156
posttest result analysis, 159
pretest result analysis, 157
pretest *vs* posttest result analysis, 158
proportional reasoning, 148
direct proportionality, 150
dynamic proportionality, 148, 149
proportionality constant, 148
proportional reasoning;*static proportionality*, 148
rate, 148
ratio, 148
students' deficiencies, 150
theoretical framework, 147
weaknesses, 146

N
National Council of Teachers of Mathematics (NCTM's), 127
Newton's law of universal gravity, 5
action-at-a-distance, 82, 83
conceptual description, 84
contemporary presentations, 84, 85
electricity and magnetism, 81, 82
field intensity at distance, 86
gravitational field intensity, 87
algebraic representation, 87
mathematical reasoning, 90
piecewise function, 89
spherical solid mass, 89
two independent variable parameters, 89
using Kepler's laws, 88
zero magnitudes, 88
historical perspective, 83
lines of force, 83
modeling type approach, 85, 86
one group mixed-method, 85
posttest result analysis, 96–98
pretest result analysis, 93–96
reconstruct law of gravitational interaction, 90
bi-numerical notation, 91
force acting at a distance, 92
parent function, 92
tangible analysis, 90
two masses and fields, 93
using cascade covariation, 92
teaching and learning, 81
Newton's second law of motion. *See* Limiting case analysis

P
Parallel-axis theorem, 13
Phenomena behavior, 4
Photoelectric effect (PE), 163
algebraic representation, 171, 173
cognitive activity, 164
Einstein's formula, 164
binding energy, 168
covariational parameters, 169
graphical and algebraic representations, 165
graphical representation, 165
kinetic energy, 167
photon's energy, 167, 168
threshold frequency, 167
work function, 166

Photoelectric effect (PE) (*cont.*)
function notation, 169
graph representation, 170, 171
piecewise function, 171, 172
proposed notation, 170
threshold energy of incoming photons, 173
using law of conservation of energy, 173
work function, 163
Physics law, 10
Physics principles, 10
Planck's constant, 168
Polynomial function, 6
Projectile motion
classifications, 104
constructivist learning theory, 105, 106
degree of movement freedom, 104
kinematics, 103
lecture component sequencing, 107
parametric equations, 104, 105
physics curriculum, 108
one-dimensional frame reference, 110
parametric equations (PE), 108
parametrization of inclined motion, 113, 114
parametrization of vertical motion, 111
two-dimensional frame reference, 110
vectorial functions, 112
physics lab, 114
algebraic representations, 116
lab logistics, 115
position functions, 117, 119
using the classroom screen, 115
velocities and acceleration functions, 119
verification process, 120
X and Y coordinates object, 117
research findings, 101
research question, 107
students' understanding, 102
study description, 106
teaching and learning, 102
treatment evaluation, 121
identify motion components, 122
sample responses, 123, 124
student responses experiments, 122
student responses lab conduct, 122
student responses observation, 122

R
Reimaging lens equation, 127
algebraic function, 131
convert into covariational function, 130, 131
expressing magnification, 136
geometric optics, 129
image characteristics
converging lens, 132
thin lens equation, 128
using graph, 132
independent students' assignment, 140, 141
lab equipment, 129
magnification function, 133, 135
mathematical and experimental representations, 136–139
mathematical background, 128
object's distance, 133, 134
scientific constraints, 132
sketching, 132

S
Speed *vs* kinetic energy graph, 190
Students comments, 182
Students' preference
algebraic representations, 197
responses favoring formulas, 197
responses favoring functions, 199

V
Velocity vs kinetic energy, 191